PLANT LIFE OF EDINBURGH
and
THE LOTHIANS

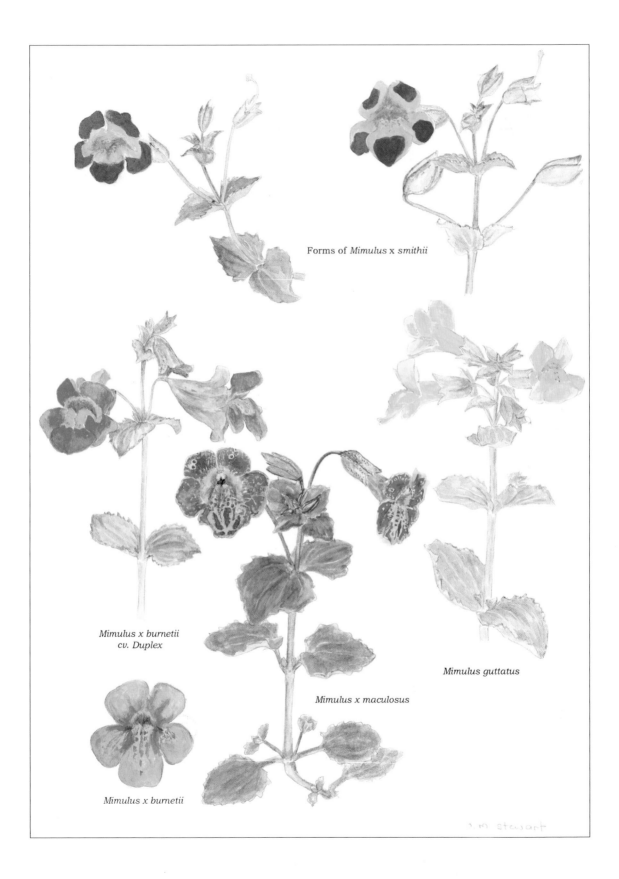

Forms of *Mimulus* x *smithii*

Mimulus x *burnetii*
cv. Duplex

Mimulus x *maculosus*

Mimulus guttatus

Mimulus x *burnetii*

PLANT LIFE OF EDINBURGH
and
THE LOTHIANS

Edited by
P. M. Smith, R. O. D. Dixon and M. P. Cochrane

Art Editor
C. E. Jeffree

Edinburgh University Press

© P. M. Smith, R. O. D. Dixon and M. P. Cochrane, 2002

Edinburgh University Press Ltd
22 George Square, Edinburgh

Typeset in Sabon and Gill Sans
by Pioneer Associates, Perthshire, and
printed and bound in Great Britain by
Bookcraft Ltd, Midsomer Norton, Bath

A CIP record for this book is available from the British Library

ISBN 0 7486 1336 6 (hardback)

The right of the contributors
to be identified as authors of this work
has been asserted in accordance with
the Copyright, Designs and Patents Act 1988.

Frontispiece: Paintings of Monkeyflowers (*Mimulus* ssp.) by Olga Stewart (1920–98),
who made a special study of alien species in the Lothians and was a diligent and enthusiastic
recorder in the Botany of the Lothians Project. The species are (clockwise from top left)
M. x smithii, *M. guttatus*, *M. x maculosus*, *M. x burnetii* cv. Duplex.
Reproduced by kind permission of the Stewart family.

Contents

Contents

Preface

Seventy-five eventful years have passed since the last Flora of Edinburgh and the Lothians appeared (Martin 1927), and that Flora was in part an updated version of Balfour's *Flora* of 1863. It is a hundred and seventy-eight years since Greville's *Flora Edinensis*, the first Edinburgh Flora, was published and that inevitably reflected a quite different scientific philosophy. In 1981, the Botanical Society of Edinburgh (now of Scotland) decided to sponsor a survey leading to a new Flora, with an accompanying broader account of local plant life, supported by maps, habitat and frequency data. It was hoped to publish near the end of the twentieth century. This is the work now completed.

Great changes in population and landscape pressures have occurred since 1927, with impacts on local plant life, as well as significant advances in scientific knowledge of it. Losses of species have, of course, taken place but, overall, the recorded flora of vascular plants is some 10 per cent richer. Change, whether of climate or land use, does not necessarily imply botanical loss, indeed it offers an opportunity for ingress and evolution.

The Society's wish was to involve as many local people as possible in the preparation of the Flora: the list of participants shows that this was realised. Many Scottish organisations have contributed to the survey and its publication: this has been a happy ship. The aim throughout has been to introduce as well as to enumerate and describe the flora. We hope the book will be found useful and will stimulate wide interest in, and further monitoring of, local botany. There is much left to discover and to understand.

We thank all contributors, all recorders and other helpers, and our publishers, for their work and we thank all who helped to fund the enterprise. Scottish Natural Heritage provided financial support for the completion of the survey and for data input.

Publication of this book would not have been possible without financial help from:

 Botanical Society of Scotland, Hon. Secretary, c/o Royal Botanic Garden Edinburgh, Inverleith Row, Edinburgh EH3 5LR

Edinburgh Natural History Society, Hon. Treasurer, J. Muscott, 1F2/69 Warrender Park Road, Edinburgh EH9 1ES

 Wild Flower Society, 82A High Street, Sawston, Cambridge CB2 4HJ

 Farming and Wildlife Advisory Group (Lothian Wildflower Seed Initiative), Lothian FWAG, Vogrie House, Vogrie Country Park, Gorebridge, Midlothian EH23 4NU

 **Historic Scotland**, Longmore House, Salisbury Place, Edinburgh EH9 1SH

Edinburgh Botanic Garden (Sibbald) Trust

The Carnegie Trust for the Universities of Scotland

Dr B. E. Sumner (proceeds of an exhibition of botanical art).

With grateful thanks to the late Miss Margaret F. Torrance who graduated from the University of Edinburgh with an Honours degree in Botany in 1940, and the generous contributions made by many individual members of the Botanical Society of Scotland.

Use of the Ordnance Survey grids for the distribution maps is with the kind permission of the Ordnance Survey, © Crown Copyright Ordnance Survey. All rights reserved N.C./02/19278.
Plate 2 and Figures 1.1 to 1.5 which are derived from Ordnance Survey-based material are reproduced by kind permission of the Ordnance Survey.
Plate 2 and Figures 1.2 and 1.4 are reproduced with kind permission of Scottish Natural Heritage.

Readers desiring guidance on the meaning of botanical terms are referred to the following works: *Ainsworth and Bisby's Dictionary of the Fungi* (Kirk et al. 2001); *Lichens: An Illustrated Guide to the British and Irish Species* (Dobson 2000); *British Mosses and Liverworts* (Watson 1981); *New Flora of the British Isles* (Stace 1997).

List of Plates

List of Authors

W. J. Baird	*formerly* of National Museums of Scotland
D. F. Chamberlain	Royal Botanic Garden Edinburgh
M. P. Cochrane	*formerly* of Scottish Agricultural College, Edinburgh
B. J. Coppins	Royal Botanic Garden Edinburgh
T. M. Darwin	Scottish Natural Heritage, Dalkeith
R. O. D. Dixon	*formerly* of Institute of Cell and Molecular Biology, University of Edinburgh
A. F. Dyer	Royal Botanic Garden Edinburgh and *formerly* of Institute of Cell and Molecular Biology, University of Edinburgh
P. A. Furley	Geography Department, University of Edinburgh
S. Helfer	Royal Botanic Garden Edinburgh
E. H. Jackson	41f Promenade, Musselburgh, Midlothian EH21 6JU
C. E. Jeffree	Institute of Cell and Molecular Biology, University of Edinburgh
G. E. Kenicer	Royal Botanic Garden Edinburgh
H. S. McHaffie	Royal Botanic Garden Edinburgh
D. R. McKean	Royal Botanic Garden Edinburgh
J. Muscott	Office of Lifelong Learning, University of Edinburgh
G. Russell	Institute of Ecology and Resource Management, University of Edinburgh
A. J. Silverside	Department of Biology, Paisley University
K. A. Smith	Institute of Ecology and Resource Management, University of Edinburgh

P. M. Smith   Institute of Cell and Molecular Biology, University of Edinburgh

N. F. Stewart   Kingfishers *at* Cholwell Cottage, Posbury, Crediton, Devon EX17 3QE

R. Watling   Caledonian Mycological Enterprises, Edinburgh *and* Royal Botanic Garden Edinburgh

M. Wilkinson   Department of Biological Sciences, Heriot-Watt University, Edinburgh

Happy the man whose lot it is to know the secrets of the earth. He hastens not to work his fellows hurt by unjust deeds but, in rapt admiration, contemplates immortal nature's ageless harmony, and how and when her order came to be.

Euripides

1

Lothian Landscapes, Geology, Climate and Soils

P. A. FURLEY and K. A. SMITH

The Character of the Lothian Environment

Even within the rich diversity of landscapes that make up the British Isles the Lothians lay claim to a unique place. The setting of Scotland's capital, Edinburgh, which lies between the Pentland Hills and the Firth of Forth, has a grandeur that is known world-wide, but the Lothians also encompass many other highly individual local landscapes. They comprise the coastal plains of the Forth estuary, the glaciated and volcanic landforms of the eastern Midland Valley and extend to the watersheds high on the Lammermuir and Moorfoot Hills in the Southern Uplands. 'Lothian is and always has been a name to ring in the ears, to challenge the imagination, an ancient name of colour and character...' (Tranter 1979). From Caerketton Hill (478m) at the southern boundary of the City of Edinburgh, the great sweep of these diverse landscapes can be appreciated, with its panorama of swelling hills, volcanic cones, coastal sand-dunes and bays, and many small streams following the undulations of the landscape of the lowlands, heading eventually towards the sea (Ogilvie 1951).

Whilst it may be argued that to some extent the Lothians are an arbitrary unit in the continuum of east-west trending landscapes of central Scotland, they have gained cohesion and unity from their historical evolution from the kingdom of Loth[1] and from the growing importance of Edinburgh. In local administrative terms, the Lothians area today is comprised of the Districts of the City of Edinburgh, West Lothian, Midlothian and East Lothian. Apart from minor boundary changes at the periphery, essentially the Districts together occupy the same area as the historic counties of West Lothian, East Lothian and Midlothian, and more latterly the Lothian Region, which was in existence between the 1970s and 1990s. The Lothians

---

1. The 'Land of Lothian' may be more of a semi-mythical kingdom (Colledge 1980) but seems to have stretched from the Forth to the Tweed, having succeeded the pre-Roman Votadini occupation and settlement. It has had a rich and independent history (Tranter 1979) and has been known as the Lothian region since at least 1633 when it figured on a Mercator Atlas map (Lothian Regional Council 1995).

cover about 2,000km² and form a shallow crescent of predominantly lowland landscape, with Edinburgh as its focus, bounded on the north by the waters of the Firth of Forth and on the south by the bulwark of the Southern Uplands. The easternmost part also includes a section of the North Sea coastline, from North Berwick to the boundary with the Scottish Borders district. Plate 1 presents a satellite image of the Lothians, and Figure 1.1 comprises a map showing the main roads and settlements.

The factors that have determined the evolution of the landscape may also be seen to play a large part in determining the character and distribution of the flora. The topography and relief of the Lothians is largely a result of the structure and composition of the underlying rocks. However, geomorphological processes caused by water, ice and even wind have shaped the surface, especially over the past 13,000 years or so (Sissons 1967). This Holocene chronology embraced periods in which profound changes took place at the soil surface, driven by marked post-glacial climatic alterations with warmer and wetter, drier and colder periods. Superimposed on these natural processes are the consequences of several thousand years of occupation, settlement and land use change.

The purpose of this introductory chapter is to set the Lothians in their environmental context by examining the physical landscape, the geology, the climatic advantages and limitations of the area and by presenting a detailed consideration of the soils which indirectly reflect all the environmental factors in their formation and which directly influence plant growth.

Lothian Landscapes

The Lothians comprise a region of strong topographic contrasts. To the south, the Southern Uplands form round, convex hills with smooth slopes and concave valleys. This dissected plateau region rises to 300–50m a.s.l. in the Lammermuir Hills and to approximately 300–650m in the Moorfoot Hills (Figure 1.2). The hills abut abruptly onto the lowland plain along a marked escarpment caused by the Southern Upland fault.

Running south-west from Edinburgh, the Pentland Hills form a striking narrow ridge some 25km long, with a series of mostly volcanic summits reaching nearly 600m a.s.l. although, at the southern end, conglomerates produce a more gentle appearance. Most striking are the igneous intrusions and volcanic hills that project above the lowland surface throughout the region. To the east lie the volcanic plugs of North Berwick Law and the Bass Rock, whilst closer to Edinburgh the laccolith intrusions of Traprain Law and the low east-west ridge of the Garleton Hills also testify to the violent volcanic activity of the past. In Edinburgh itself it is the incidence and rugged outline of igneous and extrusive volcanic rocks, that has determined the unique character of the city. These include Arthur's Seat, Calton Hill, Castle Rock, Blackford Hill and, continuing westwards, Corstorphine Hill and then the Bathgate Hills in West Lothian.

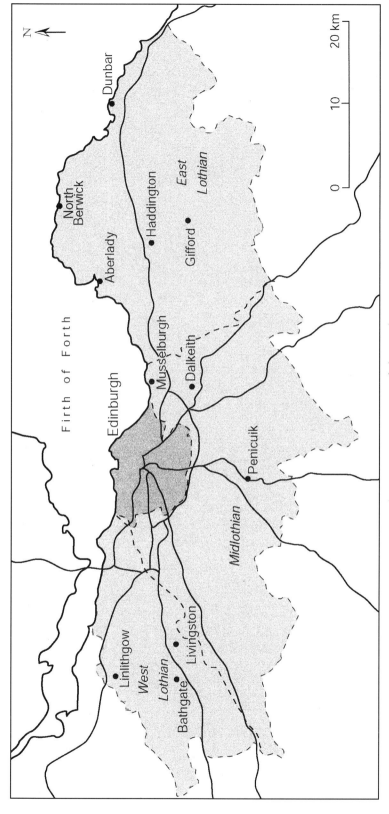

Figure 1.1 *Location of the Lothians.*

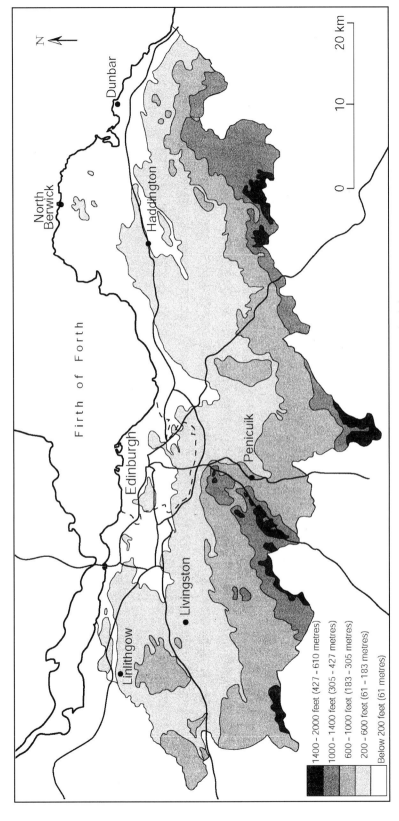

Figure 1.2 *Topography of the Lothians (modified from ASH 1998).*

Legend:
- 1400 – 2000 feet (427 – 610 metres)
- 1000 – 1400 feet (305 – 427 metres)
- 600 – 1000 feet (183 – 305 metres)
- 200 – 600 feet (61 – 183 metres)
- Below 200 feet (61 metres)

Firth of Forth

North Berwick
Dunbar
Haddington
Edinburgh
Linlithgow
Livingston
Penicuik

0   10   20 km

Much later in its geological evolution, fluxes of glacial ice have fashioned both the shape of the hills and the nature of the lowlands. The landscape today consists of 'glacially covered Devonian and Carboniferous sedimentary rocks . . . broken by numerous crags and hills of volcanic and intrusive rocks of the same ages' (McAdam *et al.* 1996). The overall eastward advances of ice and westwards retreats have shaped hillsides, infilled glacially-cut valleys and plastered a layer of till over the landscape (Figure 1.3). With the final ice retreat, the glacial meltwaters had to cut through these variable sandy and clay deposits to reach the coast, whilst periglacial processes thinned deposits over the hills and solifluction (the slow gravitational downslope movement of water-saturated, seasonally-thawed materials) shifted them to the valley bottoms. Over the lowlands, the residue of glacial retreat is everywhere discernible, with erratics (rocks or boulders carried to their present locations by ice), heavy clay tills in the hollows up to 60m thick, fluvio-glacial sand and gravel deposits, deflected water courses and incised gorges such as those of the Tyne near East Linton or the North Esk in Midlothian (Sissons 1958; Figure 1.4).

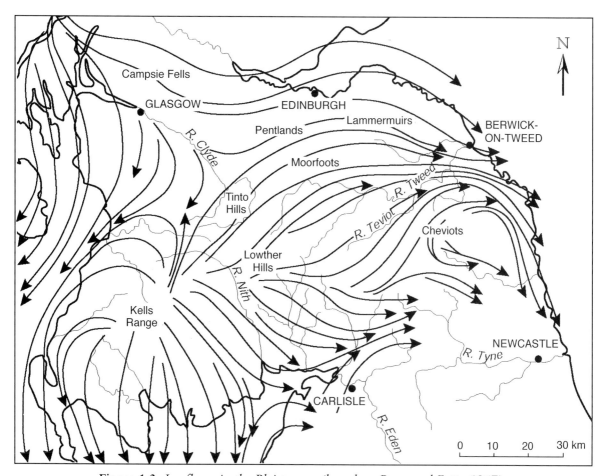

Figure 1.3 *Ice flows in the Pleistocene (based on Ragg and Futty 1967).*

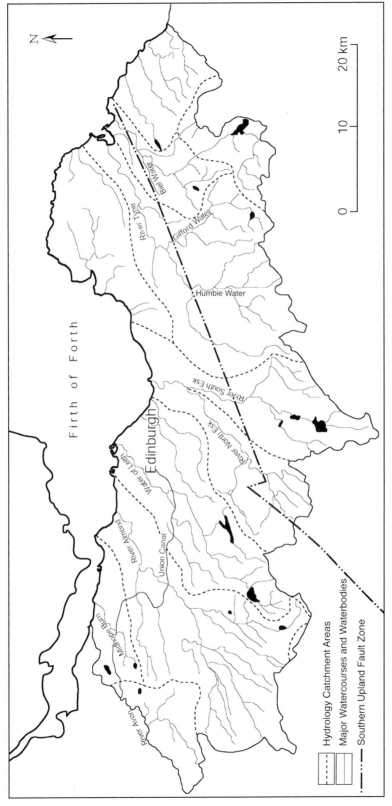

Figure 1.4 *Drainage and catchment basins in the Lothians (modified from ASH 1998).*

Tulloch and Walton (1958) reported buried channels covered by drift some 80–155m thick.

Despite these later interruptions to the water courses, drainage to the Forth in rivers such as the Almond, Water of Leith, North and South Esk and the Tyne is still conditioned by the Lower Palaeozoic structural grain which runs parallel to the Southern Upland Fault. Some of the most distinguishable features of these ice movements are the crag and tail landforms lying to the lee side of hills facing the onset of ice, some of the most notable of which are the Garleton Hills in East Lothian and the well known Castle Rock and Royal Mile in Edinburgh (Sissons 1971).

The Geological Framework

As a result of this rich legacy of glacial landforms, the Lothians are 'one of the key areas in Scotland for the elaboration of concepts related to ice sheet glaciation, glacial sediments and landforms' (Gordon and Sutherland 1993). Many of the earliest observations were made in this area, for instance on ice flow direction, orientation of clasts in till and ice-moulded striations on rocks (notably Agassiz in 1841 on Blackford Hill). Indeed it can be reasonably argued that the geology of the Lothians is of global significance because it has been 'the scene of several fundamental discoveries in the science' (McAdam *et al.* 1996), notably from the time of James Hutton (*Theory of Earth*, 1795) who is considered to be the father of modern geology.

Whilst geomorphological processes have fashioned the final shape of the land surface, the structure and composition of the underlying rocks provide the bones of the landscape (Craig and Duff 1975; Craig 1983). Each of the two main structural divisions, the Midland Valley and the Southern Uplands, has a distinguishable and individual character. The Midland Valley represents an ancient rift valley lying between the Southern Upland and Highland fault lines. Whilst most of the Lothians lie within the Midland Valley, the southern margins extend over the fault and into the rolling uplands and so possess a more typically Border character (McAdam *et al.* 1992). The geology will be considered in chronological order, starting with the oldest rocks that influence the landscape, and with the object of showing the relationship to the contemporary environment.

The solid rocks of the area range in age from Ordovician to Carboniferous but, over the lowlands and indeed spread thinly over much of the hills, there are superficial, and often unconsolidated, Quaternary drift deposits (Table 1.1). The oldest rocks in the region outcropping at the surface are the shales and greywackes of Ordovician and Silurian age that form most of the Southern Uplands. Originally marine sediments, the rocks were highly compressed during the Caledonian mountain building period (495–435 million years ago), resulting in high angles of dip and a generally NE–SW strike to the landscape. During the Devonian period (410 million years ago), much of Scotland experienced an arid or semi-desert climate, evidence of which can be seen in the contemporary landscape. Sediments from the surrounding hills at the time were deposited over the Midland Valley with coarse deposits coalescing to form conglomerates. Periodic volcanic activity unleashed an outpouring of lava, as exemplified in the interstratified nature of the Pentland Hills, which are made up of basaltic and andesitic lavas inter-bedded with Ordovician and Silurian mudstones and fine sandstones. Late Devonian processes witnessed an uplift of land with further folding and faulting (about 360 million years ago). Erosion and subsequent deposition of sandstones and finer deposits in what were desert conditions can be seen today in the red rocks and soils of the easternmost areas of the Lothians such as the land around Dunbar.

Table 1.1 The general geological succession in the Lothians

Holocene/ Recent	Peat, alluvium, colluvium, blown sand, raised beach deposits	
Pleistocene	Glacial till, fluvio-glacial sands and gravels, solifluction deposits	
		Lower Coal Measures
	Upper Carboniferous	Passage Group (Roslin Sandstone)
		Upper Limestone Group
Carboniferous		Limestone Coal Group
	Lower Carboniferous	Lower Limestone group
		Calciferous Sandstone Measures
Old Red Sandstone	Upper Old Red Sandstone	
	Lower Old Red Sandstone	
		Hawick Rocks
Silurian	Llandovery	Queensberry Grits
		Birkhill Shales
		Lowther Shales
	Ashgill and Caradoc	Hartfell Shales
Ordovician		Glenkiln Shales
	Arenig	Radiolarian Cherts and Mudstones

(Based on Ragg and Futty 1967; McAdam and Clarkson 1996)

In the late Primary Period, notably during the Carboniferous (360–280 million years ago), the Midland Valley became a large delta with a humid equatorial climate. Muds and silts were deposited there. Alternations of sea level gave rise to limestones (during marine incursions) and mudstones (with more terrestrial conditions prevailing). The decayed tropical forest vegetation ultimately metamorphosed to form the coal seams, and anoxic processes in marine sediments gave rise to oil shales, exploitation of both of which has given rise to the characteristic man-made landscapes widespread over the lowland plains. The oil shale bings of West Lothian are by far the most visible legacy of this activity. Occasional volcanic activity at this time also affected today's landscapes, notably in the Garleton and Bathgate hills.

Over the immense time between the end of the Primary and the onset of cooler conditions in the Quaternary (245 million years ago at the close of Permian to just over 2 million years ago at the Tertiary-Quaternary boundary), what is now Scotland moved northwards under the influence of continental drifting. This changed the climates yet again from tropical to more humid and temperate conditions. During the Secondary in the Jurassic and particularly in the Cretaceous, shallow seas covered parts of the lowlands and the remnants of the calcareous sands and limestones laid down at this time have a pronounced local effect on soils today – such as the gently swelling Camp Hills near Tranent.

Not surprisingly, the impact of the last geological events have had the most dramatic effect on the landscape. The cooler and wetter conditions of the late Tertiary accentuated and led to the glacial advances and retreats of the Quaternary. Numerous major and minor advances and retreats of ice occurred during the Pleistocene, which shifted materials around over the surface of the region. The last phase in the geological evolution before the present day has been the deglaciation of the late Pleistocene and Holocene, with alternating postglacial conditions from periglacial to warmer, wetter or drier regimes. Today we have a landscape that is far drier and better drained than during most of the rest of the Holocene period. At the same time, as ice melted so the sea level rose, flooding the Forth Valley to a level as much as 16m above present sea level (ASH 1998) and fine tidal sediments covered most of the river floodplains. Isostatic uplift (re-adjustments of the equilibrium between the land surface and sea level as a result of the lifting of the pressure of ice) then raised these deposits to form raised terraces and beaches, which are widespread along the Forth, its estuary and around the North Sea coast (Sissons 1962; 1963).

Climatic Controls on Plant Growth

Although climate has been changing continuously over time, affecting the palaeo-ecology, there are reasonably predictable patterns that govern present-day plant growth. The main features are the prevailing winds and pressure systems, the influence of the sea, and the altitudinal effects in the hill areas, although it is also worth remembering Kingsley Martin's apt comment that 'wherever you go, the weather is without exception, exceptional' (quoted in Hulme and Barrow 1997).

Winds
The winds are generally from the south-west, but the Forth-Clyde Valley has a major funnelling effect responsible for many of the blustery winds and gales of

autumn and early winter. Equally, at this time of year warmer winds from the south-west can produce spells of bright days in protected areas. Onshore winds from the east frequently bring in cool, misty conditions or haar, which can blanket the coastal plains and Forth Valley even in the height of summer. The higher tracts of land to the east also block wind pathways at times, backing them up over the cool North Sea and bringing chilled air conditions to the stretch of coast south of Dunbar. From April until June, the prevailing winds are often from the north-east and are frequently cold and blustery, associated with high pressure to the north. When such northerly winds reach gale force and when they occur whilst fields are bare in January–February, the soils can dry rapidly and become highly erodible. Large dust storms are a common sight at this time of the year leading to severe erosion in vulnerable types of soil (such as the Fraserburgh Series) and a reciprocal dumping of sediment elsewhere, which may also be a problem.

Precipitation

Rainfall is clearly related to the wind pattern and elevation, increasing westwards and away from the coast. Much of the lowland area below around 150m in the eastern part of the region is extremely dry, similar to the eastern coastal areas of England. The estuarine and coastal villages can have less than 650mm of rainfall annually, scattered in light showers from which the ground is rapidly dried. Annual rainfall increases with altitude to over 750m at around 300m and over 1,000mm at 525m in the Lammermuir Hills. To the west, rainfall figures of around 1,000mm can be found at lower altitudes, as over the Slamannan Plateau in West Lothian. The low averages from March until June, which are characteristic of the eastern lowlands, combined with higher evapotranspiration potentials (Figure 1.5), make the region liable to spring droughts and prolonged dry spells. Local 'rain shadow' effects to the lee sides of hills accentuate this (coupled with higher temperatures from the slight fohn wind influence where descending winds are compressed and warmed adiabatically (Meteorological Office 1982)). Within this broad pattern of dry spells, sometimes relieved by the incidence of haar and buffered by moisture-absorbent clay soils, there are also periodic wet spells. In the Edinburgh area generally and in the Borders, heavy rains followed by flooding have occurred after mid-August (as in 1948 and 1956), or earlier, as instanced by the spring floods in Edinburgh in 2000.

Temperature

Figures vary widely over space and time. Over the Midland Valley, the range in the average monthly mean temperature is approximately 12–13°C, varying from 2–3°C in January to 14–15°C in July. The influence of the sea warms coastal areas slightly in winter and tends to keep them snow-free, but can also reduce the summer temperatures along a narrow coastal fringe. As the land rises towards the Southern Uplands, so the altitudinal effect on temperature becomes more pronounced, although less distinct in the summer. For the rest of the year, mean temperatures at

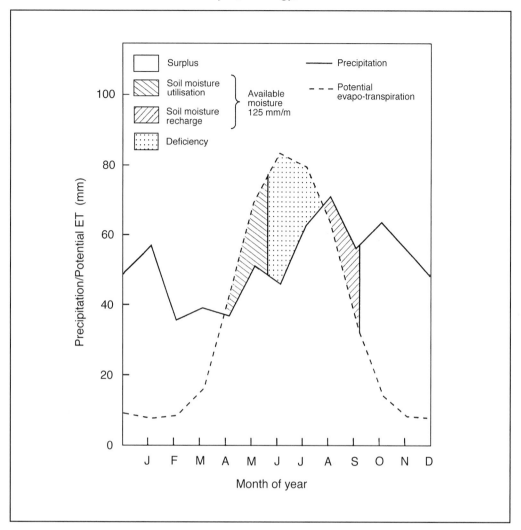

Figure 1.5 *Monthly rainfall and potential evapotranspiration in East Lothian (East Fortune) (based on Ragg and Futty 1967).*

around 150m are frequently 1°C or more below those of the coast. At higher elevations, especially facing north, the incidence of freezing in the winter months of January and February is much more marked, and snow lies for longer on the ground and in sites protected from the sun. Cold, dense air also sinks towards the valley bottoms and tends to stagnate in the Borders and in low-lying sites within the Midland Valley (Harrison 1997). Even within the Edinburgh City boundaries, there are well-known cold spots, wind tunnels and rain shadows. Indeed Edinburgh is 'one of the most topographically and climatically complex cities in the UK' (Meteorological Office 1982). These micro-variations reflect the mosaic of variation over both the lowland and upland areas and account directly and indirectly, through soil properties, for the conditions for plant growth.

Insolation

The eastern coastal areas of the Lothians claim high sunshine figures over the year whereas conditions westwards and inland become cloudier. June is generally the sunniest month followed by May, whereas July is frequently disappointingly low in sunshine. Insolation in April can be as great as that in August and more than that in September. However, the influence of fog and low cloud or haar is an important factor in the climate of the eastern Lothian and Forth Valley lowlands. These conditions are linked to easterly winds bringing warm air masses over the cold North Sea. Haar varies in its penetration inland and can often persist for several days.

Growing Period

As a consequence of these controls, the growing period, defined as the time when the mean daytime temperature is 5–6°C or higher, varies considerably over the Lothian region. In the warmer areas the temperature threshold may be reached in mid to late March and last until the middle of November, whereas at elevations of about 150m, the growing season does not start until April and will end earlier in November. Above an altitude of 150m, the length of the growth period diminishes steadily. Soil temperatures near the surface vary with the ambient temperatures, albeit with some damping of the diurnal fluctuations. At greater depths the insulating effect of the soil provides protection against surface extremes of temperature and that can be very useful for plants with deeper root systems.

There are a number of issues that override the normal climatic patterns. Extreme events such as the 1967/68 hurricane, very intense cold snaps or very marked droughts, may have an impact far beyond their periodicity and consequently may be extremely influential in shaping the pattern of vegetation at least in the short or medium term.

The Soils

The soils of the Lothians have formed as a result of several interacting factors. These comprise the climate, in particular rainfall and temperature; the geological parent materials, ranging from sedimentary and metamorphosed sandstones and shales to igneous intrusive and extrusive rocks and recent deposits such as shell-sand; topographic factors such as the steepness and length of slopes or the occurrence of water-collecting hollows; and biotic factors (vegetation, microbes and soil fauna, and human activities). The time over which these factors have operated is also an important factor in determining the stage of soil development (Ragg 1973). In pedological terms the soils of Lothian are relatively young, with development having been given a new direction by environmental changes since the end of the last Ice Age. This contrasts with uninterrupted soil development over hundreds of thousands of years in places where soils have been less disturbed, especially where they have never been swept away by moving ice sheets. On the other hand, soil-forming processes become evident within a much shorter timescale than the length of the postglacial period. One example is where plant colonisation of sand-dunes has taken place, along with rapid evolution of a soil that can assume an identifiable profile within a few hundred years or less.

Solid rocks undergo physical and chemical weathering and gradually give rise to unconsolidated surface deposits which constitute the parent materials of present-day soils. The deposit may remain *in situ*, or it may be transported elsewhere by the action of water, ice or wind. As indicated earlier, most of the Lothians is covered by such transported material – notably glacial till and fluvio-glacial deposits. The second most common deposit in the area is marine sediment left exposed along the coast by sea-level changes.

The action of rainfall and temperature on these parent materials has gradually brought about chemical changes, through the leaching of inorganic ions, and the establishment of biological life in the previously inert material. This has resulted in the creation of organic matter, the end product of decomposition being humus, which dominates the surface horizons and retains many of the nutrient ions and moisture vital for plant growth. Organic matter combines with clay particles to create water-stable aggregates of larger size, and consequently alters the water-holding capacity and the drainage characteristics, and lessens the erodibility of the mineral material. The presence of reactive compounds such as organic acids, and chemical oxidation and reduction induced by microorganisms, all contribute to the variety of properties exhibited in soil development.

The most recent overview of the soil types in the Lothian region (Bown and Shipley 1982) is largely based on the Soil Memoirs for Kelso and Lauder (Ragg 1960), and Haddington and Eyemouth (Ragg and Futty 1967) (Plates 2 and 3; Table 1.2).

Climatic and topographic influences
Leaching
Leaching is a major soil-forming mechanism in any humid region where precipitation exceeds evapotranspiration, and where the surplus water drains through the soil. The potential annual evapotranspiration (ET) in the Lothian area decreases from 500mm in the coastal fringe of East Lothian to 350mm on the highest parts of the Lammermuir and Pentland Hills (Francis 1981), so that even in the driest area there is a net surplus of precipitation over ET of more than 100mm, rising to 650mm or more at altitude. Leaching therefore occurs throughout the region to a greater or lesser degree. The principal process in leaching is the replacement of exchangeable basic ions such as those of potassium, calcium and magnesium by hydrogen and aluminium. The presence of the basic ions on exchange sites on soil mineral and organic surfaces provides a reserve for plant nutrition. However, the constant influx of hydrogen ions in rainwater (due to the reaction of atmospheric

Table 1.2 Predominant soil associations and constituent soil series found in the Lothians (after Ragg and Futty 1967; Ragg and Dent 1969)

Soil association	Parent material	Principal soil series	Soil classification
Bemersyde	Intrusive rhyolites and trachites	Bemersyde Dirrington	Freely drained brown forest soil (bfs) Freely drained iron podzol
Biel	Drifts derived from Lower Carboniferous and Upper ORS conglomerates and sandstones	Oxwell Biel	Freely drained bfs Imperfectly drained bfs
Darvel	Fluvio-glacial sands and gravels derived mainly from Lower Carboniferous igneous and sedimentary rocks	Darvel Duncrahill	Freely drained bfs Imperfectly drained bfs
Dreghorn	Raised beach sands, fine sands, silts and gravels	Dreghorn Peffer	Freely drained bfs Imperfectly drained bfs
Eckford/Innerwick	Fluvio-glacial sands and gravels derived mainly from Upper ORS conglomerates and sediments	Eckford Innerwick Skateraw	Freely drained bfs Freely drained bfs Freely drained bfs
Ettrick	Silurian and Ordovician greywackes and shales	Linhope Kedslie Minchmoor Dod Ettrick Hardlee	Freely drained bfs Imperfectly drained bfs Freely drained iron podzol Peaty podzol Non-calcareous gley Peaty gley

Association	Parent material	Series	Soil type
Fraserburgh	Raised beach and windblown shelly sand	Fraserburgh Luffness	Freely drained brown calcareous soil Imperfectly drained brown calcareous soil
Hobkirk	Till derived from Upper ORS sediments and marls, some with partially sorted upper horizons	Hobkirk Pressmennan Faw Cessford	Freely drained bfs Imperfectly drained bfs Peaty podzol, freely drained below B_1 Poorly drained non-calcareous gley
Kilmarnock	Till derived from Lower Carboniferous sediments and igneous rocks, some with partially sorted upper horizons	Kilmarnock Brownrigg	Imperfectly drained bfs Imperfectly drained bfs
Rowanhill/Giffnock/ Winton	Till derived from Carboniferous sediments and igneous rocks, some with partially sorted upper horizons	Macmerry Winton Rowanhill	Imperfectly drained bfs Imperfectly drained non-calcareous gley Poorly drained non-calcareous gley
Sourhope	Drifts derived from andesite and basalt lavas	Sourhope Bellshill Frandy	Freely drained bfs Imperfectly drained bfs Freely drained podzol
Yarrow/Fleet	Fluvio-glacial gravels derived mainly from Ordovician and Silurian greywacke	Yarrow	Freely drained bfs

ORS – Old Red Sandstone
bfs – Brown Forest Soil
B_1 – Upper part of the B (illuvial) horizon

carbon dioxide with water to form the weakly ionised carbonic acid) results in displacement of the basic ions and their removal in the drainage water. The ionisation of organic acids released from plant material also contributes to the acidity, and in recent decades an additional contribution has come from 'acid rain': deposition of sulphuric and nitric acids formed from sulphur and nitrogen oxides released by the burning of fossil fuels.

Gleying

The combination of the climate and the fact that till deposits are generally compacted compared with other types of surface deposit and have a very low hydraulic conductivity means that the till in the higher rainfall areas is commonly waterlogged or near-waterlogged for much of the year. In these conditions, the demand for oxygen to support the activity of plant roots and soil microorganisms exceeds the rate of supply from the atmosphere, and anaerobic conditions develop. Thus gleying becomes a dominant process. Gleying is the reduction of iron from the ferric (Fe(III)) state, which is the stable form in aerated conditions, to the more mobile ferrous state (Fe(II)), and likewise the reduction of manganese from Mn(IV) to the soluble Mn(II). Gleyed soils are characterised by the dominance of greenish-grey or very pale colours, with rusty-coloured patches, concretions and zones around old root channels, where re-oxidation has taken place (Bown and Shipley 1982). Gleying also occurs in more permeable material where there is a high ground-water table or where water accumulates due to movement down a hillslope. In wet hollows there may be a substantial organic surface layer developed on the gleyed material, constituting a peaty gley soil. Gleyed areas, whether in hollows or in folds in the hillside where downward movement of water is concentrated, are usually readily identified by the presence of wetland species such as rushes (*Juncus* spp.).

On sloping land where drainage is better, the lower slopes generally have a cover of brown forest soils, developed under the original vegetation cover of broadleaved woodland, but now supporting (under natural conditions) acid bent-fescue grassland. Higher up the slope, however, there is a change to humus-iron podzols, mainly under heather. The transition from brown forest soil to podzol appears to be associated with increased leaching and soil acidity (see below), lower temperatures, the absence of earthworms and generally lower biotic activity, together with plant litter that is more resistant to biological breakdown. The humus-iron podzol typically has a layer of black mor humus with an underlying grey sandy layer from which the iron has been leached away, as a result of complexing (chelation) with organic compounds released by the vegetation (Nortcliff 1988). Further down there is a dark-brown iron- and organic-rich layer where the dissolved iron and associated organic material have been redeposited. Humus-iron podzols give way in their turn to peaty podzols on higher land. Here the low temperatures and higher rainfall result in anaerobic conditions which inhibit organic matter breakdown, and a much thicker surface organic layer develops.

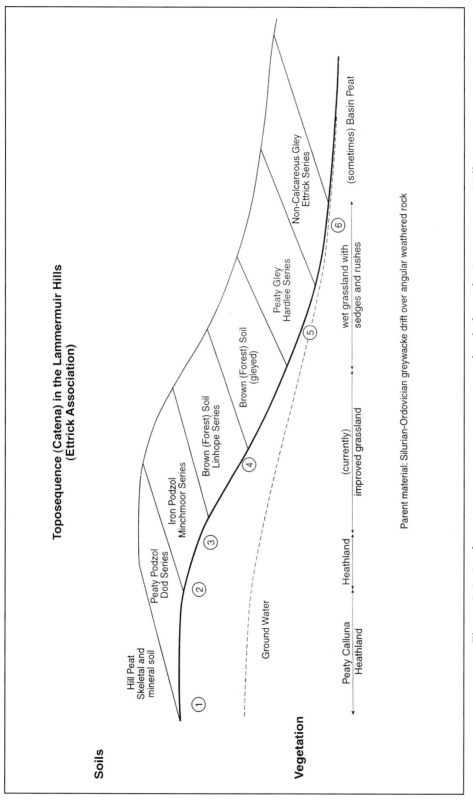

Figure 1.6 *A schematic toposequence (catena) of soils from the Lammermuir Hills.*

This spatial pattern of soils which results from profiles developing over different parts of a slope with differing moisture regimes has a marked effect on plant growth and composition. It can be illustrated for example, by a hydrological catena from the Lammermuir Hills (Figure 1.6).

Schematic Toposequence (Catena) in the Lammermuir Hills
(Ettrick Association)

1. **Hill peat/blanket bog.** Runoff is reduced over low gradients and if the subsoil is relatively impermeable moisture accumulates, encouraging the formation of peat (arbitrarily defined where organic matter is 30cm or more thick). The pH levels are low (3.5 to 4.0) and any mineral content is mostly sand.
   Vegetation: uniform throughout the profile; typically *Calluna-Eriophorum vaginatum-Trichophorum* bog communities.
2. **Peaty podzol (Dod Series).** With slightly increased gradients and greater runoff, the organic horizon is thin and overlies a podzolic profile. This happens where downward eluviation (or washing) of soluble and exchangeable ions generates a leached A horizon and an illuvial (accumulation) B horizon overlying weathered parent rock or drift (C horizon).
   Vegetation: heathland dominated by *Calluna vulgaris* with abundant *Nardus stricta*.
3. **Iron podzol (Minchmoor Series).** Profiles occur on the shoulders of rounded convex hills, where downward infiltration of water exceeds lateral runoff. This produces the most extreme form of podzol, frequently containing clearly defined very dark coloured litter (L), fermentation (F) and humus (H) organic horizons, sharply separated from grey, heavily weathered and leached A horizons (pH less than 4) and distinct reddish-brown B horizons (pH about 4.5). The latter often exhibit characteristic layering of organic matter (Bh), overlying fine-textured material (Bt) and metallic cations (or sesquioxides, Bs). In extreme cases the iron-rich material is synthesised into a thin iron-humus pan, which may inhibit drainage and give rise to localised gleying.
   Vegetation: typically heathland dominated by *Calluna vulgaris* with abundant *Erica cinerea*, *Vaccinium myrtillus*, *Deschampsia flexuosa* and *Nardus stricta*.
4. **Brown forest soil (Linhope Series).** As the slope increases, so runoff exceeds vertical percolation. The profile that results is mildly acidic (pH less than 6) with far less distinct horizon boundaries. A dark brown surface A horizon (moder and mull humus) overlies a frequently deep diffuse yellowish-brown B horizon. The profiles develop in stony drift of medium texture, and soil depth reflects the amount of material in transit downslope

---

– which may often slip as 'rafts' of surface material in wet conditions (Whitfield and Furley 1971). Further down the slope the influence of ground water may be apparent in the gleying of the lower B and C horizons.
Vegetation: mostly improved grazing with abundant semi-natural grasses such as *Agrostis tenuis*, *Anthoxanthum odoratum*, *Festuca ovina* and *Poa pratensis*.

5. **Peaty gley (Hardlee Series).** The greater proximity of ground water to the surface and more frequent downwash from the slopes above determines that these sites are usually wet. In such circumstances, minerals are present in reduced form and reflect anaerobic microbial processes. Organic matter may accumulate with moder humus (pH less than 4 at the surface). There is seasonal or periodic drying which produces a mottling effect with oxidised patches within a darker matrix of more reduced conditions (pH typically greater than 5). Occasionally a grey surface-water gley overlies a better-drained brown section, below which there is a further (ground) water-affected zone giving rise to a ground-water gley.
Vegetation: *Sphagnum* and other mosses locally dominant in a wet heath-land association of *Calluna vulgaris*, *Molinia caerulea*, *Erica tetralix* plus many typical rushes and sedges.

6. **Non-calcareous gley (Ettrick Series).** The Association takes its name from this semi-permanently wet soil. Seasonal moisture permits little drying out and the colour persists as subdued greens and blues reflecting long periods of reduction and anaerobic activity. Typically the soils are surface water gleys (pH about 5.5 at the surface and about 6.5 in the subsurface horizons). Vegetation: cultivated in places where drained, but typically semi-natural wetland plants such as *Deschampsia caespitosa* and *Holcus lanatus* with frequent sedges and rushes. In depressions lacking a stream outlet, this situation encourages the build-up of peat, forming characteristic basin peat and raised moss (for example Red Moss at Balerno).

Aspect

The effect of aspect on the soil temperature regime is very important in determining soil type. A south-facing slope of 20° can be equivalent to a southerly shift of 8–9 degrees of latitude (compared with level ground), and a similar north-facing slope to a northerly shift of 12–15 degrees of latitude (Chandler and Gregory 1976, cited in Bown and Shipley 1982). An example of the soil differences that can arise is illustrated at Boghall Glen on the south-eastern side of the Pentland Hills, about 10km south of Edinburgh. On the south-facing slope near the head of the glen, the freely drained brown forest soil (Sourhope Series) stretches up to over 450m a.s.l. before giving way to podzols, whereas on the north-facing slope the transition occurs 50–100m lower down (Ragg and Dent 1969).

In the drier parts of the lowlands, gleying and leaching are less pronounced than in the wetter uplands. There is a higher level of base saturation and a more moderate pH, which have provided a good medium for plant growth, and the natural soils here are brown forest soils, developed under broadleaved woodland. This type of environment is conducive to activity by earthworms and other soil organisms, which bring about the incorporation of plant litter into the soil and the development of a mull type of humus. The mull humus is intimately mixed into the top mineral layer to form a thick organic-rich surface horizon (in complete contrast to the upland podzols where the organic material remains as a distinct surface layer or mor humus). This type of soil development can now only be seen in very old woodland. In most of the lowlands the change in land use to arable agriculture has meant that cultivation has mixed the surface layer with the underlying material, resulting in the formation of a fairly uniform plough layer. This layer has been much modified chemically as well as physically, with regular additions of lime to control pH and mineral fertilisers to raise the nutritional status.

The Influence of Parent Materials

The parent material (the geological substrate on which a soil is formed) can have a profound effect on the soil's characteristics. A group of soils formed on a common parent material, but differing in soil type and drainage status, is known as a soil association. In the Lothians, the parent materials over much of the area are glacial drifts, mainly of Carboniferous and Old Red Sandstone origin (see section on geological framework above and Table 1.2) and fluvio-glacial sands and gravels. There is also a belt of soils along the Firth of Forth coast on raised beach deposits of marine origin. Where outcrops of more resistant igneous rocks occur, the parent material is weathered *in situ* or covered by a thin layer of local, stony drift (Ragg and Futty 1967).

The Rowanhill/Giffnock/Winton Association occupies a substantial part of the area covered by Carboniferous till (Table 1.2). Part of the till is unsorted (in other words it has not been subjected to the washing out of finer particles), and is very impermeable. This gives rise to clay loams or sandy clay loams, which are imperfectly drained brown forest soils, such as those of the Winton Series in East Lothian. In contrast, in some places for example near water courses and on lower ground, the upper layer has been subjected to water sorting, leaving a coarser and better drained material. This soil is mapped as Macmerry Series, which in spite of its better drainage is still classified as an 'imperfectly drained' soil. In the wetter parts of the area, covering much of West Lothian and Midlothian, the soils of the same association are more commonly classified as gleys (ASH 1998).

The Pentland Hills are made up of a combination of lavas and Old Red Sandstone (ORS) sediments, which have given rise to a complex mixture of different soil associations. On higher ground there are podzols, brown forest soils, rankers and peats of the Bemersyde, Mountboy and Hobkirk Associations, underlying heather moorland and rough hill grassland. The lower hill fringes in the north are covered by gleys and brown forest soils of the Sorn/Humbie/Biel Association, derived from a combination of Carboniferous and ORS sediments and lavas, and peat (ASH 1998).

The raised beach deposits (and wind-blown material) lie below 30m a.s.l. Deposits lie mainly to the east of Edinburgh, and extend from Prestonpans and Aberlady in the west to beyond Dunbar in the east. Raised beach sands, fine sands, silts and gravels give rise to coarse-textured soils of the Dreghorn Association, including the freely drained Dreghorn Series on sands and gravels and the imperfectly drained Peffer Series on silts and fine sands. Along the coastal margin the parent material consists of shelly sand, on which the Fraserburgh Series soils with a loamy sand to sand texture have formed. The presence of the calcium carbonate in the seashell fragments results in soils with a high pH, above 7.5 (Ragg and Futty 1967). A mosaic of soil types reflecting the precise combination of marine deposit, age and the presence or absence of fresh water is found along the coastal zone (Figure 1.7).

The Mosaic of Soils in the Coastal Zone: Aberlady Bay to Gullane

The coastal areas illustrate the complexity of soil profiles that result from minor differences in parent material and drainage. The stretch from Aberlady to Gullane in East Lothian provides an example of a mosaic of plant associations that can be related closely to the nature of the underlying soils. These have been formed over a raised beach and sand-dune sequence, mostly below 30m a.s.l. and with low rainfall (under 600mm per year). The shell content changes considerably and tends to be covered to varying depths by wind-blown sand.

The principal soils are the sand-dunes forming a series of parallel ridges of different age, the salt-affected (halomorphic) conditions of the saltings and zones close to onshore winds from the sea, and the dune slacks between successive dune-building phases (Kirby 1997). Over and above these variations, if one takes a line from the high water mark inland the soils may be considered as a chronosequence – that is the soils are progressively older with distance from the coast. This is reflected in, for example, the increased acidification inland (from pH 8.8 to over 6); exchangeable calcium dropping from about 80cmol kg^{-1} to over 40cmol kg^{-1} in the surface horizon; the progressive accumulation of organic matter, as evidenced by increased levels of organic carbon (0.1 to over 14%), total nitrogen (0.003 to 0.18%) and available phosphorus (13mg kg^{-1} to over 40mg kg^{-1}); and cation exchange capacities rising from 6cmol kg^{-1} at the coast to 16cmol kg^{-1} inland. It can also be seen in the increasing development of the soil profile from the embryo soils (regosols) on the most recent sand-dunes to the deeper, freely drained brown calcareous soils of the Fraserburgh Series (Furley, in preparation; Ragg and Futty 1967).

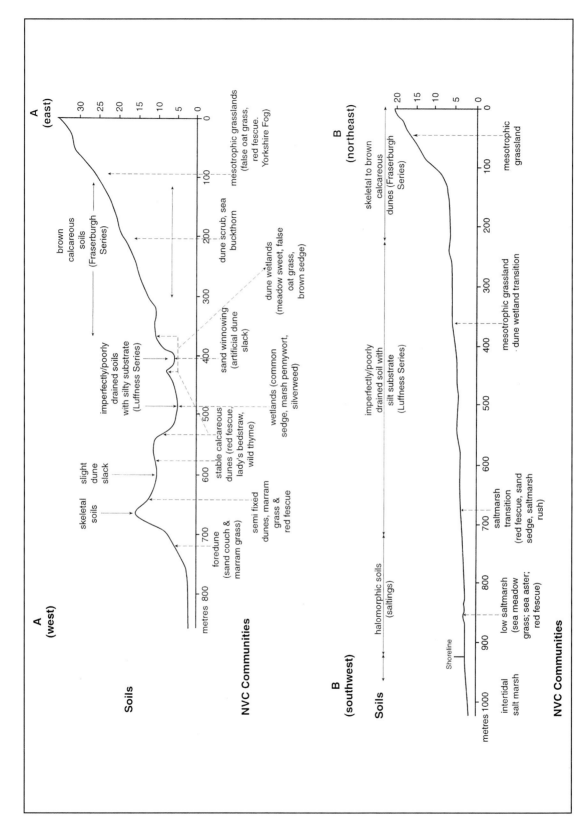

Figure 1.7 *Soils, drainage and plant communities of the coastal zone. National Vegetation Classification (NVC) communities are described in Rodwell (1991–2000).*

Another legacy of the ice age is the presence of deposits of fluvio-glacial sands and gravels along outwash channels such as the valleys of the North Esk, South Esk, Tyne and Almond, on which soils of the Darvel Association have developed: brown forest soils, gleys, podzols and alluvium. Around Dunbar, similar deposits have given rise to brown forest soils and gleys of the Eckford/Innerwick Association, while at the edge of the Lammermuir Hills around Garvald and Gifford there are freely drained soils of the Yarrow/Fleet association (Ragg and Futty 1967; ASH 1998).

The soils developed on shelly sand described above, and small areas of Winton and Darvel Series around Saltoun and Pathhead, where Carboniferous limestone occurs near the surface and the till is rich in calcium carbonate (Ragg and Futty 1967), provide an exception to the general rule that the soils of the area have become acidic through leaching.

Biotic Factors and Human Influences

Avery has summarised the relationships between biotic factors and soil characteristics as follows:

> The properties of a soil . . . are partly determined by the type of vegetation and other organisms it supports. Replacement of forest by grassland or heath, for example, causes significant changes in the moisture and temperature regimes as well as in the amount and nature of the plant remains deposited on the surface or added to subsurface horizons . . . In virgin landscapes, variations in vegetation and associated organisms generally accord with variations in climate, parent material and relief, and soil differences reflecting the independent nature of biotic factors are correspondingly limited in scope and extent . . . Wherever man has settled and managed the land, however, the natural systems have been modified or disrupted by his activities and the resulting changes are reflected in soil profiles, the upper parts in particular taking on new characteristics which depend on land use and management practices and more or less mask the effects of the primitive vegetation. (Avery 1990, p. 21)

Most of the lowland part of the Lothians is regularly ploughed, with continuous arable cultivation practised in the east, and in the wetter parts, rotation between arable and grassland or at least the ploughing and reseeding of grassland to improve the pasture. As was pointed out in the previous section, such cultivation mixes any naturally formed soil horizons (layers) to the depth of ploughing – typically 20cm. Soil organic matter in this mixed top layer typically decreases (through increased oxidation) after the initial disturbance, and reaches a new lower equilibrium level after some decades. Conversely, the organic matter content of soils returned to grass after arable use gradually increases. Loss of organic matter has caused a decline in the stability of soil crumbs and aggregates, and makes the soil more vulnerable to erosion. Regular use of nitrogenous fertilisers (up to 180 kg of nitrogen per hectare per annum on winter wheat, for example) encourages gradual acidification of these intensively used soils. Consequently, regular liming, usually to restore the pH to between 6 and 7, is practised on those soils (the great majority) which do not have natural sources of lime.

The soils of the East Lothian plain, including the Winton, Macmerry and Kilmarnock Series, are predominantly cropped with cereals (winter wheat and barley), oilseed rape and potatoes, and some of the highest wheat yields recorded in the UK have been obtained in East Lothian. The raised beach soils such as the Dreghorn Series to the east of Edinburgh were at one time much used for market gardening, but this has declined and been replaced to some extent by intensive field-scale cropping of brassicas and other vegetables. The proportion of land under grass and used for sheep or cattle grazing or silage grass production increases with increasing soil wetness towards the western part of the lowland area, particularly in West Lothian.

Much of the farmed lowland soil area has had under-drainage systems installed, to reduce the general wetness of the soil. This was first done about two centuries ago. Drainage, where it is effective, benefits crop growth in the arable areas and, importantly, it increases the number of days per year when the soils can be worked. In grassland, it reduces the 'poaching' damage done by livestock hooves. Another and much more drastic disturbance to some of the soils has been through the extraction of underlying coal, oilshale and limestone deposits, followed by the restoration of the land surface. Areas of hundreds of hectares near Tranent have been subjected to opencast coal extraction, and returned to agricultural use, while land near Dunbar was similarly restored after limestone extraction, although part has become a landfill. Much older deep mining of coal and oilshale in Midlothian and West Lothian resulted in the creation of large bings. The old shale bings of West Lothian, still with only patchy vegetation cover, remain as monuments to that era, and some opinion is against their removal. However, many of the coal bings have been removed (used in construction projects) or levelled. The latter activity has created challenging problems of recreating a soil on the top of inert and often very acid material, and much experimentation has gone on in an attempt to revegetate the ground and improve its amenity value.

Upland grazing land (especially on brown forest soils) has been subjected to improvement procedures designed to increase nutrient availability and plant and animal production. This has involved such practices as the addition of lime and phosphorus fertiliser, and grazing management. Compared with the low level of nutrient cycling under extensive hill management, surface improvements and intensified grazing have been shown to increase greatly the fractions of the nitrogen and phosphorus nutrient pools that are available to support fresh grass growth (Floate *et al.* 1973).

In the afforestation programmes of the second half of the twentieth century, previously uncultivated upland soils (those regarded as inferior to the upland pasture land) were brought under the plough in order to establish new forests. Here the disturbance was deeper than that created by ploughing for agriculture, because the objective was primarily to drain the upper layers, in a very wet environment, to provide better conditions for the establishment of young trees. Commonly, parallel strips of soil to a depth of 50cm or so were inverted by a special deep plough on top of the adjacent unploughed land (Thompson 1984). This resulted in 'buried profiles', with mineral soil at the top, then a double organic layer and then mineral soil again. After a first tree rotation of perhaps forty years, a new organic horizon has developed on the surface, making the profile even more complex. One effect of mature plantations of species such as Sitka Spruce (*Picea sitchensis*) on these wet upland soils is to lower the water table considerably, compared with that prevailing under the normal herbaceous vegetation (King *et al.* 1986). This in turn is likely to result in a loss of soil organic matter, through increased aeration and therefore oxidation. The main areas which have undergone developments of this type are in the Southern Uplands, outwith the Lothians, but examples of upland coniferous forestry within the Lothians can be found on the north side of the Pentland Hills, south-west of Balerno, and in the Silverburn-Penicuik area of Midlothian.

Conclusion

A number of issues affecting plant-soil relationships may be identified from this overview of the environment and soils of the Lothians. Gordon (1994) has emphasised several key points including critical loads on soils and the need to construct sensitivity indices for measuring disturbance and for monitoring change. Other edaphic issues that raise concern include acidification, soil pollution and rehabilitation, soil erosion, the impact of land use change on both soils and plants, changes in hydrological systems and soil moisture. There is undoubtedly a need for greater awareness of the dynamic nature of soils and thus the significance of disturbance for future generations. Amongst the possible changes which may arise in the future are:

1. The cessation of intensive agricultural activities on some hill grassland and in some wetter areas in the lowlands, brought about by changing economic circumstances and (for the wet areas) a consequent lack of investment in under-drainage. This could provide an 'environmental bonus' in that there could be an eventual restoration of botanical diversity, although this may take a long time because of the inhibitory effect of high nutrient levels, particularly phosphorus (Ford and Younie 1996), which will decline only slowly as a result of immobilisation in unavailable forms.

2. If current set-aside policies (putting at least 15 per cent of arable land out of production) should give rise to a more or less permanent reduction in cultivated land of this order, and if suitable financial incentive schemes materialise, there

could be a substantial increase in the area of broadleaved woodland on good rather than on very marginal soils. This would lead to a restoration of soil organic matter levels and improvement in soil structure, in addition to the improved amenity value. It is possible that considerations related to the Kyoto Protocol on global warming, allowing carbon sequestration by afforestation to be used to offset carbon dioxide emissions from fossil fuels, will provide an additional stimulus to move in this direction.

3. There are also climate change-related trends pointing in the opposite direction. Increased mean temperatures are likely to lead to a decline in soil organic matter. Peat and other highly organic soils are more fragile than their mineral counterparts, particularly with respect to erosion and oxidation following drying (DoE 1996). However, some projections point to an increase in precipitation in northern Britain, along with a rise in temperatures, which could offset much of the potential damage to peat soils. Also, contrary to the trends outlined in 1. and 2. above, there could be a switch of land from grassland to arable in cooler parts of the UK, such as the Lothians, to compensate for projected reductions in productivity associated with increased droughtiness further south. Under such a scenario, organic matter levels in soils affected could be reduced by 50 per cent in twenty-five years due to this land use change (DoE 1996).

4. Future trends in nitrogen deposition give some cause for concern. In 1989–92, the total deposition of nitrogen from the atmosphere was estimated to be in the range 10–15 kg nitrogen per hectare per annum (UK Review Group 1994). The future level of deposition will very much depend on the success with which agricultural emissions of ammonia and releases of nitrogen oxides from combustion are controlled. Known impacts of enhanced nitrogen inputs to natural soils include changes in species composition, increased acidification, but also increased tree growth (UK Review Group 1994); some or all of these outcomes may be detectable in years to come.

Soils are often considered to be stable, permanent features of the landscape whilst plant communities are recognised as reacting rapidly and more visibly to pressure. This chapter emphasises the continuous, variable and frequently highly dynamic character of many soil properties. Within the long-term spatial patterns and secular trends of soil and landscape evolution, there are mid- and short-term changes in soil-plant relationships that need to be taken into account. Over and above these relatively regular processes, there are less predictable catastrophic events such as storms or floods, which can set soil formation and plant growth off in a new direction.

2

Fossils and Pollen: Indications of the Past Plant Life of the Lothians up to the Dawn of Agriculture

W. J. BAIRD

A Firm Hold upon the Land

The present understanding regarding the origin of land plants is that they evolved from green algae during the Ordovician geological period 490 to 440 million years ago. By the time the Silurian rocks of the Pentland Hills were being laid down over 400 million years ago, plants already had a firm hold upon the land. Their light veil of green along braided water courses or estuaries would have been the only sign of life in an otherwise barren landscape. We know little of these earliest Lothian plants other than that they were probably still very small and that they grew in or near water. They had already developed the essential mechanisms which would enable them to continue their invasion of the land. These were a waterproof membrane (cuticle) with pores (stomata) to regulate water loss from tissues, and a system of tubes (tracheids) which allowed water and nutrients to be transported around within the plant.

Although their hold upon the soil-less Silurian landscape looked tenuous, plants had in fact consolidated earlier landings. Based upon a study of fossil plant debris including spores (Wellman and Richardson 1993), it is suggested that the flora was already well established, abundant and geographically widespread. However, as only ten species of spores were identified, it suggests that the vegetation comprised only a few different forms. Despite their small size and apparent frailty, plants were to conquer the land in succeeding geological periods.

The Missing Pages

In the following geological period, the Devonian (410 to 355 million years ago), represented in Scotland by rocks of the Old Red Sandstone, plants evolved from being the small, sparse covering in an otherwise empty land, to an occupying body of many and varied forms, obviously here to stay. Although there is little in the way of Old Red Sandstone rocks in the Lothians, and even those present are poor in fossil plant remains, we know from other sites that tremendous changes were taking place in the land plant world. This is clearly revealed by the detailed scientific study that has been carried out on the material from the hot spring silica deposits at Rhynie in Aberdeenshire. At this site there is detailed preservation of internal and external structures which has enabled reconstruction of how these Old Red Sandstone plants grew and reproduced. *Asteroxylon* seems the most advanced genus described from the site. It has leaves and lateral sporangia with features characteristic of both Lycopods and Zosterophylls. *Rhynia* is a more primitive leafless plant with terminal sporangia. It is also evident from the exquisitely preserved remains at this site that, as plants invaded the land, they took with them their camp followers from the world of the primitive arthropods and attendant fungi. Fossil material collected from around the world has opened a window into the rapid and extensive expansion of the plant world during this period. By the end of the Old Red Sandstone times, 355 million years ago, although there were still small plants, there were also tree-like forms many metres high. Several more types had appeared during this period; there were lycopods, horsetails, ferns and early gymnosperms. These were certainly growing in the Old Red Sandstone landscape of the Lothians, but the geological conditions did not favour their preservation. We have to wait until the next geological period to see, from the Lothians, what amazing plants this long period of development brought forth.

An Abundance of Green

The processes of geology were kind to this area during the Carboniferous period, 355 to 290 million years ago. Still drifting north, that part of the Earth's crust later to become the Lothians was almost on the equator during the whole period. A land mass lying to the north of the present Lothians was drained by large rivers, shallow seas occupying what is now the Lothians. Sediments were being deposited along large river channels and in the deltas and shallow seas. In the seas themselves, limestones of varying structure and origin were being formed. However, most important of all was the deposition of vast areas of swamp containing masses of fallen vegetation. These swamps would, through the processes of compaction under the weight of overlying sediments and the passing of time, become the coal basins of the Lothians. The mainly sedimentary and often fossiliferous rocks of the Carboniferous period in the Lothians are divided into groups; the Dinantian (equivalent to the Carboniferous Limestone Series of England), the Namurian (equivalent to the Millstone Grit Series) and the Westphalian (equivalent to the Coal Measures) (McAdam and Clarkson 1996). The geographical distribution of these rocks is shown in Figure 2.1.

The great advance that plants made at the beginning of the Carboniferous period was not the development of many new and different forms, but the evolution

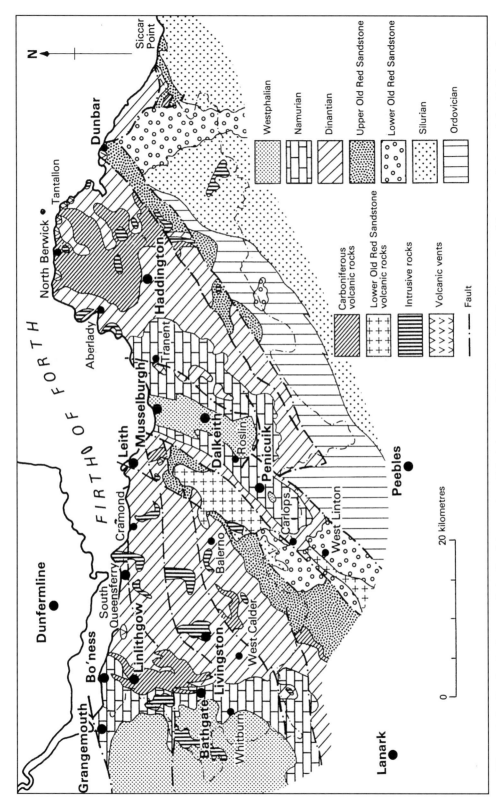

Figure 2.1 *Geological map of the Lothian region (McAdam and Clarkson 1996). Reproduced by permission of the Edinburgh Geological Society.*

and spread of those existing forms that were best fitted to the new habitat. In the Carboniferous period, conditions for plant growth were good; anyone who looks at the coal seams and the fossil evidence would concede that point. The familiar diamond-shaped scale leaf attachment pattern of *Lepidodendron* (Figure 2.2) is very common in the Coal Measures as a fossil recording the presence of branches or trunks of tree-sized plants. However, *Sphenopteris* (Plate 17, top) which probably represents more than one species of the foliage of early seed plants, and the arche-typal early seed *Genomosperma* are also common. The amount of plant material required to provide the great coal seams, all created by a chemical process powered by sunlight, is probably so huge as to be beyond accurate measurement. Fossils of the plants which formed coal were not best preserved in the coal, they were more often found as compressions in the shales or sandstones overlying the coal. On a few rare sites and under special conditions they were found as petrifactions within the coal. In other rocks where local volcanic conditions had mobilised the minerals silica or calcite, fine details of internal structures and even cells were preserved.

The early coal miners of the Lothians must have been the first to find fossil plants on a regular basis as they worked the many small mines with pick and shovel. The most common of coal fossils, often occurring in the seat earth just below the coal, is the root known as *Stigmaria*. This pitted, semi-cylindrical rock must have been the first plant fossil that many a child found on the old coal bings which dotted the landscape. The fossil had of course carried out the functions of a root for the arborescent lycopods of the Carboniferous forest. Its often crushed internal cylinder had transported water drawn from the surrounding soil through the root hairs once attached to surface pits. The miners were an inquiring and intelligent group who had an eye for anything unusual and different that they might find during their labours. In fact, a study of the acquisition registers of the National Museums of Scotland shows that some miners were so perceptive of the scientific treasures of the mines that they made a business of selling their finds, which depended on their stoneworking skills and intelligent eye.

Perhaps the first local fossil plant find that caught the imagination of the general public, was the discovery of giant tree trunks in the sandstone of Craigleith quarry. The first of these was discovered in 1826 and described by Henry Witham who gave it the name *Pitus* (Long 1979). The stem of this fossil tree was lying in a nearly horizontal position within the sandstone of the quarry and measured 36 feet long, with a diameter of 3 feet at the base of the stem. A further even larger speci-men was discovered in 1830 with a length of 59 feet, and a basal diameter of 6 feet. This specimen, now known by its full binomial of *Pitus withami*, was partially erected in the open air at the Royal Botanic Garden Edinburgh where it can be examined by those who wish to get an idea of the sheer size of these ancient plants. It is interesting to note that *Pitus* was not a tree as we would understand it in the modern botanical sense, but probably an arborescent pteridosperm. Local pteri-dosperm cupules have been much researched by A. G. Long (Plate 17, bottom).

Many people became fascinated by the study of geology and particularly that

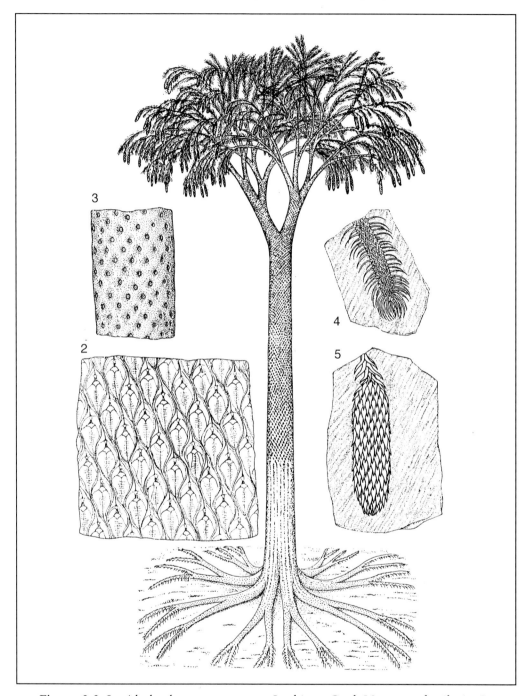

Figure 2.2 Lepidodendron, *a common Lothians Coal Measures fossil. 1. An artist's impression of the whole plant which would have been up to 50m in height. 2. Detail of trunk surface. 3. Detail of root surface. 4. Tip of branch. 5. Cone. (2–5 not to scale). Figures by courtesy of the National Museums of Scotland.*

aspect of the science which related to fossils. It is only possible to mention a few, but perhaps most people will be familiar with the name of Hugh Miller, the Cromarty stonemason, and his work in the field of palaeontology. He conveyed his knowledge to the public through the medium of several books on the subject, for example, *The Old Red Sandstone* (1841) and *Footprints of the Creator* (1849). Miller was familiar with the occurrence of fossils in the quarries from his early training as a stonemason. It may therefore have been that common bond that enabled him to get access to superb plant fossils from the limestone mines at Burdiehouse. The limestone at this location was unusual in that it preserved the foliage of the plants that had been swept into an area where the limestone was being deposited. These fossil branches and leaves could sometimes be recognised as being fertile spore-bearing fronds with occasionally the fertile cones of the early lycopod *Lepidostrobus*. Several of the specimens, which Hugh Miller described and illustrated in his books, are in the collections of the National Museums of Scotland.

Perhaps one of the most important palaeobotanists of the late nineteenth century was Dr Robert Kidston. In his address to the Royal Physical Society in Edinburgh in 1893 he summarised the results of his detailed researches on the fossil flora of the Carboniferous rocks. As a result of the studies by Kidston and others, great steps were taken in the identification, naming and classification of fossil plants. By 1910, and the publication of the second edition of the Memoirs of the Geological Survey of Scotland entitled *The Geology of the Neighbourhood of Edinburgh*, over 100 species of plant would be entered in the general list of Carboniferous fossils. One could be excused at this time for thinking that perhaps most kinds of fossil plant had been found and most of the things that could be said and done about them had already been completed. Yet in the second half of the twentieth century an even greater period of scientific discovery in the field of palaeobotany would take place in the Lothians.

Making a Peel: 3-D before Computers

During the great period of nineteenth-century research, if fossil plants were preserved in such a way that anatomical detail was present, then the palaeobotanist could always make a microscope slide. This technique consisted of first cutting a thin slice of rock containing plant material with a diamond-coated saw blade. The slice was then glued to a small sheet of glass and ground so thin, using carborundum powder, that light would shine through it. When this slide was examined under the microscope, very fine detail could be seen. However, the system required technical skill and was wasteful of specimen material, as most was simply ground away by the carborundum.

A better method of examining the finely preserved specimens was used by Dr Albert Long in his important work on the Lower Carboniferous material from sites in the south of Scotland, especially those in East Lothian. Using a technique known as the peel method, Long brought new insights to the study of fossil plants particularly to the way in which plants evolved.

The Peel Method

The peel method involved cutting across the fossil material with a diamond-toothed saw, grinding the resultant cut smooth with carborundum grit, and then etching the smoothed face with dilute acid. When the acid had etched the surface of the rock, the specimen was dried. The solvent acetone was then poured onto the etched face and immediately on top of that was added a clear cellulose sheet. When the acetone was dry the cellulose sheet could be peeled from the rock removing a thin layer of the fossil. The resultant peel could be examined under the microscope just like a glass slide. The technique was easier to learn and less wasteful of material as it could be used time and again over a short distance by simply re-grinding the etched face and repeating the process. Using this technique up to 180 peels per inch could be achieved from the material being studied.

Between 1959 and 1987 Long wrote more than thirty papers on his discoveries, many being published in the Transactions of the Royal Society of Edinburgh. Among several of those referring to fossils from the Lothians were one on a new genus and species *Tristichia ovensi* in 1961 and another, with Barnard in 1973 on petrified stems and associated seeds. Although later workers did not necessarily agree with all his views, his papers were important, and not just for his detailed observations. The impetus of his work brought together a younger dynamic group of palaeobotanists from both Britain and France to work on the material from Long's sites and other sites. From East Kirkton in West Lothian to Wardie Shore in Edinburgh and the coast at Weak Law and Oxroad Bay in East Lothian new specimens were collected and studied. In a series of papers this new gathering of scientists would refine, advance and sometimes even question Long's ideas, giving new and detailed insights into the structure and development of Carboniferous plants. In particular they would shed new light on the way in which seed plants developed during this time.

At the end of the Carboniferous period there existed a flora where many of the groups we know today had already evolved and were indeed important elements of the plant world. Some such as the calamites and arborescent lycopods had achieved structural elements such as woody tissues which their descendants would not retain. The early gymnosperms were already a major component, perhaps even more so in the unrecorded upland areas. Before we return to the ancient plants of the Lothians, ages would pass and flowering plants would come to dominate the new landscape.

What happened in the Mesozoic and the Tertiary?

There was a definite change in the climate during the time when the Carboniferous period was drawing to a close and the Permian was beginning. The conditions became drier and more continental and evidence for this can be seen in those parts of Scotland, such as Mauchline in Ayrshire, where the Permian rocks are represented by lithified desert dunes. These Permian rocks are not represented in the Lothians, other than by minor igneous intrusions. Indeed from the beginning of the Permian period 290 million years ago until the start of the Quaternary period 1.6 million years ago, fossil-bearing rocks are not found in the Lothians. We go from the Carboniferous with its abundant and occasionally exquisite fossil preservation to nothing for 300 million years. What happened? I think it is important first of all to say that rocks were being laid down and fossils preserved in other areas of Scotland. We know this from both the Annandale and Elgin areas where many tracks and some skeletal fossils of early reptiles, perhaps including ancestors of dinosaurs, have been found. These finds show that even in the desert sands of the Permian and Triassic, life was continuing. In the Jurassic rocks of north-east Scotland near Golspie the quarrying of sandstone and limestone produced further evidence of plant evolution particularly the finding of remains of cycads. The Tertiary, 65 to 1.6 million years ago, was a period in Scotland when great out-pourings of basaltic lava covered large areas in the north-west Highlands and Inner Hebrides. In the brief periods between these lava flows, soils formed on their weathered surfaces and plants grew before the next flow would engulf them. These Tertiary plant fossils are the first to show a modern flora with maples and conifers appearing with the ginkgos. But in the Lothians, we have no record of these rocks or the plant evolution which we know was taking place elsewhere. The reason is quite simple: the formation of sedimentary rock requires deposition, then protection by overlying sediments, then lithification. Even when this process is complete, if the resultant rocks are then exposed to weathering, they may be all worn away to be redeposited elsewhere, for example in the North Sea. Thus the preservation of rocks from any period is dependent on a fortuitous combination of circumstances and the inclusion of a fossil within them is a rare event. Why this process of deposition was interrupted in the Lothians is probably explained as being due to uplift and then consequent weathering of the land surface.

The Quaternary Period (1.6 million to 10,000 years ago)

The melting of the ice after the most intense period of the Devensian ice sheet reached its maximum, some 18,000 years ago, marks a period when abundant records of Lothian plant life are resumed. The effect of glaciation on the underlying rocks is to grind them away producing sand, gravel and rock flour. Although these deposits look unpromising, they are actually often relatively high in available plant nutrients and if climate conditions are equable, they can be relatively quickly colonised. It is therefore probable that by the time the Lothians area was free of ice, some 13,000 years ago, the first cleared areas were well covered in an arctic flora. The sources for these plants were probably across the land bridge that linked Britain to Continental Europe although it is possible that in certain favoured areas a few hardy plant species hung on throughout the whole period of the Devensian glaciation. The nature of this early post-glacial flora is confirmed by material collected from late glacial deposits at Corstorphine and Redhall which contains, amongst other plants, those which we would now associate with a more arctic-alpine flora e.g. *Salix herbacea, Salix polaris, Dryas octopetala, Betula nana, Empetrum nigrum, Oxyria digyna*. It is interesting that these deposits also contained remains of water plants such as *Potamogeton*, and plants such as *Viola palustris* and *Stellaria media* which are typical of open grassy areas.

The transition from an early post-glacial flora of arctic-alpine plants would have been fairly rapid in the Lothians as is shown by pollen analysis (see Chapter 3). The natural progression towards a more varied and perhaps tree-based flora may also have started in the area before the return of the ice to some highland areas for a brief 500 year period called the Younger Dryas, or more commonly in Scotland, the Loch Lomond re-advance, between 11,000 and 10,300 years ago. After this period, the climate gradually became drier and warmer leading to a Climatic Optimum between 8,000 and 5,000 years ago. During the early post-glacial period, birch and probably juniper spread rapidly throughout the area. As the climate ameliorated, other species began to replace birch wherever suitable sites became free. It is important to understand that the subsequent spread of trees such as pine, alder, oak, elm, hazel and holly took place in a gradual manner, each species occupying those places where it could establish a foothold. Tree species were also sensitive to climatic change and it is probable that alder made strong advances in the forest flora, particularly along river valleys during the warm but wetter Atlantic period 7,000 to 5,000 years ago. It is interesting to read some texts in which the appearance of oak, for example, is treated as though the trees came as an invading army driving the established but lesser native species from the land. Caution should be used in attributing ages and titles to periods within the Holocene. The description of, and explanations for, changes in the Scottish climate at this time must be regarded as at an interim stage and also in a state of flux (Whittington and Edwards 1997).

With the continued rise of sea levels after the melting of the glacial ice, it was only a matter of time before the North Sea would flood, probably around 8,000 years ago. From then on, there was no longer a land bridge to the continent of Europe.

Until the regular appearance of humans, the vegetation of the Lothians was allowed to establish itself by natural methods and maintain itself to its best advantage. The appearance of Mesolithic hunter-gatherer people around 10,000 years ago would initially make little change. True they would burn wood in their fires and use it to make shelters and tools, but it is unlikely that their numbers were such as to make a major impact on the flora. The details of what constituted that flora can be seen in such publications as *Plants and People in Ancient Scotland* (Dickson and Dickson 2000). This is a most useful compendium of detailed information on post-glacial plant materials in Scotland. From the work of the Dicksons and others, we can learn at least one thing about our early ancestors in Scotland. They were inordinately fond of hazelnuts. It would seem likely that perhaps hazel was the first tree to be deliberately coppiced by our ancestors as the resultant stems from a vigorous stool often form long, strong and straight poles ideal for spear hafts.

The Neolithic peoples, probably in the Lothians from before 6,000 BP, had a greater reliance on a settled place of habitation than did the constantly travelling Mesolithic hunters and would have brought a much more vigorous attention to the surrounding flora. It is at the change from pure livestock farming to a mixed arable farming and grazing system that we see new changes in the flora. Detailed studies of the archaeological dig at Ratho, Midlothian (Smith 1995) have shown the appearance of oats, barley and possibly wheat. The use of firewood in a settled location would very quickly create open areas where grassland would replace trees. When these peoples also started to maintain grazing animals, there would have been a deliberate policy of gathering branches for winter feed. In addition of course, the grazing stock would selectively eat young trees at the seedling stage. Once stock-keeping had commenced in an area such as the Lothians, then a chain of events had been started which would eventually end in an open farmed or grazed landscape, not woodland. How long such a process might take depends on the number of people involved, but possibly in East Lothian towards the end of the Neolithic Period, 4,500 BP, there would have been few trees left and, of those which remained, their ownership, use, and future would have been constantly disputed. Further treatment of the impact of man on the forest will be found in Chapter 3.

3

Moorland, Wetlands and Forest in the Lothians: The Growing Influence of Humans

P. M. SMITH

The previous chapter has discussed the establishment of a land flora in the Lothians, as shown by fossil evidence, through a distant time when the Forth was a tidal lagoon sedimented with peaty ooze, up to the dawn of agriculture. Chapter 18 considers, in a much more recent period, the broad impact of changing land use, especially with respect to agriculture. But what was the origin of local forest and moorland? What was the natural plant cover like when men first began to live locally? How do science and recorded history show it to have changed since?

The Beginnings of Forest: Post-glacial Vegetation and the Climatic Optimum

After the retreat of ice from the area, perhaps 12,000 years ago, rapid plant colonisation occurred. As the Pentland summits emerged from the ice sheets, there were valley glaciers in the Lammermuir Hills, while an ice-free area appeared around Penicuik. Ice retreated from the Esk Valley to the north-east and to the west. Anderson (1967) calculates that at that time the ice was receding at a rate of one mile every sixteen years.

The thin but mineral-rich soil in cracks and fissures between exposed rocks was occupied at first by mosses, then by calcicolous herbs. This was the kind of vegetation, containing arctic-alpines, that we now see as characteristic of the micaceous schist-derived soils of the Ben Lawers range. There, the soft rock continues to fragment and release minerals, maintaining a high base status in the soil. In the Lothians at that time there would have been much Mountain Avens (*Dryas octopetala*) and perhaps Dwarf Birch (*Betula nana*), until the easily removable minerals in rocks and gravels were depleted by plant growth and leaching. Only in places where calcareous sandstone outcropped, or clay and humus colloids existed in quantity, would high mineral content remain. A generally calcifugous ground flora would have swiftly succeeded the arctic-alpine phase. Post-glacial climatic amelioration would also have tended to make the climate less favourable for arctic-alpines.

In the deeper soils, the pioneer trees would then have established – birch, willow and aspen among them. In lighter, sandier soils, and perhaps in more upland sites, pine would have been frequent. Charlesworth (1957) thought that the northward advance of these trees would have been of the order of two or three hundred metres a year. More recent estimates suggest a faster expansion, taking into account advance islands of chance establishment which later coalesce.

In parts of the Lothians where the glacial rock debris was overlain by boulder clays there would have been many areas of impeded drainage (Cadell 1913; Anderson 1967). In these, a moorland flora would have dominated, perhaps with patches of forest where the till was well-drained or mineral-rich. This landscape was thus one of moor and bog, generating layers of peat. There were numerous marshes and lakes (Cadell 1913) (Figure 3.1). Rivers and streams that once were fed by the Pentland glacier continued to run in the valleys and gorges they had cut in periglacial times. The Braid Burn is an example of one of these. They ran into the Forth, the estuary of which had been formed millions of years earlier by a large river draining an extensive land surface to the north and west.

Anderson (1967) concluded that by 7,000 BP a mixed-leaf forest had formed over much of the Lothians, except where altitude, soil factors or hydrology were inimical. It was thus never a continuous forest, either in space or time. It thinned and retreated, later to expand again, as revealed so clearly by analyses of pollen taken from bog and lake cores (Figure 3.2). These analyses record the continuous rain of durable, identifiable pollen grains on to surfaces where decay was inhibited by anoxic conditions. The pollen content of different strata reveals much about the vegetation cover of an area and how it changed over time (Godwin 1975; Ingrouille 1995; Dickson and Dickson 2000). A snapshot of 7,000 BP would show a local forest probably dominated by oak, birch and hazel. Pine characterised upland sites, and lowland areas of low nutrient status, where it out-competed broad-leaved trees. Ash occurred especially where calcareous sandstone rocks outcropped. Alder was a common pioneer in wetter places, especially around lakes and rivers. Quoting work by Geikie (1866) and Hunter (1883), Anderson (1967) suggests the existence of an oak-ash-hazel woodland in the Lothians, with Goutweed (*Aegopodium podagraria*), which he clearly does not believe to have been only a Roman introduction (Godwin 1975), as a principal forest-floor herb. Hawthorn, elder and yew are found as occasional sub-fossils in the Edinburgh area. Juniper (*Juniperus communis*) was probably frequent, Rowan (*Sorbus aucuparia*) too, though the timber of the latter rots so readily that little sub-fossil material has so far been detected. There was Holly (*Ilex aquifolium*) and Bird Cherry (*Prunus padus*). These early forest remains ('bog-oaks') were often found during eighteenth- and nineteenth-century estate operations, when mosses were being cleared and

Figure 3.1 *The old lochs of Edinburgh, as presented by Cadell (1913). These old lochs, now largely drained and so disappeared, or greatly reduced in area, had this extent and distribution in immediately post-glacial times. There will have been others throughout the Lothians, perhaps particularly in the south and west. (After Cadell 1913.)*

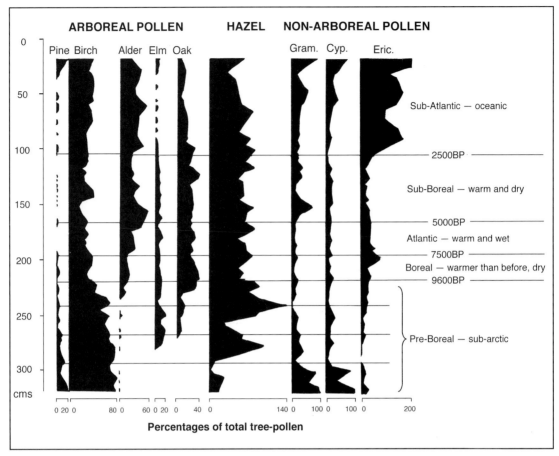

Figure 3.2 *Pollen diagrams showing the proportions of arboreal and non-arboreal pollen in a 300cm core taken in the upper Eddleston Valley, south-east Scotland. Note the decline in the proportion of elm and oak pollen in the Sub-Atlantic period as the proportion of grass (Gram.), sedge (Cyp.) and ericaceous (Eric.) pollen increases. The proportion of pollen from wetland trees (birch and alder) remains more or less constant. (Modified from the figures of Newey 1967.)*

drained for agriculture. The tree remains lie under many layers of peat, often on a clay or gravel base.

Oscillations of climate ensured that forest and moorland/bog were advancing and retreating at different times (Figure 3.2). The peat cover developed in wetter periods when bog advanced and forest shrank back. The peak of the ancient Lothian forest cover was in the Atlantic period (7,000 to 5,000 BP) as defined by Blytt and Sernander. Known as the Climatic Optimum, it was warmer and wetter than it has since been: lime, elm and hazel became more widespread – this is the time from which oak remains have been recovered from as high as the 1,000m contour of the Cairngorms (Anderson 1967).

Plant Cover in the Stone, Bronze and Iron Ages

In the beginning of this period of natural afforestation, perhaps almost 10,000 years ago, people came to the Lothians, albeit in such small numbers that forest expansion was unchecked (Anderson 1967; Piggott 1955; Dickson and Dickson 2000). These were Mesolithic people who lived by hunting the game of the forest, by gathering food plant items and by fishing. With tools of bone, horn and stone they could not make much impact on the natural vegetation except by occasional, possibly accidental, fires. Mesolithic tribes existed as small, mobile family groups who had, it should be recollected, a far wider local area over which to wander, hunt and camp than we now see around us. To the east and south, substantial areas of what is now the North Sea were dry land, in a phase of forest expansion. An illuminating indication of this period was the trawling up by the fishing boat *Colinda*, of a tree from the bed of the North Sea. Such submerged trees are well-known – on Dogger bank for example – but this one had a Mesolithic bone arrow-point embedded in its trunk.

Neolithic people occupied the Lothians from about 6,250 BP to 3,000 BP. Always in small numbers (Anderson estimates 20,000 in the whole of Britain), they nevertheless began an assault on the forest which has continued ever since. They were more or less settled farmers, increasingly successful at expanding the area in which they could grow crops or pasture animals. Coastal, estuarine and lakeside settlements would have been the bases for a slash-and-burn agriculture. Forest clearings made by axe and fire gave temporarily fertile plots for cereal cultivation. Later, more forest would be burned. Plenty of oak charcoal remains exist from that period. The gathered plant proteins would have come from acorns and hazelnuts, which Anderson (1967) speculated may even have been cultivated. Late Neolithic people had metal tools that they obtained by trade.

Expanding agriculture, inevitably followed by population increase and further encroachment upon the wild, has been the most important factor affecting the natural plant cover of the Lothians up to the present day (see Chapter 18). Early signs of agriculture are shown by pollen analysis of lake and bog cores. The pro-portion of tree pollen declines while that of herbs characteristic of open land increases. Many familiar ruderal species are represented among the latter. These incoming plants were exploiting the pastures and fields so laboriously created as the forest was removed from the better, and more accessible, lowland soils. They grew around field margins and they contaminated crops. As soil fertility was unwittingly improved by the addition of manure to pastures and middens (only later to arable plots), so there was an invasion of herbs that flourished only where soil nitrogen and phosphorus levels were augmented. Nettles began to do well.

People of the Bronze Age (about 4,000 to 2,250 BP) and Iron Age (about 2,500 to 1,900 BP) increased the rate of forest clearance. They were more permanently settled, had better metal tools for cutting, and had greater use for forest products. They used pine and oak charcoal to smelt metals, and some of them built wooden crannogs. Thinner forest, on dry, lowland sites, was attacked first. This further assault, opening up both forest and moorland, coincided with the natural decline of forest in the Sub-Boreal and Sub-Atlantic periods, following the Climatic Optimum. The forest suffered from the effects of climatic deterioration. Peat growth accelerated again as natural drainage worsened. Forest regeneration would have been inhibited by sheep grazing in upland areas, and perhaps by pig pannage in existing woodland.

Plant Cover in Roman Times (AD 79–409)

Roman occupation of the Lothians saw further forest clearance. Energetic marching from place to place needed access roads through forests and bogs. Tree trunks and branches laid over the bog (corduroy roads) achieved both purposes, and their remains are still occasionally uncovered. Cobbinshaw Bog lies partly over what looks like the remains of a birchwood: in the old branches, four feet below the surface, Roman coins have been found (Anderson 1967). 'Shaw' is Scots for a wood. Roman records of the campaign by Severus (AD 207–11) show that much forest still existed locally, but was being removed. Hanson (1997) suggests that 17–30 acres would need to be clear-felled to provide timber enough to build a single Roman fort. Forest was the haunt of the native enemy: clearance had the further merit of denying sanctuary. The Romans needed barley and oats for their horse-regiments (Ramsay and Dickson 1997) so extra agricultural acreage was in demand. Agricultural technology advanced: the Romans introduced the plough. Galen refers to the local soldiery being employed on draining marshes.

A substantial wooded area would nevertheless have remained at the end of the Roman period, though it would have been less continuous than before. Dickson and Dickson (2000) suggest that the best oak woodland would by then have been sparse. Chalmers (1887) speculates that a local West Lothian and Midlothian tribe – the Gadeni – derived their name from the many groves that 'added both strength and ornament to their various country'. Groves are isolated woodland areas, their presence implying non-continuous forest cover.

Post-Roman Changes

The pattern of an increasingly open, pastoral landscape, with forest patches ever wider apart, was reinforced by Saxon activity and influence. Angles and Saxons began to dominate the Lothians, previously the Bernicia province of Roman Northumbria, about AD 457 (Anderson 1967). Romanised North Britons, the Votadini, were displaced by Germanic invaders. After about AD 600, the Lothian area was part of the powerful Germanic kingdom of Northumbria, later subjected to incursions by the Danes (Mackie 1964). Anderson notes that there is little historical record of what happened to plant life during the subsequent Dark Age, but quotes a traditional verse (antiquity uncertain) to show local awareness of local forest decline:

> Calder Wood was fair to see,
> When it went to Camiltrie. [Camilty]
> Calder Wood was fairer still
> When it went o'er Crosswoodhill. [southern end of Pentland Hills]

'Calder' is often thought to be derived from a word for river, but it is as likely to come from 'coed', a woodland.

Agriculture became more intensive under the local version of the Feudal System (1097–1400) and there was an increase in woodland loss. Monastic houses, particularly the Cistercians, controlled huge areas of arable land, growing mainly cereals. Anderson (1967) quotes Sir John Skene as noting critically in 1641 that, unlike England, Scotland had no verderers – officers who controlled forest and punished unauthorised woodcutting. By this time, timber was clearly at a premium.

In and near Edinburgh, the most notable patch of forest, probably with the castle ridge as its northern boundary, was the Forest of Drumsheugh. This was the forest in which legend states that, in about 1128, King David was about to be attacked by a hart when a cross appeared in his hand, and the animal fled. This was on Rood Day (Good Friday) and underlies one of the theories of the origin of Holyrood Abbey (Smeaton 1904). Whatever the facts of that, the forest seems to have existed then. Royal Charter 146 of the year 1142 granted woodland for timber and pannage in the Pentland/Moorfoot forest, held by the king. Hay (1700) states that, before 1153, 'Roslin was at that time a great Forrest [sic], as also Pentland Hills, and a great part of the country about, so that there did abound in those parts great numbers of harts, hynds, deer and roes, with other wild beasts'. Legal records of grants of land with standing timber exist in East Lothian from 1160. The Statistical Accounts for Scotland show that tall oaks were swept away by a great flood at Haddington in 1358. Saltcoats, in Dirleton parish, was said to contain a large area of woodland, in which there were boars. Boece (1465?–1531) (see Boece 1938) recorded that the castle (then the town, essentially) was surrounded by 'ane gret forest'. There are many historical indications that the Lothians were still substantially forested at that time.

The early fifteenth century saw the 250-year cool period of the British climate that has been called the 'Little Ice Age' (Grove 1988). Parry (1975) shows that, by 1600, the upper arable limit in the Lammermuir Hills had dropped by over 200 metres. From about the beginning of this period, Lothian records show increasing attention to the amount of available wood, though the details are sometimes conflicting. Anderson (1967) records that Pope Pius II wrote of Midlothian in 1450: 'A sulphurous stone dug from the earth is used by its people for fuel. This stone is burned instead of wood, of which the country is destitute'. This was an exaggeration, certainly, but pressure on the remains of the Forest of Drumsheugh was great. Wood was being used for shipbuilding, charcoal, gunstocks, wheels, mills – and firewood, whatever Pius II thought. A sixteenth-century forestry record shows that sixty-two loads of wood were taken to Edinburgh Castle from Caerkettle (near Borthwick) (Anderson 1967). This would have been for charcoal for military purposes. In the early sixteenth century, timber was in demand for building Edinburgh malting houses, so that the East Lothian barley surplus might be gainfully processed. This implies a considerable cereal acreage in those drier parts of the Lothians. Though shortages of wood are reported, timber extensions to houses, bringing them forward into the street, were fashionable. John Knox's house dates from this period.

Two almost contemporaneous sources give what is probably an accurate picture of the countryside and plant cover in the Lothians before the beginning of the Industrial Revolution. In 1558, when Mary, Queen of Scots married the Dauphin, we can read that the great gun of Edinburgh Castle was fired to mark the occasion (Fraser 1976). The barrel was trained towards Granton and 'fired over the cornfields'. So the forest had gone. The gunstone, incidentally, is stated to have weighed a quarter of a ton, and to have landed on Wardie Muir, one-and-a-half miles away. Gunnery experts must evaluate the likelihood of that. In 1550, Alexander Alesius described Edinburgh and the Lothians in Sebastian Munster's *Cosmographica*, published in Switzerland. He wrote that it was 'a countryside of extremely fertile, pleasant meadows, little woods, lakes and streamlets' (Daiches 1978). Alesius probably under-records the extent of the southern marshes and moorlands, knowing little of such landscapes at home.

Conflicting records of forest cover continue. Writing of the battle of Lousie Law (Lussielaw) in 1571, the Reverend Thomas White records in his account of Liberton Parish that the Boroughmuir of Edinburgh was full of aged oaks (Anderson 1967). Yet in 1598, Fynes Moryson wrote: 'I rode to Edenborrow [Edinburgh] seated in Lodonay [Lothian], the most civill [sic] region of Scotland, being hilly and fruitful of corne [sic], but having little or no wood' (Brown 1891).

This mixture of surviving woodland patches, arable fields and marshy places is revealed by a study of Lothian place names (Wood 1989). Roslin means 'a morass at a pool'. Dalkeith means 'wood meadow'. Bathgate is 'house in the wood'. Currie is 'a boggy plain'. Berwick, as in North Berwick, means 'barley farm'.

Thus, if we ask when the local forest disappeared, there can be no simple answer. There was Stone Age and Iron Age thinning. The Romans used much timber and greatly increased forest access. The Germanic period and later the Anglo-Norman periods of influence over the Scottish monarchy saw further substantial erosion, though with some maintenance of Royal Forest for hunting. Records of timber extraction in the fifteenth and sixteenth century are accompanied by clear evidence of shortages and an increasingly open countryside.

The Recent Period (Seventeenth to Twentieth Centuries)

Timothy Pont's map of the Lothians (1636), (Plate 4), based on a somewhat earlier survey, offers one of the first views of the Lothian area in this period. It was largely farmland with interspersed estates and 'farm toun' settlements, the city of Edinburgh dominating to the north. The scene is familiar but not quite the same as today. Moorland and bog were extensive in the south. Forest was restricted to small patches of woodland. Laurie's map (1766) shows the ravine woods along the rivers particularly well (Plate 18). The old lochs of Edinburgh (Cadell 1913) (Figure 3.1) were still extant. These were the remains of post-glacial sinks of meltwater, probably smaller and fewer than they had earlier been, because of natural drainage, silting and seral succession. Nevertheless, water and waterside habitats were an important feature for local plant cover. In Edinburgh, parts of Craigcrook, Lochend and Duddingston Lochs remain, but Gogar Loch, south-west of Corstorphine, was drained in 1766. Corstorphine Loch itself, extending from Corstorphine to Haymarket, was drained, partly in 1670, completely in 1837, using a ditch – The Stank – dug for that purpose. The Water of Leith still occasionally floods into some of these old lake beds. Holyrood Loch was a large, triangular loch, now long gone: the now-ruined Abbey of Holyrood was built on the site of it. The Borough Loch lay to the south of the city. Many other lochs would have existed in the Lothians, among the glacial debris of 12–13,000 years ago. Holyrood Loch drained into the River Tumble, parts of which were finally covered up by sewerage works as recently as the 1920s. The Nor' Loch, on the site now occupied by Waverley Station, was an artificial loch dating from 1340–50, fed partly by a branch of the River Tumble, and draining out through the Dalry Burn into the Water of Leith at Roseburn. The Borough Loch was partly drained in 1722 by Thomas Hope of Rankeillor, who managed to convert it into a marsh. It ceased to be a marsh about 1840, and became Hope's Park (Hope Park), now known as The Meadows.

All around were ponds and even more 'mosses'. Old estate records reveal them. Edinburgh itself was poorly drained: the seventeenth-century citizenry often complained of evil-smelling 'dubs'. Much of this still existed when Maitland wrote his *History of Edinburgh* (1753). Botanists will note wryly that, while he says nothing of the plant cover, he gives a list of the fish most commonly caught in the Borough Loch.

The first formal plant records were now being made (see Chapter 11), and Greville's *Flora Edinensis* of 1824 reveals that an extensive wetland flora still existed in the city. Similar but unrecorded plant cover probably existed around the mosses and lochs in other lowland parts of the three vice-counties. Greville mentions '*Aira aquatica*', with decumbent stems, six feet long, in Restalrig Meadows, and in The Meadows, especially along the central walk. This was *Catabrosa aquatica*, which is certainly not there now. Nor does *Isolepis setacea* now grow in the Braid Hills Marshes, for they have now become a golf course.

'*Poa fluitans*' (*Glyceria fluitans*) grew along slow streams in The Meadows and in the King's Park (Holyrood Park). *Potamogeton natans* and *P. densus* grew in 'the water-pits of Corstorphine Hill'. *Peucedanum palustre*, then charmingly called the Marsh Milky Parsley, is gone from 'Colinton Marsh'. *Scutellaria galericulata* (Skull-cap) was common over local marshes: now it is rare and declining. *Corallorrhiza trifida* (Coral-root Orchid) grew among willows at Ravelrig Bog, Balerno (Ravelrig Toll Moss), which was also a site for the Dioecious Sedge, *Carex dioica*. There is no bog there now, just some badly drained fields. *Ceratophyllum demersum* (Rigid Hornwort) no longer 'flourishes in Canonmills Loch' – the loch has gone. But Dechmont Reservoir was built, and there today is Rigid Hornwort. Though osiers (*Salix viminalis*) are still to be found in Edinburgh, we can no longer say that they range over 'the marshes and stream banks' of Craigcrook.

Industrialisation, development, quarrying and urbanisation all required land and drainage. Musselburgh Harbour was built in 1749; the Union Canal was opened in 1822; the 'Innocent' Railway from Edinburgh to Dalkeith first ran in 1831. Leith Docks were constructed from 1806 to 1902. The Forth Rail Bridge opened in 1890. Harlaw and Threipmuir balancing reservoirs were built in 1843. Though these developments destroyed some habitats, they created new ones, and brought in new plants. The dockland aliens are reviewed in Chapter 14. One of them is *Allium paradoxum*. This striking feature of Lothian roadsides in spring, came to Leith Docks in ship ballast. It was unknown in the Lothians in 1824. The railways were accompanied by their own flora. Yellow Toadflax (*Linaria vulgaris*) is now a common sight along railway lines and in marshalling yards. We read that in 1824 it was sometimes seen on cultivated field margins, and that Rosebay Willowherb (*Chamerion angustifolium*) was rare.

The agricultural changes reviewed in Chapter 18 have had clear floristic consequences. Often these arise from better agricultural hygiene – roguing and seed-cleaning particularly. The cornfields of 1824 were contaminated with Corn Cockle (*Agrostemma githago*) and Cornsalad (*Valerianella locusta*). Corn Cockle is gone; Cornsalad is now scarce, in marginal habitats. Rye Brome (*Bromus secalinus*), a rye and wheat mimic, has gone from the fields, now appearing only as an irregular component of cheap amenity grass seed. The Corn Buttercup (*Ranunculus arvensis*), now a rare casual, grew in the Stockbridge cornfields of 1824. *Avena strigosa* (Bristle Oat) was a regular find in 1824: at least this has been replaced as an item of agrostological interest by the vigorous and wholly admirable (to a botanist) weed species, *A. fatua* – the so-called 'Wild Oat' – which was unrecorded in 1824. Corn Gromwell (*Lithospermum arvense*) was common, now is not. The 1824 cornfield poppies were 'common' and though they are less so today, they occasionally still produce a spectacular display (Plate 6). The Shepherd's Needle (or Venus' Comb) – *Scandix pecten-veneris* – now unknown locally, was 'common' in the Lothian cereal fields of the nineteenth century. In 1824, Greville congratulates local farmers on having almost completely obliterated the Corn Marigold (*Chrysanthemum segetum*) from this area. It was then generally an annoying weed.

Now it no longer annoys, so pleasing farmers, but is certainly not obliterated, so pleasing botanists.

It is essential to recognise that change is not inherently deleterious to floristic richness or habitat diversity. This is emphatically shown by the results of the present survey, which are discussed in Chapters 16 and 17. Though wetland and woodland generally have declined over the historical period here reviewed, it seems that wooded areas in the form of shelter belts have survived or even increased (Fowler 1967), offering narrow but important corridor habitats to associated woodland species. The oil shale industry bequeathed unsightly bings (see Chapters 15 and 18), but these are now rich habitat opportunities for colonising plants (Plate 15). The need for public open space, which certainly contributed from the seventeenth century onwards to wetland loss, was refined in the twentieth century to the concept and establishment of country parks such as the John Muir Country Park near Dunbar, Vogrie Country Park in Midlothian, and Almondell and Calder Wood Country Park in West Lothian. Some of these were once private estates which have now passed into enlightened public ownership. They will, over time, prove to be conserving and enriching features of the local plant cover. Old lochs were drained but new reservoirs and a canal – another important corridor habitat – were created. Dunbar limestone began to be exploited for cement as recently as 1963. Livingston – Scotland's fourth New Town, created by statute – originated in 1962. There will be botanical consequences – Livingston reduces pressure on Edinburgh's Green Belt. The plant life of the Lothians is never unchanging: it varies all the time. The future may witness its dynamic response to global climatic change, to further expansion of housing, to declining agricultural acreage and possibly even to on-shore oil extraction.

4

Microfungi of the Lothians

S. HELFER

Introduction

Microfungi constitute an integral part of any flora. Like the macrofungi they may be either beneficial or detrimental to the higher plants in any community. Very few, if any, are neutral. Unlike the macrofungi and most of the green plants, microfungi are normally not very obvious and, as a consequence are not very well documented for the Lothians. In the following article, some microfungi in the Lothians will be presented in an illustrative rather than documentary manner, the aim being to encourage all botanists to recognise, collect and record microfungi alongside plants. There is no shame in collecting a specimen with fungal leaf spots or distortion caused by a fungus. In fact, microfungi can tell their own story, and aid in the identification of plants, as well as confirming the place of plants as parts of a natural environment.

Whilst most modern authors (Winterhoff 1992) consider fungal communities separately from plant communities, there is some merit in an integrated approach, especially where there is an intimate and specific relationship between plants and fungi, such as in biotrophic, mutualistic or parasitic associations (Hirsch and Braun 1992; Helfer 1993). The potential impact of microfungi on vascular plant communities can be illustrated by the devastating effect of *Ophiostoma novo-ulmi* on elms in lowland Scotland and elsewhere (Redfern 1977).

Definitions and Scope

There is no agreed definition for what constitutes a microfungus. Most mycologists favour this term for those fungi, whose bodies are small and can only be appreciated using optical aids such as hand lenses or microscopes, as opposed to the macrofungi, which are fleshy and easily recognisable (Ellis and Ellis 1997; Hawksworth *et al.* 1995; Triebel 2001). Arnolds (1992) defines macrofungi as those 'Fungi forming reproductive structures (. . .), which are individually visible with the naked eye, that is, larger than about 1mm'. All remaining fungi are regarded as microfungi. There is therefore no systematic rank attached to the term, and this can lead to members of the same genus being treated as either microfungi or macrofungi, solely depending on the size of their reproductive structures (Arnolds 1992). However, some fungal taxa (of various ranks) can be considered as exclusively microfungal. They include the hyphomycetes, coelomycetes, rusts, smuts, downy and powdery mildews, as well as the exobasidiales and various groups of ascomycetes. There is a sizeable grey area, and some authors include the slime moulds (Jüelich 1994) and even some agaricales (Triebel 2001). Some authors have also added ecological or methodological aspects to the definition of microfungi (Arnolds 1992; Winterhoff 1992). Consequently there is a wide range of taxonomic groups represented, and the author is not sufficiently familiar with many of these to do them justice. For more detailed treatment of the individual groups please refer to a modern textbook (for example, Kendrick 1992), dictionary (for example, Hawksworth *et al.* 1995, or Holliday 1989) or the specialist literature cited in these.

Even though microfungi occur on a wide variety of substrates (Ellis and Ellis 1988; Gams 1992; Gravesen *et al.* 1994), consideration is restricted here to those occurring on plants, and the main emphasis is on examples of biotrophic parasitic microfungi. Not all microfungi, however, are directly harmful to plants, and it should be pointed out that many plants have mycorrhizal relationships with members of the microfungi (for example, *Glomus mossae*, a common vesicular arbuscular (VA) mycorrhizal species found in the roots of many plants). In addition to being of direct, nutritional benefit, this relationship also protects plants against many (microfungal!) root pathogens.

Lothian Microfungi

Baas-Becking (1934, quoted in Gams 1992) stated the general nature of microfungal distribution when he said that 'Everything is everywhere and the environment selects'. The environment, of course, is composed of the host plants (if any), external factors such as temperature and moisture, and other, competing or synergistic micro-organisms. There are no complete reports of microfungi of any particular flora known to the author. Several accounts exist for smaller areas (for example, Wicken Fen, Silverside 2001) or of several other groups of fungi (Hawksworth 1997). In the Lothians we should therefore expect to find the majority of widespread and common microfungi which normally occur in similar environments. Whilst only very rudimentary data is available at the time of writing, it is to be hoped that this expectation will be confirmed, when better surveys and checklists become available. This is no small matter, as it is expected that there are around 5,000 species of microfungi alone in the Lothians. The taxonomy of the microfungi used here follows that of Hawksworth *et al.* (1995).

Plasmodiophoromycota (slime moulds)

- A single illustration is chosen from this phylum: the cause of clubroot of Brassicaceae, *Plasmodiophora brassicae*. It infects the roots of host plants, producing extensive swelling of the root tissues (hypertrophy and hyperplasia), making them ineffective for water uptake. Infected plants show stunted growth and wilting during the day (at low relative humidity) even when the soil is wet.

Oomycota (water moulds and downy mildews)
The following group has been illustrated here: Peronosporales (white rusts and downy mildews).

- *Albugo candida* the white rust of Brassicaceae, occurs commonly on *Capsella* and *Cardamine* and other crucifers, including crop brassicas and ornamentals such as wallflowers.
- *Phytophthora infestans* the late blight of potatoes is the important agricultural pathogen that caused the Irish Potato Famine in the 1840s. It still is very common on potatoes and other Solanaceae. Other species of *Phytophthora* cause major damage to other groups of plants, for example *P. cinnamomi* which causes root rots in many woody plants.

Ascomycota (sac fungi)
The following groups have been considered here: Taphrinales, Erysiphales, Rhytismatales, Dothideales, Ophiostomatales. The Ascomycota also contain a large number of macrofungi not discussed here, from cup fungi, such as the orange peel fungus (*Aleuria aurantia*) to morels (*Morchella* spp.) and truffles (*Tuber* spp.). Many species are the fungal partners in lichens.

- The Taphrinales are represented by a single genus, *Taphrina* with some ninety-five species world wide (Kurtzman and Sugiyama 2001). In the Lothians a number of these fungi cause galls and deformations on many trees and are common on birch and cherry (witches brooms), pear and poplar (leaf blisters) and peach (leaf curl).
- The Erysiphales cause the common powdery mildews on many angiosperm species (there are no powdery mildews on gymnosperms or pteridophytes, Braun 1987). Examples from the Lothians range from powdery mildew on (*Erysiphe aquilegiae* var. *ranunculi*) on *Ranunculus* spp., to powdery mildew (*Microsphaera alphitoides*) on oak, to the widespread mildews on grasses and cereals *Blumeria graminis* and legumes *Erysiphe pisi*.
- Representatives of the Rhytismatales in the Lothians grow on leaves, *Rhytisma acerinum* on sycamore, and *Rhytisma salicinum* on willow, causing conspicuous spots, commonly known as tar spot (Plate 7).
- There is a large number of very varied microfungi in the order Dothideales. The only representatives to be mentioned here are *Venturia inaequalis* (Plate 20) and *V. pirina*, the fungi causing scab on apple and pear respectively.
- *Ophiostoma novo-ulmi* (Ophiostomatales) is the pathogen causing the current epidemic of Dutch elm disease.

Basidiomycota (club fungi)

The following groups have been considered here: Uredinales (rust fungi) and Ustilaginales (smut fungi). Arguably the largest group of Basidiomycota are macro-fungi, ranging from the puffballs (*Lycoperdon* spp.) to the fleshy mushrooms and boletes. These are discussed in Chapter 5.

- The rust fungi (Uredinales) are parasitic microfungi on many species of vascular plants. Henderson (2000) has identified 268 (plus five additional species (Henderson pers. com.) species on 327 (plus one) vascular plant genera in Britain. Most rusts are more or less host plant specific. Some rusts are restricted to one or few host genera or even species, for example *Phragmidium rosae-pimpinellifoliae* on *Rosa pimpinellifolia*. Others are dependent on two unrelated host plants to complete their life cycle, for example *Gymnosporangium cornutum* on *Sorbus aucuparia* and *Juniperus communis* (Plate 20).
- Ustilaginales (smut fungi) are parasitic on many flowering plants, especially Poaceae, Cyperaceae and Caryophyllaceae. Most smuts attack the reproductive organs of the plants, either the anthers or the seeds and many are systemic in their host plant tissue (Ingram and Robertson 1999). Some smuts such as *Ustilago striiformis* produce symptoms in the leaves of grasses.

Mitosporic fungi (true fungi with no known sexual stage)

There are an estimated 15,000 species worldwide, of microfungi which have no known sexual state (Hawksworth *et al.* 1995). They represent a very varied group, belonging to diverse taxonomic affinities. The absence of a sexual state makes their classification difficult, although it is believed that many 'really' belong to the ascomycetes (Kendrick 1992). There are many leaf spot fungi and other parasites in this group as well as the Coelomycetes, the Hyphomycetes and saprophytic rots and moulds. Two examples are below:

- *Botrytis cinerea*, the grey mould of many herbaceous plants.
- *Alternaria* sp. causing leaf spots on herbaceous plants (Plate 20).

Conclusions

The microfungi represent a mixed and varied group of organisms with equally varied ecological function. They are widespread and frequent, many of them being of common, worldwide distribution. Whilst no database exists of the microfungi occurring specifically in the Lothians, it is expected that they are represented by a large number of species. Many microfungi are parasitic on plants, others are saprophytes or mutualists. Without microfungi the flora of the Lothians (or, for that matter, any other flora) would look very different indeed.

5

Macrofungi of the Lothians

R. WATLING

History of the Study of Fungi in the Lothians

The study of the mycota (fungus flora) of the Lothians has had a surprisingly long history and can be directly related to the growing interest during the period of the Enlightenment in Edinburgh. The first records to be traced are those in 1782 by James Smith, founder of the Edinburgh Natural History Society and of the Linnean Society of London; he was a student in Edinburgh under John Hope, the then Professor of Botany and Regius Keeper. James Smith recorded a handful of larger fungi and these have been highlighted by Watling (1986) in an account of the floristics of Scotland. A rather more comprehensive study was made by Robert Kaye Greville, in *Flora Edinensis* published in 1824. Greville must be considered to be the mentor of a then little-known bryologist, the Rev. M. J. Berkeley, who later became the 'Father of British Mycology'. Thus the Lothians are linked with the very development of British mycology.

Greville recorded both macro- and microfungi and several widely known species are originally attributable to him and therefore are connected to the Lothians, including the familiar Dead Man's Fingers (*Xylaria polymorpha*). All his records, for example that of *Hydnum spathulatum*, now called *Hyphoderma radula*, were incorporated into Stevenson's *Mycologia Scotica* (1879), which covered all the records then known of fungi for Scotland, arranged according to river watersheds. Thus Stevenson's records of fungi for the Lothians are mixed with those for Fife and even for other localities further west along the Forth. However, the Lothian specialities were pinpointed by reference to a specific locality, for example *Panus stevensonii* from Penicuik, named after Stevenson by Berkeley and Broome (1879), and now known to be the rather uncommon *Phyllotopsis nidulans*. A total of 333 fungi for the Forth are listed in *Mycologia Scotica* whilst a calculation made by Watling (1986) increased this to 753 larger fungi alone. Since Stevenson, the Lothians has been host to many amateur and professional mycologists, especially Malcolm Wilson who, although he specialised in rusts and smuts, knew a good deal about forest pathogens and larger fungi in general. His student, Douglas Henderson, who also had an all-round mycological background, continued this tradition and to the records of both these mycologists can be added those on plant pathogens amassed at East Craigs and the East of Scotland College of Agriculture, as it was then known. R. W. G. Dennis, who later became Head of Mycology at the Royal Botanic Gardens, Kew, added other records now deposited in the library at Royal Botanic Garden Edinburgh and there are further notes held by the Research Branch of the Forestry Commission. Consequently the Lothians mycota, especially of the larger forms is quite well documented.

Many records have resulted from the several forays held by the Cryptogamic Section of the Botanical Society of Scotland and, since 1996, by the South-East Scotland Fungus Group. A foray held by the latter group on the first Sunday of each month has produced several unusual finds. These finds have stimulated much interest amongst a whole group of diligent collectors and, thanks particularly to the efforts of Adrian Newton and his wife Lynn Davy, a very useful database has been developed which includes many old records also. Unfortunately, nothing has been brought together as a definitive mycota as yet, although this may change in the future. However, some clues can be given here as to the diversity of the more obvious fungi in the Lothians.

Habitats

Unfortunately, fungi do not follow the strict rules that can be imposed on the recording of flowering plants as they may not fit into any particular category. Thus a pine cone or oak leaf found a great distance from a pine or oak wood would probably harbour a range of fungi specific to that substrate, which should, it would have been thought, reflect the pinewood or oakwood ecosystem. The fruiting body of a mycorrhizal fungus may be found some distance into a field when in fact the roots of the host tree with which it is associated are metres away. Also, fruiting is erratic and so a species of macromycete may not fruit every year and therefore will not be recorded. Indeed, the time period of fruiting might be so short that even though it has fructified, it may not have been recorded because the right person was not in the right place at the right time. It may also only be seen fruiting once in a lifetime, though colonising the soil or substrate in the vegetative state, out of sight. Unfortunately, it is only possible for a floristic study to record fruiting bodies. It is therefore almost impossible, except only broadly, to tie a particular fungus to a particular category of ecosystem. For instance, the Cramp Ball (*Daldinia concentrica*) is well-known on Ash (*Fraxinus excelsior*) but may be found less frequently on other hosts; thus though *Fraxinus* could be said to be its preferred host, this fungus can be isolated from the surface-sterilised buds of unrelated shrubby plants where it exists as an endophyte. However, with all these restrictions taken into account some pointers can be offered on some unusual sightings. There is little doubt that the mycota of the Lothians has changed very little over the years with some of the more interesting taxa clinging on in small scraps of woodland or ravines.

Parks and Gardens

Probably the best sites to start with are parks and gardens with their wide range of hosts and habitats, from small city gardens to the lawns of such residences as Hopetoun House. The latter has been singled out because over twenty species of *Hygrocybe* were found there in the autumn of 1998 by a group of amateurs surveying grasslands in the Lothians. This makes this particular lawn of national importance, despite the fact that in 1999 hardly any fruit-bodies were found! Holyrood Park is also an important site for the distribution of the rare *Daldinia fissa* previously called *D. vernicosa* which grows on burnt Gorse (*Ulex europaeus*); there are no other records for this fungus in the Lothians and it has been speculated that the fungus may be spread by beetles which are attracted to burnt gorse. Even the smallest garden can produce the most extraordinary records. Thus a survey by Prof. F. Last of the gardens in the city and in East Lothian which have *Eucalyptus* spp. (Last and Watling 1998) has revealed the widespread presence of *Laccaria fraterna*, an ectomycorrhizal agaric probably introduced from Australia. Eucalypts have also been associated in Midlothian with the southern hemisphere truffle *Hydnangium carneum*. Gardens make up a rather important habitat for Lothian's macromycota, including fungal pathogens of ornamental plants.

With the introduction of new horticultural procedures and indeed new species of bedding plants from abroad and the popularisation of formerly less well-known phanerogams, the park and garden mycota has changed dramaticallly. The Morel (*Morchella hortensis*) (Plate 21, bottom) has occurred in a newly landscaped bed of heathers at Liberton Hospital and the Lorel (*Gyromitra esculenta*), is a familiar inhabitant of the wood chip mulch at the Royal Botanic Garden Edinburgh. Also found recently has been *Clitocybe pruinosa*, a white agaric with decurrent gills, associated with a newly purchased *Veronica* sp. by a resident of Juniper Green. *Gymnopilus dilepis* was recorded for the first time in Britain in a pot of plants from an Edinburgh supermarket. The commonest fungus, however, in pots of plants purchased from garden centres and supermarkets is *Leucocoprinus birnbaumii*, a delicate, bright-yellow mushroom. A much larger and potentially dangerous relative was found in the indoor plantings at Ainslie Park Recreational Centre, namely *Chlorophyllum molybdites*. This also is a first UK record of this toxic parasol mushroom.

An excellent demonstration of the importance of gardens in maintaining the diversity of fungi in the Lothians resulted from the Scottish Wildlife Trust Edinburgh Garden Survey (Watling 1997). The most noteworthy find in this survey was *Geopora cervina*, a large discomycete. This is possibly the second record of this cup-fungus for the UK. Other unexpected finds in the survey were the unfamiliar *Calocybe carnea* and *Rhodocybe caelata*, generally associated with base-rich grassland, and *Melanophyllum echinatum*, a cryptic-coloured agaric which may previously have been missed. It has also been found on alluvial soils near the entrance to the Almondell and Calder Wood Country Park at Pumpherston, as has one of the less common British Eyelash fungi (*Scutellinia cejpii*). The large discomycete, *Disciotis venosa*, which smells of swimming pools, is generally associated with limestone woodland but it has appeared in some numbers in a garden in Edinburgh. This and many other records indicate how some fungi may crop up in the most unexpected places, even close to human habitation.

Country Parks also have their range of specialities. Thus Vogrie Country Park produced the Dune Stinkhorn (*Phallus hadrianii*) and some fruit-bodies of the Ballet Dancer (*Hygrocybe calyptriformis*). Although the latter is a Red Data List species, it is also known from Morningside Cemetery, the golf course at Ravelston, and several sites in the Pentland Hills and in East Lothian: at Prestonhall near Pathhead over seventy-five fruiting bodies have been recorded on a single day. This beautiful pink wax-cap has been included in Edinburgh's Biodiversity Action Plan. Visits to Beecraigs Country Park have produced the rather unusual *Paurocotylis pila* over several seasons since the mid-1990s, previously only known from a quarry in Northamptonshire. In 1992, it was found in a potato patch in Yorkshire and sites were located in Binscarth Wood on Orkney. However, now it has appeared in other parts of northern Britain including an Edinburgh garden at Corstorphine and at the edge of a car park at Lauriston Castle. It was first described in, and has been probably introduced from, New Zealand (Watling 2001).

Parkland areas in the Lothians also have distinctive mycological features for they are habitat to the familiar Fairy Ring Champignon (*Marasmius oreades*) and the Hay-maker (*Panaeolus foenisecii*), the latter often the subject of mild anxiety at the Edinburgh 'Sick Kids Hospital' in the summer months after ingestion by toddlers. *Agrocybe temulenta* has been found at Inverleith; the related *Agrocybe paludosa* is known from Yellowcraig. One exciting find from Whitekirk Hill near Sheriff Hall is the rather rare wax-cap *Hygrocybe spadicea*. Because of its rarity throughout the UK it has been chosen by Scottish Natural Heritage for further study and monitoring.

The bogs of the Pentland Hills have yielded the rather unusual looking *Sarcoleotia turficola*, an arctic-alpine discomycete previously known only from Loch Maree, Ross-shire and Unst in Shetland. It has been found recently on the moors in Yorkshire.

Plantations and Woodlands

Plantations harbour a whole range of larger fungi and at Selm Muir Wood, M. Richardson charted the appearance of a range of species of agarics over a three year period with much startling data on fresh weight, numbers and persistence of fruit-bodies (Richardson 1970). The brightly coloured *Hygrophorus bresadolae* was found at Selm Muir associated with Larch (*Larix* spp.) whilst in contrast, under Pine (*Pinus* spp.) at Woodhall Dean, East Lothian, the Ear-pick fungus (*Auriscalpium vulgare*) has been located on pine cones; incidentally this last species was recorded by Greville (1824) as occurring in the Edinburgh area. A fungus which grows on the false truffle *Elaphomyces* is *Cordyceps ophioglossoides*, a relative of the 'vegetable caterpillars', and both have been found at Beecraigs Country Park. Generally, Lothian conifer plantations have a predictable and hardly inspiring range of taxa, although they can possess a fairly high mycodiversity.

In contrast to plantations, those fungi associated with Beech (*Fagus sylvatica*) wherever they may be planted either in windbreaks, woods or as single individuals, range from highly specific *Russula* spp. to the Porcelain fungus (*Oudemansiella mucida*). The last fungus is known from Bavelaw and Buckstone within the Edinburgh City boundaries and, along with *Hygrocybe calyptriformis*, noted above, and the Golden Sock (*Phaeolepiota aurea*), has been incorporated into the Edinburgh Biodiversity Action Plan for future monitoring. *P. aurea* is a large, striking, golden mushroom and has been found amongst a pile of leaves at the Hermitage of Braid, in Ravelston Woods and in the Royal High School grounds at Barnton.

One of the best local woodlands for fungi in the Lothians is undoubtedly Saltoun Big Wood probably because there is such a wide mixture of planted trees, possibly intermingled with some volunteer specimens of original species of the area, in addition to subsequent invasive trees such as birch and willow. This mixture gives some rather uncommon taxa which one usually associates with localities further north, for example, the milk-caps *Lactarius helvus*, *L. mammosus* and *L. trivialis*. The majestic *Hebeloma radicosum*, which appears to be associated with small mammal latrines and underground passages, is also a feature of this East Lothian site.

Veteran Trees

Old trees throughout the Lothians, either ornamentals or in the woodland community, are very interesting as refuges for wood-inhabiting fungi, especially the bracket fungi. The Dalkeith Country Park, which contains some very fine specimen oaks, is characterised by a whole range of typical oak forest inhabitants such as, the Oak Dryad's Saddle fungus (*Inonotus dryadeus*), the Beefsteak fungus (*Fistulina hepatica*) and *Gymnopus fusipes* (syn. *Collybio fusipes*) – all on trees, and also *Gyroporus castaneus*, in Scotland a decidedly rare oak-woodland bolete. The Horn of Plenty (*Craterellus conucopioides*) found at Roslin is also a good indicator of old oak woodland. *Inonotus hispidus* occurs regularly on old ash trees on the north side of Corstorphine Hill and *Sparassis crispa*, a fungus generally associated with old pinewoods, occurs at a single site in Cammo Country Park. A huge fruiting of the Death Cap (*Amanita phalloides*), a rare species in Scotland anyway, occurred in East Lothian in 2000 in a troop over 100 strong, an event which was duly recorded in the press, even with a coloured photograph!

Several ornamental trees were planted in Edinburgh during historic times, the most famous probably being the Corstorphine Sycamore. This fell down in the gales in early 1999, but examination of the base of the shank shows that it had already been weakened by a fungus, as good developments of the Artist's Bracket (*Ganoderma europaeum*) soon appeared and the stain in the heartwood in the remaining shank shows that it has been a colonist for a long, long time. This species and *G. applanatum*, which differs in its more pointed margin to the bracket, can be found on a whole range of tree species: one at Inverleith is on a flowering cherry, an unusual host for *G. europaeum*. *Volvariella bombycina*, a fungus related to the Paddy Straw fungus, a mushroom very frequently used in Chinese meals, fruited on old elms in Edinburgh during 1995, an unusual sight for mycologists north of the Border because of its southern distribution in the British Isles. It only goes to show that one should always be alert and on the look out as slight shifts in climate or changes in land-management can favour the fruiting of less common taxa. *Pleurotus dryinus*, an oyster mushroom with a ring, is also rather unusual and has been found on a street-side plane tree at Raeburn, Edinburgh and on beech at Bavelaw. Of course the most dramatic species of fungus to affect the Lothians in recent years has been *Ophiostoma novo-ulmi*, the agent which causes Dutch elm disease. It is a microscopic fungus but its damage to elms has encouraged the fruiting of other larger fungi such as the *Volvariella* and *Pleurotus* noted above.

The generally northerly-distributed bracket fungus, the Tinder fungus (*Fomes fomentarius*) has been seen on old birches near Livingston, and, on a cluster of old, well-established trees of Hornbeam (*Carpinus betulus*) in Colinton Dell, *Melanconium stromaticum* and the unusual resupinate fungus *Peniophora laeta*. The *Melanconium*, although a micro-fungus, can easily be seen as it produces extensive black smears all over the branches of the trees, whereas the *Peniophora* produces bright-orange, waxy skins often high up in the canopy. The little-recorded *Flagelloscypha punctiformis* has been found in the vicinity at Craiglockhart Dell. Although shaped like a disc it is a minute relative of the mushrooms and toadstools (agarics).

Roadside Verges and Railtracks

The Lawyer's Wig or Shaggy Ink Cap (*Coprinus comatus*) is a familiar sight along our highways. At Lizzie Brice's Roundabout near Livingston the rare *Melanoleuca turrita* has been collected but sadly the willows with which it was associated have been replaced by an industrial development. The fungi are often very unpredictable organisms and sometimes fruit in the most unusual places. *Agaricus bernardii*, a relative of the field mushroom, has formed great troops in some years on the side of the carriageway at Cramond Bridge.

A relative of the Morel, *Morchella elata*, occurred on the disused railway tracks at Silverknowes, and *Mitrophora semilibera*, another morel, in the tennis courts at Inverleith. One of the earthstars *Geastrum triplex* is commonly found along the old railway line at Corstorphine. This same earthstar fruited in hundreds in 1999 on the banks of a relatively recent conversion of a railtrack at Davidson's Mains accompanying several colonies of a Coral-fungus, *Ramaria bourdotiana*, previously unrecorded for the UK.

The Ink Cap, *Coprinus flocculosus*, has been found growing on wood along the Gunpowder Trail at Roslin, emphasising that by careful survey many small woodlands, even close to the City can reveal rare and unusual finds.

Sand-dunes

The sand-dunes in a rim along the Firth of Forth are rather exciting as they are home to several uncommon fungi. In East Lothian there have been findings of *Hohenbuehelia culmicola* fruiting on Marram (*Ammophila arenaria*), *Melanoleuca cinereifolia* and the Stalked Puffball (*Tulostoma brumale*). The last species was found first in the 1960s but has been found at several sites now in and around Gullane and Aberlady. *Lepiota oreadiformis*, a small parasol mushroom, features in the sandy, grassy promontories close to the Forth Railway Bridge, South Queensferry, while the minute *Entoloma nigella* occurs on the Aberlady Bay Nature Reserve. Both the sand-dunes and the woodland immediately inshore at Yellowcraig are well known for their earthstars – *Geastrum triplex* and *G. rufescens* are regularly seen there and in the past *G. pectinatum* has been collected. Yellow-craig is also known for the beautifully coloured *Leptonia rosea* (syn. *Entoloma roseum*) and an uncommon Slippery Jack (*Suillus fluryi*). From John Muir Country Park, the recently described *Coprinus ammophilae* has been recognised. Also seen there have been *Rhodocybe popinalis*, the tiny *Lepiota melanotrichus* (now called *Leucoagaricus melanotrichus*) and a small fairy club fungus, *Typhula*, growing attached to old seaweeds and not so far documented in any books!

Unusual Locations and Substrates

Some fungi have very specific habitat requirements. One such fungus is *Xylaria carpophila* which grows only on cupules of beech, but more unusual is *Tubaria autochthona* which grows on berries of Hawthorn (*Crateagus monogyna*). The first is widespread but the latter is much rarer, although it has been seen at Woodhall Dean, East Lothian. The two elf-cups *Geopyxis carbonaria* and *Trichophaea hemisphaerioides* occur on burnt patches at Saltoun and Pressmennan Wood both in East Lothian. Some of the really small agarics include *Pleurotellus graminicola* on grass blades, and *Lentinellus tridentinus* on old stems of Rosebay Willowherb (*Chamerion angustifolium*), the latter at John Muir Country Park.

The Morel, *Morchella esculenta*, which is found regularly in April and May in the stabilised dune-systems of East Lothian has also occurred on wood-chips in a garden at Blinkbonny, the location of the uncommon *Peziza echinospora* which was found fruiting under a prostrate juniper. This species is related to *Peziza adae* (= *P. domiciliana*) which was itself described as new to science from Inverleith in 1877 by Sadler. *P. adae* is found to be not uncommon in the UK, in contrast to the closely related *P. michelii*, recently found within the City boundary. *Clitocybe sadleri*, also found at Inverleith, was named after Sadler who was the gardener at the Royal Botanic Garden Edinburgh but it is really a sterile form of the Sulphur Tuft (*Hypholoma fasciculare*), the sterility probably being the result of very cold conditions affecting the spore development.

At Inverleith, as in many old properties throughout the area, *Serpula lacrimans* has caused dry-rot and there is a story, which cannot really be substantiated, that many generations ago, old, infected timbers were brought into Edinburgh for reuse. Minor damage can be caused to house timbers by the equally widespread Wet Rot Fungus (*Coniophora puteana*). A consignment of wooden kitchen utensils destined for a shop at Stockbridge was the source of a new species of *Athelopsis*, a white, resupinate fungus with a fruiting body like a cobweb.

Special Substrates

Dung fungi range from macroscopic, although sometimes very small, agarics to microscopic moulds. In between these extremes is a group of fairly large and there-fore easily seen cup (discomycetes) and flask (pyrenomycetes) fungi. These can all be studied with relative ease as it is only really necessary to have a dissecting microscope and a student's light microscope, and of course the required texts. They are a group of organisms which continue to interest mycologists and Richardson (1998) as part of a much wider programme has recorded 120 taxa from the Lothians, including many interesting records, for example *Ascobolus carletonii* and *A. hawaiiensis*, *Anopodium ampullaceum*, *Delitschia consociata* and *D. leptospora* and *Sporormiella octonalis*.

Summary

What does all this tell us about the fungi of the Lothians? The mycota, because of the nature of the area, is a mixture of many different elements but is extremely rich and characteristic of a mesophytic vegetation with fairly cool average tempera-tures. Much more has to be done, but the Lothians contain a whole series of rather interesting taxa. Even the small remnants of original vegetation demand attention. Indeed, we have become more aware of the potential of these fragments because of the diligence of the South-East Scotland Fungus Group. Thus there is, against a background of widespread and well-known taxa, an element which contains some unusual entities, some of which may be at their most northerly limits, whilst others are approaching their southern boundary.

A calculation can be attempted to estimate the number of larger fungi we might expect in the Lothians. At the moment, a database of fungi for the Lothians exists which includes nearly 4,500 records covering 1,238 species. The largest number of records is for the mushrooms and toadstools (518) excluding the boletes and russulas and their relatives, and the cortinariaceous fungi which number 197. With the polypores and larger elf-cups and their relatives, this figure rises to a little over 775. From a knowledge of the ectomycorrhizal fungi, as they normally seem to be in a relatively constant ratio to the other basidiomycetous elements, one would expect to find at least 890. There are about equal numbers of basidiomycetes to ascomycetes in most UK mycota which have not been intensively examined by microspecialists and this would give the total number of species of fungi, including the water moulds, slime moulds, lichens etc. to be about 2,250. If Hawksworth's highly provocative calculation for the fungus/vascular plant ratio (Hawksworth 1991) is taken as six, then we would expect 7,500 different fungi in the Lothians with 1,000 of those being macrofungi. We have a long way to go but the new records are coming in at an astonishing rate.

Acknowledgements

I am extremely grateful to Lynn Davy who has kindly keyed in the records of fungi that I have made from over forty years of foraying not only in the Lothians but in Scotland as a whole, and who has supplied me with the current information held in the database. I also wish to thank Mike Richardson who has freely shared his records with me. As always I am extremely grateful to all those field mycologists who have collected for me material of fungi over the years.

A Preliminary List of Non-lichenised Fungi recorded in the Lothians

Compiled by Roy Watling and Lynn Davy

This alphabetical arrangement parallels the British Mycological Society's database from which the authorities for the taxa can be sourced. The species recorded below are in the main collections made in the last forty years where full site data supports the entries; the majority are supported by voucher material deposited in the herbarium at the Royal Botanic Garden Edinburgh. Generally, records in the literature have not been sourced as many of the records require careful re-assessment in the light of recent, radical changes in fungal systematics.

ASCOMYCOTA
Diaporthales
Melanconidaceae
Calospora platanoides
Melanconis stilbostoma
Sydowiella fenestrans
Valsaceae
Diaporthe arctii
Diaporthe velata
Gnomonia setacea
Hypospilina pustula
Ophiovalsa betulae
Valsa ambiens
Valsella adhaerens
Diatrypales
Diatrypaceae
Diatrype bullata
Diatrype disciformis
Diatrype stigma

Diatrypella favacea
Diatrypella quercina
Eutypa acharii
Eutypella prunastri
Quaternaria dissepta
Quaternaria quaternata
Dothideales
Capnodiaceae
Capnobotrys dingleyiae
Coccodiniaceae
Dennisiella babingtonii
Cucurbitariaceae
Cucurbitaria laburni
Didymosphaeriaceae
Didymosphaeria fenestrans
Hysteriaceae
Gloniopsis praelonga
Leptosphaeriaceae
Leptosphaeria acuta

Lophiostomataceae
Lophiosphaera fuckelii
Melanommataceae
Melanomma pulvis-pyrius
Mycosphaerellaceae
Mycosphaerella rhododendri
Phaeosphaeriaceae
Ophiobolus acuminatus
Ophiobolus fruticum
Ophiobolus rubellus
Phaeotrichaceae
Delitischia consociata
Delitischia leptospora
Delitischia winteri
Preusssia funiculata
Trichodelitischia bisporula
Pleosporaceae
Rhopographus filicinus
Sporormiaceae

Sporormia ambigua
Sporormia minima
Sporormiella australis
Sporormiella bipartis
Sporormiella grandispora
Sporormiella intermedia
Sporormiella lageniformis
Sporormiella leporina
Sporormiella minima
Sporormiella octonalis
Venturiaceae
Venturia inequalis
Venturia pirina
Venturia rumicis
Incertae sedis
Didymella superflua
Elaphomycetales
Elaphomycetaceae
Elaphomyces muricatus
Erysiphales
Erysiphaceae
Blumeria griminis
Erysiphe aquilegiae var.
　ranunculi
Erysiphe artemisiae
Erysiphe asperifolium
Erysiphe cruciferarum
Erysiphe heraclei
Erysiphe pisi
Erysiphe polygoni
Erysiphe trifolii
Erysiphe ulmariae
Microsphaera alni var.
　extensa
Microsphaera alphitoides
Microsphaera azaleae
Microsphaera plantaginis
Oidium ericinum
Podosphaera minor
Podosphaera leucotricha
Podosphaera oxyacanthae
Sphaerotheca humuli
Sphaerotheca mors-uvae
Sphaerotheca pannosa
Uncinula bicornis
Gymnoascales
Gymnoascaceae
Gymnoascus reesii
Hypocreales

Clavicipitaceae
Claviceps purpurea
Cordyceps militaris
Cordyceps ophioglossoides
Epichloë typhina
Hypocreaceae
Chromocrea aureovirens
Creopus gelatinosus
Hypocrea citrina
Hypomyces lateritius
Nectriaceae
Nectria cinnabarina
Nectria coccinea
Nectria episphaeria
Nectria mammoidea
Nectria peziza
Nectria radicicola
Pyxidiophoraceae
Pyxidiophora microsporus
Leotiales
Dermateaceae
Hysteropezizella valvata
Mollisia cinerea
Mollisia dilutella
Niptera excelsior
Pezicula carnea
Pezicula rubis
Ploettnera exigua
Podophacidium
　xanthomelum
Tapesia fusca
Tapesia rosae
Trochila ilicina
Trochila laurocerasi
Geoglossaceae
Cudonia confusa
Geoglossum cookeianum
Geoglossum fallax
Geoglossum glutinosum
Geoglossum umbratile
Microglossum viride
Trichoglossum hirsutum
Trichoglossum tetrasporum
Hyaloscyphaceae
Dasyscyphus apalus
Dasyscyphus bicolor var.
　rubi
Dasyscyphus clandestinus
Dasyscyphus clavigerus

Dasyscyphus diminutus
Dasyscyphus dumorum
Dasyscyphus fugiens
Dasyscyphus mollissimus
Dasyscyphus nidulus
Dasyscyphus niveus
Dasyscyphus sulphureus
Dasyscyphus virgineus
Hyaloscypha aureliella
Hyaloscypha flaveola
Hyaloscypha hyalina
Hyaloscypha leuconica
Hyaloscypha stevensonii
Lachnellula calycina
Lachnellula hahniana
Lachnellula subtillissima
Lachnellula willkommii
Polydesmia pruinosa
Scutoscypha fagi
Unguiculariopsis nov. sp.
Zoellneria rosarum
Leotiaceae
Ascocoryne cylichnium
Ascocoryne sarcoides
Bisporella citrina
Bulgaria inquinans
Calloria fusarioides
Chlorociboria aeruginascens
Claussenomyces atrovirens
Claussenomyces prasinulus
Cudoniella clavus var.
　grandis
Cyathicula coronata
Cyathicula cyathoidea
Durella atrocyanea
Encoelia furfuracea
Heyderia pusilla
Hymenoscyphus caudatus
Hymenoscyphus fagineus
Hymenoscyphus fructigenus
Hymenoscyphus herbarum
Hymenoscyphus laetus
Hymenoscyphus repandus
Hymenoscyphus rokeby-
　ensis
Hymenoscyphus scutula var.
　suspecta
Hymenoscyphus scutula var.
　solani

Hymenoscyphus
  sublateritius
Leotia lubrica
Neobulgaria pura
Pezizella chrysostigma
Sarcoleotia turficola
Verpatina rufo-alutacea
Orbiliaceae
Orbilia alnea
Orbilia leporina
Orbilia vinosa
Orbilia xanthostigma
Sclerotiniaceae
Dumontia tuberosa
Lanzia luteovirescens
Monilinia johnsonii
Poculum firmum
Poculum petiolarum
Poculum sydowianum
Rutstroemia hercynica
Rutstroemia maritima
Vibrissiaceae
Apostemidium decolorans
Incertae sedis
Phialea subhyalina
Microascales
Incertae sedis
Ceratocystis ulmi
Onygenales
Onygenaceae
Onygena corvina
Onygena equina
Ophiostomatales
Ophiostomataceae
Ophiostoma nova-ulmi
Pezizales
Ascobolaceae
Ascobolus albidus
Ascobolus brassicae
Ascobolus carbonarius
Ascobolus carletonii
Ascobolus crenulatus
Ascobolus furfuraceus
Ascobolus hawaiiensis
Ascobolus immersus
Ascobolus stictoideus
Ascobolus vinosus
Ascozonus monoascus
Ascozonus woolhopensis

Iodophanus carneus
Saccobolus versicolor
Helvellaceae
Gyromitra esculenta
Helvella crispa
Helvella elastica
Helvella lacunosa
Helvella villosa
Morchellaceae
Disciotis venosa
Mitrophora semilibera
Morchella elata
Morchella esculenta
Morchella hortensis
Verpa conica
Otideaceae
Aleuria aurantia
Anthracobia macrocystis
Cheilymenia coprinaria
Cheilymenia fimicola
Cheilymenia raripila
Cheilymenia stercorea
Cheilymenia vitellina
Coprobia granulata
Coprotus albidus
Coprotus cf. lacteus
Coprotus granuliformis
Coprotus sexdecemsporus
Geopyxis carbonaria
Humaria hemisphaerica
Lasiobolus ciliatus
Lasiobolus cuniculi
Melastiza chateri
Otidea alutacea
Otidea bufonia
Otidea onotica
Paurocotylis pila
Scutellinia cejpii
Scutellinia scutellata
Scutellinia trechispora var.
  paludicola
Pezizaceae
Peziza adae
Peziza ammophila
Peziza badia
Peziza bovina
Peziza echinospora
Peziza emileia
Peziza fimeti

Peziza michelii
Peziza petersii
Peziza repanda
Peziza vesiculosa
Pyronemataceae
Geopora cervina
Pyronema omphalodes
Trichophaea
  hemisphaerioides
Thelebolaceae
Rybarobius polysporus
Thelebolus crustaceus
Thelebolus microsporus
Thelebolus nanus
Thelebolus stercoreus
Incertae sedis
Orbicula parietina
Polystigmatales
Phyllachoraceae
Glomerella cingulata
Phyllachora junci
Protomycetales
Protomycetaceae
Protomyces macrosporus
Protomyces pachydermus
Rhytismatales
Rhytismataceae
Colpoma quercinum
Cyclaneusma minus
Hypoderma commune
Lophodermium culmigenum
Lophodermium pinastri
Rhytisma acerinum
Rhytisma salicinum
Sordariales
Chaetomiaceae
Chaetomium bostrychodes
Chaetomium globosum
Chaetomium olivaceum
Coniochaetaceae
Coniochaeta discospora
Coniochaeta hansenii
Coniochaeta leucoplaca
Coniochaeta
  polymegasperma
Coniochaeta saccardoi
Coniochaeta scatigena
Lasiosphaeriaceae
Anopodium ampullaceum

Arnium hirtum
Arnium leporinum
Arnium mandax
Chaetosphaerella
  phaeostroma
Lasiosphaeria hirsuta
Lasiosphaeria ovina
Lasiosphaeria spermoides
Podospora appendiculata
Podospora cf. bifida
Podospora curvicolla
Podospora curvula
Podospora decipiens
Podospora minuta
Podospora intestinacea
Podospora pleiospora
Podospora setosa
Podospora tavisina
Schizothecium cervinum
Schizothecium conicum
Schizothecium curvuloides
Schizothecium glutinans
Schizothecium nanum
Schizothecium tetrasporum
Schizothecium vesticola
Melanosporaceae
Sphaerodes fimicola
Nitschkiaceae
Bertia moriformis
Calyculosphaeria tristis
Nitschkia grevillei
Sordariaceae
Sordaria fimicola
Sordaria humana
Sordaria macrospora
Sphaeriales
Amphisphaeriaceae
Phomatospora coprophila
Xylariaceae
Anthostoma decipiens
Daldinia concentrica
Daldinia fissa
Hypocopra equorum
Hypocopra merdaria
Hypocopra parvula
Hypocopra planispora
Hypocopra stephanophora
Hypoxylon fragiforme
Hypoxylon multiforme

Hypoxylon serpens
Nummularia lutea
Rosellinia aquila
Rosellinia ligniaria
Ustulina deusta
Xylaria carpophila
Xylaria hypoxylon
Xylaria longipes
Xylaria polymorpha
Taphrinales
Taphrinaceae
Taphrina amentorum
Taphrina betulina
Taphrina bullata
Taphrina deformans
Taphrina padi
Taphrina populina
Taphrina pruni
BASIDIOMYCETES
Agaricales
Agaricaceae
Agaricus aestivalis
Agaricus arvensis
Agaricus bernardii
Agaricus bitorquis
Agaricus campestris
Agaricus devoniensis
Agaricus edulis
Agaricus
  haemorrhoidarius
Agaricus langei
Agaricus macrosporus
Agaricus silvaticus
Agaricus silvicola
Agaricus subperonatus
Agaricus xanthodermus
Chamaemyces fracidus
Chlorophyllum molybdites
Cystoderma amianthinum
Cystoderma granulosum
Cystoderma jasonis
Lepiota alba
Lepiota aspera
Lepiota cristata
Lepiota felina
Lepiota friesii
Lepiota fulvella
Lepiota oreadiformis
Leucoagaricus holosericeus

Leucoagaricus melanotrichus
  var. fuligineobrunneus
Leucocoprinus birnbaumii
Leucocoprinus denudatus
Leucocoprinus
  liacinogranulosus
Macrolepiota mastoidea
Macrolepiota procera
Macrolepiota rhacodes
Melanophyllum echinatum
Amanitaceae
Amanita ceciliae
Amanita citrina
Amanita citrina var. alba
Amanita crocea
Amanita excelsa
Amanita fulva
Amanita inaurata
Amanita lividopallescens
Amanita muscaria
Amanita nauseosa
Amanita phalloides
Amanita porphyria
Amanita punctata
Amanita rubescens
Amanita spissa
Amanita vaginata
Amanita vaginata var. alba
Amanita virosa
Bolbitiaceae
Agrocybe arvalis
Agrocybe brunneola
Agrocybe dura
Agrocybe erebia
Agrocybe gibberosa
Agrocybe paludosa
Agrocybe pediades
Agrocybe semiorbicularis
Agrocybe temulenta
Bolbitius reticulatus
Bolbitius titubans
Bolbitius vitellinus
Conocybe aporos
Conocybe arrhenii
Conocybe arrhenii var.
  hadrocystis
Conocybe blattaria
Conocybe dunensis
Conocybe filaris

Conocybe kuehneriana
Conocybe intrusa
Conocybe lactea
Conocybe macrocephala
Conocybe magnicapitata
Conocybe mairei
Conocybe ochracea
Conocybe pubescens
Conocybe rickenii
Conocybe semiglobata
Conocybe sordida
Conocybe subpubescens
Conocybe tenera
Coprinaceae
Coprinus acuminatus
Coprinus ammophilae
Coprinus angulatus
Coprinus atramentarius
Coprinus cineratus
Coprinus comatus
Coprinus cordisporus
Coprinus disseminatus
Coprinus domesticus
Coprinus flocculosus
Coprinus friesii
Coprinus gonophyllus
Coprinus heptemerus
Coprinus hiascens
Coprinus lagopides
Coprinus lagopus
Coprinus leiocephalus
Coprinus luteocephalus
Coprinus macrocephalus
Coprinus micaceus
Coprinus miser
Coprinus niveus
Coprinus nudiceps
Coprinus patouillardii
Coprinus pellucidus
Coprinus plicatilis
Coprinus pseudoradiatus
Coprinus radians
Coprinus radiatus
Coprinus semitalis
Coprinus spilosporus
Coprinus squamosus
Lacrymaria glareosa
Lacrymaria pyrotricha
Lacrymaria velutina

Panaeolus acuminatus
Panaeolus ater
Panaeolus campanulatus
Panaeolus fimicola
Panaeolus foenisecii
Panaeolus rickenii
Panaeolus semiovatus
Panaeolus speciosus
Panaeolus sphinctrinus
Psathyrella ammophila
Psathyrella atomata
Psathyrella candolleana
Psathyrella chondroderma
Psathyrella coprobia
Psathyrella flexispora
Psathyrella gracilis
Psathyrella gracilis var.
    substerilis
Psathyrella hydrophila
Psathyrella microrhiza
Psathyrella obtusata
Psathyrella piluliformis
Psathyrella polycystis
Psathyrella prona
Psathyrella squamosa
Psathyrella subnuda
Psathyrella trepida
Psathyrella vernalis
Entolomataceae
Clitopilus prunulus
Entoloma anatinum
Entoloma andrianae
Entoloma atromarginatum
Entoloma caesiocinctum
Entoloma catalaunicum
Entoloma cetratum
Entoloma chalybaeum var.
    lazulinum
Entoloma conferendum
Entoloma cordae
Entoloma corvinum
Entoloma cuneatum
Entoloma excentricum
Entoloma exile var.
    pyrospilum
Entoloma fuscomarginatum
Entoloma griseocyaneum
Entoloma griseorubellum
Entoloma griseorubidum

Entoloma hebes
Entoloma incanum
Entoloma jubatum
Entoloma juncinum
Entoloma lampropus
Entoloma lazulinum
Entoloma lividocyanulum
Entoloma nidorosum
Entoloma nigella
Entoloma papillatum
Entoloma poliopus
Entoloma porphyrophaeum
Entoloma prunuloides
Entoloma roseum
Entoloma scabiosum
Entoloma sericellum
Entoloma sericeonitidum
Entoloma sericeum
Entoloma sericeum f.
    nolaniformis
Entoloma serrulatum
Entoloma tenellum
Entoloma turbidum
Entoloma undatum
Entoloma xanthochroum
Nolanea radiata
Rhodocybe caelata
Rhodocybe fallax
Rhodocybe popinalis
Hydnangiaceae
Hydnangium carneum
Hygrophoraceae
Hygrocybe
    aurantiosplendens
Hygrocybe calyptriformis
Hygrocybe cantharellus
Hygrocybe ceracea
Hygrocybe chlorophana
Hygrocybe chlorophana
    var. aurantiaca
Hygrocybe citrinovirens
Hygrocybe coccinea
Hygrocybe colemanniana
Hygrocybe conica
Hygrocybe conica var.
    conicoides
Hygrocybe flavipes
Hygrocybe fornicata
Hygrocybe glutinipes

Hygrocybe insipida
Hygrocybe intermedia
Hygrocybe irrigata
Hygrocybe laeta
Hygrocybe marchii
Hygrocybe miniata
Hygrocybe mucronella
Hygrocybe nitrata
Hygrocybe persistens
Hygrocybe persistens var.
   konradii
Hygrocybe pratensis
Hygrocybe pratensis var.
   pallida
Hygrocybe psittacina
Hygrocybe punicea
Hygrocybe quieta
Hygrocybe reidii
Hygrocybe russocoriacea
Hygrocybe spadicea
Hygrocybe splendidissima
Hygrocybe strangulata
Hygrocybe virginea
Hygrocybe virginea var.
   fuscescens
Hygrocybe virginea var.
   ochraceopallida
Hygrocybe vitellina
Hygrophorus bresadolae
Hygrophorus hypothejus
Hygrophorus pustulatus
Hygrophorus subradiatus
Hygrophorus tephroleucus
Pluteaceae
Pluteus cervinus
Pluteus nanus
Pluteus olivaceus
Pluteus salicinus
Volvariella bombycina
Volvariella pubescentipes
Volvariella speciosa
Strophariaceae
Hypholoma capnoides
Hypholoma elongatum
Hypholoma fasciculare
Hypholoma marginatum
Hypholoma subericaeum
Hypholoma sublateritium
Melanotus horizontalis

Melanotus phillipsii
Pholiota adiposa
Pholiota alnicola
Pholiota apicrea
Pholiota flammans
Pholiota gummosa
Pholiota highlandensis
Pholiota lenta
Pholiota lucifera
Pholiota myosotis
Pholiota scamba
Pholiota spumosa
Pholiota squarrosa
Psilocybe coprophila
Psilocybe crobula
Psilocybe inquilina
Psilocybe merdaria
Psilocybe montana
Psilocybe muscorum
Psilocybe pratensis
Psilocybe semilanceata
Stropharia aeruginosa
Stropharia caerulea
Stropharia coronilla
Stropharia cyanea
Stropharia inuncta
Stropharia pseudocyanea
Stropharia semiglobata
Tricholomataceae
Armillaria borealis
Armillaria gallica
Armillaria mellea
Armillaria ostoyae
Arrhenia retiruge
Calocybe carnea
Calocybe gambosa
Calocybe ionides
Calyptella capula
Calyptella campanula
Cantharellula cyathiformis
Cantharellula umbonata
Clitocybe brumalis
Clitocybe candicans
Clitocybe clavipes
Clitocybe cerrusata
Clitocybe connata
Clitocybe dealbata
Clitocybe dicolor
Clitocybe ditopus

Clitocybe ericetorum
Clitocybe fragrans
Clitocybe infundibuliformis
Clitocybe langei
Clitocybe metachroa
Clitocybe nebularis
Clitocybe obsoleta
Clitocybe odora
Clitocybe parilis
Clitocybe pruinosa
Clitocybe rivulosa
Clitocybe sadleri
Clitocybe suaveolens
Clitocybe vibecina
Collybia acervata
Collybia ambusta
Collybia aquosa
Collybia asema
Collybia butyracea
Collybia cirrhata
Collybia confluens
Collybia cookei
Collybia distorta
Collybia dryophila
Collybia erythropus
Collybia fuscopurpurea
Collybia fusipes
Collybia maculata
Collybia peronata
Collybia putilla
Collybia tuberosa
Crinipellis stipitarius
Dermoloma cuneifolium
Fayodia anthracobia
Flagelloscypha punctiformis
Flammulina velutipes
Hemimycena lactea
Hohenbuehelia culmicola
Hydropus floccipes
Laccaria amethystea
Laccaria fraterna
Laccaria laccata
Laccaria proxima
Lachnella alboviolascens
Lepista flaccida
Lepista nuda
Lepista personata
Lepista saeva
Lepista sordida

Lyophyllum connatum
Lyophyllum decastes
Lyophyllum fumosum
Lyophyllum infumatum
Lyophyllum loricatum
Macrocystidia cucumis
Marasmiellus ramealis
Marasmius androsaceus
Marasmius calopus
Marasmius cohaerens
Marasmius confluens
Marasmius epiphylloides
Marasmius hariolorum
Marasmius hudsonii
Marasmius oreades
Marasmius recubans
Marasmius rotula
Marasmius setosus
Marasmius splachnoides
Melanoleuca albifolia
Melanoleuca arcuata
Melanoleuca cinereifolia
Melanoleuca cognata
Melanoleuca exscissa
Melanoleuca grammopodia
Melanoleuca melaleuca
Melanoleuca oreina
Melanoleuca polioleuca
Melanoleuca schumacheri
Melanoleuca strictipes
Melanoleuca turrita
Merismodes anomalus
Micromphale inodorum
Mycena acicula
Mycena adscendens
Mycena aetites
Mycena alba
Mycena alcalina
Mycena alphitoides
Mycena amicta
Mycena capillaris
Mycena cinerella
Mycena citrinomarginata
Mycena clavicularis
Mycena clavularis
Mycena epipterygia
Mycena filopes
Mycena flavescens
Mycena flavoalba

Mycena floridula
Mycena galericulata
Mycena galopus
Mycena galopus var.
　candida
Mycena galopus var. nigra
Mycena gypsea
Mycena haematopus
Mycena hiemalis
Mycena inclinata
Mycena leptocephala
Mycena leucogala
Mycena longiseta
Mycena metata
Mycena mirata
Mycena olivaceomarginata
Mycena oortiana
Mycena pelianthina
Mycena polygramma
Mycena pseudocorticola
Mycena pullata
Mycena pura
Mycena pura var. lutea
Mycena pura var. rosea
Mycena rorida
Mycena sanguinolenta
Mycena speirea
Mycena stylobates
Mycena tenerrima
Mycena vitilis
Myxomphalia maura
Nyctalis parasitica
Omphalina ericetorum
Omphalina galericolor
Omphalina grisella
Omphalina oniscus
Omphalina pyxidata
Ossicaulis lignatilis
Oudemansiella mucida
Panellus mitis
Panellus serotinus
Panellus stipticus
Phyllotopsis nidulans
Pseudoclitocybe
　cyathiformis
Pseudoomphalina
　compressipes
Rickenella fibula
Rickenella setipes

Ripartites metrodii
Ripartites tricholoma
Strobilurus tenacellus
Tephrocybe ambusta
Tephrocybe anthracophila
Tephrocybe atrata
Tephrocybe carbonaria
Tephrocybe rancida
Tephrocybe tesquorum
Tricholoma atrosquamosum
Tricholoma cingulatum
Tricholoma fulvum
Tricholoma imbricatum
Tricholoma pessundatum
Tricholoma psammopus
Tricholoma saponaceum
Tricholoma saponaceum
　var. squamosum
Tricholoma sciodes
Tricholoma sulphureum
Tricholoma terreum
Tricholoma ustale
Tricholoma ustaloides
Tricholoma vaccinum
Tricholomopsis rutilans
Xerula radicata
Auriculariales
Auriculariaceae
Hirneola auricula-judae
Incertae sedis
Stilbum pellucidum
Boletales
Boletaceae
Boletus aestivalis
Boletus appendiculatus
Boletus badius
Boletus calopus
Boletus chrysenteron
Boletus edulis
Boletus erythropus
Boletus impolitus
Boletus lanatus
Boletus piperatus
Boletus porosporus
Boletus pruinatus
Boletus pulverulentus
Boletus rubellus
Boletus spadiceus
Boletus subtomentosus

Leccinum aurantiacum
Leccinum holopus
Leccinum melaneum
Leccinum oxydabile
Leccinum rigidipes
Leccinum roseofracta
Leccinum scabrum
Leccinum variicolor
Leccinum versipelle
Suillus bovinus
Suillus fluryi
Suillus granulatus
Suillus grevillei
Suillus grevillei var. badius
Suillus luteus
Suillus tridentinus
Suillus variegatus
Suillus viscidus
Coniophoraceae
Coniophora arida
Coniophora puteana
Serpula lacrimans
Gomphidiaceae
Chroogomphus rutilus
Gomphidius maculatus
Gyrodontaceae
Gyroporus castaneus
Hygrophoropsidaceae
Hygrophoropsis aurantiaca
Paxillaceae
Paxillus involutus
Paxillus panuoides
Paxillus rubicundulus
Rhizopogonaceae
Rhizopogon luteolus
Cantharellales
Cantharellaceae
Cantharellus cibarius
Cantharellus tubiiformis
Clavariaceae
Clavaria acuta
Clavaria argillacea
Clavaria fumosa
Clavaria vermicularis
Clavaria zollingeri
Clavulinopsis corniculata
Clavulinopsis fusiformis
Clavulinopsis helvola
Clavulinopsis laeticolor

Clavulinopsis luteoalba
Clavulinopsis umbrinella
Macrotyphula fistulosa var.
  contorta
Ramariopsis kunzei
Clavariadelphaceae
Clavariadelphus fistulosus
Clavulinaceae
Clavulina cinerea
Clavulina cristata
Clavulina rugosa
Craterellaceae
Craterellus cornucopioides
Hydnaceae
Hydnum repandum
Hydnum rufescens
Sparassidaceae
Sparassis crispa
Typhulaceae
Typhula erythropus
Typhula phacorrhiza
Typhula quisquiliaris
Ceratobasidiales
Ceratobasidiaceae
Ceratobasidium anceps
Cortinariales
Cortinariaceae
Cortinarius acutus
Cortinarius anomalus
Cortinarius armillatus
Cortinarius aureifolius
Cortinarius betuletorum
Cortinarius bolaris
Cortinarius causticus
Cortinarius crocolitus
Cortinarius decipiens
Cortinarius delibutus
Cortinarius elatior
Cortinarius flexipes
Cortinarius glaucopus
Cortinarius hemitrichus
Cortinarius hinnuleus
Cortinarius largus
Cortinarius mucifluus
Cortinarius obtusus
Cortinarius paleaceus
Cortinarius pholideus
Cortinarius pseudocrassus
Cortinarius pseudosalor

Cortinarius purpurascens
Cortinarius rigidus
Cortinarius sanguineus
Cortinarius saturninus
Cortinarius semisanguineus
Cortinarius tabularis
Flammulaster carpophila
Flammulaster granulosa
Flammulaster subincarnata
Galerina badipes
Galerina clavata
Galerina graminea
Galerina hypnorum
Galerina laevis
Galerina mniophila
Galerina mutabilis
Galerina mycenopsis
Galerina paludosa
Galerina rubiginosa
Galerina unicolor
Galerina vittiformis
Gymnopilus dilepis
Gymnopilus hybridus
Gymnopilus junonius
Gymnopilus penetrans
Gymnopilus sapineus
Hebeloma anthracophilum
Hebeloma crustuliniforme
Hebeloma crustuliniforme
  var. minus
Hebeloma fastibile
Hebeloma fragilipes
Hebeloma leucosarx
Hebeloma mesophaeum
Hebeloma radicosum
Hebeloma sacchariolens
Inocybe eutheles
Inocybe fastigiata
Inocybe geophylla
Inocybe geophylla var.
  lilacina
Inocybe hirtella
Inocybe lacera
Inocybe lanuginella
Inocybe maculata
Inocybe napipes
Inocybe obscura
Inocybe pudica
Inocybe rennyi

Naucoria bohemica
Naucoria escharoides
Phaeolepiota aurea
Rozites caperata
Crepidotaceae
Chromocyphella muscicola
Crepidotus applanatus
Crepidotus cesatii
Crepidotus mollis
Crepidotus variabilis
Pleurotellus graminicola
Tubaria autochthona
Tubaria conspersa
Tubaria furfuracea
Dacrymycetales
Dacrymycetaceae
Calocera cornea
Calocera furcata
Calocera pallidospathulata
Calocera stricta
Calocera viscosa
Dacrymyces stillatus
Fistulinales
Fistulinaceae
Fistulina hepatica
Ganodermatales
Ganodermataceae
Ganoderma applanatum
Ganoderma europaeum
Gomphales
Gomphaceae
Ramaria bourdotiana
Ramaria formosa
Hericiales
Auriscalpiaceae
Auriscalpium vulgare
Gloeocystidiellaceae
Gloeocystidiellum lactescens
Lentinellaceae
Lentinellus cochleatus
Lentinellus tridentinus
Hymenochaetales
Hymenochaetaceae
Coltricia perennis
Inonotus dryadeus
Inonotus hispidus
Inonotus nodulosus
Inonotus obliquus
Inonotus radiatus

Phellinus ferruginosus
Phellinus hippophaecola
Phellinus pomaceus
Lycoperdales
Geastraceae
Geastrum pectinatum
Geastrum rufescens
Geastrum triplex
Lycoperdaceae
Bovista ericetorum
Bovista plumbea
Bovista nigrescens
Handkea excipuliformis
Handkea utriformis
Langermannia gigantea
Lycoperdon echinatum
Lycoperdon lividum
Lycoperdon molle
Lycoperdon nigrescens
Lycoperdon perlatum
Lycoperdon pyriforme
Vascellum pratense
Nidulariales
Nidulariaceae
Crucibulum laeve
Cyathus olla
Phallales
Phallaceae
Clathrus cibarius
Mutinus caninus
Phallus hadrianii
Phallus impudicus
Poriales
Coriolaceae
Abortiporus biennis
Antrodia xantha
Bjerkandera adusta
Ceriporia viridans
Cerrena unicolor
Cinereomyces lindbladii
Datronia mollis
Daedalea quercina
Daedaleopsis confragosa
Dichomitus campestris
Fibuloporia vaillantii
Fomes fomentarius
Gloeophyllum sepiarium
Grifola frondosa
Grifola umbellata

Hapalopilus nidulans
Heterobasidion annosum
Incrustoporia semipileata
Laetiporus sulphureus
Meripilus giganteus
Oligoporus leucomallellus
Oligoporus ptychogaster
Phaeolus schweinitzii
Piptoporus betulinus
Podoporia sanguinolenta
Postia caesia
Postia fragilis
Postia stiptica
Postia subcaesia
Skeletocutis amorphus
Trametes hirsutus
Trametes pubescens
Trametes versicolor
Trichaptum abietinum
Lentinaceae
Pleurotus cornucopiae
Pleurotus dryinus
Pleurotus ostreatus
Polyporaceae
Dichomitus campestris
Polyporus brumalis
Polyporus melanopus
Polyporus nummularius
Polyporus picipes
Polyporus squamosus
Polyporus varius
Russulales
Russulaceae
Lactarius blennius
Lactarius brittanicus
Lactarius camphoratus
Lactarius chrysorheus
Lactarius cimicarius
Lactarius deliciosus
Lactarius deterrimus
Lactarius flexuosus
Lactarius fluens
Lactarius fulvissimus
Lactarius glyciosmus
Lactarius helvus
Lactarius hysginus
Lactarius ichoratus
Lactarius mammosus
Lactarius mitissimus

Lactarius obscuratus
Lactarius pallidus
Lactarius piperatus
Lactarius pubescens
Lactarius pyrogalus
Lactarius quietus
Lactarius rufus
Lactarius serifluus
Lactarius subdulcis
Lactarius tabidus
Lactarius theiogalus
Lactarius torminosus
Lactarius trivialis
Lactarius turpis
Lactarius vellereus
Lactarius vietus
Lactarius volemus
Russula aeruginea
Russula atropurpurea
Russula betularum
Russula caerulea
Russula carminea
Russula cicatricata
Russula consobrina
Russula cyanoxantha
Russula cyanoxantha var. peltereaui
Russula delica
Russula emetica
Russula emeticella
Russula farinipes
Russula fellea
Russula fragilis
Russula fragilis var. nivea
Russula gracillima
Russula grisea
Russula laurocerasi
Russula lepida
Russula lepida var. lactea
Russula lutea
Russula mairei
Russula nigricans
Russula nitida
Russula obscura
Russula ochroleuca
Russula paludosa
Russula parazurea
Russula pectinata
Russula pectinatoides

Russula puellaris
Russula pulchella
Russula rosea
Russula sanguinea
Russula sardonia
Russula sororia
Russula turci
Russula velenovskyi
Russula vesca
Russula violeipes
Russula xerampelina
Schizophyllales
Schizophyllaceae
Plicaturopsis crispa
Schizophyllum commune
Sclerodermatales
Sclerodermataceae
Scleroderma areolatum
Scleroderma bovista
Scleroderma citrinum
Scleroderma verrucosum
Sphaerobolaceae
Sphaerobolus stellatus
Stereales
Atheliaceae
Athelia epiphylla
Atheliopsis nov. sp.
Leptosporomyces galzinii
Botryobasidiaceae
Botryobasidium subcoronatum
Hyphodermataceae
Cristina helvetica
Cylindrobasidium evolvens
Hyphoderma praetermissum
Hyphoderma radula
Hyphodontia aspera
Hyphodontia papillosa
Hyphodontia setigera
Hyphodontia subalutacea
Hypochnicium punctulatum
Lyomyces sambuci
Schizopora paradoxa
Meruliaceae
Byssomerulius corium
Ceraceocorticium confluens
Chondrostereum purpureum
Merulius tremellosus

Mycoacia aurea
Phanerochaete filamentosa
Phanerochaete laevis
Phanerochaete sordida
Phanerochaete velutina
Phlebia merismoides
Phlebia radiata
Trechispora farinacea
Peniophoraceae
Peniophora incarnata
Peniophora laeta
Peniophora limitata
Peniophora lycii
Peniophora nuda
Peniophora pithya
Peniophora quercina
Sistotremataceae
Sistrotrema brinkmanii
Steccherinaceae
Mycoleptodon ochraceum
Stereaceae
Stereum gausapatum
Stereum hirsutum
Stereum purpureum
Stereum rugosum
Stereum sanguinolentum
Vuilleminiaceae
Vuilleminia comedens
Thelephorales
Thelephoraceae
Thelephora terrestris
Tremellales
Exidiaceae
Exidia glandulosa
Exidia recisa
Tremellaceae
Pseudohydnum gelatinosum
Tremella encephala
Tremella frondosa
Tremella mesenterica
Tulostomatales
Tulostomataceae
Tulostoma brumale
TELIOMYCETES
Uredinales
Coleosporiaceae
Chrysomyxa rhododenri
Coleosporium campanulae
Coleosporium senecionis

Coleosporium tussilaginis
Melampsoraceae
Melampsora allii-populina
Melampsora capraearum
Melampsora hypericorum
Melampsora larici-populina
Melampsoridium betulinum
Pucciniastrum
   goeppertianum
Pucciniaceae
Gymnosporangium
   cornutum
Phragmidium fragariae
Phragmidium rosa-
   pimpinellifoliae
Phragmidium rubi-idaei
Phragmidium violaceum
Puccinia aegopodii
Puccinia antirrhini
Puccinia brachypodii var.
   arrhenatheri
Puccinia cnici-oleracei
Puccinia hieracii
Puccinia hordei
Puccinia lagenophorae
Puccinia luzulae
Puccinia malvacearum
Puccinia poarum
Puccinia porri
Puccinia prostii
Puccinia punctiformis
Puccinia recondita
Puccinia urticata
Tranzschelia discolor
Triphragmium ulmariae
Uromyces dactylidis
Uromyces ficariae
Uromyces muscari
Uromyces rumicis
Uromyces viciae-fabae
USTOMYCETES
Platygloeales
Platygloeaceae
Platygloea fimetaria
Ustilaginales
Tilletiaceae
Entyloma microsporum
Tilletia holci
Urocystis agropyri

Urocystis anemones
Ustilaginaceae
Microbotyron violacea
Ustilago avenae
Ustilago hypodytes
Ustilago striiformis
MITOSPORIC FUNGI –
DEUTEROMYCOTINA
Coelomycetes
Amerosporium
   patellarioides
Ceuthospora feurichii
Coryneopsis rubi
Cryptosphaeria eunomia
Cytospora oxyacanthae
Darluca filum
Dilophospora alopecuri
Dinemasporium hispidulum
Entomosporium maculatum
Fusicoccum galericulatum
Leptostroma spiraeinum
Libertella faginea
Lichenoconium parasiticum
Macrophoma fraxini
Melanconium alni
Melanconium betulinum
Melanconium bicolor
Melanconium stromaticum
Micropera cotoneastri
Phloeospora aegopodii
Phoma acuta
Phoma complanata
Phoma urticae
Phomopsis hysteriola
Phomopsis pterophila
Phomopsis vepris
Phyllosticta bellunensis
Phyllosticta ligustri
Phyllosticta maculiformis
Pycnothyrium litigiosum
Sclerophoma pythiophila
Septoria minuta
Septoria pseudoplatani
Vermicularia dematium
Vermicularia trichella
Hyphomycetes
Botryosporium
   longibrachiatum
Botrytis cinerea

Botrytis globosa
Calcarisporium arbuscula
Cladosporium
   cladosporioides
Cladosporium herbarum
Cylindrium flavovirens
Cylindrium griseum
Dendrospora erecta
Doratomyces stemonitis
Drechslera cactivorum
Drechslera iridis
Endophragmia catenulata
Epicoccum purpurascens
Fulvia fulva
Helminthosporium
fusiforme
Hobsonia christiansenii
Linodochium hyalinum
Memnoniella echinata
Mycocentrospora acerina
Oidiodendron fuscum
Oospora lactis
Ovularia obliqua
Ovularia primulana
Pullularia pullulans
Pycnostysanus azaleae
Ramularia pratensis
Rhintrichum repens
Rhopalomyces magnus
Sclerotium rhizodes
Sclerotium scutellatum
Sclerotium semen
Sepedonium
   chrysospermum
Septonema strictum
Spilocaea pomi
Spondylocladiopsis
   cupulicola
Stachybotrys lobulata
Stilbella erythrocephala
Tilachlidium tomentosum
Trichoderma viride
Trichothecium roseum
Tubercularia vulgaris
Varicosporium elodeae
Verticillium fungicola
Verticillium rexianum
Volutella ciliata
OOMYCOTA

Peronosporales
Peronosporaceae
Albugo candida
Pythiales
Pythiaceae
Phytophthora cinnamoni
Phytophthora infestans
**PLASMODIOPHORO-
MYCOTA**
Plasmodiophorales
Plasmodiophoraceae
Plasmodiophora brassicae
ZYGOMYCOTA
Glomales
Glomaceae
Glomus fasciculatus
Glomus mosseae
Mucorales
Chaetocladiaceae
Chaetocladium brefeldii
Mucoraceae
Phycomyces blakesleanus
Phycomyces nitens
Spinellus fusiger
Pilobolaceae
Pilaira anomala

Pilaira moraeui
Pilobolus crystallinus
Pilobolus kleinii
Pilobolus sphaerosporus
Pilobolus umbonatus
Zoopagales
Piptocephalidaceae
Piptocephalis cylin-
    drospora
Piptocephalis fimbriata
Piptocephalis freseniana
Piptocephalis lepidula
Piptocephalis repens
**MYXOMYCETES –
MYXOMYCOTA**
Liceales
Cribrariaceae
Dictydium cancellatum
Enteridiaceae
Enteridium lycoperdon
Reticularia olivacea
Lycogalaceae
Lycogala epidendrum
Tubiferaceae
Tubifera ferruginosa
Physarales

Didymiaceae
Diderma hemisphaericum
Mucilago spongiosa
Physaraceae
Fuligo septica
Physarum cinereum
Protosteliales
Ceratiomyxaceae
Ceratiomyxa fruticulosa
Stemonitales
Stemonitidaceae
Comatricha nigra
Stemonitis axifera
Stemonitis fusca
Stemonitis herbatica
Stemonitopsis hyperopta
Trichiales
Arcyriaceae
Arcyria ferruginea
Arcyria nutans
Trichiaceae
Trichia affinis
Trichia botrytis
Trichia contorta
Trichia varia

6

Stoneworts of the Lothians

N. F. STEWART

Stoneworts (charophytes) have received rather little attention in recent years and many of the most recent records derive from the loch survey undertaken by Scottish Natural Heritage. Undoubtedly there are sites still to be discovered and particular attention should be given to lochs and pools with low levels of nutrients but a high content of lime. The dunes around Aberlady Bay and Gullane have long been identified as an important site although Bristly Stonewort (*Chara hispida*) and Delicate Stonewort (*Chara virgata*) are the only species that have been confirmed as still present. However, there is a need for further exploration of temporary pools in the spring before they dry out.

Despite the lack of recording, it is probable that there has been a significant decline in this group over the last 100 years due to the nutrient enrichment of water bodies as a result of sewage and agricultural run-off. The listing of nearly half the species as extinct may therefore be a reality and, unless this issue is addressed, further declines can be expected.

Nomenclature in the list below follows Bryant, J. A., Stace, C. A. and Stewart, N. F. *Checklist of the Charophytes of the British Isles*, but names used in Moore, J. A. (1986) *Charophytes of Great Britain and Ireland*, B.S.B.I. Handbook No. 5 are given in brackets where appropriate.

Chara aculeolata Kütz.
Hedgehog Stonewort
(*Chara pedunculata* Kütz.)
Native. Status: extinct.
Calcareous, low-nutrient pools.

EL: (Aberlady Bay, 1961).

Chara aspera Deth. ex Willd.
Rough Stonewort
(*Chara aspera* Deth. ex Willd. var. *aspera*)
Native. Status: extinct.
Sand-dune pools and (outside the Lothians) calcareous

to faintly brackish ditches and lakes.

EL: (Gullane Links, 1897).

Chara contraria A. Braun ex Kütz.
Opposite Stonewort
(*Chara vulgaris* var. *contraria* (A. Braun ex Kütz.) J. A. Moore)
Native. Status: extinct.
Sand-dune pools and loch edges.

ML: (Gladhouse Reservoir, 1935).
EL: (Gullane Links, 1882).

Chara globularis Thuill.   Fragile Stonewort
(*Chara globularis* Thuill. var. *globularis*)
Native. Status: rare. Some pre-1920 records
may refer to *Chara virgata*.
Lochs, pools and canals and, unlike most stoneworts,
tolerant of some enrichment.

ML: west of Cramond Bridge; Union Canal
east of Lemmington Lift Bridge; Duddingston
Loch; (Braid Hills, 1849).
EL: Gosford Loch; Archerfield Loch; west of
Scoughall; (Aberlady, pre-1927; Biel Water,
West Barns, 1961).

Chara hispida L.   Bristly Stonewort
(*Chara hispida* vars *hispida* L. and *major*
(Hartm.) R. D. Wood)
Native. Status: rare.
Calcareous, low-nutrient lochs and pools.

EL: Marl Loch, Aberlady; (North Berwick,
1838; Gullane Loch, 1896).

[*Chara rudis* (A. Braun) Leonh.
          Rugged Stonewort]
(*Chara hispida* var. *rudis* A. Braun)
This species was recorded from Gullane Loch
in 1896 but this was in error for *C. hispida*.]

Chara virgata Kütz.   Delicate Stonewort
(*Chara globularis* var. *virgata* (Kütz.) R. D.
Wood)
Native. Status: rare.
Acidic to calcareous lochs, pools and ditches.

WL: Linlithgow Loch.
ML: Threipmuir Reservoir; Duddingston Loch;
(Ravelrig Toll Moss, pre-1844; Glencorse,
1872.
EL: Marl Loch, Aberlady; (ditches near
Gullane, 1831).

Chara vulgaris L.   Common Stonewort
(*Chara vulgaris* vars *vulgaris* L., *longibracteata*
(Kütz.) J. Gorves & Bullock-Webster and *papillata* Wallr. ex A. Braun)
Native. Status: rare.
Calcareous lochs, pools and ditches.

ML: Duddingston Loch, var. *longibracteata*;
(Pentland Hills, pre-1814, var. *papillata*).
EL: Gosford Loch; (Aberlady Bay/Gullane
Links, 1961, vars *vulgaris* and *papillata*)

Nitella opaca (Bruz.) Agardh
          Dark Stonewort
(*Nitella flexilis* L. var. *flexilis* pro parte)
Native. Status: occasional.
Acidic to neutral lochs, streams and ditches.

Fertile material is needed to separate this
species from Smooth Stonewort *N. flexilis*
sensu strictu and some records are therefore
given as *Nitella Flexilis* agg. These are
indicated with an asterisk in the following list.
Smooth Stonewort has not been confirmed in
the Lothians and the records are assumed to be
for Dark Stonewort. However the former may
possibly occur and it would be useful to
confirm these records.

WL: Lochcote Reservoir; Beecraigs Loch; loch
east of Addiewell.
ML: Threipmuir Reservoir; Clubbiedean
Reservoir; Dunsapie Loch*; Gladhouse
Reservoir; (ditches near Musselburgh*, pre-
1824; Harelaw Reservoir*, 1881; Hallyards,
Ratho*, 1906).
EL: Pressmennan Loch; pond by Whiteadder
Reservoir.

Nitella translucens (Persoon) Agardh
          Translucent Stonewort
Native. Status: extinct.
Acidic ditches, pools and lakes.

ML: ditches about Edinburgh, pre-1821.

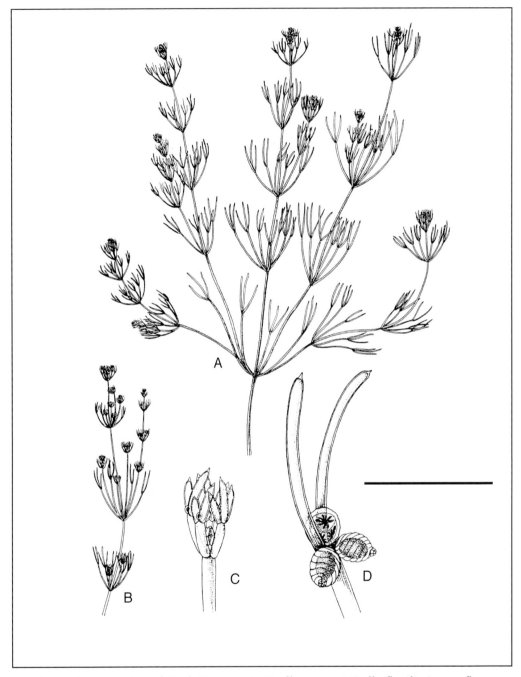

Figure 6.1 *Drawing of Dark Stonewort* (Nitella opaca (Nitella flexilis L. *var.* flexilis *pro parte)) showing A, B: habits; C: branch apex with branchlet whorls; D: gametangia with two 1-celled dactyls. Scale bar represents 6cm for A and B, 0.2cm for C and 0.15cm for D. Reproduced by kind permission of BSBI from* Charophytes of Great Britain and Ireland, *BSBI Handbook No. 5 (1986).*

7

Seaweeds of the Lothians

M. WILKINSON

What is a Seaweed?

'Seaweed' is a non-technical term meaning algae attached to the rocky seashore, both intertidal and below low-water mark (sublittoral). Algae are the simplest plants. Some recent classifications place at least some algae in other kingdoms, but here they are regarded as plants, united in the mechanism of photosynthesis. They differ from other plants in the simplicity of their reproductive organs, which lack cellular walls, and in their biochemical diversity. One aspect of this diversity is the existence of many colour groups of algae, with different pigment combinations, compared with the uniformly green land plants. Just three of the colour groups contain seaweeds, green, brown and red seaweeds, in the taxonomic divisions Chlorophyta, Phaeophyta and Rhodophyta respectively.

They range from microscopic single cells to large cartilaginous kelp plants (up to 3m in Britain but up to 70m elsewhere!). All share the absence of resistant stages such as seeds, development of the next generation being immediate. They face stresses different from those faced by land plants. Immersed in water that supplies all nutrients direct to their cells, they do not need transport systems such as xylem and phloem, nor do they need systems to regulate water loss and gas exchange, such as stomata. Buoyed up by water they have no need of support tissues. Internally they are, therefore, much simpler than land plants. The largest seaweeds have a very superficial similarity to leaves, stems and roots with flat, photosynthetic fronds, supported by a stem-like stipe, attached to the rocks by a holdfast which, unlike roots, does not penetrate and is non-absorptive. Instead, they need physical toughness of cell walls, combined with flexibility of the plant, to avoid damage by wave action or water movement.

Types of Seaweed

About 619 species of seaweed are listed for the British Isles by South and Tittley (1986). Over 100 can be found on a single shore (Wilkinson and Rendall 1985). This number, which surprises many, is partly achieved through the wide habitat range. In addition to epilithic seaweeds attached to rocks, there are endophytic, endozoic and epiphytic seaweeds. Endophytes are microscopic, filamentous forms creeping between cells of larger seaweeds. Endozoic forms are microscopic unicells and filaments living in the chitinous tests of hydroids and bryozoans, and boring through the calcareous shells of molluscs, barnacles and tube-worms. Epiphytism is especially common in seaweeds. This is the growth of plants attached to other plants. There can be several degrees of smaller plants growing on larger host plants. Many larger filamentous seaweed species may have several species of smaller forms attached to them which cannot be seen without microscopic examination.

Epilithic seaweeds have a wide range of encrusting and erect forms. Encrusting ones, only a few millimetres thick, appearing as stains on the rock surface, can be mistaken for the rock itself. The crustose corallines are red encrusters that secrete calcium carbonate, appearing whitish or pink, and even more like part of the rock. Some encrusters are not species in their own right but stages in the life-cycle of erect plants, for example, *Petrocelis* which is now known to be an encrusting stage in the life-cycle of *Mastocarpus*. Some other encrusters may be the extended basal portions of normally erect plants, for example, the calcareous filamentous plant *Corallina*, which looks like a branched string of beads, sometimes takes a flat encrusting form that is hard to distinguish from genuine crustose coralline species.

Erect forms can be filamentous, foliose or cartilaginous. Filamentous forms may be unbranched, irregularly branched, or branched in a very regular pattern, for example, *Plumaria*. Similarly, foliose plants may be irregular in outline, for example, *Ulva*, or tubular, for example, *Enteromorpha*, or have a regular shape, perhaps with midribs, as in *Phycodrys* which resembles oak leaves.

The really large, leathery cartilaginous forms are mainly brown. There are two major types. The rockweeds, also known as wracks or fucoids, dominate the intertidal zone of shores sheltered from wave-action, while the kelps, or laminarians, form dense underwater forests. These kelp forests are the most widespread habitat type in temperate and subpolar waters. Just as terrestrial forests have tree, shrub and herb layers, so kelp forests have strata of smaller plants adapted to living in the shade cast by the kelp canopy.

The life-cycles of seaweeds are more diverse than those of land plants. Many have an asexual, spore-producing generation, or sporophyte, and a sexual generation or gametophyte. Unlike higher plants, all seaweeds do not adhere to a strict life-cycle pattern. In some green seaweeds (e.g. *Cladophora*, *Enteromorpha* and *Ulva*) both generations look identical. In other seaweeds the two generations may be very distinct. In kelps, the large, leathery plant is the sporophyte and there is a microscopic filamentous gametophyte. Some green and red seaweeds have a shell-boring stage as one generation. In *Porphyra*, the foliose red plant eaten as laver bread in South Wales, the shell-boring sporophyte was classed as a separate species, *Conchocelis rosea*, until the life-cycle was elucidated by laboratory culture in 1948. We cannot know life-cycles without culture experiments. Since many seaweeds have not been cultured, it is likely that the species number may be reduced in future when existing separate species are discovered to be parts of the same life-cycle.

Seaweeds on the Intertidal Seashore

Seaweeds are not usually found on sedimentary shores unless attached to rocks. These shores have other algae, the microphytobenthos, microscopic single-celled diatoms, euglenoids and dinoflagellates, and cyanobacteria, living between sand and mud particles. Although not individually visible to the naked eye they can be abundant, forming coloured patches, and accounting for much primary production.

Shores composed of pebbles and small boulders have few algae due to movement of the rocks in the waves. Larger boulders and bedrock shores show well-marked vertical and horizontal distribution patterns of seaweeds. A brief summary of these patterns is given here. Further information is available in Wilkinson (1992).

Horizontal patterns are lateral variations along the shore, mainly in response to variations in wave action. Vertical distribution refers to zonation in relation to tidal height. This is clearest on very sheltered shores where the abundant, dominant fucoids occur in the following order going down from the high shore:

Pelvetia canaliculata	Channelled Wrack
Fucus spiralis	Spiral Wrack (Plate 8, bottom left)
Fucus vesiculosus	Bladder Wrack (Plate 8, bottom right)
Ascophyllum nodosum	Knotted Wrack or Egg Wrack (Plate 8, top)
Fucus serratus	Toothed Wrack (Plate 8, bottom right)

Subordinate species may also be zoned but these zones do not coincide with the fucoid zones, and a few tolerant species may occur throughout the whole shore, for example, *Enteromorpha*. Zonation is also apparent on wave-exposed shores but is dominated by sturdy, sessile animals such as mussels and barnacles. On shores of intermediate exposure it can be more difficult to see zonation clearly. At each level on such shores, there may be several possible communities in adjacent, small patches giving a mosaic appearance.

It used to be thought that zonation was due solely to physical tolerance to desiccation. This is supported by the decrease in species number with height on the shore, as longer emersion times will be more stressful for marine species. Since about 1960, many field experiments have shown that biotic interactions, such as competition and grazing, play a role. Limpets and periwinkles are aggressive grazers on seaweeds, while barnacles and mussels compete with them for space. Generally, on the lower shore where there are many species and conditions are strongly marine, species boundaries are set by competition and grazing. On the upper shore, where there are fewer species to interact, and conditions are harsher with longer tidal emersion, species boundaries are set by physical tolerance.

Increased wave action affects seaweed distribution patterns in several ways. Zones become wider and occur higher on the shore as a consequence of the spray wetting to a greater height on the shore. Species characteristic of shelter are replaced by those tolerant to exposure, for example, in the kelp zone on the fringe between sublittoral and intertidal, the kelp *Laminaria digitata* is replaced by *Alaria esculenta*, and on the lower shore *Himanthalia elongata* replaces *Fucus serratus*. The mosaic pattern seen on shores of intermediate exposure is enhanced by several other modifying factors that vary along the shore such as rock type, aspect and slope.

Rock pools are a specialised habitat. They provide continuous submersion, like the sublittoral, but their small volume means that, unlike the sublittoral, they undergo fluctuations in temperature and salinity. Consequently, those lower on the shore are more like the sublittoral but, with increasing intertidal height, they become more distinct. The lowest pools may contain sublittoral species, including kelps, while the highest pools are restricted to a few tolerant green seaweeds. Mid-shore pools may have a wide range of species including some from outside pools, some from open rocks lower on the shore, but also some largely restricted to pools, for example, *Halidrys*, the podweed.

Zonation patterns are not constant in time. There are seral and cyclic successions. Separate but adjacent small patches of rock may be at different stages in successions, adding to the mosaic pattern of intermediately exposed shores. An example of a cyclic succession is the regular alternation, taking several years, between fucoids and barnacles. An example of a seral succession is the recolonisation of bare shore. Sometimes when a space is cleared by removal of dead plants, replacement may be by whatever is fruiting adjacently, but sometimes a successional sequence commences with smaller, short-lived, filamentous, green forms, which are replaced progressively with foliose red and brown species, culminating in fucoids after a few years.

There is also long-term change. Geographical ranges of species spread and diminish naturally. *Codium fragile* ssp. *atlanticum* has spread northwards up the west coast of Britain during the twentieth century. Since a major factor in determining the geographical distribution of seaweeds is water temperature, it will be interesting to see if, during the twenty-first century, species composition changes in particular areas such as the Forth due to the effect of global warming on sea temperature.

Distribution of Seaweeds in the Lothians

The Firth of Forth is an arm of the North Sea. Most of it is fully marine. The Forth estuary proper starts about the Forth Bridges, where the salinity is measurably diluted below that of seawater. The inner Firth can occasionally suffer some dilution of salinity when large quantities of freshwater flow down rivers in the winter. The inner Firth may also have slightly more turbid water than the outer Firth due to estuarine influence. As a consequence, three areas of seaweed flora can be subjectively observed which correspond roughly to the three Lothian counties:

- Outer Firth of Forth – clear water – marine – East Lothian
- Inner Firth of Forth – water slightly turbid – mainly marine – Midlothian including City of Edinburgh
- Outer Forth estuary – water turbid – salinity reduced – West Lothian

The Edinburgh shoreline is additionally impacted by former severe crude sewage pollution (see later) and alteration of shores due to land reclamation. More detail of seaweeds in the Forth is given by Wilkinson *et al.* (1987).

The conventional distribution of seaweeds in an estuary (Wilkinson 1980; Wilkinson *et al.* 1995) is as follows:

1. Species numbers decline going upstream, with selective attenuation firstly of red seaweeds, then of brown seaweeds.
2. A few brackish-water species, for example, *Fucus ceranoides*, may occur in the mid- and upper reaches.
3. The outer estuary is characterised by fucoid-dominated shores, which are species-poor in comparison with such shores on the surrounding sheltered open coast.
4. The inner estuary is characterised by turfs of microscopic filamentous algae (mainly green species) and cyanobacteria with only a few species on each shore.

Some features of this are apparent if the transition from East to Mid- to West Lothian is seen as a gradient as shown by species totals in Table 7.1 (totals for each individual shore can be found in Wilkinson *et al.* 1987).

Table 7.1 Total numbers of seaweed species, by taxonomic division, for the three Lothian counties, based on unpublished surveys between 1976 and 1987 by the present author with Ian Tittley (Natural History Museum). Percentage contribution of the three divisions to each county flora is shown in brackets. Detailed species listings are given at the end of the chapter.

	East Lothian	Midlothian	West Lothian
Chlorophyta	48 (25%)	39 (34%)	26 (35%)
Phaeophyta	55 (29%)	27 (23%)	15 (20%)
Rhodophyta	86 (46%)	49 (43%)	33 (45%)
Total	189	115	74

The inner zone of the estuary is upstream of the county boundary at the River Avon so the West Lothian shores are species-poor, fucoid shores, typical of the outer estuary. In this county the lower species number can be related to the relative uniformity of the shores, predominantly shingle or muddy with boulders or rock outcrops. The full estuarine algal distribution including the turf-forming upper estuarine flora (Wilkinson *et al.* 1987; Wilkinson and Slater 1997) can be seen in the Lothians in the small sub-estuaries of the Forth, principally those of Peffer Burn and the Rivers Esk, Almond and Avon.

The greatest variation in flora is in the outer Firth, where species totals are greatest. This is seen where there is the greatest habitat variation as, for example, at Dunbar. In as little as 100m there can be a change from exposed to sheltered shores; from shores with little seaweed cover in the midshore and exposure indicators, *Alaria* and *Himanthalia* dominating the lower shore, to densely fucoid-covered shores with *Ascophyllum* and *Pelvetia* which are characteristic of shelter.

Temporal Changes in Lothian Seaweeds

Edinburgh was a centre of marine research in the nineteenth century and several famous phycologists collected in the Forth. Particularly important among these was George William Traill who lived at Joppa. He published authoritative lists for Joppa and Dunbar (Traill 1886, 1890). In the 1970s Dunbar showed broadly similar species-richness to that recorded in 1890 (Wilkinson *et al.* 1987). By contrast, Joppa, described by Traill in the 1880s as a luxuriant shore, which all phycologists should visit, had lost half its species by the 1970s. This was ascribed to discharge of crude sewage from Edinburgh. Since the inauguration of Edinburgh's sewage scheme about 1980, water quality has improved. Species-richness has only slightly increased at Joppa. It seems that the change from seaweed domination to a mussel/barnacle dominated shore, induced by the sewage, has given rise to a resistant animal-dominated climax community which is hard to dislodge. This did not occur at Granton where the sewage-induced climax was a fragile silt covering on the rocks, inhabited by small polychaete worms. Without the continual input of silt from sewage this was lost and has been replaced by a fucoid-dominated more seaweed-rich community.

Continued monitoring at Granton and Joppa since the publication of the above conclusions (Wilkinson *et al.* 1987) has shown that natural successions may also be involved. Individual rocks have been observed at Granton starting with photographic records from the 1960s (kindly provided by Prof. J. C. Smyth). Rocks, which had been seen to change from polychaete cover to fucoid cover, have now changed to greater abundance of mussels and barnacles in place of seaweeds. Similarly, the concrete wall erected around the new sewage works at Seafield in the 1970s was initially colonised by fucoid-dominated seaweeds, but by the late 1980s this had been replaced by barnacles and mussels, even though this was after introduction of the sewage scheme.

In comparing species totals on shores, it is important to consider how the data are collected. Wilkinson and Tittley (1979) suggested that seaweed species-richness remains broadly constant in the absence of environmental change, although the detailed list on successive occasions may be different. About one-third of the species may be ephemeral. This has recently been substantiated in monthly surveys at Granton and Joppa by a Heriot-Watt University Ph.D. student, Emma Wells. If successive surveys are aggregated together, the cumulative species totals increase. The total of eighty-four species at Joppa (Wilkinson *et al.* 1987) makes it seem a rich site. Yet it remains a mussel/barnacle-dominated shore to this day with only about thirty to forty species likely to be found on a single visit, including many in trace amounts. The reason it appears to have a high total is because it has been sampled more often than other shores because of its interest, so giving a high cumulative total.

Change may be less likely in estuaries where the harsh conditions induced by wide salinity fluctuations and high turbidity mean that the few species present are hardy. Nonetheless, a change observed elsewhere by the author in recent years is the migration upstream of the lower estuarine fucoid-dominated, species-poor community, linked to improvements in water quality. There has been an approximate 15 km upstream migration of the fucoid limit in both the Tyne and Tees estuaries. No such change has been seen in the Lothian sub-estuaries of the Forth but one sub-estuary just outside the geographical remit of this chapter, the Carron at Grangemouth, has undergone dramatic change. Virtually devoid of seaweeds in 1975 when it was receiving large amounts of effluent from ICI, it was initially colonised by a few species of turf-forming green seaweeds following removal of the effluents. Following further general water quality improvements, it attained dense fucoid cover in the 1990s. The only candidate sub-estuary for such change in Lothian is the Avon, also at Grangemouth. A substantial effluent from BP Chemicals was removed from this estuary about twenty years ago but the flora has not changed. Nonetheless, experiments with fucoids transplanted to the estuary, and salinity measurements within it, suggest that fucoids should be able to invade it. It may be just a matter of time for this immigration to occur.

With the rich shores of East Lothian, the outer estuarine shores of West Lothian, the shores recovering from pollution around Edinburgh, and the upper estuarine environment in the sub-estuaries of the Forth, the Lothians present a wide range of seaweed communities.

Intertidal seaweed species recorded in the Lothians

Systematic list of intertidal seaweed species recorded in the Lothians (**WL** West Lothian; **ML** Midlothian; **EL** East Lothian) in unpublished surveys made by the present author and Ian Tittley (Natural History Museum) between 1977 and 1987. Taxonomic nomenclature accords with the check-list of South & Tittley (1986).

CHLOROPHYTA
Acrochaete repens **EL**
Acrochaete wittrockii **EL**
Blidingia marginata **WL, ML, EL**

Blidingia minima **WL, ML, EL**
Bolbocoleon piliferum **EL**
Bryopsis hypnoides **ML, EL**
Bryopsis plumosa **WL, ML, EL**

81

Capsosiphon fulvescens **ML**
Chaetomorpha capillaris **ML, EL**
Chaetomorpha linum **ML, EL**
Chaetomorpha melagonium **ML, EL**
Cladophora albida **ML, EL**
Cladophora glomerata **ML**
Cladophora laetevirens **ML, EL**
Cladophora rupestris **WL, ML, EL**
Cladophora sericea **WL, ML, EL**
Codium fragile atlanticum **EL**
Enteromorpha clathrata **EL**
Enteromorpha compressa **ML, EL**
Enteromorpha flexuosa **EL**
Enteromorpha intestinalis **WL, ML, EL**
Enteromorpha linza **WL, ML, EL**
Enteromorpha prolifera **WL, ML, EL**
Enteromorpha torta **ML, EL**
Entocladia perforans **WL, ML, EL**
Entocladia viridis **ML, EL**
Epicladia flustrae **WL, ML, EL**
Eugomontia sacculata **WL, ML, EL**
Monostroma grevillei **WL, ML, EL**
Monostroma oxyspermum **WL, ML, EL**
Ostreobium queketti **EL**
Percursaria percursa **ML, EL**
Prasiola stipitata **WL, ML, EL**
Pringsheimiella scutata **ML, EL**
Rhizoclonium riparium **WL, ML, EL**
Rosenvingiella polyrhiza **WL, ML, EL**
Spongomorpha aeruginosa **EL**
Spongomorpha arcta **WL, ML, EL**
Tellamia contorta **WL, ML, EL**
Ulothrix flacca **WL, ML, EL**
Ulothrix implexa **WL, ML, EL**
Ulothrix palusalsa **WL, ML, EL**
Ulothrix speciosa **WL, ML, EL**
Ulothrix subflaccida **WL, ML, EL**
Ulva lactuca **WL, ML, EL**
Ulva rigida **EL**
Urospora bangioides **WL, ML, EL**
Urospora penicilliformis **WL, ML, EL**

PHAEOPHYTA
Acinetospora crinita **ML, EL**
Alaria esculenta **EL**
Ascophyllum nodosum **WL, ML, EL**
Asperococcus fistulosus **ML, EL**
Chorda filum **EL**
Chorda tomentosa **EL**

Chordaria flagelliformis **EL**
Cladostephus spongiosus **WL, ML, EL**
Cutleria multifida **EL**
Desmarestia aculeata **EL**
Desmarestia viridis **EL**
Dictyosiphon chordaria **EL**
Dictyosiphon foeniculaceus **EL**
Dictyota dichotoma **EL**
Ectocarpus fasciculatus **ML, EL**
Ectocarpus siliculosus **ML, EL**
Elachista fucicola **ML, EL**
Eudesme virescens **EL**
Fucus ceranoides **EL**
Fucus serratus **WL, ML, EL**
Fucus spiralis **WL, ML, EL**
Fucus vesiculosus **WL, ML, EL**
Giffordia granulosa **WL, ML, EL**
Giffordia hincksiae **ML, EL**
Giffordia sandriana **ML**
Halidrys siliquosa **EL**
Hecatonema maculans **EL**
Herponema velutinum **EL**
Himanthalia elongata **EL**
Isthmoploea sphaerophora **WL, ML**
Laminaria digitata **WL, ML, EL**
Laminaria hyperborea **WL, ML, EL**
Laminaria saccharina **WL, ML, EL**
Leathesia difformis **EL**
Litosiphon laminariae **EL**
Microspongium globosum **EL**
Mikrosyphar polysiphoniae **ML, EL**
Mikrosyphar porphyrae **EL**
Myrionema magnusii **EL**
Myrionema strangulans **EL**
Myriotrichia clavaeformis **EL**
Pelvetia canaliculata **WL, ML, EL**
Petalonia fascia **WL, ML, EL**
Pilayella littoralis **WL, ML, EL**
Protectocarpus speciosus **EL**
Pseudolithoderma extensum **EL**
Punctaria latifolia **EL**
Punctaria tenuissima **EL**
Ralfsia verrucosa **WL, ML, EL**
Scytosiphon lomentaria **WL, ML, EL**
Sphacelaria cirrosa **EL**
Sphacelaria fusca **EL**
Sphacelaria plumigera **ML**
Sphacelaria plumosa **ML, EL**
Sphacelaria radicans **ML, EL**

Spongonema tomentosum **ML**
Stictyosiphon tortilis **EL**
Waerniella lucifuga **EL**

RHODOPHYTA
Ahnfeltia plicata **EL**
Antithamnion cruciatum **EL**
Antithamnionella floccosa **ML, EL**
Antithamnionella spirographidis **EL**
Audouinella concrescens **ML**
Audouinella daviesii **ML, EL**
Audouinella endozoica **EL**
Audouinella floridula **WL, ML, EL**
Audouinella membranacea **EL**
Audouinella parvula **EL**
Audouinella purpurea **WL, ML, EL**
Audouinella secundata **WL, ML, EL**
Audouinella virgatula **ML, EL**
Bangia atro-purpurea **WL, ML, EL**
Bostrychia scorpioides **EL**
Brogniartella byssoides **EL**
Callithamnion corymbosum **EL**
Callithamnion hookeri **WL, ML, EL**
Callithamnion sepositum **ML, EL**
Callithamnion tetragonum **ML**
Callophyllis laciniata **EL**
Catenella caespitosa **WL, ML, EL**
Ceramium diaphanum **WL, ML, EL**
Ceramium flabelligerum **EL**
Ceramium rubrum **WL, ML, EL**
Ceramium shuttleworthianum **WL, ML, EL**
Chondrus crispus **WL, ML, EL**
Choreocolax polsiphoniae **EL**
Corallina elongata **EL**
Corallina officinalis **ML, EL**
Cruoria pellita **EL**
Cryptopleura ramosa **EL**
Cystoclonium purpureum **WL, ML, EL**
Delesseria sanguinea **WL, EL**
Dilsea carnosa **ML, EL**
Dumontia contorta **WL, ML, EL**
Erythrotrichia carnea **ML, EL**
Erythrotrichiopeltis boryana **EL**
Furcellaria lumbricalis **EL**
Gelidium pusillum **WL, ML, EL**
Gracillaria verrucosa **EL**

Griffithsia flosculosa **ML, EL**
Halarachnion ligulatum **EL**
Hildenbrandia rubra **WL, ML, EL**
Hypoglossum woodwardii **ML, EL**
Laurencia hybrida **ML, EL**
Laurencia pinnatifida **ML, EL**
Lithophyllum incrustans **EL**
Lithothamnion glaciale **WL, ML, EL**
Lomentaria articulata **ML, EL**
Lomentaria clavellosa **WL, ML, EL**
Lomentaria orcadensis **EL**
Mastocarpus stellatus **WL, ML, EL**
Melobesia membranacea **EL**
Membranoptera alata **WL, ML, EL**
Nemalion helminthoides **EL**
Odonthalia dentate **WL, EL**
Palmaria palmate **WL, ML, EL**
Peysonnelia dubyi **EL**
Peysonnelia harveyana **EL**
Phycodrys rubens **WL, ML, EL**
Phyllophora crispa **EL**
Phyllophora pseudoceranoides **WL, ML, EL**
Phyllophora traillii **ML, EL**
Phymatolithon lenormandii **WL, ML, EL**
Phymatolithon polymorphum **EL**
Plocamium cartilagineum **ML, EL**
Plumaria elegans **WL, ML, EL**
Polyides rotundus **EL**
Polysiphonia brodiaei **WL, ML, EL**
Polysiphonia elongata **EL**
Polysiphonia fibrata **EL**
Polysiphonia lanosa **WL, ML, EL**
Polysiphonia macrocarpa **WL, ML, EL**
Polysiphonia nigra **ML, EL**
Polysiphonia nigrescens **WL, ML, EL**
Polysiphonia urceolata **WL, ML, EL**
Porphyra leucosticta **ML, EL**
Porphyra linearis **EL**
Porphyra purpurea **WL, ML, EL**
Porphyra umbilicalis **WL, ML, EL**
Pterosiphonia parasitica **EL**
Ptilota plumosa **EL**
Ptilothamnion plumula **EL**
Rhodomela confervoides **WL, ML, EL**
Rhodomela lycopodiodes **EL**
Titanoderma pustulatum **EL**

8

Lichens of Edinburgh and the Lothians

B. J. COPPINS

The number of lichens reported from the Lothians between 1974 and 2001 totals 591 species, representing 40 per cent of the 1,470 species recorded from Scotland and 34 per cent of the 1,722 recorded from the British Isles.

Historical Background

The history of lichen recording in the Lothians prior to 1974 is essentially the same as described for Edinburgh by Coppins (1978). The great majority of species have been re-discovered since 1974, but among the notable absentees today are species that are the most sensitive to air pollution and habitat disturbance, namely the four British *Lobaria* species, *Pseudocyphellaria crocata* (Dalmahoy Hill) and *Sticta limbata* (Hermitage of Braid). All are large foliose species with cyanobacteria as their principal or secondary photobiont, and none of them have been recorded in the region since the early nineteenth century. Another notable absentee is the conspicuous, crustose *Caloplaca flavorubescens*. Listed in the Red Data Book (Church *et al.* 1996), it favours large ash trees in open, unpolluted habitats and was last found in 1805, at Arniston by George Don.

Lichen Habitats in the Lothians

Woodlands

The lichen flora of a woodland essentially comprises a 'background' component of widely occurring species, many of which will be found outside the woodland environment, and those requiring a long continuity of woodland cover. The latter are known as 'old woodland indicator species' and many of them also require a long continuity of mature or ancient trees. The success of species of both components, but especially the old woodland indicators, is further dependent on relatively clean air. Given that suitable old woodlands are few and far between in the Lothians, and that the region suffered from a century or more of high air pollution levels (especially of sulphur dioxide), the number of old woodland indicators is few (Table 8.1).

Table 8.1 *Old woodland indicator lichens recorded in the Lothians since 1974*

Arthonia anombrophila	Graphis elegans
Arthonia cinnabarina	Graphis scripta
Arthonia vinosa	Megalaria grossa
Arthopyrenia antecellens	Microcalicium ahlneri
Bacidia absistens	Nephroma parile
Bactrospora corticola	Peltigera collina
Biatora chrysantha	Pertusaria hemisphaerica
Biatora sphaeroides	Pertusaria multipuncta
Catillaria alba	Ramonia chrysophaea
Catillaria atropurpurea	Sclerophora nivea
Celothelium ischnobelum	Strigula taylorii
Cladonia parasitica	Thelotrema lepadinum
Collema subflaccidum	

All these species are very rare in the Lothians, most with very small populations at no more than two localities. The woodlands with the highest concentration of old woodland indicators are Newhall, near Carlops, and Woodhall Dean, with just one or two species present at other sites, such as Birky Bank (Crichton), Linn Dean, Roslin, Temple, East Lammermuir Deans and Humbie.

Conifer plantations are generally poor for lichens, owing to the lack of continuity with former native woodland, the very acidic bark of the trees and dense shade. However, where better lit conditions prevail because of poor tree growth or the presence of the deciduous larch, a luxuriant, albeit species-poor, lichen growth can be found. In an area of stunted pines at Monynut Forest, four of the yellow beard-lichens (*Usnea* spp.) can be seen in abundance, and larch twigs in the upper part of Pressmennan Wood have been found with *Cetraria sepincola*, a lichen largely confined to birch. Trackside clearings and banks in the aforementioned plantations are an example of where a 'respectable' list of lichens from stones and consolidated soil has been recorded.

The lignum of large tree stumps is generally colonised by common *Cladonia* species (e.g. *C. coniocraea*, *C. macilenta* and *C. polydactyla*), *Lepraria incana*, *Micarea prasina* and *Placynthiella* spp., but many other species can be found, including the less common *Cladonia digitata*, *C. ochrochlora*, *Lepraria umbricola* and *Micarea melaena*, the 'old woodland' *C. parasitica*, and the rare *C. incrassata* (Binning Wood).

The woodland habitat may also encompass substrata other than trees, and some of these are dealt with in sections below. Not dealt with, however, are niches such as bare earth banks by paths or ditches, which can be colonised by inconspicuous lichens such as *Psilolechia clavulifera*, a powdery-grey crust, and *Thrombium epigaeum*, which looks like a green-algal scum. Where more 'mossy', the banks can also support *Cladonia* species, such as *C. caespiticia*, or the inconspicuous *Bryophagus gloeocapsa* overgrowing hepatics. In the Lothians, mossy rock outcrops under tree cover are mostly devoid of lichens, unlike in the west of Scotland, although such rocks at Rockville Heughs host a large population of *Peltigera horizontalis* in its only Lothian locality. Dry rock underhangs within woodlands are generally more rewarding to the lichenologist. The species found mostly occur also on north- to east-facing outcrops in open country, but characteristic 'woodland' species on hard acidic rocks include *Bacidia trachona, B. viridifarinosa, Enterographa zonata, Opegrapha gyrocarpa, O. multipuncta* and *Verrucaria elaeina*.

Wayside and Parkland Trees

The ravages of atmospheric pollution and, in the fertile lowlands, abundant use of fertilisers and slurry, result in the lichen flora of such trees being disappointing throughout the region. In many areas, tree boles can be seen covered in an abundant 'furry' growth of *Evernia prunastri*, together with abundant grey foliose thalli of *Hypogymnia physodes* and *Parmelia sulcata*, or, where affected by fertilisers, a whitish covering of *Diploicia canescens* and common species of *Phaeophyscia, Physcia, Physconia* and, perhaps, with the orange-yellow, leafy thalli of *Xanthoria* species, and the yellow-green, shrubby *Ramalina farinacea*. Despite the cover being high, the biodiversity is generally very low, not attaining the high biodiversity frequently encountered on such trees in Angus, East Perthshire and Berwickshire. The best examples of such trees occur in well-lit but sheltered situations, where there is a margin of rough ground affording some protection to the trees from fertiliser drift, or in parkland where the grassland has not been 'improved'. It is in such parkland that *Arthonia pruinata* (at Smeaton House) and *Parmelia acetabulum* (at Middleton Hall) have been found. A favoured host for many lichens of wayside trees is elm, and as a result of Dutch elm disease many 'elm lichens' have severely declined, including the Red Data List species *Bacidia incompta* and *Caloplaca luteoalba*. In the Lothians, the future of *B. incompta* is precarious, it being now known only from a single surviving elm in open woodland below Auldhame Castle. *Caloplaca luteoalba* is no longer known on elms in the region, but it survives in one of its few, British saxicolous sites, on the wall of Innerwick Castle.

Trees standing in the open can be regarded as diaspore traps, potentially offering a home to propagules carried on the wind or by passing birds. At two sites in the Lammermuir Hills a few thalli of *Bryoria capillaris*, a species characteristic of pine-birch woods in the Central and Eastern Highlands, have been found on exposed beech trunks. Similarly, at the beech avenue near Bavelaw there is a healthy population of *Hypogymnia farinaria*, another native pinewood species. Of an opposite, southern affinity, and found in local abundance on beech and sycamore at Smeaton and Tyninghame, is *Parmelia soredians* – its nearest British localities being in East Anglia.

Worked Timber

Weathered fence-posts, railings, gates, etc., can become abundantly covered by lichens, comprising a mixed array of species normally found on bark, on naturally occurring lignum, or on exposed rocks. This habitat has greatly declined in recent decades owing to miles of fencing being replaced rapidly as a result of grant-aiding, and the use of modern timber treatments. Industrial and agri-pollution have also had a detrimental effect. However, good examples can be found in the region, especially in the foothills where there are still some locally abundant populations of characteristic species such as *Cyphelium inquinans* and *Thelomma ocellatum*.

Inland Rock Outcrops

The natural rock outcrops of the Lothians hold much more of national significance than do the woodlands. The most important are the basaltic outcrops of Lower Carboniferous origin. Of these, the jewel in the crown is Traprain Law, but diverse floras with many noteworthy species are found on the extensive outcrops at, for example, Holyrood Park, Braid Hills, Corstorphine Hill, North Berwick Law and Binny Craig, as well as on numerous smaller ones, many of which have yet to be explored lichenologically. The high lichen biodiversity (approximately 200 species) found on the Traprain Law laccolith is a result not only of the nature of the sub-stratum, but of the fact that there are outcrops in all directions and of varying slope. On and alongside the frequently flushed rocks on the north side is a strong sub-montane component, with several species not known elsewhere in south-east Scotland, such as *Catapyrenium lachneum*, *Collema glebulentum*, *Melaspilea granitophila*, *Placynthium flabellosum*, *Polyblastia cruenta*, *Porina guentheri*, *Porocyphus coccodes*, *Protothelenella sphinctrinoides*, *Pyrenopsis grumulifera* and *P. impolita*. Other sub-montane species near the summit are *Ochrolechia inaequatula*, *Ophioparma ventosa* and *Umbilicaria deusta*. Notable on the east side is the nationally scarce *Rimularia limborina*, but it is the south and west-facing outcrops that have the most obvious lichen cover to the lay eye, with an abundance of foliose (leafy) lichens, especially the yellow-green *Parmelia conspersa* and the pale brown *P. loxodes*, together with the darker brown *P. glabratula* ssp. *fuliginosa* and *P. verruculifera* and the rare, almost black, *P. disjuncta*. The ultra-basic rocks at Braid Hills are the type locality for *Lecanora andrewii*, originally collected there by James McAndrew in the early twentieth century. It is still there in abundance, as it is at the Lion's Haunch in Holyrood Park and on the south side of North Berwick Law. Lichens on rock are comparatively less affected by air pollution than are those on trees, and there are still fine saxicolous lichen communities to be found within the City of Edinburgh in Holyrood Park, Braid Hills and Corstorphine Hill. At Holyrood, noteworthy species include *Anaptychia runcinata* (normally on sea-shore rocks), *Pertusaria amarescens* with its obligate parasite *Dactylospora saxatilis*, the metalophytic *Lecanora epanora* (also at Binny Craig) and, on calcareous rocks above Duddingston Loch, *Aspicilia radiosa* and *Diplotomma venustum*. Noteworthy occurrences on smaller basaltic outcrops, all of north-facing aspect, include *Arthonia endlicheri* (otherwise known in Britain from south-west England and South Wales) and *Lecanactis latebrarum* (only site in south-east Scotland) from Rockville Heughs, a large population of *Ramalina polli-naria* at Blaikie Heugh, and the common *Lecanora rupicola*, but with its rare, obligate parasite *Opegrapha glaucomaria*, at Lawhead Hill near Tyninghame. Lawhead Hill also has the only known site in south-east Scotland for the calci-colous *Lecania cuprea*, growing in a vesicle lined with large calcite crystals. Finally, the west-facing outcrop at Hairy Craig is largely dominated by a large, fertile population of *Lecanora pannonica* in its only known Scottish locality, and is the only known British occurrence on a natural rock face. This species is common, and usually sterile, in eastern England on sandstone walls, especially the boundary walls of churchyards.

Study of outcrops of other rock types, as in the Pentland and Lammermuir Hills, have with a few exceptions, so far revealed little other than the normal, low-altitude, acidic flora found throughout much of upland Britain. However, much more careful exploration is required, and rock-faces or boulders of high interest can be very localised. The occurrences on calcareous Lower Old Red Sandstone conglomerate of *Gyalecta ulmi* (Red Data Book listed Endangered) in Woodhall Dean, and of *Polyblastia theleodes* in Sheeppath Glen, cover areas of less than a metre square, but are the only known sites in south-east Scotland. A further example is *Arthonia arthonioides* on sandstone outcropping by the River North Esk at Newhall.

Seashore Rocks

The extensive outcrops of hard intrusive rocks along the coast provide ideal sub-strata for maritime lichens, with the best examples being found between Aberlady and Dunbar. In places, the characteristic shore zonation can be seen, with a eulit-toral 'black' zone dominated by *Verrucaria maura*, together with *V. mucosa*, *V. striatula*, below the mesic-supralittoral 'orange zone' of the orange lichens, *Caloplaca marina*, *C. thallincola* and *Xanthoria parietina*, on a background grey of mostly *Lecanora helicopis*. Often present in the respective upper and lower part of these zones is the tiny black shrub-like *Lichina confinis*. Its less common relative, *Lichina pygmaea*, which was originally described from the Forth as a small seaweed by Lightfoot (1777), is found at the lower part of the 'black zone' on exposed shores, such as at The Leithies and Frances Craig. Above the 'orange zone' is seen the more species-rich 'grey zone' of the xeric-supralittoral with the conspicuous grey-green 'sea-ivory', *Ramalina siliquosa* (Plate 9), and sometimes, the related *R. cuspidata* and *R. subfarinacea*, the brown foliose *Anaptychia runcinata* and many whitish to grey crusts, that include *Aspicilia leprosescens*, *Caloplaca ceracea* (Plate 9), *Lecanora rupicola*, *Lecidella asema*, *Ochrolechia parella*, *Pertusaria pseudocorallina* and *Tephromela atra* (Plate 9).

Where there are cliffs behind the shore, a similarly diverse flora may inhabit the vertical faces, with species such as the white, often wide-spreading *Haematomma ochroleucum* var. *porphyrium*, with its finely powdery surface, the minutely lobate, orange *Caloplaca arnoldii*, and in crevices the grey, squamulose *Solenopsora vulturiensis*. The coastal rocks are sometimes calcareous in part, such as the vent agglomerates outcropping to the east of North Berwick, providing niches for characteristic, but very localised, limestone species such as *Aspicilia radiosa*, *Catapyrenium squamulosum*, *Toninia sedifolia* and *T. verrucarioides*.

Where the rocks close to the shore are low and partially surrounded by short-turf, several lichens can take advantage, and among the more notable of these are *Anaptychia ciliaris* ssp. *mamillata* (Hummell Rocks and The Leithies) and *Catapyrenium cinereum* (St Baldred's Cradle). The softer, sandstone seashore rocks are more easily eroded and mostly have a depauperate lichen flora. Appearances can be deceiving. The flat, sandstone rocks exposed at low tide on the north side of the Tyne estuary appear to be devoid of lichen cover. However, the brown coloration of these rocks is an almost total cover of *Pyrenocollema orustense* and *Verrucaria ditmarsica*, two species that have very thin thalli, minute black perithecia, and are characteristic of brackish, estuarine conditions.

Rocks in Rivers

As with the seashore rocks, rocks in rivers and streams can support a range of specialist lichens, given that the rock is not too rapidly eroded and that there is suitably high water quality. These habitats merit more study in the Lothians, but data gathered to date indicate that *Verrucaria aquatilis*, *V. hydrela* and *V. praetermissa* are the commonest species, with *Bacidia inundata* being of more local occurrence. Rarer species in the region include *Collema flaccidum* and *Dermatocarpon meiophyllizum* found at the Linn Rocks in the River Tyne at East Linton.

Small Boulders and Stones

These can be in a miscellany of situations, such as in open ground in alluvial plains, in and alongside paths and tracks, in scree slopes in open situations or within woodland, and in debris from derelict buildings and walls. They are mostly colonised by pioneer or ruderal species, many of which are easily overlooked, as is their habitat. Common species on acid rocks include *Porpidia crustulata*, *P. soredizodes*, *Rhizocarpon reductum*, *Trapelia coarctata* and *T. placodioides*. But, 'leaving no stone unturned', persistent scrutiny may reveal much less common species, such as *Catillaria atomarioides* and *Rinodina orculariopsis* (open habitats) or *Verrucaria dolosa* (damp or shaded habitats), or even nationally rare species such as *Micarea curvula* (by the headwaters of the Whiteadder Water) and *M. parva* (Newhall; the type locality). Common species on calcareous stones include *Caloplaca holocarpa*, *Verrucaria muralis* and *V. nigrescens*, but the less common *Thelidium minutulum* and *Verrucaria murina* can be found in shaded situations.

Walls, Buildings and Monuments

Old walls that are built of local stone carry a flora similar to that of the local natural outcrops, at least with regard to the dominant species. However, it is interesting to observe that the proportions of cover may vary, often with species such as the bright yellow-green 'map lichen', *Rhizocarpon geographicum*, and the white *Lecanora rupicola* and *Ochrolechia parella* assuming greater predominance on walls than on nearby outcrops. These and many other species are more common on walls and gravestones than they are on natural outcrops. This is especially true of calcicolous species that have taken advantage of the use of mortar, concrete, asbestos cement and imported marble, for example *Caloplaca decipiens*, *C. flavescens*, *C. saxicola*, *C. teicholyta*, *Collema crispum*, *Lecania erysibe*, *Protoblastenia rupestris*, *Phaeophyscia nigricans* and *Xanthoria elegans*. The ideal places to study lichens of stonework are churchyards, and many Lothian churchyards await detailed surveys. Some, such as those at Crichton, Humbie and Preston Kirk, East Linton, have been studied already and found to support in the order of 100 species. A rare British species of nutrient-enriched rocks, often associated with bird colonies and bird-perching stones, is *Ramalina polymorpha*. It is found in such habitats as a small colony near the summit of North Berwick Law, and was seen in 1998 as two small depauperate colonies on Bass Rock, where conditions for it appear to be becoming too enriched as a result of the increasing gannet population of recent years. In East Lothian, *Ramalina polymorpha* has an alternative habitat in the form of the three standing stones at Pencraig Hill, Traprain and Standingstones Farm, thus adding to the stones' archaeological importance.

Heathland and Moorland

The heathland and moorland in the Lothians has suffered greatly from past and present land use, especially muirburn, over-grazing and conversion to 'improved' pasture. Nevertheless, many heathland lichens survive in the region, albeit in small, fragmentary populations. Most characteristic of these habitats are species of *Cladonia* subg. *Cladina*, the 'reindeer lichens' (Plate 19). These are still to be found in refugia in the upland areas, and on acid grey-dunes such as Sandy Knowe, north of Gullane. Of more local occurrence is *Cetraria islandica*, 'Iceland moss', still persisting, but under threat of overgrazing by sheep, on summit ridges in the Pentland Hills. Apart from that on the acid dunes at Sandy Knowe, the most luxuriant *Cladonia*-dominated community recorded so far is at the top of Faucheldean Bing. Although *Cladonia portentosa* is the only 'reindeer lichen' present, species of subg. *Cladonia* here include *C. crispata*, *C. furcata*, *C. glauca*, *C. gracilis*, *C. ramulosa*, *C. scabriuscula* and *C. subulata*.

Grassland and Dunes

Lichens are not usually obvious in inland grasslands, but can be present where the soils are of low nutrient status and have not been 'improved' by fertiliser application. The lichens to be found can be those characteristic of heathland, but include species often present also in garden lawns, especially *Cladonia furcata* and the 'dog lichens', *Peltigera lactucifolia* and *P. membranacea*. However, it is the coastal, dune grasslands where a higher lichen biodiversity is to be found. The finest examples are along the coast (but not continuously) from Aberlady Bay to Bathan's Strand

(Tyninghame), Belhaven Bay and Barns Ness. Apart from the aforementioned three species, the commoner species are *Cladonia fimbriata*, *C. humilis*, *C. rangiformis*, *Collema tenax* var. *ceranoides*, *Peltigera canina* and *P. rufescens*, and less frequent are *Cladonia ciliata*, *C. portentosa*, *Leptogium gelatinosum* and *L. schraderi*. On short, rabbit-grazed calcareous turf, rosettes of the squamules of *Cladonia pocillum* can be noticeable and these signal that closer scrutiny is worthwhile. In such habitats some of the smaller, rarer and more demanding crustose lichens can be found, such as *Diploschistes muscorum* (which begins as a parasite on *Cladonia pocillum*), *Agonimia globulifera*, *Bacidia bagliettoana*, *Chromatochlamys muscorum*, *Polyblastia agraria*, *Rinodina conradii*, *Sarcosagium campestre*, *Thelocarpon impresellum* and *Verrucaria bryoctona*.

Post-industrial Habitats

Sites containing old mine waste and abandoned quarries can provide important habitats for lichens, although many such habitats are being further modified, to the detriment of lichens, by landscape design initiatives and landfill. In the Lothians, the finest examples of such habitats are the bings, or rather those that have escaped landscape improvement. The 'heathland' at Faucheldean has already been mentioned. In 1999, an even more intriguing find was made at this bing, namely a small population of the terricolous *Stereocaulon saxatile*, which is otherwise regarded as a montane species of the Highlands. Another, sub-montane to montane species, *Peltigera venosa*, was recorded from Philpstoun in the 1970s, but attempts to refind it have failed. Particularly impressive at Philpstoun Bing are the extensive colonies of the crustose, saxicolous *Stereocaulon* species, *S. leucophaeopsis*, *S. nanodes* and *S. pileatum*. Of these three metalophytic species, the first is so far known in the Lothians only from bings, while the last two are otherwise exceedingly rare. The road-metal quarry at Traprain, abandoned in the 1970s, has fortunately not been 'managed' and has provided some noteworthy records of species not recorded elsewhere in the Lothians. These include *Aspicilia moenium* (on concrete; only British record), and on soil, *Cladonia cariosa*, *Leptogium biatorinum* and *L. byssinum* (second British locality).

Final Remarks

This résumé is based on personal records made since the author came to the region in 1974. The records have mostly been made in an *ad hoc* manner, and do not result from a systematic survey. Lichenological colleagues who have accompanied me and provided new records are Steen Christensen, Sandy Coppins, Peter Earland-Bennett, Stefan Ekman, Alan Fryday, Tony Fletcher, Oliver Gilbert, Paul Harrold, Peter James, Sergey Kondratyuk, Colin Pope, Charles Rawcliffe, John Sheard and Ray Woods. Hundreds of suitable localities and niches remain to be explored and many more hundreds of records need to be made before the lichens of the Lothians can be considered well known. Even so, conditions are forever changing with, for example, the dramatic decline in sulphur dioxide levels reflected in the rapid decline of *Lecanora conizaeoides* from trees and sandstone walls, and the rise in ammonia levels reflected by an increased occurrence of *Xanthoria poly-carpa*. Future changes in air and water quality, agricultural policies and practices, use of building materials, and many others, will affect lichens, offering many opportunities for observation and research.

9

A Bryophyte Flora of the Lothians

D. F. CHAMBERLAIN

Introduction

About half of the bryophytes (mosses, liverworts and hornworts) that are recorded in Britain occur in the Lothians, despite a climate that is not particularly favourable. The highest ground reaches only 580m, so the truly alpine element of the British bryoflora is not represented. However, the bryophyte flora of the Lothians is rich compared to that of other predominantly lowland eastern regions in northern Britain because of the complex and varied geology of the area, and the extensive coastal dunes of East Lothian.

The Sources of Records

Literature

Greville published the earliest effective list of the bryophytes of the Lothians in 1824 in his *Flora Edinensis*. This list was based on his own records, augmented by those of Maughan, Arnott, Don, J. E. Smith and others, and contained almost 200 species. The records related largely, though not exclusively, to the Edinburgh district, to the adjacent parts of the Pentland Hills, and to the stream and river valleys at Bilston, Roslin and Auchendinny. A few West Lothian records were republished in the 1845 New Statistical Account of Scotland. Few of these early records are supported by herbarium specimens in the collections at Edinburgh. Some of the records not confirmed recently almost certainly comprise extinctions. However, there is some doubt as to the exact identity of some of the entries, especially as concepts of synonymy have changed since 1824.

Some literature records exist from the late 1860s (Sadler 1868; Bell and Sadler 1869). A more specific list relating to the species of *Grimmia* on Arthur's Seat in Edinburgh (Bell and Sadler 1870) is, for the most part, supported by herbarium specimens. The first edition of the *Moss Exchange Club Census Catalogue* (1907) presented bryophyte records of the known species for the vice-counties, those for VC 82 – East Lothian, VC 83 – Midlothian and VC 84 – West Lothian being relevant to the present account. Later lists (McAndrew 1912; Adam 1917; Evans

1917) were published, partly to update the 1907 *Catalogue* and partly to authenticate the entries, as it was not always clear on which records the *Census* was based. A list of the species of *Riccia* in the Edinburgh district (Evans 1905) provides an interesting period snapshot by which to judge modern records of the genus. Apart from new vice-county records published by the Moss Exchange Club and latterly by the British Bryological Society very little has appeared in print since 1917 that relates to the bryophytes of the Lothians.

Herbarium Records: 1800–1970

While there is an extensive historical record provided by specimens from the turn of the nineteenth and twentieth centuries in the Herbarium of the Royal Botanic Garden Edinburgh, there are relatively few extant specimens of Lothians bryophytes collected before 1880 and in the period between 1920 and 1970. There are a few specimens collected by Sadler between 1867 and 1870, and one or two specimens from the Greville Herbarium dating from the period between 1819 and 1847. The herbaria of W. and W. E. Evans, and of McAndrew, present a fairly comprehensive coverage of the bryophytes of the Lothians between 1889 and 1915. There are a few specimens collected in East Lothian by J. B. Duncan between 1927 and 1932, and by E. Beattie and U. K. Duncan in 1967. There are also a few specimens collected by A. McG. Stirling in West Lothian in the 1960s. The collectors, along with the period and vice-county or -counties in which they collected, are listed in Table 9.1.

Table 9.1 Collectors of herbarium specimens of bryophytes in the Lothians

Collectors	Period when collecting	Vice-counties in which collections were made
Adam, J. C.	1916	WL, EL
Arnott	pre-1824	
Averis, B.	post-1990	EL
Beattie, E	1967–70	EL
Bell, N. (with D. F. Chamberlain)	1999	WL
Boyd, W.	1869–70	ML, EL
Brown, J.	1864	ML
Chamberlain, D. F.	1966–present	WL, ML, EL
Crundwell, A. C.	1964–67	WL, EL
Dixon, C.	post-1990	ML
Don, G.	1806	ML
Donaldson, B. P.	1963	EL
Duncan, J. B.	1909–29	EL
Duncan, U. K. (with E. Beattie)	1967	EL
Evans, W. (Herbarium)	1868–1905	WL, ML, EL

Evans, W. E.	1904–9	ML, EL
Evans, W. W.	1847–51	WL, ML, EL
Fairley, H. A.	1955	EL
Greville, R. K. (Herbarium)	1819, 1847	ML
Harper, G. H.	1993	EL
Henderson, D. M.	1953–60	ML
Jurand, M. K.	1967	EL
Laing, J.	1847	ML
Long, D. G.	1969–present	WL, ML, EL
Lowe (Herbarium)	1855	ML
Lyall, ?W.	1860	ML
Lightowlers, P.	1978–80	ML
Maughan, R. and E. (Herbarium)	nineteenth century (1807)	ML
McAndrew, J.	1898–1913	WL, ML, EL
Monington, H. W.	1869	EL
Morton, H. (with M. K. Jurand)	1967	EL
Murray, A.	1897	ML
Nagy, L.	post-1990	EL
Nichol	1856	ML
Paton, J. A.	1971	ML
Ratcliffe, D. A.	1960–61	ML, EL
Revell, R. D.	1971	EL
Rothero, G. P.	2000	ML, EL
Sadler, J.	1867–70	WL, ML
Saville, R.	1992–present	WL, ML, EL
Sinclair, J.	1946	ML, EL
Stark, R. M.	1845	ML
Stirling, A. McG.	1964–71	WL, EL
Stewart, J.	pre-1824	
Townsend, C. C.	1971	ML
Wiggington, M.	1972, 1992	WL
Young (Herbarium)	1898	WL
Warburg, E. F. (with A. C. Crundwell)	1964	ML

Modern Records: Post-1970

By 1971, the British Bryological Society Mapping Scheme was well under way so David Long and the present author decided to undertake the first systematic survey of the Lothians, based on the 10km Ordnance Survey Grid. Specimens of the more significant finds have been incorporated into the Edinburgh Botanic Garden Herbarium. This survey now represents thirty years of fieldwork and accounts for more than 90 per cent of the modern records. These records constitute the major proportion of the species listings in this account.

The Impact of Geology on the Bryophytes of the Lothians

The most significant effect of geology in the area is to be seen in the bryoflora of the extensive outcrops of basalt that are generally associated with past volcanic activity along the south side of the Forth Estuary. These extend from Cockleroy and Binny Craig in West Lothian, through the volcanic plugs and lavas around Dalmahoy and the City of Edinburgh, to the East Lothian outcrops around Traprain and North Berwick Law, and on to the coast at Aberlady and Dirleton. Some of the species that are largely restricted to basalt in Lothians are basalt specialists, others are more ubiquitous elsewhere in Britain on hard rock surfaces and rock ledges, or owe their existence to the influence of a more base-rich form of basalt, known as trachyte, on the overlying substrates.

Species of *Grimmia* are, or have been, particularly well represented on the Edinburgh basalts. These include *G. anodon*, *G. laevigata*, and *G. orbicularis*, three species that are probably now extinct in the Lothians. Despite these losses, it is heartening that *G. decipiens*, *G. montana* and *G. longirostris* still occur within the boundaries of the City of Edinburgh and the first two of these species are still to be found, along with *G. ovalis* on the East Lothian basalts. *Coscinodon cribrosus* and *Schistidium pruinosum* are associates on the Edinburgh basalts, while *Hedwigia integrifolia* occurs only on the basalt of Traprain Law.

The dark colour of the basalt allows the rock to absorb incoming radiation, raising the temperature of the surface significantly above the ambient air temperature. This may be a factor that allows *Reboulia* and perhaps some other species to survive on basalt outcrops. Other epilithic (rock-growing) species that are largely restricted to basalts in the Lothians include: *Andraea rothii* var. *rothii*, *Cynodontium bruntonii*, *Pterogonium gracile*, *Rhabdoweisia crispata*, *R. fugax* and *Weissia controversa* var. *crispata*. All five occur on a range of different rock types elsewhere in Britain. *Porella obtusata* is known in the Lothians only from damp basalt on North Berwick Law, but it too is not confined to this rock type elsewhere. This is an interesting record for a species that generally has an Oceanic distribution in Britain.

Of the sedimentary rocks, the sandstones, mostly Carboniferous, make the most extensive impact on the bryoflora, especially where rivers have cut deep valleys through them. These sandstones are generally base-rich and may be associated with even more highly base-rich bands of limestone. In Midlothian, sandstones outcrop along the valley of the Water of Leith in Edinburgh and along the valleys of the River Esk and its tributaries, from Dalkeith, through Bilston and Roslin to Penicuik, and to Newhall on the Peebleshire border. *Fissidens crassipes*, *Rhynchostegiella teneriffae* and *Amblystegium tenax* are common associates on damp sandstone rocks along these river systems. *Tetradontium brownianum* and *Barbula sinuosa* occur more rarely on these sandstones. *Calypogeia integristipula* and the nationally endangered *Orthodontium gracile* are restricted to the sandstone cliffs along the River North Esk at Roslin. It is assumed that the single unconfirmed record for *Brachydontium trichodes* was from sandstone. In East Lothian, sandstone has a major influence on the bryoflora of the Lammermuir, Woodhall, Thornton and Dunglass Deans. *Conardia compacta* has been recorded from Dunglass, and there are also old records of this species from Roslin. The single record for *Leiocolea heterocolpos* comes from sandstone at Edgelaw. *Plagiothecium cavifolium* is also generally, though not exclusively, associated with sandstone in the Lothians.

Conglomerate outcrops are much more restricted in size and occurrence but are markedly base-rich. The most significant site on conglomerate encompasses the waterfalls on the Logan Burn at The Howe, (called Nether Habbie's Howe in old records) between Threipmuir and Loganlea in the Pentland Hills. There are modern records from this site for *Riccardia incurvata*, *Scapania aequiloba*, *S. cuspiduligera*, *Anomobryum filiforme* var. *concinnatum*, *Bryoerythrophyllum ferruginascens*, *Encalypta ciliata*, *Orthothecium intricatum*, *Philonotis arnellii*, *Plagiobryum zieri*, *Schistidium pruinosum*, *S. strictum*, *Seligeria donniana*, *S. recurvata*, *Syntrichia princeps*, and *Tortula subulata* var. *graeffii*. *Platydictya jungermannioides* has been found on conglomerate at Woodhall Dean in East Lothian. The nationally scarce *Jungermannia subelliptica* and *Schistidium platypyllum* are confined to damp base-rich rocks by streams, and *Barbilophozia atlantica* is known from boulders around Torduff in the Pentland Hills.

Thin limestone bands are often associated with the carboniferous sandstones. There are, however, significant outcrops of limestone near Bathgate in West Lothian, at Fullarton, Middleton, Newhall and Vogrie in Midlothian, and at Barns Ness, on the coast east of Dunbar. *Apometzgeria pubescens*, *Anomodon viticulosus*, *Pterygoneuron ovatum* and *Seligeria pusilla* occur on natural and man-made humid limestone rock faces in Midlothian. *Aloina rigida* is associated with limestone, but is also recorded from basalt in the Lothians. *Thuidium philibertii* and *Entodon concinnus* are limestone grassland species that occur in the Lothians.

Species restricted to more acid rocks include *Kiaeria blyttii* (on cornstone) and *Racomitrium affine* (on felsite), close to Logan Burn in the Pentland Hills.

Significant Habitats for Bryophytes in the Lothians

Coastal Habitats

The dune systems of East Lothian, at Longniddry Bents, Aberlady, Gullane, Yellow-craig and Belhaven support, or have supported, a range of different habitats that are significant for their bryofloras. *Rhynchostegium megapolitanum* has been recorded from the grey dune turf at Aberlady. The more base-rich turf associated with trachyte outcrops and/or shell sand supports *Brachythecium mildeanum*, *Distichium inclinatum*, *Ditrichum flexicaule* s.s., *Encalypta raptocarpa*, *Racomitrium elongatum*, *R. canescens* s.s., *Rhytidium rugosum* and *Thuidium abietinum* ssp. *abietinum*. *Tortella inclinata* and *Tortula protobryoides* occur on bare areas associated more directly with the trachyte. *Bryum* spp. are well represented on the dunes, especially on sand from which turf has been stripped. The nationally vulnerable *Bryum warneum*, which is probably the most significant of these, is scattered over the dunes from Aberlady to Belhaven. *B. dunense*, *B. intermedium*, and *B. knowltonii* have occurred, or still occur, as associates. Extensive dune slacks occur at Aberlady from which *Bryum calophyllum*, *Campyliadelphus elodes*, *Drepanocladus polygamus* and *D. sendtneri* are recorded. *C. elodes* is also recorded from Belhaven, associated with *Amblyodon dealbatus*.

Sea cliffs replace the soft coastal dunes around Tantallon in East Lothian. Here *Tortula atrovirens* and *Weissia longifolia* var. *longifolia* are to be found. *Riccia beyrichiana* is now apparently restricted to ledges on maritime rocks at Blackness Castle in West Lothian, though it has also been recorded from dried mud by reservoirs.

Bog and Moorland

The Lothians are well endowed with raised bogs, including, in West Lothian, Blawhorn National Nature Reserve and Easter Inch Moss; in Midlothian, Red Moss at Threipmuir, Auchencorth Moss and the bogs around Gladhouse; and in East Lothian, Fala Flow. Nationally scarce species recorded include *Cephalozia loitlesbergeri*, *Cephaloziella spinigera*, *Kurzia sylvatica*, *Hypnum imponens* and *Sphagnum flexuosum*. Blanket bog communities occur in the Pentland Hills but are more extensive in the Moorfoot Hills, especially on the Dundreich Plateau and around Eastside Heights, from where *Calypogeia azurea*, *Sphagnum austinii* and *S. fuscum* are recorded. The nationally scarce *Cephalozia catenulata*, *Diplophyllum obtusifolium* and *Campylopus pyriformis* var. *azoricus* have been found on moorland banks and exposed peat.

Base-rich springs in upland moorland and mires are comparatively infrequent in the Lothians and are usually very limited in extent. The most significant of these are to be found below East Kip and at the head of Medwin Water in the Pentland Hills, along the upper reaches of the River South Esk above Gladhouse in the Moorfoot Hills, and along the border with Berwickshire in the Lammermuir Hills. *Cephalozia pleniceps*, *Moerckia hibernica*, *Sphagnum teres* and *Tomenthypnum nitens* are known in the Pentland Hills. *Sphagnum cortortum* and *S. teres* occur along the River South Esk above Gladhouse, and *Pseudobryum cinclidioides* is known only in the Lammermuir Hills.

Woodland

Epiphytes are best represented on the branches of elder trees, the most notable being *Cryphaea heteromalla*, *Orthotrichum tenellum*, *Pylaisia polyantha* and *Zygodon conoideus*. *Dicranum montanum* has been recorded on tree trunks around Temple and at Vogrie in Midlothian. *Platygyrium repens* occurs on willow and elder in East Lothian, and *Plagiothecium laetum* has been recorded on tree bases in the southern part of Midlothian. *Antitrichia curtipendula* has occurred in the Lothians on mature tree trunks, but it also grows on rock faces.

Man-made Habitats

The reservoirs, especially those that are situated along the Pentland Hills, are the most significant of the man-made habitats The nationally vulnerable *Riccia canaliculata* is found at Threipmuir and Harlaw, the nationally threatened *Weissia rostellata* at Clubbiedean, and *Physcomitrium sphaericum* at Harperrig. Other species recorded include *Fossombronia incurva, Riccia cavernosa, Atrichum tenellum, Bryum tenuisetum, Philonotis caespitosa, Aphanoregma patens* and *Pohlia flexuosa.* Most of these species are dependent on mud on dried-up reservoir bottoms, a habitat that is threatened by changes in management practices.

Shale bings, which have been important sites for *Buxbaumia aphylla* and for *Aloina brevirostris,* are now also a threatened habitat. *Haplomitrium hookeri* has been recorded on disturbed stony soil at Loganlea. The only record of *Pohlia filum* is from bare soil in a quarry in the Moorfoot Hills, and *Anthoceros agrestis* is known only from waste ground at New Park. *Ephemerum serratum* var. *serratum,* and *Tortula acaulon* var. *pilifera* have also been recorded on disturbed ground.

Changes in the Bryoflora of the Lothians

Assessment

This account provides significant records from three periods. The first is represented by Greville's lists in the *Flora Edinensis* and covers the first quarter of the nineteenth century. Our knowledge of the second period, around the turn of the nineteenth and twentieth centuries, largely comes from the extensive collections of W. and W. E. Evans and J. McAndrew. D. Long and the author have contributed the great majority of the records for the period between 1971 and the present. These three snapshots over the past 175 years allow some tentative conclusions to be made concerning losses and gains in the Lothians bryoflora.

The authenticity of old literature records that are not backed by herbarium specimens is open to question, partly because there can be doubt arising from misunderstandings of synonymy and partly because of reliance on the quality of the naming. However, non-confirmation of some of these records in modern times probably indicates extinctions. The locality information accompanying early literature records and specimens is often not detailed enough to be certain that the taxa could be re-found. In any case, some bryophytes have very specific ecological requirements so that small changes may eliminate them or cause them to migrate. Thus, a comparison based on the relatively crude 10km grid used since 1971 has to be treated with some caution.

Some extinctions are recorded in the literature. A proportion of these are directly due to habitat destruction. Habitat modification, either directly as a result of human activity such as drainage or farming practices, or indirectly through air or water pollution, has undoubtedly resulted in the loss of some species. Some of the losses reflect national trends and are therefore to be expected; over-collection may have led to the demise of a few other species. A measure of possible extinctions may be provided by expressing the number of unconfirmed, pre-1950 records as a percentage of the total number of taxa recorded in any one 10km square (Table

9.2). While the individual species records are not necessarily significant, the over-all figures may indicate a trend.

Segregate Species Recognised Recently

Given the difficulties posed by lack of transport, even into the beginning of the twentieth century, it is not surprising that the fieldwork carried out during the first two periods was so patchy. It is therefore meaningless to quote figures of the new vice-county records published since 1950 as an indication of major change. Furthermore, a significant number of these represent a more detailed understanding of the taxonomy of some genera especially *Schistidium, Bryum, Pohlia, Plagiothecium, Kurzia* and *Calypogeia*. A number of recently recognised segregate species could not therefore have been recorded. The following list includes those newly recognised species for which the first vice-county record for the Lothians was published post-1950.

Calypogeia azurea
Fossombronia incurva
Marchantia polymorpha
   ssp. *ruderalis*
Metzgeria temperata
Plagiochila britannica
Scapania scandica
Bryum bornholmense
Bryum dunense
Bryum gemmiferum
Bryum klinggraeffii
Bryum radiculosum

Bryum ruderale
Bryum sauteri
Bryum subapiculatum
Bryum tenuisetum
Bryum violaceum
Dicranella staphylina
Ditrichum cylindricum
Ditrichum gracile
Drepanocladus cossonii
Fissidens limbatus
Hypnum andoi
Plagiothecium cavifolium

Plagiothecium curvifolium
Plagiothecium laetum
Platygyrium repens
Pohlia bulbifera
Pohlia camptotrachela
Pohlia filum
Pohlia flexuosa
Pohlia lutescens
Racomitrium elongatum
Schistidium crassipilum
Schistidium strictum

Gains

Those species that have been recorded locally for the first time within the past fifty years, but were recognised as occurring elsewhere in the British Isles, are listed below. These are genuine gains. It is possible, however, that many of them were overlooked by the early collectors. Three of the species that are recent additions to the Lothians, namely *Didymodon sinuosus*, *Tortula atrovirens* and *Weissia longifolia*, may be indicators of climate change due to global warming. All three are rare in Scotland and are at, or close to, the northern limit of their range in Britain. There is no obvious reason why the remaining species have been overlooked. They are found in a wide range of habitats, and some occur in well-worked localities. Two introduced species, *Campylopus introflexus* and *Orthodontium lineare*, are clearly spreading. *Dicranum tauricum*, which may have been introduced accidentally through forestry, is apparently also expanding its range within the Lothians.

Anthoceros laevis
Calypogeia integristipula
Calypogeia neesiana
Calypogeia sphagnicola
Cephalozia loitlesbergeri
Cephalozia macrostachya
Cephalozia pleniceps
Cephaloziella rubella
Cephaloziella spinigera
Diplophyllum obtusifolium
Haplomitrium hookeri
Hygrobiella laxifolia
Kurzia sylvatica
Leiocolea heterocolpos
Lejeunea lamacerina
Lepidozia pearsonii
Nardia compressa
Riccardia incurvata
Scapania aequiloba
Scapania cuspiduligera

Atrichum tenellum
Bryoerythrophyllum
    ferruginascens
Campylopus piriformis var.
    azoricus
Coscinodon cribrosus
Dicranodontium denudatum
Dicranum montanum
Didymodon ferrugineus
Didymodon sinuosus
Hennediella stanfordensis
Kiaeria blyttii
Philonotis caespitosa
Platydictya jungermannioides
Pseudobryum cinclidioides
Pylaisia polyantha
Racomitrium ericoides
Rhabdoweisia fugax
Rhynchostegium
    megapolitanum

Rhytidium rugosum
Seligeria pusilla
Sphagnum contortum
Sphagnum flexuosus
phagnum inundatum
Sphagnum quinquefarium
Sphagnum russowii
Sphagnum teres
Sphagnum warnstorfii
Thuidium abietinum ssp.
    abietinum
Tortula atrovirens
Weissia brachycarpa var.
    brachycarpa
Weissia longifolia var.
    longifolia
Weissia rostellata
Zygodon conoideus
Zygodon rupestris

Losses

Twenty-two species and varieties, recorded in the Lothians more than fifty years ago, have not been confirmed over the past thirty years. For some of these there are published accounts that suggest that they are true extinctions. During the development of Luffness Golf Course some of the springs were filled in with sand with the consequent loss of an important base-rich mire community. As a result, *Meesia uliginosa* and *Catascopium nigritum* were lost from their only known site in the Lothians, along with *Amblyodon dealbatus* and *Moerckia hibernica* (the last two species occur elsewhere in East Lothian and in Midlothian).

Bell and Sadler (1870) list five species of *Grimmia*: *G. anodon*, *G. laevigata*, *G. longirostris*, *G. orbicularis* and *G. ovalis* that occurred on the basalt rocks of Edinburgh. By 1912, McAndrew considered it likely that all five were extinct. While no physical event can explain their loss, air pollution resulting from the expansion of the City of Edinburgh in the late Victorian era may have been responsible. All these species grow on exposed rock faces and are therefore particularly sensitive to increased levels of air pollutants. It is also just possible that over-collecting could have been a contributory cause, especially as Arthur's Seat was a well-known site for these rarities. While both *G. longirostris* and *G. ovalis* have been found at other Lothian localities, the other three can be assumed to be extinct as extensive field-work has failed to locate them.

It seems likely that *Ptilium crista-castrensis* has gone from its only known local station, under firs near Threipmuir. *Bryum uliginosum*, for which there is a record from Dirleton Common, dated 1897, has probably become extinct, as there is now no suitable habitat in this area on which it could exist. Several attempts to find *Plagiopus oederianus* at The Howe (Nether Habbie's Howe) in the Pentland Hills have failed, so it too is presumed to be extinct. While *Cladopodiella fluitans* has not been refound on Bavelaw Marsh recently, mire communities that could support it are still there. It is therefore unwise to write this species off. Most of the remaining species that have not been confirmed recently have imprecise locality details or have not been searched for. These are:

Acaulon muticum	*Bryum pseudotriquetrum*	*Hamatacaulis vernicosus*
Andreaea rothii var. *rothii*	var. *bimum*	*Microbryum rectum*
Antitrichia curtipendula	*Cynodontium jenneri*	*Tortula subulata* var
Brachyodontium trichodes	*Fissidens limbatus*	*angustata*
	Grimmia curvata	*Anthoceros laevis*

Habitat loss and pollution have undoubtedly impoverished the bryoflora, especially around the main conurbations. It is, however, difficult to assess the losses that may be due to climate change, though this may well have been a factor in the possible demise of *Antitrichia curtipendula*. The loss of mortared walls around Edinburgh has reduced the occurrence of *Aloina rigida* and *Pterygoneuron ovatum*, though both are still present on natural substrates, particularly limestone. The man-made shale bings have been important for bryophytes. In particular, several bings have supported populations of *Buxbaumia aphylla*. Over the last twenty years bings have been exploited as a source of aggregate for road building and some no longer exist. It is not known how drastic the impact of this loss will be on *Buxbaumia*.

A more general indicator of loss of species may be seen from the statistics presented in Table 9.2. While it is probable that some of the unconfirmed taxa are still present, where the percentage of the unconfirmed records exceeds 15 per cent it is likely that there have been significant extinctions. Four 10km squares fall into this category. Square 06 has been adversely affected by the spread of Livingston, and by tree planting on Drumshoreland Muir. Square 17 has undoubtedly been affected by the extension of Edinburgh westwards and by developments at, and to the east of, South Queensferry. Square 26 has been affected by developments in south Edinburgh and square 27 has been affected by developments in central, north and east Edinburgh.

West Lothian and the environs of Edinburgh have suffered from high levels of industrial air-borne pollution. However, there is some evidence that the cleaner air of the past twenty years in these areas is having an ameliorating effect on the bryoflora. The clearest indication of this may be seen in the return of some of the epiphytic species particularly *Orthotrichum* species and *Cryphaea heteromalla*. The apparent recent spread of *Zygodon conoideus* may also reflect better air quality. The distribution of historical and recent records for four species is shown in Figures 9.1 to 9.4.

Table 9.2   *The number of bryophytes recorded in each 10km square*

Sq. no.	Landmarks	Vice-county	Post-1950	Pre-1950 (H)	Pre-1950 (nH)	%
1	*2*	*3*	*4*	*5*	*6*	*7*
86	Blawhorn Moss	WL	85	1	0	1
95	Leven Seat, Miller's Moss	ML*	77	0	0	0
96	Addiewell, Bathgate, Fauldhouse	WL, ML	171	7	4	7
97	River Avon, Cockleroy, Torphichen	WL	190	13	8	10
98	Bo'ness, Kinneil	WL	80	0	2	2
05	Cobbinshaw and Crosswood Reservoirs	ML	157	3	4	4
06	Mid Calder and West Calder, Livingston	WL, ML	187	12	8	10
07	Broxburn, Linlithgow, Hopetoun	WL	162	23	7	16
08	Blackness Castle	WL	95	3	1	4
15	Newhall	ML	232	9	0	7
16	Threipmuir, Balerno, Pentland Hills	ML	321	30	3	10
17	Cramond, Ratho, South Queensferry	WL, ML	164	41	6	2
24	Dundreich Plateau, near Gladhouse	ML	200	0	0	0
25	Penicuik Estate, Gladhouse	ML	234	10	3	5
26	South Edinburgh, Roslin, Pentland Hills	ML	285	62	7	20
27	Edinburgh	ML	229	39	5	16
34	Ladyside and Eastside, Moorfoot Hills	ML	202	0	0	0
35	Temple, Middleton	ML	239	6	0	2
36	Dalkeith, Gorebridge, Vogrie	ML*	191	8	3	5
37	Musselburgh	ML, EL	64	5	4	12
44	Stow	ML	137	0	0	0
45	Fala Moor, Soutra	ML, EL	251	0	0	0
46	Saltoun, Pencaitland	ML, EL	180	2	4	3
47	Tranent, Longniddry	EL	119	17	1	13
48	Aberlady Bay	EL	168	18	5	12
56	Hopes Reservoir, Gifford	EL	240	13	4	6
57	Haddington, Traprain, East Linton	EL	194	15	6	10
58	North Berwick, Yellowcraig	EL	157	19	2	12
66	Dunbar Common, Whiteadder Reservoir	EL	99	4	1	2
67	Woodhall Dean, Dunbar	EL	262	3	1	1
68	Tyninghame Coast	EL	91	0	2	2
76	Lammermuir Deans, Burnhope	EL	120	5	4	4
77	Barns Ness, Sheeppath, Dunglass	EL	215	1	8	4

Column **1** – 10km square numbers: **86–98** in Ordnance Survey 'NS'; **05–77** in Ordnance Survey 'NT'.

Column **2** – landmarks in 10km squares.

Column **3** – vice-counties, *indicates less than 0.5km² in another Lothians vice-county.

Column **4** – post-1950 records.

Column **5** – pre-1950 records, backed by herbarium specimens, but not confirmed for more than 50 years.

Column **6** – pre-1950 literature records, not backed by herbarium specimens or confirmed for more than 50 years.

Column **7** – pre-1950 records expressed as a percentage of the total number of records.

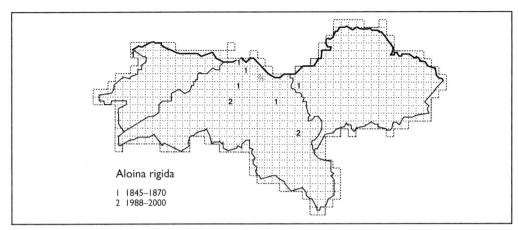

Figure 9.1 *Map of the Lothians showing the history of the recording of* Aloina rigida. *This nationally scarce species has declined following the loss of mortared walls.*

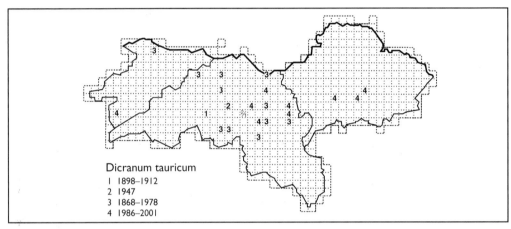

Figure 9.2 *Map of the Lothians showing the history of the recording of* Dicranum tauricum. *Generally frequent, though rare in the east, this species is probably spreading.*

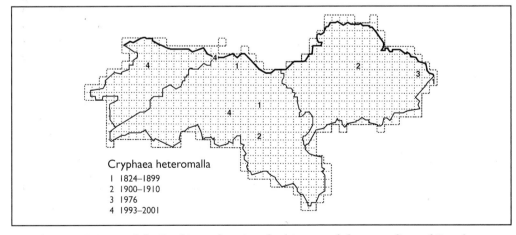

Figure 9.3 *Map of the Lothians showing the history of the recording of* Cryphaea heteromalla, *occasional, though possibly spreading as air quality improves.*

104

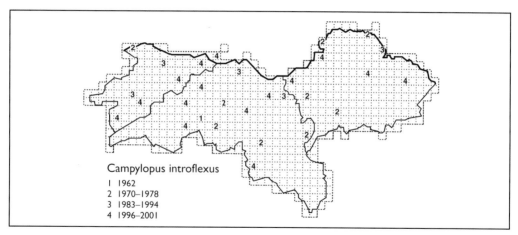

Figure 9.4 *Map of the Lothians showing the history of the recording of* Campylopus introflexus, *an introduced species, widespread and spreading.*

The Layout of the Bryophyte Flora

The Flora account that follows lists all the Lothians bryophyte genera, species, subspecies and varieties and indicates the 10km squares (Figure 9.5) from which they have been recorded. The names, authorities and reference numbers follow those used in the most recent check-list (Blockeel and Long 1998). The name is given in italics when a species, subspecies or variety is almost certainly extinct, or where all the records are considered to be mis-identifications. Synonyms are cited in brackets after the name only when confusion might arise following a recent change in the status of a name.

The vice-counties in which each taxon occurs now or has occurred are cited using a two-letter code: WL – West Lothian, ML – Midlothian, EL – East Lothian. When there are no post-1950 records, the two-letter code is enclosed in brackets. Round brackets indicate that there are records before 1950 that are confirmed by herbarium specimens; square brackets denote that there are literature records before 1950 that are not confirmed by herbarium specimens.

The status in the Lothians is given and the national status assigned by the British Bryological Society is added for those taxa that are considered to be rare or scarce in the British Isles. A brief statement of the main habitats in which each of the taxa occurs in the Lothians then follows.

Where a taxon is represented in more than eight 10km squares, the squares are included as a list of two-figure codes (see Table 9.2 and Figure 9.5). Where eight or fewer 10km squares are represented, a full citation is given of at least one record that authenticates each square, giving precedence to post-1950 records. Where the only record or records are for specimens collected before 1950, then the 10km square code is enclosed in brackets. Round brackets denote records that are confirmed by herbarium specimens and square brackets denote literature records that are not confirmed by herbarium specimens. Where three squares or fewer are represented, then either all the records are included, or a selection of the available records is presented that indicates the period of time over which the records were made. Some old records are too imprecise to allocate definitely to a 10km square. In these instances, the square number is followed by a query.

The first record for a vice-county is indicated by {VCR}; in the case of pre-1950 VCR's which have been confirmed by post-1950 collections, the first modern record {MVCR} is given. These specimens are housed in either the Edinburgh (E) or the British Bryological Society (BBSUK) Herbaria.

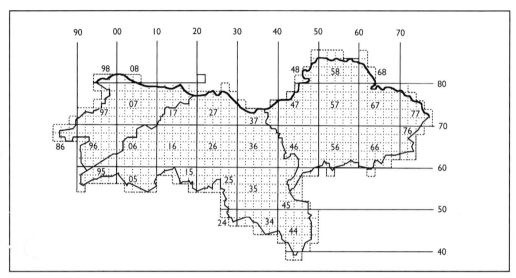

Figure 9.5 *Map of the Lothians showing the 10km squares surveyed.*

LIVERWORTS (MARCHANTIOPSIDA)

1. **Haplomitrium** Nees

1. H. hookeri (Sm.) Nees   **ML**
Once seen, nationally scarce.
Roadside.

16 - {*VCR ML*}; near Water House, Loganlea
Reservoir, Pentland Hills, 1971, J. A. Paton and
C. C. Townsend.

4. **Blepharostoma** (Dumort.) Dumort.

1. B. trichophyllum (L.) Dumort.   **ML, EL**
Frequent.
Moist base-rich rocky substrates.

15, 16, 24, (26), 35, 45, 66, 67, 77?
67/77 - {*VCR EL*}; calcareous old red sandstone
rocks, Sheeppath Dean, near Innerwick, March
1968, Stirling.

5. **Trichocolea** Dumort.

1. T. tomentella (Ehrh.) Dumort.   **ML, EL**
Occasional.
Base-rich flushes, sub-montane.

05, 16, 24, 25, 34, 66, 67, 77.
77 - {*VCR EL*}; ledge in wet shaded ravine,
Sheeppath Dean, June 1967, E. Beattie and
U. K. Duncan.

7. **Kurzia** G. Martens

1. K. pauciflora (Dicks.) Grolle (*Lepidozia
setacea* (Web.) Mitt.)
     ML, (EL)
Rare, but probably overlooked.
Damp peat in bogs, on the side of ditches and among
Sphagnum.

15 - Auchencorth Moss, September 1971,
J. A. Paton.
16 - Balerno Common, 260m, 22 July 1988,
D. G. Long 15170.
(56) - on peaty ground, Faseny Water, above
Gifford, 12 September 1903, W. Evans.
[66] - Dunbar Common, before 1910, W. Evans.

2. K. sylvatica (A. Evans) Grolle   **ML, EL**
Habitat and frequency as above, nationally
scarce.

15 - Auchencorth Moss, September 1971,
J. A. Paton.

16 - {*VCR ML*}; bog, west of Threipmuir
Reservoir, 1971, J. A. Paton.
45 - {*VCR EL*}; boggy ground, Fala Moor, 1973,
D. F. Chamberlain and D. G. Long.

3. K. trichoclados (Müll. Frib.) Grolle
     (WL), ML, EL
Frequency and habitat as above.

(97) - Crawhill, near Westfield, by River Avon,
26 September 1903, W. Evans.
15 - peat bog, Auchencorth, September 1971,
J. A. Paton.
(26) - Hawthornden, 6 June 1902, W. Evans.
34 - Rough Moss, Blackhope, near Garvald
Punks, 593m, 13 August 1977, D. G. Long.
56 - Meikle Says Law, 1991, D. G. Long.

8. **Lepidozia** (Dumort.) Dumort.

1. L. reptans (L.) Dumort.   **WL, ML, EL**
Common.
Organic soil in woods, tree stumps, amongst rocks.

95, 96, 97, 98, 05, 06, 15, 16, 17, 24, 25, 26, 27,
34, 35, 36, 44, 45, 46, 47, 56, 57, 66, 67, 76, 77.

2. L. pearsonii Spruce   **ML**
Rare.
Moorland, under *Calluna*, sub-montane.

16 - {*VCR ML*}; bank, near Logan Burn
Waterfall, 310m, 1971, J. A. Paton.
34 - Moor N of Eastside Heights, 520m, 26 June
1976, D. G. Long.

10. **Calypogeia** Raddi

1. C. fissa (L.) Raddi   **WL, ML, EL**
Common.
Woodland banks, soil in ravines, peat in bogs.

86, 96, 97, 98, 06, 07, 08, 15, 16, 17, 24, 25, 26,
27, 34, 35, 36, 44, 45, 46, 47, 56, 57, 66, 67, 77.

2. C. muelleriana (Schiffn.) Müll. Frib.
     WL, ML, EL
Common.
Moorland, peaty banks in woodland, ditch sides.

86, 95, 96, 97, 05, 06, 15, 16, 24, 25, 26, 27, 34,
35, 36, 44, 45, 46, 56, 66, 67.
96/06 - {*VCR WL*}; peaty bank amongst boggy
scrub, Easter Inch Moss, near Bathgate, 1971,
D. G. Long and D. F. Chamberlain.
66 - {*VCR EL*}; peat moor, Moss Law, 1972,
D. F. Chamberlain *et al.*

3. C. azurea Stotler & Crotz (*C. trichomanis*)
Auct., non (L.) Corda   **ML, EL**
Rare and nationally scarce.
Damp peat on moorland, sub-montane.

24 - blanket bog, 550m, Dundreich Plateau, 18
July 1988, D. F. Chamberlain.
34 - {*VCR ML*}; on wet peat slopes at 580m,
Eastside Heights, Moorfoot Hills, 26 June 1975,
D. G. Long and D. F. Chamberlain, DGL 5068.
56 - Meikle Says Law, 1991, D. G. Long.

4. C. neesiana (C. Massal. & Carestia) Müll.
Frib.   **ML**
Rare.
Damp peat on moorland and in bogs.

05 - wet moorland, Bawdy Moss, 2 June 1974,
D. G. Long 3464.
15 - Auchencorth, September 1971, J. A. Paton.
16 - {*VCR ML*}; bog, west of Threipmuir, 1971,
J. A. Paton.

5. C. integristipula Steph.   **ML**
Rare and nationally scarce.
Dry shaded sandstone rock faces.

26 - {*VCR ML*}; Roslin Glen, 2 September 1971,
J. A. Paton and D. G. Long, DGL 1625.

6. C. sphagnicola (Arnell & J. Perss.) Müll.Frib.
          WL, ML
Rare but probably overlooked.
Amongst other bryophytes in peat bogs.

86 - Blawhorn Moss, Wiggington.
05 - basic peaty flush, near the source of Medwin
Water, 363m, 21 August 1976, D. G. Long 5189.
15 - {*VCR ML*}; peat bog, Auchencorth Moss,
September 1971, J. A. Paton.
26 - flushes among *Sphagnum rubellum*, south of
Caerketton, Kirk Burn, 330m, 31 July 1977,
D. G. Long 6139.

7. C. arguta Mont. & Nees   **WL, ML, EL**
Frequent.
Woodland and moorland banks.

96, 97, 98, 06, 16, 24, 26, 27, 34, 35, 36, 45, 46,
56, 66, 67, 77.

12. Cephalozia (Dumort.) Dumort.

1. C. bicupidata (L.) Dumort. (incl. ssp. *lammer-
siana* (Huebener) R. M. Schust.)   **WL, ML, EL**
Common.
Wet soil in bogs and marshes, wet rocks.

86, 95, 96, 97, 98, 05, 06, 07, 08, 15, 16, 17, 24,
25, 26, 27, 34, 35, 36, 44, 45, 46, 47, 56, 57, 66,
67, 76, 77.

3. C. catenulata (Huebener) Lindb.   **[WL], ML**
Rare and nationally scarce.
Damp peat by ditch, upland.

05 - {*VCR ML*}; damp peat by ditch, Bawdy
Moss, above Crosswood Reservoir, 1974,
D. G. Long *et al*.
[17] - Dalmeny, 1898, Young.

4b. C. macrostachya Kaal. var. **spiniflora**
(Schiffn.) Müll. Frib.   **WL, ML**
Rare.
Bogs, bare peat, sides of ditches.

86 - {*VCR WL*}; side of ditch, Blawhorn,
21 October 1975, D. F. Chamberlain and
D. G. Long.
15 - Auchencorth, September 1971, J. A. Paton.
16 - {*VCR ML*}; Threipmuir Bog, 1971,
J. A. Paton.
16 - The Common, Threipmuir, 11 August 1973,
D. F. Chamberlain and D. G. Long.

5. C. leucantha Spruce   **ML, (EL)**
Rare.
Bogs.

(05) - Baads, Shortings, near Addiewell,
(? Baadsmill), 30 April 1904, Ewing.
16 - peaty blanket bog, near Threipmuir
Reservoir, 11 August 1973, D. G. Long 2663.
(56) - on boggy ground near head of Faseny
Water, above Gifford, 12 September 1903,
W. Evans.
(56) - Lammer Law, Lammermuir Hills, 500m,
W. Evans.

6. C. lunulifolia (Dumort.) Dumort. **WL, ML, EL**
Frequent.
Peaty banks in woods, bogs and moorland.

86, (97, 05, 06), 15, 16, 25, 26, 34, 36, 45, 46,
(66), 67, 77.

7. C. pleniceps (Austin) Lindb.   **ML**
Rare and nationally scarce.
Base-rich upland flushes.

05 - {*VCR ML*}; Garvald Syke, at the source of
the Medwin Water, 370m, 21 August 1976,
D. G. Long *et al*. DGL 5190.

8. C. loitlesbergeri Schiffn. **ML**
Rare, once seen, nationally scarce.
Raised bogs.

25 - 1 mile west of Gladhouse, 4 November 1972, D. F. Chamberlain.

9. C. connivens (Dicks.) Lindb. **WL, ML, EL**
Frequent.
Bogs.

86, 96, 05, 06, 15, 16, [17], 24, 25, 26, 34, 45, 56.
45 - {VCR EL}; on peat, Fala Moor, June 1961, Birse.

13. **Nowellia** Mitt.

1. N. curvifolia (Dicks.) Mitt. **WL, ML, EL**
Frequent.
Rotting logs, in humid ravines.

96, 97, 06, 07, 15, 16, 25, 26, 35, 36, 44, 46, 47, 56, 58, 77.
77 - {VCR EL}; rotting log in ravine, Sheeppath Dean, near Oldhamstocks, November 1967, U. K. Duncan and E. Beattie.

14. **Cladopodiella** H. Buch

1. C. fluitans (Nees) H. Buch **(ML)**
Very local, probably extinct.
Fens and mires.

(16) - Balerno Moss, October 1902, J. McAndrew.
(16) - Bavelaw Moss, 12 February 1903, W. Evans.
(16) - Bavelaw Moss, 20 September 1911, J. McAndrew.

16. **Hygrobiella** Spruce

1. H. laxifolia (Hook.) Spruce **ML**
Rocks by streams, very local.

34 - {VCR ML}; rocks by Ladyside Burn, Moorfoot Hills, 1973, D. F. Chamberlain and D. G. Long.

17. **Odontoschisma** (Dumort.) Dumort.

1. O. sphagni (Dicks.) Dumort. **WL, ML, EL**
Locally abundant.
On bare peat and amongst other bryophytes, in bogs.

86, (96), 05, 06, 15, 16, 24, 25, 34, 45, 66, 67.

2. O. denudatum (Mart.) Dumort. **(WL), ML**
As above, but much rarer.

(96) - Fauldhouse Moor, 14 March 1903, W. Evans.
15 - Auchencorth, 2 September 1971, J. A. Paton.
(16) - Bavelaw Moss, 15 August 1911, J. McAndrew.
(16) - Ravelrig Toll Moss, 1 May 1909, W. E. Evans.
16 - damp peat on blanket bog, Threipmuir, 28 August 1971, D. G. Long 1508.

18. **Cephaloziella** (Spruce) Schiffn.

1. C. spinigera (Lind.) Warnst. (*C. subdentata* Warnst.) **ML, EL**
Rare but possibly overlooked; nationally scarce.
Peat bogs, amongst other bryophytes.

15 - {VCR ML}; Auchencorth, 2 September 1971, J. A. Paton.
45 - {VCR EL}; on peat, south-east margin of Fala Moor, 10 June 1973, D. G. Long 2511.

3. C. rubella (Nees) Dumort. **ML, EL**
Rare.
Damp soil, on rocks.

26 - on soil in crevices of a rock, in a young conifer plantation, 240m, South of Torduff Hill, 23 July 1977, D. G. Long 6093.
26 - {VCR ML}; boulder on bank of River North Esk, near Auchendinny, 140m, 15 May 1977, D. G. Long.
26 - cindery bank on old railway, 103m, Colinton, 5 March 1978, D. G. Long 6524.
27 - Corstorphine Hill, field record.
67 - {VCR EL}; damp soil by track in young conifer plantation, Pressmennan Wood, 2 January 1974, D. G. Long and D. F. Chamberlain.

4. C. hampeana (Nees) Schiffn. **(WL), ML, EL**
Occasional.
Bare organic soil on banks and in bogs.

(07) - Drumshoreland Moss, 9 September 1911, J. McAndrew.
15 - Auchencorth, 1997, D. F. Chamberlain.
16 - near Loganlea Reservoir, August 1971, J. A. Paton and C. C. Townsend.
26 - Roslin, field record.
35 - field record
45 - on *Bryum capillare* on rock in ravine below

Linn Dean Cascade, 12 August 1973, D. G. Long 2678.
58 - near North Berwick, 13 October 1906, Anon.

6. C. divaricata (Sm.) Schiffn.   WL, ML, EL
Frequent.
On organic soil on banks and in woodland.
96, 97, 07, 16, 17, 25, (26), 27, 35, 36, 56, 57, [77].

21. Barbilophozia Loeske

3. B. floerckei (F. Weber & D. Mohr) Loeske
     WL, ML, EL
Common.
Banks and rocks on hillsides, moorland.
95, 96, 97, 05, 06, 07, 15, 16, 24, 25, 26, 34, 35, 36, 44, 45, 48, 56, 57, 66, 67, 76, 77.

4. B. atlantica (Kaal.) Loeske   ML
Very local and nationally scarce.
Amongst rocks on hillside.
(16/26) - Clubbiedean, March 1902, McAndrew.
(26) - {VCR ML}; head of Torduff Reservoir, 19 November 1903, J. McAndrew.
26 - {MVCR ML}; exposed rocks, south-west end of Torduff Reservoir, 23 June 1998, D. G. Long 27785.

5. B. attenuata (Mart.) Loeske   WL, ML, EL
Frequent.
Amongst *Calluna* and rocks, generally in humid situations.
96, 97, 05, 15, 16, 25, 26, 34, (36), 46, (58), 67, 77.

6. B. hatcheri (A. Evans) Loeske   WL, ML, EL
Occasional.
Shaded rock face, amongst boulders.
97, 07, 16, 17, 26, 27, 45, 57, 67.
67 - {VCR EL}; shaded conglomerate rock face, Woodhall Burn, above Spott, 1971, D. F. Chamberlain and D. G. Long.

8. B. barbata (Schmidel ex Schreb.) Loeske
     WL, ML, EL
Rare.
Generally in the lee of stone walls on moorland.
97 - Carribber Glen, post-1971, D. G. Long and D. F. Chamberlain.
16 - on base of mossy wall by a stream, 360m, by

burn east of Scald Law, Pentland Hills, 7 May 1978, D. G. Long 6712.
(17) - Craigiehill Wood, Dalmeny, 24 September 1913, J. McAndrew.
(27) - Corstorphine Hill Wood, 1902, W. Evans.
48 - Luffness Links close to the sewage works, October 2000, G. Rothero.
57 - north side of Traprain Law, 18 September 1972, D. G. Long 2124.
66 - old stone wall, Hungry Snout, east of the Whiteadder Reservoir, 4 August 1973, D. F. Chamberlain and D. G. Long.

22. Anastrepta (Lindb.) Schiffn.

1. A. orcadensis (Hook.) Schiffn.   ML
Rare.
Dry hummock in bog.
16 - {VCR ML}; dry hummock, Threipmuir Bog, 1971, D. G. Long 1505.
[26] - Bonaly, 1871, Anon.

23. Lophozia (Dumort.) Dumort.

2. L. ventricosa (Dicks.) Dumort. (incl. var. *silvicola* (H. Buch) E. W. Jones ex R. M. Schust., etc.)
     WL, ML, EL
Common.
Banks in moorland and in woods.
86, 95, 96, 97, 05, 06, 07, 15, 16, 17, 24, 25, 26, 27, 34, 35, 36, 44, 45, 47, 48, 56, 57, 66, 67, 76, 77.

5. L. sudetica (Nees ex Huebener) Grolle   ML, EL
Occasional.
Damp rocks, upland.
15, 16, 24, 25, 34, 35, 45, 57, 76.
76 - {VCR EL}; rocks by Monynut Water, Laughing Law, 1972, D. F. Chamberlain *et al.*

6. L. excisa (Dicks.) Dumort.   WL, ML, EL
Frequent.
Bare organic soil, generally in moorland and on hillsides, rarely by the sea.
96, 06, 08, 16, 17, 25, 26, (27), 35, 36, (47), 48, 56, 57, 67, [77].

11. L. incisa (Schrad.) Dumort.,   WL, ML, EL
Occasional.
Bare organic soil on moorland and in bogs.
96, 05, 15, (16), 24, 26, (27), 34, 45, 56, 57.

96 - {*MVCR* WL}; dry peaty bank of drain, Easter Inch Moss, Seafield, D. F. Chamberlain.

13. L. bicrenata (Schmidel ex Hoffm.) Dumort.
                                                    WL, ML, EL
Frequent.
Rock ledges and bare soil in moorland.

96, 97, 06, 07, 15, 16, (17, 26, 27), 44, 48, 56, (58), 67, [77].

24. Leiocolea (Müll. Frib.) H. Buch

3. L. bantriensis (Hook.) Jörg.   **ML, EL**
Occasional.
Base-rich flushes, moist base-rich rock faces.

05, 16, 24, 25, 26, 56, 66, 67, 77.

5. L. alpestris (Schleich. ex F. Weber) Isov.
(*L. muelleri* (Nees) Joerg.)   **ML, EL**
Occasional.
Moist base-rich rock ledges.

05 - Crosswood Burn, field record.
16 - Logan Burn, August 1971, J. A. Paton.
24 - damp basic soil in ravine, River South Esk, above Gladhouse, 26 July 1975, D. G. Long and D. F. Chamberlain, DGL 4603.
(26) - Roslin Glen, October 1907, J. McAndrew.
(48) - Gullane Links, June 1907, J. McAndrew.
66 - Upper Burnhope Dean, 17 August 1996, D. F. Chamberlain.
67 - calcareous rocks, Upper Sheeppath Dean, 4 June 1972, D. G. Long and D. F. Chamberlain, DGL 1983.

6. L. heterocolpos (Thed. ex C. Hartm.) H. Buch
                                                    ML
Once collected, nationally scarce.
Basic sandstone rock face.

35 - {*VCR* ML}; by river in wooded valley below Edgelaw Reservoir, Temple, 1977, D. G. Long.

7. L. badensis (Gottsche) Jörg.   **WL, ML, EL**
Occasional.
Humid base-rich rock faces, dune hollow.

07 - basic rock face on slope above Mains Burn, at Ochiltree Mill, 8 October 1975, D. G. Long 4688.
17 - quarry near Cramond Mill, field record.
26 - damp calcareous soil, near Torduff Reservoir, 27 February 1971, D. G. Long 1196.

36 - near Trotter's Bridge, River South Esk, 13 January 1974, D. G. Long 3052.
45 - damp calcareous rock, top of waterfalls, Linn Dean Water, 3 June 1973, D. G. Long 2472.
(48) - Gullane Links, 26 Sept 1908, J. McAndrew.
58 - small damp dune hollow, Yellowcraig, near Dirleton, 3m, 30 July 1977, D. G. Long 6134.

8. L. turbinata (Raddi) H. Buch   **WL, ML, EL**
Frequent.
Humid base-rich rock ledges.

96, 15, 16, 17, 25, 26, 35, 36, 46, (47), 48, 56, 57, 67, 77.

25. Gymnocolea (Dumort.) Dumort.

1. G. inflata (Huds.) Dumort.,   **WL, ML, EL**
Frequent.
Wet peat in bogs and woodland.

86, 95, 96, 97, 05, 06, 07, 15, 16, 17, 24, 25, 26, 27, 34, 35, [36], 44, 45, 56, 66.

28. Anastrophyllum (Spruce) Steph.

1. A. minutum (Schreb.) R. M. Schust.
(*Sphenolobus minutus* (Schreb.) Berggr.)   **ML, EL**
Rare.
Bare peat on stony hillsides and in bogs, in upland areas.

05 - sandstone rock face, west bank of the Medwin Water above Medwin Head, 20 April 1975, D. G. Long 4507.
24 - peat hummock in blanket bog, 550m, 20 July 1998, D. F. Chamberlain
(26) - on dyke below Glencorse Reservoir, 23 April 1909, J. McAndrew.
(26) - Caerketton Rocks, 19 June 1902, W. Evans.
45 - {*VCR* EL}; Scottish Wildlife Trust Reserve, Linn Dean, 1990, D. G. Long.

29. Tritomaria Schiffn. ex Loeske

1. T. exsectiformis (Breidl.) Loeske   **ML**
Once seen.
Soil overlying old wall.

36 - {*VCR* ML}; near the River Tyne, below Crichton Castle, 21 September 1996, D. F. Chamberlain.

3. T. quinquadentata (Huds.) H. Buch
WL, ML, EL
Frequent.
Moorland, amongst rocks, old quarries, sand-dunes.

95, (97, 07), 15, 16, 17, 24, 25, 26, (27), 34, 35, 44, 45, [48], 56, 57, 58, 66, 67, 68, 76.

31. Mylia Gray

1. M. taylorii (Hook.) Gray **(WL), ML, (EL)**
Rare and probably decreasing.
Moorland banks, bogs.

(97) - Crawhill, near Westfield, 26 September 1903, W. Evans.
(06) - moss west of Harperrig, 30 April 1902, W. Evans.
15 - near the summit of East Cairn Hill, Pentland Hills, 17 June 1973, D. F. Chamberlain and D. G. Long.
(16) - Bavelaw Moss, 3 June 1907, J. McAndrew.
(25) - north of Cowieslinn, 16 May 1903, J. McAndrew (possibly VC 78).
34 - moorland, Eastside Heights, 26 June 1976, D. F. Chamberlain and D. G. Long, DFC B337.
(56) - Lammer Law, 11 October 1902, W. Evans.

2. M. anomala (Hook.) Gray **WL, ML, EL**
Occasional but often locally abundant.
Damp peat and amongst other bryophytes, in peat bogs.

86, 96, 05, 06, 15, 16, 24, 25, (26), 34, 45, (56), 66.

32. Jungermannia L.

2. J. atrovirens Dumort. **WL, ML, EL**
Frequent.
Sreams, wet cliff faces, generally on acid substrates.

95, 96, 97, 05, 07, 15, 16, 17, 24, 25, 26, 27, 34, 35, 36, 44, 45, 46, 56, 57, 66, 67, 76, 77.

3. J. pumila With. **WL, ML, EL**
Frequent.
Streams, damp cliff faces, generally on base-rich sub-strates.

96, 97, 15, 16, 17, 24, 25, 26, 34, 35, 36, 45, 56, 67, 77.
67 - {VCR EL}; shaded conglomerate in wooded gorge, Wood Burn, near Spott, 1974, D. G. Long *et al.*

6. J. exsertifolia Steph. **ssp. cordifolia** (Dumort.) Vana **(WL), ML, EL**
Occasional.
Rocks in fast-moving upland streams.

16, (17), 24, 25, 34, 35, 45, 46, 56.
45 - {VCR EL}; partly submerged in rocky stream, ravine of Linn Dean Water, Soutra Hill, 1974, D. G. Long *et al.*

7. J. sphaerocarpa Hook. **ML**
Rare.
Wet sandstone by burns.

05 - Crosswood Burn, field record.
(26) - Bilston Burn, October 1908, J. McAndrew.
26 - on sandstone just above water level, Bilston Burn, 3 February 1972, D. F. Chamberlain and G. Argent.

9. J. gracillima Sm. **WL, ML, EL**
Frequent.
Stony tracks, compacted soil, generally on base-rich soil.

96, 05, (06), 15, 16, 24, [25], 26, 34, 35, (36), 44, 45, 46, (56), 66, 67, 77.

11. J. hyalina Lyell **(WL), ML, [EL]**
Rare.
Rocks by and in streams.

(96) - near Bathgate, 9 September 1903, W. Evans.
15 - Newhall, August 1971, BBS Excursion.
[77] - Dunglass, McAndrew.

12. J. paroica (Schiffn.) Grolle **ML, EL**
Rare.
Rocks in and by streams.

24 - damp rocks by River South Esk, above Gladhouse, 28 July 1973, D. F. Chamberlain and D. G. Long.
26 - {VCR ML}; Old Red Sandstone rocks on wooded north slope, River North Esk, Roslin, 1973, D. F. Chamberlain *et al.*
34 - rocky valley on Ladyside Burn, 24 June 1973, D. F. Chamberlain and D. G. Long.
66 - {VCR EL}; rocks by stream, Hungry Snout, east of the Whiteadder Reservoir, 4 August 1973, D. F. Chamberlain and D. G. Long.

13. J. obovata Nees **ML, EL**
Rare.
Rocks by streams, generally on at least slightly base-rich substrates.

(16) - The Howe, 18 April 1902, W. Evans.
16 - near Logan Burn, 1971, J. A. Paton.
24 - rocks by River South Esk, above Gladhouse,
28 July 1973, D. F. Chamberlain and D. G. Long.
56 - shaded wet rock face by moorland stream,
below Bleak Law, Lammermuir Hills, 1974,
D. G. Long and D. F. Chamberlain.

14. J. subelliptica (Lindb. ex Kaal.) Levier
ML, EL
Rare and nationally scarce.
Rocks by streams.

24 - Rose Cleugh, Dundreich, above River South
Esk, 5 August 1974, D. F. Chamberlain and
D. G. Long.
34 - Wolf Cleugh, foot of Dewar Burn, 350m,
13 August 1977, D. G. Long 6165.
66 - {VCR EL}; by stream below White Castle,
Lammermuir Hills, 1973, D. F. Chamberlain and
D. G. Long.

33. Nardia Gray

1. N. compressa (Hook.) Grev. **ML**
Rare.
Sandstone rocks by burn.

05 - {VCR ML}; sandstone rocks by Crosswood
Burn, above the reservoir, 2 June 1974,
D. G. Long *et al.* 3454.

2. N. scalaris S. F. Gray **WL, ML, EL**
Frequent.
Acid soils, in moorland, hillsides, tracks.

96, 97, 05, 06, 15, 16, 17, 24, 26, 27, 34, 35, 36,
44, 45, [46], 47, 56, [57], 66, 67, 76, 77.

34. Marsupella Dumort.

1. M. emarginata (Ehrh.) Dumort.
a. var. **emarginata** **WL, ML, EL**
Frequent.
Moist rocks, by streams and on cliffs.

96, 97, 05, 15, 16, (17), 24, 25, 26, 27, 34, [36],
45, [48], 56, 57, 66, 67, 77.

b. var. **aquatica** (Lindenb.) Dumort. **(ML)**
Rare.
Wet rocks by streams.

(15) - Gutterford Burn, North Esk Reservoir, 14
October 1905, W. Evans.
(16) - The Howe, 5 May 1909, W. E. Evans.

4. M. funckii (F. Weber & D. Mohr) Dumort
ML
Rare.
Rock crevice by moorland stream.

[26] - Swanston, 1856, Anon.
44 - rock crevice by moorland stream, Luggate
Water, Overshiels, 1 July 1973, D. G. Long and
D. F. Chamberlain, DGL 2574.

37. Diplophyllum (Dumort.) Dumort.

1. D. albicans (L.) Dumort. **WL, ML, EL**
Common.
Rock ledges and on the ground amongst rocks and on
peat in moorland, restricted to acid habitats.

86, 95, 96, 97, 05, 06, 07, 08, 15, 16, 17, 24, 25,
26, 27, 34, 35, 36, 44, 45, 46, 47, 48, 56, 57, 66,
67, 76, 77.

3. D. obtusifolium (Hook.) Dumort. **EL**
Once recorded, nationally scarce.
Gravelly bank, sub-montane.

56 - {VCR EL}; south of Hopes Reservoir,
Lammermuir Hills, 1974, D. G. Long and
D. F. Chamberlain.

38. Scapania (Dumort.) Dumort.

1. S. compacta (Roth) Dumort. **WL, ML, EL**
Frequent.
Rock ledges, on dry boulders.

97, 05, (07), 15, 16, 17, 24, 25, 26, 27, 34, 45,
48, 57, 67, 68.

4. S. cuspiduligera (Nees) Müll. Frib. **ML**
Rare and nationally scarce.
Base-rich conglomerate, sub-montane.

16 - {VCR ML}; by Logan Burn Waterfalls,
310m, 31 July 1976, D. G. Long 5141.

5. S. scandica (Arnell & H. Buch) Macvicar
WL, ML, EL
Frequent.
Peaty soil, on moorland, mine waste, tracks.

96, 06, 15, 24, 26, (27), 34, 35, 44, 45, 46, 48,
56, 66, 67, 76, 77.
06 - {VCR WL}; mossy bank among *Calluna*, on
damp north-facing slope, approx. 360m, Five
Sisters Bing, 1998, D. G. Long 27688.
45 - {VCR EL}; Linn Dean Ravine, Soutra Hill,
29 April 1974, D. F. Chamberlain.

10. S. umbrosa (Schrad.) Dumort. **(WL), ML, EL**
Occasional.
On shaded boulders and tree trunks, in moist woodland.

(96), 05, 15, 25, 26, [36], 66, 67, 77.

11. S. nemorea (L.) Grolle   **WL, ML**
Occasional.
Banks and shaded boulders, generally in woodland.

05, 06, 15, 16, 17, [26], 27, 35, 36, 46.

12. S. irrigua (Nees) Nees   **WL, ML, EL**
Frequent.
Damp mineral soil, by springs and in marshes.

86, 96, (97), 05, 06, 07, 16, (17), 25, 26, 27, 34, (35), 46, 48, 56, 68.

15. S. undulata (L.) Dumort.   **WL, ML, EL**
Common.
Wet rocks, on cliffs and by and in streams.

86, 95, 96, 97, 05, 06, 15, 16, 17, 24, 25, 26, 27, 34, 35, 36, 44, 45, 46, 56, 66, 67, 76, 77.

16. S. subalpina (Nees ex Lindenb.) Dumort. **ML**
Rare.
Rocks in upland streams.

[05] - Crosswood Burn, before 1910, W. Evans.
16 - near Logan Burn Waterfall, August 1971,
J. A. Paton and C. C. Townsend.
(17) - Craigiehall Bridge, Midlothian, 30 August
1906, J. McAndrew.
34 - damp rocks in burn, Wolf Cleugh, Dewar
Burn, 360m, 13 August 1977, D. G. Long 6169.
44 - Luggate Water, near Overshiels, Stow, 1 July
1973, D. G. Long 2573.

19. S. aequiloba (Schwägr.) Dumort.   **ML**
Rare and nationally scarce.
Shaded moist base-rich upland rock face.

16 - {VCR ML}; rocks by Logan Burn Waterfall,
28 August 1971, J. A. Paton.
16 - rocks by Logan Burn Waterfall, 310m,
31 July 1976, D. G. Long 5137.

20. S. aspera Bernet   **ML, EL**
Rare.
Dry, often somewhat exposed base-rich rock.

26 - {VCR ML}; rocks by bridge, top end of
Torduff Reservoir, 4 November 1973, D. G. Long
and D. F. Chamberlain, DGL 2847.

35 - Whitelaw Cleugh, near Middleton, basic
rocks in small ravine, D. G. Long 1491.
45 - {VCR EL}; shaded rock face by waterfalls,
Linn Dean Water, 24 August 1975, D. G. Long
4655.
(48) - Gullane, J. McAndrew.

21. S. gracilis Lindb.   **WL, ML, (EL)**
Occasional.
Boulders, rock faces, etc, generally under trees.

06 - {MVCR WL}; damp north-facing slope,
mossy bank under *Calluna*, 160m, Five Sisters
Bing, 1998, D. G. Long 27689.
(07) - Riccarton Hills, 18 July 1902, W. Evans.
15 - field record.
16 - Logan Burn, 4 September 1998, D. G. Long
and D. F. Chamberlain.
(17) - Dalmeny shore, 4 November 1903,
J. McAndrew.
(56) - Lammer Law, 11 October 1902, W. Evans.
(58) - North Berwick Law, June 1907, W. Evans.

40. Lophocolea (Dumort.) Dumort.

1. L. bidentata (L.) Dumort. (incl. *L. cuspidata*
(Nees) Limpr.)   **WL, ML, EL**
Common.
Amongst grasses in marshes and bogs, on fallen tree
trunks.

All Squares.

3. L. heterophylla (Schrad.) Dumort.
    WL, ML, EL
Common.
Fallen timber in woodland

96, 97, 98, 06, 07, 08, 15, 16, 17, 25, 26, 27, 35, 36, 37, 44, 46, 47, 48, 56, 57, 58, 67, 68, 77.

41. Chiloscyphus Corda

1. C. polyanthos (L.) Dumort.   **WL, ML, EL**
Frequent.
Marshy areas and rocks by streams.

86, 96, 97, 98, 05, 06, 15, 16, 17, 24, 25, 26, 27, 34, 35, 36, 44, 45, 46, 56, 66, 67, 76, 77.

2. C. pallescens (Ehrh. ex Hoffm.) Dumort.
    WL, ML, EL
Occasional.
Base-rich flushes and marshes.

(96) - near Bathgate, 9 September 1903,
W. Evans.

97 - {*MVCR* WL}; Lochcote Marsh, near
Linlithgow, 18 June 2000, D. F. Chamberlain.
[05] - Baads, (?Baadsmill), Ewing.
16 - Harlaw Reservoir, field record.
24 - River South Esk, above Gladhouse, field
record.
25 - {*MVCR* ML}; among reeds at south-west
end of Gladhouse Reservoir, 5 November 1972,
D. F. Chamberlain.
[48] - Gullane Links, 1910, J. McAndrew.
(57) - Traprain Law, 29 August 1908,
J. McAndrew.
66 - Dunbar Common, field record.

46. Plagiochila (Dumort.) Dumort.

2. P. porelloides (Torr. ex Nees) Lindenb.
                                      WL, ML, EL
Common.
Woodland banks and rocks, sheltered cliffs.

96, 97, 98, 05, 06, 07, 08, 15, 16, 17, 24, 25, 26,
27, 34, 35, 36, 44, 45, 46, 47, 48, 56, 57, 58, 66,
67, 68, 76, 77.

3. P. asplenioides (L.) Dumort.   **WL, ML, EL**
As above though less common.

97, 98, 07, 08, 15, 16, 24, 25, 26, 34, 35, 36, 45,
46, 56, 57, 58, 66, 67, 76, 77.

4. P. britannica Paton   **ML**
Rare but possibly under-recorded.
Damp base-rich shaded rock faces and banks by
streams.

45 - Scottish Wildlife Trust Reserve, Linn Dean
Water, 1990, D. G. Long.
46 - {*VCR* EL}; on damp bank by a stream,
130m, Humbie Mill Wood, 25 April 1994,
D. G. Long 25313.
57 - shady bank by woodland stream, Howden
Burn, near Newbyth, 8m, 19 February 1995,
D. G. Long 25864.
66/76 - Barn Hope, post-1990, B. Averis.
67 - Woodhall Dean, post-1990, B. Averis.
67/76 - Sheeppath Glen, post-1990, B. Averis.

6. P. spinulosa (Dicks.) Dumort.   **ML, EL**
Rare.
Sheltered rock faces in ravines.

16 - {*MVCR* ML}; rock ledges in ravine, 310m,
Logan Burn Waterfalls, Pentland Hills, 1976,
D. G. Long 5128.

24 - rocks, Bowman's Gill Ravine, Dundreich,
Upper River South Esk, above Gladhouse, 26 July
1975, D. G. Long 4604.
34 - Wolf Cleugh, Dewar Burn, 13 August 1977,
D. G. Long.
45 - shady rocks in deep ravine, below Linn Dean
Cascade, D. G. Long and D. F. Chamberlain,
DGL 2681.

51. Radula Dumort.

1. R. complanata (L.) Dumort.   **(WL), ML, EL**
Frequent though possibly decreasing.
Sheltered tree trunks, occasionally also on rock faces.

(97), 16, (17), 26, 27, 35, 36, 44, 45, 46, 56, 58,
67, 68, 76, 77.

52. Ptilidium Nees

1. P. ciliare (L.) Hampe   **WL, ML, EL**
Frequent,
Boulders and among heather on moorland.

95, 97, 05, 07, 15, 16, 17, 24, 26, 27, 34, 35, 36,
44, 45, 48, 56, 57, 58, 66.

2. P. pulcherrimum (G. Weber) Vainio
                                      [WL], ML, EL
Rare and almost certainly decreasing.
Tree trunks, and on the ground.

[97] - on wood, Carribber Glen, before 1910,
J. McAndrew.
[16] - on the ground, near Bavelaw Castle, 6 May
1905, J. Hunter.
16 - on wet sandstone shales, Dalmahoy Hill,
March 1957, D. M. Henderson.
24 - on rowan, Rose Cleugh, Dundreich, Upper
River South Esk, above Gladhouse, 5 August
1974, D. F. Chamberlain.
[26] - Hawthornden, before 1910, Anon.
(27) - Corstorphine Hill, 21 January 1904,
J. McAndrew.
(36) - on a log, Arniston Woods, 25 October
1902, W. Evans.
56 - {*VCR* EL}; on willow trunk in alder carr,
Danskine Loch, May 1967, D. F. Chamberlain.

53. Porella L.

1. P. platyphylla (L.) Pfeiff.   **WL, ML, EL**
Rare.
Shaded base-rich rock faces.

26 - in grass at base of basalt crags, above and at

south end of Torduff Reservoir, 1 September 2000, D. F. Chamberlain and G. P. Rothero.

27 - Hangman's Rock, by Duddingston Loch, Holyrood Park, Edinburgh, 1 October 1971, D. G. Long and D. F. Chamberlain, DGL 1737.

27 - quarry face, Agassiz Rock, Blackford Quarry, Edinburgh, March 1976, D. F. Chamberlain B280.

27 - {MVCR ML}; shady cliffs under trees, west end of Easter Craiglockhart Hill, 105m, 2 February 1996, D. G. Long 26039.

67 - {MVCR EL}; on conglomerate outcrop in abandoned stream course, Woodhall, 1993, G. H. Harper.

2. P. cordeana (Huebener) Moore   **WL, ML, EL**
Frequent.
Shaded rocks and tree bases.

97, 07, 08, 15, (17), 25, 26, 27, 34, 35, 45, 46, 56, 57, 58, 67, [77].

3. P. arboris-vitae (With.) Grolle (incl. var. *obscura* (Nees) M. F. V. Corley)   **ML, EL**
Rare.
Shaded rocks.

16 - rock ledge close to lower waterfall, Logan Burn, 4 September 1998, D. G. Long and D. F. Chamberlain.

(26) - Bonaly Burn, 21 May 1910, J. McAndrew.

27 - {VCR ML}; shady cliffs under trees at west end of Easter Craiglockhart Hill, 2 February 1996, 105m, D. G. Long 26041.

45 - {VCR EL}; on basic rock face in small gully in ravine, Scottish Wildlife Trust Reserve, Linn Dean, 19 May 1990, D. G. Long 18118.

(57) - Traprain Law, 1908, J. McAndrew.

4. P. obtusata (Taylor) Trevis   **EL**
Known at only one locality; nationally scarce.
Wet north-facing basaltic rock face.

(58) - North Berwick Law, 27 October 1906, J. McAndrew.

58 - at north-east end of North Berwick Law, 16 August 1997, D. F. Chamberlain.

54. Frullania Raddi

1. F. tamarisci (L.) Dumort.   **(WL), ML, EL**
Frequent in less polluted areas.
Rocks and tree trunks.

(08), 15, 16, (17), 24, 25, 26, [27], 34, 35, 45, 48, (56), 57, 58, 66, 67, 68, 76, [77].

4. F. fragilifolia (Taylor) Gottsche *et al.*   **(ML), EL**
Rare.
Base-rich rock faces, also apparently on oak.

(26) - on oak, Roslin Woods, 27 May 1902, W. Evans.

45 - {VCR EL}; rock face, Linn Dean Cascade, 1973, D. F. Chamberlain and D. G. Long.

5. F. dilatata (L.) Dumort.   **WL, ML, EL**
Common.
Generally on tree trunks, less frequently on rocks.

97, 05, 07, [17], 24, 25, 26, 27, 35, 36, (37), 44, 45, 46, 47, 56, 57, 58, 67, 68, 77.

07 - {MVCR WL}; on old elder tree, 160m, Binny Craig, near Ecclesmachan, 1998, D. G. Long 26041.

60. Lejeunea Lib.

1. L. cavifolia (Ehrh.) Lindb.   **WL, ML, EL**
Frequent.
Moist shaded rocks.

97, 07, 16, (17), 24, (26), 34, 35, 45, 56, 57, 66, 67, 77.

07 - {VCR WL}; damp shale beside burn in steep wooded valley, near Abercorn, 1971, D. G. Long.

16 - {VCR ML}; wet basic conglomerate, Logan Burn Waterfalls, Pentland Hills, 1969, D. G. Long and D. F. Chamberlain, DGL 876.

77 - {VCR EL}; shaded rock face, Sheeppath Dean, near Innerwick, 1972, D. F. Chamberlain *et al.*

2. L. lamacerina (Steph.) Schiffn.   **EL**
Rare.
Moist base-rich rocks.

45 - {VCR EL}; calcareous rocks in shaded ravine, Linn Dean, below waterfall, 1974, D. G. Long and D. F. Chamberlain.

(57) - Traprain Law, 25 July 1908, J. McAndrew.

3. L. patens Lindb.
All records of this species are referable to *L. cavifolia*.

62. Cololejeunea (Spruce) Schiffn.

1. C. calcarea (Lib.) Schiffn.   **ML, EL**
Rare.
Moist base-rich rock faces.

16 - {VCR ML}; damp conglomerate rock face, upper waterfall, Logan Burn, 1972,

D. F. Chamberlain.
45 - damp shady rock face in ravine, Kate's Cauldron, Linn Dean Cascade, 260m, 18 June 1990, D. G. Long 18253.
67 - shady basic rock face in ravine, upper part of Sheeppath Dean, 4 June 1972, D. G. Long 1980.
77 - Sheeppath Dean, field record.

64. **Fossombronia** Raddi

5. F. pusilla (L.) Nees **WL, ML, EL**
Occasional.
Damp soil, by tracks, ditches and reservoirs.

07, 15, 17, 27, 36, 46, 56, 67, 68.
15 - {VCR ML}; dried mud at mouth of Henshaw Burn, North Esk Reservoir, D. F. Chamberlain and D. G. Long.
56 - {VCR EL}; damp disturbed clay soil, Darent House Moor, 1977, D. F. Chamberlain and D. G. Long.

7. F. wondraczekii (Corda) Dumort. ex Lindb.
(WL), ML, EL
As above.

[96], 06, (07), 16, 25, 26, 35, 46, 56.

8. F. incurva Lindb. **ML**
Rare and nationally scarce.
Dried mud by drained reservoirs.

16 - {VCR ML}; roadside, Loganlea, J. A. Paton and C. C. Townsend.
25 - Gladhouse Reservoir, 19 August 1973, D. G. Long.
35 - Gladhouse Reservoir, 31 August 1971, BBS Excursion.

66. **Pellia** Raddi

1. P. epiphylla (L.) Corda **WL, ML, EL**
Common.
Shaded banks, in woodland and moorland.

86, 95, 96, 97, 98, 05, 06, 07, 08, 15, 16, 17, 24, 25, 26, 27, 34, 35, 36, 44, 45, 46, 47, [48], 56, 57, 58, 66, 67, 68, 76, 77.

2. P. neesiana (Gottsche) Limpr. **WL, ML, EL**
Frequent.
Moist acid soil, by streams, in bogs and flushes.

86, 05, 24, 25, (26), 34, 35, 45, 56, 66, 67.
67 - {VCR EL}; roadside marsh 0.5miles south of West Steel, 1971, D. G. Long and D. F. Chamberlain.

3. P. endiviifolia (Dicks) Dumort. **WL, ML, EL**
Common.
Rocks by streams, flushes, woodland banks, sand-dunes, always on base-rich substrates.

96, 97, 98, 06, 07, 08, 15, 16, 17, 24, 25, 26, 27, 34, 35, 36, 45, 46, 47, 48, 56, 57, 58, 66, 67, 68, 76, 77.

68. **Moerckia** Gottsche

1. M. hibernica (Hook.) Gottsche **ML, (EL)**
Rare and nationally scarce.
Dune slacks, sub-montane flushes.

16 - {VCR ML}; base-rich flush, 370m, valley, north-west of Scald Law, Pentland Hills, 4 September 1998, D. G. Long and D. F. Chamberlain, DGL 28034.
(48) - Gullane Links, 19 September 1910, J. McAndrew.

69. **Blasia** L.

1. B. pusilla L. **WL, ML, EL**
Frequent.
Damp gravelly areas, by streams, in quarries.

96, 06, 15, 16, 24, 25, 26, 35, [36], 56, 66, 67, 76, 77.

70. **Aneura** Dumort.

1. A. pinguis (L.) Dumort. **WL, ML, EL**
Common.
Damp rock ledges, flushes, etc., in base-rich habitats.

95, 96, 97, 05, 06, 07, 15, 16, (17), 24, 25, 26, 27, 34, 35, 36, 44, 45, 48, 57, 58, 66, 67, 76, 77.

72. **Riccardia** Gray

1. R. multifida (L.) Gray **(WL), ML, EL**
Rare.
Flushes and bogs, generally in base-rich habitats.

(96) - on wet bank, roadside near Bathgate, 9 September 1903, W. Evans.
05 - {MVCR ML}; peat in base-rich flush, source of the Medwin Water, Garval Syke, 3 September 1975, D. G. Long et al. 4668.
(16) - near Balerno, November 1907, J. McAndrew.
24 - basic flush, below Emly Bank, 4 August 1974, D. G. Long 3666.
(26) - near Glencorse Reservoir, 23 April 1904,

J. McAndrew.

[37] - Inveresk, W. Evans, before 1910.

[47] - Gosford, before 1910, J. McAndrew.

[57] - Newlands, near Gifford, before 1910, W. Evans.

77 - damp flushes in ravine, Sheeppath Glen, near Oldhamstocks, 4 June 1972, D. G. Long 1966.

2. R. chamaedrifolia (With.) Grolle (*R. sinuata* (Hook.) Trev.) **WL, ML, EL**
Frequent.
Marshes and bogs.

97, 06, 16, 24, 25, 26, 27, 34, 35, 36, 44, 46, 48, 56, 57, 67, 77.

06 - {*VCR* WL}; damp soil at edge of stubble field, Easter Inch Moss, 1971, D. G. Long and D. F. Chamberlain.

16 - {*VCR* ML}; bank in valley north of Hare Hill, Pentland Hills, 1971, J. A. Paton and C. C. Townsend.

56 - {*VCR* EL}; damp soil in wood near Gifford, 1971, D. G. Long and D. F. Chamberlain.

3. R. incurvata Lindb. **WL, ML, EL**
Rare and nationally scarce.
Damp gravel, by streams, waste ground, on base-rich substrates.

06 - {*VCR* WL}; waste ground, Easter Inch Moss, 18 September 1971, D. G. Long and D. F. Chamberlain, DGL 1688.

16 - {*VCR* ML}; roadside near Water House, Loganlea Reservoir, September 1971, J. A. Paton and C. C. Townsend.

26 - gravel by burn, south-west end of Bonaly Reservoir, 348m, 24 July 1977, D. G. Long 6123.

34 - Wolf Cleugh, field record.

35 - Gladhouse Reservoir, field record.

56 - {*VCR* EL}; damp gravel by Sting Bank Burn, north-west end of Hopes Reservoir, 11 Aug 1974, D. G. Long and D. F. Chamberlain, DGL 3695.

4. R. palmata (Hedw.) Carruth. **WL, ML, EL**
Rare.
Rotting wood, in moist, shaded situations.

97 - {*VCR* WL}; rotten log on woodland margin, Castle, by Lochcote Reservoir, D. G. Long *et al*.

15 - rotting logs by River North Esk, Newhall Estate, 20 February 2001, D. G. Long.

(16) - {*VCR* ML}; rotten wood, near Balerno Moss, 1903, J. McAndrew.

17 - shaded rotting tree trunk, southern perimeter of Ratho Park Golf Course, 20 August 2000, D. F. Chamberlain.

36 - {*MVCR* ML}; log in old woodland, 180m, Birky Bank, near Crichton Castle, 1993, D. G. Long 25053.

56 - {*VCR* EL}; on log in boggy *Betula* woodland, 180m, valley north-east of Danskine Loch, 8 January 1995, D. G. Long 25831.

5. R. latifrons (Lindb.) Lindb. **ML**
Rare.
Bogs.

15 - Auchencorth Moss, August 1971, J. A. Paton.

(16) - Bavelaw Moss, August 1909, J. McAndrew.

16 - {*MVCR* ML}; bare peat, 260m, Threipmuir bog, 26 July 1978, D. G. Long 7142.

73. Metzgeria Raddi

1. M. fruticolosa (Dicks.) A. Evans **WL, ML, EL**
Frequent.
Well-lit tree trunks and branches, especially on *Sambucus*.

05, 06, 07, 15, 16, 17, 25, 26, 35, 36, 45, 46, 47, 56, 57, 58, 67, 68, 76, 77.

06 - {*VCR* WL}; bole of willow overhanging north bank of River Almond, below Almondell, 1973, D. G. Long and D. F. Chamberlain.

2. M. temperata Kuwah. **ML**
Once found.
On *Acer* in marshy woodland.

15 - {*VCR* ML}; Newhall Estate, near Carlops, 1978, D. G. Long.

3. M. furcata (L.) Dumort. **WL, ML, EL**
Common.
Tree trunks and branches, dry rock faces.

96, 97, 05, 06, 07, 08, 15, 16, 17, 24, 25, 26, 27, 35, 36, 44, 45, 46, 47, 48, 56, 57, 58, 66, 67, 68, 76, 77.

4. M. conjugata Lindb. **ML, EL**
Occasional.
Moist shaded rock faces.

16 - Logan Burn Waterfall, 1971, J. A. Paton and C. C. Townsend.

24 - ravine by River South Esk, above Gladhouse, field record.

34 - Wolf Cleugh, field record.

45 - Kate's Cauldron, Linn Dean Water, Soutra Hill, 255m, 18 June 1990, D. G. Long 18250.

57 - north slope of Traprain Law, 120m,
12 November 1993, D. G. Long 25036.
67 - Woodhall Dean, field record.
77 - Sheeppath Dean, field record.

74. Apometzgeria Kuwah.

1. **A. pubescens** (Schrank.) Kuwah.   **ML, EL**
Occasional.
Dry base-rich rock faces.

15, 16, 24, 26, (27), 45, 46, 66, 67, 76, 77.

76. *Targionia* L.

1. *T. hypophylla* L.
The single record for **26**: Bonaly Burn, October
1855, Lowe, refers to *R. hemisphaerica*.

77. Lunularia Adams.

1. **L. cruciata** (L.) Dumort. ex Lindb.
                                          WL, ML, EL
Common.
Damp ruderal habitats, on disturbed soil, streamsides.

86, 97, 06, 07, 08, 15, 16, 17, 26, 27, 35, 36, 37,
46, 47, 48, 57, 58, 68, [77].

79. Conocephalum Hill

1. **C. conicum** (L.) Dumort.   **WL, ML, EL**
Common.
Damp rocks, especially by streams.

96, 97, 98, 05, 06, 07, 08, 15, 16, 17, 24, 25, 26,
27, 34, 35, 36, 37, 44, 45, 46, 47, 56, 57, 58, 66,
67, 68, 76, 77.

80. Reboulia Raddi

1 **R. hemisphaerica** (L.) Raddi   **WL, ML, EL**
Occasional.
Exposed basaltic and sandstone outcrops.

(07) - {*VCR WL*}; Binny Craig, Winchburgh,
19 September 1908, J. McAndrew.
07 - Binny Craig, March 1998, D. G. Long.
(26) - Bonaly Burn, Colinton, 21 May 1910,
W. Evans.
[27] - Craiglockhart, 1822, Greville.
27 - basalt, below Samson's Ribs, Holyrood Park,
30 June 1973, D. F. Chamberlain.
57 - crevices of basaltic outcrop, north slope,
Traprain Law, 120m, 12 November 1993,
D. G. Long 25033.
[67] - top of Sheeppath Dean, 1885, Anon.

81. Preissia Corda

1. **P. quadrata** (Scop.) Nees   **WL, ML, EL**
Occasional.
Moist base-rich rock ledges, often by streams.

07, 15, 16, 24, 26, 34, 35, 45, 48, 58.

82. Marchantia L.

1.* **M. polymorpha** L.
A recent revision has re-assigned material to the
three subspecies recognised here. Some records,
for which there are no vouchers, cannot be determined beyond the species.

96, 97, 05, 06, 07, 15, 16, 17, 24, 25, 26, 27, 34,
36, 44, 45, 46, 47, 48, 56, 57, 58, 66, 67, 76, 77.

a. ssp. polymorpha   **ML, EL**
Local.
Muddy stones and soil by streams.

15 - Newhall Woods, Carlops, October 1959,
F. R. Irvine.
(16) - The Howe, Pentland Hills, 7 September
1904, W. E. Evans.
26 - muddy stones in burn in open ravine, 255m,
Pentland Hills, 23 July 1977, D. G. Long 6109.
46 - damp soil by stream, Birns Water, near
Stobshiel, 28 July 1974, D. G. Long 3632.
(48) - Luffness Links, 20 July 1898, W. Evans.
56 - stone in stream, Papana Water SSSI, near
Gifford, 18 July 1999, D. F. Chamberlain.

b. ssp. ruderalis Bischl. & Boisselier   **ML, EL**
Local but often abundant in man-made habitats.
Gravel around habitations, silty rocks in a burn.

16, (26), 27, 36, 45, 56, 58, 67.

c. ssp. montivagans Bischl. & Boisselier (*M.
alpestris* (Nees) Burgeff)   **ML, EL**
Probably frequent.
Rocks by streams, marshes.

16, (17), 24, 26, 34, 45, 48, 56, 66, 67, 77.

84. Riccia L.

1. **R. cavernosa** Hoffm.   **ML**
Probably decreasing, nationally scarce.
Bare mud around empty reservoirs.

16 - Redford Bridge, Threipmuir, October 1952,
A. C. Crundwell.
(16/26) - Clubbiedean Reservoir, 28 September
1905, W. Evans.

(26) - Glencorse Reservoir, 2 October 1905,
J. McAndrew.
26 - south end of Torduff Reservoir, 13 August
1973, D. F. Chamberlain and D. G. Long.

6. R. canaliculata Hoffm.   ML
Rare and considered vulnerable nationally.
On exposed mud and floating, Pentland Hills reservoirs.

(16) - Threipmuir Reservoir, 22 September 1905,
W. E. Evans.
16 - on damp mud, Threipmuir Reservoir, 1973,
D. G. Long and D. F. Chamberlain.
16 - damp mud, Harlaw Reservoir, 1 October
1960, D. M. Henderson.

7. R. sorocarpa Bisch.   WL, ML, EL
Occasional.
Damp tracks, fields, dried mud by reservoirs, etc., on
acid soils.

98, [05], (06), 15, 16, 17, 25, 26, 27, 35, (47),
56, 57, 66.
(47) - {VCR EL}; St Germains, 13 November
1903, W. Evans.

9. R. glauca L.   WL, ML, EL
Occasional to rare.
Fields, dried mud by reservoirs.

98 - {VCR WL}; damp stubble field, near East
Kerse Mains, Kinneil, 26 January 1975,
D. G. Long and D. F. Chamberlain, DGL 4173.
[06] - Harperrig, 1905, Anon.
(15) - North Esk Reservoir, 4 October 1905,
W. Evans.
16 - Harlaw Reservoir, field record.
(16/26) - Clubbiedean Reservoir, 29 September
1905, W. Evans.
(17) - South Queensferry, 11 October 1903,
W. Evans.
(35) - Rosebery Reservoir, 4 October 1905,
W. Evans.
48 - {VCR EL}; damp clay in stubble field, near
Luffness Mains, 15 September 1974, D. G. Long
and D. F. Chamberlain, DGL 3979.
68 - stubble field, near Tyninghame Links,
2 January 1975, D. F. Chamberlain.

13. R. beyrichiana Hampe ex Lehm.
    WL, ML, (EL)
Rare, nationally scarce.
Rocks by the sea, dried mud by reservoirs.

08 - on soil on rocky slope, Blackness Castle,

10 October 1971, D. G. Long and
D. F. Chamberlain, DGL 1816.
(16) - Threipmuir, 29 September 1905,
W. E. Evans and J. McAndrew.
(17) - Dalmeny, 11 August 1906, J. McAndrew.
(17) - shore, below Forth Rail Bridge, South
Queensferry, 21 October 1906, J. McAndrew.
25 - wet mud, Gladhouse Reservoir, 31 August
1971, D. G. Long 1612.
(48) - Gullane Links, 6 July 1912, J. McAndrew.

HORNWORTS (ANTHOCEROPSIDA)

85. Anthoceros L.

2. A. agrestis Paton   ML
Once found, nationally scarce.
Damp margin of stubble field.

06 - {VCR ML}; damp margin of stubble field,
Newpark, near, Bells Quarry, 1974, D. G. Long.

86. Phaeoceros Prossk.

1. P. laevis (L.) Prossk.   (WL, ML)
Not seen recently.
Damp soil.

[96] - Bathgate, before 1910, W. Evans.
(06) - Newpark, Bells Quarry, 30 April 1902,
Anon.
(17) - South Queensferry, 11 November 1903,
Anon.

MOSSES (SPHAGNOPSIDA)

I. Sphagnum L.

1. S. austinii Sull. ex Aust. (*S. imbricatum*
Hornsch.)   ML
This is a nationally scarce species.
Undisturbed blanket bogs, rare.

[25] - Auchencorth Moss, 24 May 1902,
W. Evans.
34 - {VCR ML}; flat peat bog near Garvald
Punks, Moorfoot Hills, 13 Aug 1977, D. G. Long
6185.

3. S. papillosum Lindb.   WL, ML, EL
Frequent.
On wet peat, often dominant in the more acid bogs.

86, 95, 96, 05, 06, 15, 16, 24, 34, 45, 66, 67.

4. S. palustre L.   WL, ML, EL
Common.

Wet peaty ground, in woods, marshes and bogs, always on acid soil.

86, 96, 97, 05, 06, 07, 15, 16, 24, 25, 26, 27, 34, 35, 36, 44, 45, 46, 56, 66, 67, 76, 77.

5. S. magellanicum Brid.   WL, ML, EL
Frequent.

Bogs, on deep peat; usually an indicator of relatively undisturbed habitats.

86, 05, 06, 15, 16, 24, 25, 34, 45, 56.

6. S. squarrosum Crome   WL, ML, EL
Frequent.

Sphagnum lawns in bogs, damp peaty ground, often under trees.

96, 06, (07), [16], 24, 25, 26, 27, 45, 56, 66, 67, 77.
96 - {VCR WL}; Foulshiels Bing, May 1998, D. F. Chamberlain.
77 - {VCR EL}; bank of Aikengall Water, Sheeppath Dean, near Dunbar, September 1967, E. Beattie.

7. S. teres (Schimp.) Ångstr.   ML, EL
Occasional.

Usually associated with base-rich flushes and mire communities on peat.

05 - {VCR ML}; base-rich flush, Garval Syke, source of the Medwin Water, 21 August 1976, D. F. Chamberlain and D. G. Long.
16 - edge of marsh, at East End, Black Springs Wildlife Refuge, Harlaw Reservoir, Pentlands, 20 June 1999, D. G. Long *et al.*
24 - flushes near River South Esk, Dundreich Plateau SSSI, 18 July 1998, D. F. Chamberlain.
45 - hummocky valley mire, middle part of Longmuir Moss, 25 Jun 1995, D. G. Long 25930.
56 - {VCR EL}; boggy birch wood, near Danskine Loch, Gifford, 17 October 1971, D. G. Long and D. F. Chamberlain, DGL 1826.

8. S. fimbriatum Wilson   WL, ML, EL
Frequent.

Ditches in peat bogs, peaty ground under scrub, alder/willow carr.

86, 96, 97, 05, 06, 16, 25, 26, 27, 45, 56, 67, 77.
96 - {VCR WL}; under birches on boggy moor, Easter Inch Moss, 1971, D. G. Long and D. F. Chamberlain.

67 - {VCR EL}; bank of stream in Woodhall Dean, September 1967, E. Beattie.

9. S. girgensohnii Russow   WL, ML, EL
Frequent.

Ditches in bogs, open moorland, under trees.

05, 06, 15, 16, 24, 25, 34, 36, 45, 56, 66, 67, 76, 77.

10. S. russowii Warnst.   WL, ML, EL
Locally frequent.

Damp moorland, often sub-montane.

86, 06, 16, 24, 34, 35, 45, 56, 66, 77.
06 - {VCR WL}; boggy moor, Easter Inch Moss, 1971, D. G. Long and D. F. Chamberlain.
24 - {VCR ML}; peaty bank by tributary of River South Esk, Moorfoot Hills, 1976, D. G. Long and D. F. Chamberlain.
56 - {VCR EL}; wet grassland by stream, 385m, Faseny Water, Lammermuir Hills, August 1961, D. A. Ratcliffe.

11. S. quinquefarium (Lindb. ex Braithw.) Warnst.   WL, EL
Rare.

Steep damp slopes in upland moorland.

06 - {VCR WL}; Five Sisters Bing, D. G. Long 27683.
45 - {VCR EL}; on steep slope of open ravine, 300m, Linn Dean Water, Soutra, 19 May 1990, D. G. Long 18093.
56 - Redden Grain, 310m, 5 May 1991, D. G. Long.
66 - White Castle, 24 April 1994, D. G. Long.

12. S. warnstorfii Russow   ML

This species can be confused with *S. capillifolium* ssp. *rubellum* and may be more common than the single record implies. It always occurs in base-rich habitats.

On the edge of a base-rich flush.

34 - {VCR ML}; basic flush, approx. 400m, Wolf Cleuch, Dewar Burn, Moorfoot Hills, 13 August 1977, D. F. Chamberlain and D. G. Long, DGL 6187.

13b. S. capillifolium (Ehrh.) Hedw. **ssp. rubellum** (Wilson) M. O. Hill (*S. rubellum* Wilson)
                            WL, ML, EL

A common plant of moorland.

Dryish sites, usually associated with *Calluna*. It also occurs on hummocks in bogs and in wet woodland.

86, 95, 05, 06, 15, 16, 24, 25, 26, 34, 35, 44, 45, 56, 66, 67, 77.

14 S. fuscum (Schimp.) H. Klinggr.   **ML**
A rare species in the Lothians.
Restricted to upland blanket bogs.

24 - Dundreich Plateau, 5 August 1974,
D. G. Long and D. F. Chamberlain.
34 - {VCR ML}; Eastside Heights, Pringles Green, Moorfoot Hills, July 1960, D. A. Ratcliffe.
34 - upland blanket bog, approx. 550m, Eastside Heights, Moorfoot Hills, 26 June 1976,
D. F. Chamberlain and D. G. Long, DFC B338.
34 - on flat peat bog, near Garvald Punks, Moorfoot Hills, 593m, 13 August 1977,
D. G. Long 6184.

15a. S. subnitens Russow & Warnst. var. **sub-nitens** (*S. plumulosum* Röll.)   **WL, ML, EL**
Infrequent.
Flushes and bogs that are usually slightly base-rich.

97, 06, 24, 25, 26, 34, 45, 48, 56, 77.
97 - {VCR WL}; damp rocky flush on the north slope of Cockleroy Hill, 1974, D. G. Long *et al.*

17. S. molle Sull.   **ML, WL**
Rare but possibly overlooked.
Bogs, usually lowland; drainage ditches.

86 - old drainage ditch, Blawhorn, 20 July 1996, D. F. Chamberlain.
06 - {VCR WL}; boggy moorland, Easter Inch Moss, near Bathgate, D. G. Long and
D. F. Chamberlain, DGL 1678.
16 - Threipmuir, 26 May 1969, D. G. Long 726.

19. S. compactum Lam. & DC.   **WL, ML, EL**
Occasional.
Moorland, on damp peat.

96, 05, 06, 15, (16), 26, 45, 48 (66).

21. S. inundatum Russow   **ML, EL**
Very local but possibly overlooked.
Wet peat in bogs.

16 - {VCR ML}; marshy ground near Redford Bridge, Threipmuir Reservoir, September 1965,
J. Appleyard.
16 - flushes, south bank of Black Springs Wildlife Refuge, Harlaw Reservoir, Pentland Hills, 20 June 1999, D. G. Long *et al.*
25 - Cauldhall Moor, May 1998, D. G. Long.

77 - {VCR EL}; wet bank of stream ravine, Sheeppath Dean, November 1967, E. Beattie and U. K. Duncan.

22. S. denticulatum, Brid. (*S. subsecundum* Nees var. *auriculatum* (Schimp.) Lindb.)   **WL, ML, EL**
Scattered.
Bogs.

96, (07), 15, (16), 25, 26, 45, 66, 67, 77.
96 - {VCR WL}; wet peaty ground close to ponds, Foulshiels Bing, Stoneyburn, May 1998,
D. F. Chamberlain.
77 - {VCR EL}; wet bank of stream ravine, Sheeppath Dean, November 1967, E. Beattie and U. K. Duncan.

23. S. contortum Schultz   **ML**
Rare but possibly under-recorded.
Base-rich flushes on peat.

25 - {VCR ML}; near River South Esk, above Hirendean Castle, Moorfoot Hills, 28 July 1973,
D. G. Long and D. F. Chamberlain, DGL 2620.

25. S. tenellum (Brid.) Bory   **WL, ML**
Occasional.
On the drier parts of peat bogs, and on wet moorland.

86, (06/07), 15, 16, 24, 26, 34, 45.

26. S. cuspidatum Ehrh. ex Hoffm.   **WL, ML, EL**
Frequent.
In running water in drainage ditches and ponds in peat bogs.

86, 95, 96, [97], 05, 06, 15, 16, 24, 26, 34, 44, 45, 56, 66, (67).

30. S. fallax (H. Klinggr.) H. Klinggr. **ssp. fallax**
(*S. recurvum* P. Beauv.)   **WL, ML, EL**
Locally abundant.
Damp ground generally, in woodland, on moorland and in bogs; usually on peat but not exclusively so.

86, 95, 96, 97, 05, 06, 15, 16, 24, 25, 26, 34, 35, 36, 44, 45, 46, 56, 66, 67, 76, 77.

31. S flexuosum Dozy & Molk. (*S. recurvum* P. Beauv. var. *amblyphyllum* (Russow) Warnst.).
    ML
Once found. This is a nationally scarce species.
Raised bog.

16 - {VCR ML}; south-west corner of Red Moss, Balerno, 260m, 12 March 1998, D. G. Long.

MOSSES (ANDREAEOPSIDA)

2. Andreaea Hedw.

2a. A. rupestris Hedw. var. **rupestris**
                                        WL, ML, EL
Locally frequent.
Rock faces and boulders, sub-alpine.

97, 05, (06), 07, 15, 16, 17, 24, 25, 26, 27, 34, 35, 45, 56, 57, 66.

6. A. rothii F. Weber & D. Mohr
a. ssp. rothii   (EL)
This nationally scarce variety is rare but possibly confused with the commoner var. *falcata*. No recent records.
Rock face, on basalt.

(57) - {VCR EL}; Traprain Law, October 1908, J. McAndrew.

b. ssp. falcata (Schimp.) Lindb.   ML, (EL)
Occasional.
Rock faces, boulders, usually sub-montane.

16 - boulder near Loganlea Waterfall, 1970s, D. G. Long and D. F. Chamberlain.
24 - field record.
25 - River South Esk, above Gladhouse, south of Hirendean Castle, 26 July 1975, D. G. Long and D. F. Chamberlain, DGL 4610.
26 - Bonaly Reservoir, field record.
34 - Wolf Cleugh, Moorfoot Hills, 1977, D.G. Long and D. F. Chamberlain.
(57) - Traprain Law, East Linton, September 1928, J. B. Duncan.

MOSSES (BRYOPSIDA)

3. Pogonatum P. Beauv.

1. P. nanum (Hedw.) P. Beauv.   (WL), ML, [EL]
Rare and probably decreasing.
Banks on peat, shale bings.

06 - {MVCR ML}; Addiewell Bing, 15 May 1976, D. F. Chamberlain and D. G. Long, DFC B298.
(16) - quarry, west of Balerno, 20 October 1909, J. McAndrew.
(17) - Craigie Hill Wood, October 1906, J. McAndrew.
(25) - Penicuik Woods, 1868, Anon.
(26) - Auchendinny, 1836, Anon.
(27) - Edinburgh 1819, Greville.

[77] - Aikengall Deans, 1885, Anon.

2. P. aloides (Hedw.) P. Beauv.   WL, ML, EL
Common.
Shaded banks, often on peat, in woodland.

86, 95, 96, 97, 98, 05, 06, 07, 15, 16, 17, 24, 25, 26, 27, 34, 35, 36, 45, 46, 47, 56, 57, 66, 67, 76, 77.

3. P. urnigerum (Hedw.) P. Beauv.   WL, ML, EL
Common.
Stony soil, rock crevices, usually on slightly base-rich substrates.

96, 06, 07, 15, 16, 17, 24, (25), 26, (27), 34, 35, 36, 44, 45, 56, 57, 58, 66, 67, 76, 77.

4. Polytrichum Hedw.

1. P. alpinum Hedw.   WL, ML, EL
Scattered.
Boulders, steep rocky banks, sub-montane.

(97), 07, 16, 26, 34, 35, 45, 46, 56, [57].

2. P. longisetum Sw. ex Brid. (*P. aurantiacum* Brid., *P. gracile* Dicks.)   WL, ML, EL
Frequent.
Bare peat in moorland, soil in open ruderal situations, basifuge.

86, 95, 96, 97, 98, 05, 06, 07, 15, 16, 17, 26, 27, 45, 46, 47, 56, 57, 58, 77.

3. P. formosum Hedw.   WL, ML, EL
Common.
Banks and among rocks in woodland, rotting tree trunks.
This species can be confused with *P. longisetum*. It is much less frequent in West Lothian.

95, 96, 97, 06, 07, 15, 16, 17, 24, 25, 26, 27, 34, 35, 36, 44, 45, 46, 47, 56, 57, 66, 67, 68, 76, 77.

5. P. commune Hedw.
a. var. commune   WL, ML, EL
Locally abundant.
Wet soil, usually on peat, in bogs and wet woodland.

86, 95, 96, 97, 98, 05, 06, 07, 15, 16, 17, 24, 25, 26, 27, 34, 35, 36, 44, 45, 46, 47, 56, 57, 58, 66, 67, 76, 77.

b. var. humile Sw.
The only known specimen of this variety (**16** - Bavelaw Moss, Balerno, 1916, Anon.) has been re-identified as var. *commune*. There are therefore

no authentic records of this taxon from the Lothians.

6. P. piliferum Hedw.   **WL, ML, EL**
Common.
Dry acid soil and rocks.

86, 95, 96, 97, 05, 06, 07, 08, 15, 16, 17, 24, 25, 26, 27, 34, 36, 44, 45, 46, 48, 56, 57, 58, 66, 67, 76, 77.

7. P. juniperinum Hedw.   **WL, ML, EL**
Common.
Dry acid soil and rocks.

86, 95, 96, 97, 98, 05, 06, 07, 08, 15, 16, 17, 24, 25, 26, 27, 34, 35, 36, 37, 44, 45, 46, 48, 56, 57, 58, 66, 67, 68, 76, 77.

8. P. strictum Brid. (*P. alpestre* Hoppe)
                                             WL, ML, EL
Frequent.
Hummocks in peat bogs.

86, 95, 96, 05, 06, 15, 16, 25, 26, 34, 35, 45, 66.

5. Oligotrichum Lam. & DC.

1. O. hercynicum (Hedw.) Lam. & DC.   **ML, EL**
Occasional, often in small quantity.
Banks, usually on mineral soil, sub-montane.

05, (15), 16, 24, 25, (27), 34, 46, 56.
56 - {VCR EL}; peaty bank above Hopes Reservoir, 974m, D. G. Long and D. F. Chamberlain.

6. Atrichum P. Beauv.

2. A. tenellum (Röhl) Bruch & Schimp.   **ML**
Once seen.
Dried-up silty bed of a reservoir.

16 - {VCR ML}; Threipmuir Reservoir, west of Redford Bridge, 11 August 1975, D.G. Long and D. F. Chamberlain, DGL 2655.

3a. A. undulatum (Hedw.) P. Beauv. var.
undulatum   **WL, ML, EL**
Abundant.
Gravel by streams and on paths, woodland banks, moist soil.

86, 95, 96, 97, 98, 05, 06, 07, 08, 15, 16, 17, 24, 25, 26, 27, 34, 35, 36, 44, 45, 46, 47, 56, 57, 66, 67, 76, 77.

7. Tetraphis Hedw.

1 T. pellucida Hedw.   **WL, ML, EL**
Frequent.
Humid peaty banks, usually in the shade of trees, rotting tree trunks.

96, 97, 98, 05, 06, 07, 15, 16, 25, 26, 34, 36, 45, 46, 56, 67.
46 - {VCR EL}; shady rocks in wooded valley, Humbie Water, near Humbie Mill, 1972, D. G. Long and D. F. Chamberlain.

8. Tetradontium Schwägr.

1. T. brownianum (Dicks.) Schwägr.   **WL, ML**
Moist shaded old red sandstone, rare and possibly decreasing.

97 - {VCR WL}; damp shaded sandstone rock face, Inveravon, near Grangemouth, 1976, D. F. Chamberlain and D. G. Long.
05 - shady sandstone ledges by moorland stream, Crosswood Burn, 2 June 1974, D. G. Long 3461.
25 - old stonework by stream in wood, Hare Burn, near Penicuik, 9 April 1973, D. G. Long.
26 - under sandstone outcrop, Roslin Glen, near the Chapel, 26 October 1968, D. G. Long 568.
35 - shaded sandstone rock face by river, Edgelaw Ravine, Temple, 26 December 1977, D. G. Long 6469.
36 - damp shaded sandstone, Maggie Bowies Glen, Crichton, Pathhead, 7 March 1974, D. G. Long 3148.

9. Diphyscium D. Mohr

1. D. foliosum (Hedw.) D. Mohr   **ML**
Very local.
Peaty banks in in woods and on moorland.

15 - by River North Esk, above Carlops, 1970s, D. G. Long and D. F. Chamberlain.
25 - Penicuik Woods, 1970s, D. G. Long and D. F. Chamberlain.
34 - Ladyside Burn, below Windy Knowe, Moorfoot Hills, 24 June 1973, D.G.Long and D. F. Chamberlain, DGL 2557.
35 - Hirendean Castle, River South Esk above Gladhouse, 1970s, D. G. Long and D. F. Chamberlain.

10. Buxbaumia Hedw.

1. B. aphylla Hedw.   **WL, ML, EL**
This nationally scarce species is rare and threatened by commercial exploitation of shale.

Mine and shale waste, burnt areas in woods.

96 - {*VCR* WL}; bing, north-west of Fauldhouse, 280m, May 1966, Corner.

96 - bing, East Benham, Polkemmet, 270m, 30 May 1988, D. G. Long 15083.

06 - {*VCR* ML}; Addiewell Bing, May 1966, Corner.

06 - on peaty crust on Addiewell Bing, 1979, D. G. Long and D. F. Chamberlain.

(26) - Roslin Wood, 31 Oct. 1807, R. and E. Maughan.

46 - {*VCR* EL}; Ormiston Bing, March 1967, R. Corner.

11. Archidium Brid.

1. A. alternifolium (Hedw.) Schimp. **WL, ML, EL**
Occasional.
Damp silt and clay, in fields and on margins of reservoirs.

97, [05], 06, 15, 16, 25, 26, 35, [46] 47, 66.

12. Pleuridium Rabenh.

1. P. acuminatum Lindb. **WL, ML, EL**
Frequent.
Dry soil amongst grass, calcifuge.

(06), 07, 15, (16), 17, 25, (26), 27, 36, 44, 45, (46), 47, 56, 67, 68, 77.

2. P. subulatum (Hedw.) Rabenh. **WL, ML, EL**
Occasional.
Dry soil amongst grass, on mildly base-rich substrates.

97, (17, 27), 35, 36, 45, 56, 67, 77.

56 - {*VCR* EL}; grass ley, near Danskine Loch, Gifford, 1967, M. K. Jurand and H. Morton.

13. Pseudephemerum (Lindb.) I. Hagen

1. P. nitidum (Hedw.) Reimers **WL, ML, EL**
Frequent.
Damp silt and clay, in fields and on margins of reservoirs.

96, 97, 06, 07, 15, 16, 25, 26, 27, 35, 44, 45, 46, 47, 56, 66, 67.

66 - {*VCR* EL}; damp mud by Whiteadder Reservoir, near Millknowe, D. G. Long.

14. Ditrichum Hampe

1. D. cylindricum (Hedw.) Grout **WL, ML, EL**
Common.
Bare soil in fields.
86, 96, 98, 06, 07, 15, 16, 17, 25, 26, 27, 35, 36,

37, 44, 45, 46, 47, 48, 56, 57, 58, 66, 67, 68, 77.

06 - {*VCR* WL}; stubble field one mile north-east of Bathgate, October 1964, A. C. Crundwell, Rodway, and A. McG. Stirling.

16 - {*VCR* ML}; bank of stream below Logan Burn waterfall, 1971, C. C. Townsend.

6. D. heteromallum (Hedw.) E. Britton
(WL), ML, EL
Frequent.
Dry acid soil, especially on moorland.

[97], 05, 06, 16, 24, 25, 26, (27), 34, 44, 45, 46, 56, [66, 68], 77.

9. D. flexicaule (Schwägr.) Hampe **EL**
Rare.
Sand-dunes.

D. gracile, a species recently recognised as occurring in Britain, has been split off from *D. flexicaule*, and has generally proved to be more common. Consequently, most of the records have been transferred to the former species.

(48) - {*VCR* EL}; Gullane Links, 6 April 1907, McAndrew.

48 - on shallow sand overlying basalt, Luffness Links, near the Sewage Station, 18 October 2000, G. P. Rothero, D. G. Long and D. F. Chamberlain.

10. D. gracile (Mitt.) Kunze (*D. crispatissimum* (C. Müll.) Par) **ML, EL**
Frequent.
Rocks and turf, sand-dunes, on base-rich substrates.

16, (25), 26, 34, 35, 36, 45, 48, (56), 58, 66, 67, 76, 77.

16. Distichium Bruch, Schimp. & W. Gümbel

1. D. capillaceum (Hedw.) Bruch, Schimp. & W. Gümbel **ML, EL**
Occasional.
Base-rich rocks, sand-dunes.

(16) - near Balerno, 7 March 1894, W. Evans.

26 - near Bonaly Reservoir, 1970s.

27 - Arthur's Seat?

45 - Linn Dean Water, Soutra, at top of waterfall, 3 June 1973, D. G. Long 2476.

(48) - Gullane Links, April 1910, J. McAndrew.

2. D. inclinatum (Hedw.) Bruch, Schimp. & W. Gümbel **(ML), EL**
Rare. This is a nationally scarce species.

Sand-dunes, basalt.

(27) - Craiglockhart, Greville.
48 - Gullane Links, near Gullane Point, 13 May 1973, D. G. Long 2375.
58 - Dirleton, 1977, D. G. Long and D. F. Chamberlain.
67 - stabilised dunes, Belhaven Bay, 20 November 1975, D. F. Chamberlain.

17. Ceratodon Brid.

1. C. purpureus (Hedw.) Brid.   **WL, ML, EL**
Abundant.
Soil, rocks, rotting wood, tree trunks, always on acid substrates.

All Squares.

20. Rhabdoweisia Bruch, Schimp. & W. Gümbel

1. R. fugax (Hedw.) Bruch, Schimp. & W. Gümbel   **EL**
Rare.
Crevices in basaltic rocks.

5 - {VCR EL}; north slope of Traprain Law, 1993, D. G. Long 25040.

2. R. crispata (Dicks.) Lindb.   **WL, ML**
Rare.
Shaded or sheltered rock faces, on basalt.

[97] - {VCR WL}; Cockleroy Hill, September 1906, J. McAndrew.
97 - rocks at north end of Cockleroy Hill, 17 February 1974, C. G. C. Argent.
34 - {VCR ML}; shaded rocks, Wolf Cleugh, 1977, D. G. Long and D. F. Chamberlain.

21. Cynodontium Schimp.

1. C. bruntonii (Sm.) Bruch, Schimp. & W. Gümbel   **(WL), ML, EL**
Local.
Ledges and rock faces on igneous and conglomerate substrates, sub-montane.

[97] - Cockleroy, 1913, Anon.
[07] - Binny Craig, prior to 1916, W. Evans.
16 - rock face to north of Logan Burn waterfall, 27 August 1988, D. F. Chamberlain.
26 - Torduff Reservoir, 23 June 1998, D. G. Long.
27 - Dunsapie, Arthur's Seat, 1977, D. G. Long and D. F. Chamberlain.

57 - basalt rocks, Garleton Hills, 25 November 1978, D. G. Long 2576.
58 - ledges on north side of North Berwick Law, April 1997, D. F. Chamberlain.

3. C. jenneri (Schimp.) Stirt.   **(EL)**
Once seen. This is a nationally scarce species.
Walls, sub-montane.

(66) - {VCR EL}; Snailscleugh, Whiteadder Water, June 1929, J. B. Duncan.

23. Dichodontium Schimp.

1a. D. pellucidum (Hedw.) Schimp. var. **pellucidum**   **WL, ML, EL**
Common.
Rocks by water.
Only var. *pellucidum* has been confirmed in the Lothians.

86, 96, 97, 98, 05, 06, 07, 15, 16, 17, 24, 25, 26, 27, 34, 35, 36, 44, 45, 46, 48, 56, 57, 66, 67, 76, 77.

25. Dicranella (Müll. Hal.) Schimp.

1. D. palustris (Dicks.) Crundw. ex E. F. Warb.   **WL, ML, EL**
Frequent.
Flushes, and in bogs.

86, 97, 05, 15, 16, 24, 25, 26, 34, 35, 44, 45, 56, 66, 67, [76], 77.

2. D. schreberiana (Hedw.) Dixon   **WL, ML, EL**
Frequent.
Damp soil, sometimes near springs, usually on slightly base-rich substrates.

97, 05, 06, 15, 16, 17, 24, (25), 26, 27, 34, 35, 36, 44, 45, [46], 56, 66, 77.

5. D. subulata (Hedw.) Schimp.   **ML, EL**
Rare.
Banks, often with heather, sub-montane.

[06] - West Calder, 1871, Anon.
16 - north-west slope of Carnethy Hill, Pentland Hills, 425m, 20 August 1977, D. G. Long 6192.
35 - sandy bank, Broad Law Quarry, 31 August 1971, D. G. Long and D. F. Chamberlain, DGL 1590.
66 - {VCR EL}; White Castle above Castle Moffat, steep shady bank with *Calluna*, 24 April 1994, D. G. Long 25292.

6. D. varia (Hedw.) Schimp.　　WL, ML, EL
Frequent.
Sides of ditches, damp clay soil, on mineral base-rich substrates.

96, 97, 05, 06, (07), 15, 16, 17, 24, 25, 26, 27, 34, 35, 44, 45, 46, 47, 56, 58, 66, 67, 68, 76, 77.

7. D. staphylina H. Whitehouse　　WL, ML, EL
Widespread and common.
Disturbed ground, fields.

96, 98, 06, 07, 15, 16, 17, 25, 26, 27, 35, 36, 37, 46, 47, 48, 56, 57, 58, 66, 67, 68, 77.
07 - {VCR WL}; stubble field near Swineburn, Winchburgh, 1976, D. G. Long.
26 - {VCR ML}; bare soil in arable field, near Roslin Chapel, August 1969, D. F. Chamberlain.

8. D. rufescens (Dicks.) Schimp.　　WL, ML, EL
Locally frequent.
River banks, land slips, on base-rich soils.

86, 96, 97, (05), 06, 07, 15, 16, 24, 25, 26, [27], 34, 35, 44, 45, 46, 56, 66, 67, 76, 77.
96 - {VCR WL}; damp soil on bank of River Almond, near Blackburn, 1971, D. G. Long and D. F. Chamberlain.
77 - {VCR EL}; wet clay by stream, Sheeppath Dean, November 1967, E. Beattie and U. K. Duncan.

9. D. cerviculata (Hedw.) Schimp.　　WL, ML, EL
Occasional.
Exposed wet peat usually on sides of drainage ditches in bogs.

86, 95, 96, 05, 06, (07), 15, 16, [17], 25, 45.

10. D. heteromalla (Hedw.) Schimp. WL, ML, EL
Abundant.
Banks in woods, occasionally also in the open.

86, 95, 96, 97, 98, 05, 06, 07, 08, 15, 16, 17, 24, 25, 26, 27, 34, 35, 36, 44, 45, 46, 47, 56, 57, 58, 66, 67, 68, 76, 77.

26. Dicranoweisia Lindb. ex Milde

1. D. cirrata (Hedw.) Lindb. ex Milde
　　　　　　　　　　　　　　　WL, ML, EL
Common.
Tree trunks, rocks, walls, etc, tolerant of reasonably high levels of air pollution.

All Squares.

28. Kiaeria I. Hagen

2. K. blyttii (Bruch, Schimp. & W. Gümbel) Broth.　　　　　　　　　　　　ML
Rare. This is a nationally scarce species.
Sub-montane rock faces.

16 - acid rocks on crag to west of Logan Burn waterfall, 4 September 1998, D. G. Long and D. F. Chamberlain.
35 - {VCR ML}; exposed rock, amongst other mosses, Moorfoot Hills, C. Dixon.

29. Dicranum Hedw.

2. D. bonjeanii De Not.　　WL, ML, EL
Frequent.
Grassland, fens and mires, often on base-rich substrates.

97, 05, 16, 25, (26), 34, 45, 48, [56], (57, 58), 66, 67, 77.

4. D. scoparium Hedw.　　WL, ML, EL
Widespread and common.
Woodland floor, tree trunks, amongst grass, sand-dunes, amongst rocks and heather in moorland, bogs.

86, 95, 96, 97, 05, 06, 07, 15, 16, 17, 24, 25, 26, 27, 34, 35, 36, 44, 45, 46, 47, 48, 56, 57, 58, 66, 67, 68, 76, 77.

5. D. majus Sm.　　WL, ML, EL
Frequent.
Woodland and open upland banks, in humid situations.

96, 97, 06, 15, 16, 24, 25, 26, 34, 35, 36, 45, 46, 56, 57, (58), 66, 67, 76, 77.

8. D. fuscescens Sm.　　ML, EL
Local.
Tree trunks, rock faces.

(15), 16, 24, 25, 26, 34, 35, 45, 46, 56, (58), 67, 76.

12. D. tauricum Sapjegin (*D. strictum* Schleich.)
　　　　　　　　　　　　　　　WL, ML, EL
Frequent, though rarer in the east, probably spreading.
Tree trunks and rotting wood in humid woodland.

96, 97, 06, 07, 15, 16, 25, 26, 27, 35, 36, 37, 46, 56.
46 - {VCR EL}; birch trunk, Peter's Muir Wood, East Saltoun, 3 September 1995, D. G. Long 25829.

97 - {VCR WL}; tree trunks on south bank of the River Avon, near Torphichen, 1971, D. G. Long and D. F. Chamberlain.

13. D. montanum Hedw.   **ML**
Rare.
Tree trunks, usually exposed to the light.

35 - {VCR ML}; old quarry near Rosebery Reservoir, Temple, amongst saplings, on log, 212m, 5 February 1978, D. G. Long 6482.
36 - on tree trunk in woodland, Tyne Water, above Vogrie House, approx. 125m, 11 January 1992, D. G. Long 21658.

30. Dicranodontium
Bruch, Schimp. & W. Gümbel

3. D. denudatum (Brid.) E. Britton   **WL, ML**
Rare.
Moorland slopes, spoil heaps.

06 - {VCR WL}; mossy bank among *Calluna*, Five Sisters Bing, 1998, D. G. Long 27691.
34 - {VCR ML}; peaty moorland slope, 0.5km north of Ladyside Heights, Moorfoot Hills, 26 June 1976, D. G. Long and D. F. Chamberlain, DGL 5066.

31. Campylopus Brid.

4. C. fragilis (Brid.) Bruch, Schimp. & W. Gümbel
                              (WL), ML, EL
Occasional.
Bare ground on peaty moorland, dry peaty banks.

(97), 05, (16), 17, 24, 26, 27, 45, [56], (58), 67.

5. C. pyriformis (Schultz) Brid.
a. var. pyriformis   **WL, ML, EL**
Common.
Exposed peat in woodland, moorland and bogs.

85, 95, 96, 97, 05, 06, (07), 15, 16, 24, 25, 26, 27, 34, 44, 45, 46, (56, 57) 58, 66, 67.

b. var. azoricus (Mitt.) M. F. V. Corley   **(ML)**
This is a nationally scarce variety.
Peat in a bog.

(16) - {VCR ML}; Ravelrig Toll Moss, 5 April 1898, W. Evans.

6. C. flexuosus (Hedw.) Brid. (*C. paradoxus* Wilson)   **WL, ML, EL**
Locally common.
Wet peat on moorland and in bogs, old tree stumps.

86, 95, 97, 05, 06, 15, 16, (17), 25, 27, 34, 44, 45, 47, 56, 66, 77.

11. C. introflexus (Hedw.) Brid.   **WL, ML, EL**
Widespread and spreading.
Wet peat on moorland and in bogs, wet rotten wood, damp rock ledges.

96, 98, 06, 07, 16, 17, 25, 26, 27, 35, 36, 45, 47, 48, 57, 58, 66, 67, 68.

33. Leucobryum Hampe

1. L. glaucum (Hedw.) Angstr.   **WL, ML, EL**
Frequent.
Peaty banks in dry moorland and on hummocks in bogs, woodland banks.

86, 96, 97, 05, 06, [07], 15, 16, 25, 26, 35, 46, 56, 66, [68].

34. Fissidens Hedw.

2*. F. viridulus (Sw.) Wahlenb. **agg.** (incl. *F. limbatus* Sull., *F. pusillus* (Wilson) Milde & *F. gracillifolius* Brugg.-Nan. & Nyholm)
The species comprising this aggregate have been revised relatively recently (Corley 1981). It is therefore not always possible to place an individual record in the individual species.

96, 97, 98, 07, 08, 16 (17), 26, 27, 35, 36, 46, 47, 56, 57, 67, 68, 77.

2. F. viridulus (Sw.) Wahlenb.   **ML, EL**
Probably local, but under-recorded.
Woodland rides and banks, generally on base-rich soil. There is an unknown recent record for East Lothian in the Census Catalogue.

27 - exposed soil under trees, by quarry below Craigmillar Castle, 23 January 2001, D. F. Chamberlain.

3. F. limbatus Sull.   **(WL, ML)**
Probably rare.
Habitat unknown.

(97) - {VCR WL}; Linlithgow 1869, J. Sadler
(26) - {VCR ML}; Colinton, June 1869, J. Sadler.

4. F. pusillus (Wilson) Milde   **WL, ML, EL**
Occasional.
Base-rich rock, on limestone, sandstone, conglomerate, etc, usually near water.

97, 07, 15, 16, (17), 26, 36, 46, 56, 77.

97 - {*VCR* WL}; stone on wooded bank, East Kerse Mains, Bo'ness, 1976, D. G. Long and D. F. Chamberlain.

16 - (*MVCR* ML); on stone by waterfall in ravine, 288m, Bonaly Burn, below reservoir, 1977, D. G. Long 6107.

56 - {*VCR* EL}; stone by Danskine Loch, near Gifford, 21 June 1998, D. F. Chamberlain.

5. F. gracilifolius Brugg., Nan. & Nyholm (*F. pusillus* (Wilson) Milde var. *tenuifolius* (Boul.) Podp. **ML**
Rare.

On damp shaded limestone and sandstone.

(**17**) - River Almond, below Cramond Bridge, September 1913, J. McAndrew.

26 - on old red sandstone, Roslin, 21 Feb. 1980, P. Lightowlers.

36 - old limestone quarry, Currie Lee, Pathhead, 17 March 1974, D. G. Long.

6. F. incurvus Starke ex Röhl. **WL, ML, EL**
Rare.

Earthy banks, in woodland and in quarries, on base-rich soils.

97 - {*VCR* WL}; shady earthy bank in old lime quarries above Galabraes, near Bathgate, 16 March 1975, D. G. Long 4246.

16 - {*VCR* ML}; shaded bank above Lymphoy Burn, Currie to Balerno, February 1999, D. F. Chamberlain.

(**17**) - west of South Queensferry, Nov 1907, J. McAndrew.

27 - wooded bank, Blackford Hill, close to Agassiz Rock, 1 October 2000, D. F. Chamberlain.

77 - {*VCR* EL}; bare soil on sea banks, north of Bilsdean, 14 February 1976, D. G. Long 4713.

7. F. bryoides Hedw. **WL, ML, EL**
Common.

Shaded banks and on soil, usually in woodland, always on acid substrates.

All squares except 05.

11. F. crassipes Wilson ex Bruch, Schimp. & W. Gümbel **WL, ML, EL**
Occasional.

Rocks at about water level in streams and rivers, on base-rich, sandstone and limestone.

06, 15, 16, 17, 26, 27, 36, 37, 46, 57, 67, 77.
17 - {*VCR* WL}; rock in River Almond, north

bank above Cramond Bridge, 1972, D. F. Chamberlain.

13. F. exilis Hedw. **WL, ML, EL**
Occasional.

Bare soil under trees; limestone or sandstone substrates.

98, (07), 15, (26), 27, 36, 46, 56, 67, 77.
56 - {*VCR* EL}; bare soil in shady wood, Gifford, 1991, D. G. Long and D. F. Chamberlain.

16. F. osmundoides Hedw. **ML, EL**
Local.

On banks and amongst rocks, sub-montane.

16 - Logan Burn, 1970s, D. G. Long and D. F. Chamberlain.

24 - River South Esk above Gladhouse, 1970s, D. G. Long and D. F. Chamberlain.

25 - damp ground by hill stream, River South Esk, South of Hirendean Castle, 4 August 1974, D. G. Long and D. F. Chamberlain, DGL 3640.

35 - damp rocks by burn, Whitelaw Cleugh, 15 August 1971, D. G. Long 493.

57 - damp rocks, Traprain Law, 18 September 1972, D. G. Long 2113.

(58) - east of North Berwick, 13 October 1906, W. Evans.

67 - mapping record, 10 March 1968, A. McG. Stirling.

17a. F. taxifolius Hedw. var. **taxifolius**
WL, ML, EL
Common.

Damp acid to base-rich soil on banks, usually under trees though sometimes in the open.

96, 97, 98, 06, 07, 08, 15, 16, 17, 24, 25, 26, 27, 34, 35, 36, 44, 45, 46, 47, 48, 56, 57, 58, 66, 67, 68, 76, 77.

18. F. dubius P. Beauv. (*F. cristatus* Wilson ex Mitt.) **WL, ML, EL**
Occasional.

Wet usually shaded, rock ledges, dry grassland, on base-rich substrates.

97, 16, 26, 27, 45, 48, 66, 67, 77.

19. F. adianthoides Hedw. **WL, ML, EL**
Common.

Wet rock ledges, base-rich stony flushes.

95, 96, 97, 05, 16, 24, 25, 26, 34, 35, 44, 45, 48, 56, 58, 66, 67, 68, 76, 77.

36. Encalypta Hedw.

1. E. streptocarpa Hedw. WL, ML, EL
Frequent.
Mortar on walls, base-rich rock ledges.
95, 96, 97, 08, 15, 16, (17), 24, 25, 26, 27, 35, 45, 46, 48, 57, 58, 66, 67, 76, 77.

3. E. rhaptocarpa Schwägr. ML
Rare. This is a nationally scarce species.
Base-rich dune slacks.
(48) - Gullane Links, June 1907, J. McAndrew.
(48) - Gullane Links, August 1909, J. McAndrew and J. B. Duncan.
48 - fixed dunes near the Sewage Works, Aberlady Bay, 9 March 1975, D. G. Long 4226.

4. E. vulgaris Hedw. (WL), ML, EL
Occasional.
Mortar on walls, conglomerate rocks, sandstone and limestone.
[08], 15, 16, (17, 26), 27, 36, [37], 45, 48, 56, 57, 58, 66, 76, 77.

6. E. ciliata Hedw. ML, EL
Rare. This is a nationally scarce species.
On conglomerate and sandstone, in humid ravines, sub-montane.
(16) - The Howe, July 1845, Anon.
16 - {MVCR ML}; Logan Burn, upper waterfall, on damp conglomerate, 4 September 1998, D. G. Long and D. F. Chamberlain, DGL 28025.
(26) - Bonaly Burn, Pentland Hills, 10 May 1891, W. Evans.
(26) - ravine, west of Swanston, 13 April 1907, W. Evans.
45 - {VCR EL}; shady rocks in deep ravine, Kate's Cauldron, Linn Dean Water, 12 August 1973, D. G. Long and D. F. Chamberlain, DGL 2673.

37. Eucladium Bruch, Schimp. & W. Gümbel

1. E. verticillatum (Brid.) Bruch, Schimp. & W. Gümbel WL, ML, EL
Frequent.
Moist lime-rich rock overhangs, often forming tufts.
97, 98, 07, 15, 16, 17, 25, 26, 27, 35, 36, 37, 45, 46, 48, 56, 57, 66, 67, 68, 77.

38. Weissia Hedw.

1. W. controversa Hedw.
a. var. **controversa** WL, ML, EL

Frequent.
Base-rich soil on sandstone, limestone, or basalt, rock crevices, bare batches in base-rich grassland.
96, 97, 98, (06), 07, 08, 16, 17, 24, 25, 26, 27, 34, 35, 36, 45, 48, 56, 57, 58, 68, [76], 77.

b. var. **crispata** (Nees & Hornsch.) Nyholm (*W. crispata* (Nees & Hornsch.) C. Müll.) (ML)
This nationally scarce variety has not been recorded recently in the Lothians.
Ultra-basic rocks, very local and rare.
(27) - Arthur's Seat, 4 June 1904, J. McAndrew.
(27) - {VCR ML}; Arthur's Seat, 18 January 1907, J. McAndrew.
(27) - Blackford Hill, 4 January 1907, J. McAndrew.

c. var. **densifolia** (Bruch, Schimp. & W. Gümbel) Wilson (WL)
Wet limestone rocks.
The single record (below) matches var. *densifolia* in its growth form but the seta is significantly longer than those of this variety. This record should therefore be considered as doubtful.
(97) - wet limestone rocks, Carribber Glen, 26 February 1898, W. Evans.

3. W. rutilans (Hedw.) Lindb. WL, ML, EL
Rare but possibly overlooked.
Grass fields, exposed mud on shore of a reservoir, quarry.
(07) - grass field, Drumshoreland, 11 April 1902, W. Evans.
17 - {MVCR WL}; floor of main quarry, Craigie Hill, 31 May 1999, D. F. Chamberlain and N. Bell.
27 - bare patch on bank above road, entrance to Dalry playground, 26 January 2000, D. F. Chamberlain.
56 - {VCR EL}; exposed mud, south side of Hopes Reservoir, 12 August 1974, D. G. Long and D. F. Chamberlain.

4. *W. condensa* (Voit) Lindb. (*W. tortilis* Schwägr.)
The unconfirmed literature record for the Edinburgh area is assumed to be a misidentification.

5. W. brachycarpa (Nees & Hornsch.) Jur.
a. var. **brachycarpa** ML, EL
Rare but possibly overlooked.

130

Bare soil, in meadows and around tree roots, on base-rich substrates.

36 - {*VCR ML*}; meadow below Hagbrae, near Crichton, 1974, D. F. Chamberlain and D. G. Long.

46 - exposed soil around upturned tree root, Humbie Mill, 20 February 1972, D. F. Chamberlain *et al.*

56 - {*VCR EL*}; in grass ley, near Danskine Loch, Gifford, 1967, M. K. Jurand and H. Morton.

b. var. obliqua M.O. Hill (*W. microstoma* Hedw.) C. Müll. **WL, ML, EL**
Occasional.
In turf, on rock ledges, always on base-rich substrates.

08, 17, 26, 27, 35, 45, (47), 58, 66, 67.

7. W. rostellata (Brid.) Lindb. **ML**
This is a nationally near-threatened species.
Among grass.

16 - {*VCR ML*}; high water mark, Clubbiedean Reservoir, Pentland Hills, 23 September 1973, D. F. Chamberlain and D. G. Long.

12a. W. longifolia Mitt. var. **longifolia** **EL**
Once collected.
Grassy bank near the sea.

58 - {*VCR EL*}; grassy bank by Tantallon Castle, 24 March 1974, D. F. Chamberlain and D. G. Long.

39. Tortella (Lindb.) Limpr.

1. T. tortuosa (Hedw.) Lindb. **ML, EL**
Frequent.
Base-rich rocks, turf by the sea, grassy areas on limestone.

15, 16, 24, 25, 26, 27, 34, 45, 46, 56, 58, 66, 77.

6. T. inclinata (R. Hedw.) Limpr. **WL, EL**
Rare. This is a nationally scarce species.
Sand-dunes, soil overlying basalt rocks, always near the sea.

08 - {*VCR WL*}; Blackness Castle, 23 June 1998, D. G. Long 27793.
(48) - Gullane Links, 1912, Anon.
48 - dunes, Gullane to North Berwick, 1960, A. C. Crundwell.
48 - Gullane Links, near Gullane Point, 13 May 1973, D. F. Chamberlain and D. G. Long.

8a. T. flavovirens (Bruch) Broth. var. **flavovirens** **WL, ML, EL**
Occasional.
Rock ledges and grass, almost always close to the sea.

07), 08, 17, 47, 48, 67, 68, 77.
08 - {*MVCR WL*}; on basalt, Blackness Castle, 23 June 1998, D. G. Long 27795.
17 - {*MVCR ML*}; turf, near the sea, Cramond Island, south shore, 24 January 1999, D. F. Chamberlain.

40. Trichostomum Bruch

1. T. brachydontium Bruch **WL, ML, EL**
Rare.
Sea cliffs, basalt, ledges on base-rich rock.

08 - Blackness, 10 October 1977, D. G. Long and D. F. Chamberlain.
26 - {*VCR ML*}; basic rocky bank, Torduff Reservoir, 27 February 1971, D. G. Long 1195.
48 - grassy bank by sea, Gullane Point, 18 October 2000, D. F. Chamberlain, G. P. Rothero and D. G. Long.
27 - Arthur's Seat, 1970s, Anon.
77 - near Bilsdean, 1970s, D. G. Long.

2. T. crispulum Bruch **WL, ML, EL**
Frequent.
Rock ledges, damp soil, on base-rich substrates.

08, 16, 24, 25, 26, 27, 34, 37, 45, 48, 56, 58, 66, 77.

45. Pseudocrossidium R. S. Williams

1. P. hornschuchianum (Schultz) R. H. Zander (*Barbula hornschuchiana Schultz*) **WL, ML, EL**
Occasional.
Base-rich bare ground, often coastal.

07, 16, 17, 25, 26, 27, 35, 37, 47, 48, 58, 67, 77.

2. P. revolutum (Brid.) R. H. Zander (*Barbula revoluta Brid.*) **[WL], ML, EL**
Frequent.
Mortar in old walls, sandstone cliffs.

[98], 15, 16, 17, 25, 26, 27, 35, 36, 37, 46, (47), 48, 57, 58, 67, 68, 76, 77.

46. Bryoerythrophyllum P. C. Chen

1. B. recurvirostrum (Hedw.) P. C. Chen (*Barbula recurvirostra* (Hedw.) Dixon) **WL, ML, EL**

Common.
Mortar in walls, base-rich soil on sandstone and limestone ledges.

96, 97, 05, 06, 07, 08, 15, 16, 17, 24, 26, 27, 34, 35, 36, 37, 44, 45, 46, 47, 48, 56, 57, 58, 66, 67, 68, 77.

2. B. ferruginascens (Stirton) Giacom.　　**ML**
Once seen.
Damp soil overlying conglomerate rock.

16 - {VCR ML}; Logan Burn, between the waterfalls, 4 September 1998, D. G. Long and D. F. Chamberlain, DGL 28007.

47. Leptodontium (Müll. Hal.) Hampe ex Lindb.

1. L. flexifolium (Müll. Hal.) Hampe ex Lindb.
　　　　　　　　　　　　　　WL, ML, EL
Occasional and usually in small quantity.
Open ground on peat on moorland, especially where there has been a fire.

97, [06], (07, 15), 16, (17), 24, 25, 26, 27, 34, 35, 45, 56, 58, 66.

48. Hymenostelium Brid.

1. H. recurvirostrum (Hedw.) Dixon
(*Gymnostomum recurvirostrum* Hedw.)　**WL, EL**
Rare.
Wet base-rich rocks, sand-dunes.

07 - {MVCR WL}; dripping basic rock outcrop on bank above Mains Burn, Ochiltree Mill, 8 October 1975, D. G. Long 4691.
15 - limestone cliff, downstream from Paties Pool, Newhall, 20 February 2001, D. F. Chamberlain.
(48) - Luffness Links, 1905, Anon.
56 - dripping shaded rock face, Papana Water SSSI, near Gifford, 18 July 1999, D. F. Chamberlain.
77 - Sheeppath Dean, on dripping rocks, 210m, 28 May 1977, D. G. Long 6066.

49. Anoectangium Schwägr.

1. A. aestivum (Hedw.) Mitt.　　　**ML**
Rare.
Base-rich rock ledges, usually sub-montane.

15 - East Cairn Hill, 1970s, D. G. Long and D. F. Chamberlain.
16 - Logan Burn, 1970s, D. G. Long.

24 - ravine 2, Dundreich SSSI, River South Esk, above Gladhouse, 1970s, D. G. Long and D. F. Chamberlain.
(26) - Roslin, 1860, ?W. Lyall.

50. Gyroweisia Schimp.

1. G. tenuis (Hedw.) Schimp.　　**WL, ML, EL**
Occasional.
Damp old red sandstone, cliffs, boulders and walls.

96, 15, 16, 17, 25, 26, 35, 36, 44, 56, 57, 76, 77.
17 - {VCR ML}; damp sandstone rock by pond in old quarry, Carmelhill, Kirkliston, 1971, D. F. Chamberlain and D. G. Long.
56 - {VCR EL}; damp boulder by stream, wood near Gifford, 1971, D. G. Long.

51. Gymnostomum Nees & Hornsch.

3. G. aeruginosum Sm.　　　**WL, ML, EL**
Frequent.
Flushes and rocks, on base-rich substrates.

07, (15), 16, (17), 24, 25, 26, (27), 34, 35, 45, 48, 57, 58, 67, 68, 77.

53. Barbula Hedw.

1. B. convoluta Hedw. (incl. var. *commutata* (Jur.) Husn.)　　　　　　　　**WL, ML, EL**
Common.
A ruderal, on bare sand in dunes, on paths and walls.

86, 95, 96, 97, 98, 05, 06, 07, 15, 16, 17, 25, 26, 27, 35, 36, 37, 44, 46, 47, 48, 56, 57, 58, 66, 67, 68, 77.

2. B. unguiculata Hedw.　　　**WL, ML, EL**
Sometimes abundant.
A ruderal on bare soil in fields, on paths and walls.

86, 95, 96, 97, 98, 05, 06, 07, 08, 15, 16, 17, 25, 26, 27, 35, 36, 37, 44, 46, 47, 48, 56, 57, 58, 66, 67, 68, 76, 77.

54. Didymodon Hedw.

1. D. rigidulus Hedw. (*Barbula rigidula* (Hedw.) Mitt.)　　　　　　　　**WL, ML, EL**
Common.
Base-rich rock ledges, old mortar on walls.

86, 96, 97, 98, 05, (06), 07, 08, 15, 16, 17, 25, 26, 27, 34, 35, 36, 37, 44, 46, 47, 48, 56, 57, 58, 66, 67, 77.

8. D. vinealis (Brid) R. H. Zander (*Barbula vinealis* Brid.)   **WL, ML, EL**
Commonly on walls, occasional in other dry ruderal habitats.

[06], 08, (15), 16, 26, 27, 37, 45, 47, 48, 56.

9. D. insulans (De Not.) M. O. Hill (*Barbula cylindrica* (Tayl.) Schimp.)   **WL, ML, EL**
Common.
On soil, in woodland, on banks and by streams.

96, 97, 98, 06, 07, 08, 15, 16, 17, 24, 25, 26, 27, 34, 35, 36, 37, 44, 45, 46, 47, 48, 56, 57, 58, 66, 67, 68, 76, 77.

10. D. luridus Hornsch. ex Spreng. (*Barbula trifaria* Auct., non (Hedw) Mitt.)   **WL, ML, EL**
Occasional.
Stonework by water.

05, 15, 17, 26, 27, 35, 36, (37), 57, 67, 76, 77.

12. D. sinuosus (Mitt.) Delogne (*Barbula sinuosa* (Mitt.) Garov. *Oxystegus sinuosus* (Mitt.) Hilpert)   **ML, EL**
Rare in Scotland, commoner in Southern England.
Soil-covered sandstone rocks by streams.

16 - retaining wall by stream, Water of Leith Walkway at Currie, 6 February 1999,
D. F. Chamberlain.
26 - {VCR ML}; retaining wall, Redhall Lade, Colinton Dell, Edinburgh, 29 November 1998,
D. F. Chamberlain.
27 - concrete by stream above Hermitage of Braid Offices, 21 February 2001,
D. F. Chamberlain.
46 - {VCR EL}; on shaded sandstone, Humbie Woods, 25 April 1994, D. G. Long.
57 - rocks, Hailes Castle, approx. 30m, 4 November 1994, D. G. Long 25716.

13. D. tophaceus (Brid.) Lisa (*Barbula tophacea* (Brid.) Mitt.)   **WL, ML, EL**
Common.
Damp soil and rocks on base-rich substrates, tufa-forming.

95, 96, 97, 98, 07, 08, 15, 16, 17, 26, 27, 35, 36, 37, 45, 46, 47, 48, 58, 67, 68, 77.

14. D. spadiceus (Mitt.) Limpr. (*Barbula spadicea* (Mitt.) Braithw.)   **(WL), ML, EL**
Occasional.
Amongst base-rich rocks by upland streams.

15, 16, (17), 24, (26), 35, 45, [58], 66, 67, 77.

15. D. fallax (Hedw.) R. H. Zander (*Barbula fallax* Hedw.)   **WL, ML, EL**
Frequent.
Base-rich soil, on limestone, sandstone, or on the coast.

95, 96, 97, 05, 06, 16, 17, 25, 26, 27, 34, 35, 36, 46, 48, 58, 66, 67, 68, 76, 77.

17. D. ferrugineus (Schimp. ex Besch.) M. O. Hill (*Barbula reflexa* (Brid.) Brid.)   **EL**
Rare.
Base-rich sand-dunes or soil, generally near the coast.

48 - Gullane Links, near Gullane Point, on soil-covered concrete block, 15 September 1974,
D. G. Long and D. F. Chamberlain, DGL 3878.
67 - edge of saltings, Belhaven Bay 20 November 1975, D. G. Long.
77 - {VCR EL}; basic soil in ravine, Sheeppath Glen, near Oldhamstocks, November 1967,
E. Beattie and U. K. Duncan.

57. Pterygoneuron Jur.

1. P. ovatum (Hedw.) Dixon   **ML, (EL)**
This nationally scarce species is rare and decreasing in the Lothians.
Limestone quarry, mortar in walls, sand-dunes.

25 - dry soil on limestone in old quarry, Fullarton Water, 4 August 1974, D. G. Long 3636.
(26) - near Colinton, Edinburgh, March 1868, W. Evans.
(27) - near Duddingston, Edinburgh, 1867, Anon.
(27) - near Granton, 4 April 1867, J. Sadler.
(27) - Craiglockhart, East Dyke, July 1906, J. McAndrew.
(27) - old quarry, Craigmillar, November 1902, J. McAndrew.
36 - crumbling bank on low limestone outcrop, quarry in Crichton Glen, 24 January 1988,
D. G. Long 14852.
(48) - near Gullane Point, 17 November 1946, J. Sinclair 2695.

58. Aloina Kindb.

1. A. brevirostris (Hook. & Grev.) Kindb.   **WL**
Rare.
Spoil on pit bing.

Records of this nationally scarce species from walls in the Edinburgh district apply to *A. rigida*.

06 - {*VCR* WL}; on heap of shale agglomerate, on old bing, Drumshoreland Muir, approx. 110m, 3 February 1989, D. G. Long 15467.

2. A. rigida (Hedw.) Limpr.   [WL], ML, (EL)
This nationally scarce species has declined following the loss of mortared walls.
Mortar on old walls, limestone in a disused quarry.

[97] - Inveravon, 1845, Anon.
26 - spoil heap to south-west of the main pond, Torphin Quarry, South Edinburgh, 1September 2000, G. Rothero and D. F. Chamberlain.
(27) - wall, Meadowbank, 1847, R. M. Stark.
(27) - near Granton, April 1870, J. Sadler.
(36) - near Dalkeith, October 1845, R. M. Stark.
36 - crumbling bank on low limestone outcrop, Kilnwood Quarry, near Crichton, 10 January 1988, D. G. Long 14843.
(37) - walls at Elphinstone, 1847, R. M. Stark.

3. A. aloides (Schultz) Kindb.   WL, ML, EL
Occasional.
Soil overlying limestone, old red sandstone, walls, always on base-rich substrates.

96, 97, (17), 26, (27), 35, 48, 58, 77.
96 - {*VCR* WL}; soil overlying limestone outcrop in quarry, Petershill, near Bathgate, 1975, D. G. Long and D. F. Chamberlain.

60. Tortula Hedw.

1. T. subulata Hedw.
a. var. **subulata**   WL, ML, EL
Frequent.
Base-rich soil on banks.

96, 07, (15), 16, 17, 24, 25, 26, 27, 34, 35, 36, 44, 45, 46, 48, 57, 66, 67, 76, 77.

b. var. **angustata** (Schimp.) Limpr.   (ML, EL)
Habitat unknown, no recent records.

(16) - Water of Leith, Balerno, 9 May 1910, J. McAndrew.
(26) - Water of Leith, Hailes, 1909, J. McAndrew.
(27) - in a field, south of Craigmillar Castle, Edinburgh, March 1902, J. McAndrew.
(36) - Heriot, 29 April 1911, J. McAndrew.
[48] - Gullane Links, June 1907, J. McAndrew.
(57) - Haddington, 24 November 1906, W. Evans.

d. var. **graeffii** Warnst.   ML
Once found. This is a nationally scarce variety.

Base-rich sub-montane rocks.

16 - {*VCR* ML}; waterfalls, Logan Burn, 4 September 1998, D. G. Long and D. F. Chamberlain, DGL 28009.

8a. T. muralis Hedw. var. **muralis**   WL, ML, EL
Very common.
Occurring most commonly on stonework on walls and on concrete, but also occurring naturally on a range of natural stone surfaces.

All Squares.

11. T. atrovirens (Sm.) Lindb. (*Desmatodon convolutus* (Brid.) Grout)   EL
Rare. This is a nationally scarce species.
Earthy banks, on the coast.

58 - {*VCR* EL}; earthy bank by cliff path, Oxroad Bay, Tantallon, 19 May 1974, D. G. Long *et al.*, DGL 3431.
77 - Bilsdean, post-1971, D. G. Long.

12. T. lanceola R. H. Zander (*Pottia lanceolata* (Hedw.) Müll. Hal.)   (WL), ML, EL
Occasional.
Base-rich soil, on limestone.

(07), [16], (17), 27, 48, 58, 66, 77.

15. T. modica R. H. Zander (*Pottia intermedia* (Turner) Fürnr.)   ML, EL
Rare but possibly overlooked.
Soil, usually slightly base-rich.

[26] - Liberton, December 1878, W. Evans.
(27) - Craigleith (former) Station, 31 December 1910, J. McAndrew.
27 - gully by Agassiz Rock, Blackford Glen, Edinburgh, 13 March 1978, D. G. Long 6539.
27 - soil above crag, Arthur's Seat, Edinburgh, 17 January 1978, P. Lightowlers.
48 - {*VCR* EL}; sandy bank, south-west end of Gullane Beach, 1974, D. F. Chamberlain and D. G. Long.

16. T. truncata (Hedw.) Mitt. (*Pottia truncata* Bruch & Schimp.)   WL, ML, EL
Common.
Acid soil, in fields and in meadows.

96, 97, 98, 06, 07, 08, 15, 16, 17, 25, 26, 27, 35, 36, 37, 46, 47, 48, 56, 57, 66, 67, 68, 77.

17. T. protobryoides R. H. Zander (*Pottia bryoides* (Dicks.) Mitt.)   EL

Rare. This is a nationally scarce species.
On sand in sand-dunes.

(47) - shore, Longniddry, May 1907, J. Hunter.
47 - sandy track, Gosford Bay , 10 March 1974,
D. F. Chamberlain.
(48) - Gullane Quarry, near Gullane Point,
24 October 1908, J. McAndrew.
48 - track leading to Gullane Quarry, Gullane
Bents, February 1999, D. F. Chamberlain.

18. T. acaulon (L. ex With.) R. H. Zander
a. var. **acaulon** (*Phascum cuspidatum* Hedw.)
WL, ML, EL
Frequent.
On soil in fields, especially where it is base-rich.

96, 97, 06, 07, 08, 17, 25, 26, 27, 35, 36, 37, 46,
47, 48, 57, 58, 67, 68, 77.

b. var. **pilifera** (Hedw.) R. H. Zander (*Phascum
cuspidatum* Hedw. var. *piliferum* (Hedw.) Hook
& Taylor) **(WL), ML, EL**
This nationally scarce variety occurs in dry
habitats.

17 - gravel path near the Clubhouse, Ratho Park
Golf Course, 20 August 2000,
D. F. Chamberlain.
(17) - South Queensferry, 11 November 1903,
W. Evans.
(27) - Arthur's Seat, Samson's Ribs, Edinburgh,
April 1847, Anon.
(27) - Craiglockhart Hill, February 1872,
W. Evans.
58 - cliff top, Oxroad Bay, near Tantallon Castle,
1974, D. F. Chamberlain and D. G. Long.

61. Microbryum Schimp.

2. M. davallianum (Sm.) R. H. Zander (*Pottia
starckeana* (Hedw.) Müll .Hal. ssp. *conica*
Schleich. ex Schwägr.) D. F. Chamb. & ssp.
minutula (Schleich. ex Schwägr.) Bouvet)
WL, ML, EL
Local.
Soil on limestone and other calcareous substrates.

96 - soil overlying limestone outcrop in quarry,
Petershill, near Bathgate, 1975, D. G. Long and
D. F. Chamberlain.
16 - Balerno, 1955, Anon.
(26) - Straiton, 15 December 1946, J. Sinclair
2708.
(27) - Arthur's Seat, 10 March 1870, J. Sadler.
35 - Middleton Limestone Quarries, 31 August

1971, D. G. Long 1609.
36 - soil overlying limestone, below Crichton
Castle, 21 November 1996, D. F. Chamberlain.
(67) - Belhaven, Dunbar, 7 September 1928,
E. N. Stevens.
77 - limestone sea cliff, Skateraw, 20 March
1976, D. F. Chamberlain and D. G. Long, DFC
B263.

3. M. rectum (With.) R. H. Zander (*Pottia recta*
(With.) Mitt.) **[ML]**
Rare, not seen recently.
Base-rich soil.

[27] - Holyrood Park, prior to 1824,
R. K. Greville.
[27] - flower pot in greenhouse, Morningside
Park, Edinburgh, September 1904, W. Evans.

62. Hennediella Paris

1. H. stanfordensis (Steere) Blockeel (*Tortula
stanfordensis* Steere) **ML, EL**
Rare and possibly ephemeral. This is a nationally
scarce species.
Exposed earth in a field and beside a river.

27 - {VCR ML}; bare compacted soil, Water of
Leith, near Canonmills, Edinburgh, 7 March
1976, D. G. Long.
77 - {VCR EL}; sloping pasture shaded by woodland, dean opposite Thornton Castle, 20 March
1976, D. G. Long and D. F. Chamberlain.

2. H. heimii (Hedw.) R. H. Zander (*Pottia heimii*
(Hedw.) Fürnr.) **WL, (ML), EL**
Occasional.
Banks, generally by the sea.

[98], 08, (17, 27), 47, 48, 58, 67, 68, 77.

63. Acaulon Müll. Hal.

1. A. muticum (Brid.) Müll. Hal. **(ML)**
Only once seen and then possibly introduced.
Open ground.

(27) - Experimental Gardens, Royal Botanic
Garden, Midlothian, 1847, Anon.

65. Syntrichia Brid.

1. S. ruralis (Hedw.) F. Weber & D. Mohr
(*Tortula ruralis* (Hedw.) P. Gaertn.) **WL, ML, EL**
Frequent.
Base-rich rocks, walls, sand-dunes.

15, 16, 17, 25, 26, 27, 35, 36, (37), 46, 47, 48, 57, 58, 66, 67, 68, 77.

2. S. ruraliformis (Besch.) Cardot (*Tortula ruralis* (Hedw.) P. Gaertn. ssp. *ruraliformis* (Besch.) Dixon) **WL, ML, EL**
Frequent.
Sand-dunes and in turf, near the sea.

[07], 08, (17), 37, 47, 48, 58, 67, 68, 77.

4. S. intermedia Brid. (*Tortula intermedia* (Brid.) Berk.) **WL, ML, EL**
Occasional.
Dry base-rich rock faces, concrete.

07, 16, 26, 27, 35, (47), [57], 58, 67.
07 - {VCR WL}; concrete wall, south-east end of Faucheldean Bing, 31 May 1999,
D. F. Chamberlain and N. Bell.

5. S. princeps (De Not.) Mitt. (*T. princeps* De Not.) **ML**
Rare. This is a nationally scarce species.
Base-rich rock faces, usually sub-montane.

16 - {VCR ML}; agglomerate rock ledge, water-fall, Logan Burn, 4 September 1998,
D. F. Chamberlain and D. G. Long.
25 - wall, Gladhouse, Anon.
26 - sandstone outcrop, above Flotterstone, 1970s, D. F. Chamberlain.
27 - Agassiz Rock, Blackford Glen, 1984,
R. F. O. Kemp.
(27) - Craiglockhart, Edinburgh, February 1898,
J. McAndrew.

7a. S. laevipila Brid. var. **laevipila** (*Tortula laevipila* (Brid.) Schwägr.) **ML, EL**
Frequent.
Trunks of trees, especially ash and sycamore.

25, 26, 27, 35, 36, 44, 46, (47), 56, 57, (58), 67, 77.

8. S. papillosa (Wilson) Jur. (*Tortula papillosa* Wilson) **ML, EL**
Frequent.
Epiphytic on elder, ash, willow and sycamore.

17, (26), 27, 35, 36, 44, 45, 46, 47, (56), 57, 58, 67, 77.

9. S. latifolia (Bruch ex Hartm.) Huebener (*Tortula latifolia* Bruch ex Hart) **WL, ML, EL**
Occasional.

Silted tree trunks and roots, occasionally also on boulders, usually close to water.

17, 26, 37, 46, 47, 57, 67, 68.
17 - {VCR WL}; above Cramond Bridge, 1972,
D. F. Chamberlain.
36 - {MVCR ML}; on silty stonework by burn, 105m, Lion's Lodge Bridge below Ford, 1994,
D. G. Long 25803.

66. Cinclidotus P. Beauv.

1. C. fontinaloides (Hedw.) P. Beauv. **WL, ML, EL**
Occasional.
Base-rich rocks around water level in streams and rivers.

97 - River Avon, near Torphichen, Anon.
07 - Mains Burn, Anon.
17 - River Almond, Cramond Bridge, Anon.
26 - rocks by Glencorse Reservoir, 30 July 2000,
D. F. Chamberlain.
27 - sandstone by Water of Leith, below Dean Bridge, 27 July 2001, D. F. Chamberlain.
45 - Linn Dean Water, Anon.
57 - Hailes Castle, near East Linton, approx. 25m, 4 November 1994, D. G. Long 25717.

67. Coscinodon Spreng.

1. C. cribrosus (Hedw.) Spruce **ML**
Rare. This is a nationally scarce species.
Basalt rocks.

27 - {VCR ML}; south-facing slab, Agassiz Rock, Blackford Glen, 1978, D. G. Long.
27 - Samson's Ribs, Arthur's Seat, 30 September 1971, D. G. Long 1734.
27 - erratic basalt boulder to west of Duddingston Loch, Holyrood Park, 1997,
D. F. Chamberlain.

68. Schistidium Brid.

1. S. maritimum (Turner) Bruch & Schimp. **WL, ML, EL**
Occasional.
Rock faces just above the high tide mark.

[07], 08, 17, 27, 47, 48, [57], 58, 68.

2. S. rivulare (Brid.) Podp. (*S. alpicola* (Hedw.) Limpr. var. *rivulare* (Brid.) Limpr.) **WL, ML, EL**
Occasional.
Rocks by rivers.

97, 06, 07, 16, 17, 26, 27, 35, 36, 44, 45, 46, 56, 57.

3. S. platyphyllum (Mitt.) H. Perss. (*S. alpicola* (Hedw.) Limpr., *S. rivulare* (Brid.) Podp. ssp. *latifolium* (Z. E. Zetterst.) B. Bremer)   **WL, ML**
Rare. This is a nationally rare species.
Base-rich usually moist rocks, sub-alpine.

(**97**) - River Avon, Carribber Glen, 10 April 1909, W. Evans, (conf. Orange).
(**17**) - River Almond, Craigiehall Bridge (ML), 1903, McAndrew.
17 - River Almond, Craigiehall (WL), 1977, D. G. Long.
25 - {VCR ML}; River South Esk, near Hirendean Castle, Moorfoot Hills, 28 July 1973, D. G. Long 2615.

5. S. apocarpum (Hedw.) Bruch & Schimp.
                                          WL, ML, EL
Recent revisions of *Schisitidium* have resulted in the recognition of several distinct species that are related to *S. apocarpum*. It is therefore likely that a significant number of the records refer to other species, especially *S. crassipilum*. *S. apocarpum* in the strict sense is clearly rarer than the records suggest. An asterisk indicates recently confirmed records.
Base-rich rocks, mortar and cement in walls.

86, 95, 96, 97, 05, 06, 07, 15, 16, 24, 25, 26, 27*, 34, 35, 36, 44, 45, 46, 48, 56, 57, 58, 66, 67*, 76, 77.

8. S. pruinosum (Wilson ex Schimp.) Roth   **ML**
Rare.
Agglomerate and basalt rocks.

16 - {MVCR ML}; base-rich south-facing agglomerate rocks by lower waterfall, Logan Burn, 4 September 1998, D. G. Long and D. F. Chamberlain, DGL 28002.
(**26**) - Braid Hills, March 1869, W. Evans.
(**26**) - Bonaly, 26 March 1880, W. Evans.
(**27**) - Arthur's Seat, June 1847, J. Laing.
(**27**) - St. Anthony's Chapel, Holyrood Park, January 1870, J. Sadler.
27 - above Duddingston Loch, Holyrood Park, Edinburgh, 24 November 1973, D. G. Long 2872.

9. S. strictum (Turner) Loeske ex Martensson
                                                    ML
Rare. Once seen.
Base-rich rocks, sub-montane.

16 - {VCR ML}; upper waterfall, Logan Burn, 4 September 1998, D. G. Long and D. F. Chamberlain, DGL 28004.

12. S. confertum (Funck) Bruch & Schimp.   (**ML**)
Rare.
Base-rich rocks.

(**26**) - above Boghall Wood, 1 March 1902, W. Evans.
(**27**) - Holyrood Park, Edinburgh, May 1847, Greville.
(**27**) - Arthur's Seat, October 1869, J. Sadler.
(**27**) - Arthur's Seat, June 1871, W. Evans.

16a. S. crassipilum Blom   **ML, EL**
This species has only recently been recognised in Britain and is apparently widespread, especially in man-made habitats. It is therefore likely to be commoner than *S. apocarpum* in the Lothians.
Rocks, concrete, walls.

(**27**) - rocks at Craigmillar Castle, 6 April 1909, W. E. Evans.
27 - limestone outcrop below Craigmillar Castle, 23 January 2001, D. F. Chamberlain.
48 - basalt outcrops on sand-dunes, slope above the sewage works, Luffness Links, 18 October 2000, D. G. Long *et al.*

69. **Grimmia** Hedw.

1. G. anodon Bruch & Schimp.   (**ML**)
This nationally endangered species has not been seen since the nineteenth century and is probably extinct.
Basalt rocks.

(**27**) - Arthur's Seat, March 1869, J. Sadler.
(**27**) - Arthur's Seat 1871, W. Evans.

4. G. laevigata (Brid.) Brid.   (**ML, EL**)
Not seen since 1907 and possibly extinct. This is a nationally scarce species.
Basalt rock faces.

(**27**) - Arthur's Seat, 7 March 1902, W. Evans.
(**27**) - Arthur's Seat, 11 May 1830, Mrs. H. Lyall.
(**27**) - Arthur's Seat, 1906, Anon.
(**58**) - east side of North Berwick Law, June 1907, J. McAndrew.

5. G. montana Bruch & Schimp.   **ML, EL**
Rare. This is a nationally scarce species.
Basalt rock faces.

27 - {*VCR* ML}; Agassiz Rock, Blackford Quarry, Edinburgh, 13 March 1978, D. G. Long 6540.
(57) - Traprain Law, October 1908, J. McAndrew.
57 - exposed basalt rocks, west summit of Traprain Law, 210m, 4 November 1994, D. G. Long 25715.

7. G. donniana Sm.   **ML, EL**
Occasional.
Acid rocks, stone walls.

05 - Cobbinshaw Reservoir, 13 May 1972, D. F. Chamberlain.
16 - Scald Law, field record.
(25) - Leadburn, 17 March 1894, W. Evans.
26 - wall, near Cairn Burn, South of Carnethy Hill, Pentland Hills, 280m, 20 August 1977, D. G. Long 6187.
27 - Arthur's Seat, 8 February 1978, P. Lightowlers.
45 - on old wall, Longmuir Moss, approx. 275m, 20 April 1990, D. G. Long 17947.

10. G. longirostris Hook. (*G. affinis* Hornsch.)
                                                 ML
Rare. This is a nationally scarce species.
Basalt rock faces.

27 - Agassiz Rock, Blackford Quarry, 1971, D. F. Chamberlain.
27 - sloping rocks above Agassiz Rock, Blackford Quarry, Edinburgh, 1978, D. G. Long.
(27) - Blackford Hill, April 1869, J. Sadler.
(27) - Arthur's Seat, 1868, W. Evans.
(27) - Corstorphine Hill, Hb. Maughan.

11. G. ovalis (Hedw.) Lindb.   **(ML), EL**
This nationally vulnerable species is rare in the Lothians.
Basalt rocks.

(27) - {*VCR* ML}; Arthur's Seat, 1870, W. Boyd.
(27) - Arthur's Seat, June 1871, W. Evans.
57 - basalt rock slabs, north slope of Traprain Law, 120m, 12 November 1993, D. G. Long 25039.
58 - north side of North Berwick Law, 12 November 1993, D. G. Long.
(58) - North side of North Berwick Law, June 1869, W. Boyd.

15. G. pulvinata (Hedw.) Sm. var. **pulvinata**
                                            WL, ML, EL
Common.
Base-rich rocks, mortar in walls.

All Squares except 95.

16. G. orbicularis Bruch ex Wilson   **(ML), [EL]**
Rare; not seen recently. This is a nationally scarce species.
Basalt and other rocks.

(27) - Holyrood Park, 1864, John Brown.
(27) - Holyrood Park, above Duddingston Loch, 4 April 1870, J. Sadler.
[67] - Dunbar, June 1869, H. W. Monington.

19. G. trichophylla Grev. (incl. var. *stirtonii* (Schimp.) Moell. and var. *subsquarrosa* (Wilson) A. J. E. Sm.)   **WL, ML, EL**
Basalt and other rocks.
Rock faces.

96, 97, 05, 08, 16, 17, 24, 26, 27, 45, 46, 48, 56, 57, 58, 67, 76.

23. G. decipiens (Schultz) Lindb.   **ML, EL**
Rare. This is a nationally scarce species.
Basalt rocks.

27 - on a boulder in a marsh, Whinny Hill, Arthur's Seat, 20 Jan 1978, P. Lightowlers.
57 - dry acid rocks, Traprain Law, 18 September 1972, D. G. Long 2121.
57 - Traprain Law, April 1955, H. A. Fairley.
(57) - Traprain Law, 3 October 1908, J. McAndrew.
(57) - on rocks, Hailes Castle, north bank of the River Tyne, 19 September 1908, W. Evans.

25. G. curvata (Brid.) De Sloover (*Dryptodon patens* (Hedw.) Brid.)   **(EL)**
Rare; not seen recently.
Damp rocks.

(57) - Traprain Law, 20 June 1908, J. McAndrew.
(57) - Traprain Law, September 1928, J. B. Duncan.

70. Racomitrium Brid.

2. R. aciculare (Hedw.) Brid.   **WL, ML, EL**
Common.
Damp rocks, often by water, common.

86, 95, 96, 97, 05, 06, 07, 15, 16, 17, 24, 25, 26, 27, 34, 35, 36, 44, 45, 46, 56, 57, 66, 67, 76, 77.

3. R. aquaticum (Schrad.) Brid.   **WL, ML, EL**
Rare.
Rocks by water.

97 - damp shaded rock face above Union Canal, Linlithgow Golf Course, 16 May 1999, D. F. Chamberlain.
16 - The Howe, field record.
34 - Ladyside Heights, field record.
57 - {*VCR EL*}; wet basalt rock slabs, north side of Traprain Law, approx. 120m, 12 November 1993, D. G. Long 25041.

4. R. fasciculare (Hedw.) Brid.   **WL, ML, EL**
Common.
Acid rocks and boulders.

86, 95, 96, 97, 98, 05, 06, 07, 15, 16, 17, 24, 25, 26, 27, 34, 35, 44, 45, 46, 48, 56, 57, 66, 67, 76.

7. R. affine (Schleich. ex F. Weber & D. Moore) Lindb.   **ML, [EL]**
Rare. This is a nationally scarce species.
Sub-montane rocks.

16 - {*VCR ML*}; Logan Burn above waterfalls, 4 September 1998, D. G. Long and D. F. Chamberlain, DGL 27982.
[57] - Traprain Law, September 1928, J. B. Duncan *et al.*

8. R. heterostichum (Hedw.) Brid.   **WL, ML, EL**
Common.
Acid rocks.

95, 96, 97, 05, 06, 07, 08, 15, 16, 17, 24, 25, 26, 27, 34, 35, 36, 44, 45, 46, 48, 56, 57, 58, 66, 67, 76, 77.

10. R. lanuginosum (Hedw.) Brid.   **WL, ML, EL**
Frequent.
Acid rocks, sub-montane.

95, 96, 97, 05, 06, 07, 15, 16, 24, 25, 26, 34, 35, 57, (67), 76.

11. R. ericoides (Brid.) Brid. (*R. canescens* (Hedw.) Brid. var. *ericoides* (Brid.) Hampe)
                                        WL, ML, EL
Frequent.
Gravelly ground, tracks.

96, 05, 06, 07, 15, 24, 25, 26, 34, 35, 44, 57, 66, 67, 76.
06 - {*VCR WL*}; slope on Philpstoun South Bing, 1988, D. G. Long 15091.
57 - {*VCR EL*}; gravelly ground, Traprain Law Quarry, approx. 360m, 8 May 1995, D. G. Long 25909.

12. R. elongatum Frisvol   **ML, EL**

Rare. This is a nationally scarce species.
Rock ledges, sandy ground by the sea.

16 - {*VCR ML*}; rock ledge by lower waterfall, Logan Burn, 4 September 1998, D. G. Long and D. F. Chamberlain, DGL 27980.
67 - {*VCR EL*}; sandy ground near mouth of Hedderwick Burn, West Barns, approx. 4m, 12 November 1993, D. G. Long 25031.

13. R. canescens (Hedw.) Brid.   **EL**
Rare.
Sand-dunes by the sea.

48 - {*VCR EL*}; dune pasture, Aberlady, 1963, B. P. Donaldson.
48 - bare basalt soil, near Gullane Point, 18 October 2000, D. G. Long, G. P. Rothero and D. F. Chamberlain.
58 - thin soil on basaltic outcrop in fixed dunes, Yellowcraig, near Dirleton, 5 February 1994, D. G. Long 25218.

71. **Ptychomitrium** Fürnr.

1. P. polyphyllum (Sw.) Bruch & Schimp.
                                        WL, ML, EL
Occasional.
Base-rich rocks.

07 - {*VCR WL*}; on rocks in hollow, approx. 80m, Philpstoun Bing, 12 March 1998, D. G. Long 27457.
16 - Harperrig, 1979, field record.
24 - River South Esk, above Gladhouse, field record.
(26) - near Glencorse Reservoir, September 1905, J. McAndrew.
27 - Water of Leith, Dean Village, field record.
48 - Aberlady Bay, 1971, R. D. Revell.
58 - basaltic rock outcrop on dunes, 3m, 5 February 1994, D. G. Long 25219.

74. **Blindia** Bruch, Schimp. & W. Gümbel

1. B. acuta (Hedw.) Bruch, Schimp. & W. Gümbel   **ML, EL**
Occasional.
Rocks in streams, sub-montane.

05, 16, 24, 34, 35, 56, 57, 67, 77.

75. **Seligeria** Bruch, Schimp. & W. Gümbel

1. S. pusilla (Hedw.) Bruch, Schimp. & W. Gümbel   **ML**

Rare. This is a nationally scarce species.
Limestone rock face.

15 - limestone bluffs along River North Esk,
Newhall Estate, 20 February 2001,
D. F. Chamberlain.
36 - {VCR ML}; side of stream in wooded valley
to north of Hagbrae Quarries, 1974, D. G. Long
and D. F. Chamberlain.

6. S. recurvata (Hedw.) Bruch, Schimp. &
W. Gümbel   **WL, ML**
Occasional.
Damp shaded sandstone and limestone rock faces.

96 - {VCR WL}; limestone blocks in stream,
Petershill Quarry, north-east of Bathgate,
21 October 1975, D. G. Long 4696.
(07) - rock in wood, south bank of Mains Burn
at Ecclesmachan, 8 April 1916, J. C. Adam.
16 - conglomerate rock face, above lower waterfall, Logan Burn, 4 September 1998, D. G. Long
and D. F. Chamberlain.
(26) - Bilston Burn, April 1903, J. McAndrew.
27 - sandstone boulder by Water of Leith,
Slateford, Edinburgh, 15 November 1998,
D. F. Chamberlain.
35 - limestone above a damp bank, Middleton
Quarry, 31 August 1971, D. G. Long *et al.* 1607.

9. S. donniana (Sm.) Müll. Hal.   **ML**
Rare. This is a nationally scarce species.
Damp calcareous rock.

16 - damp overhanging calcareous rock, Loganlea
Waterfall, 19 May 1972, D. F. Chamberlain.
(26) - Bilston Glen, 4 April 1902, W. Evans.

76. Brachydontium Fürnr.

1. B. trichodes (F. Weber) Milde   **(ML)**
Not seen recently.
Moist sandstone rock faces.

(16) - south of Currie, 15 March 1904, W. Evans.
(26) - Hawthornden, 4 April 1903, J. McAndrew.

77. Discelium Brid.

1. D. nudum (Dicks.) Brid.   **[ML]**
The single literature record from 47 - Adniston,
March 1853, A. O. Black has not been confirmed
since. Even if the original determination was correct it seems likely that this species does not now
occur in the Lothians.

78. Funaria Hedw.

1. F. hygrometrica L.   **WL, ML, EL**
Common.
Disturbed ground, especially after fires.

95, 96, 97, 98, 05, 06, 07, 08, 15, 16, 17, 25, 26,
27, 34, 35, 36, 37, 44, 45, 46, 47, 48, 56, 57, 58,
66, 67, 68, 77.

79. Entoshodon Schwägr.

1. E. attenuatus (Dicks.) Bryhn   **(ML)**
Peaty banks in moorland, rare.
There is some doubt as to the locality for the single specimen known as it is recorded as having
been collected in the Pentland Hills in 1897.

2. E. fascicularis (Hedw.) Müll. Hal. (*Funaria fascicularis* (Hedw.) Lindb.)   **[ML]**
The single literature record for this species (26,
Auchendinny, 1871) has not been subsequently
confirmed.

3. E. obtusus (Hedw.) Müll. Hal. (*Funaria obtusa*
(Hedw.) Lindb.)   **ML**
Rare.
Peaty banks.

26 - Roslin, 1960s, D. F. Chamberlain, field
record.
(26) - Allermuir Glen, 1 March 1902, W. Evans.
(27) - Holyrood Park, Edinburgh, Anon.

80. Physcomitrium (Brid.) Brid.

1. P. pyriforme (Hedw.) Brid.   **WL, ML, EL**
Occasional.
Damp ground, amongst grass.

07, (17, 26, 27), 36, (37), 45, (47, 57).
07 - {MVCR WL}; meadow east of Linlithgow
Loch, 16 May 1999, D. F. Chamberlain.
36 - {MVCR ML}; on soil in marshy field, Ford,
near Pathhead, 28 June 1987, D. G. Long 14158.
(37) - near Musselburgh, 16 May 1907,
J. McAndrew.
45 - {MVCR EL}; boggy field trampled by cattle,
approx. 320m, Upper Linn Dean Water, Soutra,
21 May 1990, D. G. Long 18126.

3. P. sphaericum (Ludw.) Brid.   **ML**
Only known from one locality. This is a national-
ly near-threatened species.
Mud by dried-out reservoir.

06 - Harperrig Reservoir, 1979, D. G. Long and D. F. Chamberlain.

81. Aphanoregma Sull.

1. A. patens (Hedw.) Lindb.   **(WL), ML, EL**
Rare.
Reservoir margins, muddy stream banks.

16 - dried out silt, west end of Clubbiedean Reservoir, 1973, D. F. Chamberlain and D. G. Long, DGL 2192.
(17) - west of South Queensferry, 7 October 1911, J. McAndrew.
(26) - south end of Torduff Reservoir, Pentland Hills, 1973, D. G. Long and D. F. Chamberlain.
67 - {VCR EL}; ditch, near East Links, Belhaven, 1977, D. G. Long and D. F. Chamberlain.

83. Ephemerum Hampe

1. E. serratum (Hedw.) Hampe
a. var. serratum   **(WL), ML, EL**
Occasional. This is a nationally scarce variety.
Damp mud on paths and in fields.

(07), 15, 16, 25, 26, 35, 46, 56.
56 - {VCR, EL}. Damp mud by path, wood near Gifford, 1971, D. G. Long.

b. var. minutissimum (Lindb.) Grout   **ML**
Rare.
An ephemeral in fields.

(27) - Craiglockhart, 18 February 1897, W. Evans.
36 - meadow below Hag Brae, Currie Lea, D. G. Long.

86. Tetraplodon Bruch, Schimp. & W. Gümbel

1. T. mnioides (Hedw.) Bruch & Schimp. **ML, EL**
Rare.
On dung and animal remains in moorland.

16 - Threipmuir, on dead sheep, 1967, M. K. Jurand.
24 - Dundreich Plateau, Moorfoot Hills, 1970s, D. G. Long.
[26] - Swanston and Bonaly Hills, prior to 1917, W. Evans.
34 - Eastside Heights, field record.
35 - {MVCR ML}; West Heath, Moorfoot Hills, September 1997, C. Dixon.

45 - south-east margin of Fala Moor (EL), D. G. Long and D. F. Chamberlain, DGL 2508.

88. Splachnum Hedw.

1. S. sphaericum Hedw.   **(WL), ML**
Occasional.
On dung on acid moorland.

(86) - Drumelzie Moss, adjoining Blawhorn Moss, 27 May 1916, W. Evans.
(05) - Cobbinshaw Moss, 11 June 1910, W. Evans.
15 - field record.
16 - field record.
24 - Dundreich, field record.
(26) - Bonaly Hill, 31 August 1890, W. Evans.

2. S. ampullaceum Hedw.   **ML, EL**
Rare.
On dung and animal remains, in base-rich mires.

16 - sheep dung on blanket bog, Ravelrig Toll Moss, 27 January 1974, DGL 3104.
16 - dung in marsh, near Logan Burn, Pentland Hills, September 1977, D. F. Chamberlain.
45 - flushes amongst willows, south-east margin of Fala Moor, 2 September 1996, DGL 26680.
56 - Gifford, 1969, E. P. Beattie.

91. Orthodontium Schwägr.

1. O. lineare Schwägr.   **WL, ML, EL**
Frequent to common. This introduced species is spreading in suitable habitats.
Rotting wood, damp peaty banks.

96, 97, 98, 06, 07, 15, 16, 17, 24, 25, 26, 27, 35, 36, 37, 45, 46, 47, 48, 58, 66, 68, 77.
07 - {VCR WL}; stump in spruce plantation, side of pond, near Dechmont, 1971, D. G. Long and D. F. Chamberlain.
16 - {VCR ML}; on wood, by Threipmuir Reservoir, August 1964, E. F. Warburg and A. C. Crundwell.

2. O. gracile Schwägr. ex Bruch, Schimp. &
W. Gümbel   **ML**
Rare.
Shaded sandstone rock faces.

This native, nationally endangered species is decreasing in Britain. Recent observations suggest that this species is more conspicuous in the winter; in summer it may survive in the protonemal state.

(26) - Roslin Glen, 1898, A. Murray and
C. Scot.
26 - sandstone outcrop above the River Esk,
Roslin, 1992, D. F. Chamberlain; 27 September
2000, G. Rothero and D. G. Long.

92. Leptobryum Wilson

1. L. pyriforme (Hedw.) Wilson   **WL, ML, (EL)**
Stone walls, rocks, cinder heaps. It also occurs
frequently in plant pots.

[96], 08, 16, (17), 26, 27, (48, 57).

93. Pohlia Hedw.

3. P. cruda (Hedw.) Lindb.   **WL, ML, EL**
Frequent, especially in upland areas.
On shaded banks and ledges.

97, 98, 07, 08, 15, 16, 17, 24, 25, 26, 27, 34, 35,
36, 45, 56, 57, 66, 67, 77.
98 - {*MVCR* WL}; basic rocks on wooded bank,
near East Kerse Mains, Bo'ness, 26 January 1975,
D. G. Long and D. F. Chamberlain, DGL 4185.

4. P. nutans (Hedw.) Lindb.   **WL, ML, EL**
Common.

Peaty banks in bogs and on moorland; rotting tree
trunks.

All Squares.

6. P. drummondii (Müll. Hal.) A. L. Andrews
          WL, ML, EL
Occasional.
Stony and gravelly ground on banks and tracks.
This species has been recently recognised as dis-
tinct from the much rarer *P. rothii*. All Lothian
records refer to *P. drummondii*.

97, 06, 15, 16, 26, 45, 56, 66, 67.
06 - {*VCR* WL - as *P. rothii*}; edge of field near
Easter Inch Moss, Bathgate, 1971, D. G. Long
and D. F. Chamberlain.
16 - {*VCR* ML - as *P. rothii*}; heathy track near
Logan Burn, Pentlands Hills, 1971, J. A. Paton
and C. C. Townsend.
67 - {*VCR* EL - as *P. rothii*}; shaded loamy bank,
Burn Hope Dean, near Innerwick, 1972,
D. F. Chamberlain *et al.*

8. P. filum (Schimp.) Mårtensson (*P. gracilis*
(Bruch *et al.*) Lindb.)   **ML**
Once found. This is a nationally scarce species.
Gravelly quarry bottom; sub-montane.

35 - {*VCR* ML}; Broad Law Quarry, Moorfoot
Hills, 1971, A. C. Crundwell and J. A. Paton.

10. P. bulbifera (Warnst.) Warnst.   **WL, ML, EL**
Frequent.
Damp tracks, marsh margins.

86, 96, 97, 05, 06, 15, 16, 25, 26, 27, 34, 35, 44,
45, 46, 56, 67.
96 - {*VCR* WL}; damp roadside ditch, near
Fauldhouse, 1972, D. G. Long and
D. F. Chamberlain.
45 - {*VCR* EL}; disturbed soil, Fala Moor, 1973,
D. F. Chamberlain *et al.*

11. P. annotina (Hedw.) Lindb.   **WL, ML, EL**
Frequent.
Rock ledges, stony ground.

96, [97], 05, 06, 07, 08, 15, 16, 17, 24, 26, 27,
34, 35, 46, 47, (48), 56, (57), 66, 67, 68, 76, 77.

13. P. camptotrachela (Renauld & Cardot) Broth
          WL, ML, EL
Rare.
Damp ground.
07 - grass ley, east of Linlithgow Loch, 16 May
1999, D. F. Chamberlain.
07 - {*VCR* WL}; damp pasture, Ochiltree Hill,
8 October 1975, D. G. Long 4687.
25/35 - Gladhouse Reservoir, 1971, J. A. Paton.
45 - {*VCR* EL}. Linn Dean Water, Soutra Hill,
7 June 1973, D. G. Long 2490.

14. P. flexuosa Harv. (*P. meyldermansii*
R. Wilczek & Demaret)   **ML**
Once found. This is a nationally scarce species.
Reservoir margin.
16 - {*VCR* ML}; Loganlee Reservoir, 1971,
J. A. Paton.

16. P. lutescens (Limpr.) H. Lindb.   **ML, EL**
Rare.
Woodland banks, bare soil on tracks and in pasture.

16 - damp soil on side of ditch, near Threipmuir
Reservoir, 2 September 1973, D. G. Long 2727.
35 - {*VCR* ML}; track to Broad Law Quarry,
Moorfoot Hills, 1971, J. A. Paton.
66 - {*VCR* EL}; damp sandy river bank in
wooded ravine, 4 August 1973, D. G. Long
2640.
67 - bare soil in rough pasture, above Weartherly,
near Spott, 13 October 1974, D. G. Long and
D. F. Chamberlain, DGL 4069.

18. P. melanodon (Brid.) A.L. Shaw (*P. carnea* (Schimp.) Lindb., *P. delicatula* (Hedw.) Grout) **WL, ML, EL**
Widespread.
Shaded banks, on damp soil.

96, 97, 05, 06, 07, 08, 15, 16, 17, 24, 25, 26, 27, 35, 36, 37, 45, 46, 47, 56, 57, 58, 66, 67, 76.

19a. P. wahlenbergii F. Weber & D. Mohr var. **wahlenbergii** **WL, ML, EL**
Frequent to common.
Damp ground, on tracks and in marshes and in woodland.

96, 97, 06, 07, 15, 16, 17, 24, 25, 26, 27, 34, 35, 36, 44, 45, 46, 47, 48, 56, 57, 58, 66, 67, 76, 77.

95. Plagiobryum Lindb.

1. P. zieri (Hedw.) Lindb. **ML, EL**
Rare.
Base-rich rocks by streams, sub-montane.

16 - waterfalls, Logan Burn, 4 September 1998, D. F. Chamberlain and D. G. Long.
24 - ravine above River South Esk, above Gladhouse, 450m, D. G. Long and D. F. Chamberlain.
(26) - west of Swanston, 1898, Anon.
45 - Linn Dean Water, Soutra, D. G. Long.
(57) - Traprain Law, 1908, Anon.
66 - Burn Hope, field record.

96. Anomobryum Schimp.

1. A. julaceum (Gaertn., Meyer & Schreb.) Schimp.
Occasional.
Base-rich rocks by streams, generally sub-montane.

a. var. julaceum (*A. filiforme* (Dicks.) Solms) **ML, EL**
16, 24, 25, 26, 34, 35, 44, 45, 77.
77 - {VCR EL}; damp rocks below Sheeppath Dean, 1972, D. F. Chamberlain *et al.*

b. var. concinnatum (Spruce) J. E. Zettererst. (*A. filiforme* (Dicks.) Solms var. *concinnatum* (Spruce) Loeske) **ML, EL**

15 - River North Esk, above Carlops, 17 June 1973, D. F. Chamberlain and D. G. Long.
16 - rock ledges, Logan Burn waterfalls, 310m, 31 July 1976, D. G. Long 5134.
26 - wet basic rocks in small gully, Bonaly ravine,

243m, 23 July 1977, D. G. Long 6115.
34 - dry basic rocks, Wolf Cleugh, Dewar Burn, 360m, 13 August 1977, D. G. Long 6172.
45 - {VCR EL}; damp basic rocks, Linn Dean Water, Soutra, 1990, D. G. Long 18113.

97. Bryum Hedw.

2. B. warneum (Röhl.) Blandow ex Brid. **EL**
Rare. This is a nationally vulnerable species.
Damp sandy soil on sand-dunes.

(48) - Gullane Links, 2 July 1897, W. Evans.
48 - damp sand stripped of turf, Gullane Links, August 1978, D. F. Chamberlain.
58 - disturbed sand on bank below Marine Villas, West Links, Yellowcraig, 15 August 2001, D. F. Chamberlain.
67 - damp ground on edge of salt marsh, Belhaven Bay, 20 March 1976, D. G. Long 4779.

4. B. calophyllum R.Br. **EL**
This nationally vulnerable species has only been recorded from one East Lothian dune system.
Upper part of salt marsh.

(48) - Gullane Links, July 1897, W. Evans.
(48) - Luffness Links, 7 November 1908, W. E. Evans.
48 - Aberlady Bay, marshy ground by mouth of Peffer Burn, near the wooden bridge, 15 September 1974, D. G. Long and D. F. Chamberlain, DGL 3961.

7. B. uliginosum (Brid.) Bruch & Schimp. **(EL)**
This nationally critically endangered species is probably extinct in the Lothians.
Dune slacks.

(58) - Dirleton Common, 11 August 1897, W. Evans.

8. B. pallens Sw. **WL, ML, EL**
Widespread.
Damp base-rich open ground, edge of marshes, sand-dunes, by streams.

95, 96, 97, 05, 06, 07, 08, 15, 16, 24, 25, 26, 34, 35, 44, 45, 47, 48, 56, 57, 58, 66, 67, 76, 77.

13. B. algovicum (Sendtn. ex Müll. Hal. var. **ruthenicum** (Warnst.) Crundw. (*B. pendulum* (Hornsch.) Schimp.). **WL, ML, EL**
Occasional.
Mortar on walls, sand-dunes.

86, (17, 26, 27), 48, 57, (58), 66, 67, 68.

86 - {*MVCR* EL}; mortar on stone podium at entrance to Blawhorn Moss NNR, 20 July 1996, D. F. Chamberlain.

15. B. knowltonii *Barnes*   (EL)
The only known specimen of this nationally vulnerable species (**48** - Gullane, 1909, Anon.) is sterile. There is therefore some doubt as to the authenticity of this record.

17. B. imbricatum (Schwägr.) Bruch & Schimp.
(*B. inclinatum* (Brid.) Blandow)   **ML, EL**
Occasional.
Sand-dunes, old walls.

16, 26, 36, 37, 47, 48, 58, 66, 67, 77.
48 - {*VCR* EL}; on shell sand stripped of turf, Gullane Links, 1978, D. F. Chamberlain.

18. B. intermedium (Brid.) Barnes   **(ML), EL**
Rare. This is a nationally scarce species.
Sand-dunes.

(26) - south side of Roslin Glen, 30 Oct. 1897, W. Evans.
48 - shell sand stripped of turf, Gullane Links, August 1978, D. F. Chamberlain.
77 - old lime kiln, Skateraw, 20 March 1976, D. F. Chamberlain and D. G. Long, DFC B271.

20. B. capillare Hedw.
a. var. capillare   **WL, ML, EL**
Common.
Tree trunks, rock faces, walls.

All Squares.

b. var. rufifolium (Dixon) Podp.   **ML, EL**
Rare.
Base-rich rocks.

26 - {*VCR* ML}; dry basic rock , gully of Bonaly Burn, 1977, D. G. Long.
58 - {*VCR* EL}; stone of cliff path, Oxroad Bay, near Tantallon Castle, 1974, A. C. Crundwell *et al.*

23. B. subelegans Kindb. (*B. flaccidum* Brid.)
    WL, ML, EL
Occasional.
Epiphytic on shrubs and trees.

97, 17, 25, 26, 35, 36, 45, 46, 47, 48, 56, 57, 58, 67, 68, 77.
97 - {*VCR* WL}; shady bank in old lime quarry above Galabraes, Bathgate Hills, 1975, D. G. Long.

27. B. pallescens Schleich. ex Schwägr.   **EL**
Rare but possibly overlooked.
Damp places on sand-dunes.

(48) - damp places, Gullane Links, 12 June 1916, J. C. Adam.
58 - disturbed sand on bank below Marine Villas, West Links, Yellowcraig, 15 August 2001, D. F. Chamberlain.

28. B. pseudotriquetrum (Hedw.) Gaertn, Meyer & Schreb.
a. var. pseudotriquetrum   **WL, ML, EL**
Common.
Bogs and marshes, by streams.

95, 96, 05, 06, 07, 15, 16, 24, 25, 26, 27, 34, 35, 36, 44, 45, 46, 48, 56, 57, 58, 66, 67, 76, 77.

b. var. bimum (Brid.) Lilj.   **[ML]**
This taxon is only known from a single literature record from [26] - Allermuir Glen, Bonaly Burn, 1900, Anon. This record should be treated with some caution.

30. B. caespiticium Hedw.   **WL, ML, EL**
Widespread.
Rocks, walls, cinder on bings.

95, 07, 17, 24, 25, 26, 27, 34, 36, 37, 48, 56, 58, 66, 67, 68, 76, 77.

31. B. argenteum Hedw. (incl. var. *lanatum* Bruch *et al.*)   **WL, ML, EL**
Abundant in built-up areas.
Tarmac, tracks and man-made habitats generally.

All squares.

34. B. gemmiferum R. Wilczek & Demaret
    WL, EL
Rare but probably overlooked.
Mud and bare soil.

06 - {*VCR* WL}; mud on a bing, Drumshoreland Muir, 110m, 30 May 1988, D. G. Long 15072.
47 - {*VCR* EL}; bare soil, sloping bank of ditch, Cuddie Wood, near Gladsmuir, 9 October 1977, D. G. Long 6350.

36. B. bicolor Dicks.   **WL, ML, EL**
Common.
Man-made habitats generally.

86, 96, 97, 98, 05, 06, 07, 08, 15, 16, 17, 25, 26, 27, 34, 35, 36, 37, 44, 45, 46, 47, 48, 56, 57, 58, 66, 67, 68, 77.

37. B. dunense A. J. E. Smith & H. Whitehouse
WL, EL
Rare. This is a nationally scarce species.
Sandy ground by the sea.

17 - {*VCR* WL}; sandy ground by the sea, near
Dalmeny House, 1981, D. G. Long.
48 - {*VCR* EL}; damp sandy ground by sea, at
footbridge, Luffness Links, September 1974,
D. G. Long and D. F. Chamberlain.
67 - grass flats at Belhaven Bay, 20 November,
1975, D. F. Chamberlain B243.

38. B. radiculosum Brid. **WL, ML, EL**
Occasional.
Mortar in old walls.

06, 17, 26, 27, 35, 36, 37, 46, 57, 58.
17 - {*VCR* ML}; crevices in sandstone wall of
farm building, East Craigs, Corstorphine, 1975,
D. G. Long.
77 - {*VCR* EL}; damp brickwork by stream,
Thornton Burn, 1976, D. G. Long.

39. B. ruderale A. C. Crundwell & Nyholm
WL, ML, EL
Occasional but probably under-recorded.
Disturbed ground, fields.

97, 05, 25, 26, 27, 35, 46, 58.
97 - {*VCR* WL}; bare soil in field on south-east
side of River Avon, Muiravonside, 1971,
A. C. Crundwell and A. McG. Stirling.

40. B. violaceum A. C. Crundwell & Nyholm
WL, ML, EL
Rare but probably under-recorded.
Bare soil in fields.

97 - {*VCR* WL}; bare soil in field on south-east
side of River Avon, Muiravonside, 1971,
A. C. Crundwell and A. McG. Stirling.
26 - {*VCR* ML}; on bare soil in arable field, near
Roslin Chapel, August 1969,
D. F. Chamberlain.
27 - Canonmills, Edinburgh, 1977,
D. G. Long.
68 - Tyninghame, 1970s, Anon.

41. B. klinggraeffii Schimp. **(WL), ML, EL**
Rare.
Dried-up ponds and reservoirs.

(07) - {*VCR* WL}; west of South Queensferry,
7 October 1911, J. McAndrew.
26 - Torduff Reservoir, 13 Sept 1973, D. G. Long
and D. F. Chamberlain.

56 - {*VCR* EL}; damp soil on woodland track,
near Gifford, August 1971, D. G. Long.
57 - Whiteadder Reservoir, September 1996,
D. F. Chamberlain.
58 - pond, North Berwick Law, 20 January 1974,
D. G. Long 3083._

42. B. sauteri Bruch *et al.* **ML**
Rare but probably overlooked.
Disturbed ground.

[26] - {*VCR* ML}; Glencorse, Pentland Hills,
October 1952, A. C. Crundwell.

43. B. tenuisetum Limpr. **ML**
Once found. This is a nationally scarce species.
Dried-up reservoir.

16 - {*VCR* ML}; mud by Threipmuir Reservoir, at
Redford Bridge, Balerno, 1975, D. G. Long and
D. F. Chamberlain.

44. B. subapiculatum Hampe
(*B. microerythrocarpum* Müll. Hal. & Kindb.)
WL, ML, EL
Occasional but almost certainly under-recorded.
Disturbed soil, generally on acid substrates.

[96], 06, 07, 17, 47, 48, 56, 58, 67.
56 - {*VCR* EL}; stream bank, Hopes Water, above
West Hopes, south-east of Gifford, 1971,
D. F. Chamberlain and D. G. Long.

45. B. bornholmense Wink. & R. Ruthe **ML**
Once recorded.
Reservoir.

35 - {*VCR* ML}; Gladhouse Reservoir, 1971,
A. C. Crundwell.

46. B. rubens Mitt. **WL, ML, EL**
Common.
Disturbed soil, in fields and gardens.

95, 97, 98, 06, 07, 16, 17, 25, 26, 27, 34, 35, 36,
37, 45, 46, 47, 56, 57, 58, 67, 68, 77.
97 - {*VCR* WL}; bare soil in field on south-east
side of River Avon, Muiravonside, 1971,
A. C. Crundwell and A. McG. Stirling.
26 - {*VCR* ML}; bare soil in arable field near
Roslin Chapel, August 1969, D. F. Chamberlain.

51. B. alpinum With. **(WL), ML, EL**
Local and rare.
Damp basalt.

(97) - Cockleroy, 12 September 1908,

J. McAndrew.
(16) - Dalmahoy Hill, 29 April 1907,
J. McAndrew.
16 - basalt dykes Kaimes Hill, Balerno, 12 March
1953, D. M. Henderson.
(57) - Traprain Law, 12 September 1908,
W. Evans.
57 - Traprain Law, 1970s, field record.

98. **Rhodobryum** (Schimp.) Limpr.

1. **R. roseum** (Schimp.) Limpr.　　　**WL, ML, EL**
Occasional.
Base-rich grassland and woodland banks.

(08), 16, 17, 25, [26], (27, 36, 47), 48, 58, 67,
68, 77.
16 - {*MVCR* ML}; short turf, near Logan Burn
Waterfall, September 1977, D. F. Chamberlain.
17 - {*MVCR* WL}; on sandy slope by sea, 5m,
Hound Point, Dalmeny Estate, 1981, D. G. Long
10065.

99. **Mnium** Hedw.

1. **M. hornum** Hedw.　　　**WL, ML, EL**
Common.
Acid substrates generally, in woodland and on rocks.

All Squares.

2. **M. marginatum** (Dicks.) P. Beauv.
　　　　　　　　　　　　　(WL), ML, EL
Occasional.
Shaded base-rich rocks.

(97), 16, (17), 24, (26), [37], 45 (58), 67.
67 - {*VCR* EL}; on shady rocks in gorge, Boonslie
Burn, below Boonslie, Lammermuir Hills, 1974,
D. G. Long 4070.

3. **M. stellare** Hedw.　　　**WL, ML, EL**
Frequent.
Base-rich rock ledges.

97, 07, 08, 15, 16, 26, 27, 35, 36, 45, 46, 48, 56,
57, 66, 67, 77.

101. **Rhizomnium** T. J. Kop.

1. **R. punctatum** (Hedw.) T. J. Kop.　**WL, ML, EL**
Common.
**Wet rock ledges, damp ground under trees and in
marshes.**

86, 96, 97, 98, 05, 06, 07, 08, 15, 16, 17, 24, 25,

26, 27, 34, 35, 36, 37, 44, 45, 46, 47, 48, 56, 57,
58, 66, 67, 76, 77.

3. **R. pseudopunctatum** (Bruch & Schimp.)
T. J. Kop.　　　　　　　　　**(WL), ML**
Occasional.
Base-rich flushes and mires.

(97) - Cockleroy Hill, 8 September 1906,
J. McAndrew.
(06) - near Harburn, West Calder, May 1856,
Nichol.
15 - mire, by River North Esk, above Carlops,
17 June 1973, D. F. Chamberlain and
D. G. Long.
16 - Logan Burn, 1970s, D. G. Long and
D. F. Chamberlain.
26 - by outflow stream, Bonaly Reservoir, 325m,
24 July 1977, D. G. Long 6118.
27 - mapping record.
34 - basic flush on side of gully, Wolf Cleugh,
Dewar Burn, 360m, 13 August 1977, D. G. Long
6166.

102. **Plagiomnium** T. J. Kop.

1. **P. cuspidatum** (Hedw.) T. J. Kop.　**WL, ML, EL**
Occasional.
Basalt, lawns, wood.

07, 17, 27, 47, (48), 57, (58), 67.
07 - {*MVCR* WL}; shady boulders in wood,
below Abercorn Church, 6 January 1974,
D. G. Long 3043.
27 - {*MVCR* ML}; lawn, Royal Botanic Garden
Edinburgh, 21 April 1976, D. F. Chamberlain.

2. **P. affine** (Blandow) T. J. Kop.　　**(WL), ML, EL**
Local but probably under-recorded as it can be
confused with the commoner *P. rostratum*.
Sandy soil on dunes, walls.

(97) - Carribber Glen, May 1916,
J. C. Adam.
(06) - foot of wall by road, Morton, 4 March
1898, W. Evans.
(07) - Riccarton Hills, 18 January 1902,
W. Evans.
(17) - {*VCR* WL}; Carlowrie, near Kirkliston,
March 1898, W. Evans.
26 - {*VCR* ML}; Clubbiedean Reservoir, Pentland
Hills, 10 September 2000, G. P. Rothero.
27 - shaded basalt cliff above Blackford Pond,
16 April 2001, D. F. Chamberlain.
67 - {*VCR* EL}; sandy ground at margin of pine

plantation, on dunes, 4m, near mouth of the Hedderwick Burn, West Barns, 1993, D. G. Long 25030.

4. P. elatum (Bruch & Schimp.) T. J. Kop. (*Mnium seligeri* Lindb.) **WL, ML, EL**
Occasional though locally frequent.
Base-rich, mires and dune slacks.

97, 05, 15, 16, 24, 25, 26, 34, 35, 45, 48, 56, 66.

5. P. ellipticum (Brid.) T. J. Kop. (*Mnium rugicum* Laur.) **ML, EL**
Occasional.
Neutral to slightly acid mires.

05, 16, 26, (27), 35, 45, 46, 48, 56, 67.

6. P. undulatum (Hedw.) T. J. Kop. **WL, ML, EL**
Common.
Amongst grass, woodland banks.

95, 96, 97, 98, 05, 06, 07, 08, 15, 16, 17, 24, 25, 26, 27, 34, 35, 36, 37, 44, 45, 46, 47, 48, 56, 57, 58, 66, 67, 68, 76, 77.

7. P. rostratum (Schrad.) T. J. Kop. **WL, ML, EL**
Common.
Woodland banks, by rivers.

95, 96, 97, 06, 07, 08, 15, 16, 17, 25, 26, 27, 34, 35, 36, 37, 45, 46, 47, 48, 56, 57, 58, 67, 68, 76, 77.

103. Pseudobryum (Kindb.) T. J. Kop.

1. P. cinclidioides (Huebener) T. J. Kop. **ML, EL**
Rare. This is a nationally scarce species.
Upland mires.

45 - {VCR ML}; middle part of Longmuir Moss, 275m, 25 June 1995, D. G. Long 25932.
56 - {VCR EL}; foot of Rotten Cleugh, upper Dye Water, 455m, 1991, D. G. Long 19562.

104. Aulacomnium Schwägr.

1. A. palustre (Hedw.) Schwägr. **WL, ML, EL**
Frequent.
Bogs.

86, 95, 96, 05, 06, 07, 15, 16, 24, 25, 26, 34, 35, 44, 45, 48, 56, 66.

3. A. androgynum (Hedw.) Schwägr. **WL, EL**
Occasional.

Epiphytic on tree branches.

96, 97, 06, 07, 16, 17, 25, 26, 27, (35), 36, 37.

106. Meesia Hedw.

1. M. uliginosa Hedw. (EL)
This nationally scarce species is probably extinct in the Lothians.
Flushes on sand-dunes.

(48) - Gullane Links, June 1907, J. McAndrew.
(48) - Gullane Links, October 1907, J. McAndrew.

107. Amblyodon Bruch & Schimp.

1. A. dealbatus (Hedw.) Bruch & Schimp. **ML, EL**
This nationally scarce species is rare in the Lothians.
Base-rich flushes, dune slacks.

26 - by outflow stream, Bonaly Reservoir, 330m. 23 July 1977, D. G. Long 6101.
(27) - Holyrood Park, Edinburgh, 27 April 1806, G. Don.
(48) - Luffness Links, 19 September 1910, J. McAndrew.
67 - ditches in sand-dune, Belhaven Bay, 20 March 1976, D. F. Chamberlain and D. G. Long, DFC B272.

108. Catascopium Brid.

1. C. nigritum *(Hedw.) Brid.* (EL)
This nationally scarce species is probably extinct in the Lothians.
Flushes and dune slacks.

(48) - Gullane Links, July 1897, W. Evans.
(48) - Luffness Links 16 June 1909, W. E. Evans.
(48) - west side of Gullane Hill, 16 June 1909, W. Evans.

109. Plagiopus Brid.

1. P. oederianus *(Sw.) H. A .Crum & L. E. Anderson* (ML)
This species has not been seen for ninety years and is probably extinct.
Base-rich agglomerate rocks.

(16) - The Howe, 1867, J. Sadler.
(16) - The Howe, 5 May 1909, W. E. Evans.

110. **Bartramia** Hedw.

2. **B. pomiformis** Hedw. (incl. var. *elongata* Turn.)   WL, ML, EL
Frequent.
Banks and rock ledges, usually upland.

97, 07, 15, 16, 24, 25, 26, 34, 35, 44, 45, 46, 56, 57, 58, 66, 67, 76, 77.

3. **B. ithyphylla** Brid.   WL, ML, EL
Frequent.
Bank and rock ledges, usually on slightly base-rich substrates.

06, 07, 15, 16, (17), 24, (25, 26), 27, 34, 45, 56, 57, (58), 66, 76.
07 - {*MVCR* WL}; on side of block of conglomerate ash on north-facing slope of Philpstoun South Bing, 12 March 1998, D. G. Long 27445.

112. **Philonotis** Brid.

3. **P. arnellii** Husn.   ML, EL
Rare.
Reservoir margins, bare soil by streams, moist base-rich rock ledges.

16 - north end of Harperrig Reservoir, Winter 1979, D. F. Chamberlain.
35 - east end of Gladhouse Reservoir, 19 August 1973, D. G. Long 2689.
45 - bare sandy soil by stream, lower part of Linn Dean ravine, 28 April 1974, D. G. Long 3393.
56 - damp mud, Stobshiel Reservoir, 19 August 1973, D. G. Long 3629.
(57) - Traprain Law, September 1928, J. B. Duncan.

4. **P. caespitosa** Jur.   WL, ML
Rare. This is a nationally scarce species.
Pasture, bare mud by reservoir.

97 - {*VCR* WL}; pasture near Bathgate, 1992, Wiggington.
16 - {*VCR* ML}; Threipmuir, 1964, A. C. Crundwell and E. F. Warburg.

5. **P. fontana** (Hedw.) Brid.   WL, ML, EL
Common.
Wet ground, flushes, marshes and by streams.

86, 96, 97, 05, 06, 15, 16, 17, 24, 25, 26, 27, 34, 35, 44, 45, 46, 56, 57, 66, 67, 76, 77.

8. **P. calcarea** (Bruch & Schimp.) Schimp.   ML, EL

Occasional.
Calcareous flushes, generally sub-montane.

15, 16, 24, 25, 26, 34, (48), 66, 67, [76], 77.

113. **Breutelia** Schimp.

1. **B. chrysocoma** (Hedw.) Limpr.   [WL], ML
Local and probably decreasing.
Moorland banks.

[06] - Drumshoreland, 1917, Anon.
(16) - The Howe, Pentland Hills, March 1847, W. Evans.
24 - River South Esk, near Gladhouse, 20 July 1998, D. F. Chamberlain.
25 - River South Esk, near Gladhouse, 1970s, field record.
(26) - Bonaly Glen, 3 October 1904, W. Evans.
34 - tributary of the Dewar Burn, 15 August 1971, D. G. Long 1501.

115. **Amphidium** Schimp.

2. **A. mougeotii** (Bruch & Schimp.) Schimp.
           WL, ML, EL
Frequent.
Shaded rock ledges, banks.

96, 97, 08, 15, 16, 24, 25, 26, 34, 35, 45, 56, 57, 66, 67, 76, 77.

116. **Zygodon** Hook. & Taylor

1. **Z. viridissimus** (Dicks.) Brid.
a. var. **viridissimus**   WL, ML, EL
Common.
Rock faces, bark of trees.

86, 96, 97, 06, 07, 08, 15, 16, 17, 25, 26, 27, 35, 36, 45, 46, 47, 48, 56, 57, 67, 68, 77.

b. var. **stirtonii** (Schimp. ex Stirt.) I. Hagen
           WL, ML, EL
Occasional.
Rock ledges, bark of trees.

97, (17), 26, 27, 36, 45, 48, 57, 58, 68.
45 - {*VCR* EL}; calcareous ledge in ravine, Linn Dean Water, Soutra Hill, 1974, D. G. Long and D. F. Chamberlain.

2. **Z. rupestris** Schimp. ex Lor. (*Z. baumgartneri* Malta; *Z. viridissimus* Schimp. ex Lor. var. *vulgaris* Malta)   ML, EL
Rare.
Base-rich rock faces.

16 - {*VCR ML*}; agglomerate rock face, between the waterfalls, Logan Burn, 4 September 1998, D. G. Long and D. F. Chamberlain, DGL 28014.
45 - {*VCR EL*}; calcareous ledge in ravine, Linn Dean Water, Soutra Hill, 1970s, D. G. Long and D. F. Chamberlain.

3. Z. conoideus (Dicks.) Hook & Tayl.   ML, EL
Rare.
Old elder and apple trees.

17 - elder, ruins at Cammo Estate, West Edinburgh, 4 March 2001, D. F. Chamberlain.
26 - {*VCR ML*}; elder, River North Esk, opposite Hawthornden, 27 September 2000, D. G. Long and G. P. Rothero.
27 - elder, below Corbie's Crag, Blackford Hill, 16 April 2001, D. F. Chamberlain.
46 - house with arches, Ormiston Hall, 16 April 2000, D. F. Chamberlain.
67 - Woodhall Dean, 1992, R. Saville.

117. Orthotrichum Hedw.

1. O. lyellii Hook. & Tayl.   ML, EL
Occasional.
Trunks of trees.

16, 25, (26), 35, (36), 44, 45, 46, 47, 56, 57, 67.

2. O. striatum Hedw.   ML, (EL)
Rare. The apparent decline in this species may be as a result of pollution.
Epiphytic on elder and other trees.

(25) - trees in Penicuik Woods, September 1868, W. Evans.
26 - on elder, filter beds, above Flotterstone, Pentland Hills, 30 July 2000, D. F. Chamberlain.
(35) - Arniston, 12 May 1898, W. Evans.
(46) - near Humbie Station, to the east of Saltoun Woods, 14 May 1904, W. Evans.
(56) - Bolton, 24 May 1904, W. Evans.
[58] - literature record, no further information.

4. O. affine Brid.   WL, ML, EL
Common except in the most polluted areas.
An epiphyte on a variety of trees.

86, 96, 97, 05, 06, 07, 15, 16, 17, 25, 26, 27, 35, 36, 37, 45, 46, 47, 48, 56, 57, 58, 66, 67, 76, 77.
07 - {*MVCR WL*}; on elder tree on north-facing slope, approx. 80m, of Philpstoun South Bing, 12 March 1998, D. G. Long 27443.

5. O. rupestre Schleich ex Schwägr.   ML, EL
Occasional.
Rock faces, quarries.

(16), 25, 26, 27, 48, 57, (58), 67.

8. O. anomalum Hedw.   WL, ML, EL
Frequent.
Base-rich rocks, walls and concrete.

96, 97, 05, (07), 25, 26, 27, 35, 36, 37, 44, 45, 46, 48, 57, 58, 66, 67.

9. O. cupulatum Brid.   (WL), ML, EL
Frequent.
Base-rich rocks, walls, concrete, often near water.
It is not always clear to which variety an individual record belongs.

(07), 16, (17), 25, 26, 27, 34, 35, 36, 46, 48, 56, 57, 58, 66.

a. var. **cupulatum**

16, 25, (26), 27, 35, 36, 56, 58, 66.

b. var. **riparium** Hueb. (var. *nudum* (Dicks) Braithw.)

[97], 16, (17), 25, 26, 27, 57.
25 - {*MVCR ML*}; River South Esk, Gladhouse, 28 May 1975, D. F. Chamberlain.

10. O. rivulare Turner   WL, ML, EL
Occasional.
Silt-covered tree trunks, roots and rocks, generally by rivers and streams.

(97), 16, 17, 25, 26, 27, 45, 46.
17 - {*MVCR WL*}; wet rocks by River Almond, Grotto Bridge, Craigiehall, 1977, D. G. Long 5528.

12. O. stramineum Hornsch. ex Brid.
   WL, ML, EL
Occasional though widespread.
Epiphyte on trees, especially elder, ash and sycamore.

07, (16), 17, 27, 35, 36, 44, 45, 46, 47, 56, 57, 67, 77.
07 - {*VCR WL*}; willow trunk, north side of Philpstoun Bing, March 1998, D. G. Long 27451.

13. O. tenellum Bruch ex Brid.   ML, EL
Rare.
Epiphyte on sycamore and willow.

35 - {*VCR ML*}; old sycamore, Heriot Churchyard, 26 June 1976, D. F. Chamberlain and D. G. Long.

57 - {*VCR EL*}; on old willow on bank of River Tyne opposite Brae Heads, south-west of East Linton, 24m, 19 February 1995, D. G. Long 25849.

15. O. pumilum *Sw. (O. schimperi* Hammar)
[*ML*]
The single unconfirmed literature record from the Edinburgh district is probably a misidentification.

16. O. diaphanum Brid. WL, ML, EL
Frequent and widespread.
Epiphyte, also on rocks.
This is the most tolerant of pollution of all the species of *Orthotrichum*.

86, 96, 97, 05, 06, 07, 16, 17, 25, 26, 27, 35, 36, 37, 44, 45, 46, 47, 48, 56, 57, 58, 67, 68, 77.

17. O. pulchellum Brunton WL, ML, EL
Frequent.
Epiphyte, especially on elder.

[06], 07, 15, 16, 17, 25, 26, 27, 35, 36, 46, 47, 48, 56, 57, 58, 67, 77.
07 - {*MVCR WL*}; elder trunk in wood, near Abercorn Church, 6 January 1974, D. G. Long 3045.

118. Ulota D. Mohr

2. U. drummondii (Hook. & Greville) Brid.
ML, EL
Occasional to rare.
Epiphyte, on ash, rowan, and hazel, usually in upland areas.

24 - {*VCR ML*}; on branch of rowan, Dundreich Plateau SSSI, Bowman's Gill ravine, August 1998, D. F. Chamberlain.
(25) - on hazel, Edgelaw Reservoir, 14 June 1902, W. Evans.
45 - near Linn Dean Water, 3 June 1973, D. F. Chamberlain.
46 - on hazel, wooded valley below Humbie Kirk, 130m, 25 April 1994, D. G. Long 25309.
56 - on rowan, Papana Water SSSI, near Gifford, 18 July 1999, D. F. Chamberlain.
66 - Burn Hope Dean, near Innerwick, 4 June 1972, D. F. Chamberlain.
67 - Woodhall Dean, post-1989, B. Averis.

3. U. crispa Brid. WL, ML, EL
Occasional.
An epiphyte that is not tolerant of air pollution.

96, 97, 06, 15, 25, 26, [27], 34, 35, 36, 45, 46, 56, 67, 76, 77.

4. U. bruchii Hornsch. ex Brid. (*U. crispa* Brid. var. *norvegica* (Grönvall) A. J. E. Sm. & M. O. Hill) WL, ML, EL
Frequent outside the most polluted areas.
An epiphyte.

96, 97, 06, 07, 15, 24, 25, 26, 34, 35, 36, 46, 56, 66, 67, 76, 77.
07 - {*MVCR WL*}; willow trunk, 80m, Philpstoun North Bing, March 1998, D. G. Long 27449.

5. U. phyllantha Brid. WL, ML, EL
Frequent.
An epiphyte in areas with clean air.

96, 07, 16, 17, 24, 26, 27, 35, 47, 48, 56, 57, 58, 66, 67, 68, 77.
26 - {*MVCR ML*}; on elder tree on dry roadside bank, approx. 210m, below Glencorse Reservoir, 31 July 1977, D. G. Long 6146a.

119. Hedwigia P. Beauv.

2. H. stellata Hedenäs WL, ML, EL
Occasional.
Rocks, usually sub-montane.
This species has recently been separated from *H. ciliata* (Hedw.) P. Beauv. which is now considered to be a rare species in Britain. All the Lothians records of *H. ciliata* are referable to *H. stellata*.

15, 16, 17, 24, 26, [27], 45, 48, 56, 57.

3. H. integrifolia P. Beauv. EL
Rare. This is a nationally scarce species.
Dry basalt rock faces.

(57) - Traprain Law, September 1928, J. B. Duncan.
57 - on basalt at east end of Traprain Law, 1997, D. G. Long.

120. Fontinalis Hedw.

1. F. antipyretica Hedw.
a. var. **antipyretica** WL, ML, EL
Common.
Rocks in fast-moving water.

96, 97, 05, 06, [07], 15, 16, [17], 24, 25, 26, 27, 34, 35, 44, 45, 46, 56, 57, 66, 67, 76, 77.

b. var. **gracilis** (Hedw.) Schimp.   **ML**
Rare.
Rocks in streams, sub-montane.

(15) - Gutterford Burn, above North Esk Reservoir, 14 October 1905, W. Evans.
(16) - The Howe, Pentland Hills, 5 May 1909, W. E. Evans.
16 - {VCR ML}; in Logan Burn, above upper waterfall, 320m, 31 July 1976, D. G. Long 5143.

c. var. **gigantea** (Sull.) Sull.   **ML**
Rare.
Rocks in streams.

36 - {VCR ML}; on stones in burn, Tyne Water, Lothian Bridge, near Ford, 90m, 28 September 1986, D. G. Long 13941.

2. F. squamosa Hedw.   *[ML]*
The single literature record from Redhall, Water of Leith, 1824, requires confirmation.

121. **Climacium** F. Weber & D. Mohr

1. C. **dendroides** (Hedw.) F. Weber & D. Mohr
          WL, ML, EL
Frequent.
Damp base-rich ground, fens, mires, dune slacks.

96, 97, 05, 06, [07], 15, 16, 24, 25, 26, 27, 34, 35, 36, 44, 45, 46, 47, 48, 56, 57, 58, 66, 67, 68, 76, 77.

122. **Cryphaea** D. Mohr

1. C. **heteromalla** (Hedw.) D. Mohr
          WL, ML, EL
Occasional though possibly now spreading as air quality improves.
An epiphyte, especially on elder.

07 - {VCR WL}; on elder in sheltered area below embankment at edge of loch, Beecraigs Country Park, 1993, R. Saville.
17 - Cramond North Wood, 1993, R. Saville.
26 - {VCR ML}; near Glencorse Reservoir, 1993, R. Saville.
27 - Water of Leith, near St. Marks Park, 1993, R. Saville.
(35) - Rosebery, 1905, W. Evans.
(57) - Hailes, 1908, W. Evans.
77 - on elder, Thornton Quarry, 1970s, D. G. Long and D. F. Chamberlain.

123. **Leucodon** Schwägr.

1a. L. **sciuroides** Schwägr. var. **sciuroides**
          ML, EL
Rare and clearly decreasing.
Rocks and epiphytic on tree trunks.

15 - old chestnut tree, Newhall Estate, near the house, 20 February 2001, D. G. Long.
(27) - rocks, Arthur's Seat, June 1871, W. Evans.
(35) - old elm in Park, Rosebery, 18 February 1905, W. Evans.
(36) - on old plane tree, Arniston, 12 May 1898, W. Evans.
(56) - Yester Woods, 23 January 1904, W. Evans.
(57) - on rocks, Hailes, East Linton, 19 September 1908, W. Evans.
58 - exposed rock outcrop, south-east end of North Berwick Law, 20 January 1974, D. G. Long 3081.
67 - old sycamores, Tyninghame House, 12 July 1990, D. G. Long.

124. **Antitrichia** Brid.

1. A. **curtipendula** (Hedw.) Brid.   **(ML), EL**
Rare and decreasing; possibly now extinct in the Lothians.
Tree trunks, rock faces.

(26) - Penicuik Woods, April 1868, W. Evans.
(26) - Penicuik House grounds, April 1921, W. Evans.
[27] - Arthur's Seat, pre-1826, R. K. Greville.
[56] - old tree, Yester, August 1902, W. Evans.
66 - post-1950 mapping record.

125. **Pterogonium** Sw.

1. P. **gracile** (Hedw.) Brid.   **ML, EL**
Rare and probably decreasing.
Basalt rock faces.

(26) - rocks, Braid Hills, February 1869, W. Evans.
(27) - Holyrood Park, Edinburgh, April 1867, J. Sadler.
27 - crags below Radio Station, Blackford Hill, 16 April 2001, D. F. Chamberlain.
(27) - Craiglockhart Hill, 5 March 1902, W. Evans.
(27) - Craigmillar prior to 1850, R. Maughan.
57 - {MVCR EL}; rocks on low slope, 130m, west end of Traprain Law, 10 February 1998, D. G. Long 27424.

57 - low basalt outcrop, north bank of the River Tyne, near Hailes Castle, 19 February 1995, D. G. Long 25853.

128. **Neckera** Hedw.

2. **N. crispa** Hedw.   ML, EL
Rare.
Moist base-rich rocks.

16 - rocks in ravine, Logan Burn, 370m, 31 July 1976, D. G. Long 5139.
25 - near Penicuik, 31 August 1971, Anon.
26 - Bonaly Ravine, 1970s, D. G. Long.
45 - {VCR EL}; dry shady rocks in steep gorge, below Linn Dean, Cascade, 12 August 1973, D. G. Long and D. F. Chamberlain, DGL 2667.

4. **N. complanata** (Hedw.) Huebener
WL, ML, EL
Frequent.
Shaded base-rich rock faces, sycamore and ash trunks.

97, 07, 15, 16, 17, 25, 26, 27, 34, 35, 36, 45, 46, 56, 57, 58, 66, 67, 68, 76, 77.

129. **Homalia** (Brid.) Bruch, Schimp. & W. Gümbel

1. **H. trichomanoides** (Hedw.) Bruch, Schimp. & W. Gümbel   WL, ML, EL
Frequent.
Tree bases and boulders, generally in woodland on base-rich substrates.

97, 98, 06, 08, 15, 16, (17), 26, 27, 35, 36, 45, 46, (47), 56, 57, 58, 67.

130. **Thamnobryum** Nieuwl.

1. **T. alopecurum** (Hedw.) Gangulee
WL, ML, EL
Common.
Shaded rocks by water.

97, 98, 07, 08, 15, 16, 17, 24, 25, 26, 27, 34, 35, 36, 45, 46, 56, 57, 58, 66, 67, 77.

133. **Hookeria** Sm.

1. **H. lucens** (Hedw.) Sm.   WL, ML, EL
Occasional.
Woodland and moorland banks.

97, (15), 16, 24, 25, 26, 34, 35, 36, 46, (57), 67, 77.

138. **Leskea** Hedw.

1. **L. polycarpa** Hedw.   WL, ML, EL
Occasional.
Silt-covered boulders and tree trunks by water.

17, 26, (46), 47, 56, 57, 58, 68.

143. **Anomodon** Hook. & Tayl.

1. **A. viticulosus** (Hedw.) Hook. & Tayl.   ML, EL
Occasional.
Wet base-rich rock faces.

15 - Newhall Woods, October 1959, F. R. Irvine.
(16) - Kinleith Glen, Currie, 22 March 1911, J. McAndrew.
27 - Easter Craiglockhart Hill, 13 January 1999, D. F. Chamberlain.
58 - east of Carperstane, 1970s, D. G. Long.
67 - in steep ravine, Woodhall Dean, near Spott, 19 May 1974, D. G. Long *et al.* 3437.
77 - Thornton Dean, 1970s, D. G. Long and D. F. Chamberlain.

144. **Heterocladium** Bruch, Schimp. & W. Gümbel

1. **H. heteropterum** (Bruch ex Schwägr.) Bruch, Schimp. & W. Gümbel
a. var. **heteropterum**   WL, ML, EL
Frequent.
Moist woodland and moorland banks.

97, 15, 16, 25, (26, 27), 34, 35, 45, 57, 67.

b. var. **flaccidum** Bruch, Schimp. & W. Gümbel
WL, (ML), EL
Rare.
Moist woodland and moorland banks.

97 - {VCR WL}; wood near Linlithgow, 1992, Wiggington.
(26) - {VCR ML}; Bilston Glen, near Polton, 6 February 1903, W. Evans.
45 - damp shaded rock face, below Linn Dean Cascade, Kate's Cauldron, approx. 255m, 18 June 1990, D. G. Long 18282.
66 - Hungry Snout, 1970s, D. G. Long and D. F. Chamberlain.

145. **Thuidium** Bruch, Schimp. & W. Gümbel

1a. **T. abietinum** (Hedw.) Bruch, Schimp. & W. Gümbel ssp. **abietinum**   EL

Once recorded. This is a nationally scarce species. Knoll in sand-dunes.

67 - {VCR EL}; sandy knoll, 2m, mouth of the Hedderwick Burn, 6 August 1977, D. G. Long 6153.

2. T. tamariscinum (Hedw.) Bruch, Schimp. & W. Gümbel **WL, ML, EL**
Common.
Grassland, woodland and moorland banks.

95, 96, 97, 98, 05, 07, 15, 16, 17, 24, 25, 26, 27, 34, 35, 36, 44, 45, 46, 47, 56, 57, 58, 66, 67, 68, 76, 77.

3. T. delicatulum (Hedw.) Mitt. **WL, ML, EL**
Occasional.
Grassy banks, boulders in woodland and moorland.

05, 06, 08, 15, 16, (27), 34, 58.

4. T. philbertii Limpr. **WL, ML, EL**
Occasional.
Base-rich grassland, on limestone and on sand-dunes.

97 - {MVCR WL}; turf on limestone, lime workings, Galabraes, above Bathgate, 16 March 1975, DGL 4243.
(07/17) - west of South Queensferry, 24 April 1909, J. McAndrew.
25 - grassy bank, quarry near Fullarton Water, 4 August 1974, D. G Long 3634.
36 - Currie Lee, near Gorebridge, 13 January 1974, D. F. Chamberlain and D. G. Long, DGL 3072.
(47) - Gosford Bay, November 1906, W. Evans.
48 - south-west end of sand-dunes, by the sea, Gullane, 24 February 1974, D. G. Long 3117.

5. T. recognitum (Hedw.) Lindb. **(ML)**
The single specimen from **25** - south of Fullarton, 1905, Anon. is unconvincing. The presence of this species in the Lothians therefore requires confirmation.

147. Palustriella Ochyra

1. P. commutata (Hedw.) Ochyra (*Cratoneuron commutatum* (Hedw.) Roth)
a. var. **commutata** **WL, ML, EL**
Frequent.
Wet base-rich rock faces.

96, 97, 98, 07, 15, 16, 24, 25, 26, 27, 34, 36, 45, 46, 48, 56, 66, 67, 76, 77.

b. var. **falcata** (Brid.) Ochrya **WL, ML, EL**
Frequent.
Wet peaty or stony ground, in base-rich habitats.

96, (97/ 07), 05, 16, 24, 25, 26, 34, 35, 45, 48, 66, 67, 76, 77.
96 - {MVCR WL}; wet peaty ground close to ponds, Foulshiels Bing, Stoneyburn, 1998, D. F. Chamberlain.

148. Cratoneuron (Sull.) Spruce

1a. C. filicinum (Hedw.) Spruce var. **filicinum** **WL, ML, EL**
Common.
Base-rich flushes, mires, wet rocks, stream sides.

96, 97, 98, 05, 06, 07, 15, 16, 17, 24, 25, 26, 27, 34, 35, 36, 44, 45, 46, 47, 48, 56, 57, 58, 66, 67, 68, 76, 77.

150. Campylium (Sull.) Mitt.

1. C. stellatum (Hedw.) J. Lange & C. E. O. Jensen
a. var. **stellatum** **WL, ML, EL**
Frequent.
Wet grassy areas, in base-rich mires and fens.

95, 96, 97, 05, 15, 16, (17), 24, 25, 26, 27, 34, 35, 44, 45, 48, 58, 67, (77).

b. var. **protensum** (Brid.) Bryhn **WL, ML, EL**
Occasional.
Habitat as above.

96, 97, [06], (16, 25, 26, 27), 34, 35, 45, 67.
97 - {MVCR WL}; basic flush, old quarries above Galabraes, Bathgate Hills, 16 March 1975, D. G. Long 4247.

151. Campyliadelphus (Kindb.) R. S. Chopra

1. C. chrysophyllus (Brid.) Kanda (*Campylium chrysophyllum* (Brid.) J. Lange) **ML, EL**
Occasional.
Dry base-rich grassland.

(26) - banks of Allermuir Burn, 29 December 1900, W. Evans.
27 - seepage line on hillside to east of Agassiz Rock, Blackford Quarry, Edinburgh, 16 April 2001, D. F. Chamberlain.
35 - Middleton Quarry, 1970s D. G. Long and D. F. Chamberlain.
48 - by footbridge, Luffness, Aberlady, 9 March

1975, D. G. Long 4223.
58 - damp sand, Yellowcraig, near Dirleton,
31 July 1977, D. F. Chamberlain and D. G. Long.
68 - St Baldred's Cradle, near Tyninghame,
20 November 1975, D. F. Chamberlain B242.

2. C. elodes (Lindb.) Kanda (*Campylium elodes*
(Lindb.) Kindb.)   **EL**
Rare. This is a nationally scarce species.
Dune slacks.

(48) - Luffness marshes, 7 November 1908,
W. E. Evans.
48 - near golf course, Aberlady Bay, 13 March
1975, D. G. Long 2377.
67 - in ditches at margin of pine plantation, on
dunes, approx. 3m, Belhaven Bay near Dunbar,
12 November 1993, D. G. Long 25029.

153. **Amblystegium** Bruch, Schimp. & W. Gümbel

1. A. serpens (Hedw.) Bruch, Schimp. &
W. Gümbel
a. var. **serpens**   **WL, ML, EL**
Common.
Shaded stonework and tree trunks.

86, 96, 97, 98, 05, 06, 07, 08, 15, 16, 17, 25, 26,
27, 35, 36, 37, 44, 45, 46, 47, 48, 56, 57, 58, 66,
67, 68, 76, 77.

b. var. **salinum** Carringt.   **EL**
Rare.
Turf in salt marshes.

48 - Aberlady, 18 October 2000, G. P. Rothero *et
al.*
(48) - Gullane Links, 19 September 1910,
J. McAndrew.

2. A. fluviatile (Hedw.) Bruch *et al.* **WL, ML, EL**
Occasional.
Base-rich rocks in or by streams, principally sandstone.

06, 15, 17, 26, 27, 45, 56, 66.
06 - {VCR WL}; Calder River, Almondbury Park,
Livingston, (?Almondell), 1972, Wiggington.

3. A. tenax (Hedw.) C. E. O. Jensen
                                                     WL, ML, EL
Frequent.
Habitat as above.

97, 15, 16, 17, 25, 26, 27, 35, 36, 44, 45, 46, 57,
66, 67.

97 - {VCR WL}; boulder in River Avon, near
Torphichen, 1971, D. G. Long and
D. F. Chamberlain.
67 - {VCR EL}; rock in Woodhall Dean, above
Spott, 1971, D. G. Long and D. F. Chamberlain.

4. A. varium (Hedw.) Lindb.   **ML**
Rare.
Wet mud, tree trunks.

16 - fallen tree trunk, Kinleith, 18 April 1976,
D. F. Chamberlain B289.
25 - on wet mud, beside Gladhouse Reservoir,
1971, D. G. Long and D. F. Chamberlain.
27 - Edinburgh, 1970s, field record.

154. **Leptodictyum** (Schimp.) Warnst.

1. L. riparium (Schimp.) Warnst. (*Ambylestegium
riparium* (Hedw.) Bruch *et al.*)   **WL, ML, EL**
Frequent.
On fallen trunks and boulders in and around streams.

96, [97], 06, 07, 15, 16, 17, 25, 26, 27, 35, 36,
37, 46, 48, 56, 57, 58, 67.

155. **Conardia** H. Rob.

1. C. compacta (Müll. Hal.) C. Rob.
(*Amblystegium compactum* (Müll. Hal.) Aust.
                                                    (ML), EL
Rare. This is a nationally scarce species.
Shaded base-rich rock ledges.

(26) - Roslin Glen, July 1932, J. B. Duncan.
77 - Dunglass Dean, 1970s, field record.

156. **Warnstorfia** Loeske

1. W. fluitans (Hedw.) Loeske (*Drepanocladus
fluitans* (Hedw.) Warnst., incl. var. *falcatus* (Sanio
ex C. Jens.) Roth)   **WL, ML, EL**
Frequent.
Margins of bog pools and runnels on peat moors.

86, [96], 05, 15, 16, 24, 25, 26, 34, 44, 45, [46],
56, 66, 67, [76].

2. W. exannulata (Bruch, Schimp. & W. Gümbel)
Loeske (*Drepanocladus exannulatus* (Bruch *et al.*)
Warnst., incl. var. *rotae* (De Not.) Loeske)
                                                    WL, ML, EL
Occasional.
Marshes, mires and dune slacks, usually in base-rich
habitats.

86, 96, 05, 06, 15, 16, 35, 48, 66, 67.
05 - {MCVR ML}; basic flush by moorland
stream, near the source of the Medwin Water,
Pentland Hills, 3 September 1975, D. G. Long
4673.
67 - {MVCR EL}; drainage ditch, Dunbar
Common, 1978, D. F. Chamberlain.

157. Drepanocladus (Müll. Hal.) Roth

1. D. polygamus (Bruch et al.) Hadenäs
(*Campylium polygamum* (Bruch et al.) J. Lange
& C. Jens.)   **WL, (ML, EL)**
Rare. This is a nationally scarce species.
Base-rich ground in mires and dune slacks.

06 - {VCR WL}; mineral soil in marsh, Easter
Inch Moss, May 1998, D. F. Chamberlain.
(16) - Redford Bridge, Threipmuir, Balerno,
31 July 1913, J. McAndrew.
(48) - Gullane Links, May 1916, J. C. Adam.

2. D. aduncus (Hedw.) Warnst.   **WL, ML, EL**
Occasional to frequent.
Reservoir margins, mineral soil in marshes.
Some of the queried field records require further
confirmation.

96, 06, 07?, 16?, 25?, (26), 27, 35?, (47), 48, 56,
66?, 67?, 77.

3. D. sendtneri (Schimp. ex H. Muell.) Warnst.
                                            EL
Rare. This is a nationally scarce species.
Dune slacks.

(48) - North Lagoon, Gullane Links, June 1913,
J. McAndrew.
(48) - Luffness marshes, 7 November 1908,
W. E. Evans.
48 - {MVCR EL}; overgrown pond by salt marsh,
Aberlady Bay, October 1971, D. F. Chamberlain
et al.

5. D. revolvens (Sw.) Warnst. agg. (incl. *D. revolvens* (Sw.) Warnst. *s.s.* and *D. cossonii* (Schimp.)
Loeske)   **WL, ML, EL**
Occasional.
Base-rich flushes, usually sub-montane.
D. cossonii has only recently been recognised as
distinct from *D. revolvens*. While there are no
confirmed specimens of *D. revolvens* in the strict
sense, some of the following records may refer to
the latter species.

95, 15, 16, 34, 35, 47.

6. D. cossonii (Schimp.) Loeske (*D. revolvens*
(Sw.) Warnst. var. *intermedius* (Lindb.) P. W.
Richards & E. C. Wallace)   **ML, EL**
Rare.
Base-rich flushes, dune slacks, generally lowland.

16 - flush below East Kip, Pentland Hills,
4 September 1998, D. F. Chamberlain and
D. G. Long.
24 - {VCR ML}; basic flush on valley side, River
South Esk, South of Hirendean Castle, Moorfoot
Hills, 4 August 1984, D. G. Long 3643.
48 - {VCR EL}; dune slack, Aberlady, 1960,
A. C. Crundwell.

158. Hamatacaulis Hadenäs

1. H. vernicosus (Mitt.) Hadenäs (*Drepanocladus
vernicosus* (Mitt.) Warnst.)   **(ML)**
Once found. This species is considered to be rare
nationally but data deficient.
Upland mire.
A record from Aberlady is referrable to
Drepanocladus cossonii (q.v.).

(16) - {VCR ML}; Bavelaw Moss, Balerno,
October 1909, J. McAndrew (comm.
T. Blockeel).

159. Tomenthypnum Loeske

1. T. nitens (Hedw.) Loeske (*Homalothecium
nitens* (Hedw.) Robins.)   **ML**
Rare. This is a nationally scarce species.
Upland base-rich flushes.

05 - base-rich flush by a stream, source of the
Medwin Water, 3 September 1975, D. G. Long
4670.
16 - {MVCR ML}; base-rich flush, north-west of
Scald Law, 355m, 13 November 1977,
D. G. Long 6432.
[25] - Fullarton Water, April 1870, W. Evans.
[26] - above Swanston, 1824, Anon.

160. Sanionia Loeske

1. S. uncinata (Hedw.) Loeske (*Drepanocladus
uncinatus* Hedw.) Warnst.)   **WL, ML, EL**
Frequent.
Rocks, heathery banks, margins of flushes.

96, 97, 05, 06, 07, 15, 16, 17, 24, 25, 26, 27, 34,
35, 36, 45, 56, 57, 66, 67.

161. **Hygrohypnum** Lindb.

1. H. ochraceum (Turn. ex Wilson) Loeske
WL, ML, EL
Occasional.
Base-rich rocks and boulders in streams.

96, 05, 16, 24, 25, 26, (27), 34, 56, 67, 77.
96 - {VCR WL}; rocks in stream, Breich Water, near Addiewell, 15 May 1976, D. F. Chamberlain and D. G. Long.

3a. H. luridum (Hedw.) Jenn. var. **luridum**
WL, ML, EL
Frequent.
Acid to base-rich boulders, usually associated with running water.

96, 05, 06, 07, 15, 16, 17, 24, 25, 26, 27, 34, 35, 44, 45, 46, 56, 57, 67, 77.

5. H. eugyrium (Bruch *et al.*) Broth.
[WL], ML, EL
Occasional to rare.
Base-rich rocks by water.

[07] - Riccarton, 1916, Anon.
16 - Logan Burn Waterfall, 1970s, field record.
(26) - Clubbiedean Reservoir, 21 March 1898, W. Evans.
36 - stone in River South Esk, Carrington Barns, 13 January 1974, D. G. Long and D. F. Chamberlain.
46 - stones by stream, Church Wood, Humbie, 29 October 1995, D. G. Long and D. F. Chamberlain.

163. **Scorpidium** (Schimp.) Limpr.

1. S. scorpioides (Hedw.) Limpr.　　(ML), EL
Rare.
Stony base-rich flushes, dune slacks.

(16) - Ravelrig Toll Moss, prior to 1826, R. K. Greville.
(25) - Pomathorn Moor, May 1869, W. Evans.
48 - fen near Marl Loch, 15m, Aberlady Bay, 4 November 1994, D. G. Long 25733.

164. **Calliergon** (Sull.) Kindb.

1. C. stramineum (Brid.) Kindb.　　WL, ML, EL
Frequent.
Bogs.

86, 95, (96), 05, 06, (07), 16, 24, 25, 26, 34, 35, 45, 56, 66, 77.

3. C. cordifolium (Hedw.) Kindb.　　WL, ML, EL
Occasional.
Fens and mires.

96, 97, 05, 06, [07], 16, 25, 27, 48, 56.

4. C. giganteum (Schimp.) Kindb.　　WL, ML, EL
Occasional.
Fens.

97, 05, 15, 16, 24, (25), 26, 48, 56.

165. **Calliergonella** Loeske

1. C. cuspidata (Hedw.) Loeske (*Calliergon cuspidatum* (Hedw.) Kindb.)　　WL, ML, EL
Common to abundant.
Damp grassland, mires, flushes.

86, 95, 96, 97, 05, 06, 07, 08, 15, 16, 17, 24, 25, 26, 27, 34, 35, 36, 44, 45, 46, 47, 48, 56, 57, 58, 66, 67, 68, 76, 77.

166. **Isothecium** Brid.

1a. I. myosuroides Brid. var. **myosuroides**
WL, ML, EL
Common.
Tree trunks and boulders generally in woodland.

97, 06, 07, 08, 15, 16, 17, 24, 25, 26, 27, 34, 35, 36, 44, 45, 46, 47, 56, 57, 58, 66, 67, 77.

2. I. alopecuroides (Lam. ex Dubois) Isov. (*I. myurum* Brid.)　　WL, ML, EL
Frequent to common.
Habitat as above.

97, 98, 05, 06, 07, 08, 15, 16, 17, 24, 25, 26, 27, 34, 35, 36, 44, 45, 46, 47, 56, 57, 58, 66, 67, 76, 77.

168. **Homalothecium** Bruch, Schimp. & W. Gümbel

1. H. sericeum (Hedw.) Bruch, Schimp. & W. Gümbel　　WL, ML, EL
Common.
Exposed base-rich rock faces, walls and on concrete, ash and sycamore trunks.

All squares except 95.

2. H. lutescens (Hedw.) H. Rob.　　WL, ML, EL
Frequent to occasional.
Grassland, gravel and among rocks, on base-rich substrates.

97, 07, 25, 26, (27), 35, 36, 48, 57, 58, 67, [76], 77.

169. Brachythecium Bruch, Schimp. & W. Gümbel

1. B. albicans (Hedw.) Bruch, Schimp. &
W. Gümbel   **WL, ML, EL**
Frequent.
Beside tracks, quarries, dry sandy places.

96, 97, 05, 07, 08, 26, 27, 35, 36, 37, 44, 45, 46, 47, 48, (56), 57, 58, 67, 68, 77.

3. B. glareosum (Spruce) Bruch, Schimp. &
W. Gümbel   **ML, EL**
Occasional.
Base-rich rocks, especially limestone, walls.

(26), 36, 46, (47), 48, (58), 67, 68.

5. B. mildeanum (Schimp.) Milde   **ML, EL**
Rare. This is a nationally scarce species.
Damp ground, sand-dunes.

98 - {VCR WL}; damp sandy ground by the sea, Carriden, Bo'ness, D. G. Long and
D. F. Chamberlain.
(16) - {VCR ML}; Threipmuir Reservoir, 9 December 1899, W. Evans.
47 - {VCR EL}; Gosford dunes, 1974, D. F. Chamberlain.
(48) - Gullane Links, May 1908, J. McAndrew.

6. B. rutabulum (Hedw.) Bruch, Schimp. &
W. Gümbel   **WL, ML, EL**
Abundant.
Amongst grass, on fallen trunks, stones and concrete, woodland banks.

All squares except 24.

7. B. rivulare Bruch, Schimp. & W. Gümbel
    WL, ML, EL
Frequent.
Rocks by streams, flushes.

96, 97, 05, 06, 07, 15, 16, 24, 25, 26, (27), 34, 35, 36, 44, 45, 46, 56, (57, 58), 66, 67, 76, 77.

11. B. velutinum (Hedw.) Bruch, Schimp. &
W. Gümbel   **WL, ML, EL**
Frequent.
Shaded, generally base-rich, rocks, concrete, on walls.

95, 96, 97, 98, 05, 06, 07, 08, (15), 16, 17, 26, 27, 35, 36, 37, 45, 46, 48, 56, 57, 67, 77.

14. B. populeum (Hedw.) Bruch, Schimp. &
W. Gümbel   **WL, ML, EL**
Frequent.
Habitat as for *B. velutinum*.

95, 96, 97, 98, 06, 07, 08, 16, 17, 25, 26, 27, 35, 36, 44, 45, 46, 47, 56, 57, (58), 66, 67, 77.

15. B. plumosum (Hedw.) Bruch, Schimp. &
W. Gümbel   **WL, ML, EL**
Frequent.
Rocks by streams, more common in upland areas.

97, 05, 06, 07, 15, 16, 17, 24, 25, 26, 27, 34, 35, 36, 44, 45, 46, (47), 56, 57, 66, 67, 76, 77.

170. Scleropodium Bruch, Schimp. & W. Gümbel

1. S. purum (Hedw.) Limpr.
(*Pseudoscleropodium purum* (Hedw.) Fleisch.)
    WL, ML, EL
Common.
Banks in grassland, sand-dunes.

96, 97, 05, 06, 07, 08, 15, 16, 24, 25, 26, 27, 34, 35, 36, 37, 44, 45, 46, 47, 48, 56, 57, 58, 66, 67, 68, 76, 77.

171. Cirriphyllum Grout

1. C. piliferum (Hedw.) Grout   **WL, ML, EL**
Frequent.
Woodland banks, on base-rich substrates.

96, 97, 98, 06, 07, 15, 16, 17, 25, 26, 27, 35, 36, 45, 46, 56, 57, 58, 66, 67, 76, 77.

172. Rhynchostegium Bruch, Schimp. & W. Gümbel

1. R. riparioides (Hedw.) Cardot   **WL, ML, EL**
Common.
Rocks in and by streams and lakes.

86, 96, 97, 98, 05, 06, 07, 08, 15, 16, 17, 24, 25, 26, 27, 34, 35, 36, 37, 44, 45, 46, 47, 56, 57, 58, 66, 67, 76, 77.

2. R. murale (Hedw.) Bruch, Schimp. &
W. Gümbel   **WL, ML, EL**
Frequent.
Shaded base-rich rocks.

96, 97, 98, 05, 06, 07, 08, 15, 16, 17, 25, 26, 27, 35, 36, 44, 45, 46, 47, 56, 57, 67, 68, 77.

4. R. confertum (Dicks.) Bruch, Schimp. &
W. Gümbel WL, ML, EL
Frequent.
Trunks, boulders, walls, often associated with man-made habitats.
97, 98, 06, 07, 08, 16, 17, 25, 26, 27, 35, 36, 45, 46, 47, 48, 56, 57, 66, 68, 77.

5. R. megapolitanum (F. Web. & D. Mohr)
Bruch, Schimp. & W. Gümbel EL
Once recorded.
Grey sand-dunes.
48 - {(VCR EL}; mossy turf, Aberlady Bay, 1990s, Nagy (comm. D. G. Long).

173. Eurhynchium Bruch, Schimp. & W. Gümbel

1. E. striatum (Hedw.) Schimp. WL, ML, EL
Frequent.
Woodland banks, walls and rocks, on base-rich sub-strates.
96, 97, 98, 06, 07, 08, 15, 16, 17, 25, 26, 27, 35, 36, 44, 45, 46, 47, 48, 56, 57, 58, 66, 67, 76, 77.

5. E. pumilum (Wils.) Schimp. WL, ML, EL
Occasional.
Soil on woodland banks and by streams, rocks, on base-rich substrates.
08, 15, 16, 17, 27, 36, 46, 56, 57, 67, 76, 77.
36 - {VCR ML}; soil overlying limestone, Currie Lee, 1974, D. G. Long and D. F. Chamberlain.

6. E. praelongum (Hedw.) Bruch, Schimp. &
W. Gümbel (incl. var. *stokesii* (Turner) Dixon)
WL, ML, EL
Abundant.
Woodland, open grassland, rocks and on the ground, in woods and by rivers, fallen tree trunks.
All squares except 24.

7. E. hians (Hedw.) Sande Lac (*E. swartzii* (Turn.)
Curn. and var. *rigidum* (Boul.) Thér.)
WL, ML, EL
Common.
Rocks and bare ground by rivers and in woodland, gen-erally on at least mildly base-rich substrates.
96, 97, 98, 06, 07, 08, 15, 16, 17, 25, 26, 27, 35, 36, 37, 45, 46, 47, 48, 56, 57, 58, 66, 67, 68, 77.

10. E. crassinervium (Wilson) Schimp.
(*Cirriphyllum crassinervium* (Tayl.) Loeske &

Fleisch.) WL, ML, EL
Occasional.
Base-rich boulders in and by streams.
06, 17, 26, 27, 36, 46, 57, (58), 77.
36 - {MVCR ML}; tree trunk by River South Esk, Dalkeith Park, 29 October 1978,
D. F. Chamberlain.

174. Rhynchostegiella (Bruch, Schimp. &
W. Gümbel) Limpr.

1. R. tenella (Dicks.) Limpr. WL, ML, EL
Occasional.
Limestone old mortar on walls, calcareous rocks, gen-erally in relatively dry habitats.
96, 07, 15, 17, 26, 27, (47), 68, 77.
07 - {VCR WL}; calcareous rock face by stream, Midhope Burn, near Abercorn, 1974,
D. G. Long.

4. R. teneriffae (Mont.) Dirkse & Bouman
(*R. teesdalei* (Bruch, Schimp. & W. Gümbel)
Limpr.) ML, EL
Occasional.
Moist shaded base-rich especially sandstone rock faces and boulders, generally by streams.
16, 26, 27, 45, 46, 56, 57, 67, 77.
77 - {VCR EL}; basic rock face in shady ravine, near the foot of Sheeppath Glen, Oldhamstocks, 1972, D. G. Long and D. F. Chamberlain.

175. Entodon Müll. Hal.

1. E. concinnus (De Not.) Paris ML, EL
Rare.
Base-rich grassland on limestone and on sand-dunes.
36 - {VCR ML}; grassy slopes by old lime quar-ries, Currie Lee, 17 March 1974, D. G. Long and D. F. Chamberlain, DGL 3139.
48 - {VCR EL}; dune grassland, Gullane Links, May 1967, A. McG. Stirling.
77 - limestone grassland by the sea, Catcraig, Barns Ness, approx. 5m, 8 May 1995,
D. G. Long 25905.

176. Pleurozium Mitt.

1. P. schreberi (Brid.) Mitt. WL, ML, EL
Common to abundant.
Acid grassland, moorland among heather.
95, 96, 97, 05, 06, 07, (08), 15, 16, (17), 24, 25,

26, 27, 34, 35, 36, 44, 45, 46, 47, 48, 56, 57, 58, 66, 67, 68, 76, 77.

178. Platydictya Berk.

1. P. jungermannioides (Brid.) H. A. Crum
Once seen. This is a nationally scarce species.
Ledge on conglomerate outcrop.

67 - Tinker's Leap, Woodhall Dean, SSSI, near Innerwick, 19 August 2001, D. F. Chamberlain.

179. Orthothecium Bruch, Schimp. & W. Gümbel

1. O. intricatum (Hartm.) Bruch, Schimp. & W. Gümbel   ML
Rare.
Shaded base-rich agglomerate and sandstone rock ledges.

(16) - The Howe, 1856, Nichol.
16 - dry rock ledges in ravine, Loganlea Waterfall, 310m, 31 July 1976, D. G. Long 5123.
(26) - near Torduff Reservoir, 26 March 1898, W. Evans.
26 - waterfall near Torduff Reservoir, 23 September 1973, D. G. Long and D. F. Chamberlain, DGL 2732.

180. Plagiothecium Bruch, Schimp. & W. Gümbel

3a. P. denticulatum (Hedw.) Bruch, Schimp. & W. Gümbel var. **denticulatum**   WL, ML, EL
Frequent to common.
Tree bases, moist shaded rocks in woodland.

86, 95, 97, 98, 05, 06, 07, (15), 16, 17, 24, 25, 26, 27, 35, 36, 45, 46, 47, 48, 56, 58, 66, 67, 68, 77.

5. P. curvifolium Schlieph. ex Limpr.
                        WL, ML, EL
Frequent.
Habitat as above.

96, 98, 06, 08, 16, 25, 26, 27, 35, 36, 46, 67.
98 - {VCR WL}; base of tree trunk in wooded dean, near Carriden House, Bo'ness, 1974, D. G. Long and D. F. Chamberlain.
16 - {VCR ML}; cut tree branch, by Threipmuir Reservoir, August 1964, E. F. Warburg and A. C. Crundwell.
77 - {VCR EL}; decaying wood in deep shady ravine, Sheeppath Glen, near Oldhamstocks, 1972, D. G. Long *et al.*

6. P. laetum Bruch, Schimp. & W. Gümbel   ML
Rare. This is a nationally scarce species.
Habitat as above.

25 - tree base in mixed woodland, Cornton Burn, Penicuik Woods, 5 February 1978, D. G. Long and D. F. Chamberlain, DGL 6495.
26 - {VCR ML}; tributary of the River North Esk, Kirkettle, below Auchendinny, 8 December 1974, D. G. Long and D. F. Chamberlain, DGL 412.7

9. P. cavifolium (Brid.) Z. Iwats.   ML, EL
Rare. This is a nationally scarce species.
Base-rich rocks.

05 - shaded sandstone, Crosswood Burn, below Reservoir, 2 June 1974, D. F. Chamberlain *et al.*
36 - {VCR ML}; limestone rock in wooded valley, Hagbrae Quarries, Currie Lee, 17 March 1974, D. G. Long and D. F. Chamberlain, DGL 3138.
45 - basic rocks, above Cascade, Linn Dean, Soutra, 24 August 1975, D. F. Chamberlain and D. G. Long.
56 - Castle Wood, Yester, field record.
67 - damp old red sandstone on rocky slope in ravine, Woodhall Dean, below Weatherly, 13 October 1974, D. G. Long 4063.

10. P. succulentum (Wils.) Lindb.   WL, ML, EL
Frequent to common.
Shaded rocks, woodland banks.

96, 97, 98, 06, 07, 08, 15, 16, 17, 24, 25, 26, 27, 34, 35, 36, 37, 44, 45, 46, 47, 56, 57, 58, 66, 67, 77.
26 - {VCR ML}; on rock, Roslin Glen, September 1968, D. F. Chamberlain.
77 - {VCR EL}; wet bank of stream in ravine, Sheeppath Dean, Oldhamstocks, November 1967, E. Beattie and U. Duncan.

11. P. nemorale (Mitt.) Jaeg.   WL, (ML)
Occasional to rare.
Dry base-rich rocks.

97 - basaltic cliff face, Cockleroy Hill, 17 February 1974, D. G. Long et al., DGL 3110.
98 - basic rocks on woodland bank, near. East Kerse Mains, Bo'ness, 26 January 1975, D. G. Long 4174.
07 - Binny Craig, field record.
08 - on limestone, Carriden, 6 January 1974, D. F. Chamberlain and D. G. Long.
(27) - Craiglockhart Rocks, 8 September 1910, J. McAndrew.

12. P. undulatum (Hedw.) Bruch, Schimp. &
W. Gümbel   **WL, ML, EL**
Frequent to common.
Woodland and moorland banks.

86, 95, 96, 97, 06, 07, 15, 16, 17, 24, 25, 26, 27,
34, 35, 36, 44, 45, 46, 47, 56, 57, 66, 67, 68, 76,
77.

181. Isopterygiopsis Z. Iwats.

1. I. pulchella (Hedw.) Z. Iwats. (*Isopterygium
pulchellum* (Hedw.) Jaeg.)   **ML, EL**
Rare.
Rock ledges, woodland banks.

16 - The Howe, 1971, J. A. Paton &
C. C. Townsend.
(26) - Auchendinny Woods, Anon.
(57) - Traprain Law, June 1908, J. McAndrew.
67 - woodland bank, Woodhall Dean, near Spott,
19 May 1974, D. F. Chamberlain.

182. Pseudotaxiphyllum Z. Iwats.

1. P. elegans (Brid.) Z. Iwats. (*Isopterygium ele-
gans* (Brid.) Lindb.)   **WL, ML, EL**
Common.
Shaded rock faces, woodland banks.

86, 95, 96, 98, 05, 06, 07, 08, 15, 16, 17, 24, 25,
26, 27, 34, 35, 36, 37, 44, 45, 46, (47), 48, 56,
57, 66, 67, 76, 77.

184. Taxiphyllum M. Fleisch.

1. T. wissgrillii (Garov.) Wijk. & Margad.
(*Isopterygium depressum* (Brid.) Mitt & Lindb.)
                                          WL, ML, EL
Occasional.
Moist base-rich rocks, usually by streams.

98, 15, 16, (17, 26), 36, 67, 77.

185. Rhytidium (Sull.) Kindb.

1. R. rugosum (Hedw.) Kindb.   **EL**
Rare. This is a nationally scarce species.
Sand-dunes.

67 - {VCR EL}; on landward slope of old dune
ridge, Belhaven Bay, April 1968,
P. J. Myerscough.
67 - edge of pine plantation, Belhaven, 20 March
1976, D. G. Long 4778.

187. Pylaisia Schimp.

1. P. polyantha (Hedw.) Schimp.   **ML, EL**
Occasional to rare. This is a nationally scarce
species.
Epiphytic, especially on elder.

(26) - Bonaly Reservoir, Anon. (Hb. Glasgow)
36 - {VCR ML}; on old elder, near Pathhead,
1993, D. G. Long.
56 - {VCR EL}; elder on river bank, by Hopes
Reservoir, near Yester Castle, 27 October 1974,
D. F. Chamberlain and D. G. Long.
57 - by the River Tyne, Westfield, Haddington,
48m, 9 April 1978, D. G. Long 6644.
77 - on elder, Thornton Quarry, 20 March 1976,
D. F. Chamberlain and D. G. Long, DFC 258.

188. Platygyrium Bruch, Schimp. & W. Gümbel

1. P. repens (Brid.) Bruch, Schimp. & W. Gümbel
                                                  EL
Rare. This is a nationally scarce species.
Willow and elder branches, close to water.

47 - {VCR EL}; on elder in marshy wood, River
Tyne, near Westfield, Haddington, 9 April 1978,
D. G. Long 6649.
58 - on willow in marshy wood, Binning Wood,
near Whitekirk, 16 April 1978, D. G. Long 6459.

189. Homomallium (Schimp.) Loeske

1. H. incurvatum (Brid) Loeske
The only record - 17, on a wall, Dalmeny,
December 1898, W. Evans, is for a specimen that
has been redetermined as *Hypnum resupinatum*
Tayl. This species should therefore be deleted
from the Lothians List.

190. Hypnum Hedw.

3. H. cupressiforme Hedw.   **WL, ML, EL**
Abundant.
Tree branches and trunks, rocks, stone walls.

All squares.

4. H. lacunosum (Brid.) Hoffm.   **WL, ML, EL**
Frequent to common.
Rocks and walls, sand-dunes.
The distinction between the two varieties has
only recently been clarified. It is not therefore
certain as to which of them some individual
records refer.

95, 96, 97, 98, 15, 16, 17, 24, 25, 26, 27, 34, 35,
36, 44, 45, 48, 56, 57, 58, 66, 67, 68, 76, 77.

a. var. lacunosum
This is the common variety on walls and is under-recorded.

97, 15, 16, 17, 25, 27, 35, 45, 48, 56, 58.

b. var. tectorum (Brid.) J. P. Frahm
(*H. cupressiforme* Hedw. var. *tectorum* Brid.) **EL**

97 - {VCR WL}; on stone in field on south-east side of the River Avon, Muiravonside, 1971, A. C. Crundwell and A. McG. Stirling.

5. H. resupinatum Taylor (*H. cupressiforme* Hedw. var. *resupinatum* (Taylor) Schimp.)
WL, ML, EL
Frequent.
Shaded tree trunks.

97, 06, 07, 08, 15, 16, 17, 26, 27, 34, 35, 36, 44, 45, 46, (47), 48, 56, 57, 58, 67, 68, 76.

6. H. andoi A. J. E. Sm. (*H. mammilatum* (Brid.) Loeske)
ML, EL
Occasional though probably under-recorded.
Tree trunks.

06, 15, 25, 34, 35, 36, 56, 67, 77.
36 - {VCR ML}; hazel by river, River South Esk, above Trotter's Bridge, Arniston, 1974, D. G. Long and D. F. Chamberlain.
77 - {VCR EL}; trunk of birch, in stream ravine, Sheeppath Dean, near Oldhamstocks, November 1967, E. Beattie and U. K. Duncan.

8. H. jutlandicum Holmen & E. Warncke (*H. cupressiforme* Hedw. var. *ericitorum* Bruch, Schimp. & W. Gümbel)
WL, ML, EL
Often locally abundant.
Moorland, usually amongst heather.

86, 95, 96, 97, 05, 06, 07, 08, 15, 16, 17, 24, 25, 26, 27, 34, 35, 36, [37], 44, 45, 46, 47, 48, 56, 57, 58, 66, 67, 68, 76, 77.

9. H. imponens Hedw.
ML, EL
Rare. This is a nationally scarce species.
Peat in moorland and on edge of bogs.

(16) - Red Moss, 7 September 1911, J. McAndrew.
25 - Auchencorth Moss, post-1990, B. Averis.
45 - {VCR ML}; hollow in *Callunetum* on raised bog, Fala Flow, 2 September 1996, D. G. Long 26685.
56 - {VCR EL}; Lowran's Law, 480m, 5 May 1991, D. G. Long.

11. H. lindbergii Mitt.
WL, ML, (EL)
Occasional.
Stony tracks and banks.

96, 97, [06], (07?), 16, 26, 27, [56], (66).

191. **Ptilium** De Not.

1. P. crista-castrensis (Hedw.) De Not.
(ML)
Not seen for over ninety years and probably now extinct.
Under firs.

(16) - Redford Wood, Bavelaw, 16 November 1903, J. McAndrew.
(16) - Bavelaw Fir Wood, South of Balerno, 18 April 1902, W. Evans.

192. **Ctenidium** (Schimp.) Mitt.

1. C. molluscum (Hedw.) Mitt.
a. var. molluscum
WL, ML, EL
Frequent to locally common.
Dry base-rich rocks.

95, 96, 97, 05, 06, 15, 16, 24, 25, 26, 34, 35, 36, 44, 45, 46, 47, 48, 56, 57, 58, 66, 67, 76, 77.

b. var. condensatum (Schimp.) Britt.
ML
Rare.
Habitat as above.

(16) - Crummie Burn, Logan Burn, 3 January 1897, J. Murray.
16 - Logan Burn, 4 September 1998, D. G. Long and D. F. Chamberlain.
(26) - Bilston Glen, Polton, 11 November 1902, J. McAndrew.
44 - {MVCR ML}; basic damp soil at edge of track, Torsance Woods, near Stow, 26 June 1976, D. G. Long 5054.

194. **Rhytidiadelphus** (Limpr.) Warnst.

1. R. triquetrus (Hedw.) Warnst.
WL, ML, EL
Common.
Woodland banks, sand-dunes.

97, 06, 07, 15, 16, [17], 24, 25, (26), 34, 35, 36, 44, 45, 46, 47, 48, 56, 57, 58, 66, 67, 68, 76, 77.

2. R. squarrosus (Hedw.) Warnst.
WL, ML, EL
Abundant.
Amongst grass in lawns, on moorland and sand-dunes.

All Squares.

4. R. loreus (Hedw.) Warnst.   **WL, ML, EL**
Frequent.
Woodlands and moorland banks, cliffs.

96, 97, 06, 15, 16, 24, 25, 26, 27, 34, 35, 44, 45, 47, 56, 57, 66, 67, [76], 77.

195. Hylocomium Bruch, Schimp. & W. Gümbel

1. H. splendens (Hedw.) Bruch, Schimp. &
W. Gümbel   **WL, ML, EL**

Common to locally abundant.
Rough acid grassland, open moorland, bogs.

95, 96, 97, 05, 06, 07, 08, 15, 16, 17, 24, 25, 26, 27, 34, 35, 36, 44, 45, 46, 47, 48, 56, 57, 58, 66, 67, 68, 76, 77.
07 - {*MVCR* WL}; on north-facing slope of Philpstoun South Bing, by canal, approx. 80m, 12 March 1998, D. G. Long 27447.

10

Ferns and Fern Allies of the Lothians

A. F. DYER and H. S. McHAFFIE

The Lothians Pteridophyte Flora in Context

Although they have evolved independently for some 350 million years, the ferns and fern allies are usually grouped together, as the **pteridophytes**, because of shared characteristics that distinguish them from all other plant groups (Table 10.1). **Ferns** form the largest group in the pteridophytes (Table 10.2). Most ferns are characterised by sporangia aggregated into sori on the undersurface of fronds that appear as coiled croziers while they extend. The fern allies include the **club-mosses**, with sporangia in the axils of small leaves towards the end of creeping stems, the **quillworts**, with sporangia embedded in the bases of quill-like leaves grouped in tufted rosettes, the **horsetails**, with jointed stems, reduced scale leaves in whorls, and sporangia in terminal cones, and the tropical **fork-ferns**, with large two- or three-celled sporangia on rootless dichotomous stems. The British pterido-phyte flora is small but four of these groups are represented. Full details are in Chapter 13. The Lothians contain 60 per cent of the British pteridophyte flora, including all the British, and more than half the world's, list of horsetail species. In addition, there are in the Lothians four of the twenty hybrid ferns recorded in Britain and four of the eight British hybrid horsetails.

Among the twenty-eight fern species native to, and currently occurring natu-rally in, the Lothians, taxonomic diversity is inevitably limited, but ten families are represented. Eight of these belong to the so-called Modern Leptosporangiate group of ferns that arose within the last 200 million years, and two of these eight, Woodsiaceae and Dryopteridaceae, with seven species each, account for half the native fern flora of the Lothians. The two remaining families have a longer evolu-tionary history. The Hymenophyllaceae, represented by a Filmy Fern (*Hymenophyllum wilsonii*), belongs to the more ancient Early Leptosporangiate group. The Ophioglossaceae, with two species in the Lothians, *Botrychium lunaria* and *Ophioglossum vulgatum*, belongs to the even older Primitive or Eusporangiate fern group. A separate evolutionary history stretching back 250 mil-lion years or more has resulted in plants that are strikingly different from all other ferns.

Table 10.1 *Fundamental characteristics of the major groups of land plants*

Character	Pteridophytes	Bryophytes	Gymnosperms and Angiosperms
Phase of the life-cycle capable of free-living existence	Gametophyte and Sporophyte	Gametophyte only	Sporophyte only
Presence of differentiated vascular tissue (xylem and phloem)	Present	Absent	Present
Dispersal unit for colonising new site	Single-celled spore	Single-celled spore	Seed, containing embryo

Table 10.2 *Numbers of native species in each of the pteridophyte groups*

	World	British Isles	Scotland	Lothians
Ferns	approx. 12,000	50 (20)	44 (14)	28 (4)
Clubmosses	approx. 1,150	6	6	4
Quillworts	approx. 130	3	2	—
Horsetails	15	8 (8)	8 (6)	8 (4)
Fork Ferns	approx. 12	—	—	—
TOTAL	approx. 13,300	67 (28)	60 (20)	40 (8)

numbers of interspecific hybrids in brackets

Three additional fern species recorded in the Botany of the Lothians survey (see Chapters 12 and 13) are considered to be introductions to the Lothians. The Royal Fern (*Osmunda regalis*), a popular plant with gardeners, which probably once grew naturally in the Lothians and is still native elsewhere in Britain, belongs to another ancient family, Osmundaceae. Like the twenty-eight naturally occurring species, *O. regalis* is homosporous, producing just one type of spore. Pillwort (*Pilularia globulifera*) (Marsileaceae), recorded in recent years in the Lothians only in re-introduced populations, is an example of the small minority of ferns world-wide that are heterosporous. Such ferns produce large megaspores and small microspores enclosed within protective pill-like sporocarps for dispersal. These ferns are restricted to more or less aquatic habitats. The Water Fern (*Azolla filiculoides*) (Azollaceae), an introduced exotic, is another heterosporous species; all heterosporous ferns belong among the Modern Leptosporangiate species.

As with the angiosperms, the present pteridophyte flora of the Lothians is a legacy in part of the impact of the changing climate on the vegetation that developed after the last Ice Age. Taken together with the geography and topography of the area, this has affected the balance of the floristic elements recognised within the British pteridophyte flora by Page in his *Natural History of Britain's Ferns* (1988) (Table 10.3). All but one of the elements recognised by Page are equally represented within the British fern flora. The Lothians flora, however, is dominated by two elements, the Sub-Atlantic and the Widespread Northern-Continental. All the British ferns and most of the British fern allies included within these two elements are found in the Lothians. By contrast, the Arctic-Alpine and the Atlantic elements, associated with opposite extremes of climate, are very poorly represented. For two of the other elements, the Mediterranean-Atlantic and the Continental, about half the British species occur in the Lothians. The Lothians flora thus reflects the absence there of both the low temperatures of the higher mountains of northern Britain and of the mild oceanic conditions of the coast of southern and western Britain.

Climate also partly determines the relative abundance of the individual species within these elements although there has been the additional influence of the superimposed effects, both direct and indirect, of human activity. The ten most abundant pteridophytes in the present Lothians flora (as indicated by the number of survey records) are, in order of decreasing abundance: *Equisetum arvense*, *Dryopteris filix-mas*, *D. dilatata*, *Pteridium aquilinum*, *Athyrium filix-femina*, *Dryopteris affinis*, *Blechnum spicant*, *Equisetum palustre*, *E. fluviatile*, and *Oreopteris limbosperma*. The rarest native species, with three or fewer records in the survey, although in several cases other sites are known, are: *Ceterach officinarum*, *Dryopteris oreades* and *Phegopteris connectilis*, three records each; *Botrychium lunaria*, *Asplenium septentrionale*, *Equisetum variegatum* and *Polystichum setiferum*, two records each; and *Asplenium marinum*, *Cryptogramma crispa*, *Diphasiastrum alpinum* and *Ophioglossum vulgatum*, one record each. *Pilularia globulifera* is declining within its Western Europe range and has significant populations in Britain. It became extinct in the Lothians but, due to a Species Action Programme, has now been re-introduced at five sites around Edinburgh.

Post-glacial History of the Pteridophyte Flora of the Lothians

The pteridophyte flora re-established in what is now the Lothians as the ice retreated about 14,000 years ago. The initial tundra-like vegetation, similar to that found in Iceland today, featured species that flourished in the abundant lochs and grew on the unleached mineral soil of freshly exposed moraines. Spores found in soil cores throughout Scotland suggest a wide distribution then of Clubmosses

Table 10.3 *Representation in the pteridophyte flora of Britain and the Lothians of the floristic elements recognised by Page (1988)*

Floristic element	Number of species native in the British Isles		Number of species native in the Lothians	
	Ferns	*Fern Allies*	*Ferns*	*Fern Allies*
Mediterranean-Atlantic	9	2	5	1
Atlantic	9	—	3	—
Sub-Atlantic	8[a]	4	7[a]	1
Continental	8[a]	—	4[a]	—
Arctic-Alpine	7[b]	2	1	2
Widespread Northern-Continental	9	5	9	5
Northern Montane	1[c]	4	—	3

a. *Asplenium trichomanes* is included in the table twice; in the Sub-Atlantic element as *A. trichomanes* ssp. *quadrivalens* and in the Continental element as *A. trichomanes* ssp. *trichomanes*.

b. Page (1988) includes *Athyrium flexile* as an additional species in this element but in this table it is treated as a variety of *A. distentifolium*.

c. Page (1988) also includes *Cystopteris dickieana* in this element but because its specific status has been questioned and its relationship with plants under this name from continental Europe is unclear, in this table it is not distinguished from *C. fragilis* (a widespread Northern-Continental species).

(*Lycopodium* spp.), Lesser Clubmoss (*Selaginella selaginoides*), Moonwort (*Botrychium lunaria*) (Plate 10, top left) and Adder's-tongue (*Ophioglossum vulgatum*) (Plate 10, top right), species that are now found less frequently. As the climate changed through successive warmer and cooler, drier and wetter phases, different species would have periodically become more abundant, and from 5,000 years ago the human impact became increasingly important (Newey 1967; 1970).

The rarest British arctic-alpine species such as *Woodsia* are no longer found in the Lothians where the hills reach only moderate elevations, but in places the altitude is sufficient to retain small populations of some upland or northern species that would have been more widespread during the early post-glacial period. A few sites for Mountain Male-fern (*Dryopteris oreades*) and Parsley Fern (*Cryptogramma crispa*) occur on the southern margin of the area where it enters the Southern Uplands. The Forked Spleenwort (*Asplenium septentrionale*), usually found in a more continental climate with cold winters and hot summers, typical of

Scandinavia and central Europe, occurs on south-facing rocks on Arthur's Seat. The Shady Horsetail (*Equisetum pratense*) is at the southern edge of a northern range and is now very rare in Midlothian. By contrast, other species are relics of wider distributions during warmer and wetter periods about 5,000 years ago. The Soft Shield-fern (*Polystichum setiferum*), a relatively frequent species in the west of Scotland and in the south of Britain, grows in a few sheltered woodland sites in the Lothians where it is on the northern edge of its range. The Southern Polypody (*Polypodium cambricum*) grows on Arthur's Seat and is also at the northern edge of a more southern Mediterranean-Atlantic range. Similarly, the Great Horsetail (*E. telmateia*) (Plate 23) too is at the northern edge of a southern distribution, although it extends further north on the West Coast. It is now very uncommon in the Lothians. Another relic from a warmer wetter past is Wilson's Filmy-fern (*Hymenophyllum wilsonii*), now found in only one site in a shady ravine at Soutra. Sea Spleenwort (*Asplenium marinum*) has been recorded in this area only from the Bass Rock. It requires the high humidity and frost protection which it derives from close proximity to the sea. As an Atlantic species, it is more common on the West Coast. The Royal Fern (*Osmunda regalis*) also reached its peak during the Climatic Optimum but is now only present in the Lothians as an introduced species. In this case, however, human influences through land drainage and even collecting might also have contributed to its loss from the area.

Human activity has resulted in the increase of some species and the reduction of others. Forest clearance would have led to a reduction in most woodland ferns but an increase in Bracken (*Pteridium aquilinum*), although intensive cultivation in the Lothians has restricted its spread on the lower ground. Adder's-tongue (*Ophioglossum vulgatum*) (Plate 10, top right), in the Lothians now near its north-ern limit inland in Britain, would once have been much more widespread in open woodland and unimproved pasture but it was lost when fields were ploughed. There are now very few sites, although an expanding population of several thousand individuals on Arthur's Seat has been found since the sheep were removed. Widespread draining and agricultural improvement have probably also led to a decline of the Narrow Buckler-fern (*Dryopteris carthusiana*) which grows in boggy places. However, the Broad Buckler-fern (*Dryopteris dilatata*) can grow in the dense shade of forestry plantations or policy woodlands and is still abundant.

Several other man-made habitats have created new opportunities for natural colonisation by pteridophytes in the Lothians. Perhaps the most obvious is old walls built with lime mortar. Introduced by the Romans, the use of lime mortar increased from the Middle Ages until the twentieth century when it was largely replaced by cement mortar. After it has weathered, it is the mortar, rather than the brick or stone, that provides a substrate for various base-tolerant ferns. This has allowed several species to re-enter areas previously lost to urbanisation. Edinburgh New Town has a fern flora on walls of at least nine species, and at least three more species occupy similar situations outside the city. All twelve species are also found in natural habitats within the Lothians. Some are opportunists, usually found in terrestrial habitats such as hedgerows and woodland margins. Bracken (*Pteridium aquilinum*), Golden Male Fern (*Dryopteris affinis*) and Lady Fern (*Athyrium filix-femina*) are examples. Among the others are small calcicole rock ferns including the Brittle Bladder Fern (*Cystopteris fragilis*), Black Spleenwort (*Asplenium adiantum-nigrum*), Common Maidenhair Spleenwort (*Asplenium trichomanes* ssp. *quadri-valens*) (Plate 10, bottom) and, as the name implies, the Wall Rue (*Asplenium ruta-muraria*). Over the last two millennia, these species have become more abundant than they would have been if restricted to their natural habitats. *Asplenium ruta-muraria* exemplifies this most clearly. Populations of this species on natural rock substrates are rare, one being on Arthur's Seat, but populations are common on old walls throughout Edinburgh and many towns in the Lothians. Of the ninety records in the present survey, seventy-seven are on walls. *A. ruta-muraria* seems to be particularly tolerant of urban atmospheres and the xeric conditions of exposed walls and sometimes it is the only vascular plant present or it is accompanied only by another plant particularly associated with walls, the naturalised Ivy-leaved Toadflax (*Cymbalaria muralis*). Fern and fern-allies have also colonised semi-natural rock faces, such as quarries and road, rail and canal cuttings, and man-made scree, such as waste heaps and disused railway track beds. The Stag's-horn Clubmoss (*Lycopodium clavatum*) and Alpine Clubmoss (*Diphasiastrum alpinum*) are found on the shale bings of West Lothian. *Lycopodium clavatum* also occasionally occurs on newly-exposed surfaces on roadsides and quarries. *Equisetum arvense* thrives on railtrack beds, aided by its tolerance of herbicides.

Pillwort (*Pilularia globulifera*) grew in pond margins kept open by disturbance from stock, particularly cattle. Populations may therefore be naturally short-lived, relying on opportunistic colonisation of newly disturbed sites as they become available, and thus particularly vulnerable to changes in farming practice. Most, if not all, of the old sites recorded in the nineteenth century have been lost, especially in West Lothian, but a reintroduction programme has been initiated and some of the re-introduced populations have survived and are spreading. Man-made reservoirs have provided additional habitats for other aquatic species, such as the Water Horsetail (*Equisetum fluviatile*).

Rarities, Hybrids, Apomicts, Weeds and Aliens

Rarities

The three rarest pteridophytes occurring naturally in the Lothians are the Sea Spleenwort (*Asplenium marinum*), known from only one locality, and the Variegated Horsetail (*Equisetum variegatum*) and the Forked Spleenwort (*A. septentrionale*), each found at only two sites. Only two Lothian species can be considered to be uncommon nationally, *Equisetum variegatum* and *Asplenium septentrionale*. *E. variegatum*, recorded in the Lothians from two coastal sites at Aberlady Bay and Yellowcraig, is widespread in bare coastal and upland sites in northern and western Britain but is nowhere abundant. In higher latitudes elsewhere in its almost circumpolar range, it is one of the first plants to colonise after glaciers retreat. This fact, and its present ecology in Britain, suggest that it might have been one of the first pteridophytes to establish after the post-glacial retreat of the ice, and more recently it became restricted to isolated open habitats as the forest developed. *A. septentrionale* has been recorded from two sites within the Edinburgh city boundary, Arthur's Seat and Blackford Hill. It is scarce and very local throughout Britain where it is at the edge of its continental distribution. It is restricted to dark volcanic rock faces with a southerly aspect, and occupies a microhabitat that reaches higher temperatures than most fern habitats in Britain. Perhaps allied to this is the observation that in laboratory experiments this species had the highest minimum temperature (about 15°C) for spore germination of several British ferns tested (Dyer 1995). The population on Arthur's Seat has been recognised for over 200 years and is mentioned by James Bolton in his *Filices Britannicae* of 1785. This is probably the earliest published account of a fern locality in the Lothians. The Blackford Hill site is mentioned in Greville's *Flora Edinensis* of 1824. Though small and isolated, these two populations are clearly persistent.

Hybrids

Interspecific hybrids are relatively common among pteridophytes and, in Britain, all five genera represented by more than two native species include hybrids (Table 10.4). Some of the British hybrids from all five genera are found in the Lothians.

Table 10.4 Hybrids

Genus	British Isles		Lothians	
	Species	Hybrids	Species	Hybrids
Equisetum	8	8	8	4
Asplenium (including related genera *Ceterach* and *Phyllitis*)	12	8	5	1
Dryopteris	9	6	5	1
Polypodium	3	3	3	1
Polystichum	3	3	2	1

Some *Equisetum* hybrids form extensive vegetative clones, sometimes maintained in the absence locally of one or both parents. Three of the *Equisetum* hybrids in the Lothians (*E. x dycei*, *E. x rothmaleri* and *E. x mildeanum*) are rare nationally, having been first recognised within only the last forty years and previously known only from localities further north. Fern hybrids are usually scattered sterile individuals accompanied by both parents. Three of the fern hybrids in the Lothians are widespread in Britain in scattered localities where the parents grow close together. The fourth, *Asplenium x murbeckii*, is extremely rare, reported from only about three localities throughout Britain where the parents grow together. At least one individual occurs among plants of *A. ruta-muraria* and close to a population of *A. septentrionale* on Arthur's Seat. Another very rare hybrid involving *Asplenium septentrionale* as one parent, *A. x alternifolium* Wolf., was recorded (as *A. germanicum*) from Blackford Hill by Balfour and Sadler in their *Flora of Edinburgh* in 1863. Superficially similar to *A. x murbeckii*, it can be distinguished by the lack of glandular hairs on the stipes. Future searches may again discover it at either of the *A. x septentrionale* sites.

Apomicts and Clones

Two Lothian ferns, *Dryopteris affinis* and *Phegopteris connectilis*, are apomictic, producing viable spores that have the same genetic complement, both qualitatively and quantitatively, as the parent sporophyte. Although completely unrelated, both ferns have the same modifications of the sexual life cycle to restore spore fertility to plants that would otherwise be sterile due to meiotic irregularities. This has allowed the propagation by spores of plants with a complex hybrid history. Sporelings develop from sporophytic buds developed mitotically from the prothallus cushion, a process known as apogamy, and not from fertilised egg cells in archegonia. Sporelings thus have the same chromosome complement as the parent gametophyte. The absence of chromosome doubling by fertilisation is compensated for by the elimination of meiotic reduction; chromosome separation at the premeiotic mitosis is suppressed so that the spore mother cell has twice the normal chromosome complement, each chromosome having a sister chromosome produced at the previous mitotic replication. This allows regular bivalent pairing in a mechanically normal meiosis, as a result of which spores have a complete but unreduced chromosome complement that is the same as the parent sporophyte. This cycle of apogamy and diplospory can be repeated indefinitely and produces natural genetic clones similar to those from vegetative reproduction but with the added characteristic of spore dispersal to allow long-distance colonisation. The Scaly Male Fern (*D. affinis*) encompasses several morphologically different forms. Some are abundant and widespread as a result of clonal propagation by spores. There have been several attempts to apply names to the most conspicuous clones but, despite much investigation, there is still heated debate about the appropriate taxonomic treatment. All forms of this species are of interspecific hybrid origin and most are triploid, though diploids and higher ploidies are known. Further confusion in identification results from the fact that although no archegonia are produced, the gametophytes form antheridia and can act as the male parent in forming morphologically intermediate hybrids with other related Male Ferns. The Beech Fern (*P. connectilis*) has no close relatives in Britain but it too is triploid and probably of ancient hybrid origin and produces viable spores by the same combination of apogamy and diplospory. Although the taxonomy of *P. connectilis* in Britain appears to be straightforward, more research into the species and its relatives worldwide might reveal a more complicated history.

Weeds

Although several pteridophyte species are 'weedy' in that they colonise open or disturbed soil surfaces by means of spore dispersal, only two species, Bracken (*Pteridium aquilinum*) and Field or Corn Horsetail (*Equisetum arvense*), could be considered serious weeds of cultivated ground in the Lothians, as in Britain as a whole. Both have become more abundant since forest clearance, colonising open habitats by spores and then spreading locally by means of deeply subterranean spreading rhizomes. Although not as serious a pest as in the west of Scotland, *Pteridium aquilinum* encroaches on upland pasture, heather moorland and forestry plantations and is difficult to control even with chemicals. Dense stands can suppress the other plant species. It is eliminated by regular ploughing or mowing, and so is absent from the more intensively managed areas or restricted to hedgerows and waste ground. *E. arvense* is familiar as a pernicious weed of horticultural ground including gardens where only the most persistent and determined measures can eliminate it. As with bracken, dormant buds on the rhizomes survive most onslaughts on the aerial parts and garden cultivation simply serves to spread rhizome fragments over a wider area.

Aliens

It is interesting to note that, as in Britain as a whole, there are few naturalised alien species of pteridophytes and none are aggressive colonisers such as the flowering plants Giant Hogweed (*Heracleum mantegazzianum*), Few-flowered Leek (*Allium paradoxum*) or Oxford Ragwort (*Senecio squalidus*). This may be in part due to the fact that, unlike the angiosperms, there are no weedy ephemeral or annual pteridophytes that complete their life cycle within one growing season. (The Annual Jersey Fern (*Anogramma leptophylla*), restricted within the British Isles to the Channel Islands, comes closest to an annual life cycle. It is almost unique among ferns in having an annual sporophyte, but the gametophyte is perennial.) Moreover, most of the invasive perennial pteridophytes are native to warmer parts of the world and almost none of the temperate species that can be grown outside in British gardens has escaped from cultivation. The only examples of established aliens in the Lothians are the fern ally Mossy Clubmoss (*Selaginella kraussiana*), native to southern Africa and the Atlantic islands, which has established itself locally as a garden weed in the sheltered habitat of Edinburgh New Town, and the Water Fern (*Azolla filiculoides*), native to North America, which appears fitfully in canals and ponds in warmer years. The edible Ostrich Fern (*Matteuccia struthiopteris*), a native of northern North America and Europe, has become naturalised in the Borders, but not apparently within the Lothian boundary. These and other alien species were popular introductions, alongside varieties of native species, to gardens during the Victorian fern craze of the nineteenth century.

Looking to the Future

The pteridophyte species list for the Lothians has changed little since records began some 200 years ago. Whether the same will be said when the next *Plant Life of Edinburgh and the Lothians* is written in 50 or 100 years time depends critically on the extent and direction of climate change. Some southern species such as *Polypodium cambricum* or *Polystichum setiferum* might be lost or some montane species such as *Asplenium viride* or *Polystichum lonchitis* might return, but most species will probably survive, albeit with increased or reduced abundance. Even a rarity such as *Asplenium septentrionale* has shown great resilience at a site within the City of Edinburgh throughout the environmental changes of the industrial revolution and does not appear to be threatened. Even if there is no substantial change in the climate, there is still the possibility of adding the Killarney Fern (*Trichomanes speciosum*) to a future list. Like some other species in the Hymeno-phyllaceae, this species has the ability to survive as independent, perennial, clonal gametophytes propagating vegetatively by gemmae and by branching of the filaments to form mats. By this means it survives, bryophyte-like, in shady rock crevices in conditions where the sporophyte cannot live. Thus, while the sporophyte is known from only a handful of sites within Britain, it is becoming increasingly apparent that the gametophyte colonies are surprisingly widespread with more than 100 sites now known. It could yet be found in suitable habitats in the Lothians.

There is also the possibility in the future of new introductions becoming established in the wild. Horticultural interest in ferns has revived in recent years and increasing numbers of foreign temperate species are being planted, creating a potential for the naturalisation of additional aliens.

11

The Earlier Study of the Plants of Edinburgh and the Lothians

P. M. SMITH

Introduction

There was a time when the plant cover of Edinburgh and the Lothians went for-
mally unrecorded. What is known of the flora in this pre-historical phase is
discussed in Chapters 2 and 3. Until the late seventeenth century, plants were
regarded in intellectual circles, if they were regarded at all, as the handmaidens of
Medicine. Though a Professor of Botany was appointed in Edinburgh University in
1676, it was not until the eighteenth-century Scottish Enlightenment that plants
began to be looked at scientifically for their own sake. Then, as Sir Kenneth Clark
has it, the Smile of Reason appeared – in Edinburgh. Even then, though lists of
plants encountered locally at last existed, the indications of plant cover were
accidental, always incomplete or patchy. In the nineteenth century, a formal listing
and description began and, rather later, developed into the current, we hope never-
ending, evaluation, analysis and discussion of local floristics.

As in many other parts of the British Isles, knowledge of the flora of the
Lothians benefited from the growing enthusiasm for natural history in the later
nineteenth century. From that time, the spread of education also meant that the
enjoyable tasks of exploring and cataloguing local botanical riches were no longer
the preserve of an intellectual elite. Professional botanists themselves began to
increase in numbers. In Edinburgh, fortunately, there was a concentration and a
continuity of interested and active botanists, because of the long-established university and Royal Botanic Garden (RBGE). The Directorship (Regius Keepership)
of the RBGE and the Regius Chair of Botany in the University of Edinburgh were,
remarkably, a single incumbency until 1956. Successive holders of this joint post
showed a concern for research into the local flora until well into the twentieth
century. The excellence of the herbarium and living collections of the RBGE were
a powerful stimulus to floristic enquiry. The founding of the Botanical Society of
Edinburgh (now the Botanical Society of Scotland) in 1836 was another enormous
spur to local botanical investigation, involving both amateur and professional
botanists – a spur which continues to the present day, as this current survey and

publication attest. Other local societies also contributed to the growing interest in, and investigation of, the plant life of the Edinburgh area, notably among them the Edinburgh Field Naturalist and Microscopical Society which amalgamated with the Scottish Natural History Society in 1923 to form the Edinburgh Natural History Society. The establishment of Vice-county Recorderships by the Botanical Society of the British Isles (BSBI) was a further, very helpful step forward. It is pleasant to note that BSBI Recorders have, since their inception, been actively engaged in local floristic inventories, and that they played a major role in the management and production of the present work.

It is the general case that knowledge of the floristics of lower plants (crypto-gams) has been slower to accumulate and to be published than has that of the flowering plants. This is also true of the cryptogamic flora of the Lothians. The slowness is attributable to the relatively specialised interests and skills needed to sustain the research, to the difficulties of collection and identification, and even to the requirement for good magnification of the often minute features needed for recognition. Fewer people have been willing or able to research the cryptogams, even though the physical attractions of these plants, their ecological significance and inherent botanical interest are enormous and undoubted. Local cryptogam floristics have been covered only partially or by scattered texts and articles, often published as part of a greater whole. The present work attempts to correct that as far as possible.

This chapter reviews the past study of the plant life of Edinburgh and the Lothians, and briefly considers future monitoring and prospects for extending what we now know.

History of Floristic Study in Edinburgh and the Lothians

The first Flora of Edinburgh and its environs was Robert Kaye Greville's *Flora Edinensis*, a very considerable achievement, published by Blackwoods in Edinburgh and by T. Cadell in the Strand, London, on Thursday 18th March, 1824 (Plate 27). Before this we have few published details of local plants. Old maps give some information (see Chapter 3). Maitland's otherwise fascinating *History of Edinburgh* (1753) tells us nothing of the local plants: historians are perhaps inevitably anthropocentric in their evaluations.

Dr Greville's book is very substantial, covering most groups of plants, and is arranged by the Linnaean system. He was a highly gifted Edinburgh botanist, though he was not professional in the sense of being employed as such. A Fellow of the Royal Society, Greville (1794–1866) rather loftily expresses surprise that 'no previous attempt prior to the present has been made to describe or even to enumerate the plants growing around the Metropolis of Scotland'. Perhaps the professional botanists up to that date were as busy, or as pre-occupied, as their present counterparts. He thought the omission the more peculiar because of the world-wide reputation of the University Medical School. At that time, we should remind ourselves, the subjects of botany and medicine were scarcely separated. Until quite recently, educated medical practitioners knew their plants and, in Edinburgh, the Regius Chair of Botany remained within the Faculty of Medicine until 1958. Greville, who was also the author of the *Scottish Cryptogamic Flora* (1823–8), dedicates *Flora Edinensis* to his friend Robert Graham M.D., who was then Professor of Botany in Edinburgh.

In fact, as Greville later notes, a list of plants growing in Edinburgh *had* been assembled earlier by Sir Robert Sibbald (1641–1722), an Edinburgh physician who was at various times Professor of Medicine and President of the College of Physicians. His *Catalogue of Plants growing in the King's Park* [Holyrood Park] was compiled in 1684. This, the first reliable record of Edinburgh's plants, lists 381 species and varieties. His *Scotia Illustrata*, also of 1684, listed about 500 plants growing in the Botanic Garden, by then on a site at the eastern shore of the Nor' Loch (now part of Waverley Station). Some of these plants were local to Edinburgh, some to other parts of Scotland, and some were introductions of actual or potential medical significance. It is not always clear which were which. Sibbald was the energetic doctor who had helped to found the Edinburgh Physic Garden, originally at Holyrood (later to become the Royal Botanic Garden).

Other early lists of plants undoubtedly existed. Balfour (1863) refers to one made by a Mr Woodforde, about the time that Greville was producing his Flora. Woodforde had won the prize for the best herbarium of plants submitted by an Edinburgh medical student. This prize (a gold medal) had originally been offered by John Hope (1725–86), Professor of Botany from 1760, to stimulate interest in botany. In this, he built well. An earlier winner had been the illustrious Sir James Edward Smith, later President of the Linnean Society of London, who far-sightedly purchased the herbarium of Linnaeus. An earlier list is John Hope's *List of plants growing in the neighbourhood of Edinburgh, collected in flower 1765, as a sketch of the CALENDARIUM FLORA OF EDINBURGH*. This resides in a notebook at the Royal Botanic Garden Edinburgh, and will have been known to Greville. One other enumeration was by Robert Maughan, who published *A List of the Rarer Plants Observed in the Neighbourhood of Edinburgh* in the first volume (1811) of the Wernerian Natural History Society.

Greville's coverage was a stated radius of ten miles from the city as centre. Within this area it was comprehensive – the flowering plants, ferns and other pteridophytes, mosses, liverworts, fungi, lichens, freshwater and marine algae. An illustrated introduction is given to the mysterious Cryptogamia. The few native gymnosperms are interred in the classification of angiosperms. *Pinus* comes close to *Corylus*, in the Monoecia Polyandria, while *Juniperus* (said to be abundant in the Pentland Hills) follows *Hydrocharis* in the Dioecia Monadelphia. Such is the artificiality of the Linnaean system. The author acknowledges expert contributors – Mr Yalden and the famous George Don for angiosperms, Mr George Walker-Arnott (later Professor at Glasgow) for freshwater algae, Dr Richardson for marine algae and

Captain Wauch of Foxhall for fungi. Like the present work, Greville's was enriched by input from many sources.

Flora Edinensis starts with Linnaeus' first Class, the Monandria, and its first Order, Monogynia. For the bulk of the record, largely of flowering plants, he gives names, species descriptions, localities, collectors and flowering times. He cites illustrations. Like most natural history works of the time, the book is clearly inspired by the teaching of Linnaeus (1707–78), but fortunately is written in English, despite its Latin title. It incorporates a fulsome account of the Linnaean system, which Greville greatly admired. His philosophy may be indicated by his flyleaf quotation, from Linnaeus' *Critica Botanica*, which appears as follows:

> Omnes species originam familiae suae primam ab ipsissima Omnipotentis Creatoria manu numerant : creatis enim specibus aeternam legem generationis et multiplicationis intra speciem propriam imposuit Naturae Aucto rebus.

A free translation of this would be: All species take their origin from the very hand of the All-powerful Creator: indeed the eternal laws of generation and multiplication within species were imposed by Nature's Author.

Greville's great *Flora* was the only authority on plants local to the Edinburgh area for almost forty years. It has never been surpassed for size and detail, though it naturally became outdated as the local flora changed, as knowledge advanced, and as academic fashion progressed. It was not written for a mass or popular readership – in 1824 there was none. It was bulky and expensive. In the early Victorian period which followed, there were more people, more of them interested in the local flora. Increasingly, they had time and better means to travel more cheaply and further to see and collect plants. Roads and railways were rapidly extending and improving. Like human society, the flora itself was changing fast. A new plant handbook was needed.

John Hutton Balfour's *Flora of Edinburgh* appeared in 1863 (Plate 27). The author, an Edinburgh physician who became Regius Keeper of the Botanic Garden and Regius Professor, was assisted by John Sadler, a diligent Fifer who was Demonstrator at the Royal Botanic Garden and Vice-President of the Botanical Society of Edinburgh. By the time of the second edition (1871), Sadler had become the first ever Lecturer in Botany at the Royal High School, Edinburgh, a post he held until 1879. Balfour himself was a founder member of the Botanical Society of Edinburgh, indeed it was at his home in Dundas Street on the 8th of February 1836 that the Society effectively began. Among his many published works was a *Manual of Botany* (1848) that he dedicated to Greville. His *Flora* was a small handbook, about 17 x 10 x 1.5cm, which was obviously designed for the pocket and the field excursion. A folded map at the front (which retreated to the safer, less often disturbed end papers in the second edition) indicated that the geographical coverage was more extensive than that of Greville's *Flora*. Balfour notes that railways were making day excursions possible over a larger area and that a twenty mile radius was now the extent of country to be embraced by those submitting herbaria for the Hope Prize. The map shows that the northern limit was marked by Milnathort and Anstruther, while the southern margins included Lanark, Peebles and Greenlaw. Thus the list of plants assembled related not only to Edinburgh and the Lothians, but also to much of Fife, Clackmannan and the Borders. Balfour was writing in Edinburgh almost a century after, in his own happy phrase, 'Botany was cultivated in this city'. Much had changed since then, and since Greville wrote. Saving space for a pocket book format meant that the pedagogical material of *Flora Edinensis* had to go. But there were now plenty of available textbooks, apart from his own, where the elements of botany could easily be acquired. Balfour's Flora gave no keys or descriptions – these too could now be accessed elsewhere – and it is not much more than a list of plants found locally, their habitats, localities and flowering times. Flowering plants (969 species in 1863; 1,008 by 1871), ferns and fern allies (forty species) and 520 mosses and other lower plants are included. As a basis of the record, Balfour used all the previous lists, plus Greville's work, and added more records of his own, augmented by those made by his students. 'The Botany class' of that time made weekly excursions in summer, as a formal part of their university course. Those must have been happy days.

Effectively a third edition of Balfour's book, C. O. Sonntag's *A Pocket Flora of Edinburgh and the Surrounding District* (1894), (Plate 27) was conceived as a handy guide for identification of local plants. Sonntag included a glossary, keys to Natural Orders and to species of flowering plants, ferns and fern allies. The bulk of the book is species descriptions, with extensive citing of localities, the latter largely as in Balfour (1871) but partly also from his own plant collection, which he began in 1884. Sonntag's own records are mostly marked with an asterisk. The area covered is indicated by a reduced version of Balfour's map (back at the front of the book again), and is stated to have a diameter of twenty to thirty miles. It still included territory north of the Forth, extending from Kincardine to Kirkcaldy. In the south, it reached Lanark and Peebles. The John Sadler connection with the Royal High School is continued: Sonntag also was a teacher there.

Isa Martin's *Field-Club Flora of the Lothians* (first edition, 1927, second edition, 1934) (Plate 27) was produced by the Botanical Committee of the Edinburgh Natural History Society. It drew extensively from records published before, but also included data accrued during fieldwork by the Society members themselves. The *Field-Club Flora* is based on Sonntag's book, which it closely resembles in content and format. The Society had purchased the Sonntag copyright, and the work is best regarded as an extended and modernised version of that earlier *Flora*. Thus, at one remove, it is an updated Balfour. The book was intended to be a general introduction to field botany, for amateurs and students particularly, using the local flora as a vehicle. Consequently it included a note on the taxonomic hierarchy, a general key, keys to Natural Orders (now called families) and to genera within them. Species were numbered according to the then current (eleventh) edition of *The*

London Catalogue of British Plants. There were brief descriptions of the species, localities, and an illustrated botanical glossary. A brief, defining explanation of ecology, then a new concept to many, and a listing of plants in ecological groups – Loch Plants, Seashore Plants, Dockyard Casuals etc. Gymnosperms and Pteridophytes were included but lower plants were not. The total number of species listed was 1,018. The geographical coverage, revealed by the map (now safely ensconced in the back once more) is much the same as that of Sonntag's *Flora*. No attempt was made at a historical evaluation of floristic change, nor was there any treatment of the physical or climatic background to the distribution of species. The Field Club Flora was in print at least until comparatively recently, and is a valuable historical document.

Isa Martin was the Convener and Secretary of the Edinburgh Natural History Society's Botanical Committee, and she edited the *Flora* they produced. As additions to her own record of the members' findings on regular Saturday morning and Wednesday evening excursions to nearby Edinburgh sites, she included data from lists submitted by several prominent members. Among them were W. Edgar Evans and W. G. Smith, whose work in Scottish botany resonates clearly today. Another contributor was George Taylor of Cockburnspath, who later became Director of the Royal Botanic Garden, Kew. Sir George was, until his death in 1993 at Belhaven, East Lothian, Patron of the Botany of the Lothians Project. He attended several of the field trips that were run as part of this survey and is well remembered presiding robustly over tables of specimens – for instance in the Public Hall, Stow, and in the Black Bull Inn, Lauder. When an undergraduate student, Sir George had, as an exercise, compiled the first complete survey of plants from Duddingston Loch.

Completing the history of the study of the flora of the Lothians up to the present day requires mention of the hard work of the Vice-county Recorders appointed by the Botanical Society of the British Isles (BSBI). The first Recorder was the indefatigable Sir George, who assembled and maintained the record for what were then Linlithgowshire, Edinburghshire and Haddingtonshire. The BSBI and the Recorder organisation had themselves grown out of the great Hewett Cottrell Watson's Botanical Exchange Club. It is interesting to recollect that, when an Edinburgh student, Watson had been inspired into botany by Professor Robert Graham, who had awarded him a gold medal for an essay which was the basis for his later *Botanical Geography* (see Fletcher and Brown 1970). It *is* a small world. The next significant figure was Miss E. P. (Betty) Beattie, a BSBI member since 1955, who lectured on mapping schemes. In 1956, she was Scottish Regional Officer for the BSBI and secretary of its Committee for the Study of the Scottish Flora. She took up the Recordership in 1958, keeping a careful card index of finds and localities. Betty Beattie had hoped for the production of a new county Flora, but she did not live to see the completion of the present work, to which she had given her blessing. She handed over the Midlothian Recordership to D. R. McKean in 1980. McKean, together with the current Recorders for West Lothian (J. Muscott)

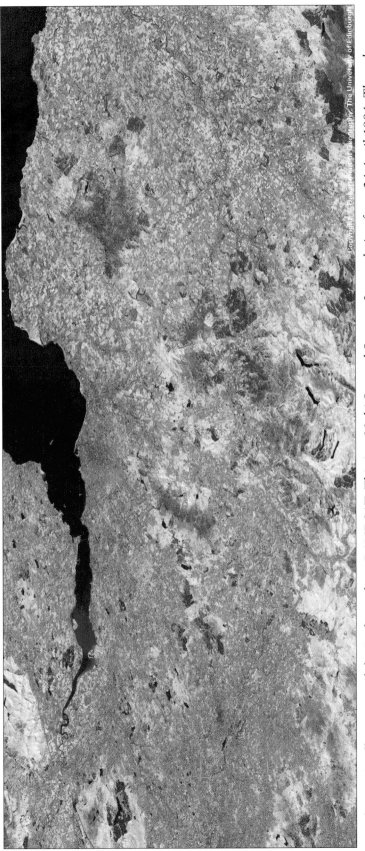

Plate 1 *Satellite image of the Lothians from LANDSAT. This is a Multi-Spectral Scanner Image dating from 24 April 1984. The colours represent the following: Forth Estuary and North Sea – dark blue; reservoirs – nearly black; upland areas mostly wooded – dark brown; open moorland – olive green; grazed land – light green; agricultural areas – pink; urban areas/bare soil – blue/grey. Copyright: The Department of Geography, The University of Edinburgh.*

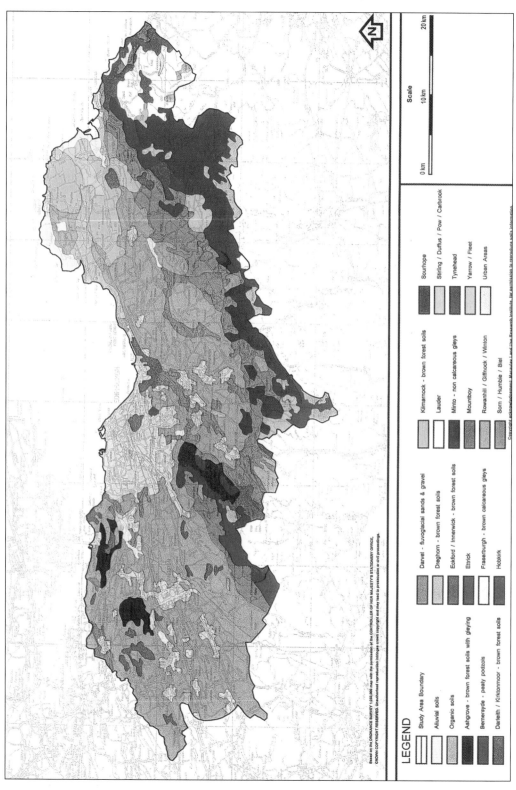

LEGEND

Study Area Boundary	Darvel - fluvoglacial sands & gravel	Kilmarnock - brown forest soils	Sourhope
Alluvial soils	Dreghorn - brown forest soils	Lauder	Stirling / Duffus / Pow / Carbrook
Organic soils	Eskford / Innerwick - brown forest soils	Minto - non calcareous gleys	Tynehead
Ashgrove - brown forest soils with gleying	Ettrick	Mountboy	Yarrow / Fleet
Bemersyde - peaty podzols	Fraserburgh - brown calcareous gleys	Rowanhill / Giffnock / Winton	Urban Areas
Darleith / Kirktonmoor - brown forest soils	Hobkirk	Sorn / Humbie / Biel	

Scale

0 km   10 km   20 km

Plate 2 *Soils map of the Lothians (from ASH 1998, reproduced by permission of the Macaulay Land Use Research Institute, Aberdeen, copyright owner).*

Plate 3 Top: *'Prairie' Lothian – the drier and sunnier East Lothian with its flat land and fertile soils is a grain-producing area. Reproduced by permission of the Scottish Agricultural Science Agency.* **Bottom:** *Characteristic soil profiles: 1. Kilmarnock Series, 0–70 cm; 2. Skateraw Series, 0–60 cm; 3. Winton series, 0–60 cm; 4. Fraserburgh Series, 0–60 cm (from Ragg and Futty 1967, reproduced by permission of the Macaulay Land Use Research Institute, Aberdeen, copyright owner).*

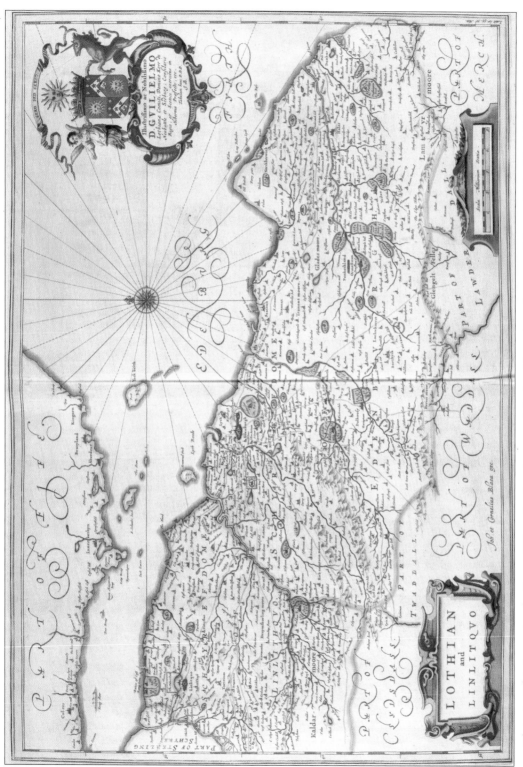

Plate 4 This is a reproduction of Timothy Pont's map of the Lothians, based on an edition published in 1636. The original survey work for it may have been completed before the end of the sixteenth century. It is plain that woodlands are sparse, except in the clearly demarcated great estates of that time. Much of the area is moor or arable land, with scattered farm toun settlements. Reproduced by kind permission of the Trustees of the National Library of Scotland.

Plate 5 *Large reed mats of* Phragmites australis *and birch/alder carr are encroaching on the margins of Duddingston Loch. Duddingston was once a commercial source of reed for thatch – see Chapter 19. C. E. Jeffree.*

Plate 6 Top: *Sheep grazed Holyrood Park and Arthur's Seat until the early 1980s. Large areas once grassland are now dominated by Gorse* (Ulex europaea) *which provides spectacular displays of colour in early May.* **Bottom:** *Field Poppies* (Papaver rhoeas) *regularly splash the fields of Inveresk with red. This year (2001) the crop was oilseed rape. C. E. Jeffree.*

Plate 7 Top: *The Tar-spot Fungus* (Rhytisma acerinum) *(Chapter 4) forms jet-black spots on leaves of Sycamore* (Acer pseudoplatanus) *in late summer.* Rhytisma *has been used as an air pollution indicator because of its sensitivity to sulphur dioxide. The incidence of this species has increased within the city boundary as air pollution by coal smoke has declined.* **Bottom left:** *The distinctive scaly brackets of Dryad's Saddle* (Polyporus squamosus) *are edible when young and fresh. The species, a parasite of deciduous trees such as beech and sycamore, is common in Lothian woodlands and gardens, adding to the woes of elms stricken by Dutch elm disease.* **Bottom right:** *The Candle-snuff Fungus* (Xylaria hypoxylon), *common on dead wood, is named for its resemblance to extinguished candle wicks. C. E. Jeffree.*

Plate 8 *Seaweeds.* **Top:** *Egg or Knotted Wrack* (Ascophyllum nodosum) *with attached tufts of the red seaweed* (Polysiphonia lanosa). **Bottom left:** *Spiral Wrack* (Fucus spiralis). **Bottom right:** *Bladder Wrack* (Fucus vesiculosus) *lying over Toothed Wrack* (Fucus serratus). *Top and bottom right: M. Wilkinson; bottom left: C. E. Jeffree.*

Plate 9 *A gallery of lichens on coastal rocks at Tyninghame and North Berwick.* **Top:** Tephromela atra, *syn.* Lecanora atra, *is common on siliceous rocks at the coast and also inland on mortared walls.* **Bottom left:** Xanthoria ectanoides *with* Ramalina siliquosa. **Bottom right:** Caloplaca ceracea, *syn.* C. caesiorufa, *is common on coastal rocks but rare inland. Top: C. E. Jeffree; bottom, right and left: D. White, courtesy of RBGE.*

Plate 10 *Ferns. Two distinctive ferns, both members of the primitive Ophioglossaceae, are now rare in the Lothians. Adder's-tongue* (Ophioglossum vulgatum) (**Top right**) *occurs in Holyrood Park, and Moonwort* (Botrychium lunaria) (**Top left**) *is infrequent in dunes and dune slacks at Aberlady and Yellowcraig.* **Bottom:** *The evergreen Maidenhair Spleenwort* (Asplenium trichomanes) *growing in a drystone wall. C. E. Jeffree.*

Plate 11 Top: *Wood Sage* (Teucrium scorodonia) *is frequent and widespread throughout the Lothians.* **Bottom left:** *Bogbean* (Menyanthes trifoliata) *is recorded as occasional or rare in eleven localities in the region.* **Bottom right:** *Harebell or Bluebell* (Campanula rotundifolia) *seen here below Samson's Ribs on Arthur's Seat, is common and locally abundant, recorded in 370 squares. C. E. Jeffree.*

Plate 12 *Coastal flora.* **Top left:** *Bloody Cranesbill* (Geranium sanguineum) *at Aberlady Bay.* **Top right:** *Common Centaury* (Centaurium erythraea) *is an occasional in coastal grassland and also on dry calcaereous soils elsewhere.* **Bottom left:** *young plants of Glasswort* (Salicornia europaea), *abundant in salt-marsh at Tyninghame, is in decline in the Lothians as saltmarsh area diminishes.* **Bottom right:** *Restharrow* (Ononis repens), *here photographed in dune pasture at Yellowcraig, is common in a wide range of other habitats including roadsides.* C. E. Jeffree.

Plate 13 *Aliens: Both are garden escapes. White Butterbur* (Petasites albus) (**Top**) *grows in profusion on roadsides, pollution notwithstanding, as here at Bush, Midlothian, while Giant Hogweed* (Heracleum mantegazzianum) (**Bottom**), *photographed at Sheriffhall, spreads by windblown and waterborne seed along roads, railways and rivers (Chapter 17). C. E. Jeffree.*

Plate 14 *The Union Canal and Lothian's riversides provide a significant area of semi-wild habitats for plants even in the midst of urban development.* **Top:** *A view of the Union Canal at Craiglockhart in autumn.* **Bottom left:** *Unbranched Bur-reed* (Sparganium emersum), *recorded in seven squares, with Common Duckweed* (Lemna minor) *beneath.* (**Bottom right**) *Tufted Loosestrife* (Lysimachia thyrsiflora) *with a close-up of an inflorescence* (**inset**). *Both occur in canal edges and ditches. J. Muscott.*

Plate 15 *Bing floras can be surprisingly species-rich.* **Top:** *this mixed-flower meadow on Broxburn Bing includes Ragwort* (Senecio jacobaea), *Rosebay Willowherb* (Chamerion angustifolium), *Ox-eye Daisy* (Leucanthemum vulgare) *and the leaves of Coltsfoot* (Tussilago farfara). **Bottom left:** *Mouse-ear Hawkweed* (Pilosella offic-inarum) *showing its stolons.* **Bottom right:** *Alpine Clubmoss* (Diphasiastrum alpinum) *was photographed on Faucheldean Bing. Top and bottom right: J. Muscott; Bottom left: C. E. Jeffree.*

Plate 16 *Ethnobotany* (Chapter 19). *Sea Buckthorn* (Hippophae rhamnoides) (**Top**) *is common on stable dunes along the coast of the Forth estuary. The inset shows the abundant orange fruits.* **Bottom left:** *Woad* (Isatis tinctoria) *was once a commercial source of blue dye.* **Bottom right:** *The Burry Man of South Queensferry parades the town on the second Friday in August coated in the hooked fruits of Lesser Burdock* (Arctium minus). *Top: J. Muscott; bottom left and right: G. J. Kenicer.*

and East Lothian (E. H. Jackson) have all been intimately involved in the present survey and in the preparation of the Flora. A. J. Silverside has also been a major contributor, and has particularly contributed to the East Lothian survey. Details will be found in Chapter 12.

Future Local Floristic Studies

To maintain the record requires continuity of interest and effort by as many local botanists – amateur and professional – as possible. The BSBI Vice-county Recorders, the Botanical Society of Scotland, the Edinburgh Natural History Society, the BSBI, the University of Edinburgh, the Royal Botanic Garden Edinburgh, the National Museum of Scotland, the Scottish Wildlife Trust and Scottish Natural Heritage – all made important contributions to the present endeavour. They have collective and individual responsibility for future studies of floristic botany in the Lothians. We must all play our part in keeping botany alive for the delight as well as the edification of students yet to come. We cannot know them, but we think of them with affection.

12

The Botany of the Lothians Project
and the compilation of
A Flora of the Lothians

M. P. COCHRANE, R. O. D. DIXON,
J. MUSCOTT and P. M. SMITH

The Project

Origins and History

The Council of the Botanical Society of Edinburgh (since renamed the Botanical Society of Scotland), having expressed a wish for the Society to undertake some large-scale practical field research, accepted, in October 1981, a proposal put forward by P. M. Smith that a study should be made of the botany of the Lothians. The Council considered that this proposal would fulfil three objectives: (a) to engage the Botanical Society of Edinburgh in some fundamental research activity that would be of value to Scottish society and would increase the knowledge of the plant life of Scotland, (b) to involve as many members of the Society as possible in the field project, regardless of initial knowledge or skill and (c) to disseminate as widely as possible, knowledge of, delight in and respect for, the flora. The project was given the title 'Botany of the Lothians' to indicate that in the final publication a county Flora would, if possible, be presented in the context of authoritative accounts of the geology and land use of the area, of the history of the flora, and of the algae, fungi, lichens and bryophytes of the region.

The Botanical Society of Edinburgh also decided that the area to be surveyed would not be the four recently re-organised Districts that together made up Lothian Region, but the three Watsonian vice-counties that constitute the Lothians. These vice-counties are used for all biological recording, and were first devised in 1852 by H. C. Watson. They have remained unchanged over the years, so neither their internal nor their external boundaries coincide with the modern political entities of the same names. The areas covered in the survey, VC 82 East Lothian, VC 83 Midlothian and VC 84 West Lothian are shown in the gazetteer map. The use of the vice-county boundaries to define 'The Lothians' also gave us access to the expertise of the three Vice-county Recorders appointed by the Botanical Society of the British Isles (BSBI) to keep botanical records for their areas, E. H. Jackson (VC 82), D. R. McKean (VC 83), and J. Muscott (VC 84). It would have been more difficult to obtain such expertise if the vice-county boundaries had not been adhered to.

Earlier 'Botanies' followed the general plan of a physical and climatic background supporting an account of the local flora within the boundaries of the area under study. P. M. Smith's conceptual model for the Botany of the Lothians was Amphlett and Rea's *The Botany of Worcestershire* (1909), and his aim throughout was to enlarge and extend local appreciation of botany in its *broadest* sense, at a time of absurd reductionism. These earlier 'Botanies' were part of the fashion for new County Floras in the late nineteenth and early twentieth centuries, following the seminal work of Hewett Cottrell Watson, for example, in *Cybele Britannica* (1847) and *Topographical Botany* (1873). In Ireland, the accompanying work was Moore and More's *Contributions towards a Cybele Hibernica* (1866).

The nineteenth-century 'Botanies' included some agricultural information and it was considered essential to record agricultural land use in the Lothians survey. Six of the habitat types formed a group entitled 'Agricuse' for convenient reference. It is a time of change for local agriculture: it was thought that the pattern of present-day commercial cultivation in an area that shows considerable variation in soil-type and climate should be recorded, and that relationships with the weed flora should be investigated.

Relatively few of the early county Floras or 'Botanies' included phytogeographical or ecological studies. Towards the latter part of the nineteenth century, as these disciplines became more defined and scientific, this type of analysis began to appear in footnotes and increasingly detailed commentaries. It was thought essential that the survey data of the 'Botany of the Lothians' should be capable of generating valid information about the ecology and distribution of individual species. This meant that the whole area would have to be covered evenly and that concentration on 'hot spots' where rare or unusual plants had traditionally been noted, would have to be avoided. To this end a rigorous sampling procedure was set up (see section below on Data Collection). This was based on a grid of tetrads (2km x 2km) on the Ordnance Survey maps, and would allow well-based distribution maps and tables of habitat particulars to be prepared, which could be scanned for ecological and phytogeographical purposes. It was decided that estimations of the frequency of each species within squares would provide valuable additional information.

Mapping distributions in county Floras became the norm in the late twentieth century following the notable success of the pioneering work of the BSBI in producing the *Atlas of the British Flora* (Perring and Walters 1962). A new version of the Atlas is currently being prepared. The Botanical Society of Edinburgh was adamant that every taxon of angiosperms (flowering plants), gymnosperms (conifers), ferns and fern-allies in the Lothians should be potentially mappable. It was assumed that advances in computers and the growing expertise of local enthusiasts would make map production possible by the time the data were collected. This expectation was realised, R. O. D. Dixon being responsible for map production. From the late-1980s onwards, M. P. Cochrane acted as project secretary. She subsequently took charge of the organisation of the writing of this Flora of the Lothians, compiling most of the draft text and then incorporating all the additional material supplied by the Vice-county Recorders.

It would not be right to end this brief account of the origins and history of the project without recollecting the Society event that marked its public genesis. Programmes that depend on wide public interest and participation need a resounding start and plenty of publicity to keep them going. It had been the experience of being marginally involved in the production of a computer-mapped local Flora (Cadbury, Hawkes and Readett 1971) that had further inspired P. M. Smith with notions of what could be done, if sufficient local interest could be harnessed and sustained. It was no accident that the public launch of the Botany of the Lothians project in 1981, involved a major address by Professor J. G. Hawkes of Birmingham, who had been chief pilot of the Warwickshire endeavour. He had been P. M. Smith's Ph.D. supervisor, and mentor.

As one project can stimulate new intellectual effort in different places, so it was and is the Society's intention that whatever schemes and devices were developed to generate the Botany of the Lothians project would be available to members in any part of its Scottish demesne. It was the earnest hope of the Society that other local Scottish surveys, 'Botanies', or Floras would, in due time, appear.

Finances

Any local Flora project needs supporting funds to meet the costs of communications and postage, the duplicating of working documents and the hiring of local halls for meetings, talks and field gatherings. The Botany of the Lothians project has been run on a shoestring, quite deliberately. It was not intended to be a major drain on the finances of the sponsoring Society, or of its members. In the event, the project has been fortunate to receive periodical grants-in-aid from the Botanical Society of Scotland and some useful small gifts of materials and cash from individuals – by no means all of them members of the Society. Early in the life of the project, a generous donation was made by the Edinburgh Natural History Society. A kind donation to general expenses was made by the late Sir George Taylor of Belhaven, and a bequest was made from the estate of the late Dr John Gregor, previously Director of the Scottish Plant Breeding Station, Roslin.

The University of Edinburgh and the Royal Botanic Garden Edinburgh (RBGE) have helped enormously by giving free room hire and storage facilities. In addition, they perceived that the work done toward the survey by P. M. Smith and D. R. McKean, respectively, was a legitimate part of their professional activity. Participating members of the Society, Botany of the Lothians Committee members, and all recorders of squares have individually increased the resources available in many ways, not least by their contributions to transport and communications costs.

During the later stages of the survey (1992–4), Scottish Natural Heritage (SNH) provided grants to cover travel costs for a 'mopping up' of more distant and inaccessible squares. This 'mopping up' was largely undertaken by J. Muscott, but in 1994, C. Dixon and D. R. McKean were also involved, and the RBGE provided additional travelling expenses.

Later, the Society received a generous grant of £6,300 from SNH to help finance the input of the survey data into computer files. It had been hoped that there would be a sufficient number of Society members able to do this work at home, and an input program had been especially written with this in mind. However, in these days of increased workloads there proved to be too few computer-literate botanists with time to spare for this task. Thus, the SNH grant was used to pay the Scottish Wildlife Trust (SWT) to transfer a large part of the data to computer files. This was done with commendable speed and accuracy.

Costs of publication of academic work are considerable and are found burdensome by all societies and institutions. Edinburgh University Press were willing to publish the book but asked for an indemnity both to protect them from commercial loss and to enable a realistic cover price to be set. For a time it was hoped that a local commercial sponsor would subsidise publishing costs, but this proved illusory. Most large commercial sponsors are attracted only to projects with mass appeal. A publishing fund was set up and due to the generosity of several organisations and societies and of many individual members of the Botanical Society of Scotland, the requisite sum was raised.

Professional Backing

The Botany of the Lothians project was fortunate to secure the general advice of members of the BSBI Panel of Referees for identifications of critical genera. Their aid was essential in determining and confirming local identifications and in guiding the collection of specimens in suitable condition. It is not possible to acknowledge all of them individually here, but special thanks go to those listed in Table 12.1. The Editors place on record their special debt to D. R. McKean (RBGE) for overseeing identifications. Thanks are due also to A. J. Silverside (University of Paisley), who helped to plan the layout of the Flora and also helped to prepare the checklist for, and evaluation of material from, East Lothian. R. Saville (SWT) provided valuable professional support in managing data transfer to computer files. After the contentious matter of how many 1km x 1km squares to sample, and how, had been resolved by the Botany of the Lothians Committee, J. Grace (University of Edinburgh) produced an entirely independent, randomised list of squares to be recorded, one from each of the tetrads in the grid. J. Grace also defined the habitat categories that have been used in the survey.

It was considered essential to interest young people in the work and, at the outset, school groups were involved in the recording. Teacher and school groups were energetically organised by W. McNab, formerly Principal Lecturer in the Department of Biology at Moray House College of Education. However, this was a time of change in education and increasing workloads for teachers and pupils meant that few were able to continue to take part in the project.

Table 12.1 A list of referees who identified material from critical genera

Equisetum	C. Dixon	*Cotoneaster*	J. Fryer
Aster	P. F. Yeo	*Salix*	R. D. Meikle
Fumaria	P. D. Sell, M. G. Daker	*Potentilla*	B. Harold
Ranunculus	N. T. H. Holmes	*Rosa*	G. G. Graham, A. L. Primavesi
Cochlearia	T. C. G. Rich, D. H. Dalby	*Symphytum*	F. H. Perring, T. W. J. Gadella
Erophila	T. T. Elkington, T. C. G. Rich	*Hieracium*	D. J. McCosh
Chenopodium	J. M. Mullin	*Taraxacum*	
Atriplex	P. M. Taschereau	and *Epipactis*	A. J. Richards
Spiraea			
Mimulus	A. J. Silverside	*Carex*	A. O. Chater
Euphrasia		*Agrostis*	A. D. Bradshaw
Rubus	G. H. Ballantyne	*Potamogeton*	C. D. Preston

Flora of the Lothians

Data Collection

The total area of the three vice-counties that constitute the Lothians is approximately 2,200 square kilometres. For sampling purposes, 578 2km x 2km squares covering the Lothians on the Ordnance Survey maps were identified. These squares are here referred to as tetrads, each tetrad being made up of four 1km x 1km squares. One square in each tetrad was then chosen at random as the square to be surveyed. Twenty-three of the tetrads were marginal, with less than 10 per cent of their area within the Lothians, and these were largely ignored, though islands in the Firth of Forth were included regardless of their area. Of the 555 non-marginal tetrads, all but seventeen had their squares surveyed, giving 97 per cent cover (Figure 12.1). The squares are listed by eastings and northings map references in Table 12.2.

Table 12.2 Map references of the 1km x 1km squares randomly selected for recording in the survey

NS8567	NS9866	NT0673	NT1656	NT2476	NT3263
NS8568	NS9869	NT0755	NT1663	*NT2555*	NT3265
NS8667	NS9872	NT0757	NT1664	NT2559	NT3267
NS8768	NS9875	NT0761	NT1668	NT2562	NT3269
NS8868	NS9879	NT0765	NT1670	NT2569	NT3347
NS8967	NS9963	NT0774	NT1676	NT2572	NT3349
NS8970	NS9970	NT0777	NT1678	NT2575	NT3350
NS9067	NS9976	NT0779	NT1759	NT2656	NT3353
NS9069	NS9980	NT0855	**NT1760**	NT2663	NT3354
NS9155	NT0066	NT0863	NT1766	NT2665	NT3359
NS9157	NT0070	NT0864	NT1772	NT2666	NT3361
NS9158	NT0077	NT0868	NT1775	NT2668	NT3370
NS9161	NT0157	NT0873	NT1859	NT2672	NT3373
NS9163	NT0159	NT0875	NT1864	**NT2675**	NT3447
NS9164	NT0161	NT0957	NT1868	NT2676	NT3450
NS9170	NT0163	NT0958	NT1873	*NT2678*	NT3460
NS9172	NT0164	NT0961	NT1877	*NT2748*	NT3462
NS9257	NT0168	NT0966	NT1955	NT2751	NT3465
NS9261	NT0172	NT0971	NT1957	NT2753	**NT3467**
NS9263	NT0175	NT0977	NT1961	NT2755	NT3470
NS9269	NT0179	NT0979	NT1963	NT2759	NT3473
NS9271	NT0180	NT1059	NT1966	NT2760	*NT3545*
NS9358	NT0257	NT1063	NT1971	NT2770	NT3548
NS9364	NT0260	NT1069	NT1974	NT2853	NT3552
NS9366	NT0262	NT1071	NT1978	NT2861	NT3554
NS9372	NT0266	NT1073	NT2055	NT2864	NT3557
NS9457	NT0268	NT1160	NT2056	NT2866	NT3559
NS9458	NT0274	NT1164	NT2063	NT2871	NT3569
NS9560	NT0281	NT1166	NT2064	NT2872	NT3650
NS9562	*NT0355*	NT1175	NT2066	NT2875	**NT3656**
NS9564	NT0359	NT1177	NT2080	*NT2876*	NT3661
NS9567	NT0365	NT1178	NT2158	NT2947	NT3662
NS9568	NT0370	NT1259	NT2161	NT2949	NT3665
NS9570	NT0373	NT1260	NT2169	NT2950	NT3668
NS9572	NT0377	NT1263	NT2171	NT2955	NT3745
NS9574	NT0379	NT1267	NT2173	NT2957	NT3747
NS9579	NT0460	NT1270	NT2174	NT2958	NT3748
NS9580	NT0471	NT1273	NT2177	NT2962	NT3752
NS9657	NT0476	NT1278	NT2255	NT2968	NT3755
NS9667	NT0480	NT1365	NT2260	NT3055	NT3759
NS9668	NT0555	**NT1369**	NT2263	NT3057	NT3767
NS9672	NT0556	NT1375	NT2268	NT3058	NT3770
NS9674	NT0559	NT1377	NT2273	NT3062	NT3772
NS9759	**NT0563**	**NT1459**	NT2275	**NT3069**	NT3847
NS9761	NT0565	NT1463	NT2356	NT3074	NT3851
NS9763	NT0566	NT1468	NT2358	*NT3147*	NT3853
NS9764	NT0569	NT1471	NT2365	NT3149	NT3858
NS9770	NT0573	NT1557	NT2366	NT3151	NT3862
NS9776	NT0575	NT1561	NT2371	NT3152	NT3865
NS9779	NT0578	NT1565	NT2377	NT3160	NT3868
NS9780	NT0659	NT1566	NT2457	NT3165	NT3870
NS9857	NT0662	**NT1573**	NT2461	NT3166	NT3945
NS9858	NT0667	NT1575	NT2464	NT3170	NT3948
NS9861	NT0668	NT1577	NT2467	NT3172	NT3954
NS9864	NT0671	NT1578	NT2471	NT3257	NT3956

NT3960	NT4442	*NT4846*	NT5381	NT5982	NT6666
NT3967	NT4444	*NT4848*	NT5383	NT5985	NT6668
NT3972	NT4450	*NT4859*	NT5384	NT6063	**NT6671**
NT3975	*NT4452*	NT4863	NT5386	NT6066	NT6676
NT4049	NT4455	NT4864	NT5461	NT6073	NT6678
NT4051	NT4459	NT4867	NT5462	NT6074	NT6773
NT4054	NT4460	NT4871	NT5465	NT6076	NT6775
NT4057	NT4464	NT4874	NT5466	NT6080	NT6871
NT4062	NT4469	NT4879	NT5475	NT6087	NT6878
NT4066	NT4471	NT4880	NT5481	NT6160	NT6963
NT4143	NT4476	NT4882	NT5483	NT6165	NT6965
NT4144	NT4541	NT4961	NT5569	NT6168	NT6967
NT4147	NT4547	NT4968	NT5570	NT6171	NT6969
NT4152	NT4549	NT4973	NT5572	NT6179	NT6972
NT4158	NT4557	NT4976	NT5577	NT6182	NT6975
NT4160	NT4563	NT4985	NT5579	NT6184	NT6977
NT4165	NT4566	NT4986	NT5585	*NT6259*	NT7063
NT4169	NT4572	NT5061	NT5586	NT6263	NT7066
NT4171	NT4574	**NT5065**	NT5661	NT6271	NT7071
NT4172	NT4578	NT5069	NT5662	NT6273	NT7073
NT4175	NT4580	NT5072	NT5670	NT6274	NT7078
NT4242	NT4641	NT5075	NT5675	*NT6283*	NT7164
NT4250	NT4649	NT5076	NT5677	**NT6360**	NT7168
NT4257	NT4651	NT5079	NT5678	NT6365	NT7174
NT4263	NT4667	NT5080	NT5681	NT6367	NT7176
NT4267	NT4671	NT5162	NT5683	**NT6368**	NT7263
NT4273	NT4681	NT5167	NT5764	NT6377	*NT7264*
NT4274	NT4683	NT5171	NT5766	NT6379	NT7266
NT4339	NT4743	NT5182	NT5768	NT6380	NT7270
NT4340	NT4745	NT5185	NT5772	NT6463	NT7276
NT4345	NT4747	NT5186	NT5785	NT6465	NT7368
NT4347	NT4759	NT5264	NT5865	NT6466	NT7373
NT4348	NT4761	NT5271	NT5872	NT6469	NT7374
NT4352	NT4763	NT5275	NT5877	NT6474	NT7469
NT4355	NT4764	NT5278	NT5879	NT6479	NT7472
NT4359	NT4769	*NT5359*	NT5961	NT6561	NT7475
NT4361	NT4772	NT5360	NT5963	**NT6570**	**NT7570**
NT4364	NT4775	NT5362	NT5966	NT6573	NT7671
NT4368	NT4777	NT5367	NT5968	NT6577	NT7672
NT4370	**NT4778**	NT5368	NT5971	*NT6660*	
NT4376	*NT4843*	NT5373	NT5974	NT6663	
NT4439	*NT4845*	NT5377	NT5980	NT6665	

Un-recorded squares are in **bold.**

Edge squares (un-recorded) are in *italics*.

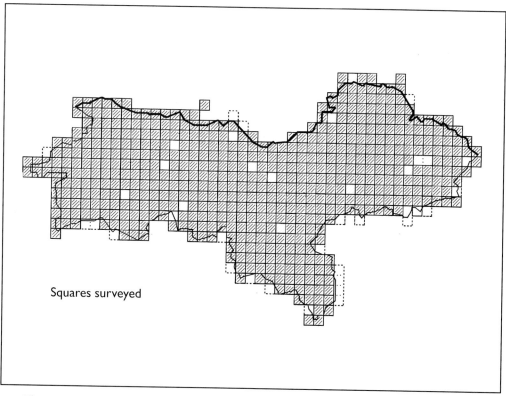

Figure 12.1 *Map of the Lothians showing the tetrads in which 1km x 1km squares were surveyed (stippled) or not surveyed (white).*

The survey squares were then assigned to recorders. Recorders came from a wide range of backgrounds, some were professional botanists, some were experienced and knowledgeable amateurs, some were enthusiastic beginners. Over the course of the project more than 300 people recorded plants. Recorders worked either individually or in groups.

Under the chairmanship of P. M. Smith, a committee that included the Vice-county Recorders compiled a set of instructions for recorders and produced a recording sheet. This recording sheet was geared towards card input, and made use of the standard university card input forms of the time. The layout of the card is shown in Figure 12.2 and the habitat categories are listed in Table 12.3.

Table 12.3 The 41 habitat categories used in the survey, and the symbols used on recording sheets to represent these habitat categories

WO	**Woodland**		**WS**	**Waterside**
WO-C	conifer woodland		WS-C	canalside
WO-M	mixed woodland		WS-D	ditchside
WO-D	deciduous woodland		WS-L	lakeside (lochside)
			WS-M	marsh
			WS-P	pondside
			WS-R	riverside
WA	**Immersed in Water**		**M**	**Marginal**
WA-C	canal		M-HR	hedgerow
WA-D	ditch		M-SC	scrub
WA-L	lake (loch)			
WA-P	pond		**HB**	**Heathland and Bogs**
WA-R	river		HB-B	bog
			HB-D	dry heath, moor
			HB-G	heath grassland
R	**Ruderal**			
R-RO	roadsides			
R-RY	railway embankment/cutting		**S**	**Stony ground**
R-BR	brick or marl pit		S-WA	walls
R-FM	farmyard		S-QU	quarries
R-WS	waste ground, tips, refuse dumps			
			MA	**Maritime**
OC	**Other Cultivated Land**		MA-D	dunes
OC-A	allotments		MA-C	cliffs
OC-G	gardens		MA-B	beach
OC-P	public park		MA-M	saltmarsh
			MA-S	coastal scrub/turf
Agricuse				
A	**Agricultural**			
A-CE	cereal fields			
A-RO	rootcrops			
A-VE	other vegetable crops			
A-GI	intensive grassland			
A-GR	rough grassland			
A-GL	leys			
A-OR	orchard			

The recorders were expected to fill in the name of each species in shortened form using the first four letters of the genus name and the first four letters of the species name, for example, Prim vulg for *Primula vulgaris*. Then they were asked

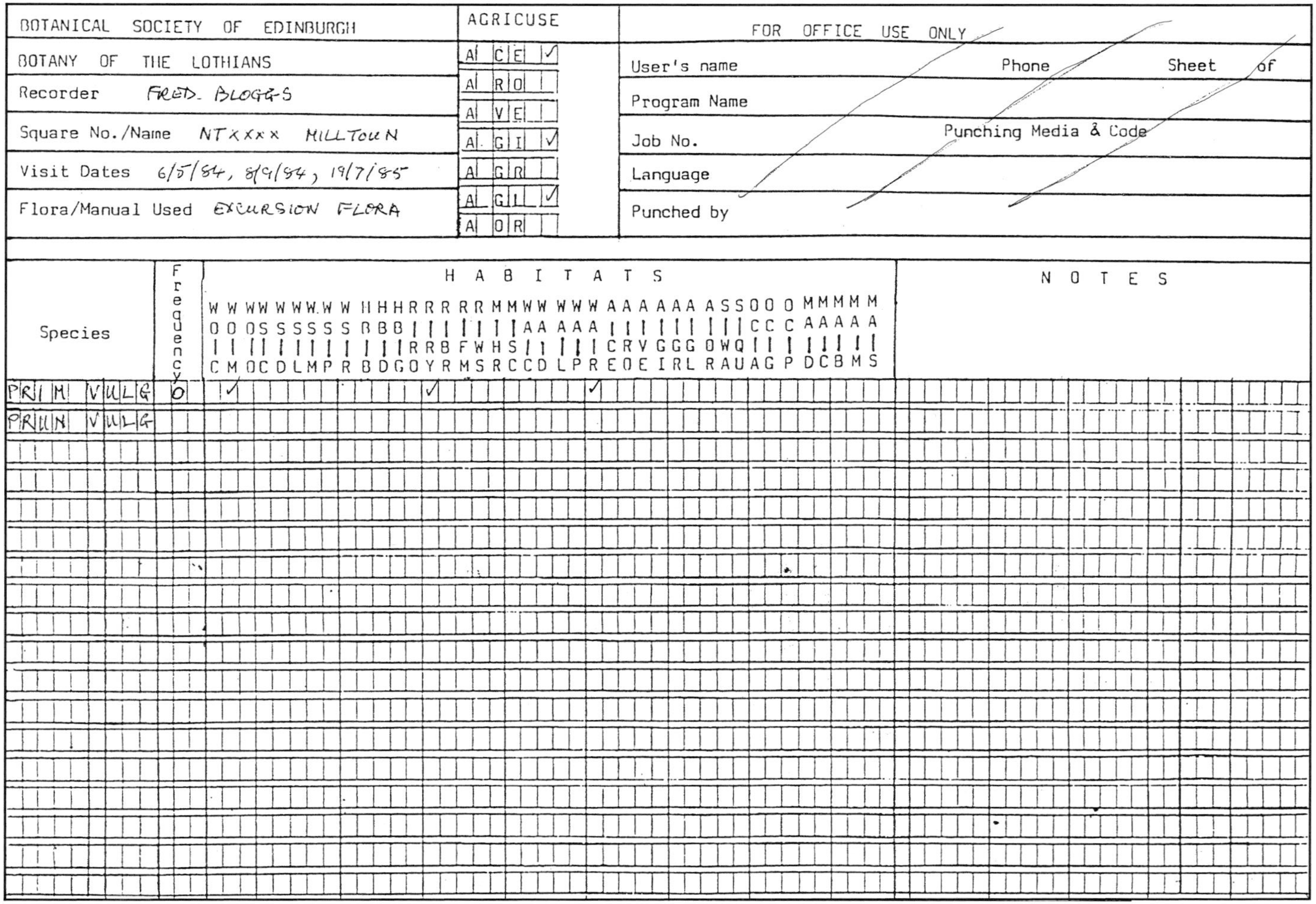
BOTANICAL SOCIETY OF EDINBURGH
BOTANY OF THE LOTHIANS
Recorder FRED. BLOGGS
Square No./Name NTxxxx MILLTOWN
Visit Dates 6/5/84, 8/9/84, 19/7/85
Flora/Manual Used EXCURSION FLORA
AGRICUSE
A CE
A RO
A VE
A GI
A GR
A GL
A OR
FOR OFFICE USE ONLY
User's name
Phone
Sheet of
Program Name
Job No.
Punching Media & Code
Language
Punched by
Species
Frequency
HABITATS
NOTES
PRIM VULG
PRUN VULG

Figure 12.2 *Part of a recording card of the type used during the first few years of the survey.*

to mark in the appropriate column(s) the habitat(s) where the plant was found, and to insert an estimate of its distribution density over the square as a whole, using the letters R (rare), O (occasional), F (frequent), A (abundant). Recorders were allowed to qualify these terms using the prefix 'L' (locally) to indicate that a species was not uniformly distributed in the square. The recorders were asked to visit the square in spring, summer and autumn, and to record the dates of their visits, the agricultural land use and the name of the Flora used for identifications. Field and laboratory meetings were held to explain and standardise procedures and to help beginners and 'lapsed' botanists to identify specimens. Lists of difficult genera were drawn up and recorders were asked to collect and press specimens of these at the appropriate stage of plant development so that identification could be checked. Much of this checking work was carried out by D. R. McKean at the Royal Botanic Garden Edinburgh, but where necessary, specimens were sent to appropriate national authorities.

Recorders were instructed to include all ferns, fern allies, gymnosperms (conifers), and flowering plants growing wild in the Lothians. Garden plants were generally recorded only if growing wild (escapes), but some, particularly trees and shrubs, were included when planted in public places such as parks and roadsides for amenity purposes. Garden, allotment and agricultural weeds were included, and crop plants noted.

Filling in the recording sheet was quite laborious as the plant names had to be entered. It was also somewhat prone to error because it was quite easy to tick the wrong habitat column. Later, a new, more flexible sheet was produced with the abbreviated plant names printed in alphabetical order with space for the recorder to add species density and habitat codes in any order. The abbreviated plant names caused some problems, as not all were distinct. The coincidence between *Rumex acetosa* and *Rumex acetosella* was quickly recognised and brought to the attention of recorders, but the problem with *Salicornia fragilis* and *Salix fragilis* was only recognised and corrected when the first maps were produced.

Completed sheets were sent to the appropriate Vice-county Recorder who then checked for unlikely identifications or obvious errors in the ticking of boxes and/or species. Where appropriate, recorders were asked to 'look again' or to provide specimens to confirm identifications. Some of the squares considered to have been inadequately recorded were re-visited.

It was thought initially that the recording phase of the project might take ten years to complete. That proved to be an under-estimate, recording was spread over fourteen years. The survey work, however, was only one aspect of the Botany of the Lothians project. The Vice-county Recorders collated the plant records from the first seven years of the survey with those obtained during the BSBI Monitoring scheme of 1987–8, and other records in their possession, to produce checklists. The Botanical Society of Edinburgh published *A Checklist of the Flowering Plants and Ferns of East Lothian* (Silverside and Jackson 1988), *A Checklist of the Flowering Plants and Ferns of Midlothian* (McKean 1988), and *A Checklist of the Flowering*

Plants and Ferns of West Lothian (Muscott 1989). All were reprinted. These check-lists proved to be very valuable to recorders and were used in the compilation of the second type of recording sheet. The checklists have also been used extensively in the writing of the Flora.

Access

We must record our grateful thanks to the numerous private landowners and local authority and other public agencies who have agreed to recorders entering land to observe the plant cover. Fortunately no crises such as Foot-and-Mouth epidemics impeded progress during the recording phase of the survey. It was a feature of the advice given to all recorders that 'permission to visit' always be secured before botanising on private land. Though in Scotland there is no criminal law of trespass, it is always courteous to gain formal permission and in some cases it proved very rewarding. During visits to private land, the attention of recorders was often drawn to plants that had intrigued or puzzled the landowner for many years. Several farmers and farmers' wives were subsequently recruited as recorders themselves. People *are* attracted by local plants: the interest is always there to be informed, fostered and increased.

Data Input and Analysis

The total number of tetrads covered was 557, including two 'edge' tetrads. The highest number of species recorded from any one square was 360, and there were few squares with under 100 species. Entry of this amount of data (149,398 records!) was a considerable task. It was originally hoped that volunteers would do much of the work on home computers. This was not possible using large databases such as Recorder and Access and so data input and analysis software was written by R. O. D. Dixon specifically for the project. The data input program could be distributed to helpers on a floppy disk and it enabled data to be entered speedily, with typing and scrolling reduced to a minimum. The only typing necessary was the entry of the Ordnance Survey co-ordinates of the square being entered and, when the entries on the record sheet were not in alphabetical order, the first two or three letters of the genus. To avoid the need for scrolling, the abbreviated species names were presented on screens of fifty-four species names that could be selected using a mouse. Paging between screens involved the page up and page down keys or clicking with the mouse on PgUp or PgDn on the screen. A typical screen is shown in Figure 12.3.

After the species was selected, a separate screen was shown from which any of the forty-one habitats could be selected together with the density distribution of the species. The data from each square input was saved onto a floppy disk and passed on for entry into the main database. For each square, the input program produced two files, one containing habitat information, the other containing distribution density information. In order to reduce memory requirements and usage of disk space, species names, habitat names and distribution density were represented by

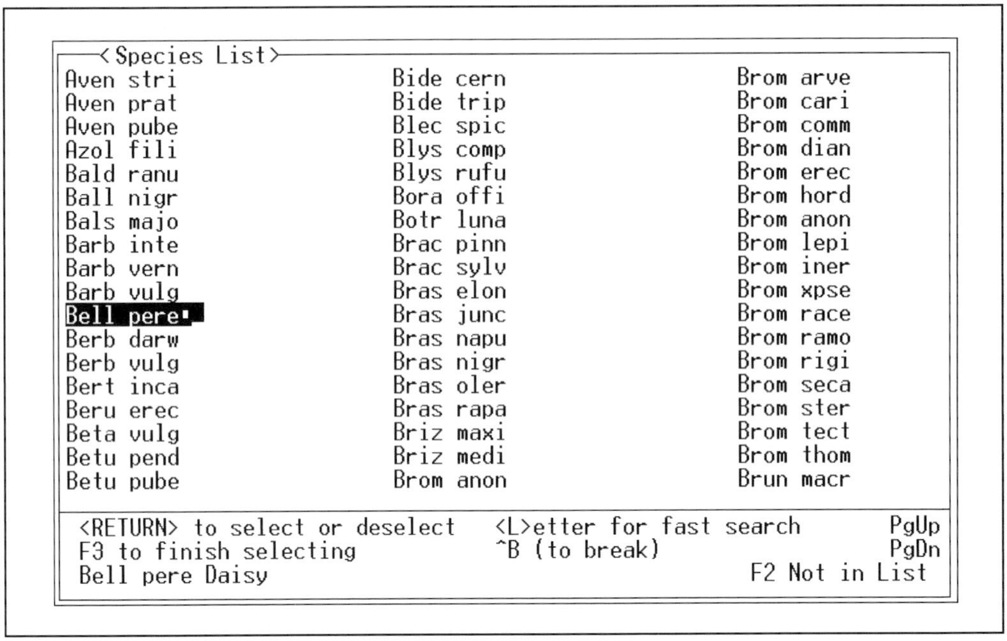

```
┌─<Species List>──────────────────────────────────────────────┐
│ Aven stri          Bide cern          Brom arve             │
│ Aven prat          Bide trip          Brom cari             │
│ Aven pube          Blec spic          Brom comm             │
│ Azol fili          Blys comp          Brom dian             │
│ Bald ranu          Blys rufu          Brom erec             │
│ Ball nigr          Bora offi          Brom hord             │
│ Bals majo          Botr luna          Brom anon             │
│ Barb inte          Brac pinn          Brom lepi             │
│ Barb vern          Brac sylv          Brom iner             │
│ Barb vulg          Bras elon          Brom xpse             │
│ Bell pere▪         Bras junc          Brom race             │
│ Berb darw          Bras napu          Brom ramo             │
│ Berb vulg          Bras nigr          Brom rigi             │
│ Bert inca          Bras oler          Brom seca             │
│ Beru erec          Bras rapa          Brom ster             │
│ Beta vulg          Briz maxi          Brom tect             │
│ Betu pend          Briz medi          Brom thom             │
│ Betu pube          Brom anon          Brun macr             │
│─────────────────────────────────────────────────────────────│
│ <RETURN> to select or deselect   <L>etter for fast search  PgUp│
│ F3 to finish selecting           ^B (to break)             PgDn│
│ Bell pere Daisy                              F2 Not in List │
└─────────────────────────────────────────────────────────────┘
```

Figure 12.3. *The contents of a screen showing part of the species list in the data input program.*

integers and later related to the proper names for output to screen or print. The amount of data to be entered into the main database was about three megabytes and so, in order to be able to use extended memory on the computer, a compiler that was compatible with a DOS extender was used to produce the main database program. The two files from each square were amalgamated into two large files that formed the basis of two tables in the main database program. The database program produced maps of the distribution of each species (Figure 12.4) and also information with regard to the habitats in which each species was recorded (Figure 12.5). The program was distributed to the writers of the Flora and to chapter authors.

It is clear from the data in Figure 12.5 that *Ranunculus flammula* (Lesser Spearwort) is a water plant, the vast majority of the habitats being wetlands of some sort. However one or two habitats are, at first glance, rather surprising. For a discussion of these apparent anomalies, see section below on Interpretation of Data.

The Writing of the Flora

Earlier generations of county Floras usually included descriptions of plants and their flowering times and even keys to their identification. These have gradually become less necessary as inexpensive and accurate national Floras and other

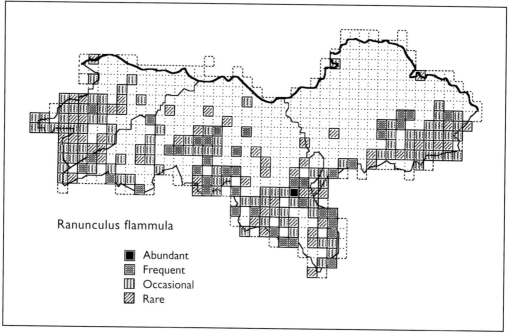

Figure 12.4 *Map showing the distribution of* Ranunculus flammula *in the Lothians. The key shows the type of shading used to indicate distribution density in all the maps in the Flora.*

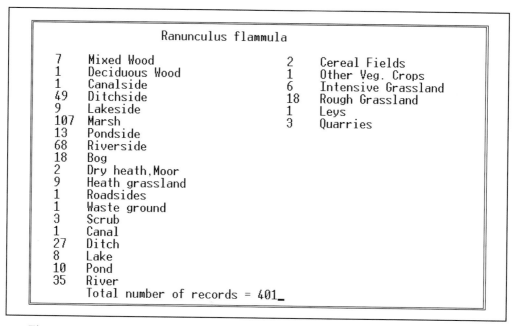

Figure 12.5 *The database list of the habitats in which* Ranunculus flammula *was recorded during the survey.*

manuals, often very well illustrated, have become available. This account of the survey omits such material, with a great saving of space. Historical information, such as the date of the first record of every species and the collector, is also routinely omitted to save space. Where such information seems significant it is included.

Maps

In the Flora of the Lothians, presented in the following pages, the maps are those resulting from the survey work of the project. A map was produced for each species recorded, though not all are included in the text. The text is based on the information from the survey supplemented by information supplied by the Vice-county Recorders whose records include areas not covered in the survey. Thus, some of the locations mentioned in the text do not appear on the maps. The squares on the map represent tetrads and are shaded to represent the distribution density of the species as estimated in the 1km x 1km square recorded in that tetrad (Figure 12.4). In general, maps showing very dense or very sparse distribution have not been included in the Flora unless they show distribution patterns that might be related to topography or land use.

Taxa Included

The text is more broadly based than the maps and all taxa recorded in the Lothians are included whether or not they were recorded in the survey. In the case of rare taxa, sites are usually identified by name as they do not necessarily appear on the maps. Pre-1981 records are in brackets where there is a later record in at least one of the vice-counties. Where there is no post-1981 record anywhere in the Lothians, an entry is included in small print at the end of the genus. The historical records in the text are records held by the Vice-county Recorders or those in published Floras. The terms 'pre-1894', 'pre-1927' and 'pre-1934', refer to Sonntag (1894), Martin, (1927), and Martin (1934), respectively. Other texts from which historical records were obtained include Balfour and Sadler (1863) and Greville (1824).

Nomenclature

The scientific nomenclature used for species, subspecies and hybrids is generally that followed in *New Flora of the British Isles* (Second Edition), (Stace 1997). It was not followed in the case of the genus *Bromus*, where the limits of the genus are controversial. Thus the genera *Anisantha*, *Bromopsis* and *Ceratochloa* are here recorded as sections of, and so incorporated within, *Bromus*. In some cases, often of apomictically reproducing plants, it was recognised that accurate identification of species or microspecies could not in general be expected, and so identification had to be more broadly based. For example, dandelions were generally recorded as '*Taraxacum officinale*' agg., that is, an aggregate species in both the scientific and folk senses. Recorders were not encouraged to send in their own identifications of segregate species within such aggregates. As many collected specimens as possible were authoritatively identified and added to the record of authenticated specimens maintained by the Vice-county Recorders. In some cases, where a large number of taxa are contained within these aggregate species, as in '*Taraxacum officinale*', *Hieracium* and *Rubus*, the Flora entry classifies the separate taxa in sections or subsections.

In other cases, there is a genuine taxonomic controversy about how broad the concept of a particular species may be. It may be recognised *sensu lato* (in a wide sense) or *sensu stricto* (in a narrow sense). This is not the same situation as is covered by use of an aggregate concept. *Sensu lato* (*s.l.*) and *sensu stricto* (*s.s.*). meanings are identified in the text where they have been used. Subspecies recorded in the Lothians are included in the text. In some species, the type subspecies is absent from the record, while other subspecies are commonly found.

Changes in nomenclature adopted since 1997 have not been used because it was felt that it was important for our readers to be able to refer to a single work. Well aware of the confusion and frustration that name changes can engender, we have included a list of pre-1997 synonyms at the end of the book. Information about name changes since 1997 may be found in the BSBI Database (Leicester). Common names in the Flora are those currently used by botanists working in the Lothians.

Native and Introduced Species

The terms 'introduced' and 'native' are used to describe the origin of a taxon in the Lothians. It is interesting to note how often this differs from the origin of the same taxon in other parts of the British Isles, for example, *Aconitum napellus* (Monkshood) and *Fagus sylvatica* (Beech) are not considered to be native in the Lothians, but native populations of these species occur in other parts of Great Britain. Our definition of 'native plant' includes those plants that have been known in the area over the period of recorded history and that are not known to have been introduced by humans since they became ecologically more significant than other animals. The native element of the British flora was largely defined by the severance of the land bridge between Great Britain and continental Europe about 8,300 years ago. Its content is confirmed by sub-fossil records and pollen deposits of known age. Most pre-Roman arrivals are effectively native plants. In the absence of written or other historical evidence we cannot know, locally, when introductions were made. Evidence of introductions being 'archaeophytes' or (curiously) 'neophytes' (Dahl 1998) has no local basis – and this is a work about local botany.

'Introductions' are mainly plants known to have been introduced, consciously or unconsciously, by humans: some of these plants are crop or garden escapes, some are relics of herb gardens. Casuals, such as *Hordeum jubatum*, are a special case of introduction: they are occasional, not persisting, and are here because of repeated introduction only. Not quite asylum-seekers, they are perhaps the van of an army of potential future colonists, which may become established as the climate changes. Introductions that are persistent and spreading are described as well naturalised, while those that are persistent but not spreading are described as established. A good example of the former is *Allium paradoxum* (Few-flowered Garlic) which has shown itself to be extremely invasive. An example of the latter is *Maianthemum bifolium* (May Lily) which has been known from the same site since 1868, but has never spread. Some plants such as *Hyoscyamus niger* (Henbane) are more difficult to classify. At one site it has become naturalised but elsewhere it behaves as a casual.

Status

The 'status' of the taxon refers to its distribution in the whole of the Lothians area. In almost all cases, one of five terms (rare, scarce, local, widespread, common) is used to describe the status of the taxon. 'Rare' is used when three or fewer filled squares appear on the map, and 'scarce' is used when four to ten filled squares appear on the map. 'Widespread' is used when 40 per cent or more of the squares are filled, the distribution of the squares throughout the whole of the Lothians being more or less uniform and the taxon being seldom if ever recorded as abundant. 'Local' is used to describe distributions intermediate between 'scarce' and 'widespread'. The term 'local' covers a wide range of distribution densities. At the lower end, the taxon can almost be considered as 'scarce', while at the higher end 'local' is used in preference to 'widespread' when the distribution of the taxon is markedly non-uniform. 'Common' is used when more than 70 per cent of the squares are filled and at least five squares are recorded as abundant and twenty squares are recorded as frequent. The terminology used to describe the distribution of a taxon within a vice-county is more loosely based and not tied to the survey, for Vice-county Recorders had information about other sites, only one quarter of the total area having been surveyed in the project. 'Rare' or 'scarce' in this context means something between five and ten known sites, depending upon the vice-county. Where a taxon is regarded as rare or scarce in a particular vice-county, a list of sites is usually included. The list may not be completely comprehensive, but it gives some indication of where the taxon may be found. More importantly, if the taxon is found at some other site, the Vice-county Recorders would like to be informed! Readers are reminded that the locations mentioned in the text do not necessarily appear on the distribution maps because the maps record only the data collected in the survey.

Habitats

The habitats in which a taxon is found are listed in the text in rank order as recorded in the project. Modifications such as 'damp woodland' or 'acid soils' were added on the advice of the Vice-county Recorders. In the survey, woodland habitats were classified as conifer, deciduous or mixed and the database has records for each taxon in each woodland type. To avoid repetition in the Flora text, the term 'woodland' has been used except where there was a characteristic distribution of the taxon in one of the three woodland types. It should be noted that the number of records for conifer woodlands was very much lower than that for the other two types of woodland. The terms 'lake' and 'lakeside' include reservoirs and small lakes but not ponds. Place names in the Lothians use 'loch' instead of 'lake' and so the term loch has been used throughout the text. Much of the land in the Lothians is or was in the form of large estates. These are seldom known as parks, though the term parkland is used occasionally to describe large areas of rough pasture containing widely-spaced mature trees. Woodland in estates was originally planted for either commercial or recreational purposes. The former is referred to in the text as estate woodland, the latter as policy woodland or policies. Some estates such as Vogrie and Dalkeith now contain areas designated as Country Parks and as such are open to the public. The term 'public parks' in the habitat list refers only to parks created in urban areas and in villages.

Interpretation of Data

The Botany of the Lothians project was a very ambitious one and it achieved many of its aims. It involved about 300 people in field botany, surveyed 98 per cent of the tetrads in the Lothians and produced a substantial body of data. The grid-based sampling system meant that some well-known rarities in botanical 'hot spots' were not recorded in the survey, but they are all included in the text of the Flora. It also meant that many previously unexplored areas were surveyed, producing rewards and surprises – at one level the realisation that *Fallopia convolvulus* (Black Bindweed) was much more widespread than previously believed, at another, the discovery of *Epilobium alsinifolium* (Chickweed Willowherb) in the Lammermuir Hills.

Many recorders involved in the project were, or became, quite experienced field botanists, but inevitably there was considerable variation in knowledge, diligence, observational skills and physical fitness. Recorders received full instructions and plenty of support was available, but strict training and monitoring regimes for recorders were considered to be inappropriate. The variation between recorders gave rise to some variation in (a) the ratio of the number of species recorded to the number of species actually present in the square, (b) the accuracy with which species were identified, (c) the interpretation of the habitat classifications, and (d) the interpretation of the terms used to record the distribution density of the species. While the grosser errors were picked up by Vice-county Recorders, some less obvious errors inevitably slipped through. Errors of omission are hard to quantify. Even the most experienced field botanists fail to record all the species present in an area (Rich and Smith 1996), and squares differ greatly in their habitat diversity and thus in species content. Re-visiting of squares in which recording had been obviously deficient was undertaken, but it is inevitable that some species are under-recorded. Where this is suspected it is noted in the text. Therefore, while it is reasonable to conclude from the presence of a filled square on the map that a given species is present in the area represented, it is not valid to conclude that the absence of shading in a particular square on the map indicates the absence of the species. The species may have been present in the square but the recorder failed to record it, or the species may have been absent from the surveyed square but present in one of the other three 1km x 1km squares in the tetrad.

Density of Distribution

The aspect of the survey work that gave rise to the most unease among recorders and Vice-county Recorders was the estimation of the density of distribution of a taxon within a square. Recorders were asked to classify the distribution density as 'rare', 'occasional', 'frequent' or 'abundant'. The first and last of these terms are true measures of density, but the middle two terms, though well used by field biologists, refer to events in time and not in place. Botanists walking through their squares will describe the distribution density of a taxon as 'occasional' if they occasionally come across that taxon. In a square of uniform habitat it is the taxon that is sparsely distributed, whereas in a square composed of many patches of a number of habitats, the taxon may be abundant in a particular habitat, but the habitat is sparsely distributed in the square. Recorders were asked to record the latter type of situation using the terms locally frequent or locally abundant but they were not asked to estimate the proportion of the area of their square occupied by each type of habitat. The representation on the maps therefore indicates the density of species in the habitats in which they grow. In some cases it indicates their density over the whole one kilometre square. Despite the fact that all estimations were subjective and there seems to have been a tendency among recorders to over-estimate plant density, useful data were obtained and useful lessons were learnt.

Habitats

Variation in the recorders' interpretations of habitat types was difficult to detect or quantify. Advice was given to recorders on the meaning of the terms used to define the habitats and obvious errors such as recording 'bog' instead of 'marsh' were sorted out. However, it is unlikely that complete consistency was achieved in the distinction between roadside and hedgerow, roadside and ditch, or roadside and rough grassland. Recorders did not always report accurately whether a plant was in or merely beside a pond, river or ditch and in any case how close to the edge of a pond does a plant have to be to be considered as being 'pondside'? Other habitats likely to have been confused are woodland and riverside, since most semi-natural woodland in the Lothians is beside rivers. Many roads also run past woodland, and the Union Canal intersects a wide variety of habitats. The terms 'river' and 'riverside' covered all flowing water and did not allow the recorder, and hence do not enable the reader, to distinguish between a river and a stream or burn. There was no term for an upland flush. This may account for the apparent presence of *Ranunculus flammula* on 'dry heather moorland' in the example in Figure 12.5.

The distinction between 'moorland' and 'heath grassland' was blurred in the minds of many recorders, and sometimes also on the ground. The term 'rough grassland' was widely used for habitats ranging from roadside verges and coarse grasses on wasteland in lowland areas to unimproved pasture and heath grassland in upland areas. Habitat types not on the recorders' list included tracks and footpaths, old railway lines such as those converted to cycle tracks, and old industrial sites and bings that had been restored. Most of the habitats recorded for a given species were in keeping with the known ecology of that species, but outliers in the habitat lists occurred and had to be interpreted with care. Returning to the *Ranunculus flammula* example, the 'mixed woodland' habitat probably implies a stream, the 'quarry', a pond, the 'rough grassland', a marsh, and the 'cereal field', a ditch. Habitat information is further discussed in Chapters 16 and 17.

13

Flora of the Lothians

EDITED BY
J. MUSCOTT (VC 84 West Lothian)
D. R. McKEAN (VC 83 Midlothian)
E. H. JACKSON (VC 82 East Lothian)

Text compiled by M. P. Cochrane and P. M. Smith
Maps produced by R. O. D. Dixon

The key to the maps in the Flora can be found in Figure 12.4 page 193.
For an explanation of the format and terms used in the writing of the Flora,
please refer to Chapter 12.

LYCOPODIACEAE

Huperzia Bernh.

H. selago (L.) Bernh. ex Schrank & Mart.
Fir Clubmoss
Native. Status: rare.
Moorland, upland heath grassland.

WL: rare, a single plant on Foulshiels Bing, 1998;
(Knock Hill, 1903).
ML: rare, Moorfoot Hills; (Tynehead and
Balerno, pre-1934).
EL: rare, Linn Dean; Traprain Law; (Aberlady
Bay, 1963; Faseny Water, 1972).

Lycopodiella Holub

L. inundata (L.) Holub   Marsh Clubmoss
ML: Glencorse parish, 1845.

Lycopodium L.

L. clavatum L.   Stag's-horn Clubmoss
Native. Status: scarce.
Upland heath grassland, moor, conifer woodland, quarries, slag and shale bings.

WL: local, very locally frequent on Foulshiels and
Faucheldean Bings; scattered records from near
Blackridge; Fauldhouse and Polkemmet Moor;
(Carmelhill and Carriden, 1971).
ML: scarce, West Calder; Penicuik; Pentland
Hills; Gladsmuir; Bavelaw; Dundreich Plateau.
EL: rare, Humbie; Hopes area; Pressmennan;
Hedderwick; Monynut Water; (East Saltoun,
1966).

Diphasiastrum Holub

D. alpinum (L.) Holub   Alpine Clubmoss
Native. Status: rare.
Damp upland grassland, shale bings, forestry clearings.

WL: rare, Faucheldean Bing; (Broxburn, 1974).
ML: rare, Moorfoot Hills, 1995; near Penicuik,
1994; (Black Hill, 1863).

SELAGINELLACEAE

Selaginella P. Beauv.

S. selaginoides (L.) P. Beauv.   Lesser Clubmoss
Native. Status: rare.

Moist, mainly upland habitats.

ML: rare, Moorfoot and Pentland Hills.
EL: rare, coastal at Aberlady Bay, Yellowcraig and Belhaven Bay; inland at Pressmennan; upland at Linn Dean; Papana Water; Killmade Burn; Yearn Hope.

S. kraussiana (Kunze) A. Braun
Mossy Clubmoss, Krauss's Clubmoss
Introduced. Status: local.
Gardens.

ML: local, a horticultural weed in the New Town.
EL: (Biel Estate, 1957).

EQUISETACEAE

Equisetum L.

E. hyemale L. Rough Horsetail
Native. Status: rare.
River banks.

WL: (Carriden and Hopetoun, pre-1934).
ML: rare, Roslin.

E. variegatum Schleich. ex F. Weber & D. Mohr
Variegated Horsetail
Native. Status: rare.
Dune-slacks and marshes.

EL: thriving at Aberlady Bay; rare at Yellowcraig.

E. fluviatile L. Water Horsetail
Native. Status: widespread.
Marshes, in and beside rivers, ditches and ponds.

WL: widespread, especially in S. and W.
ML, EL: quite widespread in upland areas, locally frequent, occasional elsewhere.

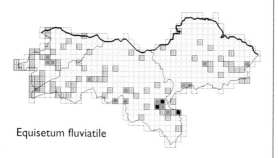

Equisetum fluviatile

E. x dycei C. N. Page (*E. fluviatile* x *E. palustre*)
Native. Status: rare.
Dune-slacks, beside water.

ML: rare, Harperrig Reservoir.
EL: rare, Aberlady Bay, 1987.

E. x litorale Kühlew. ex Rupr. (*E. fluviatile* x *E. arvense*) Shore Horsetail
Native. Status: scarce, probably overlooked.
Beside rivers and ditches.

ML: locally frequent, Carlops; Vogrie; Penicuik; Hillend; Addiewell; Loanhead; Arthur's Seat.
EL: rare, Aberlady; Linn Dean; (Saltoun, 1980).

E. arvense L.
Common Horsetail, Field Horsetail
Native. Status: common.
Roadsides, waste ground, beside rivers and ditches, woodland, railway embankments, arable fields.

WL, ML, EL: common.

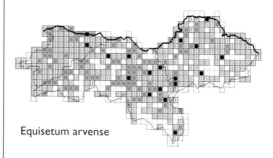

Equisetum arvense

E. x rothmaleri C. N. Page (*E. arvense* x *E. palustre*)
Native. Status: rare.
Dune-slacks, upland marsh.

ML: rare, east of Mount Main.
EL: rare, Aberlady Bay, 1987.

E. pratense Ehrh. Shady Horsetail
Native. Status: rare.
Beside rivers and ditches, usually in upland areas.

WL: (Carribber, pre-1934).
ML: rare, Penicuik Estate; Moorfoot Hills; (Mid Calder, pre-1927).

E. x mildeanum Rothm. (*E. pratense* x *E. sylvaticum*)
Native. Status: rare.
Wet woodland.

ML: rare, Penicuik Estate.

E. sylvaticum L.   Wood Horsetail
Native. Status: local.
Waterside habitats, wet woodland and grassland.

WL: quite widespread, mainly in upland areas.
ML: local, mainly in upland areas.
EL: scarce, Linn Dean; Humbie; Gosford House;
Stobshiel Reservoir; Danskine; Papana Water;
Blaik Law; Sheeppath Burn; Dod Hill.

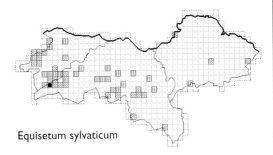

Equisetum sylvaticum

E. palustre L.   Marsh Horsetail
Native. Status: widespread, especially in S.
Marshes, waterside habitats, wet grassland.

WL: quite widespread, especially in S. and W.
ML: common in upland areas, occasional elsewhere.
EL: quite widespread in upland areas, occasional
elsewhere.

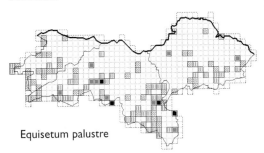

Equisetum palustre

E. telmateia Ehrh.   Great Horsetail
Native. Status: scarce.
Waterside habitats, marshes, damp woods.

WL: rare, Hopetoun.
ML: local, Roslin; Edgelaw; Carrington;
Mavisbank, Loanhead.
EL: rare, Keith Water; Humbie; Danskine Loch;
Papana Water; Dunglass Dean; (Oldhamstocks
Water, 1968).

OPHIOGLOSSACEAE

Ophioglossum L.

O. vulgatum L.   Adder's-tongue
Native. Status: local.
Scrub, damp grassland.

WL: rare, Birkhill; (Knock Hill, 1897; Lochcote,
1901; Carlowrie, Hopetoun, Dalmeny and
Linlithgow, pre-1934).
ML: local, Kirknewton; Linhouse Glen;
Murieston; locally abundant on Arthur's Seat;
(Dalhousie and Arniston, 1863).
EL: rare, Aberlady Bay, 2000; Tyninghame;
(Dunglass, pre-1934).

Botrychium Sw.

B. lunaria (L.) Sw.   Moonwort
Native. Status: rare, population sizes fluctuate.
Conifer woodland, short turf, dunes, upland heath.

WL: rare, Easter Inch Moss; Faucheldean Bing;
Dalmeny.
ML: rare, Bonaly; Middleton; Cobbinshaw; West
Calder; North Esk Reservoir.
EL: rare, Linn Dean; Aberlady Bay; Gullane;
Yellowcraig; Belhaven Bay; (Canty Bay, 1933;
two inland sites, pre-1962).

OSMUNDACEAE

Osmunda L.

O. regalis L.   Royal Fern
Introduced. Status: rare, often planted.
Damp woodland in policies and country parks.

ML: rare, Beeslack; (Hermitage of Braid, 1968).
EL: (Pencaitland, pre-1894).

ADIANTACEAE

Cryptogramma R. Br.

C. crispa (L.) R. Br. ex Hook.   Parsley Fern
Native. Status: rare.
Acid rocks and scree.

ML: rare, Pentland and Moorfoot Hills.

MARSILEACEAE

Pilularia L.

P. globulifera L.   Pillwort
Native. Status: rare.
Marshes.

WL: attempted re-introduction at Newliston, 1998; (Philpstoun, 1874; Winchburgh, 1885; Drumshoreland, 1890; Broxburn, pre-1934).
ML: rare, re-introduced at Bawsinch, 1997, and at Meadows Yard Local Nature Reserve, 1998; (Threipmuir, Logan Cottage and Braid Hills, pre-1863).

HYMENOPHYLLACEAE

Hymenophyllum Sm.

H. wilsonii Hook.   Wilson's Filmy Fern
Native. Status: rare.
Wet, shady, rocky gorge.

EL: rare, Linn Dean.

POLYPODIACEAE

Polypodium L.

P. vulgare agg.   Polypody
Native. Status: local.
Walls, quarries, moorland, heath grassland, woodland.

WL: scarce.
ML: quite widespread.
EL: scattered distribution, mainly in lowland areas.

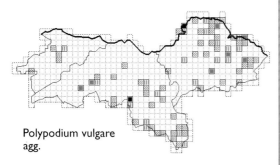

Polypodium vulgare
agg.

Includes the following three taxa that were not always distinguished in the survey.

P. vulgare L. *s.s.*   Common Polypody
Native. Status: local.
Walls, quarries, moorland, heath grassland, woodland.

WL: scarce, Linlithgow; Hopetoun and Dundas Estates; Wester Ochiltree.
ML: quite widespread.
EL: scattered distribution, rarely specified in the survey.

P. x mantoniae Rothm. & U. Schneid. (*P. vulgare* x *P. interjectum*)
Native. Status: rare, probably under-recorded.
Heath grassland, dunes.

ML: rare, Pentland Hills.
EL: (Traprain Law, 1972; Gullane, 1973).

P. interjectum Shivas   Intermediate Polypody
Native. Status: scarce, probably under-recorded.
Walls, quarries, woodland and moorland.

WL: rare, Bathgate; Binny Craig.
ML: scattered, perhaps under-recorded, Penicuik; Roslin; Blackford Hill; Newbattle; Inveresk; Cramond Bridge; Torphin.
EL: scarce and scattered, mainly in sheltered sites, Hailes Castle; Woodhall Dean; Dunglass Dean.

P. cambricum L.   Southern Polypody
Native. Status: rare.

ML: rare, only in Holyrood Park where it is locally frequent on base-rich rocks.

DENNSTAEDTIACEAE

Pteridium Gled. ex Scop.

P. aquilinum (L.) Kuhn   Bracken
Native. Status: common.
Woodland, heath grassland, moorland, scrub, rough grassland, beside roads and rivers.

WL: widespread in upland areas in W., scarcer elsewhere.
ML: common, especially in upland areas.
EL: common in upland areas, occasional in arable areas and along the coast.

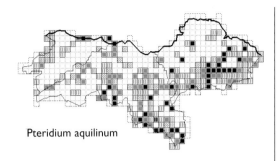

Pteridium aquilinum

THELYPTERIDACEAE

Phegopteris (C. Presl) Fée

P. connectilis (Michx.) Watt   Beech Fern
Native. Status: rare.
Woods, heath grassland, scrub, rocky outcrops.

WL: (Cockleroy, 1974; Lochcote, 1977).
ML: rare, Auchendinny; Roslin; Moorfoot Hills.
EL: rare, Linn Dean; cleughs below Lothian Edge;
Burn Hope; (Sheeppath Dean, 1908; Traprain
Law, pre-1934; Pressmennan, 1966; Bladdering
Cleugh, 1967).

Oreopteris Holub

O. limbosperma (Bellardi ex All.) Holub
          Lemon-scented Fern, Mountain Fern
Native. Status: local.
Heath grassland, moorland, upland streamsides, damp
woodland.

WL: local, mainly in wet, acid areas in SW.
ML: widespread and locally frequent in SE., local
elsewhere.
EL: widespread in the Lammermuir Hills; rare in
lowland areas, Saltoun Forest; Gladsmuir.

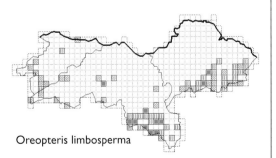

Oreopteris limbosperma

ASPLENIACEAE

Phyllitis Hill

P. scolopendrium (L.) Newman
          Hart's-tongue Fern
Native. Status: local, and spreading.
Damp walls, including wells, quarries, on rocks in
woodland.

WL: local, mainly in N.
ML: local, mainly in lowland areas.
EL: quite widespread along the coast, local
inland, mainly in lowland areas.

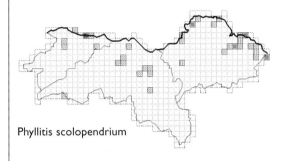

Phyllitis scolopendrium

Asplenium L.

A. adiantum-nigrum L.   Black Spleenwort
Native. Status: local.
Walls, rocky outcrops, quarries, shale banks.

WL: rare, by Mains Burn near Ochiltree Mill;
Hopetoun; (Knock Hill and east of Lochcote
Reservoir, 1977; Carribber Glen, 1978; Hiltly
Crags and Parkley Craigs, 1982).
ML: widespread, under-recorded in this survey.
EL: local, mainly in NE., Gullane; Yellowcraig;
Hailes; Stenton; Innerwick; (ravines along
Monynut Water, 1970).

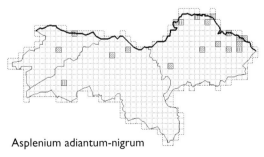

Asplenium adiantum-nigrum

A. marinum L.   Sea Spleenwort
Native. Status: rare.
Rocks and walls by the sea.

EL: rare, old ruins, Bass Rock, 1985.

A. trichomanes L. *s.l.*   Maidenhair Spleenwort
Native: Status: local, under-recorded.
Walls, quarries, upland ravines.

WL: local, mainly in N., locally frequent along
the coast.
ML: scattered.
EL: widely scattered, quite widespread near the
north-east coast, occasional in upland areas.

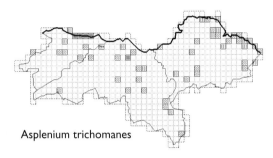

Asplenium trichomanes

ssp. trichomanes
Lime-free rocks.

ML: rare, Calton Road Crags; Blackford Hill;
Loganlea Reservoir.

ssp. quadrivalens D. E. Mey.
Old walls, lime-rich rocks, quarries, upland ravines.

WL: local.
ML: common, especially in the city.
EL: the common ssp., but few recent records
specify it.

A. ruta-muraria L.   Wall-rue
Native. Status: local.
Almost wholly on walls, occasionally in quarries.

WL: local, locally frequent in N.
ML: widespread, sometimes frequent on old
walls; on rocks on Arthur's Seat and Blackford
Hill.
EL: rare in upland areas, quite widespread
elsewhere, especially in coastal areas.

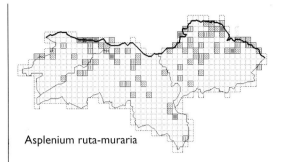

Asplenium ruta-muraria

A. x murbeckii Dörfl. (*A. ruta-muraria* x *A.
septentrionale*)
Native. Status: rare, with parents.
On basaltic rock.

ML: rare, Arthur's Seat.

A. septentrionale (L.) Hoffm.
                              Forked Spleenwort
Native. Status: rare.
On basaltic rock.

ML: rare, Blackford Hill; colonies on rocks on
Arthur's Seat.

A. viride Huds.   Green Spleenwort
ML: lime-rich rocks, Medwin, 1871.

Ceterach Willd.

C. officinarum Willd.   Rustyback
Native. Status: rare.
Old walls.

WL: (wall of factor's garden, Abercorn, destroyed
by repointing, 1973.)
ML: rare, Hallheriot, 1989.
EL: rare, Garleton; garden walls in Gullane;
Spott; Woodhall Dean.

WOODSIACEAE

Athyrium Roth

A. filix-femina (L.) Roth   Lady Fern
Native. Status: widespread.
Wet woodland, waterside habitats, damp roadsides.

WL: widespread, locally frequent.
ML, EL: local in N., widespread and locally
frequent elsewhere.

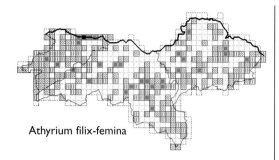

Athyrium filix-femina

Gymnocarpium Newman

G. dryopteris (L.) Newman   Oak Fern
Native. Status: scarce.
Heath grassland, quarries, shaded areas in woodland.

WL: rare, Philpstoun; (Woodcockdale, 1863;
Bowden Hill, 1867).
ML: scarce, Moorfoot Hills; Carberry; Roslin;
Stoneyburn; Auchendinny; Arniston; Bilston
Burn.
EL: rare, Hope Hills; cleughs below Lothian
Edge; Woodhall Dean; deans above
Oldhamstocks.

Cystopteris Bernh.

C. fragilis (L.) Bernh.   Brittle Bladder Fern
Native. Status: scarce.
Walls, quarries, shale banks, upland ravines.

WL: local.
ML: local, mainly in S.
EL: scarce, Linn Dean; Traprain Law; Woodhall
Dean; Aikengall Water area.

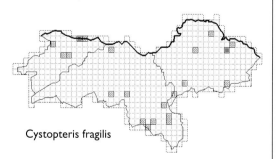

Cystopteris fragilis

DRYOPTERIDACEAE

Polystichum Roth

P. setiferum (Forssk.) T. Moore ex Woyn.
          Soft Shield Fern

Native. Status: rare.
Woodland, ravines.

WL: rare, beside canal near Craigton; Dalmeny
Village; Cockleroy.
ML: rare, Balerno; Dreghorn; Roslin; Penicuik.
EL: rare, woodland west of Yellowcraig, 2000;
(Dunglass Dean, 1834; Yester Estate, 1957).

P. aculeatum (L.) Roth   Hard Shield Fern
Native. Status: local.
Rocks in woodland, ravines, steep river banks.

WL: scattered.
ML: local.
EL: occasional, mainly inland, Humbie; Garleton
Hills; Garvald; Bladdering Cleugh; Bilsdean.

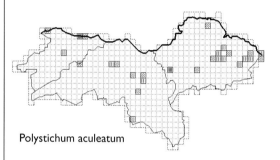

Polystichum aculeatum

P. x bicknellii (H. Christ) Hahne (*P. setiferum* x
P. aculeatum)
Native. Status: rare.
Beside streams.

WL: rare, Birdsmill; west of Port Edgar.

Dryopteris Adans.

D. oreades Fomin   Mountain Male Fern
Native. Status: rare.
Upland scree.

ML: rare, Stow; Moorfoot Hills.

D. filix-mas (L.) Schott   Male Fern
Native. Status: common.
Woodland, scrub, beside roads, rivers and ditches,
walls.

WL, ML, EL: common.

D. affinis complex   Scaly Male Fern
Native. Status: local.
Woodland, beside rivers and roads, heath grassland,
scrub.

WL: widespread, seldom locally frequent.
ML: local, locally frequent, seemingly absent from a corridor from Edinburgh to the SW.
EL: widespread, locally frequent.

The segregates of this species are too imperfectly known to be included separately.

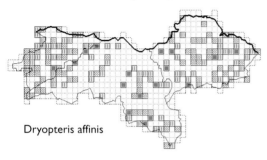

Dryopteris affinis

D. carthusiana (Vill.) H. P. Fuchs

Narrow Buckler Fern

Native. Status: local.
Woodland, bogs and marshes, waterside habitats.

WL: local, mainly in SW.
ML: local, West Calder; Penicuik; Roslin; Red Moss; Stow; Gladhouse Reservoir; Woodmuir.
EL: scarce, mainly in S., Gifford; Luggate Burn; a few sites in the Lammermuir Hills; (Humbie, 1975; Pressmennan, 1976).

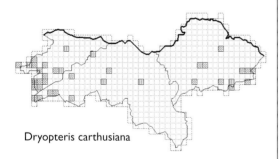

Dryopteris carthusiana

D. x deweveri (J. T. Jansen) Wacht. (*D. carthusiana* x *D. dilatata*)
Native. Status: rare.
Woodland, waterside habitats.

WL: rare, north of Drumcross, 1983; probably under-recorded.
ML: rare, Hermand; West Calder; Stow; (Balerno, 1902).
EL: rare, near New Winton, 1988.

D. dilatata (Hoffm.) A. Gray

Broad Buckler Fern

Native. Status: common.
Woodland, scrub, waterside habitats, roadsides, heath grassland, moorland.

WL: common.
ML: common in S., local in N.
EL: local in NW., common elsewhere.

BLECHNACEAE

Blechnum L.

B. spicant (L.) Roth   Hard Fern
Native. Status: quite widespread.
Dry heath grassland, moorland, woodland, waterside habitats.

WL: widespread in S. and W., rare elsewhere.
ML: widespread in most upland areas, occasional in lowland areas.
EL: widespread and locally frequent in upland areas, occasional elsewhere.

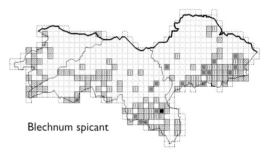

Blechnum spicant

AZOLLACEAE

Azolla Lam.

A. filiculoides Lam.   Water Fern
Introduced. Status: rare.
Ponds, canals.

WL: rare, appeared in Kinneil Lake and in the Union Canal in 1992. Seems to have disappeared from Kinneil Lake and to be declining in the canal.
ML: rare, Bawsinch, 1981, not there in 1983 having been destroyed by the exceptionally cold winter 1981/82.

PINACEAE

Abies Mill.

A. alba Mill.   European Silver Fir
Introduced. Status: rare, planted.
Estate woodland, public amenity areas.

ML: rare, Stow.
EL: rare, Gosford Estate.

A. concolor (Gord.) Hildebrand   Colorado Fir
Introduced. Status: rare, planted.
Policies.

WL: rare, Hopetoun.

A. grandis (Douglas ex D. Don) Lindl.   Giant Fir
Introduced. Status: rare, planted.
Coniferous and mixed woodland, estate woodland.

WL: local.
EL: scarce, Butterdean Wood; Humbie;
Clerkington; Yester; Bara; Tyninghame;
Hedderwick; Monynut Water.

A. procera Rehder   Noble Fir
Introduced. Status: local.
Coniferous and mixed woodland, estate woodland.

WL: scarce.
ML: local plantings.
EL: local plantings, Butterdean Wood; Humbie;
Yester; Markle; Monynut Water; Philip Burn;
Stottencleugh.

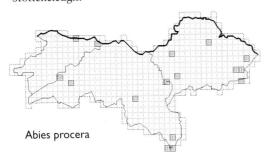

Abies procera

A. delavayi Franch   Delavay's Silver Fir
Introduced. Status: rare, planted.
Policy woodlands.

WL: rare, Hopetoun.

Pseudotsuga Carrière

P. menziesii (Mirb.) Franco   Douglas Fir

Introduced. Status: local.
Mixed and coniferous woodland in estates and
plantations, occasionally in public parks.

WL: local, poorly recorded in survey.
ML: local, Stow; Edgelaw; Dalkeith and Cammo
Country Parks.
EL: local, Gosford; Humbie area; Yester;
Morham; Whittingehame; Monynut Water;
Dunglass.

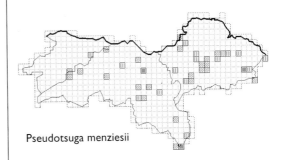

Pseudotsuga menziesii

Tsuga (Antoine) Carrière

T. heterophylla (Raf.) Sarg.   Western Hemlock
Introduced. Status: local, planted.
Coniferous and mixed woodland in estates and
plantations.

WL: quite widespread, poorly recorded in survey.
ML: local, mainly in centre, Hermitage of Braid;
Dalkeith Country Park; Roslin Glen;
Corstorphine Hill.
EL: local, mainly in centre, Pencaitland;
Butterdean Wood; Saltoun; Humbie; Yellowcraig;
Lennoxlove; Yester; Whittingehame.

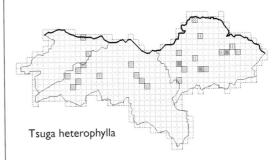

Tsuga heterophylla

Picea A. Dietr.

P. sitchensis (Bong.) Carrière   Sitka Spruce
Introduced. Status: widespread, planted and
sometimes self-set.

Coniferous and mixed plantation woodland, scrub, roadsides.

WL: common.
ML: local in N., widespread but patchy in S., locally abundant.
EL: rare in N., locally frequent in W., locally abundant in SE.

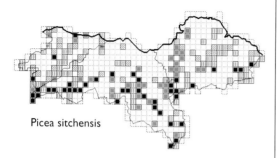

Picea sitchensis

P. jezoensis (Sieb. & Zucc.) Carrière
var. **hondoensis** (Mayr) Rehd.   Hondo Spruce
Introduced. Status: rare.
Estate woodland.

WL: rare, Hopetoun.

P. abies (L.) H. Karst.   Norway Spruce
Introduced. Status: widespread, planted.
Mixed and coniferous plantation woodland, gardens.

WL, ML, EL: widespread but patchy, locally abundant.

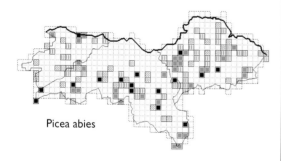

Picea abies

P. smithiana (Wall.) Boiss.
                    West Himalayan Spruce
Introduced. Status: rare, planted.
Policy woodland.

WL: rare, Hopetoun, where it was first introduced to Britain.

Larix Mill.

L. decidua agg.   Larch
Introduced. Status: widespread, much planted, sometimes self-set.
Mixed and coniferous plantation and estate woodland, scrub.

WL, ML, EL: widespread, locally frequent.

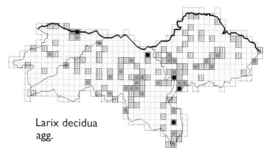

Larix decidua
agg.

Includes the following three taxa that were not always distinguished in the survey.

L. decidua Mill. *s.s.*   European Larch
Introduced. Status: widespread, planted, sometimes self-set.
Plantation and estate woodlands.

WL, ML, EL: widespread but patchy.

L. x marschlinsii Coaz (*L. decidua* x *L. kaempferi*)
                    Hybrid Larch
Introduced and native. Status: local, much planted for commercial forestry, occasionally spontaneous in scrub and near roads.
Mixed and coniferous commercial woodland.

WL: widespread.
ML: scattered distribution, mainly in S., locally abundant.
EL: scattered distribution, locally frequent.

L. kaempferi (Lindl.) Carrière   Japanese Larch
Introduced. Status: local, planted.
Plantations and estate woodland, public parks.

WL, ML, EL: scattered distribution, locally frequent.

Cedrus Trew

C. deodara (Roxb. ex D. Don) G. Don   Deodar
Introduced. Status: rare, planted.
Policy woodland.

WL: rare, Hopetoun.
ML: rare, Cammo Country Park and probably elsewhere.

C. libani A. Rich.   Cedar of Lebanon
Introduced. Status: rare, planted.
Policy woodland.

WL: rare, Hopetoun; Polkemmet Country Park.
ML: rare, Dalkeith and Cammo Country Parks; Craigcrook Castle.
EL: rare, Seton House; Bowerhouse; Dunglass.

C. atlantica (Endl.) Carrière   Atlas Cedar
Introduced. Status: rare, planted.
Public parks, gardens, policy woodland.

WL: rare, Hopetoun.
ML: local, Dalkeith and Cammo Country Parks.
EL: rare, Yester House grounds.

Pinus L.

P. sylvestris L.   Scots Pine
Native and introduced. Status: common, usually planted.
Coniferous and mixed woodland, scrub, shelterbelts, roadsides, sometimes self-set.

WL: very widespread.
ML, EL: common.

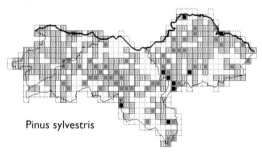

Pinus sylvestris

P. nigra J. F. Arnold   Black Pine
Introduced. Status: local, planted.
Policy woodland, scrub, public parks, often self-set.

WL: widespread.
ML: local.
EL: local, Fountainhall; Yellowcraig; Elmscleugh.

ssp. nigra   Austrian Pine
ML: local, Cammo Country Park.

ssp. laricio Maire   Corsican Pine
ML: rare, Selm Muir Wood.

EL: local, large area planted in Saltoun Wood; rare in Hedderwick Hill plantation in John Muir Country Park.

P. mugo Turra   Dwarf Mountain Pine
Introduced. Status: scarce.
Public amenity areas.

ML: scarce, roundabouts at Cameron Toll; motorway plantings.

P. contorta Douglas ex Loudon
                          Lodgepole Pine, Shore Pine
Introduced. Status: local, planted, becoming naturalised.
Coniferous and mixed woodland, scrub, roadsides.

WL: widespread in coniferous plantations, rarely in policy woodland or in NE.
ML: scattered distribution, mainly in S., locally abundant.
EL: local, Gosford; Papana Water; Pefferside; becoming naturalised at Yellowcraig.

P. pinaster Aiton   Maritime Pine
Introduced. Status: rare, planted.
Policy woodland, public amenity areas.

ML: rare, Silverknowes.
EL: rare, near Gosford.

P. strobus L.   Weymouth Pine
Introduced. Status: rare, planted.
Policy woodland.

WL: rare, Dundas Estate.

P. peuce Griseb.   Macedonian Pine
Introduced. Status: rare, planted.
Policy woodland.

WL: rare, Hopetoun.

P. pinea L.   Stone Pine
Introduced. Status: rare, planted.
Public amenity areas.

ML: rare, Edinburgh Zoo.
EL: planted along a railway embankment near Dunglass.

TAXODIACEAE

Sequoiadendron Buchholz

S. giganteum (Lindl.) Buchholz
                          Wellingtonia, Big Tree

Introduced. Status: scarce, planted.
Policy woodland, public parks.

WL: rare, Hopetoun; Lochcote; Carriden; Newliston; Almondell.
ML: rare, Colinton Dell; Cammo Country Park.
EL: rare, Pencaitland; Gosford; Huntington; Yester; Dunglass.

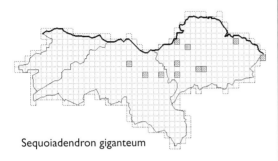

Sequoiadendron giganteum

Metasequoia Miki

M. glyptostroboides Hu & W. C. Cheng
Dawn Redwood
Introduced. Status: rare, planted.
Policy woodland.

WL: rare, Hopetoun.
ML: rare, Vogrie Country Park.
EL: rare, Whittingehame.

CUPRESSACEAE

Cupressus L.

C. macrocarpa Hartw. ex Gordon
Monterey Cypress
Introduced. Status: rare.
Policies.

EL: rare, Dunglass Estate.

x Cupressocyparis Dallim.
(*Cupressus* x *Chamaecyparis*)

x C. leylandii (A. B. Jack. & Dallim.) Dallim.
(*Cupressus macrocarpa* x *Chamaecyparis nootkatensis*)　　　Leyland Cypress
Introduced. Status: frequently planted as a hedge or for screening, rarely escaping.
Public amenity areas, gardens.

WL: few records, possibly under-recorded.
ML: self seeds near town gardens.

Chamaecyparis Spach

C. lawsoniana (A. Murray bis) Parl.
Lawson's Cypress
Introduced. Status: local, planted.
Mixed woodland, hedgerows, public parks.

WL: local.
ML: local, Hermitage of Braid; Corstorphine Hill; Cammo Country Park.
EL: local, Saltoun; Drem; Pressmennan; Dunglass.

C. obtusa (Siebold & Zucc.) Endl.
Honoki Cypress
Introduced. Status: rare, planted.
Policy woodland.

WL: rare, Hopetoun.

C. pisifera Siebold & Zucc.　　　Sawara Cypress
Introduced. Status: rare, planted.
Policy woodland, cemeteries, public parks.

WL: rare, Hopetoun.
ML: occasional.
EL: rare, in park at Yester House.

C. nootkatensis (Lamb.) Spach　　Nootka Cypress
Introduced. Status: rare, planted.
Mixed woodland.

EL: rare, Dunglass Estate.

Thuja L.

T. plicata Donn ex D. Don
Western Red Cedar
Introduced. Status: scarce, planted.
Policy and plantation woodland.

WL: scarce, Carriden; Torphichen; Hopetoun; Bangour Reservoir; woodland near South Queensferry.
ML: rare, Cammo Country Park.
EL: rare, Saltoun Wood; Lennoxlove; Yester.

Thujopsis Sieb. & Zucc.

T. dolabrata (L.f.) Sieb. & Zucc.　　　Hiba
Introduced. Status: rare, planted.
Policies.

EL: rare, Yester.

Juniperus L.

J. communis L.　　　Common Juniper
ssp. communis

Native. Status: local.
Heath grassland, moorland, woodland, scrub, rough grassland, quarries.

WL: rare, planted at Hopetoun and Kinneil; (recorded for 'Linlithgowshire', pre-1934).
ML: local, mainly in Pentland and Moorfoot Hills.
EL: local, mainly in Lammermuir Hills; introduced in dunes at Yellowcraig.

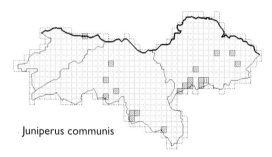

Juniperus communis

GINKGOACEAE

Ginkgo L.

G. biloba L.   Maidenhair Tree
Introduced. Status: rare, planted.
Policies.

EL: rare, one tree in the grounds of Yester House.

ARAUCARIACEAE

Araucaria Juss.

A. araucana (Molina) K. Koch   Monkey-puzzle
Introduced. Status: rare, planted.
Policy woodland, public parks.

WL: scarce, Carriden; Hopetoun; Lochcote; Dundas; Dalmeny; Polkemmet Country Park.
ML: rare, Lauriston Castle; Dalkeith, Vogrie and Cammo Country Parks; Craigcrook Castle.
EL: rare, Whittingehame; Bowerhouse; Dunglass.

TAXACEAE

Taxus L.

T. baccata L.   Yew
Introduced. Status: widespread, mostly planted, rarely naturalised.

Policy woodland, scrub, roadsides, hedgerows, gardens, public parks, graveyards.

WL, ML: widespread in lowland areas, seldom locally frequent, rare in upland areas.
EL: widespread, mainly in lowland areas; small bushes along coastal path, Gullane and in dunes at Yellowcraig.

cv. **Fastigiata** Irish Yew, was frequently planted in churchyards and has a widespread distribution in the Lothians.

NYMPHAEACEAE

Nymphaea L.

N. alba L.   White Water-lily
Introduced. Status: rare, sometimes planted.
Ponds.

WL: rare, planted in Livingston Reservoir; (Linlithgow and Newliston, pre-1934).
ML: rare, Penicuik Estate, 1983; Bawsinch, 1993; (Lochend, 1764–1824; Meadowbank and Tynehead, pre-1934).
EL: rare, Spilmersford; Gosford House; Dunglass Estate.

Nuphar Sm.

N. lutea (L.) Sm.   Yellow Water-lily
Native and introduced. Status: rare.
Lochs, ponds.

WL: rare, planted in Livingston Reservoir; (formerly in Linlithgow Loch but disappeared *c.* 1990, probably as a result of eutrophication; Craigiehall, 1963).
ML: rare, near Cobbinshaw; Mavisbank, Loanhead; (Lochend, 1764–1824; Ravelston, 1894; Elf Loch, pre-1927).
EL: (Archerfield pond and stream, 1884; Prestonpans and Dunglass, pre-1934; East Saltoun, 1957).

N. x spenneriana Gaudin (*N. lutea* x *N. pumila*)
                              Hybrid Water-lily
Introduced. Status: rare, planted.
Lochs.

ML: rare, Elf Loch, 1937 to *c.* 1990, now gone.

CERATOPHYLLACEAE

Ceratophyllum L.

C. demersum L.   Rigid Hornwort
Native. Status: rare.
Lochs and ponds rich in nutrients.

WL: rare, Bangour Reservoir; (Linlithgow Loch, 1838).
ML: rare, Duddingston and Dunsapie Lochs; (Myreside, 1764; Canonmills Loch, 1809–24; Union Canal, pre-1934).
EL: rare introduction first recorded in 1988, mainly in estate ponds, Hoprig; Gosford House; Drem Pool; Markle Quarry; Smeaton; Newbyth; Scoughall; Dunglass.

RANUNCULACEAE

Caltha L.

C. palustris L.   Marsh Marigold, Kingcup
Native. Status: widespread.
Marshes, in and beside rivers, ditches, lochs and ponds.

WL: widespread, locally frequent.
ML: occasional in N., widespread elsewhere, locally frequent in SE.
EL: rare in N. and W., widespread elsewhere, seldom locally frequent.
Small-flowered upland plants, var. **minor** DC., have been recorded from Deuchrie; Sheeppath Burn; (Hopes Water, 1977).

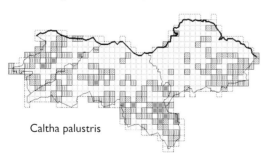

Caltha palustris

Trollius L.

T. europaeus L.   Globeflower
Native. Status: rare.
Marshes, damp hill pastures.

WL: rare, near Beecraigs Country Park;

(Carribber Glen, pre-1934).
ML: rare, Stow; Addiewell; Cockmuir; (Borthwick, Auchendinny, Crichton Castle and Nine Mile Burn, pre-1900; Penicuik, Currie and Mid Calder, pre-1934).
EL: (Garvald and Aikengall, pre-1934; near Bothwell Water, 1973).

Helleborus L.

H. foetidus L.   Stinking Hellebore
Introduced. Status: rare.
Woodland, scrub.

ML: rare, Corstorphine Hill, 2001; (Warriston and Lasswade, pre-1934).
EL: rare, West Saltoun; Balgone; Pressmennan.

H. viridis L.   Green Hellebore
Introduced. Status: rare.
Woodland.

WL: (Duntarvie, 1931).
ML: Balerno, 2000; (Turnhouse, pre-1934).
EL: (Dunglass, pre-1934; Pressmennan, 1956; Gifford, 1960).

Eranthis Salisb.

E. hyemalis (L.) Salisb.   Winter Aconite
Introduced. Status: scarce, garden escape, long persisting, under-recorded in survey due to early flowering.
Woodland, hedgerows, scrub.

WL: rare, Dalmeny.
ML: local, Dean Village; Hermitage of Braid; Lauriston Castle; Craiglockhart Dell; Oxenfoord Castle.
EL: local, Pencaitland; Luffness Friary; Dirleton; Tyninghame.

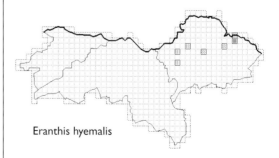

Eranthis hyemalis

Nigella L.

N. damascena L.   Love-in-a-mist
Garden escape.

ML: near Edinburgh, 1840; Leith Docks, 1904 and 1909.

Aconitum L.

A. napellus agg.   Monk's-hood
Introduced. Status: local, garden escape.
Woodland, scrub, roadsides, waterside habitats.

WL: rare, Birkhill; Carribber; Craigie; Cramond
Bridge; (Dalmeny and Linlithgow, 1973).
ML: local, Auchendinny; (Craigmillar Quarry,
1946; Middleton, 1971).
EL: occasional, mainly in estate woodland,
Pencaitland; Humbie; Luffness; Colstoun;
Whittingehame; Dunglass. Most specimens
checked have turned out to be *A. x cammarum*.

A. napellus L. ssp. **napellus**
EL: roadside verge near Elmscleugh, 1972.

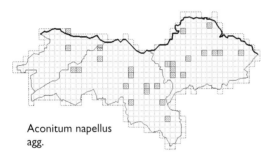

Aconitum napellus
agg.

A. x cammarum L. (*A. napellus* x *A. variegatum*)
                    Hybrid Monk's-hood
Introduced. Status: rare.
Woodland; roadsides.

ML: (Balerno, 1953; Ravelston Dykes Quarry,
1967; Crichton, 1969; Ratho; Colinton;
Musselburgh).
EL: rare, Saltcoats Castle; Mungoswells;
(Tantallon, 1921; Humbie, 1956; Archerfield,
1960; Luffness, 1973).

Consolida (DC.) Gray

C. regalis S. F. Gray   Forking Larkspur
Introduced. Status: rare, casual.
Waste ground.

ML: rare, Leith Docks, 1989 and 1992.
EL: (Dirleton, 1963).

C. ajacis (L.) Schur   Larkspur
ML: the semi-double form of this garden escape was
recorded from Leith, 1910.

C. orientalis (J. Gay) Schroedinger   Easter Larkspur
Grain alien.
ML: Leith Docks, 1903–1908; Slateford rubbish tip,
1905–1908.

Actaea L.

A. spicata L.   Baneberry
Introduced. Status: rare, possibly casual.
Scrub.

ML: rare, one plant only at Fala, 1990, presumably bird-sown.
EL: rare, near Whittingehame, 1993; Gosford.

Anemone L.

A. nemorosa L.   Wood Anemone
Native. Status: quite widespread.
Mixed and deciduous woodland, rough grassland,
moorland, riversides.

WL: quite widespread, seldom locally frequent.
ML: local, locally abundant.
EL: rare in N., quite widespread elsewhere,
locally frequent.

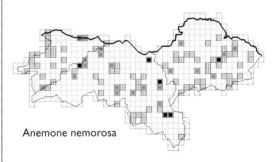

Anemone nemorosa

A. apennina L.   Blue Anemone
Introduced. Status: rare.
Woodland.

ML: rare, Almondell, 1982; (Comiston House,
1963).

Hepatica Mill.

H. nobilis Schreb.   Liverleaf
Introduced. Status: rare, garden escape.
Waste ground, scrub.

ML: rare, Lennox Tower, 1996; (Granton, 1836).

Clematis L.

C. vitalba L.   Traveller's-joy, Old Man's Beard
Introduced. Status: scarce.
Hedgerows, woodland, scrub, roadsides.

WL: rare, near The Binns; (Kinneil area, 1957).
ML: rare, Dalkeith; Calton Hill; near Roslin
Castle; Bawsinch.
EL: rare, near Pencaitland, 2000; Gosford; A198
at Craigielaw; near Yarrow; Broxburn;
(Haddington, Dirleton and North Berwick, pre-
1934; Biel, 1973; Ormiston Hall, 1976).

Ranunculus L.

R. acris L.   Meadow Buttercup
Native. Status: common.
Rough grassland, beside roads and rivers, marshes,
intensive grassland, waste ground, scrub, woodland.

WL, ML, EL: common.

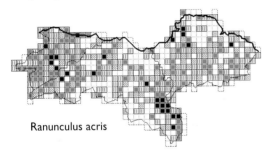

Ranunculus acris

R. repens L.   Creeping Buttercup
Native. Status: very common.
Roadsides, rough grassland, woodland, marsh, intensive
grassland, waterside habitats, waste ground, arable
fields.

WL, ML, EL: very common.

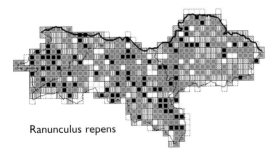

Ranunculus repens

R. bulbosus L.   Bulbous Buttercup
Native. Status: local.

Rough grassland, public parks, coastal habitats,
roadsides, unimproved grassland on dry soils.

WL: local, mainly in N.
ML: scattered.
EL: widespread along the coast, locally abundant,
occasional inland.

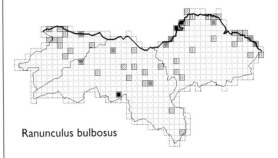

Ranunculus bulbosus

R. arvensis L.   Corn Buttercup
Native. Status: rare, casual, formerly a cornfield
weed, later a grain alien.
Waste and cultivated ground.

WL: (fields around Linlithgow, pre-1934).
ML: rare, Leith, 1990; Blackford Hill, 1985;
(formerly frequent, 1809–1934).
EL: (North Berwick, 1884; Gullane and Dunbar,
pre-1934).

R. auricomus L.   Goldilocks Buttercup
Native. Status: scarce.
Damp woodland and scrub.

WL: scarce, Almondell; Pepper Wood; scattered
sites along the River Avon from Westfield to the
Union Canal Viaduct.
ML: local, Dalkeith Country Park; Colinton and
Craiglockhart Dells; Vogrie; West Calder;
Cramond; Lady Lothian Wood; (Newbattle;
Mortonhall).
EL: rare, Saltoun; Pencaitland; several sites below
Lothian Edge; (Biel, 1881; Gifford and Danskine,
pre-1962).

R. sceleratus L.   Celery-leaved Buttercup
Native. Status: local.
Marshes, in and beside water.

WL: quite widespread but patchy in N., with one
cluster of records south-west of Linlithgow, and
another cluster around Broxburn; occasional in S.
ML: rare, Duddingston Loch; Lochend Pond.
EL: scarce, Aberlady Bay; Markle; Smeaton;
Belhaven Bay.

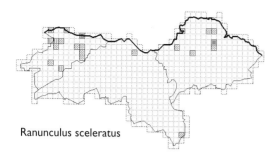

Ranunculus sceleratus

R. lingua L.   Greater Spearwort
Native. Status: rare, often planted.
Marshes, ponds, lochs.

WL: rare, Polkemmet Country Park; Half Loaf
Pond; pond near Easter Inch; (near The Binns,
1863; Linlithgow, pre-1934).
ML: rare, Duddingston Loch; Vogrie Pond;
(Lochend Pond, Prestonhall and Craigmillar,
pre-1934).
EL: rare, garden escape, beside stream,
Longniddry Bents, 1997.

R. flammula L.   Lesser Spearwort
ssp. flammula
Native. Status: widespread.
Marshes, waterside habitats, ditches, streams, bogs.

WL: very widespread in centre and S., occasional
in N.
ML: occasional in N., widespread and locally
frequent elsewhere.
EL: widespread and locally frequent in SE., rare
elsewhere.

Variants have been recorded but further research
on them is required.

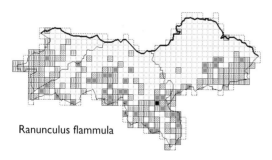

Ranunculus flammula

R. x levenensis Druce ex Gornall (*R. flammula* x
R. reptans)   Loch Leven Spearwort
Native. Status: rare, spontaneous hybrid.
Shores of lochs and reservoirs.

WL: (un-named site, 1963).
ML: rare, Gladhouse Reservoir, 1996.

R. ficaria L.   Lesser Celandine
Native. Status: common.
Woodland, beside roads, rivers and ditches, scrub,
marshes, rough grassland, public parks.

WL, ML, EL: common, especially in lowland
areas.

ssp. ficaria
Open habitats.

WL: distribution frequency unknown.
ML: common.
EL: a few scattered records. Greatly under-
recorded due to the difficulty of making sure the
plants are not going to produce bulbils at a later
date.

ssp. bulbilifer Lambinon
Shady places, deans.

WL: widespread.
ML: local, Duddingston Golf Course;
Craiglockhart Dell; Dalkeith Country Park;
Bilston Glen; Newhailes.
EL: scattered across lowland areas, few records
from upland areas.

R. hederaceus L.   Ivy-leaved Crowfoot
Native. Status: local.
Marshes, in and beside muddy springs and ditches.

WL: local, mainly in SW.
ML: widespread and locally abundant in SE.,
local elsewhere.
EL: scarce, mainly in upland areas.

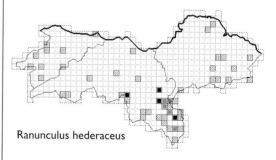

Ranunculus hederaceus

R. omiophyllus Ten.   Round-leaved Crowfoot
Native. Status: rare, possibly overlooked.
Shallow ponds and streams, on wet mud.

ML: rare, Tweedale Burn, 1999.

R. baudotii Godr. Brackish Water Crowfoot
Native. Status: rare.
Ponds and ditches near the sea, base-rich waters inland.

WL: (South Queensferry, 1882).
ML: rare, Duddingston Loch; (Braid Hill Marshes, 1889; Lochend, 1919).
EL: (Aberlady Bay, 1972).

R. aquatilis agg. Water Crowfoot
Native. Status: local.
In and beside rivers, ponds and ditches, marshes.

WL: rare.
ML: local.
EL: scarce.

Includes the following six taxa that were not always separated in the survey.

R. aquatilis L. *s.s.* Common Water Crowfoot
Native. Status: rare.
Ponds, ditches, marshes.

ML: rare, near Rosewell; Bawsinch; (Threipmuir, 1967).
EL: (records from North Berwick, 1835, Gullane, 1860 and Prestonpans, pre-1863, all appear to have been *R. aquatilis* L. *s.s*).

R. trichophyllus Chaix
Thread-leaved Water Crowfoot
Native. Status: scarce.
Marshes, ponds, lochs, reservoir mud.

WL: rare, Drumshoreland; Hopetoun, pre-1999, lost when the pond was drained; (South Queensferry, 1905; Carriden *c.* 1970).
ML: occasional, Cobbinshaw, Harperrig, Glencorse and Gladhouse Reservoirs; Duddingston Loch.
EL: scarce, Keith; Aberlady Bay; north of Drem; Donolly Reservoir; White Castle; Dunbar Common area; Faseny Water; Bothwell Water; Catcraig.

R. peltatus Schrank Pond Water Crowfoot
Native. Status: local.
In and beside lochs, rivers and reservoirs, marshes.

WL: rare, Bathgate; Beecraigs; Tailend Moss; Bangour Reservoir; Knock Pond; (Linlithgow, pre-1900); probably under-recorded.
ML: local, quite widespread on the east side of the Pentland Hills.
EL: rare, Stobshiel Reservoir; Kippielaw Ponds;

Whiteadder Reservoir; Bothwell Water area; Black Loch; (Gullane, 1860; Pressmennan Lake and Little Spott, 1871; Gosford, pre-1934).

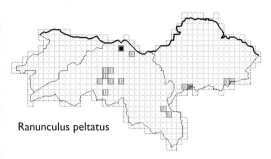

Ranunculus peltatus

R. penicillatus (Dumort.) Bab.
Stream Water Crowfoot
ssp. pseudofluitans (Syme) S. D. Webster
Native. Status: local.
Rivers.

ML: scarce, locally frequent, Water of Leith; Gogarburn; Braidburn.
EL: (Dunbar Links, 1872).

R. x bachii Wirtg. (*R. fluitans* x *R. trichophyllus* & *R. fluitans* x *R. aquatilis*)
Wirtgen's Water Crowfoot
Native. Status: rare.
Rivers.

ML: Burdiehouse Burn, 1999. This may be the furthest north this hybrid has been recorded.

R. fluitans Lam. River Water Crowfoot
Native. Status: rare.
Rivers.

EL: rare, River Tyne at Haddington and East Linton.

R. sardous Crantz Hairy Buttercup
Rare grain alien.
WL: Linlithgow, pre-1894.
ML: Dalkeith, pre-1894; Leith intermittently from 1903 to 1934, and again in 1980.

R. aconitifolius L. var. *flore pleno*
Aconite-leaved Buttercup
Garden escape.
WL: Kirkliston, 1944.

R. circinatus Sibth. Fan-leaved Water Crowfoot
Native.
ML: Duddingston Loch, 1960; Lochend Pond, 1965.

Adonis L.

A. annua L.   Pheasant's Eye
Grain casual.
ML: Leith Docks, 1906; Duddingston Station, 1914;
Seafield, 1934.

Aquilegia L.

A. vulgaris L.   Columbine
Introduced. Status: local, garden escape.
Roadsides, waste ground, scrub, walls, usually near
habitation.

WL, ML, EL: widely scattered distribution, small
numbers of plants, including some garden
cultivars.

Thalictrum L.

T. aquilegiifolium L.   French Meadow Rue
Introduced. Status: rare.
Waste ground.

EL: rare, derelict area near a garden,
Pressmennan, *c.* 1987.

T. minus L. *s.l.*   Lesser Meadow Rue
Native. Status: local.
Dunes, waste ground.

WL: rare, Carriden; Dalmeny; Bedlormie.
ML: rare, Roslin Chapel; near Balerno.
EL: scattered sites along the coast.

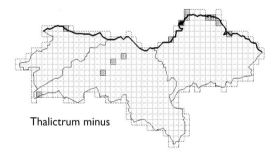

Thalictrum minus

ssp. arenarium (Butcher) A. R. Clapham
                          Sand Meadow Rue
Native. Status; local.
Sandy soil by the coast.

WL: rare, Dalmeny dunes.
ML: (from Cramond to Newhaven, pre-1934;
previously abundant).
EL: quite frequent along the north coast on dunes
and cliffs.

T. flavum L.   Common Meadow Rue
EL: Dunglass Dean, nineteenth century.

BERBERIDACEAE

Berberis L.

B. vulgaris L.   Barberry
Introduced. Status: local, usually planted in
hedges, long naturalised.
Hedgerows, roadsides.

WL, ML: local.
EL: scattered distribution, mainly in lowland
areas, Garleton; Drem; Gifford.

B. wilsoniae Hemsl.   Mrs Wilson's Barberry
Introduced. Status: rare.
Coastal scrub, dunes.

EL: rare, Hummell Rocks.

B. darwinii Hook.   Darwin's Barberry
Introduced. Status: scarce, planted, sometimes a
garden escape.
Roadsides, waste ground, woodland, coastal grassland.

WL: rare, west of Livingston; north-east of
Blawhorn; Hopetoun.
ML: rare, Blackford Hill; West Calder; (Crichton,
1956).
EL: rare, Gullane Hill; Garleton; Yellowcraig;
Dunglass Estate.

B. x stenophylla Lindl. (*B. darwinii* x *B. empetri-*
folia Lam.)   Hedge Barberry
Introduced. Status: rare.
Grassland.

EL: rare, Gullane Hill.

Mahonia Nutt.

M. aquifolium (Pursh) Nutt.   Oregon Grape
Introduced. Status: local, occasionally
naturalised.
Woodland, hedgerows, roadsides, public parks, policies.

WL, ML: local, mainly in N.
EL: rare, confined to lowland areas, Fountainhall;
West Saltoun; Luffness Friary; Morham; East
Fortune; Biel Estate; Dunglass Estate.

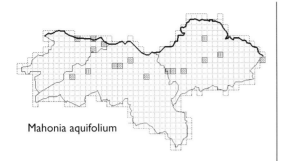

Mahonia aquifolium

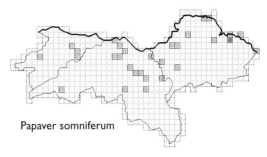

Papaver somniferum

Epimedium L.

E. alpinum L.   Barrenwort
Garden escape.
WL: Kirkliston 1871.
ML: Prestonfield, pre-1934; Dalhousie, 1938.

PAPAVERACEAE

Papaver L.

P. pseudoorientale (Fedde) Medw.
                         Oriental Poppy
Introduced. Status: rare, garden escape.
Waste ground, roadsides.

WL: rare, Bo'ness; The Binns; Greendykes Bing;
old quarry at Gowanbank.
ML: rare, near Balerno.
EL: rare, Seton Sands; Saltcoats, Gullane; North
Berwick; Markle. The **EL** records should be
considered as *P. pseudoorientale* agg. as material
collected there shows characters of *P. bracteatum*
Lindl. and requires further work.

P. atlanticum (Ball) Coss.   Atlas Poppy
Introduced. Status: rare, casual.
Waste ground, cliffs.

WL: rare, rubbish tip near Philpstoun Bing.
ML: rare, Dalmahoy; (Kingsknowe, 1945; Wester
Hailes, 1972).
EL: rare, North Berwick, 1994.

P. somniferum L.   Opium Poppy
Introduced. Status: local, casual garden escape,
many cultivated forms.
Waste ground, roadsides, farmyards, gardens, bings.

WL, ML: local.
EL: scattered distribution in lowland areas.

P. rhoeas L.   Common Poppy, Corn Poppy
Native. Status: widespread in N.
Cereal fields, waste ground, roadsides.

WL: widespread in lowland areas, very occasionally frequent.
ML: widespread and locally frequent in lowland
areas, occasional in S.
EL: widespread and sometimes locally frequent in
lowland areas, rare in upland areas.

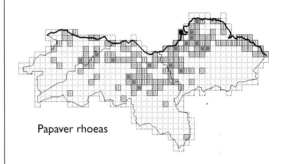

Papaver rhoeas

P. x hungaricum Borbás (*P. rhoeas* x *P. dubium*)
Native. Status: rare, with both parents.
Disturbed roadside bank.

EL: rare, West Barns, 1981.

P. dubium L.
ssp. dubium   Long-headed Poppy
Native. Status: widespread in lowland areas.
Waste ground, cereal fields, roadsides, bings.

WL, ML, EL: widespread in lowland areas, rare
in upland areas.

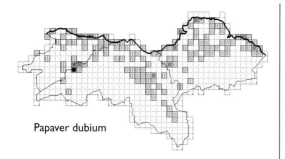

Papaver dubium

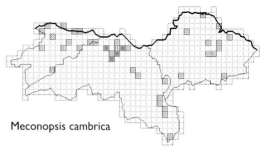

Meconopsis cambrica

ssp. lecoqii (Lamotte) Syme　　　　Yellow Juice Poppy
ML: Leith, 1905; Ratho, 1914; Granton, 1972.
EL: North Berwick, 1872.

P. hybridum L.　　　　　　　　Rough Poppy
Introduced. Status: rare, casual grain alien.
Waste ground.

ML: rare, Leith Docks, 1989; (Bilston, 1864;
Craigmillar Quarry and Comiston, 1902;
Slateford rubbish tip, 1902–6).
EL: (Dunbar and Skateraw, pre-1934).

P. argemone L.　　　　　　　　Prickly Poppy
Introduced. Status: rare, casual, possibly native.
Roadsides, cereal fields, disturbed ground.

WL: (South Queensferry and Linlithgow,
pre-1934).
ML: (Granton. 1827; Dalkeith, 1866; Buckstone,
1869; Leith, 1903; Fushiebridge, 1963; Inveresk,
1975).
EL: rare, North Berwick; Crowhill; (Broxburn
and Thorntonloch, 1972; West Barns, 1980;
formerly more widespread in lowland areas).

Meconopsis Vig.

M. cambrica (L.) Vig.　　　　　Welsh Poppy
Introduced. Status: local, garden escape.
Roadsides, waste ground, gardens, woodland.

WL: local, mainly in N.
ML: local, mainly in urban areas in N. and in the
valleys of Heriot Water and Gala Water.
EL: scattered distribution.

Argemone L.

A. mexicana L.　　　　　　　　Mexican Poppy
Wool or birdseed casual.
ML: Slateford rubbish tip, 1905; Leith Docks, 1906.

Glaucium Mill.

G. flavum Crantz　　　Yellow Horned Poppy
Native. Status: rare.
Sandy shores.

WL: (South Queensferry, 1824; Blackness, 1894).
EL: rare and sporadic at Skateraw. About forty
plants were counted at the top of the beach at
Skateraw Harbour in 1989. This was one of three
places that received plants rescued from the
nearby site of the Torness nuclear power station
in 1975. Previously a scattered distribution along
the coast, (North Berwick, pre-1894; Canty Bay,
1932; Gosford, pre-1934; Ravensheugh area,
probably 1960s; Aberlady Bay, 1968).

G. corniculatum (L.) Rudolph
　　　　　　　　　　Red Horned Poppy
Introduced. Status: rare, grain casual or garden
escape.
Waste ground, sometimes at the site of maltings.

ML: rare, Fisherrow, 1990; (Leith Docks and
Slateford rubbish tip, 1887–1921; Borthwick
railway tip, 1965; Granton, 1972.)
EL: (West Barns, 1973).

Chelidonium L.

C. majus L.　　　　　　　　Greater Celandine
Introduced. Status: scarce.
Roadsides, waste ground, woodland, scrub.

WL: rare, Carriden; Hopetoun; (Blackness, 1956;
Linlithgow, 1973).
ML: scarce, Lady Lothian Wood, *c.* 1982; Trinity,

1999; (Colinton, 1961; formerly a widespread casual).
EL: rare, Tranent; Seton (since before 1863); Longniddry Dean; Gosford; Stenton; Dunglass.

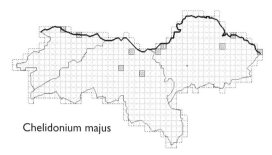

Chelidonium majus

Eschscholzia Cham.

E. californica Cham.　　　Californian Poppy
Introduced. Status: rare, casual, garden escape.
Roadsides, garden rubbish tips.

WL: (Carriden, 1960).
ML: (Leith Docks, Slateford rubbish tip and Craigmillar Quarry, 1889–1906).
EL: rare, Prestongrange, 1998; West Barns, 1980 and 1996; (Gullane, 1957 and 1958).

FUMARIACEAE

Dicentra Bernh.

D. spectabilis (L.) Lem.　　　Bleeding Heart
Introduced. Status: rare, garden escape.
Waste ground, riverbanks.

ML: rare, Canonmills, 1996.

Corydalis DC.

C. solida (L.) Clairv.　　　Bird-in-a-bush
Long-persisting garden escape.
WL: Carlowrie, 1908.
ML: Colinton Woods, 1834; Dalkeith Woods, 1847; Comiston Woods, 1914–46.

Pseudofumaria Medik.

P. lutea (L.) Borkh.　　　Yellow Corydalis
Introduced. Status: scarce.
Walls, mainly garden walls.

WL: rare, Abercorn; Carriden; Overton; (Craigton, 1954; Inveravon, 1961).

ML: local, Cramond; Telford Road; Canonmills; Braidburn; (Stow, 1973).
EL: rare, Leaston; Dirleton (since 1831); North Berwick; Belhaven; Innerwick; Dunglass.

Ceratocapnos Durieu

C. claviculata (L.) Lidén　　　Climbing Corydalis
?Native. Status: local.
Woodland, scrub, quarries, cliffs.

WL: scarce, Binny Craig; Lampinsdub; Pepper Wood; Torphichen; Craigie Hill; near Pardovan House.
ML: rare, Corstorphine; Dalmahoy; Stow; formerly common on rubbish dumps and on thatched roofs.
EL: rare, Bothwell Water; Thurston Mains Burn; (Bolton Muir Wood, 1969).

Fumaria L.

F. capreolata L.　　　White Ramping Fumitory
ssp. babingtonii (Pugsley) P. D. Sell
Native. Status: rare.
Cereal fields, rough grassland, railway embankments, quarries, disturbed ground.

WL: (Ecclesmachan and Kirkliston, pre-1934; Broxburn and Carriden, 1972).
ML: rare, Roslin, 2000; (Redhall, Craigmillar, Ratho and Currie, pre-1934; formerly widespread in fields and waste ground).
EL: rare, West Barns; Dunbar; Broxmouth; (Dirleton, 1844; Gosford, pre-1934; Thorntonloch, 1972).

F. muralis Sond. ex W. D. J. Koch
　　　　　　Common Ramping Fumitory
ssp. boraei (Jord.) Pugsley
Native. Status: widespread.
Arable fields, waste ground, roadsides.

WL, ML, EL: rare in upland areas, widespread elsewhere.

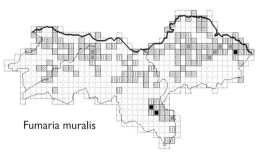

Fumaria muralis

F. purpurea Pugsley   Purple Ramping Fumitory
Native. Status: rare.
Rough grassland.

WL: (Kirkliston, 1915; Carlowrie, 1971).
ML: rare, Cockpen; Musselburgh; (Currie, 1835; Granton, 1972).
EL: rare, Dunglass, 1992; (Tyninghame, 1866; Gullane, pre-1934; North Berwick, 1914).

F. officinalis L. *s.l.*   Common Fumitory
Native. Status: widespread.
Arable fields, waste ground, roadsides.

WL, ML, EL: rare in upland areas, widespread elsewhere.

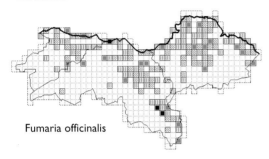

Fumaria officinalis

ssp. officinalis
Most records refer to this subspecies.

ML: rare in upland areas, widespread elsewhere.
EL: rarely recorded, Prestongrange; East Linton; Innerwick.

ssp. wirtgenii (W. D. J. Koch) Arcang.
ML: rare, perhaps casual, Roslin, 1996; (Musselburgh, 1966).
EL: rarely recorded, Setonhill; Markle; Pinkerton; East Barns, 1985 and 1999; (Luffness, 1966).

F. densiflora DC.   Dense-flowered Fumitory
?Native. Status: rare, previously more widespread.
Arable fields, roadsides, disturbed ground.

WL: (Linlithgow, 1970).
ML: rare, Bawsinch, 1990; Stow, 2000; (Granton, 1955; Cramond, 1956; Inveresk, 1975).
EL: rare, scattered distribution in lowland areas, Prestonpans; Pencaitland; Longniddry; Drem; Markle; Whitekirk; West Barns; Skateraw; (North Berwick, 1971).

F. bastardii Boreau   Tall Ramping Fumitory
WL: Kirkliston, pre-1934.

ML: Leith Docks, 1885; Colinton, 1914.
EL: Aberlady Bay, 1953; Bothwell Water, pre-1970.

F. parviflora Lam.   Fine-leaved Fumitory
Grain alien.
ML: Leith and Slateford, 1833–1906.
EL: Longniddry and Dirleton, pre-1934.

PLATANACEAE

Platanus L.

P. x hispanica Mill. ex Münchh.
(*P. occidentalis* L. x *P. orientalis* L.)
    London Plane
Introduced. Status: rare, planted.
Streets, policy woodland, public parks.

WL: rare, Kinneil Estate.
ML: local, Princes Street Gardens, Raeburn Place, Charlotte Square, Pilrig Park etc. in Edinburgh.

ULMACEAE

Ulmus L.

Most elms are susceptible to Dutch elm disease causing a general decline in frequency. Even the suckers eventually succumb to the disease.
Many dead elm trees can be seen throughout the Lothians.

U. glabra Huds.   Wych Elm
Native. Status: widespread but decreasing due to Dutch elm disease.
Mixed and deciduous woodland, roadsides, hedgerows, riversides, scrub.

WL, ML, EL: widespread and locally frequent except in some upland areas.

ssp. glabra and **ssp. montana** Hyl. have been recorded from **EL**.

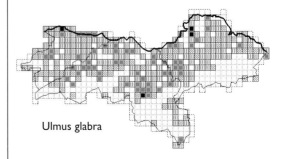

Ulmus glabra

U. x vegeta (Loudon) Ley (*U. glabra* x *U. minor*)
Huntingdon Elm
Introduced. Status: rare, planted.
Public parks, policies.

ML: rare, The Meadows, Edinburgh.
EL: rare, Woodhall Burn.

U. x elegantissima Horw. (*U. glabra* x *U. plotii*
Druce)
Introduced. Status: rare, planted.
Public amenity areas.

ML: rare, by the Water of Leith, Stockbridge.

U. procera Salisb.　　　　　English Elm
Introduced. Status: local, mainly planted.
Mixed and deciduous woodland, roadsides, hedgerows,
scrub.

WL: scattered distribution.
ML: occasional, mainly in N.
EL: rare, Ormiston, 1992; Haddington area,
1994; North Berwick, 1994; Woodhall Burn,
1988.

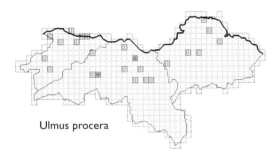

Ulmus procera

U. minor Mill.
ssp. sarniensis (C. K. Schneid.) Stace　Jersey Elm
Introduced. Status: rare, planted.
Streets, public parks.

ML: local, Ravelston, Comiston Road, Ferry
Road etc. in Edinburgh.

CANNABACEAE

Cannabis L.

C. sativa L.　　　　　　　Hemp
Introduced. Status: rare birdseed casual, formerly
an agricultural seed casual.
Waste ground.

ML: rare, Marchmont, *c.* 1985; (Leith Docks,
1887–1907).
EL: (Spittal, 1880).

Humulus L.

H. lupulus L.　　　　　　　Hop
Introduced. Status: rare.
Waste ground, hedgerows, scrub, woodland.

WL: rare, Hopetoun; Dalmeny; (Kirkliston, pre-
1934; Kinneil, 1969).
ML: rare, Craiglockhart Pond; Penicuik; Straiton;
Pathhead; (Levenhall, 1953; East Calder, 1973).
EL: rare, Tranent; Pencaitland; Gullane; Luffness;
Haddington; Pefferside; Bilsdean; (East Linton,
1973).

MORACEAE

Ficus L.

F. carica L.　　　　　　　Fig
Introduced. Status: rare, long persisting.
Walls, gardens, riverbanks.

ML: rare, Bonnington; Warriston; (Dean Bridge,
1962).

URTICACEAE

Urtica L.

U. dioica L.　　　Stinging Nettle, Common Nettle
Native. Status: common.
Roadsides, waste ground, mixed woodland, rough
grassland, particularly on well-manured soil.

WL: widespread and locally frequent.
ML, EL: common.

U. urens L.　　　　　　　Small Nettle
Probably native. Status: local.
Waste ground, roadsides, cereal fields, gardens.

WL: occasional in N.
ML: local, mainly in N.
EL: widespread in N., rare elsewhere.

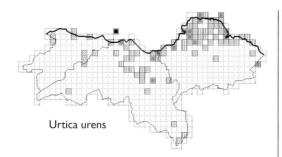

Urtica urens

U. pilulifera L.   Roman Nettle
Introduced. Status: rare, casual.
ML: Slateford and Leith, pre-1934 .
EL: North Berwick, pre-1934.

Parietaria L.

P. judaica L.   Pellitory-of-the-wall
Native. Status: local.
Old walls, often those of castles, rocks.

WL: rare, Niddry; (Dundas and Linlithgow,
pre-1934; Midhope, 1963).
ML: local, Borthwick, Roslin, Craigmillar and
Hawthornden Castles.
EL: local and only in N., scattered colonies,
mostly near habitation, Redhouse and Saltcoats
Castles; River Tyne wall, Haddington; North
Berwick sea front; West Barns.

Soleirolia Gaudich.

S. soleirolii (Req.) Dandy
  Mind-your-own-business, Mother of Thousands
Introduced. Status: rare, garden escape.
Shady walls and banks.

ML: rare, established at Marchmont and in the
New Town, Edinburgh.
EL: rare, established at Innerwick.

JUGLANDACEAE

Juglans L.

J. regia L.   Walnut
Introduced. Status: rare, planted.
Estate woodlands.

WL: rare, Hopetoun.
ML: rare, Edinburgh Zoo.
EL: rare, Luffness; near Innerwick; (Biel, 1962;
Dunglass, 1955).

MYRICACEAE

Myrica L.

M. gale L.   Bog Myrtle, Sweet Gale
Native. Status: rare.
Bogs.

ML: rare, Thriepmuir.
EL: (Blinkie Burn Moor, north of Lammer Law,
pre-1934).

FAGACEAE

Fagus L.

F. sylvatica L.   Beech
Introduced. Status: widespread, mainly planted,
naturalised in some areas.
Woodland, hedgerows.

WL, ML, EL: widespread, locally frequent.

Nothofagus Blume

N. obliqua (Mirb.) Blume
                          Roble Southern Beech
Introduced. Status: rare, planted.
Estate woodland, public parks.

WL: rare, public planting at Bo'ness.
ML: rare, planted in estate woodlands at
Dalkeith and Vogrie.
EL: rare, Fountainhall; Saltoun Forest.

N. nervosa (Phil.) Krasser   Rauli Southern Beech
Introduced. Status: rare, planted.
Estate and other woodlands.

ML: rare, Vogrie Country Park.
EL: rare, Saltoun Forest.

N. antarctica (Forst. f.) Oerst.   Antarctic Beech
Introduced. Status: rare, planted.
Estate woodland.

WL: a single tree, Kinneil Estate.

Castanea Mill.

C. sativa Mill.   Sweet Chestnut
Introduced. Status: local, planted.
Estate and other woodlands, public parks.

WL, ML: occasional.
EL: occasional, Saltoun Forest; Humbie;
Stoneypath Tower; Dunglass.

Quercus L.

Q. cerris L.   Turkey Oak
Introduced. Status: local, planted.
Estate woodland, public parks.

WL, ML: occasional.
EL: occasional, Ferny Ness; Luffness Friary;
Dunglass.

Q. ilex L.   Holm Oak, Evergreen Oak
Introduced. Status: rare, planted.
Estate woodland, public parks.

WL, ML: rare.
EL: rare, Gosford House; Seacliff; Dunbar.

Q. petraea (Matt.) Liebl.   Sessile Oak
Native. Status: local, probably mainly planted.
Woodland.

WL, ML: occasional.
EL: locally frequent. There are remnants of native
woodland in the Lammermuir Hills.

Q. x rosacea Bechst. (*Q. petraea* x *Q. robur*)
Native. Status: local, probably under-recorded.
Woodland, hedgerows, scrub.

WL, ML: occasional.
EL: rare, Saltoun Forest; North Berwick;
Pressmennan.

Q. robur L.   Pedunculate Oak, Common Oak
Native. Status: common, probably mainly
planted.
Woodland, hedgerows, scrub.

WL, ML, EL: widespread, except in S.

Q. rubra L.   Red Oak
Introduced. Status: rare, planted.
Estate woodland, public parks.

WL: rare, Hopetoun; Philpstoun; Livingston.
ML: rare, Warriston.
EL: rare, near Tranent.

Q. frainetta Ten.   Hungarian Oak
Introduced. Status: rare, planted.
Estate woodland.

WL: rare, Hopetoun; Philpstoun; Dundas.

BETULACEAE

Betula L.

B. pendula Roth   Silver Birch
Native. Status: widespread.
Woodland, scrub, hedgerows. Coloniser of waste
ground and bings. Also planted as an amenity tree.

WL: widespread.
ML: widespread except in S., locally abundant.
EL: local, few records in N. and E., probably
mostly planted.

B. pubescens Ehrh.   Downy Birch
Native. Status: widespread.
Woodland, scrub, hedgerow, shale bings. Recorded less
often than *B. pendula*. The subspecies have not been
distinguished in WL and ML.

WL: widespread, especially in S. and W.
ML: widespread.
EL: locally frequent but few records in N., proba-
bly mostly planted, except in upland areas.

ssp. pubescens
EL: rare, Whitekirk, 1985; (Humbie, 1970;
Bothwell Water, 1971).

ssp. tortuosa (Ledeb.) Nyman
EL: rare, Harestone Hill; White Castle; Woodhall
Dean.

Alnus Mill.

A. glutinosa (L.) Gaertn.   Alder
Native. Status: widespread, sometimes planted.
Beside rivers, ponds, marshes and railway lines, and in
scrub and mixed woodland.

WL: widespread.
ML: widespread except in S.
EL: local, few records in NW. and SE.,
increasingly planted.

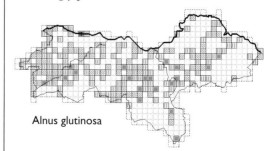

Alnus glutinosa

A. x hybrida A. Braun ex Rchb. (*A. glutinosa* x
A. incana)
Native. Status: rare.
Found where the parents grow in proximity.

WL: rare, Foulshiels Bing.
ML: rare, south of Livingston.
EL: rare, Prestongrange, planted; (Haddington,
1847).

A. incana (L.) Moench   Grey Alder
Introduced. Status: local, planted.

WL: widespread, planted on waste ground and
bings.
ML: local, Warriston; West Calder; Duddingston.
EL: local in W., first recorded in 1992 from
Ormiston; Tranent area; Luffness; Athelstaneford;
North Berwick Law car park.

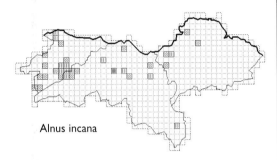

Alnus incana

A. rubra Bong.   Red Alder
Introduced. Status: rare, planted.
Public amenity areas.

WL: rare, near Uphall Station.
ML: rare, single tree near Silverknowes.

A. cordata (Loisel.) Duby   Italian Alder
Introduced. Status: rare, planted.
Public amenity areas.

WL: rare, Blackridge; Fauldhouse; Linlithgow;
Livingston.
ML: rare, Holyrood Park; Stow; West Calder.

A. hirsuta (Spach) Ruprecht
Introduced. Status: rare, planted.

ML: rare, Cramond Bridge.

Carpinus L.

C. betulus L.   Hornbeam
Introduced. Status: local, planted.
Hedgerows, mixed woodlands.

WL, ML: local.
EL: scarce, Pencaitland; Athelstaneford; Dunglass
Dean.

Corylus L.

C. avellana L.   Hazel
Native. Status: widespread.
Woodland, scrub, hedgerows.

WL: scattered distribution in old woodlands,
frequently planted on waste ground for amenity.
ML: widespread except in S.
EL: widespread except in N., though generally in
small quantites, sometimes planted.

CHENOPODIACEAE

Chenopodium L.

C. pumilio R. Br.   Clammy Goosefoot
Introduced. Status: rare, casual.
Waste ground, rubbish tips.

ML: rare, Borthwick railway tip, 1987.

C. bonus-henricus L.   Good King Henry
Introduced. Status: scarce.
Roadsides, waste ground.

WL: rare, on a verge near Uphall; (Bo'ness, South
Queensferry and Kirkliston, pre-1934;
Woodcockdale, 1958; Cramond Bridge, 1960).
ML: rare, Cousland.
EL: occasional, often close to farms, on beach at
Dunglass.

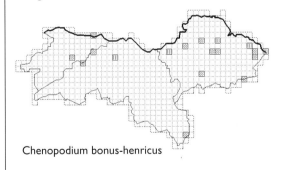

Chenopodium bonus-henricus

C. rubrum L.   Red Goosefoot
Native. Status: rare.
Waste ground, rubbish tips, brackish hollows.

WL: rare, Carmelhill Loch shore and nearby
verges; (Bo'ness, 1961; South Queensferry, 1837).

ML: rare, Leith, 1987.
EL: rare, Aberlady Bay; occasional casuals inland; (Gullane rubbish tip, 1977).

C. vulvaria L.   Stinking Goosefoot
Introduced. Status: rare.
Waste ground.

ML: rare, Granton, 1996; (Musselburgh and other sites, pre-1934).
EL: (Prestonpans and Preston Castle, pre-1934).

C. murale L.   Nettle-leaved Goosefoot
Introduced. Status: rare, casual.
Waste ground, gardens.

ML: rare, Leith, 1987.

C. album L.   Fat Hen
Native. Status: widespread.
Arable ground, roadsides, waste ground, farmyards, gardens.

WL: widespread.
ML, EL: widespread except in upland areas.

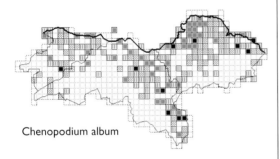

Chenopodium album

C. polyspermum L.   Many-seeded Goosefoot
ML: Hailes rubbish tip, 1971.

C. hybridum L.   Maple-leaved Goosefoot
WL: Kinneil, pre-1934.
EL: Dunbar, pre-1894.

C. urbicum L.   Upright Goosefoot
EL: Gullane, pre-1904.

Bassia All.

B. scoparia (L.) Voss   Summer-cypress
Introduced. Status: rare, casual.
Waste ground.

ML: rare, Leith, 1982.

Atriplex L.

A. prostrata agg.   Orache
Native. Status: widespread on the shores of the Firth of Forth, rare inland.
Beaches, dunes and other coastal habitats, waste ground, arable fields.

WL: locally frequent along the coast.
ML: occasional.
EL: locally frequent along the coast but fewer sites on north-east facing North Sea coast.

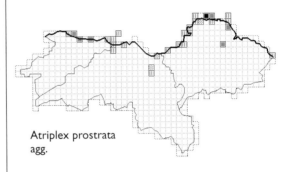

Atriplex prostrata agg.

Includes the following two taxa that were not always distinguished in the survey.

A. prostrata Boucher ex DC.
                    Spear-leaved Orache
Native. Status: widespread on the shores of the Firth of Forth, rare inland.
Beaches, dunes and other coastal habitats, waste ground, arable fields.

WL: locally frequent along the coast.
ML: occasional.
EL: locally frequent along the coast.

A. glabriuscula Edmondston
                    Babington's Orache
Native. Status: occasional along the coast but possibly under-recorded because of confusion with *A. prostrata*.
Beaches, dunes and other coastal habitats.

WL: occasional.
ML: rare, Levenhall; Leith.
EL: scarce, Prestonpans; Aberlady Bay; Seacliff; east of Dunbar; Barns Ness.

A. littoralis L.   Grass-leaved Orache
Native. Status: local, confined to coastline.
Beaches, dunes and other coastal habitats.

WL, ML: local.

EL: a few plants recorded at numerous sites, abundant at Aberlady Point; frequent at Lawrie's Den, near Dunbar.

A. patula L.   Common Orache
Native. Status: widespread except in upland areas.
Found in a wide range of habitats, predominantly roadsides, waste ground and arable fields.

WL: widespread, locally frequent.
ML: widespread except in upland areas, locally abundant.
EL: widespread except in upland areas, seldom frequent.

A. laciniata L.   Frosted Orache
Native. Status: local.
Strandline on sandy beaches.

WL: rare, Blackness Bay; Wester Shore; Dalmeny.
ML: rare, Musselburgh; Cramond.
EL: frequent along the coast.

A. hortensis L.   Garden Orache
ML: Hailes Quarry tip, 1971.

A. littoralis x A. patula
ML: Leith Docks, 1977, with both parents. Possibly the first record of this hybrid in Britain.

Beta L.

B. vulgaris L.
ssp. maritima (L.) Arcang.   Sea Beet
Native. Status: rare.
Beaches.

WL: (South Queensferry, pre-1934).
ML: rare, Cramond, last seen in 1981.
EL: rare, Prestongrange; north of Gullane; Bass Rock; Pefferside; (Gosford, 1957).

ssp. vulgaris   Root Beet
Introduced. Status: rare.
Waste ground, rough grassland.

ML: rare, casual, Granton, 1992; two other waste ground sites.

Sarcocornia A. J. Scott

S. perennis (Mill.) A. J. Scott   Perennial Glasswort
EL: Aberlady Bay, 1969, det. P. W. Ball.

Salicornia L.

S. europaea agg.   Glasswort
Native. Status: very local. A difficult genus

needing more research.
Saltmarsh, coastal mud.

WL: rare, Bo'ness; Blackness; Hopetoun.
EL: local, Aberlady Bay; abundant at Tyninghame Bay; Belhaven Bay; (Prestonpans, 1824; Gosford, 1973; North Berwick, pre-1934).

S. ramosissima Woods   Purple Glasswort
Native. Status: very local.
Saltmarsh, among other vegetation.

EL: locally abundant at Aberlady Point, 1993; (North Berwick, pre-1934; Aberlady Bay, 1969, det. P. W. Ball). A variable species, probably the same as *S. europaea* (Stace 1997).

S. europaea L.   Common Glasswort
Native. Status: very local.
Saltmarsh and coastal mud.

WL: recorded from Bo'ness, det. N. Stewart.
EL: (recorded from Aberlady Bay and Tyninghame Bay, 1969, det. P. W. Ball).

S. fragilis P. W. Ball & Tutin   Yellow Glasswort
Native. Status: rare.
Saltmarsh, bare mud.

EL: frequent very locally at Aberlady Point, 1993, det. R. Learmonth; (Aberlady Bay, 1969, det. P. W. Ball).

S. dolichostachya Moss   Long-spiked Glasswort
Native. Status: rare.
Saltmarsh.

WL: rare, Bo'ness, det. N. Stewart.
EL: rare, Tyninghame Bay, 1987; (Aberlady Bay and Tyninghame Bay, 1969, det. P. W. Ball; Belhaven Bay, 1980).

Suaeda Forssk. ex J. F. Gmel.

S. maritima (L.) Dumort.   Annual Sea-blite
Native. Status: local, locally abundant in EL.
Saltmarsh.

WL: rare, Bo'ness; Wester Shore; (Abercorn and Dalmeny, pre-1934).
EL: scattered sites along the coast, locally plentiful at Aberlady Bay and Pefferside.

Salsola L.

S. kali L.
ssp. kali   Prickly Saltwort
Native. Status: local.

Sandy beaches.

WL: rare, Blackness; Dalmeny; (Abercorn and South Queensferry, pre-1934).
ML: rare, Fisherrow Sands.
EL: established at Tyninghame Bay; occasional and sporadic elsewhere.

ssp. ruthenica (Iljin) Soó   Spineless Saltwort
Introduced. Status: rare, casual.
Waste ground, rubbish tips, cultivated fields.

ML: rare, Gogar, 1987; (occasional, docks and rubbish tips, 1980).

AMARANTHACEAE

Amaranthus L.

A. retroflexus L.   Common Amaranth
Introduced. Status: rare, casual.
Waste ground, rubbish tips.

ML: rare, Leith Docks, 1989; (Hailes rubbish tip, 1971).

PORTULACACEAE

Claytonia L.

C. perfoliata Donn ex Willd.
                    Miner's Lettuce Spring Beauty
Introduced. Status: local.
Roadsides, cultivated and waste ground, sand-dunes, scrub.

WL: rare, Abercorn; Dechmont; Kirkliston; (Carriden, pre-1934).
ML: in N. only, with several sites in Edinburgh, local.
EL: local, frequent under coastal scrub at Longniddry Bents and Gullane; rare inland.

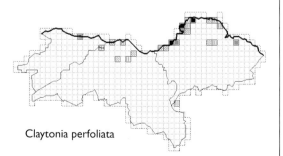

Claytonia perfoliata

C. sibirica L.   Pink Purslane
Introduced. Status: widespread.
Riversides, damp woodland, often in quite dense shade.

WL: widespread in new and old woodland.
ML: widespread, inland only, locally frequent.
EL: local, inland only, locally abundant at a site near Haddington.

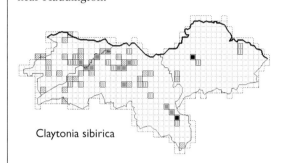

Claytonia sibirica

Montia L.

M. fontana L.   Blinks
Native. Status: widespread.
Beside streams and rivers, marshes, flushes, wet places in habitats such as quarries, roadsides and unimproved grassland, mainly in upland areas.

WL: widespread in SW.
ML: widespread and locally frequent in S.
EL: widespread in the Lammermuir Hills, rare in lowland areas.

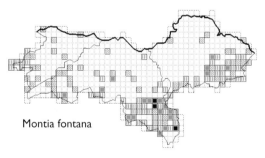

Montia fontana

The following subspecies have been recorded.

ssp. fontana
ML: (Pentland and Moorfoot Hills).
EL: Rook Law; Sheeppath Burn.

ssp. variabilis Walters
ML: (Pentland and Moorfoot Hills).
EL: Sheeppath Burn; Woodhall Burn.

ssp. chondrosperma (Fenzl.) Walters
EL: Markle; Traprain Law; Whitekirk; White Castle.

CARYOPHYLLACEAE

Arenaria L.

A. serpyllifolia L.   Thyme-leaved Sandwort
Native. Status: widespread.
Roadsides, waste ground, bare ground in numerous other habitats.

WL: widespread.
ML: widespread, locally common.
EL: widespread in lowland areas.

ssp. serpyllifolia
Native. Status: widespread. Probably most records of the species refer to this subspecies.
Roadsides, waste ground, bare ground in numerous other habitats.
WL, ML: widespread.
EL: widespread in lowland areas.

ssp. leptoclados (Rchb.) Nyman
                              Slender Sandwort
Native. Status: rare, probably under-recorded.
Roadsides, waste ground, dunes.
ML: rare, Monktonhall Colliery; Borthwick Bank; (Leith Docks; Portobello; Granton, 1972).
EL: occasional in dunes, Longniddry Bents and elsewhere on the coast, and at sites just inland.

A. balearica L.   Mossy Sandwort
Introduced. Status: rare.
Gardens.

EL: rare, naturalised at Smeaton, near East Linton.

Moehringia L.

M. trinervia (L.) Clairv.   Three-nerved Sandwort
Native. Status: widespread.
Woodland, scrub.

WL: widespread in N. only.
ML: widespread in centre only.
EL: widespread.

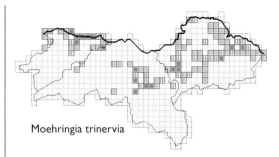

Moehringia trinervia

Honckenya Ehrh.

H. peploides (L.) Ehrh.   Sea Sandwort
Native. Status: local.
Sandy shore.

WL: local, Carriden; Blackness; Hopetoun; Dalmeny.
ML: rare, Cramond; Granton.
EL: frequent along the coast.

Minuartia L.

M. verna (L.) Hiern   Spring Sandwort
Native. Status: rare (declining).
Wet upland grassland, rocky outcrops.

ML: rare, Arthur's Seat; Craiglockhart Hill; (Blackford Hill, pre-1934).
EL: (Traprain Law, pre-1835).

Stellaria L.

S. nemorum L.   Wood Stitchwort
ssp. nemorum
Native. Status: local.
Damp woodland, often beside watercourses.

WL: (Abercorn, Carribber and Philpstoun, pre-1934).
ML: locally frequent, mainly in centre but also well established at Musselburgh.
EL: local, in centre only, Colstoun; River Tyne; Gifford; Woodhall Burn.

S. media (L.) Vill.   Common Chickweed
Native. Status: common.
Cultivated and waste ground, rough grassland, roadsides.

WL, ML, EL: widespread, except in some upland areas, locally abundant.

S. pallida (Dumort.) Crép.   Lesser Chickweed
Native. Status: local, under-recorded.
On bare, sandy soil or in short grass.

WL: rare, mainly coastal, Bo'ness; south of Blackness; Ochiltree; Hopetoun; Dalmeny.
ML: inland at Colinton Dell, rare.
EL: numerous sites, habitats mainly coastal turf and volcanic outcrops.

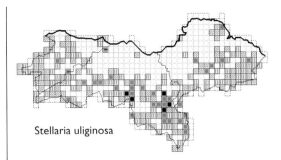

Stellaria uliginosa

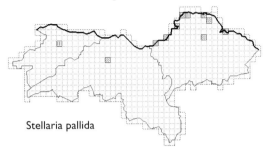

Stellaria pallida

S. holostea L.   Greater Stitchwort
Native. Status: widespread.
Damp woodland, scrub, hedgerows, rough grassland.

WL: widespread except in SE.
ML: widespread except in N. and in upland areas.
EL: widespread except in N. (from where, however, there are old records).

S. palustris Retz.   Marsh Stitchwort
Native. Status: rare (declining).
Damp ground, riverside, woodland.

WL: rare, Bathgate; (Winchburgh, 1885; Linlithgow and Philpstoun, pre-1934).
ML: (Duddingston Loch, 1957; Ratho; Kirknewton; Glencorse; Brunstane).
EL: rare, Aberlady Bay.

S. graminea L.   Lesser Stitchwort
Native. Status: common.
Rough and heath grassland, roadsides, riversides.

WL: widespread, locally frequent.
ML: widespread, locally abundant in SE.
EL: widespread.

S. uliginosa Murray   Bog Stitchwort
Native. Status: widespread, except in N.
Marshes, bogs and wet areas beside rivers and ditches.

WL: widespread.
ML: widespread except in N., locally abundant in S.
EL: widespread except in NW. (from where, however, there are old records).

Cerastium L.

C. arvense L.   Field Mouse-ear Chickweed
Native. Status: scarce.
Roadsides, waste ground, dunes, coastal turf, volcanic outcrops.

ML: local, inland only, Borthwick; Holyrood Park.
EL: local, in coastal turf, and inland mainly on volcanic outcrops, Longniddry Bents; Gullane Hill; Longyester; Black Loch.

C. arvense x C. tomentosum
Native. Status: rare, near both parents.
Railway embankment.

ML: rare, Borthwick Bank, 1983; recorded once only.

C. tomentosum L.
                    Snow in Summer, Dusty Miller
Introduced. Status: local, garden escape or throw-out, often well established.
Roadsides, waste ground, rubbish tips.

WL, ML, EL: occasional near habitation.

C. fontanum Baumg.
                    Common Mouse-ear Chickweed
ssp. vulgare (Hartm.) Greuter & Burdet
Native. Status: common.
Found in almost all habitats but most frequently on roadsides and waste ground and in unimproved grassland.

WL, ML, EL: common.

ssp. holosteoides (Fr.) Salman, Ommering & de Voogd
EL: recorded from Markle and East Barns.

C. glomeratum Thuill.
                    Sticky Mouse-ear Chickweed
Native. Status: common.

Roadsides, waste ground, all types of lowland pasture, usually by paths or trampled areas.

WL, ML, EL: widespread and frequent except in upland areas.

C. diffusum Pers.   Sea Mouse-ear Chickweed
Native. Status: frequent on parts of the **EL** coastline, scarce elsewhere.
Coastal habitats, inland on rock outcrops, walls and in quarries.

WL: local in N.
ML: on volcanic rocks, scarce, Cramond Island; Inchmickery.
EL: locally frequent on coast, scarce inland.

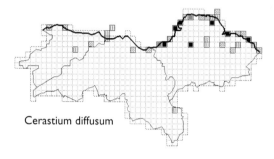

Cerastium diffusum

C. semidecandrum L.
                    Little Mouse-ear Chickweed
Native. Status: locally abundant on the coast, scarce inland.
Coastal habitats, dry open habitats on sandy or calcareous soils.

WL: rare, coastal, Carriden; Blackness; Dalmeny.
ML: locally abundant on coast, rare inland.
EL: locally frequent on coast, scarce inland.

Sagina L.

S. nodosa (L.) Fenzl   Knotted Pearlwort
Native. Status: scarce.
Coastal and damp upland habitats.

WL: rare, on damp gravel south of Blackridge; south of Bathgate.
ML: one coastal site, the Roman Camp, Cramond; a few sites in the Pentland and Moorfoot Hills.
EL: rare, along the coast, Aberlady Bay; Gullane; Belhaven Bay; inland, Papana Water.

S. subulata (Sw.) C. Presl   Heath Pearlwort
Native. Status: rare (declining).
Waste ground, dry soils.

WL: rare, Bo'ness; Hopetoun; Dalmeny.
ML: rare, Arthur's Seat; (Dalmahoy, pre-1934; Balerno, 1972).
EL: (Traprain Law, 1974).

S. procumbens L.   Procumbent Pearlwort
Native. Status: common.
Mainly ruderal, on paths and in short vegetation.

WL, ML: widespread except in arable areas, locally abundant.
EL: widespread, less so in arable areas.

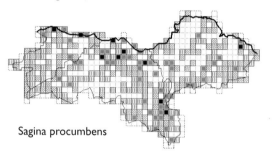

Sagina procumbens

S. apetala Ard.   Annual Pearlwort
ssp. apetala
Native. Status: scarce or under-recorded.
Quarries, walls, roadsides.

ML: occasional.
EL: occasional, mostly near the coast, Prestonpans; Gosford Bay; Gullane Hill; West Barns; Dunbar; East Barns.

ssp. erecta F. Herm.
Native. Status: local.
Ruderal, bare ground, walls, roadsides.

WL: occasional.
ML: widespread.
EL: widely scattered distribution, Prestongrange; Nether Brotherstone; Garleton Hills; North Berwick; Dunbar; Catcraig.

S. maritima Don   Sea Pearlwort
Native. Status: scarce.
Coastal habitats, roadsides inland.

WL: rare, Bo'ness; Dalmeny; (South Queensferry, 1958).
ML: occasional, Seafield; Musselburgh; Leith; Inchmickery; Cramond Island; Granton.
EL: rare on rocks along the coast; locally abundant in sandy saltmarsh, Belhaven Bay, 1993; recorded since 1993 on the verges of salted roads inland, Soutra; Broxburn; The Bell, 2000.

Scleranthus L.

S. annuus L.   Annual Knawel
Native. Status: rare (declining).
Volcanic outcrops, dry grassland.

WL: rare, Mochries Craig Hill; (South
Queensferry, 1836; Kirkliston, pre-1934).
ML: rare, Musselburgh; Corstorphine; (Lasswade
and Colinton, pre-1934).
EL: rare, mainly on volcanic outcrops, East
Linton area; Whitekirk Hill; Woodhall Dean
Reserve; (Ormiston, 1881).

Herniaria L.

H. hirsuta L.   Hairy Rupturewort
Introduced. Status: rare, transient.
Waste ground.

WL: rare, verge at Bonhard Farm, *c.* 1980.
ML: (Leith, pre-1912; weed at Royal Botanic
Garden Edinburgh, 1955 and 1970).

Polycarpon L.

P. tetraphyllum (L.) L.   Four-leaved Allseed
Introduced. Status: rare, casual.
Waste ground.

ML: rare, Leith Docks, 1986.

Spergula L.

S. arvensis L.   Corn Spurrey
Native. Status: widespread but somewhat patchy,
possibly under-recorded because of access
problems.
Arable fields, waste ground, roadsides.

WL: very widespread.
ML: locally abundant in S., scarce in N., formerly
common throughout on cultivated ground.
EL: widely scattered distribution, but scarce in
intensively farmed arable land in N., formerly
common in arable fields.

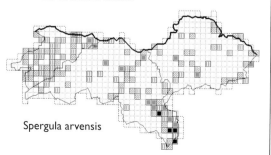

Spergula arvensis

Spergularia (Pers.) J. & C. Presl

S. media (L.) C. Presl   Greater Sea Spurrey
Native. Status: scarce.
Saltmarsh, beaches.

WL: rare, Bo'ness; Blackness; Wester Shore.
ML: rare, Cramond; Seafield.
EL: scarce, Gosford Bay; Aberlady Bay; The
Leithies; Tyninghame Bay; Dunbar area; Catcraig.

S. marina (L.) Griseb.   Lesser Sea Spurrey
Native. Status: local.
Beaches and saltmarsh, inland beside roads that are
salted in winter, and on waste ground.

WL: quite widespread, along the coast from
Bo'ness to Wester Shore and Dalmeny; recorded
from frequently salted roadside verges at twelve
inland sites.
ML: local, Musselburgh; Gorebridge; A7 at
Heriot; Dean Bridge, Edinburgh; locally frequent.
EL: rare on coast, Aberlady Bay; The Leithies;
Belhaven; Catcraig; recorded since 1992 from six
roadside sites.

S. rubra (L.) J. & C. Presl   Sand Spurrey
Native. Status: local.
Waste ground, roadsides, quarries, bare ground, bings,
dry places.

WL: local.
ML: local, locally frequent.
EL: scarce, scattered distribution, Prestonpans
marina; Macmerry; Garleton Hills; Whitekirk
Hill; Dunbar; Priestlaw Hill.

S. bocconei (Scheele) Graebn.   Greek Sea Spurrey
Introduced.
ML: Leith, pre-1908.

Lychnis L.

L. coronaria (L.) Murray   Rose Campion
Introduced. Status: rare, garden escape.
Waste ground.

WL: rare, Bo'ness; Port Edgar.
EL: rare, in the garden of an abandoned cottage
near East Linton.

L. flos-cuculi L.   Ragged Robin
Native. Status: widespread in S., scarce in N.
Marshes.

WL: widespread except in N.
ML: widespread and locally frequent in S., rare
in N.

EL: clustered in S. and SE. upland areas, scattered distribution in lowland areas.

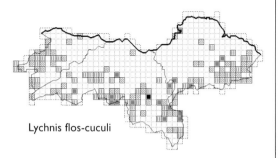

Lychnis flos-cuculi

L. viscaria L.   Sticky Catchfly
Native. Status: rare (declining).
Cliffs and rocky places.

WL: (Dundas Hill, 1824; on a railway embankment, Linlithgow, 1955, possibly a garden escape).
ML: rare, Arthur's Seat, a dwindling population reinforced by planting; re-introduced on Castle Rock, Edinburgh; (Blackford and Corstorphine Hills).
The main threat to this species is shade caused by the encroachment of scrub species, especially gorse.

Agrostemma L.

A. githago L.   Corn Cockle
Introduced. Status: rare, previously a cornfield weed, now deliberately sown in 'wild flower seed', not persisting.
Cultivated and waste ground.

ML: (Ratho, 1878; cornfields and docksides, pre-1934).
EL: rare, beside path, East Links, North Berwick, 1998; Prestonpans, 2001; (Longniddry, Aberlady and Dirleton, pre-1934).

Silene L.

S. vulgaris Garcke   Bladder Campion
Native. Status: widespread except in S.
Roadsides, waste ground, railway embankments, some coastal habitats.

WL: local, but only in N.
ML: widespread except in SW. and SE. where it is rare or absent.
EL: widespread in lowland areas.

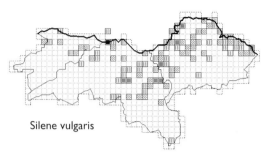

Silene vulgaris

S. uniflora Roth   Sea Campion
Native. Status: scarce.
Coastal habitats.

WL: rare, one plant at Dalmeny, 2000; (Abercorn and Blackness, pre-1934).
ML: local, Cramond Island; Inchmickery.
EL: local, Craigleith Island; Bass Rock; NE coast; no longer on NW coast.

S. armeria L.   Sweet William Catchfly
Introduced. Status: rare, garden escape, casual.
Waste ground, rough grassland.

WL: ('Linlithgowshire', 1863).
EL: rare, close to shore, St Baldred's Cradle, 1982.

S. noctiflora L.   Night-flowering Catchfly
Introduced. Status: rare, casual.
Waste ground, rough grassland.

ML: rare, Leith, 1982; Pilton, 1997.
EL: (occasional on sandy soils up to 1907; Humbie, 1956; Dunglass, 1968; West Barns, 1973).

S. latifolia Poir.   White Campion
ssp. **alba** (Mill.) Greuter & Burdet
Native. Status: widespread except in upland or boggy areas.

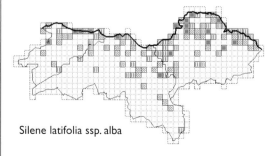

Silene latifolia ssp. alba

Roadsides, railway embankments, waste ground, arable fields, some coastal habitats.

WL: widespread in N., scarce in S.
ML: widespread in N. and E., scarce in S. and W.
EL: widespread and locally frequent in lowland areas.

S. x hampeana Meusel & K. Werner (*S. latifolia* x *S. dioica*)
Native. Status: frequent where the parents occur in the same area. Possibly over-recorded as *S. dioica* may also have pale pink flowers.

S. dioica (L.) Clairv.   Red Campion
Native. Status: widespread except in upland areas.
Woodland, scrub, roadsides, riversides.

WL: widespread in N., scarce in S.
ML: widespread and locally frequent in lowland areas, scarce in upland areas.
EL: common in lowland areas, rare or absent in upland areas.

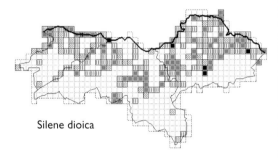

Silene dioica

S. gallica L.   Small-flowered Catchfly
Introduced. Status: rare, casual.
Waste ground.

WL: (South Queensferry, *c.* 1840).
ML: rare, Leith Docks, 1989.
EL: (Dirleton, 1834; Gullane, 1872 and 1888).

S. conica L.   Sand Catchfly
ML: Craigmillar and Slateford, pre-1934.
EL: Dirleton Common, 1835; Gullane, 1863.

Saponaria L.

S. officinalis L.   Soapwort
Introduced. Status: rare casual.
Riverside, woodland, waste ground.

WL: rare, Bathgate; Harthill.
ML: rare, Newbattle; Addistoun; Loanhead;

(Colinton Dell, 1975).
EL: rare, Prestongrange; Skid Hill quarry; North Berwick; Binning Wood.

Vaccaria Wolf

V. hispanica (Mill.) Rauschert   Cowherb
Introduced. Status: rare, casual.
Waste ground, rubbish tips.

WL: (Carriden, 1968).
ML: (Leith Docks, 1978; Granton, 1980).
EL: rare, North Berwick, 1996; (Dunbar, 1903; North Berwick, 1960).

Dianthus L.

D. deltoides L.   Maiden Pink
Native. Status: scarce.
Rough grassland on volcanic outcrops and dunes.

WL: (Niddry Castle, pre-1934).
ML: scarce, Arthur's Seat; West Craiglockhart Hill; West Calder; Heriot; Craigmillar; (formerly on Blackford, Braid and Pentland Hills).
EL: rare, Aberlady Bay; North Berwick Law and several sites south of North Berwick.

D. barbatus L.   Sweet William
Introduced. Status: rare, garden escape.
Hedgerows, walls, woodland.

EL: rare, Tranent; (Gullane, 1868; near North Berwick, 1903).

POLYGONACEAE

Persicaria Mill.

P. campanulata (Hook. f.) Ronse Decr.
                Lesser Knotweed
Introduced. Status: rare, garden escape.
Damp shady places.

ML: rare, Ratho.
EL: rare, Oldhamstocks.

P. wallichii Greuter & Burdet
                Himalayan Knotweed
Introduced. Status: rare.
Roadsides, rough grassland, beside ponds.

WL: rare, west of Port Edgar.
ML: (Carberry, 1960).
EL: rare, Dunglass Estate.

P. bistorta (L.) Samp.   Common Bistort
Native. Status: local.
Damp areas beside roads and rivers.

WL, ML: local.
EL: rare, Prestongrange; Yester; Balgone; Biel; possibly garden escapes or planted at these sites.

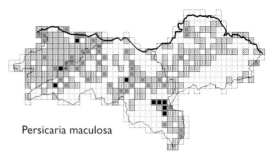

Persicaria maculosa

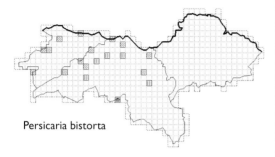

Persicaria bistorta

P. amplexicaulis (D. Don) Ronse Decr.
          Red Bistort
Introduced. Status: rare, garden escape or throw-out.
Rubbish tips, rough grassland.

WL: rare, Dalmeny village.
ML: (Dalmahoy, 1975).
EL: rare, Gullane, 1999; (Gullane, 1953).

P. vivipara (L.) Ronse Decr.   Alpine Bistort
Native. Status: scarce.
Wet upland pasture.

ML: very local, Carlops; Leadburn.

P. amphibia (L.) Gray   Amphibious Bistort
Native. Status: local.
In and beside water, also on roadsides and waste ground, arable fields. Possibly planted in new ponds.

WL: widespread in N., local elsewhere.
ML: local, locally frequent.
EL: occasional in lowland areas and in upland reservoirs.

P. maculosa Gray   Redshank
Native. Status: common.
Arable fields, pastures, roadsides, waste ground.

WL: common.
ML: locally abundant, widespread except in some upland areas.
EL: occasional in some intensively farmed areas in N. and upland areas in S., widespread elsewhere.

P. lapathifolia (L.) Gray   Pale Persicaria
Native. Status: scarce.
Waste ground, arable fields.

WL: rare, south-west of Bo'ness; west of Livingston; (Broxburn; Carriden; Linlithgow, *c.* 1970).
ML: (Levenhall, 1955; Granton, 1958; Gorebridge, 1959; Currie, Roslin and Temple, 1960–71; Dalmahoy, 1971).
EL: scattered sites, Keith; Gullane; Drem; Gifford; North Berwick; Spott.

P. hydropiper (L.) Spach   Water-pepper
Native. Status: local.
In or beside water.

WL: occasional, mainly in SW., (Linlithgow Loch, 1980).
ML: local, Harperrig; Gladhouse, Bonaly and Glencorse Reservoirs; West Calder.
EL: rare, Gosford; Danskine; (Humbie, 1957).

Fagopyrum Mill.

F. esculentum Moench   Buckwheat
Introduced. Status: rare, casual.
Cultivated ground, scrub.

ML: (rare casual, pre-1934).
EL: rare, Pressmennan, 1987; (Archerfield, in sunflower crop, 1960).

Polygonum L.

P. oxyspermum C. A. Mey. & Bunge ex Ledeb.
          Ray's Knotgrass
Native. Status: rare.
Sandy beaches.

ML: rare, Musselburgh; (Granton, 1894).
EL: rare, Prestonpans, 1988; (North Berwick, 1872; Longniddry, 1975). Mechanical beach cleaning now prevents recolonisation on tourist beaches.

Though usually regarded as **ssp. raii** (Bab.)
D. A. Webb & Chater, East Lothian specimens,
like others from E. Scotland, differ markedly
from those of western Britain in their narrower,
paler fruits and more uniformly narrow and more
acute leaves, and closely resemble **ssp.
oxyspermum** of eastern Scandinavia.

P. arenastrum Boreau   Equal-leaved Knotgrass
Native. Status: local, possibly under-recorded.
Roadsides and waste ground.

WL: local.
ML: locally frequent.
EL: local, a widely scattered distribution, frequent
in N.

P. aviculare L.   Knotgrass
Native. Status: common in lowland areas.
Records may include *P. arenastrum*.
Roadsides, waste and cultivated ground, dry soils,
gravel.

WL, ML, EL: common in lowland areas.

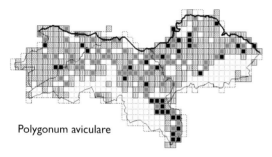

Polygonum aviculare

var. **littorale** Koch
Native. Status: rarely recorded.
Sandy beaches.

ML: Fisherrow, 1988.
EL: Prestonpans, 1988.

P. boreale (Lange) Small   Northern Knotgrass
Native. Status: rare.
Open ground.

WL: (South Queensferry, 1903).
ML: rare, Inchmickery; Pilton.

P. rurivagum Jord. ex Boreau
Cornfield Knotgrass
Introduced. Status: rare, casual.
Waste ground, paths.

WL: rare, Torphichen, 1983.
ML: rare, Craigmillar, 1999.

EL: rare, Saltoun, 1983; near North Berwick,
1996.

Fallopia Adans.

F. japonica (Houtt.) Ronse Decr.
Japanese Knotweed
Introduced. Status: local.
Woodland, beside rivers, roads and railways, waste
ground.

WL: widespread.
ML: widespread in N. where it is locally abun-
dant.
EL: scattered distribution, in estate woodlands
and on waste ground, Pencaitland; St Germains;
North Berwick; Haddington; Thornton.

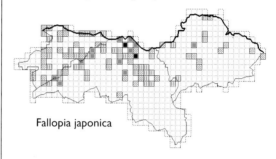

Fallopia japonica

F. x bohemica (Chrtek & Chrtková) J. P. Bailey
(*F. japonica* x *F. sachalinensis*)
Introduced. Status: rare.

WL: rare, several colonies around Port Edgar.
ML: rare, Craiglockhart Dell.

F. sachalinensis (F. Schm. ex Maxim) Ronse Decr.
Giant Knotweed
Introduced. Status: rare.
Policies, beside water.

WL: rare, possibly recorded in error for *F. x
bohemica*.
ML: local, Currie; Silverknowes; Penicuik; Roslin;
Musselburgh.
EL: rare, Winton; Saltoun; Keith Water; Yester;
Dunglass.

F. baldschuanica (Regel) Holub   Russian Vine
Introduced. Status: rare, garden escape.
Waste ground, hedgerows, scrub surrounding garden
rubbish tips.

ML: rare, Balerno; Drylaw; Bawsinch.
EL: rare, but increasing, Seton Sands;
Longniddry; Gullane; Haddington.

F. convolvulus (L.) Á. Löve   Black Bindweed
Native. Status: widespread.
Cultivated and waste ground.

WL: widespread except in S.
ML, EL: widespread except in upland areas.

Rheum L.

R. x hybridum Murray (*R. rhaponticum* L. x *R. palmatum* L.)   Rhubarb
Introduced. Status: local, garden escape.
Waste ground, garden rubbish tips, often by water.

WL: occasional.
ML: scarce.
EL: garden escape at Aberlady Bay; Tyninghame; Bothwell Water; also at several sites as a relic of cultivation.

Rumex L.

R. acetosella L.   Sheep's Sorrel
Native. Status: common.
Dry grassland, roadsides, waste ground, quarries.

WL, ML: common.
EL: occasional in N., widespread elsewhere.

ssp. acetosella var. **tenuifolius** Wallr.
Recorded from Faucheldean Bing (**WL**), and from **ML** and **EL**.

ssp. pyrenaicus (Pourr.) Akeroyd
Recorded from **EL**.

R. acetosa L.   Common Sorrel
Native. Status: very common.
Recorded frequently in almost every habitat.
Comparatively few records from arable land.

WL, ML: very common.
EL: common.

R. pseudoalpinus Höfft   Monk's Rhubarb
Introduced. Status: scarce.
Riversides and roadsides.

WL: local in SW., near Blackridge and at a few other sites.
ML: rare, Addiewell.

R. longifolius DC.   Northern Dock
Native. Status: local.
Roadsides, waste ground, rough grassland.

WL: mainly in S. where it is widespread but sparse.

ML: not in N., local in S.
EL: rare, perhaps overlooked, Linn Dean; Spittal; Danskine Loch; Nether Hailes; North Berwick.

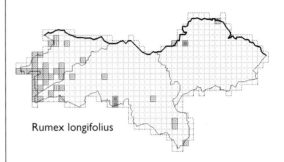

Rumex longifolius

R. x propinquus Aresch. (*R. longifolius* x *R. crispus*)
Native. Status: rare, possibly under-recorded.
Roadside.

ML: rare, south of Fountainhall.
EL: rare, Balgone, 1988; (Pilmuir, 1969).

R. x hybridus Kindb. (*R. longifolius* x *R. obtusifolius*)
Native. Status: local, where parents grow together.
Roadsides, waste ground, pasture.

WL: occasional.
ML: (Threipmuir, 1878).

R. crispus L.   Curled Dock
ssp. crispus
Native. Status: common.
Roadsides, waste and cultivated ground, riversides, coastal habitats.

WL: common.
ML, EL: common except in some upland areas.

ssp. littoreus (J. Hardy) Akeroyd.
Native. Status: under-recorded, probably widespread along the coast.
Coastal habitats.

WL: few records, under-recorded.
EL: scattered sites along the coast.

R. x pratensis Mert. & W. D. J. Koch (*R. crispus* x *R. obtusifolius*)
Native. Status: rare, probably under-recorded.
Roadsides, rough grassland, field edges, scrub.

EL: scarce, recorded from eight lowland sites.

R. conglomeratus Murray   Clustered Dock
Native. Status: scarce.
Wet places.

WL: rare, Carmelhill Loch; Dundas Loch.
ML: rare, Duddingston, *c.* 1984; (recorded as common *c.* 1824).
EL: scattered distribution in lowland areas, Butterdean Wood; Gullane; Whitekirk.

R x abortivus Ruhmer (*R. conglomeratus* x *R. obtusifolius*)
Native. Status: rare.
Disturbed woodland edge.

EL: rare, Dunglass Estate, 1983.

R. sanguineus L.   Wood Dock
Native and introduced. Status: widespread in N.
Damp, shady places in woodland and scrub.

WL: locally frequent in N., rare in S.
ML: locally frequent except in W. and S.
EL: frequent in wooded areas.

The common native plant is var. **viridis** (Sibth.) W. D. J. Koch.

var. **sanguineus**
WL: has been planted at Hopetoun and naturalises occasionally nearby.
EL: naturalised beside the lake, Gosford House, 2000.

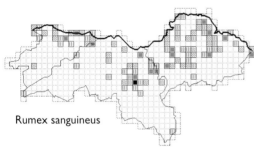
Rumex sanguineus

R. x dufftii Hausskn. (*R. sanguineus* x *R. obtusifolius*)
Native. Status: rare.
Scrub.

EL: rare, West Saltoun, 1983.

R. obtusifolius L.   Broad-leaved Dock
Native. Status: common.
Roadsides, waste and cultivated ground, pasture, woodland, scrub, riversides.
WL, ML, EL: common.

R. scutatus L.   French Sorrel
Last recorded from the walls of Craigmillar Castle (**ML**), 1971. It is possible that the introduction of this species dates from the time of Mary Queen of Scots.

R. maritimus L.   Golden Dock
ML: Meadowbank, 1840.

Oxyria Hill

O. digyna (L.) Hill   Mountain Sorrel
ML: Pentland Hills, pre-1934.

PLUMBAGINACEAE

Limonium Mill.

L. vulgare Mill.   Common Sea Lavender
Introduced. Status: transient.
Saltmarsh.
EL: first found at Aberlady Bay, *c.* 1973, subsequently lost during sewerage works.

Armeria Willd.

A. maritima Willd.   Thrift
Native. Status: common in **EL**, scarce elsewhere.
Marine turf, saltmarsh, rocks and cliffs by the sea.

WL: rare, Bo'ness; Blackness; Dalmeny.
ML: local, Cramond Island; Musselburgh; inland at Calder Road, 2000–2001.
EL: scattered distribution along the coast, abundant in places.

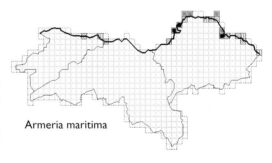
Armeria maritima

PAEONIACEAE

Paeonia L.

P. officinalis L.   Garden Peony
Introduced. Status: rare, garden escape.
Waste ground, rubbish tips.
WL: rare, near Uphall Station.
ML: rare, Davidson's Mains, 1996.

CLUSIACEAE

Hypericum L.

H. calycinum L.   Rose of Sharon
Introduced. Status: rare, garden escape.
Roadsides, waste ground, scrub, woodland.

ML: (Arniston, Roslin and Ratho, pre-1934).
EL: rare, Athelstaneford; Pressmennan; Dunglass
Estate.

H. androsaemum L.   Tutsan
Introduced. Status: local, garden escape.
Woodland, roadsides, waste ground.

WL: occasional.
ML: local.
EL: occasional, Gosford House; Colstoun Wood;
Gifford; Deuchrie; Bilsdean.

H. xylosteifolium (Spach) N. Robson
                    Turkish Tutsan
Introduced. Status: rare, garden escape.
Shady places on waste ground.

ML: rare, Drylaw, 1999.

H. perforatum L.   Perforate St John's Wort
Native. Status: local.
Roadsides, railway embankments, waste ground, scrub.

WL: locally frequent.
ML: locally frequent, not in S.
EL: quite widespread in lowland areas.

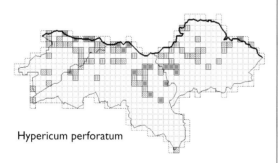

Hypericum perforatum

H. x desetangsii Lamotte (*H. perforatum* x *H. maculatum*)   Des Etangs' St John's Wort
Native. Status: rare.
Roadsides, riversides.

ML: rare, Morton Reservoir; Leadburn;
Duddingston Loch; (Dalhousie, 1838).

H. maculatum Crantz
                    Imperforate St John's Wort
Native. Status: scarce.
Riversides, railway embankments, waste ground.

WL: rare, Polkemmet Country Park; Craigton
Quarry; Dundas.
ML: scarce.
EL: rare, near Gifford, *c.* 1985.

H. tetrapterum Fr.
                    Square-stalked St John's Wort
Native. Status: local.
Marshes, riverbanks and other wet areas.

WL, ML, EL: scattered distribution, but never
frequent

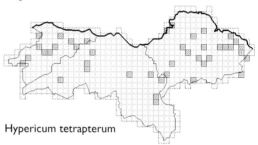

Hypericum tetrapterum

H. humifusum L.   Trailing St John's Wort
Native. Status: rare.
Dry heath grassland, woodland tracks, waste ground.

WL: rare, Blackburn; Seafield.
ML: rare, Rosewell; Roslin Glen; Leadburn;
(Cobbinshaw and Crookston, 1971).
EL: rare, Keith Glen; Binning Wood; Hedderwick;
Innerwick area; (Whittingehame, 1848; Dunglass,
pre-1934; Wester Pencaitland, 1957).

H. pulchrum L.   Slender St John's Wort
Native. Status: widespread.
Heath grassland, riverside, woodland, scrub.

WL: widespread, especially in W.
ML, EL: widespread in S., local in N.

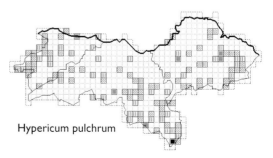

Hypericum pulchrum

H. hirsutum L. Hairy St John's Wort
Native. Status: scarce.
Woodland, scrub, waste ground.

WL: rare, Dalmeny; Hopetoun; Midhope.
ML: scarce, Roslin Glen; Bawsinch; Vogrie;
Polton.
EL: rare, Longniddry; Biel; (Gosford, 1955;
Tantallon, 1957; Papple, 1968; Gullane, 1973).

TILIACEAE

Tilia L.

T. platyphyllos Scop. Large-leaved Lime
Introduced. Status: scarce, planted.
Estate woodland and policies, public parks.

WL: scattered sites.
ML: rare, Astley-Ainslie Hospital grounds;
Hermitage of Braid; Lauriston Castle.
EL: rare, Seton; Aberlady Bay; Yester; Dunglass.

T. x europaea L. (*T. platyphyllos* x *T. cordata*)
Lime
Introduced. Status: widespread, often planted,
producing seedlings in recent years.
Woodland, policies, scrub, hedgerows, public parks.

WL: widespread.
ML, EL: widespread, except in upland areas.

T. cordata Mill. Small-leaved Lime
Introduced. Status: rare, planted.
Woodland, policies, hedgerows, public parks.

ML: rare.
EL: rare, Seton; wood near Spott.

MALVACEAE

Malva L.

M. moschata L. Musk Mallow
Probably introduced. Status: rare.
Roadsides, canal sides, edges of cultivated fields, garden
rubbish tips.

WL: scarce, Bo'ness; Linlithgow; Craigton
Quarry; north of Kirkliston.
ML: rare, Netherton; (Roslin and Auchendinny,
pre-1934).
EL: rare, most records appear to be garden
escapes, Gullane rubbish tip; Whittingehame.

M. sylvestris L. Common Mallow
Native. Status: local.
Hedgerows, roadsides, scrub, waste ground.

WL: (Kirkliston, 1955; Winchburgh, 1965;
Bo'ness, 1970s).
ML: occasional except in upland areas.
EL: quite widespread except in upland areas.

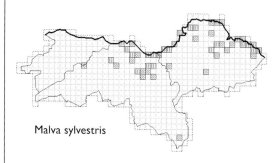

Malva sylvestris

M. nicaeensis All. French Mallow
Introduced. Status: rare.
Waste ground, rubbish tips.

ML: rare, Leith; Borthwick railway rubbish tip.

M. pusilla Sm. Small Mallow
Introduced. Status: rare, casual.
Waste ground.

ML: rare, Newhaven, 1982.
EL: rare, West Barns, 1995 to present.

M. neglecta Wallr. Dwarf Mallow
Native. Status: rare.
Waste ground, roadsides.

WL: (Linlithgow and Dalmeny, pre-1934;
Bo'ness, 1961).
ML: rare, Fisherrow.
EL: occasional on light sandy soils and rocky
outcrops in N., Port Seton; North Berwick;
Hailes Castle; Skateraw; Oldhamstocks.

M. alcea L. Greater Musk Mallow
ML: Hailes Quarry, 1971.

M. parviflora L. Least Mallow
ML: Borthwick railway rubbish tip, 1968.

Malope L.

M. trifida Cav. Mallow Wort
ML: Murieston, 1910.

Lavatera L.

L. arborea L.   Tree Mallow
Introduced. Status: rare.
Coastal habitats, waste ground.

ML: rare, ?planted at Blackford Hill and Joppa;
(Inchmickery, 1684–1824).
EL: local, abundant on Craigleith Island and the
Bass Rock (known there from before 1684);
rarely taking root along the mainland shore;
planted on North Berwick Law.

Althaea L.

A. hirsuta L.
      Hispid Mallow, Rough Marsh Mallow
Introduced. Status: rare, casual.
Waste ground.

ML: Leith and Balerno, 1982.
EL: (West Barns, 1973).

Sidalcea A. Gray ex Benth.

S. malviflora (DC.) A. Gray ex Benth.
            Greek Mallow
Introduced. Status: rare, garden escape.
Waste ground.

ML: rare, Leith; Niddrie; Bilston.

S. candida A. Gray   Prairie Mallow
Garden escape.
WL: Carriden, 1971.

DROSERACEAE

Drosera L.

D. rotundifolia L.   Round-leaved Sundew
Native. Status: local.
Peat bogs, heath grassland.

WL: local, mainly in remnant bogs in SW.
ML: locally frequent in Pentland and Moorfoot
Hills.
EL: (formerly rare in the Lammermuir Hills,
Aikengall Moss, pre-1934; Gifford area sometime
between 1950 and 1962; Faseny Water, 1972).

D. anglica Huds.   Great Sundew
ML: Auchencorth, 1972.

CISTACEAE

Helianthemum Mill.

H. nummularium (L.) Mill.   Common Rock-rose
Native. Status: local.
Heath grassland, quarries, rocky outcrops.

WL: rare, north of Blackridge.
ML: locally frequent in SE., rare elsewhere,
except in Holyrood Park where it is abundant.
EL: occasional, scattered sites, Linn Dean;
Markle; Stenton; Dry Burn.

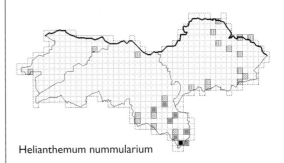
Helianthemum nummularium

VIOLACEAE

Viola L.

V. odorata L.   Sweet Violet
Probably introduced. Status: scarce.
Woodland, scrub, roadsides, railway embankments.

WL: rare, Bangour Reservoir; (Carriden and
Kinneil, 1969; Carlowrie, pre-1934).
ML: scarce, mainly in estate woodlands.
EL: widely scattered sites, mainly in estate wood-
lands, Gosford; Whitekirk; Broxmouth.

V. hirta L.   Hairy Violet
Native. Status: rare, (declining).
Sand-dunes, dry grassland.

WL: rare, Carriden; (Carribber and Hopetoun,
pre-1934; Dalmeny, 1955).
ML: (Blackford Hill, Roslin, Cramond and
Currie, pre-1934).
EL: rare, very small numbers of plants at
scattered sites along the coast, Ferny Ness;
Aberlady Bay; west of Yellowcraig; Tyninghame;
(Catcraig, 1956; Yellowcraig, 1975).

V. riviniana Rchb.   Common Dog Violet
Native. Status: common.

Woodland, scrub, riverside, roadside, heath and other rough grassland.

WL: common.
ML, EL: common except in areas of intensive arable farming.

V. canina L.   Heath Dog Violet
ssp. canina
Native. Status: rare.
Dunes, coastal turf, dry heath, rocky outcrops.

WL: (Dalmeny, 1955; Broxburn, 1974).
ML: rare, forest ride at Gladsmuir.
EL: coastal sites, rare, declining or overlooked, Gullane Point to Yellowcraig; Pefferside to Belhaven Bay; Barns Ness.

V. palustris L.   Marsh Violet
Native. Status: widespread in S., rare in N.
Marshes, bogs, beside rivers and ditches, woodland.

WL, ML: locally frequent in upland areas, rare below 200m.
EL: locally frequent in upland areas, rare in lowland areas where it occurs in wet woodland.

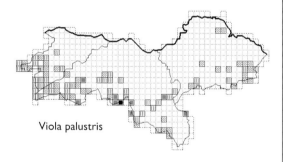

Viola palustris

V. cornuta L.   Horned Pansy
Introduced. Status: rare, garden escape.

ML: rare, Tynehead; Gala Water; Bowland Bridge.

V. lutea Huds.   Mountain Pansy
Native. Status: local.
Pasture slopes in upland areas and on volcanic outcrops.

WL: local, on volcanic hills.
ML: locally frequent, upland areas only.
EL: rare, Linn Dean Water; Garleton Hills quarry; (Traprain Law, 1873; East Linton, 1955; Lammermuir Hills above Humbie, 1970).

V. tricolor L.   Wild Pansy
ssp. tricolor
Native. Status: local, (declining).
Waste ground, arable fields, roadsides, railway embankments, rough grassland.

WL: local, Bo'ness Golf Course; Hillhouse Quarry.
ML: local, Granton; (Roslin, 1962; Inveresk, 1971; Tynehead, 1978.
EL: rare, small numbers at scattered sites, Linn Dean Water; Longyester; Markle; Traprain Law; West Barns; Catcraig; (Garleton Hills quarry, 1970).

V. x wittrockiana Gams ex Kappert (*V. tricolor* x *V. arvensis*)   Garden Pansy
Introduced. Status: rare, garden escape.
Waste ground.

WL: rare, Bo'ness rubbish tip.
EL: rare, Prestonpans marina; Humbie.

V. arvensis Murray   Field Pansy
Native. Status: widespread.
Arable fields and other cultivated ground, waste ground, roadsides.

WL, ML, EL: widespread except in upland areas, locally frequent.

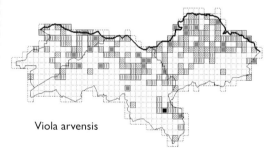

Viola arvensis

V. reichenbachiana Jord. ex Boreau   Early Dog Violet
?Native
WL: Linlithgow, pre-1934.
ML: Arniston, pre-1934.
EL: Gosford, pre-1934. This is now considered to be a doubtful record.

V. contempta Jord. (*V. arvensis* x *V. tricolor*)
Native.
WL: Linlithgow, 1847.

CUCURBITACEAE

Bryonia L.

B. dioica Jacq.   White Bryony
Introduced. Status: rare, naturalised escape.
Roadsides, railway embankments, dunes, coastal turf
and scrub, hedgerows.

ML: rare, Warriston.
EL: rare, Aberlady Bay; East Linton; Tyninghame;
Belhaven Bay; (Luffness, 1906; Gosford,
pre-1934).

SALICACEAE

Populus L.

P. alba L.   White Poplar
Introduced. Status: local, mainly planted.
Woodland, scrub, public parks.

WL, ML: local.
EL: local, Luffness; Bara; Dry Burn.

P. x canescens (Aiton) Sm. (*P. alba* x *P. tremula*)
                    Grey Poplar
Introduced. Status: local.
Woodland, scrub, roadsides, waste ground.

WL: local.
ML: widespread, planted.
EL: occasional, planted, Longniddry; Keith;
Luffness; Athelstaneford; Belhaven.

P. tremula L.   Aspen
Native. Status: local.
Woodland, scrub, riversides.

WL, ML: local.
EL: occasional, mostly in planted woodland,
Gladsmuir; Papana Water; Woodhall Dean.

P. nigra L.   Black Poplar
ssp. betulifolia (Pursh) Dippel.
Introduced. Status: rare.
Mixed woodland.

EL: rare, Dunglass.

cv. 'Italica'   Lombardy Poplar
Introduced. Status: widespread, planted.
Woodland, scrub, roadsides, riversides.

WL, ML: widespread, planted.
EL: rare, Prestonpans; Colstoun; formerly widely
planted.

P. x canadensis Moench (*P. nigra* x *P. deltoides*)
                    Hybrid Black Poplar
Introduced. Status: local.
Woodland, wet places.

WL, ML: local in N., planted.
EL: quite widespread, planted, Cockenzie;
Pencaitland; Whitekirk; Pefferside.

P. x jackii Sarg. (*P. deltoides* x *P. balsamifera*)
                    Balm of Gilead, Balsam Poplar
Introduced. Status: local, planted. Not critically
separated from the other balsam poplars.
Wet places, woodland.

WL, ML: local.
EL: rare, Keith; Danskine Loch; Whittingehame;
Tyninghame.

Salix L.

S. pentandra L.   Bay Willow
Native. Status: local.
Beside rivers, ditches, lakes and ponds, marshes.

WL: widespread in S. but in small numbers.
ML: local, Threipmuir; Duddingston; West
Calder; Dalkeith; Stow; Ratho; Cockmuir;
Arniston.
EL: occasional, possibly planted at some sites,
Bara; Balgone; Elmscleugh; Stottencleugh.

S. fragilis L. *s.l.*   Crack Willow
Native. Status: widespread, often planted.
Beside rivers, ditches, lakes and ponds, woodland,
scrub, marshes.

WL: widespread but possibly over-recorded due
to confusion with *S. x rubens*.
ML: widespread in lowland areas, rare elsewhere.
EL: widespread in lowland areas, rare elsewhere,
Saltoun Forest; Humbie area; Morham;
Tyninghame.

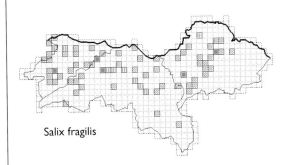

Salix fragilis

var. **decipiens** (Hoffm.) W. D. J. Koch
White Welsh Willow
Introduced. Status: rare, planted.
Wet places.

ML: rare, Gala Water; Inveresk.
EL: rare, Bara; Waughton Castle.

var. **russelliana** (Sm.) W. D. J. Koch   Bedford Willow
EL: beside the River Tyne, Haddington, 1978.

S. x rubens Schrank (*S. alba* x *S. fragilis*)
Hybrid Crack Willow
Possibly introduced. Status: rare.
Riversides, waste ground.

WL: planted on bings and waste ground, possibly under-recorded because of confusion with *S. fragilis*.
ML: rare, Inveresk; Breich.
EL: rare, River Tyne at Samuelston, with both parents.

S. alba L.   White Willow
Native. Status: widespread, mostly planted.
Riversides and other wet places, woodland, roadsides.

WL: widespread.
ML: widespread in lowland areas, rare elsewhere.
EL: widespread in lowland areas, rare elsewhere, Samuelston; Luffness; Gifford; Pefferside.

S. x sepulcralis Simonk. (*S. alba* var. *vitellina* (L.) Stokes x *S. babylonica* L.)   Weeping Willow
Introduced. Status: local, planted, not naturalised.
Policies and public parks.

ML: local, Hermitage of Braid; Lochend; Dalkeith.

S. triandra L.   Almond Willow
Introduced. Status: rare.
Riverside.

ML: rare, Trinity, 1999, one tree only; (Granton, Saughton and Craigcrook, pre-1934).
EL: rare, River Tyne at Pencaitland and Haddington.

S. purpurea L.   Purple Willow
Native. Status: scarce, mainly planted.
Riversides, lakesides, marsh, woodland, scrub.

WL: rare, on the banks of the Avon and Almond Rivers; Hopetoun.
ML: scarce, West Calder; Stow; Threipmuir;

Addiewell; Tynehead.
EL: rare, scattered sites, mainly on River Tyne, probably planted.

S. x pontederiana Willd. (*S. purpurea* x *S. cinerea*)
Native. Status: rare.
Wet places.

ML: rare, Gala Water.

S. x rubra Huds. (*S. purpurea* x *S. viminalis*)
Green-leaved Willow
Native. Status: rare.
Wet places.

ML: rare, Duddingston; Harburn.
EL: rare, River Tyne at Samuelston, probably planted.

S. daphnoides Vill.   European Violet Willow
Introduced. Status: rare.
Riversides, railway embankments, waste ground.

WL: rare, Bo'ness.
ML: rare, Granton; (Roslin, 1933, planted).
EL: rare, Archerfield Estate; Haddington; North Berwick rubbish tip; Whittingehame Water.

S. acutifolia Willd.   Siberian Violet Willow
Introduced. Status: rare, planted.
Railway embankments.

ML: rare, Granton, 1985.
EL: rare, near Drem.

S. viminalis L.   Osier
Native. Status: widespread, often planted.
Riversides and other wet places in scrub, beside roads and railways and on waste ground.

WL: widespread, locally frequent.
ML: widespread except in upland areas.
EL: quite widespread, often planted.

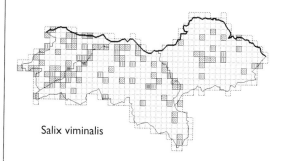

Salix viminalis

S. elaeagnos Scop. Olive Willow
Introduced. Status: rare, planted.
Waste ground.

WL: rare, Bo'ness.
ML: rare, foreshore at Granton.

S. udensis Trautv. & C. A. Mey.
Sachalin Willow
Introduced. Status: rare, planted.
Farm steading.

EL: rare, one tree near Gullane.

S. x sericans Tausch ex A. Kern. (*S. viminalis* x *S. caprea*) Broad-leaved Osier
Native. Status: local, mainly planted.
Riversides, waste ground.

WL: rare, Almondell; Bathgate; Blackridge;
Philpstoun.
ML: widespread, Hillend; Balerno; Cousland;
Roslin.
EL: rarely recorded, Bangly Hill; Latch; Papana
Water; Stoneypath; Belhaven.

S. x smithiana Willd. (*S. viminalis* x *S. cinerea*)
Silky-leaved Osier
Native. Status: rarely recorded.
Waste ground, riverside.

WL: rare, Uphall Station, 1987; (Carriden,
pre-1934).
ML: rare, Gala Water; Threipmuir; Bell's Quarry.
EL: rare, Gifford, both parents also recorded.

S. x fruticosa Döll (*S. viminalis* x *S. aurita*)
Shrubby Osier

Native. Status: rare.
Wet places.

ML: rare, Stow.

S. caprea L. Goat Willow
Native. Status: common.
Woodland, scrub, roadsides, waste ground, railway
embankments, riversides, bings.

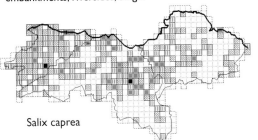

Salix caprea

WL: common.
ML: common except in upland areas.
EL: widespread.

S. x reichardtii A. Kern. (*S. caprea* x *S. cinerea*)
Native. Status: local, probably under-recorded.
Woodland, wet places, waste ground, bings.

WL: rare, probably under-recorded, Greendykes
Bing.
ML: (Threipmuir, 1973; Niddrie, 1975).
EL: rare, Fountainhall Plantation; Saltoun Forest;
Gosford; Colstoun; (Wester Pencaitland, 1957).

S. cinerea L. Grey Willow
ssp. oleifolia Macreight
Native. Status: widespread.
Riversides and other wet places, woodland, scrub,
roadsides, railway embankments, waste ground.

WL: widespread, particularly in SW.
ML: widespread, locally frequent.
EL: quite widespread.

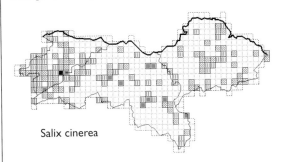

Salix cinerea

S. x multinervis Döll (*S. cinerea* x *S. aurita*)
Native. Status: rare, possibly under-recorded.
Marshes, riversides, waste ground.

WL: rare, River Avon, east of Avonbridge;
Bo'ness.
ML: rare, Burnhouses; Gladhouse and Glencorse
Reservoirs; Penicuik.
EL: rare, Monynut Water near Nether Monynut.

S. cinerea x **S. myrsinifolia**
Native. Status: rare.
Wet places.

ML: rare, Gladhouse and Threipmuir Reservoirs.
EL: (Harehead near Bothwell Water, 1973).

S. x laurina Sm. (*S. cinerea* x *S. phylicifolia*)
Laurel-leaved Willow
Native. Status: rare.
Pond margins, railway embankments.

WL: (South Queensferry, 1894; Bathgate, 1977).
ML: rare, Roslin, 1997.

S. aurita L.   Eared Willow
Native. Status: widespread in upland areas and
heathland, rare elsewhere.
Riversides, marsh, woodland, scrub, roadsides, heath
grassland.

WL, ML, EL: scattered distribution throughout
most upland areas, seldom locally frequent, rare
in lowland areas.

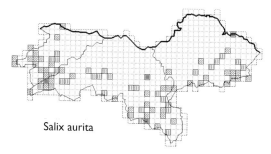

Salix aurita

S. x ambigua Ehrh. (*S. aurita* x *S. repens*)
Native. Status: rare.
Upland heath, with parents.

ML: rare, Carlops, 1999.

S. myrsinifolia Salisb.   Dark-leaved Willow
Native. Status: rare.
Bogs, ditches, railway embankments.

WL: rare, Blackburn.
ML: rare, Granton; Musselburgh; Heriot Station;
Trinity.
EL: (Linn Dean Water, 1973).

S. x tetrapla Walker (*S. myrsinifolia* x *S. phylicifolia*)
Native. Status: rare, usually with parents, often
confused with parents.
Riversides, railway embankments, roadsides.

ML: rare, Trinity; Heriot; Craighall; River
Almond.

S. phylicifolia L.   Tea-leaved Willow
Native. Status: scarce.
Riversides and other wet areas, heath grassland, waste
ground.

WL: rare, Faucheldean.
ML: occasional, mainly in upland areas,
Leadburn; West Calder; Threipmuir.
EL: rare, Linn Dean Water, 1993.

S. repens L.   Creeping Willow
var. **repens**
Native. Status: rare.
Bogs, riversides, roadsides, coastal heath.

WL: rare, Blawhorn Moss; (Drumshoreland Muir,
pre-1934).
ML: rare, Morton; Leadburn; Mount Lothian;
Longmuir Rig; Cockmuir; East Calder.
EL: rare, Pefferside; Tyninghame; (Lammer Law,
pre-1934; Bolton crossroads, 1955).

S. x mollissima Hoffm. ex Elwert (*S. triandra* x *S.
viminalis*)   Sharp-stipuled Willow
ML: River Esk, Musselburgh, 1809.

S. x calodendron Wimm. (*S. viminalis* x *S. caprea* x *S.
cinerea*)   Holme Willow
ML: Musselburgh and Colinton, pre-1927.

S. x laschiana Zahn (*S. caprea* x *S. repens*)
WL: Drumshoreland Muir, 1955.

BRASSICACEAE

Sisymbrium L.

S. altissimum L.   Tumbling Mustard, Tall Rocket
Introduced. Status: local, declining because of
habitat loss.
Waste ground.

WL: rare, Bo'ness; (Newliston and South
Queensferry, pre-1934; Carriden, 1955).
ML: recorded from a small number of mainly
coastal or urban sites, Borthwick railway tip;
Leith.
EL: recorded from a small number of mainly
coastal sites, population size varies but is
generally small, Prestongrange to Seton Sands;
Tranent; Luffness; Pefferside; West Barns;
Dunbar.

S. orientale L.   Eastern Rocket
Introduced. Status: local.
Waste ground, roadsides, beaches.

WL: rare, Bo'ness; Winchburgh; (Carriden and
South Queensferry, 1976; Linlithgow, 1977).
ML: local, frequent in some sites in Edinburgh,
Leith; Granton; Warriston; Calton Hill.
EL: quite widespread along the coast, including
Fidra and Bass Rock; scattered sites inland,
increasing.

S. officinale (L.) Scop.   Hedge Mustard

Native. Status: common except in upland areas. Roadsides, waste and cultivated ground, hedgerows, railway embankments.

WL: widespread.
ML, EL: common except in upland areas where it is rare.

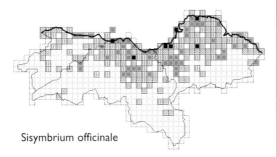

Sisymbrium officinale

var. **leiocarpum** DC.
EL: North Berwick, 1955; West Barns, 1972.

S. loeselii L.   False London Rocket
Grain casual.
WL: South Queensferry,1906.
ML: Leith Docks, 1887–1922.

Descurainia Webb & Berthel.

D. sophia (L.) Webb ex Prantl.   Flixweed
Introduced. Status: rare, casual.
Waste ground, arable fields, docks, coastal turf.

WL: rare, Greendykes Bing and Craigton Quarry, 1982; field north of Linlithgow, 1995; (Bo'ness, 1959).
ML: rare, Leith; Loanhead; Granton.
EL: an occasional plant found at scattered sites, mainly along the coast, Cockenzie; Luffness; Drem; North Berwick; West Barns; Dunbar; Catcraig.

Alliaria Heist. ex Fabr.

A. petiolata (M. Bieb.) Cavara & Grande
                    Garlic Mustard, Jack-by-the-hedge
Native. Status: common.
Roadsides, hedgerows, waste and cultivated ground, woodland, riversides, scrub.

WL, ML, EL: common except in upland areas where it is rare or absent.

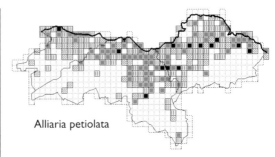

Alliaria petiolata

Arabidopsis (DC.) Heynh.

A. thaliana (L.) Heynh.   Thale Cress
Native. Status: local.
Waste ground, roadsides, walls, gardens, railway tracks.

WL: widespread.
ML: local, locally abundant.
EL: scattered distribution throughout lowland areas.

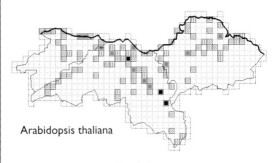

Arabidopsis thaliana

Isatis L.

I. tinctoria L.   Woad
Introduced. Status: rare, not persisting.

ML: rare, planted in nature reserve, Bawsinch, 1986.
EL: (Aberlady, 1863; Prestonpans, 1894).

Bunias L.

B. orientalis L.   Warty Cabbage
Introduced. Status: rare.
Railway embankments, waste ground.

WL: rare, Bo'ness.
ML: rare, Kingsknowe; Catcune; Granton; Ravelston.
EL: rare, Keith, 1993.

Erysimum L.

E. cheiranthoides L.   Treacle Mustard
Introduced. Status: rare, casual.

Waste and cultivated ground.

WL: rare, Bonhard Farm area; north-west of
Cauldcoats; (South Queensferry, pre-1934).
ML: rare, Leith; Craigmillar; near Stow.
EL: rare, Gullane Hill, 1997; sites around the
village of West Barns since 1958; in crops near
Philip Burn, 1994; (Seton Mains, pre-1934).

E. cheiri (L.) Crantz　　　　　　Wallflower
Introduced.　Status: local, garden escape, often
naturalised.
Quarries, walls, cliffs, waste ground.

WL: rare, Port Edgar; (Linlithgow, pre-1934).
ML: occasional, locally frequent on Edinburgh,
Craigmillar and Roslin Castles; Calton Hill.
EL: rare, Seton Sands; Gullane; Tantallon Castle
since before 1927; Belhaven; Dunbar.

Hesperis L.

H. matronalis L.　　Sweet Rocket, Dame's Violet
Introduced.　Status: widespread, garden escape.
Waste ground, riversides, roadsides, woodland,
hedgerows.

WL: quite widespread, well naturalised in places.
ML: widespread, confined to lowland areas.
EL: scattered sites in lowland areas, Seton Sands;
Pencaitland; River Tyne below Hailes.

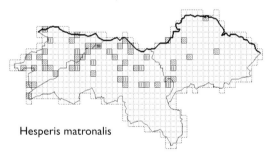

Hesperis matronalis

Malcolmia W.T. Aiton

M. maritima (L.) W. T. Aiton　　　　Virginia Stock
EL: North Berwick, 1914.

Barbarea W.T. Aiton

B. vulgaris W. T. Aiton
　　　　　　　Winter Cress, Yellow Rocket
Native.　Status: widespread except in upland
areas.
Roadsides, riversides, waste ground, rough grassland.

WL: quite widespread except in S.
ML: local, mainly in lowland areas.
EL: quite widespread in lowland areas.

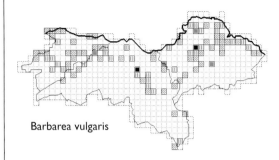

Barbarea vulgaris

var. **arcuata** Fries
EL: East Linton, pre-1927.

B. intermedia Boreau
　　　　　　Medium-flowered Winter Cress
Introduced.　Status: scarce.
Cereal fields, roadsides, riversides.

WL: local, often on or near bings.
ML: rare, Blackford; Currie; Symington, 1994.
EL: rare, Prestonpans, 1988.

B. verna (Mill.) Asch.　　American Winter Cress
Introduced.　Status: rare.
Roadsides, railway embankments.

WL: (Kirkliston, pre-1934).
ML: (occasional, pre-1934; Hailes Quarry, 1972).
EL: rare, Haddington, 1993.

Rorippa Scop.

R. nasturtium-aquaticum agg.　　　　Watercress
Native.　Status: widespread.
In and beside streams and ditches, marshes.

WL: widespread, locally frequent.
ML: widespread, locally abundant.
EL: widespread.

Includes the following three taxa that were not
always distinguished during the survey.

R. nasturtium-aquaticum (L.) Hayek　Watercress
Native.　Status: scarce.
In and beside streams, ditches and ponds, marshes.

WL: no definite records.
ML: rarely recorded.
EL: rarely recorded, Ware Road, Tyninghame;
Hedderwick; Catcraig.

R. x sterilis Airy Shaw (*R. nasturtium-aquaticum* x *R. microphylla*)   Hybrid Watercress
Native. Status: rare.
Marshes.

ML: rare, Niven's Knowe, 1999.
EL: rare, Haddington; Luggate Burn; Hedderwick; West Barns; (Tantallon Castle, 1932; Longniddry, 1954).

R. microphylla (Boenn.) Hyl. ex Á. & D. Löve
                    Narrow-fruited Watercress
Native. Status: widespread, under-recorded in the survey. The most common of the three taxa.
In and beside streams and ditches, marshes.

WL: widespread, locally frequent.
ML: fairly widespread.
EL: carefully recorded across most of the vice-county, fairly widespread, locally frequent.

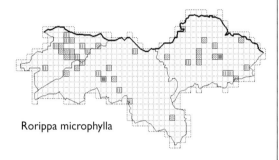

Rorippa microphylla

R. islandica (Oeder ex Gunnerus) Borbás
                    Northern Yellow-cress
Native. Status: rare.
Beside open water, goose roosts.

ML: rare, locally frequent at Harlaw Reservoir.

R. palustris (L.) Besser   Marsh Yellow-cress
Native. Status; scarce.
Beside open water, waste ground, marshes.

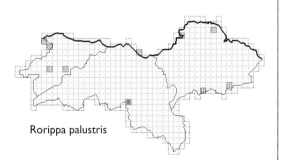

Rorippa palustris

WL: scattered distribution along the Avon and Almond Rivers; also found beside lochs.
ML: rare, Figgate Burn; Duddingston Loch; Threipmuir and Rosebery Reservoirs.
EL: rare, Gladsmuir; Aberlady Bay; Hopes Reservoir; Whitekirk; Whiteadder Reservoir; (Luffness and Dirleton, 1835; Drem, 1871).

R. sylvestris (L.) Besser   Creeping Yellow-cress
Native. Status: scarce.
Beside lakes, ponds and streams.

WL: rare, Petershill; (South Queensferry and Linlithgow, 1960s).
ML: local, Musselburgh; Bowshank; Threipmuir.
EL: rare, briefly beside a new pond, Luffness, *c.* 1985.

R. x armoracioides (Tausch) Fuss (*R. sylvestris* x *R. austriaca*)   Walthamstow Yellow-cress
Introduced. Status: rare, bird seed alien, possibly increasing, overlooked.
Waste ground.

ML: rare, Craigmillar, 2000; Blackhall.

R. amphibia (L.) Besser   Great Yellow-cress
Introduced. Status: rare.
In and beside ponds and ditches, damp grassland.

ML: rare, Musselburgh; Holyrood Park, 1997.

R. austriaca (Crantz) Besser   Austrian Yellow-cress
WL: South Queensferry, 1959.
ML: Davidson's Mains, 1972.

Armoracia P. Gaertn., B. Mey. & Scherb.

A. rusticana P. Gaertn., B. Mey. & Scherb.
                    Horse-radish
Introduced. Status: scarce.
Roadsides, waste ground.

WL: rare, Blackness area; Armadale; (Bo'ness, 1973; Ryal, 1974).
ML: rare, Slateford; Musselburgh; Leith; West Calder.
EL: rare, Prestonpans; Luffness; Gullane rubbish tip; Dirleton; Gifford; Stenton.

Cardamine L.

C. amara L.   Large Bittercress
Native. Status: local.
In and beside streams, marshes.

WL: occasional in N. only.

ML: occasional except in N. and SE.
EL: occasional along River Tyne; Biel Water; Woodhall Burn; Dunglass Burn and their tributaries.

var. **erubescens** Peterm.
EL: East Linton, pre-1927.

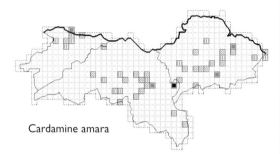

Cardamine amara

C. raphanifolia Pourr.   Greater Cuckoo Flower
Introduced. Status: rare, but possibly increasing.
Riversides.

ML: rare, Glencorse; Gorebridge; Roslin.

C. pratensis L.   Cuckoo Flower, Lady's Smock
Native. Status: common.
Marshes, beside rivers and ditches, damp rough grassland, damp roadside verges.

WL: common.
ML: common except in urban and intensively farmed areas.
EL: widespread and locally frequent, rarer in intensively farmed areas

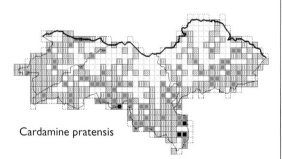

Cardamine pratensis

C. flexuosa With.   Wavy Bittercress
Native. Status: common.
Beside rivers and ditches, woodland, marshes, roadsides, damp rough grassland.

WL: widespread.
ML: widespread, locally frequent.
EL: widespread, locally frequent except in N.

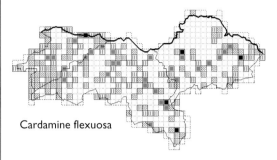

Cardamine flexuosa

C. hirsuta L.   Hairy Bittercress
Native. Status: common.
Roadsides, waste ground, gardens, public parks, farmyards, railway embankments, walls, pasture, woodland.

WL: widespread.
ML: widespread and locally abundant except in upland areas where it is rare.
EL: common in central lowland areas, occasional in N., rare in upland areas.

C. corymbosa Hook. f.   New Zealand Bittercress
Introduced. Status: rare, garden weed.
Gardens.

ML: a weed at Royal Botanic Garden Edinburgh, from 1975 onwards, quite well established.
EL: a weed in bought-in potted-up plants in a nursery near Dunbar, 1992. Liable to be introduced to East Lothian soil.

C. trifolia L.   Trefoil Cress
ML: Craigmillar Quarry, 1946–81.

Arabis L.

A. caucasica Willd. ex Schltdl.   Garden Arabis
Introduced. Status: rare, casual garden escape.
Waste ground, walls.

WL: (Blackridge and Carriden, 1970; Bathgate, 1976).
ML: rare, Little Cathpair, 1995.
EL: (East Linton, 1960; Haddington, 1970; Thurston, 1971; Gosford, 1976).

A. hirsuta (L.) Scop.   Hairy Rock-cress
Native. Status: scarce.
Roadsides, quarries, walls, volcanic outcrops, dune grassland.

WL: (Carriden, pre-1934).
ML: scarce, Blackford; Esperston; Habbie's Howe.

251

EL: scarce, Ferny Ness; Aberlady Bay; Hailes; Stenton; Belhaven Bay.

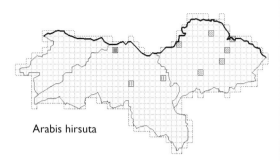

Arabis hirsuta

A. turrita L.   Tower Cress
WL: Linlithgow and Kinneil, pre-1934.

Aubrieta Adans.

A. deltoidea (L.) DC.   Aubretia
Introduced. Status: rare, garden escape.
Walls, quarries.

WL: rare, Carriden; (Midhope Castle, 1956).
ML: rare, Roseburn, 1996.
EL: rare, Markle Quarry; Innerwick; (quarry at Auldhame, 1957).

Lunaria L.

L. annua L.   Honesty, Silver Shekels
Introduced. Status: local, garden escape.
Roadsides, waste ground, rubbish tips, hedgerows.

WL: local.
ML: widespread in urban areas.
EL: scattered sites in lowland areas, Longniddry Bents; Keith; Dirleton; Whittingehame.

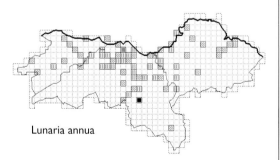

Lunaria annua

Alyssum L.

A. alyssoides (L.) L.   Small Alison
ML: Arthur's Seat, 1853; Leith Docks, 1906.
EL: Dirleton Common, 1855; North Berwick, pre-1863; Gullane, 1903.

Berteroa DC.

B. incana (L.) DC.   Hoary Alison
EL: beside the railway at Seton Mains Halt, 1915.

Lobularia Desv.

L. maritima (L.) Desv.   Sweet Alison
Introduced. Status: rare, garden escape, apparently ephemeral.
Waste ground, walls, pavements, dunes.

WL: rare, Bathgate, formerly elsewhere.
ML: rare, pavements, Edinburgh.
EL: rare escape from gardens near the coast.

Draba L.

D. muralis L.   Wall Whitlowgrass
Introduced. Status: local.
Waste ground, walls, bings, rocky outcrops.

WL: scarce, Bridgend; Craigton; Hopetoun; Woodend; abundant on parts of Philpstoun Bing.
ML: rare casual, Cairntow, 1998; previously widespread.
EL: rare, near Whitekirk; Broxmouth; (Port Seton pre-1934; near West Saltoun, 1971).

Erophila DC.

E. verna agg.   Whitlowgrass
Native. Status: widespread.
Roadsides, waste ground, rough grassland, quarries, dunes. Usually on dry, open ground.

WL: quite widespread.
ML: local, abundant in places.
EL: quite widespread, locally frequent.

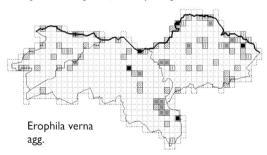

Erophila verna agg.

Includes the following three taxa.

E. majuscula Jord.   Hairy Whitlowgrass
Native. Status: rare.
Volcanic outcrops.

ML: rare, St. Leonards, 1999; (Blackford Hill,
Arthur's Seat, Craiglockhart Hill, Mortonhall and
Colinton, pre-1915).
EL: rarely recorded, Markle; Hailes; Whitekirk
Hill; (Gosford, 1889; Aberlady, 1933).

E. verna (L.) DC.   Common Whitlowgrass
Native. Status: widespread, most records refer to
this species.
Roadsides, waste ground, rough grassland, quarries,
dunes. Usually on dry, open ground.

WL: quite widespread.
ML: local.
EL: rarely recorded, Cockenzie; Longniddry
Bents; Markle; Hailes Castle; St Baldred's Cradle.

E. glabrescens Jord.   Glabrous Whitlowgrass
Native. Status: rare or under-recorded.
Open habitats, gardens, coastal turf, volcanic outcrops,
walls.

WL: rare, Dalmeny.
ML: rare, Granton Harbour, 1998; Blackford Hill
and St. Leonards 1999; (Swanston, Arthur's Seat,
Inverleith, Craigmillar and Corstorphine Hill,
pre-1910).
EL: rare, Longniddry Bents; Aberlady; Hailes
Castle; (Biel Estate, 1956).

Cochlearia L.

C. megalosperma (Maire) Vogt   Tall Scurvygrass
Introduced. Status: naturalised on one site.
Old stone wall and verge of lane.

EL: near Humbie Kirk, already well established in
1994.

C. officinalis L.   Common Scurvygrass
Native. Status: along the coast, common in **EL**,
local elsewhere.
Beaches, coastal turf, saltmarsh, cliffs, coastal rocks,
dunes, rough grassland.

WL: locally frequent along the coast.
ML: local, Seafield; Cramond Island; Fisherrow;
Inchmickery.
EL: common along the coast, locally abundant.

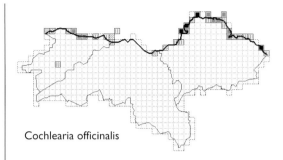

Cochlearia officinalis

ssp. **scotica** (Druce) P. S. Wyse Jacks.
ML: Musselburgh, 1868.
EL: North Berwick, 1911, specimens determined recently
by P. S. Wyse Jackson.

C. officinalis x C. danica
Native. Status: rare.
Dunes, coastal rocks, beaches, where both parents
occur.

EL: rare, Prestonpans, 1987; Archerfield, 1982;
North Berwick, 1982; Bass Rock, 1985; Dunbar
Castle, 1987.

C. danica L.   Danish Scurvygrass
Native. Status: local.
Coastal turf, cliffs, coastal rocks, dunes, roadsides.

WL: local, Port Edgar, rare elsewhere on the
coast; now colonising the M9 Motorway.
ML: local, locally frequent, Cramond Island;
Inchmickery; salted trunk roads.
EL: occasional and locally frequent on islands in
the Firth of Forth and on north-facing rocky
shores, also Cockenzie Power Station; Dunbar
Castle.

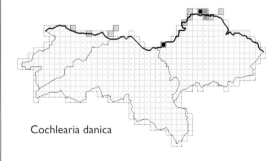

Cochlearia danica

C. micacea E. S. Marshall   Mountain Scurvygrass
ML: Pentland Hills, 1966; Medwinhead, det. T. C. G. Rich.

Camelina Crantz

C. sativa (L.) Crantz   Gold-of-pleasure
WL: Newliston, pre-1934.
ML: Leith and Pathhead, 1963.
EL: North Berwick, 1872; Haddington, 1888.

C. microcarpa Andrz. ex DC.
                    Lesser Gold-of-pleasure
Introduced. Status: rare, casual.
Waste ground, quarries, rubbish tips.

ML: rare, Granton, 1982; (Leith Docks,
sporadically, 1887–1908; Craigmillar Quarry,
1961; Gorebridge, 1963; Hailes Quarry, 1971).

Neslia Desv.

N. paniculata (L.) Desv.   Ball Mustard
Introduced. Status: rare, casual.
Roadsides.

ML: rare, Camilty, 1992.
EL: (roadside between Gullane and Dirleton,
1912).

Capsella Medik.

C. bursa-pastoris (L.) Medik.   Shepherd's Purse
Native. Status: common.
Roadsides, cultivated and waste ground, pasture.

WL, ML, EL: common except in some upland
areas.

Teesdalia W. T. Aiton

T. nudicaulis (L.) W. T. Aiton   Shepherd's Cress
ML: Braid Hills and Dalmahoy, pre-1934.
EL: between Tyninghame and East Linton, pre-1934.

Thlaspi L.

T. arvense L.   Field Pennycress
Introduced. Status: local.
Waste ground, arable fields, roadsides.

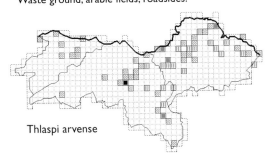

Thlaspi arvense

WL, ML: local.
EL: scattered distribution in lowland areas.

Iberis L.

I. sempervirens L.   Perennial Candytuft
Introduced. Status: rare, garden escape.
Railway tracks.

ML: rare, Kinleith, 1996.

I. umbellata L.   Garden Candytuft
Introduced. Status: rare, non-persistent garden
escape.
Rubbish tips, vicinity of gardens.

WL: rare, Bo'ness; (Bathgate, 1973).
ML: rare, Hailes Quarry; Borthwick Bank;
(Musselburgh, 1969).
EL: rare, East Linton, 1986; lane at Dunglass,
1998; (North Berwick, 1965; east of East Linton,
1966).

I. pruitii Tineo
Introduced. Status: rare, casual.
Waste ground.

ML: rare, Craigmillar, 1999; (Leith Docks,
1970–2).

I. crenata Lam.
Grain casual.
WL: South Queensferry, 1907.
ML: Leith Docks, pre-1900.

Lepidium L.

L. sativum L.   Garden Cress
Introduced. Status: rare, casual.
Waste ground.

ML: rare, Leith Docks, 1986.
EL: (West Barns, 1969).

L. campestre (L.) W. T. Aiton   Field Pepperwort
? Native. Status: rare.
Quarries, rocky outcrops, rubbish tips, waste ground.

WL: rare, Uphall Station; (Linlithgow and
Hopetoun, pre-1934).
ML: rare, Sighthill; Granton; (Borthwick, 1963).
EL: rare, Stottencleugh; (Dunglass Dean, 1834;
Biel; fields at Seton Mains, pre-1934).

L. heterophyllum Benth.   Smith's Pepperwort
Native. Status: scarce.

Waste ground, roadsides, quarries, rocky outcrops, rough grassland, railways.

WL: rare, Bo'ness; old railway embankment north of Winchburgh; (Carriden and Linlithgow, 1957; Broxburn, 1967).
ML: scarce, Dalmahoy; Rosewell; Borthwick; Musselburgh; Milton Bridge, 1988.
EL: rare, Ormiston; Markle; Whitekirk; (Garleton Hills, 1970; Tranent, 1971).

L. latifolium L.   Dittander
Introduced. Status: rare.
Dunes, beaches, cliffs, waste ground.

WL: (Kinneil, pre-1934).
EL: rare, at Tantallon Castle for more than a century; West Barns, since 1973.

L. draba L.   Hoary Cress
ssp. draba
Introduced. Status: scarce, but established.
Roadsides, waste ground, coastal habitats, public parks, rough grassland.

WL: rare, Bo'ness; Blackness Castle; Niddry; (Carriden, 1960; Linlithgow, *c.* 1980).
ML: local, locally frequent, Seafield; Ratho; Granton; Fisherrow.
EL: scarce, Prestonpans; West Barns; Dunbar; old A1 trunk road in Skateraw area; (Aberlady, pre-1934; Seton Sands, 1965; Gullane, 1973).

ssp. chalepense (L.) Thell.
Introduced. Status: rare.
Waste ground, verges.

ML: rare, Musselburgh Lagoons and Seafield, 1997.

L. ruderale L.   Narrow-leaved Pepperwort
Casual.
WL: Kirkliston, pre-1934.
ML: Leith Docks and Portobello, pre-1934.
EL: Prestonpans, pre-1894.

Coronopus Zinn

C. didymus (L.) Sm.   Lesser Swinecress
Introduced. Status: rare.
Waste and cultivated ground.

WL: rare, north of Winchburgh; near Uphall Station; (Linlithgow, pre-1934).
ML: rare, Blackford Hill; King's Buildings, 1985; Inverleith; Granton.
EL: rare, near Aberlady; known since 1900 in the

Luffness area; (Haddington and Dunbar, pre-1934).

C. squamatus (Forssk.) Asch.   Swinecress
ML: Musselburgh, pre-1934.
EL: Dirleton, Gullane, Tantallon and Dunbar, pre-1934.

Subularia L.

S. aquatica L.   Awlwort
WL: Linlithgow Loch, pre-1894.

Conringia Heist. ex Fabr.

C. orientalis (L.) Dumort.   Hare's-ear Mustard
WL: South Queensferry, pre-1934.
ML: Inveresk and Leith, pre-1934.
EL: Ormiston, 1883; Haddington, 1897.

Diplotaxis DC.

D. tenuifolia (L.) DC.   Perennial Wall-rocket
Introduced. Status: rare, casual.
Waste ground.

WL: rare, track near Dalmeny village; (Abercorn, pre-1934).
ML: rare, Leith; Granton.
EL: rare, Prestonpans; Aberlady; (Morrison's Haven, Port Seton, and Longniddry, pre-1934).

D. muralis (L.) DC.   Annual Wall-rocket
Introduced. Status: rare, often casual.
Farmyards, waste ground, quarries, public parks, roadsides.

WL: (Bo'ness, 1956; Broxburn, 1973).
ML: rare, formerly occasional, Leith Docks; Levenhall, 1994; (Granton, 1960).
EL: occasional, possibly increasing, established in the Prestongrange/Cockenzie area and at North Berwick; casual elsewhere.
The perennial growth form found at Cockenzie, 1998, resembles *D. tenuifolia* in general appearance.

Brassica L.

B. oleracea L.   Cabbage
Introduced. Status: rare, escape from cultivation.
Waste ground, roadsides.

WL, ML, EL: rare.

B. napus L.   Rape, Swede
Introduced. Status: widespread except in upland

areas, escape from, or relic of, cultivation.
Arable fields, waste ground, roadsides.

WL, ML, EL: widespread except in upland areas, locally frequent.
Most recent records are probably:

ssp. oleifera (DC.) Metzg. Oil-seed Rape
This has become a standard crop over the last twenty years and is now an extremely common and widespread escape.

B. rapa L. Turnip
Introduced. Status: local, casual.
Waste ground, rubbish tips, roadsides.

WL, EL: rare.

ssp. rapa
Introduced. Status: local, relic of cultivation. Most records for the species are probably this subspecies.

ssp. oleifera (DC.) Metzg. Turnip Rape
Introduced. Status: unknown. Now being grown as a crop in the Lothians, may occur in habitats in which *B. napus* ssp. *oleifera* is found.

B. nigra (L.) W. D. J. Koch Black Mustard
Probably introduced. Status: rare.
Waste ground.

WL: (Bo'ness, pre-1934).
ML: rare, Leith, 1982; Cockpen, 1983; (Granton, 1972).
EL: rare, Dirleton; West Barns; (North Berwick rubbish tip, 1958; Craigleith Island, 1965; Prestonpans, 1972 and 1976).

B. juncea (L.) Czern. Chinese Mustard
Introduced, casual.
Quarry tips, waste ground.
ML: Hailes rubbish tip, 1964 and 1971; Granton, 1972.

B. elongata Ehrh. Long-stalked Rape
Introduced, casual.
Quarry tips.
ML: Hailes, 1971.

Sinapis L.

S. arvensis L. Charlock
Native. Status: common.
Arable fields, waste ground, roadsides, rough grassland.

WL: widespread.

ML: widespread except in upland areas, locally frequent.
EL: widespread and locally frequent in lowland areas, rare in upland areas.

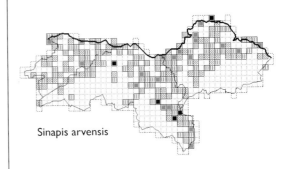

Sinapis arvensis

S. alba L. White Mustard
Introduced. Status: rare.
Arable fields, waste ground.

WL: (Carriden and Linlithgow, pre-1934).
ML: rare, Polton; Fala; Leith Docks, 1989; (Slateford rubbish tip, 1945).
EL: rare, Tranent; near Soutra; Tyninghame; West Barns; (Tranent, pre-1934; Pencaitland, 1956; Dirleton, 1957; Danskine, 1970).

Eruca Mill.

E. vesicaria (L.) Cav. Garden Rocket
ssp. sativa (Mill.) Thell.
Introduced. Status: rare, casual, garden escape.
Waste ground.

ML: rare, Craigmillar, 1999.

Erucastrum C. Presl

E. gallicum (Willd.) O. E. Schultz Hairy Rocket
Casual.
ML: Leith Docks, 1972.

Coincya Rouy

C. monensis (L.) Greuter & Burdet
ssp. monensis Isle of Man Cabbage
Introduced. Status: rare, casual.
Waste ground.

WL: rare, Broxburn; (Bo'ness, 1972).
ML: (Currie, 1953).
EL: (Archerfield, 1959).

ssp. cheiranthos (Vill.) Aedo, Leadlay & Muñoz Garm.   Wallflower Cabbage
Introduced. Status: rare, casual.
Waste ground, rubbish tips.

ML: rare, Borthwick and Craigmillar, 1999; (Granton, from 1976 to 1979).
EL: (Gullane, 1853).

Hirschfeldia Moench

H. incana (L.) Lagr.-Foss.   Hoary Mustard
Introduced. Status: rare, casual.
Waste ground.

ML: rare, Leith Docks, 1985; (Levenhall, *c.* 1900; recorded as long-established at Eskbridge, 1908).

Cakile Mill.

C. maritima Scop.   Sea Rocket
ssp. integrifolia (Hornem.) Hyl. ex Greuter & Burdet
Native. Status: local.
Beaches, dunes.

WL: rare, Blackness; Dalmeny.
ML: locally frequent along the coast from Cramond Island to Levenhall.
EL: occasional along the whole coastline.

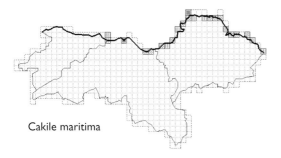

Cakile maritima

Rapistrum Crantz

R. rugosum (L.) J. P. Bergeret   Bastard Cabbage
ssp. linnaeanum (Coss.) Rouy & Foucaud
ML: Fushiebridge, 1958; Gorebridge, 1964; Blackford Quarry and rubbish tips at Leith Docks, 1969.

Crambe L.

C. maritima L.   Sea Kale
WL: Dalmeny and South Queensferry, pre-1934.
ML: Cramond, pre-1934.
EL: Bass Rock, probably in the nineteenth century; The Lamb, pre-1934.

Raphanus L.

R. raphanistrum L.
ssp. raphanistrum   Wild Radish
Introduced. Status: local.
Cultivated and waste ground, rough grassland.

WL, ML: local.
EL: rare, Tranent; North Berwick area; Monynut; (widespread 1958–71, mainly in fields).

ssp. maritimus (Sm.) Thell.   Sea Radish
? Native. Status: rare.
Maritime cliff top.

EL: rare, Bass Rock.

R. sativus L.   Garden Radish
ML: Leith Docks, 1959; Bilston, 1960.

RESEDACEAE

Reseda L.

R. luteola L.   Dyer's Rocket, Weld
Native. Status: local.
Waste ground, roadsides, rocks, walls, open grassland, dry ground.

WL: widespread.
ML: widespread in N. and centre, rare elsewhere.
EL: widespread in lowland areas.

R. lutea L.   Wild Mignonette
? Introduced. Status: local.
Waste ground, beside roads and railways.

WL: local, mainly around Bo'ness.
ML: local, Monktonhall and Seafield, 1988; Duddingston, 1995; (Borthwick, 1950; Esperston, 1973).
EL: rare, only in N. and mainly near the coast; most easily seen beside the road to Yellowcraig and by Linkfield car park, West Barns.

EMPETRACEAE

Empetrum L.

E. nigrum L.   Crowberry
ssp. nigrum
Native. Status: widespread in upland areas.
Moors, coastal heath.

WL: rare, north of Blackridge including

Blawhorn Moss; Livingston Reservoir; one plant
on Faucheldean Bing.
ML: widespread and locally frequent in upland
areas.
EL: widespread and locally frequent in upland
areas, rare along the coast, Gullane Hill; St
Baldred's Cradle; Bilsdean.

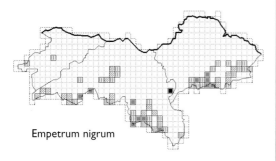

Empetrum nigrum

ERICACEAE

Rhododendron L.

R. ponticum L.   Rhododendron
Introduced. Status: widespread.
Woodland, scrub, policies, public parks, roadsides,
riversides.

WL: widespread and locally frequent.
ML: widespread and locally frequent except in
SE.
EL: widespread and locally frequent, less so in
SE.

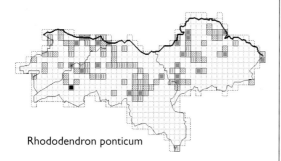

Rhododendron ponticum

R. luteum Sweet   Yellow Azalea
Introduced. Status: local, planted on old estates.
Woodland, scrub.

WL: local, Hopetoun; Almondell; old estate near
Bathgate.

Andromeda L.

A. polifolia L.   Bog Rosemary
ML: Auchencorth Moss, 1971.

Gaultheria L.

G. shallon Pursh   Shallon
Introduced. Status: rare.
Scrub in old estates.

WL: rare, Almondell; Polkemmet; near
Blackridge.
ML: rare, Kirknewton; Penicuik and Roslin
Estates; near Glencorse House.

Calluna Salisb.

C. vulgaris (L.) Hull   Ling, Heather
Native. Status: common in upland areas.
Moors, heath grassland, upland pasture, bogs,
woodland, scrub, waste ground, some coastal habitats.

WL: common in SW., scattered remnants
elsewhere.
ML: abundant in upland areas, occasional
elsewhere.
EL: abundant in upland areas, occasional in
heathy woods and on the coast.

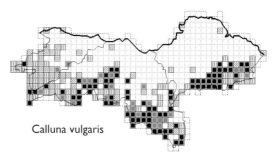

Calluna vulgaris

Erica L.

E. tetralix L.   Cross-leaved Heath
Native. Status: widespread in S. and W., rare
elsewhere.
Bogs, moors, wet heath grassland, wet woodland edges.

WL: widespread and locally frequent in SW.
ML: widespread and locally abundant in
Moorfoot Hills, occasional in some other upland
areas.
EL: widespread and locally frequent in southern
areas of the Lammermuir Hills, rare in lowland
areas and on the coast, Saltoun Forest;
Gladsmuir; St Baldred's Cradle.

E. cinerea L.   Bell Heather
Native. Status: widespread in most upland areas.
Dry moors, upland and coastal heath grassland,
heathy woodland edges, volcanic outcrops.

WL: scarce, mainly in the Bathgate Hills.
ML: widespread and locally frequent in upland
areas in S. and E., local in upland areas elsewhere.
EL: widespread in upland areas, rare elsewhere;
summit of Gullane Hill; Petersmuir Wood; Bolton
Muir Wood; St Baldred's Cradle.

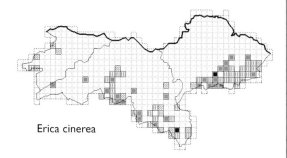

Erica cinerea

Vaccinium L.

V. oxycoccos L.   Cranberry
Native. Status: local.
Bogs, usually on *Sphagnum*.

WL: local, in remnant bogs in SW. There are
seventeen sites, most of which are small.
ML: local, Bawdy Moss; Toxside; Auchencorth;
Cobbinshaw; Levenseat.

V. microcarpum (Turcz ex Rupr.) Schmalh.
                     Small Cranberry
Native. Status: rare.
Wet moorland.

ML: rare, Bawdy Moss.

V. vitis-idaea L.   Cowberry
Native. Status: local.
Dry moors.

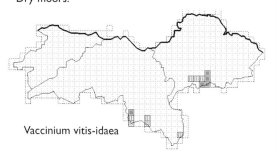

Vaccinium vitis-idaea

ML: local, east and west sides of Cairn Hill;
Craigengar; Blackhope Scar.
EL: local, concentrated in the south of the upland
area, locally frequent, Lammer Law; Hopes
Reservoir.

V. myrtillus L.   Blaeberry, Bilberry
Native. Status: common in upland areas, local
elsewhere.
Dry moors, upland pasture, woodland, bogs, roadsides,
quarries, rocky ledges.

WL: widespread, most frequent in SW.
ML: common in upland areas, occasional in
woods in lowland areas.
EL: common in upland areas, occasional in lowland areas, rare on the coast.

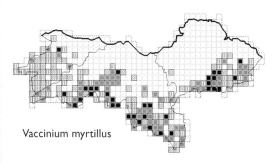

Vaccinium myrtillus

PYROLACEAE

Pyrola L.

P. minor L.   Common Wintergreen
Native. Status: scarce (declining).
Woodland, often under birch, bings.

WL: local, mainly in SW., large colony on
Foulshiels Bing.
ML: scarce, Penicuik to Roslin, 1984; Howgate,
1987; Threipmuir, 1993; previously more wide-
spread.
EL: rare, Linn Dean Water; three sites in the
Gifford area; (Whittingehame, 1873).

P. media Sw.   Intermediate Wintergreen
Native. Status: rare, (declining).
Woodland.

ML: (Bilston, 1863; Tynehead and Juniper Lee,
pre-1934).
EL: rare, one site in the Gifford area.

259

P. rotundifolia L.   Large or Round-leaved Wintergreen
ML: Auchendinny, pre-1863.

Orthilia Raf.

O. secunda (L.) House
                    Serrated Wintergreen, Yavering Bells
WL: Bo'ness, recorded by Sonntag, 1894.

PRIMULACEAE

Primula L.

P. vulgaris Huds.   Primrose
Native. Status: widespread.
Woodland, scrub, riversides, rough grassland, roadsides, hedgerows, dunes.

WL: widespread, except in S.
ML: widespread, except in N.
EL: widespread.

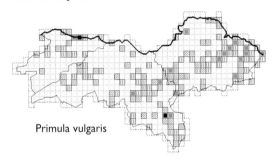

Primula vulgaris

P. x polyantha Mill. (*P. vulgaris* x *P. veris*)
                    False Oxlip
Native. Status: rare, where parents occur together.
Rough grassland, dunes, coastal turf/scrub.

WL: rare, Wester Shore; Hopetoun; (Dalmeny, *c*. 1980).
ML: scarce, locally frequent, Borthwick Bank; Tynehead.
EL: rare, some good populations along cliff path east of North Berwick; Tantallon; Seacliff.

P. elatior (L.) Hill   Oxlip
Introduced. Status: rare.
Woodland.

ML: rare, Dalkeith Country Park, 1996.

P. veris L.   Cowslip
Native. Status: local.
Woodland, policies, rough grassland, dunes, roadsides.

WL: local, good population on the Dalmeny Estate; scattered sites elsewhere.
ML: local, Chalkieside 1983; Dalkeith Country Park, 1996; (Borthwick, 1956; Oxenfoord Estate, 1965).
EL: scattered sites in lowland areas and along the coast, enhanced by planting, frequent at Longniddry Bents and Gosford Bay.

Lysimachia L.

L. nemorum L.   Yellow Pimpernel
Native. Status: widespread in S. and E., occasional elsewhere.
Marshes, beside rivers and ditches, woodland.

WL: scarce.
ML, EL: widespread in S. and E. where it is locally frequent, occasional elsewhere.

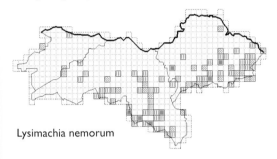

Lysimachia nemorum

L. nummularia L.   Creeping-Jenny
Introduced. Status: rare.
Woodland, roadsides, ditchsides.

WL: Auchenhard Woods; (Linlithgow and Philpstoun, pre-1934; Pepper Wood until *c*. 1980).
ML: rare, Kirknewton Estate, 1982; (Carberry, 1978).
EL: rare, Gladsmuir; Yester; Papana Water; near Elmscleugh.

L. vulgaris L.   Yellow Loosestrife
Native. Status: rare, rarely casual.
Marshes, beside ditches and ponds.

WL: (Linlithgow and Carlowrie, pre-1934).
ML: rare, Musselburgh (casual); Penicuik Estate (doubtful if native here).
EL: (in dunes at Bilsdean, 1957).

L. punctata L.   Dotted Loosestrife
Introduced. Status: local, garden escape.
Waste ground, roadsides, riversides, scrub, dunes.

WL: scattered.
ML: local.
EL: rare, Longniddry Bents; Garvald; Belhaven;
(Bilsdean).

var. **verticillata** (M. Bieb.) Boiss.
ML: Cramond Bridge, 1959.

L. thyrsiflora L.   Tufted Loosestrife
Native. Status: scarce.
Marshes, canal.

WL: scarce, Union Canal.
ML: rare, Union Canal; Meggetland; Long
Hermiston; Cobbinshaw Marshes; (Duddingston
Loch).

Trientalis L.

T. europaea L.   Chickweed Wintergreen
Native. Status: rare, (declining).
Woodland.

ML: rare, Woodmuir Plantation, 1993; (Pentland
and Moorfoot Hills, 1965; Quarrel Burn
Reservoir and Kirknewton Estate, 1976).
EL: (Sheeppath Dean, 1908).

Anagallis L.

A. tenella (L.) L.   Bog Pimpernel
Native. Status: rare, (declining).
Marshes, damp places.

WL: (South Queensferry, pre-1934).
ML: (Hunter's Bog and River Esk near Penicuik,
pre-1934).
EL: rare, Aberlady Bay; Yellowcraig; (Gullane,
pre-1934; Tyninghame; Dunbar).

A. arvensis L.
ssp. arvensis   Scarlet Pimpernel
Native. Status: rare.
Roadsides, waste ground, public parks, dunes and other
sandy places, scrub.

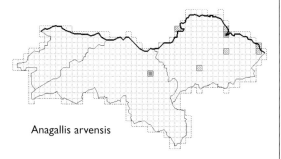

Anagallis arvensis

WL: rare, one plant at Port Edgar, 1997;
(Hopetoun and Linlithgow, 1977).
ML: rare, casual, Borthwick Bank; Sighthill;
Leith; Levenhall; (Granton, 1972).
EL: scarce, occasional plants along the coast;
inland at Spittal; Gifford; Pressmennan.

ssp. foemina (Mill.) Schinz & Thell.
                                            Blue Pimpernel
Native. Status: rare, casual.
Cultivated ground.

WL: (Linlithgow, pre-1934).
ML: rare, one plant only at Craigmillar, 1999;
(Portobello, Leith and Slateford, pre-1934).
EL: (Gullane, pre-1934; North Berwick).

Glaux L.

G. maritima L.   Sea Milkwort
Native. Status: local.
Beaches, saltmarshes, coastal rocks.

WL: local, along the coast east of Bo'ness.
ML: rare, Granton; Cramond Island, 1998.
EL: scattered along the coast, abundant in places.

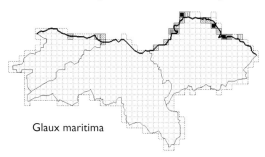

Glaux maritima

Samolus L.

S. valerandi L.   Brookweed
Native. Status: rare, (declining).
Marshes in coastal areas.

WL: (South Queensferry, pre-1934).
ML: (Roslin, 1896).
EL: rare, Tyninghame; Bilsdean; Aberlady Bay,
1989, possibly now extinct at this site;
(Broxmouth ?1857; Seton Sands, pre-1824).

HYDRANGEACEAE

Philadelphus L.

P. coronarius L.   Mock Orange
Introduced. Status: scarce, planted.

Waste ground, scrub, hedgerows.

WL: occasional.
ML: rare, Corstorphine Hill.
EL: rare, Luffness; Whittingehame.

GROSSULARIACEAE

Ribes L.

R. rubrum L.   Red Currant
Introduced. Status: local, never frequent.
Woodland, scrub, riversides, hedgerows, waste ground.

WL: quite widespread.
ML: scattered sites.
EL: scattered sites in lowland areas.

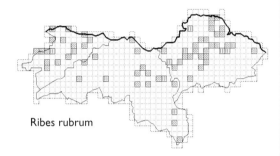

Ribes rubrum

R. nigrum L.   Black Currant
Introduced. Status: local.
Woodland, scrub, riversides, roadsides.

WL, ML: scattered distribution, not frequent in any location.
EL: very thinly scattered distribution, some upland sites.

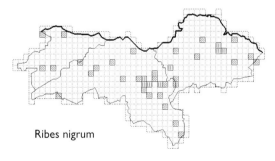

Ribes nigrum

R. sanguineum Pursh   Flowering Currant
Introduced. Status: local, planted and escaped.
Woodland, hedgerows, scrub, roadsides, railway embankments.

WL: scattered, usually single plants.
ML: widespread in N., not locally frequent, rare elsewhere.
EL: widely scattered, several plants near Gullane rubbish tip and a well-naturalised colony at East Linton, otherwise single plants at most sites.

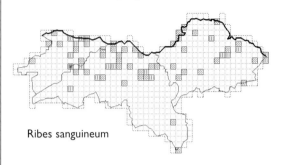

Ribes sanguineum

R. alpinum L.   Mountain Currant
Introduced. Status: local.
Woodland, hedgerows, usually in estates.

WL: in old estates, quite widespread.
ML: local, New Hailes; Carrington; Almondell; Penicuik; Linhouse Valley.
EL: rare, Ormiston; Haddington; Dunglass Dean; (Seton Glen, 1968).

R. uva-crispa L.   Gooseberry
Introduced and also possibly native. Status: widespread, usually an escape.
Woodland, hedgerows, scrub, roadsides, riversides.

WL, ML, EL: widespread except in upland and intensively farmed areas.

CRASSULACEAE

Crassula L.

C. helmsii (Kirk) Cockayne
          New Zealand Pigmyweed
Introduced. Status: rare, escape.
Ponds.

WL: rare, Carriden area.
ML: local, locally frequent, Bawsinch; Craigentinny.
EL: rare, near Whitekirk.

C. tillaea Lest.-Garl.   Mossy Stonecrop
Garden weed.
ML: Royal Botanic Garden Edinburgh, 1978.

Sempervivum L.

S. tectorum L.   House-leek
Introduced. Status: rare.
Tops of walls.

ML: (rockery in Craigmillar, 1943).
EL: rare, Gullane; (Gullane, Tantallon and East
Linton, pre-1934).

Sedum L.

S. telephium L.   Orpine
Introduced. Status: rare.
Beside canals and ditches, hedgerows, quarries, rocky
outcrops.

WL: rare, several sites, usually single plants.
ML: rare, Borthwick Bank; Ravelston; Balerno.
EL: rare, near North Berwick; Hailes Castle area;
Traprain Law; (Ormiston, 1969).

ssp. telephium
EL: North Berwick, Traprain Law and East Linton, pre-
1934.

S. spurium M. Bieb.   Caucasian Stonecrop
Introduced. Status: scarce, garden escape.
Roadsides, waste ground, walls, quarries.

WL: rare, several sites, usually single plants.
ML: rare, Gorebridge; Blackford Quarry.
EL: rare, Markle Quarry; near East Linton; near
West Barns; (Whitekirk, 1970; on walls beside
River Tyne at Haddington and East Linton,
1972).

S. rupestre L.   Reflexed Stonecrop
Introduced. Status: rare, garden escape.
Waste ground, gardens.

WL: rare, one plant, small quarry near Kinneil.
ML: rare, Balerno; Granton; Slateford, 2000.
EL: rare, roadside at Prestongrange, 1985;
(house-tops in the village of Preston, pre-1824;
Hailes Castle and Yarrow, 1873; Longniddry,
1888; Dirleton Castle, pre-1934).

S. forsterianum Sm.   Rock Stonecrop
Introduced. Status: rare, garden escape.
Waste ground.

WL: rare, verge, Ecclesmachen, 1987.
ML: (Kingsknowe, pre-1934).
EL: (Dirleton Castle, 1870).

S. acre L.   Biting Stonecrop
Native. Status: widespread.
Walls, waste ground, roadsides, coastal habitats,
quarries, rocky outcrops.

WL: local.
ML: quite widespread in N., occasional elsewhere.
EL: widespread and locally frequent in lowland
areas, rare in Lammermuir Hills.

S. album L.   White Stonecrop
Introduced. Status: local.
Walls, waste ground, gardens, quarries, railways.

WL: local, around Hopetoun; Blackfaulds;
Armadale; Torphichen.
ML: local, Gorebridge and Heriot stations; West
Calder; Roslin; Stow; Blackford Quarry, 1999.
EL: occasionally established, usually close to
human habitation, coast road at Seton Sands and
Longniddry; West Barns; Innerwick.

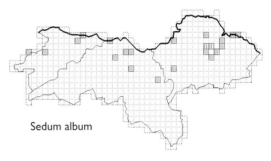

Sedum album

S. anglicum Huds.   English Stonecrop
Introduced. Status: scarce.
Walls, gardens, rocky outcrops.

ML: a few scattered locations, Dreghorn; Niven's
Knowe; Cramond Island.
EL: rare, Whitekirk Hill, known since 1969;
Bowerhouse; Broxburn; (Dunbar, pre-1934;
Traprain Law, 1967). Recent coastal records are
now considered to have been in error for *S. acre*.

S. villosum L.   Hairy Stonecrop
Native. Status: local.
Wet, upland areas.

WL: (near Knock Hill, pre-1934).
ML: occasional, upland areas only.
EL: occasional, upland areas only, Hopes
Reservoir; Faseny Water; Sheeppath Burn;
Monynut Water.

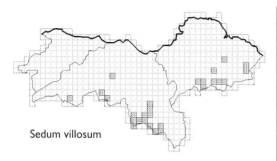

Sedum villosum

S. rosea (L.) Scop.   Roseroot
ML: upland cliffs, Habbie's Howe, pre-1934.

SAXIFRAGACEAE

Darmera Voss ex Post & Kuntze

D. peltata (Torr. ex Benth.) Voss ex Post &
Kuntze   Indian Rhubarb
Introduced. Status: rare.
Riversides, marshes.

ML: rare, naturalised at Penicuik, 1983–99.

Saxifraga L.

S. hirculus L.   Marsh Saxifrage
Native. Status: rare.
Base-rich flushes and bogs.

ML: rare, near Bawdy Moss.

S. cymbalaria L.   Celandine Saxifrage
Introduced. Status: rare, garden weed.
Gardens.

ML: rare, Marchmont, 1998 to present.
EL: rare, Belhaven, 1992.

S. stellaris L.   Starry Saxifrage
Native. Status: rare.
Wet rocks and stony places, in flushes by streams in
upland areas.

ML: rare, Moorfoot Hills south of Gladhouse.

S. x urbium D. A. Webb (*S. umbrosa* L. x *S.
spathularis* Brot.)   London Pride
Introduced. Status: local, garden escape.
Roadsides, woodland, beside canals and rivers.

WL: occasional.
ML: occasional and long persisting.
EL: rare, mainly in policy woodlands.

S. hirsuta L.   Kidney Saxifrage
Introduced. Status: rare.
Beside ponds and rivers, hedgerows.

WL: rare, Bedlormie House; (Pepper Wood, until
c. 1980).
ML: rare, Dreghorn; Kirknewton; Balerno.
EL: rare, Gosford House; (Ormiston Hall, 1911).

S. granulata L.   Meadow Saxifrage
Native. Status: scarce.
Woodland, grassland, scrub, coastal turf, riversides,
volcanic outcrops.

WL: scarce.
ML: local, scattered in small colonies.
EL: occasional across lowland areas, mainly in
well-drained grassland, Ferny Ness; Waughton
cross roads; Traprain Law; Dunglass.

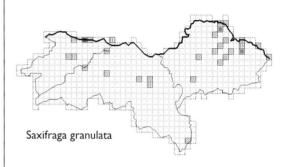

Saxifraga granulata

S. hypnoides L.   Mossy Saxifrage
Native. Status: rare, (declining).
Damp upland areas.

ML: (Dalmahoy, Auchendinny and Arthur's Seat,
pre-1934; Habbie's Howe, 1970).
EL: rare, Linn Dean Water.

S. tridactylites L.   Rue-leaved Saxifrage
Native. Status: rare, (declining).
Walls, rocks, dunes.

WL: (Blackness and Linlithgow, pre-1934).
ML: rare, Craiglockhart and Blackford Hills;
(Duddingston, pre-1934).
EL: rare, Ferny Ness; Aberlady Bay; dunes north
of Gullane; Yellowcraig.

S. x geum L. (*S. umbrosa* L. x *S. hirsuta*)
                          Scarce London Pride
EL: Ormiston Hall, pre-1934, record probably refers to *S.
hirsuta*.

Tolmiea Torr. & A. Gray

T. menziesii (Pursh) Torr. & A. Gray
Pick-a-back Plant
Introduced. Status: scarce, locally well-established garden escape.
Wooded riversides.

WL: local, frequent along the River Avon, scarce elsewhere.
ML: local, locally frequent, Cramond; Colinton; Roslin; Dalkeith.
EL: rare, Pencaitland; Gifford; Yester; (beside Newlands Burn, upstream from Yester, 1970).

Tellima R. Br.

T. grandiflora (Pursh) Douglas ex Lindl.
Fringecups
Introduced. Status: scarce but locally frequent.
Woodland, riversides.

WL: rare, occurs in small numbers, Dundas; Hopetoun; near Linlithgow Golf Course; (Polkemmet, 1978).
ML: occasional, The Drum; Roslin; Dean Village; Dalkeith Park; Cowpits; Craiglockhart Dell.
EL: rare, policy woodland, Tyninghame; (Thurston, 1959).

Chrysosplenium L.

C. oppositifolium L.
Opposite-leaved Golden Saxifrage
Native. Status: widespread.
Beside rivers and ditches, in streams, marshes, damp woodland.

WL: widespread, except in SW.
ML: widespread and locally frequent except in urban areas and on the west side of the Pentland Hills.
EL: widespread and locally frequent in S. and E., rare in N. and W.

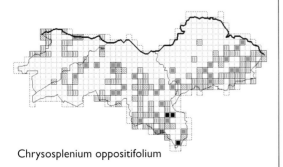
Chrysosplenium oppositifolium

C. alternifolium L.
Alternate-leaved Golden Saxifrage
Native. Status: scarce.
In and beside rivers and streams, damp woodland in base-rich areas.

WL: rare, along the River Avon.
ML: local, Musselburgh; Arniston and Roslin Glens; Balerno; Gorebridge.
EL: scarce, Keith; Humbie; Yester; East Linton; Stoneypath; Biel; Woodhall; Dunglass.

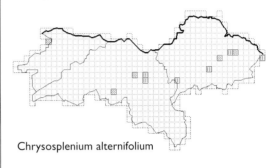
Chrysosplenium alternifolium

Parnassia L.

P. palustris L.
Grass of Parnassus
Native. Status: local.
Marshes, damp grassland beside rivers and ditches, dune slacks.

ML: quite widespread in upland areas in SE., rare or absent elsewhere.
EL: scattered sites in the Lammermuir Hills; occasional in coastal locations, Aberlady Bay; Yellowcraig; Belhaven Bay; Bilsdean.

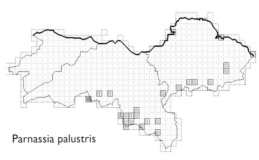
Parnassia palustris

var. **condensata** Travis & Wheldon
EL: Luffness Links, 1907; coast, north of Dirleton, 1958.

ROSACEAE

Physocarpus (Cambess.) Maxim.

P. opulifolius (L.) Maxim.   Ninebark
Introduced. Status: rare, planted.
Riversides, gardens.

WL: rare, Hopetoun; (Dalmeny).
ML: rare, (Gala Water, 1970).
EL: (Saltoun Big Wood, 1967).

Spiraea L.

S. salicifolia agg.   Bridewort
Spiraea spp. other than *S. japonica* and *S. x arguta* were aggregated under the above heading during the survey.
Introduced. Status: local.
Waste ground, scrub, riversides, canal-sides.

WL: scattered distribution, locally frequent.
ML: local, locally frequent.
EL: rare.

The following taxa have been identified.

S. salicifolia L.   Bridewort
No longer thought to occur in the Lothians; all RBGE herbarium specimens have been reclassified as *S. x pseudosalicifolia*.

S. x pseudosalicifolia Silverside (*S. salicifolia* x *S. douglasii*)   Confused Bridewort
Introduced. Status: local.
Waste ground.

WL: scattered distribution, forming large patches.
ML: occasional, Gladsmuir; Bonnyrigg; Smeaton; Balerno.
EL: rare, in hedge near Hailes Castle.

S. x billardii Hérincq (*S. alba* Du Roi x *S. douglasii*)   Billard's Bridewort
Introduced. Status: rare.
Hedgerow.

EL: rare, Luffness.

S. douglasii Hook.
ssp. douglasii   Steeple-bush
Introduced. Status: rare.
Beside canal and rivers, scrub.

WL: rare, Linlithgow; Stoneyburn.
ML: local, locally frequent, Corstorphine Hill; West Calder; Craigmillar Castle.

S. tomentosa L.   Hardhack
Introduced. Status: rare.
Waste ground.

WL: rare, near Stoneyburn.

S. japonica L. f.   Japanese Spiraea
Introduced. Status: rare, planted.
Hedgerows, public amenity areas.

WL: rare, Linlithgow; old Bangour Hospital.
ML: rare, Smeaton.

S. x arguta Zabel (*S. thunbergii* Sieb. ex Blume x *S x multiflora* Zabel)   Bridal Spray
Introduced. Status: rare.
Hedgerows.

WL: rare, planted as a hedge in an industrial estate, North Livingston.

Aruncus L.

A. dioicus (Walter) Fernald   Buck's-beard
Introduced. Status: rare, garden escape.
Rough grassland.

WL: rare, recorded in 1983 from a building site near Whitburn.
ML: (Tynehead, 1969).

Filipendula Mill.

F. vulgaris Moench   Dropwort
Native. Status: rare, (declining).
Dry grassland.

WL: (Linlithgow, pre-1934).
ML: rare, Arthur's Seat; Moorfoot Hills; (Blackford Hill; Penicuik).

F. ulmaria (L.) Maxim.   Meadowsweet
Native. Status: common.
Riversides, marshes, ditchsides, damp woodland and scrub, roadsides.

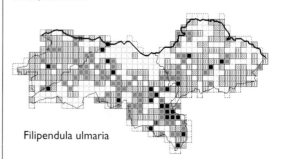

Filipendula ulmaria

WL, EL: common.
ML: common except in urban areas and on the western side of the Pentland Hills.

Kerria DC.

K. japonica (L.) DC.   Kerria
Introduced. Status: rare.
Old gardens, policies.

WL: rare, planted at Hopetoun.
EL: rare, Whittingehame.

Rubus L.

R. chamaemorus L.   Cloudberry
Native. Status: scarce.
Upland heath, peaty moorland, usually above 450m.

ML: locally frequent high up in the Pentland and Moorfoot Hills.
EL: rare, Lammermuir Hills, above Gifford.

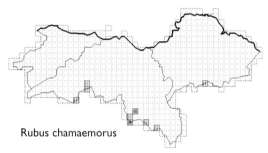

Rubus chamaemorus

R. tricolor Focke   Chinese Bramble
Introduced. Status: local, planted.
Roadsides, public amenity areas.

WL: rare, west of Livingston.
ML: local, locally frequent, Granton; Warriston; Buckstone, 1992; Leith, 2000.

R. saxatilis L.   Stone Bramble
Native. Status: rare, (declining).
Walls, quarries, base-rich upland rocks.

WL: (Linlithgow, pre-1934).
ML: rare, Moorfoot Hills, 1990; (Roslin; Auchendinny; Mid Calder; pre-1934).
EL: rare, Linn Dean; Woodhall area; East Lammermuir Deans.

R. parviflorus Nutt.   Thimbleberry
Introduced. Status: local, planted.
Public amenity areas e.g. landscaped areas beside motorways.

ML: occasional, Sighthill.

R. idaeus L.   Raspberry
?Native. Status: common.
Woodland, roadsides, scrub, riversides, hedgerows, railway embankments, waste ground.

WL: common.
ML: common.
EL: widespread and locally frequent in centre, occasional near coast and in the hills, probably often a naturalised escape from cultivation.

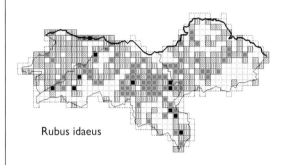

Rubus idaeus

R. spectabilis Pursh   Salmonberry
Introduced. Status: local.
Woodland, scrub, roadsides, riversides.

WL: rare, planted, beside Linlithgow Loch; Carriden Woods.
ML: scattered, locally frequent, naturalised in policy woodland, spreading and being considered as a troublesome weed.
EL: rare, Keith Marischal; Yester; Stoneypath Tower; Stenton area; (Nunraw and Dunglass Dean, 1960).

R. cockburnianus Hemsl.
                              White-stemmed Bramble
Introduced. Status: local, planted.
Roadsides, public amenity areas.

WL: rare, west of Livingston; Beecraigs.
ML: local, Warriston; Sighthill, 2000; Gogar.

R. loganobaccus L. H. Bailey   Loganberry
Introduced. Status: rare, escape.

ML: rare, Blackford Hill; Warriston; Stow.

'Rubus fruticosus agg.'   Blackberry, Bramble
Native. Status: common.
Hedgerows, scrub, waste ground, roadsides, woodland.

WL: common.
ML, EL: common in lowland areas, occasional in upland areas.

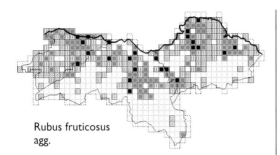

Rubus fruticosus agg.

Rubus fruticosus agg. is an aggregate name used by botanists to cover the many microspecies in the British Isles. Most recorders in the survey did not collect specimens for naming by specialists, but simply used the well-known aggregate name. However, specimens that have been authoritatively named are listed below. We were fortunate to be able to rely on Mr George Ballantyne, the noted *Rubus* specialist, who supplied a number of records and gave assistance with the list. By classifying what we have so far found by section and series, it is hoped to stimulate local interest in these fascinating plants and to promote further collection and perhaps the emergence of local expertise. Such indications of frequency as appear below are tentative – we need **more** authenticated records. The format used to indicate whether a record is recent or historic has not been applied.

section **Rubus** subsection **Rubus**

R. fissus Lindley
Native.

ML: rare, Arniston; near Hawthornden, 1876.

R. nessensis Hall
Native.

ML: rare, Colinton; near Penicuik, 1873.

R. plicatus Wiehe & Nees
ML: Musselburgh and Dalkeith, pre-1934.

R. scissus W. C. R. Watson
Native.

ML: rare, Dalmahoy, 1900; Balerno, 1945; Briestonhill, 1994.
EL: rare, south of Ormiston, 1989; West Barns, 1976.

subsection **Hiemales** E. H. L. Krause

series **Sylvatici** (P. J. Müll.) Focke

R. errabundus W. C. R. Watson
Native.

WL: scattered distribution, Beecraigs; Linlithgow; Totleywells.
ML: local, Granton Harbour, 1970; Bavelaw; Roseburn; New Craighall.

R. laciniatus Willd.
Introduced. Bird-sown escape. The usual culinary species.

ML: local, Currie; Braid Hills; Blackford Hill.
EL: North Berwick, 1966.

R. leptothyrsos G. Braun
Native.

WL: common, Linlithgow; Hopetoun; Dalmeny.
ML: rather local, Gogar; Blackford Hill; Newbattle Abbey.
EL: rare, Longniddry.

A trifoliate newcomer in this section, rare in **EL**, seemingly spreading into our area from Northumberland (where it is widespread) and Berwickshire, is shortly to be named as a new species in honour of Alan Newton.

series **Rhamnifolii** (Bab.) Focke

R. elegantispinosus (Schumach.) H. E. Weber
Introduced. Escape from gardens or allotments, often found near railway lines.

WL: widespread and increasing, Carriden; Linlithgow; Hopetoun; Dalmeny.
ML: common, Balerno; Dalkeith; Warriston.
EL: rare, Bielmill.

R. lindebergii P. J. Müll.
Native.

EL: rare, Keith area (map, Edees and Newton 1988).

R. nemoralis P. J. Müll.
Native.

ML: rare, north of Turnhouse; Blackford Hill; Newtongrange; Melville Castle.

R. polyanthemus Lindeb.
Native.

ML: rare, Inverleith; New Craighall; Niddrie Marischal, 1966.

R. septentrionalis W. C. R. Watson
Native.

ML: rare, Blackford Hill; Breich.
EL: common, Colstoun; Papple; Binning Wood; Bielmill; The Brunt; Oldhamstocks.

R. subinermoides Druce
Native. Usually a west coast species.

EL: rare, by shore, Dunbar swimming pool, 1983.

series **Sprengeliani** Focke

R. sprengelii Weihe
Native.

ML: rare, Cramond Island, 1987.

series **Discolores** (P. J. Müll.) Focke

R. armeniacus Focke (*R. procerus* P. J. Müll. ex Boulay)
Introduced as cv. **Himalayan Giant**. Status: local, spreading.

WL: rare, escape, increasing, Carmelhill; Dalmeny.
ML: local, locally frequent, Cramond Island; Turnhouse; Granton; Warriston; Craigleith; Roseburn.
EL: rare, Ormiston; old railway cycle path, Haddington; woodland at Luffness, 1966.

R. ulmifolius Schott
Introduced.

ML: rare, Dalkeith, 1999; Newcraighall.

series **Vestiti** (Focke) Focke

R. vestitus Weihe
Native.

ML: rare, a 'weed' in the grounds of St Mary's Cathedral, Palmerston Place; Roseburn; Newcraighall.
EL: rare, John Muir Country Park; Links Wood, Tyninghame, 1995.

series **Mucronati** (Focke) H. E. Weber

R. mucronulatus Boreau
Native.

WL: rare, south of Bo'ness; Birkhill; Linlithgow, 1988.

ML: rare, Seafield; Newbattle Abbey; Vogrie Country Park.
EL: occasional, Ormiston; Gladsmuir; Aberlady; Jerusalem, Haddington.

series **Anisacanthi** H. E. Weber

R. drejeri Jensen ex Lange
Native.

EL: rare, Bielmill; The Brunt; near Oldhamstocks.

R. infestus Weihe ex Boenn.
Native.

WL: rare, near Linlithgow, 1988, one bush.
ML: rare, west of Newtongrange, 1986.
EL: rare, The Brunt; Oldhamstocks area.

series **Radulae** (Focke) Focke

R. echinatoides (Rogers) Dallman
Native.

ML: common, Blackford Hill; Duddingston Loch; Arthur's Seat; Cousland; Cramond Brig; Craigleith; Southfield; Vogrie Country Park.
EL: rare, Renton Hall, Morham; Papple.

R. radula Weihe ex Boenn.
Native.

WL: very common, Linlithgow; Dechmont; Blackness; Dalmeny; Kirkliston.
ML: very common, Craiglockhart; Portobello; Temple; Musselburgh; Cramond Island; Blackford Hill, 1966.
EL: common in lowland areas, Prestongrange; Longniddry; Keith; Dirleton; Dunglass.

R. scoticus (Rogers & Ley) Edees
Native.

WL: rare, Dalmeny.
ML: rare, Warriston.

R. subtercanens W. C. R. Watson
Native.

ML: rare, Roseburn; Braid Hills and Calton Hill, 1986.

series **Hystrices** Focke

R. dasyphyllus (Rogers) E. S. Marshall
Native.

ML: local, Seafield; Hermitage of Braid; Colinton Dell.

EL: rather local, Ormiston; Saltoun; Longyester; North Berwick; wood near Black Loch; Spott.

section **Corylifolii** Lindley

R. conjungens (Bab.) Rogers
Native.

EL: (Longniddry, 1972, to be confirmed), (map, Edees and Newton 1988).

R. eboracensis W. C. R. Watson
Native.

ML: rare, Temple; Carberry; Roslin, Auchendinny and Craigcrook, pre-1934.
EL: local, West Saltoun; near Tyninghame; Oldhamstocks; Bilsdean.

R. latifolius Bab.
Native. Possibly the commonest of the microspecies. Originally described from Cramond Bridge.

WL: very common, Carriden; Blackness; Uphall.
ML: very common, Cramond; West Calder; Granton.
EL: common in lowland areas, Samuelston; Gullane; Dunbar; rare in upland areas.

R. pruinosus Arrh.
Native.

EL: (railway embankment at Saltoun, 1980, to be confirmed).

R. tuberculatus Bab.
Native.

WL: local, Linlithgow; Hopetoun; Dalmeny; Kirkliston.
ML: common, Gogar; Blackford; Breich; Leith Docks; Dalkeith.

section **Caesii** Lej. & Courtois

R. caesius L.   Dewberry
Native. Status: rare, possibly overlooked or confused with section *Corylifolii.*
Disturbed ground.

WL: (Bo'ness and Hopetoun, pre-1934).
ML: rare, Davidson's Mains, 1999; (Dalkeith, Liberton, and Cramond Bridge, pre-1934).
EL: rare, in walled garden, Pilmuir House, 1996; (Luffness, 1902).

Potentilla L.

P. fruticosa L.   Shrubby Cinquefoil
Introduced. Status: rare, garden escape.
Roadsides, scrub, public parks, waste ground.

WL: occasional amenity plantings in public places.
ML: rare, Redhall, 1996.
EL: (Gosford, pre-1927).

P. palustris (L.) Scop.   Marsh Cinquefoil
Native. Status: local.
Marshes, beside rivers, ditches and lochs, bogs.

WL: widespread in S., rare in N.
ML: mainly in S., local.
EL: scarce, Keith Water; Aberlady Bay; Belhaven Bay; sites in the Lammermuir Hills.

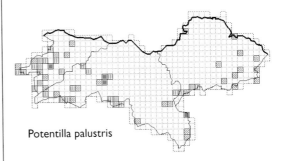

Potentilla palustris

P. anserina L.   Silverweed
Native. Status: widespread.
Roadsides, waste ground, rough grassland, sand, shingle.

WL, ML, EL: widespread, locally frequent.

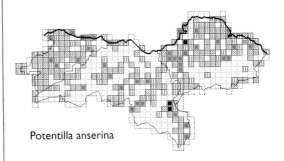

Potentilla anserina

P. argentea L.   Hoary Cinquefoil
Native. Status: rare, (declining).
Waste ground, quarries and rocky outcrops, dry grassland.

WL: (Binny Craig and Kirkliston, pre-1934; Bo'ness, 1976; Carriden, 1977).

ML: (Borthwick sidings, 1970; Blackford Hill).
EL: rare, three current sites only; (Dirleton and Tyninghame, pre-1934; introduced casual at West Barns, 1973–5).

P. recta L.   Sulphur Cinquefoil
Introduced. Status: rare.
Waste ground, cereal fields.

ML: rare, Cramond; Carberry; Musselburgh Lagoons, 1988.
EL: rare, West Barns, 1991; Catcraig, 1996; (Dirleton, 1870; West Barns, 1974, 1976, 1977).

P. intermedia L.   Russian Cinquefoil
Introduced. Status: rare.
Waste ground, rubbish tips, mainly at ports.

WL: rare, Bo'ness.
ML: rare, Leith; (Seafield, 1968).
EL: (near Prestonpans, 1972).

P. norvegica L.   Ternate-leaved Cinquefoil
Introduced. Status: rare.
Rubbish tips, waste ground, canalside.

WL: (Bo'ness, 1962; Carriden, 1975).
ML: rare, Borthwick, 1987; Granton, 1996.
EL: (Dunbar, 1936; West Barns, 1973–8).

P. neumanniana Rchb.   Spring Cinquefoil
Native. Status: rare.
Coastal turf, basalt outcrops.

ML: rare, Arthur's Seat; Blackford Hill.
EL: rare, mainly coastal, Ferny Ness car park, Longniddry; Hailes area; (Craigleith Island, 1973).

P. erecta (L.) Raeusch.   Tormentil
Native. Status: common in upland areas, local elsewhere.
Upland pasture, dry heath, woodland, roadsides, riversides, bogs, coastal grassland.

WL: common in heathland and remnants of heathland.
ML, EL: common in upland areas, scarce elsewhere.

ssp. erecta
EL: three lowland and four upland records.

ssp. strictissima (Zimmeter) A. J. Richards
EL: with ssp. *erecta* on the Dunbar Common area of the Lammermuir Hills.

P. anglica Laichard.   Trailing Tormentil
Native. Status: rare.
Woodland, waste ground, dry banks, grassy tracks.

WL: rare, old railway track south of Armadale; Dalmeny; (Bo'ness, 1970).
ML: (Bavelaw, pre-1934, unconfirmed).
EL: rare, Aberlady Bay; Yellowcraig; Bearford Burn. Possibly under-recorded due to difficulty in distinguishing it from *P. x mixta*.

P. x mixta Nolte ex Rchb. (*P. erecta* x *P. reptans* and *P. anglica* x *P. reptans*)   Hybrid Cinquefoil
Native. Status: rare, less rare than *P. anglica* and often found in the absence of the parents.
Dry banks, grassy tracks.

WL: rare, track near Bonhard Farm; (track near Linlithgow Golf Course, 1970).
ML: rare, Tynehead.
EL: rare, track through Binning Wood. Probably under-recorded.

P. reptans L.   Creeping Cinquefoil
Native. Status: local.
Roadsides, railway embankments, dry ground.

WL, ML: local.
EL: local, mainly in lowland areas, common in W.

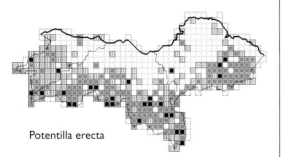

Potentilla erecta

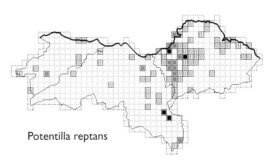

Potentilla reptans

P. sterilis (L.) Garcke   Barren Strawberry
Native. Status: widespread in upland areas, local elsewhere.
Woodland, dry heath, upland pasture, scrub, riversides, roadsides, old railways, dunes.

WL: local, rare in centre, scattered distribution elsewhere.
ML: widespread and locally frequent in S., occasional elsewhere.
EL: widespread and locally frequent in S. and E., rare and apparently decreasing in N. and NW.

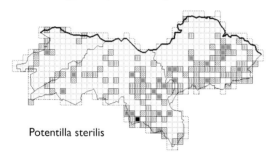

Potentilla sterilis

Fragaria L.

F. vesca L.   Wild Strawberry
Native. Status: widespread.
Woodland, roadsides, railway embankments, waste ground, scrub, hedgerow, old bings, dunes.

WL, ML, EL: widespread, locally frequent.

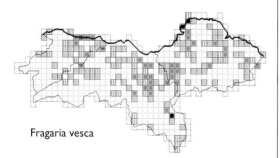

Fragaria vesca

F. x ananassa (Duchesne) Duchesne
          Garden Strawberry
Introduced. Status: rare, garden origin.
Railway embankments, roadsides, waste ground.

WL: rare, Bathgate; Bo'ness; Faucheldean Bing; old railway at Stoneyburn.
ML: (Heriot station, 1969; Ratho, 1972; Musselburgh, 1975).
EL: rare, railway sites near Winton, Huntington, Gullane and North Berwick.

Geum L.

G. rivale L.   Water Avens
Native. Status: widespread.
Beside rivers, ditches and roads, marshes, damp woodland.

WL, ML: widespread except in urban areas, locally frequent.
EL: rare and decreasing in N., widespread elsewhere, locally frequent in W.

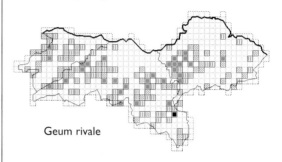

Geum rivale

var. **pallidum**
EL: near Gifford, 1973.

G. x intermedium Ehrh.(*G. rivale* x *G. urbanum*)
Native. Status: local, usually with one or both parents.
Roadsides, woodland.

WL, ML: local.
EL: local, mainly in Ormiston/Saltoun/Humbie area, rare elsewhere.

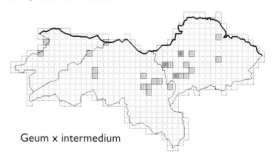

Geum x intermedium

G. urbanum L.   Herb Bennet, Wood Avens
Native. Status: common, except in some upland areas.
Woodland, roadsides, scrub, hedgerow.

WL: common except in boggy upland areas in SW.
ML: common except in some upland areas.
EL: widespread, mainly in lowland areas, locally frequent.

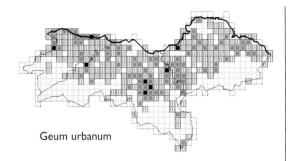

Geum urbanum

Agrimonia L.

A. eupatoria L.   Agrimony
Native. Status: local.
Roadsides, railway embankments, rough grassland, hedgerows.

WL: scarce, only in N. and E.
ML: rare, Gorebridge, 1982; (Middleton, 1976; Stow, 1977; Arthur's Seat).
EL: scattered sites in dry places in lowland areas.

A. procera Wallr.   Fragrant Agrimony
ML: Hallyards, 1940.
EL: Prestonpans and Gullane, pre-1894; six other sites, 1972.

Aremonia Neck. ex Nestl.

A. agrimonioides (L.) DC.   Bastard Agrimony
Introduced. Status: rare, naturalised.
Woodland.

WL: rare, Hopetoun East Lodge; (Carlowrie, 1869).
ML: rare, Dalkeith, 1996; (Silverknowes, 1961).

Sanguisorba L.

S. minor Scop.
ssp. minor   Salad Burnet
Introduced. Status: rare, casual.
Dry rocky places, railway embankments.

ML: (Gorebridge, 1978; Inveresk, Dalkeith and Straiton; pre-1934).
EL: rare, Aberlady Bay; (Aberlady, 1903; North Berwick and Tyninghame, pre-1934; railway embankment at West Barns, 1977).

ssp. muricata (Gremli) Briq.   Fodder Burnet
Introduced. Status: rare, persisting.
Verges beside disused railways, bings.

WL: rare, Kinneil.

ML: rare, Arthur's Seat, 1998; Innocent Railway, 1995; Calder Bing; (Currie, Inveresk and Leith, pre-1934).
EL: rare, Prestongrange Mining Museum.

Acaena Mutis ex L.

A. novae-zelandiae Kirk   Pirri-pirri-bur
Introduced. Status: rare.
Dunes.

EL: rare, Yellowcraig.

A. anserinifolia (J. R. & G. Forst.) Druce
              Bronze Pirri-pirri-bur
Introduced. Status: rare.
Bare ground in grassland.

WL: rare, well-established in grassland at Hopetoun.

A. ovalifolia Ruiz & Pav.   Two-spined Acaena
Introduced. Status: rare, garden escape.
Dunes. waste ground.

ML: (disused railway, Kingsknowe, 1970).
EL: rare, two sites in dunes west of Gullane.

A. inermis Hook. f.   Spineless Acaena
Introduced. Status: rare, well naturalised.
Rough grassland, bare ground, river shingle.

WL: rare, abandoned garden centre, Hopetoun, 1999.
ML: rare, abundant by riverside, Heriot, 1983-present; Halltree, 1991; Carcant Burn, 1998.

Alchemilla L.

A. conjuncta Bab.   Silver Lady's Mantle
Introduced. Status: rare, garden escape.
Waste ground, rubbish tips.

WL: rare, quarry near Kinneil; (near Philpstoun Bing, 1974).
EL: (Dirleton, 1962).

A. vulgaris L. agg.   Lady's Mantle
Native. Status: widespread.
Roadsides, hedgerows, waste ground, rough grassland.

WL, ML, EL: widespread.

The following segregates were recorded:

A. xanthochlora Rothm.
Native. Status: local.
Roadsides, riversides, rough grassland.

WL, ML: quite widespread.
EL: rare, Ormiston; Keith; Yester; Woodhall Burn; Thurston; (Bothwell Water and Philip Burn, 1970; not recorded in NW. since 1960).

A. filicaulis Buser.
ssp. vestita (Buser) M. E. Bradshaw
Native. Status: local.
Damp grassland.

WL: occasional, possibly under-recorded.
ML: occasional, mainly in upland areas, Pentland and Moorfoot Hills; Kirkliston; Cockpen; Gladhouse and Rosebery Reservoirs.
EL: scarce, mainly in upland areas, Linn Dean; Haddington; Yester; Danskine; Whiteadder Water; Woodhall Burn; Wide Hope.

A. glabra Neygenf.
Native. Status: widespread.
Beside rivers, ditches and roads, rough grassland, marshes.

WL: very widespread but seldom locally frequent.
ML: widespread.
EL: grading from widespread in upland areas in S. to rare and apparently decreasing in N.

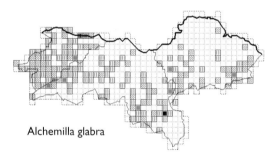

Alchemilla glabra

A. mollis (Buser) Rothm.
Introduced. Status: scarce, but increasing, garden escape.
Waste ground, roadsides, disused railways.

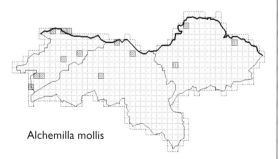

Alchemilla mollis

WL: scattered distribution, locally abundant.
ML: scarce, Granton; Wolf Cleuch; Craiglockhart Dell; Trinity.
EL: rare, Pencaitland; Ferny Ness; Gullane; Tyninghame.

A. glaucescens Wallr.
ML: Leith, 1973.

Aphanes L.

A. arvensis agg.   Parsley Piert
Native. Status: local.
Roadsides, cereal fields, waste ground, rough grassland, dry ground.

WL, ML: local.
EL: occasional.

Includes the following two taxa that were not always distinguished in the survey.

A. arvensis L.   Parsley Piert
Native. Status: local.
Roadsides, cereal fields, waste ground, open grassland.

WL: scarce.
ML: local.
EL: occasional, widely scattered sites.

A. australis Rydb.   Slender Parsley Piert
Native. Status: local.
Dry grassland, arable fields, roadsides, quarries, dunes.

WL: scarce, but less so than *A. arvensis*.
ML: occasional, Craigmillar; Wester Craiglockhart Hill; Roslin; West Calder; Heriot.
EL: occasional, widely scattered sites, Luffness quarry; Hailes area; Philip Burn; Catcraig.

Rosa L.

R. multiflora Thunb. ex Murray
                                        Many-flowered Rose
Introduced. Status: rare, garden escape.
Waste ground, derelict gardens.

ML: rare, Marchbank, 1999; (towpath near Ratho, 1968).

R. arvensis Huds.   Field Rose
?Native. Status: rare, (declining).
Hedgerow, scrub.

ML: rare, Dreghorn, 1987; (scattered distribution, pre-1934).
EL: rare, Sheriff Hall; (on the links between

Cockenzie and Gosford, pre-1824; woodland and hedges in Cockenzie and Gosford area, pre-1934).

R. pimpinellifolia L.   Burnet Rose
Native. Status: local, (declining).
Hedgerow, scrub, roadsides, rock outcrops, dunes.

WL: local, often planted; the main natural population is at Dalmeny.
ML: scarce, Stow; Gogar; Loganlea; Arthur's Seat; Habbie's Howe; (previously widespread).
EL: local, coastal between Seton Sands and Aberlady Bay, most frequent at Longniddry Bents, but the population there includes hybrids with pink in the flowers which otherwise are hard to distinguish; previously more widespread along the coast; occasional at scattered sites inland, sometimes planted.

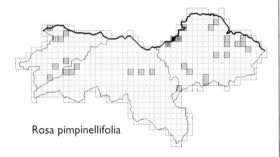

Rosa pimpinellifolia

R. pimpinellifolia x R. canina = R. x hibernica
Templeton
EL: (near Ormiston, 1907).

R. pimpinellifolia x R. caesia = R. x margerisonii (Wolley-Dod) Wolley-Dod
EL: (near Ormiston, 1907; Port Seton, 1912).

R. pimpinellifolia x R. sherardii = R. x involuta Sm.
EL: (Longniddry, 1978).

R. pimpinellifolia x R. mollis = R. x sabinii Woods
EL: Longniddry Bents, 1995.

R. pimpinellifolia x R. rubiginosa = R. x biturigensis Boreau
EL: near parents at Longniddry Bents, 1995, and Seton Sands, 1996.

R. rugosa Thunb. ex Murray   Japanese Rose
Introduced. Status: local, garden escape, often planted.
Scrub, roadsides, dunes.

WL: local, frequently planted in public places such as roadsides and bings where it becomes well-established and naturalises; first record Dalmeny Estate, 1969.
ML: local, Cramond; Warriston.
EL: local, naturalised in dunes, Aberlady Bay; Gullane; (Thorntonloch, 1967); planted and established elsewhere.

R. 'Hollandica'   Dutch Rose
Introduced. Status: rare, possibly confused with *R. rugosa*.
Waste ground.

WL: rare, Bathgate.
ML: rare, Leith Docks, 1982.

R. ferruginea Vill.   Red-leaved Rose
Introduced. Status: rare, planted.
Public amenity areas.

ML: rare, Roslin.

'R. canina agg.'   Wild Rose
Native. Status: common.
Roadsides, hedgerows, woodland, scrub, riversides.

WL, ML, EL: widespread, locally frequent, mainly in lowland areas.

Individual taxa were rarely noted during the survey, but the following have been definitely identified.

R. canina L.   Dog Rose
WL: (Bo'ness, Kirkliston and Midhope, pre-1934).

Although rarely recorded, records exist in **ML** and **EL** for the four informal groups within this species (Graham and Primavesi 1993):

group **Lutetianae**
ML: rare, Ravelrig; (Balerno, 1978).

group **Dumales**
EL: rare, Gullane Bents.

group **Transitoriae**
ML: rare, Niddry; Juniper Green; Catcune, 1982.
EL: rare, Gullane Bents; Gifford; Barns Ness.

group **Pubescentes**
ML: rare, Cramond Bridge; Currie; Gorebridge.
EL: rare, Gifford.

R. x dumalis Bechst. (*R. canina* x *R. caesia* ssp.
glauca).
Native. Status: quite widespread.
Hedgerows, scrub.

ML: quite widespread, Colinton Dell, 1999;
Stow; Pirn House; Cowpits; (Blackford Hill,
1891; Curriehill, 1914).
EL: Longniddry Bents, 1995; Gullane Bay, 1995.

R. canina x **R. sherardii = R. x rothschildii** Druce
EL: Gullane Bents, 1999; Gifford, 1986; (North
Berwick, 1912).

R. canina x **R. rubiginosa = R. x nitidula** Besser
EL: Longniddry Bents, 1995.

R. caesia Sm.
Native. Status: local.
Roadsides, hedgerow, railway embankment.

WL: rare, Hiltly Crags; Longmuir Plantation;
Wyndford Woods, det. Stewart, 1982.
EL: rare, Keith; Longyester.

ssp. caesia   Hairy Dog Rose
Native. Status: local.
Roadsides, hedgerow, railway embankment, scrub.

WL: rare, Faucheldean; Midhope; west of
Drumcross; (Abercorn, 1973).
ML: local, Ratho; Blackford Glen; Colinton;
Stow; (Dalmahoy, 1978).
EL: rare, Tranent; Haddington area; near
Kingston; Woodhall Burn; (Ormiston and
Longniddry, pre-1934).

ssp. glauca (Nyman) G. G. Graham & Primavesi
                            Glaucous Dog Rose
Native. Status: local.
Roadsides, hedgerows, scrub, dunes.

WL: rare, Bo'ness; Broxburn; Hillhouse;
Linlithgow.
ML: scattered, locally frequent, Musselburgh;
Blackford Glen; Cousland; Balerno; Auchendinny.
EL: scarce, several widely scattered sites, Tranent;
Longniddry; Gullane; Papana Water; Dry Burn.

R. caesia ssp. glauca x **R. sherardii**
WL: rare, west of Drumcross, 1983; (Linlithgow,
1973).

ML: rare, Hawthornden; (near Ratho, 1978;
Auchendinny, 1978; Easter Newton).
EL: (Dirleton, 1876; Longniddry, 1978).

R x glaucoides Wolley-Dod (*R. caesia ssp. glauca*
x *R. mollis*)
Native. Status: rare.
Roadsides.

ML: rare, Dalkeith, 1999; Cousland, 1985.
EL: (Townhead, near Gifford, 1973).

R. sherardii Davies   Sherard's Downy Rose
Native. Status: local, possibly under-recorded.
Hedgerows, roadsides, inland and coastal scrub.

WL: rare, Midhope; Linlithgow.
ML: (Stow, Carlops, Gorebridge and Balerno,
1978).
EL: quite widespread in lowland areas.

R. sherardii x **R. mollis = R. x shoolbredii**
Wolley-Dod
WL: beside canal at Linlithgow.
ML: (Roslin, 1900; north of Middleton, 1978).
EL: Gullane Bay, 1995; (Longniddry, 1868).

R. sherardii x **R. rubiginosa = R. x suberecta**
(Woods) Ley
WL: on north side of Linlithgow Loch.
EL: near Gifford, 1986; (Seton Sands and
Longniddry dunes, 1978).

R. mollis Sm.   Soft Downy Rose
Native. Status: local.
Hedgerows, roadsides, railway embankments, scrub,
dunes.

WL: occasional, Midhope; Linlithgow.
ML: local, locally frequent, Borthwick; Stow;
Balerno; Flotterstone; Edgelaw.
EL: scarce, Ormiston; Pencaitland; Keith;
Longniddry Bents; Gullane Bents;
Newmains/Stoneypath; Bilsdean.

R. mollis x **R. rubiginosa = R. x molliformis**
Wolley-Dod
EL: Longniddry Bents, 1995; Gifford, 1986;
(Seton Sands, 1978).

R. rubiginosa L.   Sweet Briar
Native. Status: local.
Scrub, hedgerows, roadsides, railway embankments,
dunes.

WL: occasional, Bo'ness; Linlithgow; Hopetoun; Dalmeny; Petershill.
ML: local, Balerno; Ratho; Craiglockhart; Dalkeith; Stenhouse; frequent along the Union Canal.
EL: occasional, on coastal dunes and at scattered sites inland, Longniddry Bents; Gullane Bents; walkway along disused railway near Haddington; Whitekirk; Bilsdean.

Prunus L.

P. dulcis (Mill.) D. A. Webb   Almond
Introduced. Status: rare, planted.
Policies.

EL: rare, Seton House.

P. cerasifera Ehrh.   Cherry Plum
Introduced. Status: scarce, often planted.
Hedgerows, public parks, woodland.

WL: scarce.
ML: rare, Roslin; Bawsinch; New Craighall; Blackford Quarry; Corstorphine Hill.
EL: rare, Longniddry Bents; Keith; Luffness; Dunbar; Dunglass.

P. spinosa L.   Blackthorn, Sloe
Native. Status: widespread.
Scrub, hedgerows, woodland, roadsides.

WL: widespread.
ML: widespread, locally frequent.
EL: quite widespread, especially in central lowland areas, locally frequent, sometimes planted.

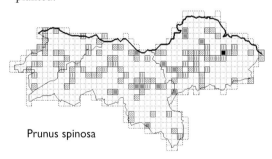
Prunus spinosa

P. domestica L. *s. l.*   Wild Plum
Introduced. Status: local.
Scrub, hedgerows, public parks, farmyards, roadsides.

WL: scarce, Abercorn; Bo'ness; Carriden; Linlithgow; Bathgate; Easter Dalmeny.
ML: local, Stow; Musselburgh; Temple.

EL: rare, derelict orchard at Gladsmuir; Hummell Rocks; Longyester; (Gifford, 1960; Hedderwick 1961; Humbie, 1965; Longniddry, 1968).

ssp. domestica   Plum
EL: rare, planted in hedge, Whitekirk.

ssp. insititia (L.) Bonnier & Layens
                Bullace, Damson
Hedgerows, woodland.

ML: scattered sites, Gogar station; Duddingston; Hallyards.
EL: rare, derelict orchard, Luffness; roadside between Luffness and Mungoswells; Morham; Dunglass.

P. avium (L.) L.   Gean, Wild Cherry
Native. Status: widespread, frequently planted.
Woodland, scrub, roadsides, hedgerows, riversides, public parks, policies.

WL: very widespread, often planted.
ML: widespread in N., rare in upland areas.
EL: widespread and locally frequent in centre, rare in N. and S.

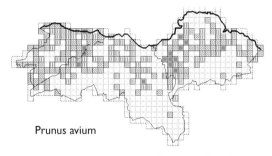
Prunus avium

P. cerasus L.   Dwarf or Morello Cherry
Introduced. Status: rare.
Scrub, hedgerows, derelict gardens.

WL: rare, Dalmeny; Linlithgow Golf Course.
ML: (Colinton Dell, 1963; Arniston, Dalmahoy and Roslin, pre-1934).
EL: rare, Pressmennan; (Dunglass, pre-1927).

P. padus L.   Bird Cherry
Native. Status: local, sometimes planted.
Woodland, scrub, hedgerows, roadsides, on fairly rich soils.

WL: local, usually planted.
ML: much more widespread than map indicates.
EL: scarce, scattered distribution, at least two-thirds of the records are in planted woodland.

277

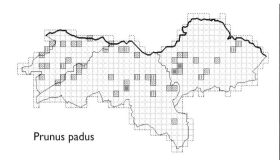

Prunus padus

P. serotina Ehrh. Rum Cherry
Introduced. Status: rare.
Beside footpaths, possibly planted.

ML: rare, Kinleith; Colinton.

P. lusitanica L. Portugal Laurel
Introduced. Status: local, planted.
Policies, public parks.

WL: quite widespread.
ML, EL: local in N., rare or absent in S.

P. laurocerasus L. Cherry Laurel
Introduced. Status: widespread, planted.
Policies, public parks, scrub, beside roads and rivers.

WL, ML, EL: widespread in N., rare or absent in S.

Chaenomeles Lindl.

C. speciosa (Sweet) Nakai Chinese Quince
Introduced. Status: rare, garden escape.
Hedgerow.

WL: rare, woodland near Cauldcoats Holdings.

Pyrus L.

P. communis L. Pear
Introduced. Status: rare.
Railway embankment, hedgerows.

WL: (Dalmeny, pre-1934).
ML: rare, Warriston, 1998; (Duddingston and Craigmillar; pre-1934).

Malus Mill.

M. sylvestris agg. Apple
Introduced. Status: local, escaped or planted in hedgerows.
Hedgerows, roadsides, riversides, woodland, disused railways.

WL: widespread in N., rare in upland areas.
ML: widespread occurrence but sparse in all locations.
EL: widespread in lowland areas.

Includes the following two species and hybrids between them.

M. sylvestris (L.) Mill. Crab Apple
Introduced. Status: uncertain owing to confusion with *M. domestica*. No definite records.

EL: Intermediates between *M. sylvestris* and *M. domestica* occur.

M. domestica Borkh. Cultivated Apple
Introduced. Status: widespread.
Hedgerows, roadsides, woodland, disused railways, riversides.

WL, ML: most records for M. sylvestris agg. probably refer to this species.
EL: occasional, widely scattered distribution, Prestonpans; Gullane; Gifford; Dunglass.

Sorbus L.

S. aucuparia L. Mountain Ash, Rowan
Native. Status: very widespread, often planted.
Woodland, scrub, beside roads and rivers, hedgerows, heath grassland, public parks, waste ground.

WL: very widespread but never in large numbers.
ML: common.
EL: rare in N., common elsewhere.

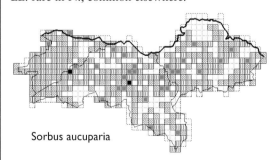

Sorbus aucuparia

S. 'x pinnatifida' auct. (*S. aucuparia* x ?*S. intermedia*)
Native. Status: spontaneous hybrid, rare, but may increase as *S. intermedia* becomes more widely planted.
Waste ground, disused railways.

ML: rare, Granton to Roseburn, 1991.
EL: (Aberlady, 1847).

S. x thuringiaca (Ilse) Fritsch (*S. aucuparia* x *S. aria*)
Native. Status: rare, possibly confused with *S. 'x pinnatifida'*.
Woodland, public parks.

WL: rare, Kinneil.
ML: rare, Holyrood, planted; Ravelston area of the Roseburn to Leith old railway cycle path; (Gorebridge, 1977).

S. intermedia agg. Swedish Whitebeam
Introduced. Status: local, mainly planted.
Scrub, roadsides, woodland, public parks, policies.

WL: widespread.
ML: local, Edinburgh Castle Rock; Dalkeith Estate; Newhouse; Leith Docks.
EL: rare, coast at Gullane; railway cycle path at Longniddry; some sites in policies.

S. aria agg. Common Whitebeam
Introduced. Status: local, often planted.
Woodland, roadsides, public parks, scrub.

WL: local.
ML: local in N., rare in S.
EL: scarce, mainly in lowland areas.

S. rupicola (Syme) Hedl. Rock Whitebeam
Native. Status: rare.
Crags.

ML: rare, Arthur's Seat; (Stenhouse, Arniston, Musselburgh and Liberton, pre-1934).

S. croceocarpa P. D. Sell.
Broad-leaved Whitebeam
Introduced. Status: rare, planted.
Policy woodlands, public places.

WL: rare, Dalmeny Estate.
ML: rare, Orchard Brae, 1993.

S. torminalis (L.) Crantz Wild Service Tree
Introduced. Status: rare, planted.
Public places.

ML: rare, Bawsinch; Royal Edinburgh Hospital.

Amelanchier Medik.

A. lamarkii F. G. Schroed. Juneberry
Introduced. Status: rare.
Policy woodland, scrub.

WL: rare, Hopetoun, planted.

Cotoneaster Medik.

C. frigidus Wall. ex Lindl. Tree Cotoneaster
Introduced. Status: rare, escape.
Public parks, disused railways, woodland.

ML: rare, Blackford Hill; Mortonhall; Ravelston; Craigmillar; Warriston.

C. salicifolius Franch.
Willow-leaved Cotoneaster
Introduced. Status: rare.
Waste ground.

WL: rare, near South Queensferry.
EL: rare, Prestongrange, 1997.

C. dammeri C. K. Schneider
Bearberry Cotoneaster
Introduced. Status: rare, escape.
Waste ground.

ML: rare, Saughton; Musselburgh; Granton, 1997; Slateford.
EL: rare, Prestongrange Mining Museum, 1997.

C. x suecicus G. Klotz (*C. ?dammeri* x *C. conspicuus*) Swedish Cotoneaster
Introduced. Status: rare, garden origin.
Waste ground.

ML: rare, Fushiebridge, 1997.

C. lacteus W. W. Sm. Late Cotoneaster
Introduced. Status: rare, escape.
Scrub.

ML: rare, Arthur's Seat, 1990.

C. cochleatus (Franch.) G. Klotz
Yunnan Cotoneaster
Introduced. Status: rare.
Walls.

ML: rare, Colinton Road, 1995.

C. cashmiriensis G. Klotz Kashmir Cotoneaster
Introduced. Status: rare.
Walls.

ML: rare, near Gorebridge, 1995.

C. integrifolius (Roxb.) G. Klotz
Entire-leaved Cotoneaster
Introduced. Status: local.
Roadsides, walls.

WL: (gateway to Midhope Castle, 1960).

ML: local, locally frequent, Ratho Quarry; Fushiebridge; Hadfast Castle.
EL: rare, Morham; North Berwick.

C. nitidus Jacques   Distichous Cotoneaster
Introduced. Status: rare, planted.
Public amenity area.

WL: rare, Port Edgar.

C. horizontalis Decne.   Wall Cotoneaster
Introduced. Status: local.
Walls, hedgerows, roadsides, railway embankments.

WL: scarce.
ML: local.
EL: rare, Linn Dean; coast at Gullane; a few records near habitation.

C. atropurpureus Flinck & B. Hylmö
                    Purple-flowered Cotoneaster
Introduced. Status: rare.
Dunes.

EL: rare, Hummell Rocks, Gullane, 1993.

C. divaricatus Rehder & E. H. Wilson
                          Spreading Cotoneaster
Introduced. Status: rare.
Public amenity areas, wall tops.

ML: rare, Loanhead; walkway at Warriston, 1999.

C. simonsii Baker   Himalayan Cotoneaster
Introduced. Status: local, often used as a hedging plant.
Woodland, waste ground, roadsides, hedgerows, public parks.

WL, ML: occasional.
EL: rare, Gullane Hill and coast; Whitekirk Hill; a few records near habitation.

C. bullatus Bois   Hollyberry Cotoneaster
Introduced. Status: rare.
Hedgerows, public parks, old railway paths.

WL: rare, Five Sisters Bing; Port Edgar.
ML: rare, Leith, 1982; Trinity, 1994.

C. rehderi Pojark   Bullate Cotoneaster
Introduced. Status: rare.
Policy woodlands.

WL: rare, Hopetoun, 1992.

C. franchetii Bois   Franchet's Cotoneaster
Introduced. Status: rare, planted.
Public amenity areas.

WL: rare, Port Edgar.

C. sternianus (Turrill) Boom   Stern's Cotoneaster
Introduced. Status: rare, escape.
Walls, sandy soil.

ML: rare, Levenhall, 1988; Dean Village, 1997.

C. dielsianus E. Pritz. ex Diels
                              Diels' Cotoneaster
Introduced. Status: rare.
Waste ground.

WL: rare, near South Queensferry.
ML: rare, Musselburgh, 1996; Corstorphine Hill, 1999; Salisbury Crags, 2000.

C. rotundifolius Wall. ex Lindl.   Round-leaved Cotoneaster
EL: Gullane, 1960.

C. juranus Gandoger
ML: Craiglockhart, 1836; Colinton Dell, 1903; Colinton, 1933.

Crataegus L.

C. monogyna Jacq.   Hawthorn
Native. Status: common.
Hedgerows, roadsides, woodland, scrub, beside rivers and ditches, waste ground, railway embankments, rough grassland.

WL, **ML**, EL: local in upland areas, common elsewhere, planted in hedges.

C. x media Bechst. (*C. monogyna* x *C. laevigata*)
Introduced. Status: local.
Policy woodlands, old hedgerows.

ML: occasional, Blackford Hill; Roslin; Loanhead; (Duddingston Loch, 1962; Arniston, 1980).

C. laevigata (Poir.) DC.   Midland Hawthorn
Probably introduced. Status: rare.
Roadsides, woodland.

WL: rare, Linlithgow Loch; (near Bonnytown, 1877).
ML: the cv. 'Paul's Scarlet' is often planted.

C. submollis Sarg.   Hairy Cockspurthorn
EL: Yester Estate, 1957.

FABACEAE

Robinia L.

R. pseudoacacia L. False Acacia
Introduced. Status: rare.
Policies.

EL: rare, Dunglass Estate.

Galega L

G. officinalis L. Goat's Rue
Introduced. Status: rare, casual.
Waste ground, disused railway.

ML: rare, Lochend, 1988.

Astragalus L.

A. cicer L. Chick-pea Milk Vetch, Wild Lentil
Introduced. Status: rare.
Hedgebank.

ML: rare, naturalised at Fushiebridge near Gorebridge since 1930.

A. danicus Retz. Purple Milk Vetch
Native. Status: local.
Dunes, coastal turf, volcanic outcrops, rough grassland.

WL: rare, Dalmeny.
ML: local, Arthur's Seat; Cramond Island; Blackford and Craiglockhart Hills; railway track at Granton.
EL: locally frequent along the coast, rare inland, North Berwick Law; Traprain area; Whitekirk Hill.

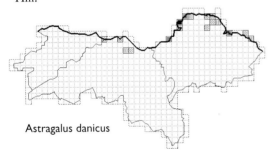

Astragalus danicus

A. glycyphyllos L. Wild Liquorice
Native. Status: rare, (declining).
Dunes, coastal turf, rough grassland.

WL: (beside River Almond near Broxburn, 1841; Dalmeny, 1954).
ML: railway track at Lochend, 1988; (occasional along the coast, pre-1934).

EL: (along the coast at Prestonpans, Gosford, Gullane, Catcraig and Skateraw, pre-1934).

A. hamosus L.
Casual grain alien.

ML: Granton, 1972.

Onobrychis Mill.

O. viciifolia Scop. Sainfoin
Introduced. Status: rare, casual.
Rough grassland, railway embankments.

WL: (Carriden and Kinneil, pre-1934).
ML: rare, Bawsinch, 1995; Dalry station, 1999.
EL: (Belhaven, pre-1934).

Anthyllis L.

A. vulneraria L. Kidney Vetch
Native. Status: scarce.
Dry coastal habitats, waste ground, sandy pastures, quarries, rocky outcrops. Sometimes sown with 'wildflower seed'.

WL: abundant at Bo'ness due to sowing on a reclaimed bing, scarce elsewhere.
ML: scarce, Arthur's Seat; Drylaw; Pilton; Sighthill; Cobbinshaw; north of Middleton; (Blackford Hill).
EL: scarce and declining along the coast, Seton Sands; Yellowcraig; North Berwick; Tantallon; Dunbar; (Seacliff, 1953; Aberlady Bay and Dunglass, 1967); rare inland, Hailes area; Whitekirk Hill; Newmains.

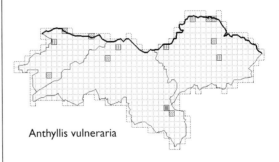

Anthyllis vulneraria

The **ML** records are for **ssp. vulneraria**, this ssp. has also been recorded in **EL**, but coastal plants are generally intermediate between this and **ssp. lapponica**.

ssp. lapponica (Hyl.) Jalas
EL: North Berwick area, pre-1968.

Lotus L.

L. corniculatus L.　　Common Bird's-foot Trefoil
Native and introduced.　Status: common.
Roadsides, rough grassland, waste ground, heath
grassland, riversides, coastal habitats.

WL, ML, EL: common except in intensively
farmed areas.

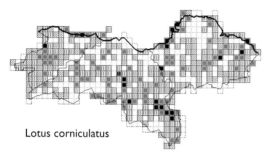

Lotus corniculatus

var. **sativus** Chrtková
Introduced with grass seed.　Status: rare,
probably under-recorded.

ML: rare, Musselburgh Lagoons *c.* 1990; Dalry
railway yard, 1999.
EL: Prestongrange, where it is established;
Gullane, *c.* 1987.

L. pedunculatus Cav.　　Greater Bird's-foot Trefoil
Native.　Status: widespread.
Marshes, beside rivers and ditches, roadsides, rough
grassland.

WL: common in SW., widespread and locally fre-
quent elsewhere.
ML: occasional in N., widespread elsewhere.
EL: rare in NW. and N., widespread elsewhere.

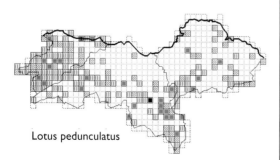

Lotus pedunculatus

L. glaber Mill.　　Narrow-leaved Bird's-foot Trefoil
Casual.

WL: Hopetoun and Linlithgow, 1872.
ML: dry pastures, Dalmahoy, 1863.
EL: Prestonpans, pre-1871; wasteground in West Barns
from 1958 until the site was bulldozed in 1962.

Ornithopus L.

O. perpusillus L.　　　　　　Bird's-foot
Introduced.　Status: rare, casual.
Dry sandy ground.

WL: (Belsyde, 1855; Carriden and Linlithgow,
pre-1934).
ML: (Musselburgh; Portobello, 1863)
EL: rare, Yellowcraig, 1994; (Stenton, 1872).

Securigera DC.

S. varia (L.) Lassen　　　　Crown Vetch
Introduced.　Status: rare.
Old railway lines, waste ground, riversides.

WL: rare, Carriden; (Dalmeny, pre-1934; Pepper
Wood, 1963).
ML: scarce, mainly on old railways, cycle track,
Craigleith; Drylaw; Warriston; Gorgie;
(Newcraighall, 1978).
EL: rare, originally planted to stabilise railway
embankments, now well-established on railway
cycle path Pencaitland/Saltoun and Longniddry;
West Barns; (embankment south of Gullane, 1968).

Vicia L.

V. orobus DC.　　　　Wood Bitter Vetch
Native.　Status: rare, (declining).
Steep grassy banks.

WL: (near Pepper Wood, 1916).
ML: rare, Stow; (Braid Hills, Blackford Hill and
Tynehead, pre-1934).
EL: (near Newmains Smithy, Whittingehame,
1872).

V. cracca L.　　　　　　Tufted Vetch
Native.　Status: common.
Roadsides, hedgerows, waste ground, rough grassland,
riversides, railway embankments, scrub.

WL: common.
ML, EL: rare in some upland areas, common
elsewhere.

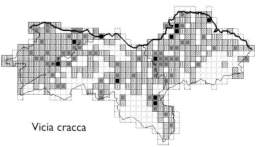

Vicia cracca

V. tenuifolia Roth   Fine-leaved Vetch
Introduced. Status: rare, casual, sometimes
naturalised.
Dry grassland.

WL: rare, on verge of A90 near Cramond Bridge.
ML: rare, Fushiebridge, near Gorebridge, 1960 to
present; Pilton, 1997.

V. sylvatica L.   Wood Vetch
Native. Status: scarce.
Woodland, rough grassland, coastal turf, dunes.

WL: (River Almond, 1878; Carribber Glen,
1956).
ML: rare, Arthur's Seat, 2000; (Roslin, pre-1934).
EL: occasional, sometimes planted, Longniddry to
Haddington cycle path; Tantallon; coast at
Bilsdean; East Lammermuir Deans.

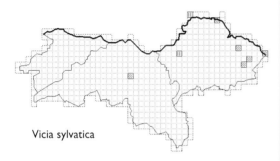

Vicia sylvatica

V. villosa Roth   Fodder Vetch
Introduced. Status: rare.
Waste ground, railway embankments.

ML: rare, Leith; Craighall; Borthwick.

V. hirsuta (L.) Gray   Hairy Tare
Native. Status: widespread.
Roadsides, waste ground, rough grassland, railway
embankments, scrub.

WL, ML, EL: widespread and locally frequent in
lowland areas, rare elsewhere.

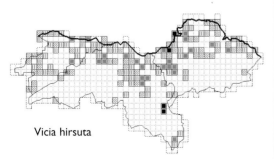

Vicia hirsuta

V. tetrasperma (L.) Schreb.   Smooth Tare
Introduced. Status: rare.
Waste ground, railway embankments.

WL: rare, Bo'ness; Broxburn; Linlithgow;
(Kirkliston, 1906).
ML: rare, Bawsinch, 1997; Duddingston Road,
1998; (Arthur's Seat, 1894; Hillend and Inveresk,
pre-1927).
EL: rare, Archerfield, Dirleton, 1989;
Longniddry/Haddington railway cycle path;
(Dirleton Station, 1914).

V. sepium L.   Bush Vetch
Native. Status: common.
Roadsides, woodland, rough grassland, hedgerow,
riverside, scrub, waste ground, railway embankments.

WL, ML, EL: widespread, locally frequent.

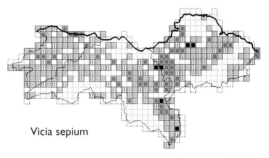

Vicia sepium

V. sativa L.   Common Vetch
Native and introduced. Status: widespread.
Roadsides, waste ground, rough grassland, scrub,
railway embankments.

WL, ML, EL: widespread.

Subspecies have not been consistently recorded,
but the following data are accurate.

ssp. sativa
Introduced. Status: rare. Plants identified in the
past as ssp. *sativa* are now recognised as mainly
belonging to ssp. *segetalis*.
Waste ground, roadsides, rough grassland.

WL: no definite records.
ML: rare, casual, Balerno; Craigmillar; Pilton.
EL: rare, possibly recorded in error for ssp.
segetalis.

ssp. nigra (L.) Ehrh.
Native. Status: local.
Roadside grassy banks, waste ground, rough grassland,
reclaimed bings, coastal turf, volcanic outcrops.

WL: quite widespread.
ML: occasional.
EL: occasional, mostly coastal grassland and volcanic outcrops.

ssp. segetalis (Thuill.) Gaudin
Introduced. Status: local.
Roadsides, rough grassland, waste ground, dunes, beside railways.

WL: local.
ML: rare, Leith Docks.
EL: occasional, Prestongrange; Cockenzie; Seton Sands; Macmerry; North Berwick; Scoughall.

V. lathyroides L.   Spring Vetch
Native. Status: scarce.
Coastal turf, rough grassland, dunes, volcanic outcrops.

WL: (Binny Craig, 1866; Linlithgow, pre-1934).
ML: local, Arthur's Seat; Blackford Hill; Cramond Island where it was locally frequent after a fire; New Hailes.
EL: locally frequent along the coast, rare inland, Ferny Ness; Hailes Castle; Tantallon; Burn Hope; Ling Hope; Catcraig.

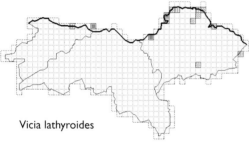

Vicia lathyroides

V. lutea L.   Yellow Vetch
Introduced. Status: rare, casual.
Waste ground.

WL: (Bo'ness, 1894).
ML: rare, Russell Road, 1988; Pilton, 1997.

V. monantha Retz.
ML: Borthwick, 1966; Granton, 1972.

Lathyrus L.

L. linifolius (Reichard) Bässler   Bitter Vetch
Native. Status: widespread in upland areas, scarce elsewhere.
Upland pastures, beside streams and roads.

WL: quite widespread in upland areas.
ML, EL: widespread in upland areas, rare elsewhere.

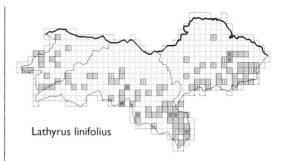

Lathyrus linifolius

var. tenuifolius Druce
ML: (Gorebridge and Moorfoot Glen, pre-1927).
EL: rare, Whiteadder area, 1993; (Hopes Water, 1972).

L. pratensis L.   Meadow Vetchling
Native. Status: common.
Roadsides, rough grassland, riversides, marsh, waste ground, scrub, hedgerows, woodland.

WL, ML, EL: common.

L. tuberosus L.   Tuberous Pea
Introduced. Status: rare.
Waste ground, railway embankments.

ML: (Granton, 1963).
EL: rare, Saltoun; Yellowcraig, on road beside Linkhouse Wood; West Barns.

L. grandiflorus Sm.
                    Two-flowered Everlasting Pea
Introduced. Status: scarce, garden escape.
Railway embankments, waste ground, hedgerows.

WL: rare, Bo'ness.
ML: scarce, Lothian Bridge; Lochend; Slateford, 2000.
EL: rare, New Winton; Innerwick; (Scoughall, 1970; Elphinstone, 1974).

L. sylvestris L.   Narrow-leaved Everlasting Pea
Native. Status: rare.
Scree.

ML: rare, below Salisbury Crags.

L. latifolius L.   Broad-leaved Everlasting Pea
Introduced. Status: rare, garden escape.
Waste ground, hedgerows.

WL: rare, edge of informal car park, Fawnspark.
ML: rare, Bawsinch, 1995; (Mortonhall, 1962).
EL: rare, beside mainline railway, Longniddry; Gullane Links; (railway line, Gullane, 1968).

L. aphaca L.   Yellow Vetchling
Introduced. Status: rare, casual.
Waste ground.

ML: rare, Salamander St., Leith, 1983; Currie, 2001; (Leith Docks, 1971).
EL: (Dunbar, 1903).

Pisum L.

P. sativum L.   Garden Pea
Introduced. Status: rare, crop or garden escape.
Roadsides, rough grassland.

WL: rare, streamside, Inveravon.
ML: rare.
EL: rare, in cereal fields, Tranent and near Spott.

var. **arvense** (L.) Poiret   Field Pea
Introduced. Status: rare, casual, grown for fodder.
Waste ground.

ML: rare, Colinton Dell, 1999.

Cicer L.

C. arietinum L.   Chick Pea
Introduced. Status: rare, casual.
Waste ground.

ML: rare, Granton, 1990.

Ononis L.

O. repens L.   Common Restharrow
Native. Status: local.
Roadsides, rough grassland, dunes, coastal turf.

WL: local.
ML: occasional, Millerhill; Lawhead; Salisbury Crags; (Newhaven, 1966; Colinton 1972; Cobbinshaw, 1977).
EL: quite widespread in lowland areas, locally abundant on coast.

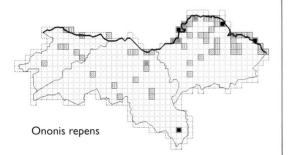

Ononis repens

var. **horrida** Lange.   with spines
ML: Arthur's Seat and Newcraighall, pre-1934.
EL: Longniddry, pre-1934.

Melilotus Mill.

M. altissimus Thuill.   Tall Melilot
Introduced. Status: rare.
Waste ground.

WL: rare, bing near Drumshoreland.
ML: rare, Leith; Millerhill; (Roslin, 1957).
EL: rare, Prestongrange; West Barns; (Aberlady and Gullane, pre-1934).

M. albus Medik.   White Melilot
Introduced. Status: scarce.
Waste ground.

WL: (Carriden, 1965; River Almond near Kirkliston, 1977).
ML: scarce, Leith; Granton; Silverknowes; Duddingston; Polton; Pilton; (Ingliston, 1975).
EL: rare, Prestongrange; West Barns.

M. officinalis (L.) Pall.   Ribbed Melilot
Introduced. Status: local.
Waste ground, bings, disturbed ground.

WL, ML: local.
EL: rare, Prestongrange; Gullane; North Berwick; Hedderwick; West Barns; Torness.

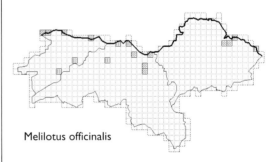

Melilotus officinalis

M. indicus (L.) All.   Small Melilot
Introduced. Status: scarce, casual.
Waste ground, dunes, disturbed ground.

WL: rare, near Uphall Station, 1989; rubbish dump north of Winchburgh, 1990; (Carriden, 1964; Bathgate, 1973).
ML: rare, Bawsinch; Seafield; Leith Docks; Fisherrow; (Borthwick, 1964).
EL: rare, Tranent; Aberlady Bay; West Barns; Dunbar; (Thorntonloch, 1975).

Trigonella L.

T. caerulea (L.) Ser.   Blue Fenugreek
Introduced. Status: rare, casual arable weed.
Waste ground.

ML: rare, Granton, 1989; (Leith Docks, 1973).

Medicago L.

M. lupulina L.   Black Medick
Native. Status: common in lowland areas, local elsewhere.
Roadsides, waste ground, rough grassland, railway embankments, dry ground.

WL, ML, EL: common in lowland areas, local elsewhere.

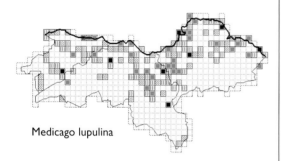

Medicago lupulina

M. sativa L.
ssp. sativa   Lucerne, Alfalfa
Introduced. Status: scarce, casual.
Waste ground, grassy banks, dunes.

WL: rare, Fauldhouse Moor, 1992; Armadale and Dalmeny, 1993; (near Kirkliston, pre-1934).
ML: rare, Dalry station, 1999; (Seafield, 1977; Turnhouse, 1977).
EL: occasionally established, mainly coastal, Prestongrange, apparently sown on coal spoil heap; Seton Sands; Haddington; Dunbar; Torness; (Yellowcraig and Thorntonloch, 1967; Scoughall, 1968).

ssp. falcata (L.) Arcang.   Sickle Medick
Introduced. Status: rare.
Waste ground.

ML: rare, Granton Harbour, 1997.
EL: rare, established at West Barns from 1963 until 1992 when the site was resurfaced to make a lorry park.

ssp. varia (Martyn) Arcang.   Sand Lucerne
EL: Prestongrange, 1980, apparently sown on a landscaped coal spoil heap as part of a grass/legume mix.

M. polymorpha L.   Toothed Medick
Introduced. Status: rare.
Waste ground.

ML: rare, Leith; Granton; (Borthwick rubbish tip, 1968).
EL: (Prestongrange, 1893).

M. arabica (L.) Huds.   Spotted Medick
Introduced. Status: rare, casual.
Waste ground.

WL: (South Queensferry, pre-1856; Linlithgow, pre-1934).
ML: rare, Leith; verge in Holyrood Park after road works, one year only; (Musselburgh, Liberton and Portobello, pre-1934).
EL: rare, landscaped mound at Linkfield car park, West Barns; (Longniddry, pre-1934).

M. truncatula Gaertn.   Strong-spined Medick
ML: Granton, 1972.

Trifolium L.

T. ornithopodioides L.   Bird's-foot Clover
Native. Status: rare.
Sandy, open ground, coastal.

ML: rare, Fisherrow Links, 1988, sown after pipe-laying, did not become established; (Musselburgh Links, pre-1934).

T. repens L.   White Clover
Native. Status: very common, frequently a crop plant. Viviparous plants often occur towards the end of the season.
Roadsides, rough grassland, intensive grassland, waste ground, riversides, leys, arable fields, public parks.

WL, ML: very common, often abundant.
EL: common, locally abundant, often sown in areas such as roadsides.

T. hybridum L.   Alsike Clover
Introduced. Status: local.
Waste ground, roadsides, rough grassland, disturbed ground.

WL: quite widespread but never locally frequent.
ML: local.
EL: scattered distribution in lowland areas, rare in upland areas.

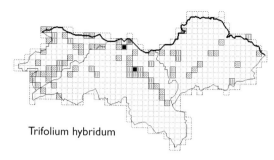

Trifolium hybridum

T. fragiferum L.   Strawberry Clover
Native. Status: rare.
Rough grassland, saltmarsh.

ML: rare, Musselburgh, 1999; (Leith Links,
1824).
EL: rare, Longniddry; Aberlady Bay; West Barns.

T. campestre Schreb.   Hop Trefoil
Native. Status: local.
Waste ground, roadsides, rough grassland, railway
embankments, dry ground, dunes.

WL, ML: quite widespread.
EL: quite widespread and locally frequent in N.,
rare in southern half of vice-county.

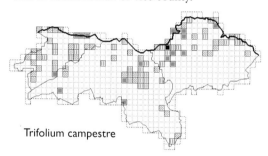

Trifolium campestre

T. dubium Sibth.   Lesser Trefoil
Native. Status: widespread.
Roadsides, waste ground, rough grassland, public parks,
railway embankments, dry ground.

WL: widespread.
ML, EL: widespread and locally frequent in lowland areas, local elsewhere.

T. micranthum Viv.   Slender Trefoil
Native and introduced. Status: rare. sometimes
difficult to distinguish from *T. dubium*.
Waste ground, lawns.

ML: rare, possibly under-recorded, Carrington;
Craigcrook; Loanhead.
EL: rare, Thurston, 1994; (Gullane, pre-1863).

T. pratense L.   Red Clover
Native. Status: common, sometimes sown.
Roadsides, rough grassland, waste ground, riversides,
intensive grassland and leys, railway embankments.

WL: very widespread.
ML, EL: common.

T. medium L.   Zigzag Clover
Native. Status: widespread.
Roadsides, rough grassland, railway embankments.

WL: widespread, locally frequent.
ML: local.
EL: quite widespread.

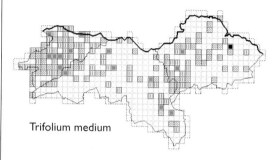

Trifolium medium

T. striatum L.   Soft or Knotted Clover
Native. Status: local.
Dry grassland, quarries, volcanic outcrops, roadsides.

WL: rare, Parkley Craigs; Greendykes Quarry,
1992; Dalmeny, 1996; (Bo'ness, 1894).
ML: local, Arthur's Seat; Blackford, Braid and
Craiglockhart Hills; Craigmillar; Fisherrow Links;
(Ratho, pre-1934).
EL: rare, Luffness; Garleton Hills; Traprain area;
North Berwick area; Tantallon; Dunglass;
(Longniddry, pre-1934).

T. scabrum L.   Rough Clover
Native. Status: rare.
Quarries, rocky outcrops, public parks, grassland.

WL: (Linlithgow, pre-1934).
ML: rare, Arthur's Seat; (Inveresk and Leith;
pre-1934).
EL: rare, Tantallon; Dunbar.

T. arvense L.   Hare's-foot Clover
Native. Status: scarce and often casual in disturbed ground.
Sandy grassland, roadsides, dunes, coastal turf, quarries,
rocky outcrops, waste ground.

WL: rare, mainly single plants, Bo'ness;

Broxburn; a more permanent colony on dry ground under the Forth Road Bridge; (Carriden, 1972; Bathgate, 1977).
ML: rare, Musselburgh; Granton; Cockpen; Leith Docks; (Borthwick, 1958).
EL: scarce, coastal and on rocky outcrops inland; a prostrate form in grass at Prestongrange was presumably sown.

T. lappaceum L.   Bur Clover
Introduced. Status: rare, casual.
Waste ground.

ML: rare, Leith Docks, 1989; (Craigmillar Castle, 1965).

T. incarnatum L.   Crimson Clover
Rare casual.
WL: Bo'ness, pre-1934.
EL: near Aberlady and near Dirleton, 1836; Gullane and North Berwick, pre-1934. The record published in the Wild Flower Society Magazine, Spring, 1988 was unfortunately an error.

T. resupinatum L.   Reversed Clover
ML: Leith and Musselburgh, *c.* 1900.
EL: Belhaven, 1892.

T. aureum Pollich   Large Trefoil
ML: railway tip at Borthwick, 1968.
EL: waste ground at West Barns 1973–5.

T. retusum L., **T. parviflorum** Ehrh and **T. spumosum** L.
ML: Craigmillar Castle, 1965. They were thought to have been introduced with grass seed imported from Italy.

Lupinus L.

L. arboreus Sims   Tree Lupin
Introduced. Status: rare, mainly casual.
Waste ground, railway embankments, dunes.

WL: rare, Whitburn; South Queensferry.
ML: rare, Monktonhall; (Craiglockhart, 1977).
EL: rare, established in and near Gullane; (Gosford, 1976; North Berwick, 1977; Prestongrange, 1978).

L. polyphyllus Lindl.   Garden Lupin
Introduced. Status: ?local, probably over-record-ed in error for *L. x regalis*.
Waste ground, railway embankments.

WL: no definite records, probably rare.
ML: no definite records.
EL: no definite records, further identification needed of plants by railway at Longniddry and in a few other sites.

L. x regalis Bergmans (*L. arboreus* x *L. poly-phyllus*)   Russell Lupin
Introduced. Status: local, garden escape.
Waste ground, verges, old bings.

WL: local, occasionally in some quantity.
ML: rare, Leith, 1982; Levenhall, 1989; Fillyside, 2000.
EL: rare, casual, Prestongrange, 1989.

Laburnum Fabr.

L. anagyroides Medik.   Laburnum
Introduced. Status: fairly widespread, often planted, sometimes naturalising.
Mixed woodlands, estates, roadsides, gardens, public places.

WL, ML, EL: widespread but occurs locally in small numbers, scarce in upland areas.

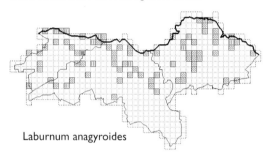

Laburnum anagyroides

The map includes the following two taxa which may be more widespread than indicated here.

L. x watereri (Wettst.) Dippel (*L. anagyroides* x *L. alpinum*).
Introduced. Status: rare, planted.
Policies, public parks.

WL: rare, Kinneil Estate.

L. alpinum (Mill.) J. Presl   Scottish Laburnum
Introduced. Status: rare.
Woodland.

WL: (Carribber, 1955; Linlithgow, 1904).
ML: rare, Bilston Burn, 1998; Corstorphine Hill, 1999.
EL: rare, Seton.

Cytisus Desf.

C. multiflorus (L'Hér. ex Aiton) Sweet
                                                    White Broom
Introduced. Status: rare.

Policies.

EL: rare, Dunglass.

C. striatus (Hill) Rothm.   Hairy-fruited Broom
Introduced. Status: rare, usually planted.
Public amenity areas, scrub.

WL: rare, Bo'ness, planted.
ML: rare, Calton Hill.
EL: rare, Prestonpans.

C. scoparius (L.) Link   Broom
Native. Status: widespread.
Roadsides, waste ground, scrub, railway embankments,
hedgerows, woodland, riversides.

WL: common.
ML, EL: rare in some upland and arable areas,
widespread and locally frequent elsewhere,
sometimes planted.

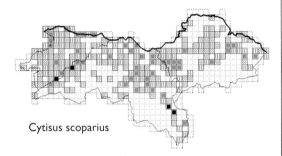

Cytisus scoparius

Spartium L.

S. junceum L.   Spanish Broom
Introduced. Status: rare, planted.
Public amenity areas.

ML: rare, beside A720 near Colinton.
EL: rare, Prestongrange Mining Museum.

Genista L.

G. tinctoria L.   Dyer's Greenweed
Introduced. Status: rare.
Rough grassland.

WL: rare, different areas on towpath near
Winchburgh, 1970 and 1987; (Drumshoreland,
1825–1934).
ML: rare, Granton, 1999, possibly planted.
(Pentland Hills, pre-1934).

G. anglica L.   Petty Whin
Native. Status: rare.
Dry heath.

WL: (Knock Hill, pre-1850).
ML: rare, Moorfoot and Pentland Hills;
Tynehead.
EL: rare, Lammermuir Hills, south and south-east
of Gifford; Bothwell Water; (Aikengall, 1885;
Faseny Water, 1968; Killpallet, 1972).

Ulex L.

U. europaeus L.   Whin, Gorse
Native. Status: common.
Rough grassland, roadsides, woodland, scrub,
hedgerows, riversides, waste ground, railway
embankments.

WL: common.
ML, EL: occasional in some upland areas,
common elsewhere.

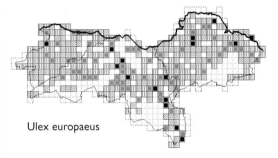

Ulex europaeus

U. gallii Planch.   Western Gorse
Native. Status: rare, planted.
Heaths, public amenity areas.

ML: (Pentland and Corstorphine Hills, pre-1900).
EL: rare, in landscaped thicket with *Ulex
europaeus* and *Cytisus* spp., Prestonpans.

U. minor Roth   Dwarf Gorse
Introduced. Status: rare.
Public amenity areas.

ML: rare, Granton, 1988, planted.
EL: rare, beside railway line, Prestongrange
Mining Museum.

ELAEAGNACEAE

Hippophae L.

H. rhamnoides L.   Sea Buckthorn
?Native and introduced. Status: local, often
originally planted, but readily becoming invasive.
Coastal scrub, dunes, roadsides, waste ground.

WL: local, Bo'ness; Hopetoun; South Queensferry; planted inland, Bathgate; Uphall Station.

ML: local, Musselburgh; Cramond; some inland sites, Crichton; Bawsinch; Penicuik; Tynehead.

EL: said to be native at Dunglass, otherwise introduced; locally abundant along the coast, rare inland.

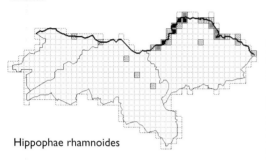

Hippophae rhamnoides

HALORAGACEAE

Myriophyllum L.

M. spicatum L.   Spiked Water Milfoil
Native. Status: local, but may have been con-fused with *M. alterniflorum* in the survey, as the two are not easily distinguished when not in flower.
Base-rich ponds, lochs, reservoirs, slow-flowing streams, ditches.

WL: scarce, mainly in lowland areas.
ML: local, West Calder; Duddingston; Elf Loch; Cobbinshaw Reservoir.
EL: scarce, widely scattered sites, Aberlady Bay; Donolly Reservoir; Black Loch; Catcraig.

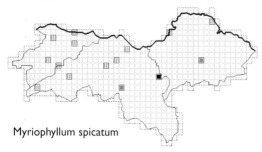

Myriophyllum spicatum

M. alterniflorum DC.   Alternate Water Milfoil
Native. Status: rare.
Streams, base-poor ponds and lochs.

WL: (Linlithgow Loch, pre-1934; Half Loaf

Pond, 1972; Union Canal, 1976).
ML: rare, Clubbiedean and Harlaw Reservoirs; River North Esk, Penicuik.
EL: rare, Luffness; Whiteadder Water; (Gullane, pre-1934).

M. verticillatum L.   Whorled Water Milfoil
ML: two doubtful records from Duddingston Loch, 1764 and 1837.

GUNNERACEAE

Gunnera L.

G. tinctoria (Molina) Mirb.   Gunnera
Introduced. Status: rare, garden escape.
Marshy areas.

ML: rare, Roslin Glen.

LYTHRACEAE

Lythrum L.

L. salicaria L.   Purple Loosestrife
?Native and introduced. Status: rare.
Beside ponds and lochs, marshes.

WL: rare, mainly planted, Glendevon; Kinneil; east of Whitburn; Dalmeny village; Midhope; Easter Inch.
ML: rare, Bawsinch; Carberry; Camilty; Boghall; Hunter's Bog; Craiglockhart Pond.
EL: rare, Aberlady Bay, 1990; (Binning Wood).

L. portula (L.) D. A. Webb   Water Purslane
Native. Status: local.
Beside lochs and reservoirs.

WL: (Linlithgow Loch, pre-1934).
ML: local, Harperrig, Threipmuir and Rosebery Reservoirs.
EL: rare, Stobshiel Reservoir; pond near Gifford; Black Loch; (Bass Rock, 1873; Bothwell Water).

THYMELAEACEAE

Daphne L.

D. laureola L.   Spurge Laurel
Introduced. Status: local, usually planted.
Estate woodlands.

WL: rare, occasional plants at Hopetoun and Inveravon; naturalised at Dalmeny.
ML: local, Dalkeith; Musselburgh; Dreghorn; Arniston.
EL: occasional plants in estate woodlands, Gosford; near Whitekirk; naturalised at Biel.

ONAGRACEAE

Epilobium L.

E. hirsutum L.   Great Willowherb
Native. Status: widespread.
In and beside ditches and rivers, marshes, wet grassland, roadsides and waste ground.

WL: widespread and locally frequent.
ML, EL: widespread and locally frequent in lowland areas, rare elsewhere.

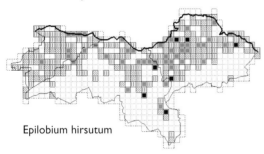
Epilobium hirsutum

E. x erroneum Hausskn. (*E. hirsutum* x *E. montanum*)
Native. Status: rare.
Roadside verge with parents.

ML: rare, Fountainhall, 1990.

E. parviflorum Schreb.   Hoary Willowherb
Native. Status: local.
Marshes, beside ditches, streams and the canal.

WL, ML: local.
EL: local, Aberlady Bay; Donolly Reservoir.

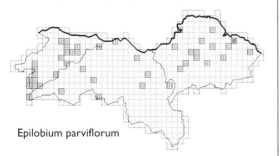
Epilobium parviflorum

E. montanum L.   Broad-leaved Willowherb
Native. Status: common.
Woodland, roadsides, waste ground, scrub.

WL, ML, EL: common.

E. x aggregatum Celak. (*E. montanum* x *E. obscurum*)
Native. Status: rare, probably under-recorded.
Roadsides, gardens.

ML: rare, garden weed.
EL: rare, Dunglass, 1993.

E. montanum x E. ciliatum
Native. Status: rare, possibly under-recorded.
Roadsides, old railway lines.

EL: rare, Ormiston, 1992; Winton, 1993.

E. tetragonum L.   Square-stalked Willowherb
All Scottish specimens formerly identified as *E. tetragonum* have been found to be *E. obscurum*.

E. obscurum Schreb.   Short-fruited Willowherb
Native. Status: widespread. Some records may be in error for *E. ciliatum*.
Beside streams, ditches and roads, marshes, damp waste ground and grassland.

WL: widespread especially in W. and S., not locally frequent.
ML: widespread but patchy.
EL: rare in N., widespread elsewhere.

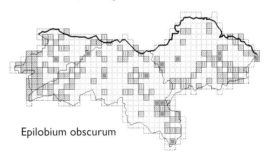
Epilobium obscurum

E. roseum Schreb.   Pale Willowherb
Native. Status: local.
Beside ditches, ponds, streams and lochs, marshes.

WL: occasional.
ML: scarce, possibly overlooked, Duddingston; Granton; Gogar; Tynehead; Warriston.
EL: rare, West Saltoun; south of Haddington; Dunglass; (North Berwick, pre-1934; Blackcastle Hill, 1967; Dunbar, 1972).

E. ciliatum Raf.   American Willowherb
Introduced. Status: widespread and becoming, or
may already be, the commonest species of
Epilobium in the Lothians.
Beside roads, rivers and ditches, waste ground,
marshes, cereal fields, woodland, scrub.

WL, EL: widespread, seldom locally frequent.
ML: widespread, locally frequent.

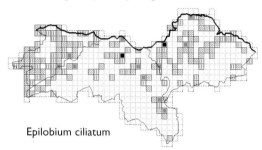

Epilobium ciliatum

E. palustre L.   Marsh Willowherb
Native. Status: widespread.
Marshes, beside rivers and ditches, bogs.

WL, ML, EL: widespread and locally frequent in
S., occasional in the better-drained N.

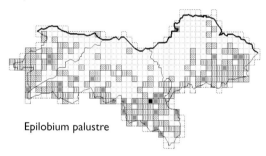

Epilobium palustre

E. alsinifolium Vill.   Chickweed Willowherb
Native. Status: rare.
Upland flushes.

EL: rare, in one flush in the Lammermuir Hills.

E. brunnescens (Cockayne) P. H. Raven &
Engelhorn   New Zealand Willowherb
Introduced. Status: local, well naturalised.
Beside streams and ditches, marshes, quarries, hill
ravines, gardens.

WL: rare, old bing near Fauldhouse; wood,
Drumshoreland.
ML: quite widespread in a small area in the S.,
rare elsewhere. First recorded in Britain at
Craigmillar in 1904.
EL: scattered sites in the Lammermuir Hills.

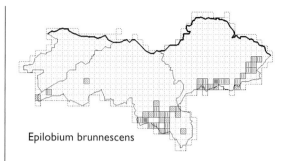

Epilobium brunnescens

E. pedunculare A. Cunn.   Rockery Willowherb
Introduced. Status: rare.
Gardens.

ML: long-established weed at Royal Botanic
Garden Edinburgh.

Chamerion (Raf.) Raf.

C. angustifolium (L.) Holub
     Rosebay Willowherb
Native. Status: very common.
Roadsides, woodland, waste ground, scrub, riversides,
rough grassland, railway embankments, hedgerows.

WL, ML: common, often locally abundant.
EL: common and often locally abundant, but
known only from Yester and two Lammermuir
deans in the nineteenth century.

Oenothera L.

O. glazioviana P. Micheli ex Mart.
     Large-flowered Evening Primrose
Introduced. Status: rare, garden escape.
Rubbish tips, dunes.

ML: rare, Roslin Glen; Leith Docks; Blackford
Hill Observatory, 1999; (Levenhall, 1966).
EL: rare, Aberlady Bay; long-established on
North Berwick sea-front; (Gullane rubbish tip,
1958 and 1979; Port Seton, 1960 and 1972).

O. biennis L.   Common Evening Primrose
WL: Kirkliston and Linlithgow, pre-1934.
ML: Leith, 1957.

O. renneri H. Scholz
ML: railway tip at Borthwick, 1966.

Fuchsia L.

F. magellanica Lam.   Fuchsia
Introduced. Status: scarce, garden escape.

Beside roads and rivers, hedgerows, railway embankments, waste ground.

WL: rare, Carriden; near Faucheldean.
ML: rare, Cramond; Warriston; Trinity.
EL: rare, naturalised in Dunglass Dean; (Gullane rubbish tip, 1970).

Circaea L.

C. lutetiana L.   Enchanter's Nightshade
Native. Status: local.
Mixed and deciduous woodland on nutrient-rich soils.

WL: occasional in N., rare or absent elsewhere.
ML: widespread and locally frequent in centre, rare elsewhere.
EL: scattered distribution, locally frequent in centre, rare elsewhere, Humbie area; Woodhall Dean; Dunglass.

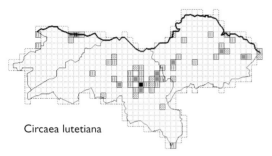

Circaea lutetiana

C. x intermedia Ehrh. (*C. lutetiana* x *C. alpina* L.)   Upland Enchanter's Nightshade
Native. Status: rare.
Woodland, riversides.

WL: near Ochiltree Mill; (Linlithgow, 1877; Hopetoun, 1903).
ML: rare, Kirkhill; Rosebank; Bilston Burn, 1999; (Colinton, Costorphine Hill, Cramond and Newbattle, pre-1934).
EL: rare, estates south of Haddington; Hopes Reservoir; Woodhall Dean; (Yester and Sheeppath Dean, pre-1934).

CORNACEAE

Cornus L.

C. sanguinea L.   Dogwood
Introduced. Status: local.
Mixed and deciduous woodland, scrub.

WL: some recent plantings beside roads and in public amenity areas.

ML: widely scattered distribution, never locally frequent.
EL: rare, planted on several estates; path on disused railway south of Longniddry; Danskine Loch.

C. sericea L.   Red-osier Dogwood
Introduced. Status: local, often planted.
Mixed and deciduous woodland, roadsides, scrub, public parks, policies.

WL: local, well naturalised on old railway line near Kirkliston.
ML: occasional, West Calder, 1993; (Dreghorn 1957; Carberry, 1969).
EL: rare, on old estates, Bara; Broxmouth; Dunglass; (Gosford, pre-1927; Yester, 1966).

C. alba L.   White Dogwood
Introduced. Status: rare, planted.
Lakeside.

EL: rare, estate near Whitekirk.

C. mas L.   Cornelian Cherry
Introduced. Status: rare, planted.
Policy woodland, public amenity areas.

WL: rare, Hopetoun; Port Edgar.

C. suecica L.   Dwarf Cornel
ML: doubtful record from the Pentland Hills, 1764.

VISCACEAE

Viscum L.

V. album L.   Mistletoe
Introduced. Status: rare, scarcely naturalising.
Woodland.

ML: rare, mainly on lime and hawthorn trees, Dean Cemetery; Ravelston Terrace; Gallery of Modern Art; on poplar at Bawsinch, 1993 until recently.

CELASTRACEAE

Euonymus L.

E. europaeus L.   Spindle
Introduced. Status: rare.
Mixed woodland, roadsides.

WL: rare, planted, Hopetoun; Duddingston

Wood; Dalmeny.
ML: rare, Dalhousie Castle; Cockpen; Bawsinch; Bonnington; Cramond; Hermiston.
EL: rare, Gosford; Pressmennan; (Dunglass, pre-1934; Wester Pencaitland, 1957; Luffness, 1981).

E. latifolius (L.) Mill.   Large-leaved Spindle
Introduced. Status: rare, planted.
Mixed woodland.

WL: rare, Hopetoun Estate.

AQUIFOLIACEAE

Ilex L.

I. aquifolium L.   Holly
Native. Status: widespread, sometimes planted.
Woodland, hedgerow, roadsides, scrub.

WL, ML, EL: widespread and locally frequent except in upland areas.

I. x altaclerensis (Loudon) Dallim. (*I. aquifolium* x *I. perado* Aiton)   Highclere Holly
Introduced. Status: rare, planted.
Woodland, hedgerow, public amenity areas.

ML: rare, Roseburn; Balerno.
EL: rare, Dunglass.

BUXACEAE

Buxus L.

B. sempervirens L.   Box
Introduced. Status: local, planted, rarely naturalised.
Woodland.

WL, ML: local, mainly in estate woodlands.
EL: scattered sites in lowland areas, mainly in estate woodlands, Links Wood, Tyninghame; Gifford.

EUPHORBIACEAE

Mercurialis L.

M. perennis L.   Dog's Mercury
Native. Status: local.
Old woodland, and remnant woodland.

WL, ML: local, locally abundant.
EL: quite widespread, locally abundant.

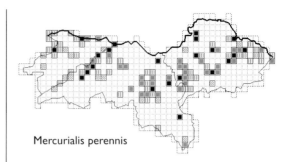

Mercurialis perennis

M. annua L.   Annual Mercury
Introduced. Status: casual.
ML: clay pit Portobello, 1909; Leith Docks, 1888–1907.
EL: Tranent, 1836.

Euphorbia L.

E. helioscopia L.   Sun Spurge
Native. Status: widespread.
Arable fields, waste ground, roadsides, disturbed ground, gardens.

WL: widespread in N., rare in S.
ML: widespread and locally frequent in E., very local in W.
EL: widespread in lowland areas.

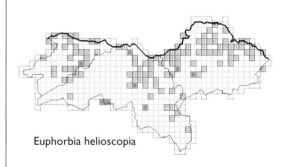

Euphorbia helioscopia

E. lathyris L.   Caper Spurge
Introduced. Status: rare.
Waste ground, beside railways.

WL: rare, Abercorn.
ML: (Prestonhall, pre-1934).
EL: (Seton Mains Halt, pre-1934).

E. peplus L.   Petty Spurge
Native. Status: local, mainly a horticultural weed.
Waste ground, roadsides, gardens.

WL: scarce.
ML: local in lowland areas, rare elsewhere.
EL: occasional in lowland areas.

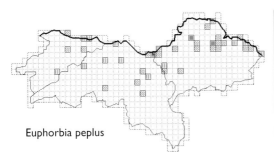

Euphorbia peplus

E. x pseudovirgata (Schur) Soó (*E. waldsteinii* (Soják) Czerep. x *E. esula*) Twiggy Spurge
Introduced. Status: scarce.
Roadsides, waste ground, railway embankments.

WL: (South Queensferry, 1958).
ML: local, Fushiebridge; Leith Docks.
EL: rare, bank above the River Tyne, Haddington; West Barns; (East Linton, 1973).

E. cyparissias L. Cypress Spurge
Introduced. Status: rare.
Rough grassland, public parks.

WL: rare, South Queensferry; (Port Edgar, pre-1934; Philpstoun, 1980).
ML: rare, Carberry; Turnhouse; Duddingston.
EL: (West Barns, 1981).

E. amygdaloides L. Wood Spurge
ssp. robbiae (Turrill) Stace
Introduced. Status: rare, garden throw-out.
Woodland.

EL: rare, naturalised in woodland at Gifford.

ssp. amygdaloides
EL: Bara Wood, 1965.

E. characias L. *s.l.* Mediterranean Spurge
Introduced. Status: rare, garden escape.
River shingle.

ML: rare, beside Water of Leith at Stockbridge 'colonies', 1999.

E. exigua L. Dwarf Spurge
WL: once abundant in fields north of Carlowrie; last record from Kirkliston, 1934.
ML: Musselburgh, pre-1934.
EL: St. Germains, 1824; Longniddry, 1908; Yester, pre-1934.

E. esula L. Leafy Spurge
WL: Abercorn, 1799.
ML: Duddingston, 1835; Slateford, 1902.
EL: lane near Gladsmuir Kirk, 1825.

RHAMNACEAE

Frangula Mill.

F. alnus Mill. Alder Buckthorn
Introduced. Status: rare, planted.
Estate woodlands.

WL: rare, Hopetoun Wood.
ML: rare, Bawsinch.

VITACEAE

Vitis L.

V. vinifera L. Grape Vine
Introduced. Status: rare.
Hedgerow, scrub, waste ground.

ML: rare, Craiglockhart Pond; Wardie railway bridge.

LINACEAE

Linum L.

L. usitatissimum L. Flax
Introduced. Status: rare, casual, bird seed alien and agricultural relic.
Arable fields, leys, beside canal and lochs, gardens, waste ground, pavements.

WL: rare, by the Union Canal, Linlithgow, 1992; (South Queensferry, pre-1934; sown in the 1980s at Carriden, now extinct there).
ML: rare, Dundas Street, 1995; Craigmillar, 1999; (Leith, 1959).
EL: rare, Dirleton, 1996; North Berwick, 1998; Wamphray, 1994; Sandy's Mill, 1996; Pressmennan, 1986; (Whittingehame, 1873; Ormiston, 1883; Aberlady, 1950; Archerfield, 1960).

L. catharticum L. Fairy Flax, Purging Flax
Native. Status: local.
Rough grassland, heath grassland, coastal turf, quarries, railway embankments, waste ground, old bings, tracks.

WL: widespread.
ML: rare in N., local elsewhere, locally frequent in SE.
EL: local, mainly along the coast and in SE.

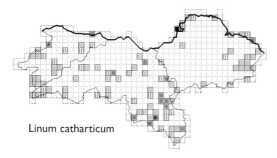

Linum catharticum

Radiola Hill

R. linoides Roth   Allseed
Native. Status: rare.
Bare peaty or sandy soil, usually coastal.

WL: rare, Bo'ness.
ML: rare, Crichton, 1995; Polton; (Fisherrow, 1765).
EL: (Gullane Links, pre-1894).

POLYGALACEAE

Polygala L.

P. vulgaris L.   Common Milkwort
Native. Status: local.
Moorland, heath grassland, rocky outcrops, quarries, coastal turf.

WL: rare, Blackridge; Bathgate; Torphichen; Riccarton Hills.
ML: local, Cobbinshaw; Loganlea; Fala.
EL: scarce, mainly on coast, Aberlady; Gullane; Hedderwick; Barns Ness; Lammermuir Hills.

P. serpyllifolia Hosé   Heath Milkwort
Native. Status: local.
Upland pasture, moorland, beside streams and roads, coastal heath.

WL, ML, EL: widespread in upland areas, rare elsewhere.

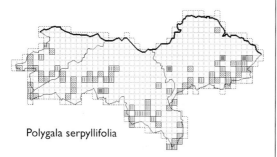

Polygala serpyllifolia

STAPHYLEACEAE

Staphylea L.

S. pinnata L.   Bladdernut
Introduced. Status: rare, planted.
Estate woodlands.

WL: rare, Hopetoun.

HIPPOCASTANACEAE

Aesculus L.

A. hippocastanum L.   Horse Chestnut
Introduced. Status: widespread, planted, occasionally self-set.
Woodland, roadsides, scrub, public parks, policies.

WL, ML, EL: widespread in lowland areas.

A. carnea J. Zeyh.   Red Horse Chestnut
Introduced. Status: rare, planted.
Policies.

WL: rare, Carriden; Midhope; Hopetoun; Dundas.
EL: rare, Yester.

A. flava Salander   Sweet Buckeye
Introduced. Status: rare.
Estate woodland.

WL: rare, Midhope.

A. indica (Cambess.) Hook.
       Indian Horse Chestnut
Introduced. Status: rare, planted.
Estate woodland.

ML: rare, Vogrie Country Park.

ACERACEAE

Acer L.

A. platanoides L.   Norway Maple
Introduced. Status: local, widely planted, becoming naturalised.
Woodland, roadsides, public parks, policies, scrub.

WL: widespread.
ML: local.
EL: scattered distribution, in lowland areas only.

296

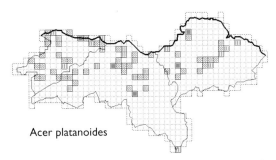

Acer platanoides

A. cappadocicum Gled.   Cappadocian Maple
Introduced. Status: rare, planted.
Estate woodlands, public parks.

WL: rare, Dundas Estate; Livingston.
ML: rare, Mavisbank.
EL: rare, Yester.

A. campestre L.   Field Maple
Introduced. Status: local, planted.
Scrub, woodland, roadsides, public parks, land
reclaimed from mining or railways.

WL: quite widespread.
ML: occasional, mainly in N.
EL: occasional and increasing, mainly in centre.

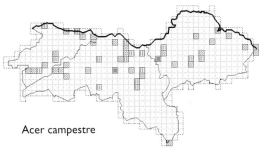

Acer campestre

A. monspessulanum L.   Montpelier Maple
Introduced. Status: rare, planted.
Policies, amenity areas.

ML: rare, Calton Hill, 2001.
EL: (Yester Estate, 1902).

A. pseudoplatanus L.   Sycamore
Introduced. Status: common, naturalised.
Woodland, roadsides, scrub, hedgerows, riversides,
waste ground, public parks.

WL, ML, EL: common.

A. saccharinum L.   Silver Maple
Introduced. Status: rare, planted.
Policies, public amenity areas.

ML: rare, Easthouses; Vogrie Country Park.

A. macrophyllum Pursh   Oregon Maple
Introduced. Status: rare, planted.
Estate woodland.

ML: rare, Vogrie Country Park.

A. palmatum Thunberg   Japanese Maple
Introduced. Status: rare, planted.
Policies.

WL: rare, Hopetoun.

A. cissifolium (Siebold & Zucc.) K. Koch
Vine-leaf Maple
Introduced. Status: rare, planted.
Policies.

WL: rare, Hopetoun.

OXALIDACEAE

Oxalis L.

O. corniculata L.   Procumbent Yellow Sorrel
Introduced. Status: rare.
Woodland, scrub, quarries, waste ground, gardens
(pernicious weed).

WL: (Dalmeny Estate, 1960).
ML: rare, Craigcrook Road; Royal Botanic
Garden Edinburgh; (Craiglockhart, 1975).
EL: rare, Dirleton Castle; Smeaton, East Linton;
(Biel Estate, 1963).

O. exilis A. Cunn.   Least Yellow Sorrel
Introduced. Status: rare.
Waste ground, gardens, (pernicious weed).

ML: rare, Eden Lane; Belgrave Crescent;
Inverleith Row; Royal Botanic Garden
Edinburgh.

O. articulata Savigny   Pink Sorrel
Introduced. Status: rare, garden escape or garden
relic.
Roadsides, scrub, gardens, waste ground.

ML: (Musselburgh Lagoons, 1975).
EL: rare, Pencaitland; Aberlady; Yellowcraig;
North Berwick; Whitekirk; Garvald; West Barns.

O. acetosella L.   Wood Sorrel
Native. Status: common.
Mixed and coniferous woodland, dry heath, scrub,
heath grassland, rough grassland, streamsides.

WL: locally abundant in N., less common in SW.

ML, EL: occasional in N., common elsewhere, locally abundant.

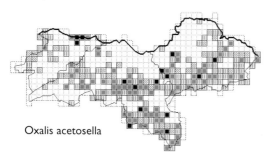

Oxalis acetosella

O. stricta L.   Upright Yellow Sorrel
WL: Broxburn.
ML: Slateford, pre-1934.
EL: Wester Pentcaitland, 1957.

GERANIACEAE

Geranium L.

G. endressii J. Gay   French Cranesbill
Introduced. Status: rare, garden escape.
Waste ground, roadsides, woodland.

WL: rare, Almondell; Broxburn; Blackness; Longridge.
ML: rare, Colinton Dell; Penicuik; Craigpark Quarry; (Newhouse, 1973).
EL: rare, established near Blance; Gullane, 1997; North Berwick, 1996.

G. x oxonianum Yeo (*G. endressii* x *G. versicolor*)   Druce's Cranesbill
Introduced. Status: rare.
Dunes, waste ground.

ML: rare, old mill site at Juniper Green, 1996.
EL: (Gosford, 1978).

G. versicolor L.   Pencilled Cranesbill
Introduced. Status: rare.
Waste ground.

EL: rare, Ormiston.

G. rotundifolium L.   Round-leaved Cranesbill
Introduced. Status: rare.
Walls, grassy banks, roadside verges.

ML: (Leith Walk, 1833–7).
EL: rare, Prestongrange; two sites at North Berwick; Hailes Castle.

G. sylvaticum L.   Wood Cranesbill
Native. Status: local.
Woodland, riversides, roadsides.

WL: local, mainly in SW.
ML: rare in N., local in centre and S., notably in river valleys.
EL: occasional around Humbie and in the Lammermuir Hills; absent from agricultural land in lowland areas.

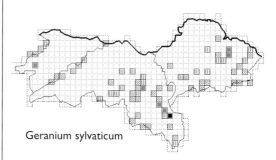

Geranium sylvaticum

G. psilostemon Ledeb.   Armenian Cranesbill
Introduced. Status: rare.
Rough grassland.

ML: rare, Granton foreshore, 1999.

G. pratense L.   Meadow Cranesbill
Native. Status: local.
Roadsides, riversides, hedgerows, rough grassland.

WL: widespread in N., rare elsewhere.
ML: local, mainly in centre.
EL: widespread in centre, rare elsewhere.

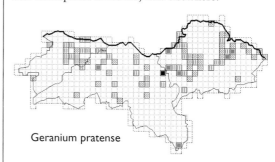

Geranium pratense

G. sanguineum L.   Bloody Cranesbill
Native. Status: local, sometimes a garden escape.
Waste ground, quarries, lime-rich grassland, dunes, cliffs.

WL: (Carriden, 1972; Linlithgow, pre-1934).
ML: rare, Arthur's Seat; (Blackford Hill; Penicuik).
EL: local, mainly coastal, frequent at Ferny

Ness/Gosford Bay; rare elsewhere, Seton;
Longniddry Bents; Aberlady Mains; Traprain
Law; Tantallon; Ravensheugh Sands; (Tyne
Estuary, 1909; Aberlady Bay, 1955).

G. dissectum L. Cut-leaved Cranesbill
Native. Status: widespread.
Roadsides, waste ground, rough grassland, scrub, cereal
fields.

WL: widespread but nowhere frequent.
ML: widespread in E., rare elsewhere except in
the valley of the River Almond.
EL: widespread in lowland areas, rare in the
Lammermuir Hills.

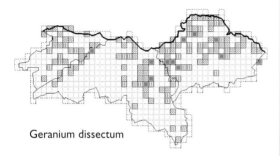
Geranium dissectum

G. x magnificum Hyl. (*G. ibericum* x *G. platype-
talum* Fisch. & C. A. Mey.) Purple Cranesbill
Introduced. Status: rare, garden escape, sterile.
Roadsides, waste ground, disused railways.

WL: rare, single plants, Bo'ness; Port Edgar;
Cauldcoats Holdings; Fauldhouse; (Blackridge,
1972).
ML: local, Leith, 1988 and 1994; south of
Livingston; (Currie, 1975; Nine Mile Burn,
1977).
EL: rare, Gifford; (Pencaitland, 1973).

G. pyrenaicum Burm. f. Hedgerow Cranesbill
Introduced. Status: local.
Roadsides, waste ground, disused railways.

WL: rare, Bathgate; (Bo'ness, 1973).
ML: occasional, Drylaw; Duddingston; Pilton;
Holyrood Park; Blackford Glen.
EL: scattered distribution, mainly in lowland
areas, Seton Sands; Gullane; Garleton Hills
quarry; Spott Mill.

G. pusillum L. Small-flowered Cranesbill
Native. Status: scarce, possibly under-recorded.
Rough grassland, roadsides, public parks, coastal
habitats, bare dry soils.

WL: (Linlithgow, pre-1934).
ML: scarce, Hallyards; Craigmillar Castle;
Salisbury Crags; Leith Docks; Musselburgh;
Duddingston.
EL: occasional in N, mainly near the coast,
Longniddry Bents; Kilspindie; Yellowcraig;
Whitekirk.

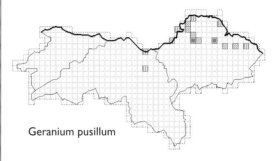
Geranium pusillum

G. molle L. Dove's-foot Cranesbill
Native. Status: widespread.
Waste ground, roadsides, rough grassland, cereal fields,
public parks.

WL: widespread in N., occasional in S.
ML: widespread and locally frequent in N. and
E., local in S.
EL: widespread and locally frequent in lowland
areas, scarce in Lammermuir Hills.

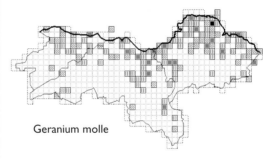
Geranium molle

G. macrorrhizum L. Rock Cranesbill
Introduced. Status: rare, garden throw-out.
Waste ground.

EL: rare, in a bramble patch at Prestongrange,
1989–91.

G. lucidum L. Shining Cranesbill
Probably introduced. Status: scarce, often a
garden escape.
Walls, rocks, rough grassland, gardens.

WL: scarce, Carribber; Blackness; Hopetoun;
Parkley; Cauldcoats.

ML: local, Lasswade; Craiglockhart Hill; Midmar; Liberton; Valleyfield, Penicuik; Lady Lothian Wood.
EL: rare, Humbie; Haddington; Dirleton; Waughton Castle; Balfour Monument; Hailes Castle; Belhaven; (Traprain Law, 1848; North Berwick, pre-1934; East Linton, 1975).

G. robertianum L.   Herb Robert
Native. Status: common. White varieties occur.
Woodland, beside roads and rivers, scrub, hedgerows, walls, waste ground.

WL: rare in S., common in N.
ML: local in S., common elsewhere.
EL: common in lowland areas, rare in upland areas.

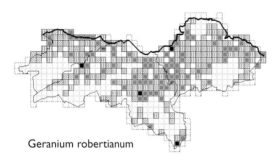

Geranium robertianum

G. phaeum L.   Dusky Cranesbill
Introduced. Status: rare, garden escape.
Scrub, rough grassland, roadsides.

WL: (Kinneil, South Queensferry and Linlithgow, pre-1934; Carriden, 1957; Pepper Wood, 1975).
ML: rare, Hallyards; (Inveresk, 1973, 1976 and 1978).
EL: rare, Elvingston; Bara; Balgone; Hailes; Innerwick; (Whittingehame, 1873).

G. nodosum L.   Knotted Cranesbill
WL: Carribber, Carriden, Dalmeny and Ecclesmachan, pre-1934.
ML: Prestonhall and Colinton, pre-1934.

G. columbinum L.   Long-stalked Cranesbill
ML: Leith, Roslin and Slateford, pre-1934.
EL: Dunbar, pre-1934.

G. ibericum Cav.   Caucasian Cranesbill
Garden escape.
EL: dunes at Gosford, 1955–71.

G. x monacense Harz (*G. phaeum* L. x *G. reflexum* L.)
     Munich Cranesbill
Garden escape.

WL: Kirkliston, 1929.
ML: Ratho Station.

Erodium L'Hér.

E. moschatum (L.) L'Hér.   Musk Storksbill
Introduced. Status: rare, casual.
Roadside.

EL: rare, Dunbar, 1991; (Prestonpans, pre-1824; Dunbar, 1903).

E. cicutarium (L.) L'Hér.   Common Storksbill
Native. Status: local.
Waste ground, quarries, rocky outcrops, public parks, coastal habitats, bare slopes, railway embankments.

WL: rare, Bo'ness; Broxburn; near Duntarvie; Dalmeny; (Linlithgow, 1977).
ML: occasional in N. and centre.
EL: local, mainly along coast, scarce inland.

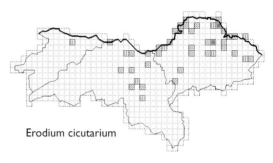

Erodium cicutarium

E. malacoides (L.) L'Hér.   Soft Storksbill
WL: Overton, 1909.

LIMNANTHACEAE

Limnanthes R. Br.

L. douglasii R. Br.   Meadow-foam
Introduced. Status: rare, casual, garden escape or throw-out.
Waste ground, rubbish tips, roadsides.

WL: rare, Faucheldean; Kinneil; Uphall Station.

TROPAEOLACEAE

Tropaeolum L.

T. majus L.   Nasturtium
Introduced. Status: rare, garden escape.
Garden rubbish tips, roadsides.

WL: rare, Bo'ness; Blackburn; Harthill.
ML: rare, Musselburgh; Warriston.
EL: rare, Longniddry; Luffness.

BALSAMINACEAE

Impatiens L.

I. parviflora DC.   Small Balsam
Introduced. Status: local, naturalised.
Woodland, waste ground, scrub.

ML: local, Easter Craiglockhart and Corstorphine
Hills; Granton; Polton.
EL: local, a few sites, plentiful in some,
Pencaitland; Bolton Muir Wood; Haddington
(since 1908); Stevenson; East Linton; Stoneypath
Tower; River Tyne; Tyninghame Links Wood;
(Saltoun Hall, 1972).

I. glandulifera Royle
                    Indian or Himalayan Balsam
Introduced. Status: local, strongly naturalised.
Beside rivers, streams and lochs.

WL: scattered sites in N., frequent only along
parts of the River Almond.
ML: locally abundant/frequent in urban areas in
N. and along the River Almond, the Water of
Leith and the River Esk, occasional elsewhere.
EL: rare, small populations at Seton Sands;
Cuddie Wood; Colstoun; Bara; West Barns;
Innerwick; (Dirleton, 1970; Aberlady Bay, 1978).

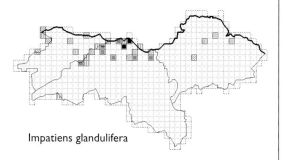

Impatiens glandulifera

ARALIACEAE

Hedera L.

H. colchica (K. Koch) K. Koch   Persian Ivy
Introduced. Status: local, persistent, not
naturalising.
Walls, roadsides.

ML: local, Slateford Aqueduct; Dean Bridge;
Craigmillar; Warriston Road; Dreghorn.
EL: rare, wooded roadside near Pencaitland.

H. helix L.
ssp. helix   Common Ivy
Native. Status: common.
Woodland, walls, roadsides, hedgerows, scrub.

WL, ML, EL: common, locally abundant in
lowland areas, occasional elsewhere.

ssp. hibernica (G. Kirchn.) D. C. McClint.
cv. Hibernica   Irish Ivy
Introduced. Status: local, probably under-
recorded.
Woodland, scrub, walls, railway embankments.

WL: rare, Kirkliston; Bo'ness.
ML: locally frequent at a few sites, Hermitage of
Braid; Corstorphine Hill; Marchbanks; Colinton
Woods.
EL: local, dominating large areas of woodland at
some sites, Prestongrange; Setonhill; North
Berwick; East Linton; Whittingehame.

APIACEAE

Hydrocotyle L.

H. vulgaris L.   Marsh Pennywort
Native. Status: scarce.
Marshes, ponds, bogs.

WL: rare, Balgornie, 1982; South Couston, 1984.
ML: locally frequent at a small number of sites,
Duddingston Loch; Thriepmuir, Harlaw, Bonaly
and Loganlee Reservoirs.
EL: local, coastal and upland sites, Gosford Pond;
Aberlady Bay; Gullane; Yellowcraig; St Baldred's
Cradle; Belhaven Bay; Hopes area; Danskine; Star
Wood; Whiteadder Water; (Innerwick, 1969).

Sanicula L.

S. europaea L.   Sanicle
Native. Status: local.
Woodland, on quite rich soils.

WL: locally frequent in N., rare elsewhere.
ML: locally frequent in centre, rare elsewhere.
EL: locally frequent in woods and deans.

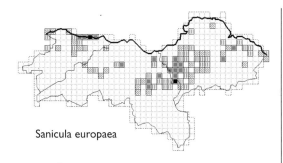

Sanicula europaea

Astrantia L.

A. major L. Astrantia, Great Masterwort
Introduced. Status: rare, garden escape.
Waste ground.

WL: rare, Dalmeny; (Carlowrie and Carribber,
pre-1934).
ML: (Muirhouse, 1965).

Eryngium L.

E. maritimum L. Sea Holly
ML: Musselburgh, pre-1894.
EL: between Musselburgh and Prestonpans, 1764; North
Berwick, pre-1927; Dunbar, pre-1934, where it was first
recorded as being 'most abundantly near Dunbar',
pre-1684.

Chaerophyllum L.

C. temulum L. Rough Chervil
Native. Status: local.
Roadsides, woodland, hedgerows, scrub, cereal fields,
waste ground.

WL: scattered in N., rare elsewhere.
ML: local, mainly in centre.
EL: widespread and locally frequent in lowland
areas, rare in Lammermuir Hills.

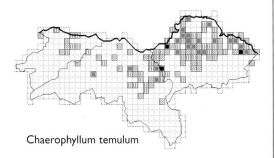

Chaerophyllum temulum

C. aureum L. Golden Chervil
Introduced. Status: rare, naturalised.
Rough grassland.

ML: rare, Calton Hill, 1884 to present;
(Corstorphine Hill, 1894).

Anthriscus Pers.

A. sylvestris (L.) Hoffm. Cow Parsley
Native. Status: common.
Roadsides, woodland, hedgerows, waste ground,
riversides, rough grassland, scrub, cereal fields.

WL, ML, EL: common except in some upland
areas.

A. caucalis M. Bieb. Bur Chervil
Native. Status: local.
Dunes, coastal scrub and other coastal habitats, rough
grassland, waste ground, cereal fields, roadsides,
volcanic outcrops.

WL: (Kirkliston, 1949; Carlowrie, Carribber and
Hopetoun, pre-1934).
ML: rare, Arthur's Seat.
EL: locally frequent along the coast, rare inland,
Longniddry Bents; Whitekirk Hill.

Scandix L.

S. pecten-veneris L. Shepherd's Needle
Field weed, formerly widespread.

WL: near Linlithgow and near South Queensferry,
pre-1934.
ML: Portobello, Dalhousie, Leith and Slateford pre-1934.
EL: Haddington, 1847; East Linton, 1857; Longniddry,
1860; Whitekirk, 1866; North Berwick, 1870; Dirleton,
pre-1934.

Myrrhis Mill.

M. odorata (L.) Scop. Sweet Cicely
Introduced. Status: widespread.
Damp roadsides, riversides, woodland, waste ground,
hedgerows, scrub.

WL: widespread and locally frequent in N., rare
in S.
ML: rare in upland areas, widespread and locally
frequent elsewhere.
EL: widespread and locally frequent in centre,
rare in intensively farmed land in N. and in
upland areas.

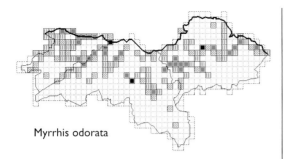
Myrrhis odorata

Bifora Hoffm.

B. radians M. Bieb.
Introduced. Status: rare, grain casual.
Waste ground.

ML: rare, Leith Docks, 1989.

Smyrnium L.

S. olusatrum L.   Alexanders
Introduced. Status: scarce.
Waste ground, edge of woodland.

ML: locally frequent at Craigmillar Castle;
Fillyside, 1999; Swanston, 2000.
EL: rare, Dirleton; Yellowcraig; (Dunglass Dean,
pre-1934).

Conopodium W. D. J. Koch

C. majus (Gouan) Loret   Pignut
Native. Status: widespread.
Rough grassland, woodland, roadsides, riversides, scrub,
heath grassland.

WL: widespread, locally frequent.
ML: widespread, locally abundant in SE., less
common in N.
EL: quite widespread, less so in N., locally
frequent.

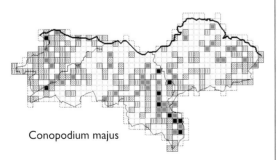
Conopodium majus

Pimpinella L.

P. major (L.) Huds.   Greater Burnet Saxifrage
Native. Status: rare.
Grassland, disused railways.

WL: (Linlithgow, pre-1934).
ML: rare, Arthur's Seat; locally frequent at
Warriston.

P. saxifraga L.   Burnet Saxifrage
Native. Status: local.
Rough grassland, quarries, rocky outcrops, roadsides.

WL: local.
ML: locally frequent in SE., rare elsewhere.
EL: rare and declining in W., occasional
elsewhere, Waughton crossroads; Stenton;
Bothwell Water area; Thornton.

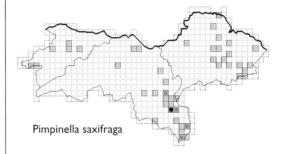
Pimpinella saxifraga

Aegopodium L.

A. podagraria L.
          Ground Elder, Goutweed, Bishop's Weed
Introduced. Status: common.
Roadsides, woodland, waste ground, riversides,
hedgerows, scrub, railway embankments, rough
grassland, gardens.

WL: common, locally abundant.
ML, EL: rare in upland areas, common
elsewhere.

Sium L.

S. latifolium L.   Greater Water Parsnip
WL: Kinneil, pre-1934.
ML: Duddingston, 1897.
EL: Dunbar, pre-1927.

Berula Besser ex W. J. Koch

B. erecta (Huds.) Coville   Lesser Water Parsnip
Native. Status: rare.
Beside rivers, lochs, ponds and ditches, marshes.

ML: rare, Dalmahoy; Duddingston Loch.

EL: rare, Gosford Estate; Aberlady Bay;
Pressmennan; Whiteadder Water; (Dunbar, 1886;
Seton, pre-1894; Dirleton, nineteenth century;
Longniddry, pre-1934).

Oenanthe L.

O. lachenalii C. C. Gmel.

Parsley Water Dropwort

Native. Status: rare.
Saltmarsh.

WL: (Carriden, pre-1934).
EL: rare, Aberlady Bay, re-found in 1990, having
previously been thought to be extinct; (Dirleton,
1836; Dunbar, pre-1934).

O. crocata L.   Hemlock Water Dropwort
Native. Status: local.
Marshes, beside rivers, ditches, lochs, ponds and
reservoirs, coastal habitats.

WL: local, in fresh water by the coast.
ML: rare, Cobbinshaw, Morton and Threipmuir
Reservoirs; Duddingston Loch; Niddrie Burn.
EL: locally frequent, mainly coastal sites and sites
in E., Seton; Longniddry Bents; Aberlady Bay;
Heckies Hole; Broxmouth; Catcraig; Dunglass;
Humbie; Pressmennan; Spott; Elmscleugh;
Thornton.

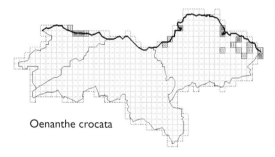

Oenanthe crocata

O. aquatica (L.) Poir.   Fine-leaved Water Dropwort
ML: Corstorphine, pre-1927.
EL: Dunbar, pre-1927.

Aethusa L.

A. cynapium L. *s.s.*   Fool's Parsley
Native. Status: local.
Waste ground, cereal fields, roadsides, hedgerows,
gardens.

WL: local, mainly in N.
ML: local, mainly in NE.
EL: occasional in lowland areas.

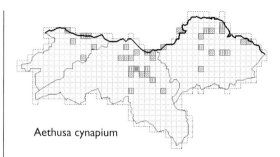

Aethusa cynapium

Foeniculum Mill.

F. vulgare Mill.   Fennel
Introduced. Status: rare, garden
throw-out/escape.
Waste ground, pavements, graveyards, dunes.

ML: (scattered occurrences, pre-1934).
EL: rare, but increasing, Gullane, 1999; North
Berwick (first recorded in 1764); Innerwick,
1999.

Anethum L.

A. graveolens L.   Dill
Introduced. Status: rare, garden escape,
sometimes persisting in allotments.
Waste ground.

ML: rare, Midmar; Leith Docks, 1989.

Silaum Mill.

S. silaus (L.) Schinz & Thell.   Pepper Saxifrage
Native. Status: rare.
Roadsides.

ML: (Dalkeith, pre-1894).
EL: rare, Blance; Bolton crossroads;
(Herdmanston, 1841; Tranent, pre-1894; Biel,
1894).

Meum Mill.

M. athamanticum Jacq.   Spignel
WL: recorded from Breich Water by Sibbald, 1684 and
Lightfoot, 1782; Linlithgow, pre-1934.
ML: Penicuik, pre-1934.
EL: Dunbar, pre-1894.

Conium L.

C. maculatum L.   Hemlock
Native. Status: local.
Roadsides, waste ground, woodland, arable land, rough
grassland, coastal habitats.

WL: occasional.
ML: locally frequent in N., rare in W. and upland areas, occasional elsewhere.
EL: widespread and locally abundant in lowland areas, especially along the coast.

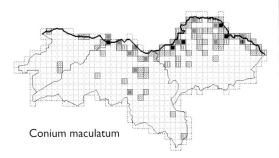
Conium maculatum

Bupleurum L.

B. rotundifolium L.   Thorow-wax
Introduced. Status: rare, casual.
Waste ground.

ML: rare, Leith Docks, 1989.

B. subovatum Link ex Spreng.   False Thorow-wax
Bird-seed alien.

ML: Dalmahoy, 1974.

Apium L.

A. nodiflorum (L.) Lag.   Fool's Watercress
Native. Status: rare.
Ditches, marshes.

WL: (Almondell and Drumshoreland, pre-1934).
ML: rare, Duddingston Loch; (Braid Hills and Myreside, pre-1934).
EL: rare, Aberlady Bay, 1986; Whiteadder Water, 1988; (Dirleton, 1836; Haddington, 1910).

Old records of *A. nodiflorum* x *A. repens* and *A. repens* are now considered to have been variants of *A. nodiflorum*.

A. inundatum (L.) Rchb. f.   Lesser Marshwort
Native. Status: rare.
Ponds, marshes.

WL: rare, Drumshoreland, 2001, previous record from this area, 1885; (Winchburgh, late 1800s; Linlithgow Loch, pre-1934).
ML: rare, Threipmuir; Bawsinch; Roslin; Straiton Pond.

EL: rare, Tyninghame; (Tranent, 1868; Luffness, 1906; Gullane, pre-1934).

A. graveolens L.   Wild Celery
ML: Musselburgh, Leith and Slateford, pre-1934.

Petroselinum Hill

P. crispum (Mill.) Nyman ex A.W. Hill   Garden Parsley
WL: Linlithgow, pre-1863.
ML: casual, Ratho, Gogar, Craigmillar and Leith, pre-1934.
EL: Prestonpans, 1851; Cockenzie, pre-1863.

Cicuta L.

C. virosa L.   Cowbane
ML: Lochend Loch, pre-1927.

Carum L.

C. carvi L.   Caraway
WL: South Queensferry, 1863; Abercorn, 1871; Kirkliston, pre-1934.
ML: casual, Roslin and Leith, pre-1934.

Ligusticum L.

L. scoticum L.   Scot's Lovage
Native. Status: local.
Cliffs and coastal rocks, dunes, coastal scrub.

WL: (formerly 'very frequent' between Hopetoun and Dalmeny, still at Dalmeny, 1934).
ML: rare, Cramond Island; (Seafield, 1966).
EL: occasional along the coast, locally frequent.

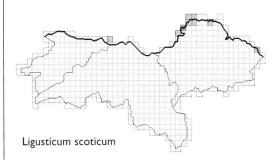
Ligusticum scoticum

Angelica L.

A. sylvestris L.   Wild Angelica
Native. Status: widespread.
Beside rivers, ditches and roads, woodland, marshes, scrub.

WL: widespread, seldom frequent.
ML: rare in some urban areas in N. and in some

upland areas in SE., widespread and locally
frequent elsewhere.
EL: widespread and locally frequent in centre,
local elsewhere.

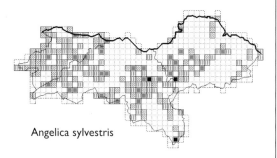

Angelica sylvestris

A. archangelica L.   Garden Angelica
EL: Longniddry, 1904–8.

Levisticum Hill

L. officinale W. D. J. Koch   Lovage
Introduced. Status: rare, garden escape.
Waste ground.

WL: rare, one plant near Old Philpstoun, 2000.
ML: rare, Holyrood Park.

Peucedanum L.

P. ostruthium (L.) W. D. J. Koch   Masterwort
Introduced. Status: rare.
Walls, paths.

ML: rare, Stow.
EL: (mouth of Biel Burn, 1933).

Pastinaca L.

P. sativa L.   Parsnip
Introduced. Status: local, casual, naturalised
or sown.
Roadsides, hedgerow, railway embankments.

WL: rare, near M8 motorway south of Broxburn,
1993.
ML: local, locally frequent, Granton; Seafield;
Newton.
EL: rare, throughout 'wild' landscaped area,
Cockenzie, 2000; Soutra, 1998 to present;
Gifford; West Barns since 1996; Skateraw;
Dunglass; (Dirleton, 1889; North Berwick,
pre-1934; Gullane, 1972).

Heracleum L.

H. sphondylium L.   Hogweed
Native. Status: common.
Roadsides, woodland, rough grassland, hedgerows,
scrub, beside rivers and ditches, arable fields.

WL: common.
ML, EL: rare in upland areas in SE., common
elsewhere.

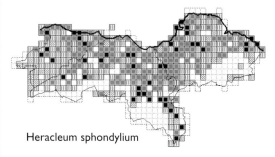

Heracleum sphondylium

H. sphondylium x H. mantegazzianum
Spontaneous hybrid with the parents. Status:
scarce, possibly overlooked.
Riversides, wooded slopes, waste ground.

ML: scarce, Princes Street gardens; Melville
Castle; Dalkeith Country Park.
EL: rare, with parents at Prestongrange, 1986,
but the whole population has since been
destroyed.

H. mantegazzianum Sommier & Levier
                              Giant Hogweed
Introduced. Status: local.
Riversides, waste ground, woodland, roadsides.

WL: occasional in N.
ML: locally frequent in N., rare elsewhere.
EL: occasional along River Tyne below East
Linton and along Biel Water; (earliest record is
'plentiful below Whittingehame', 1904);
controlled and rare elsewhere.

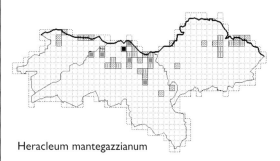

Heracleum mantegazzianum

Torilis Adans.

T. japonica (Houtt.) DC.   Upright Hedge Parsley
Native. Status: widespread.
Roadsides, hedgerows, woodland, scrub, waste ground,
cereal fields.

WL: widespread especially in N., seldom
frequent.
ML: patchy distribution, locally frequent in
centre, rare in urban and upland areas.
EL: rare in upland areas, common elsewhere.

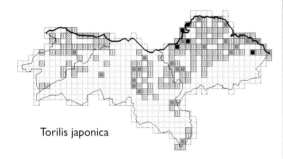

Torilis japonica

T. nodosa (L.) Gaertn.   Knotted Hedge Parsley
Native. Status: rare.
Rough grassland.

WL: (Linlithgow, pre-1934).
ML: rare, Duddingston; (Craigmillar and
Seafield, pre-1934; Arthur's Seat, 1977).
EL: rare, Aberlady; near Whitekirk; (near
Prestonpans, pre-1824; Tantallon, 1869; Gullane,
pre-1934).

T. arvensis (Huds.) Link   Spreading Hedge Parsley
ML: casual ruderal, Slateford, Leith and Levenhall,
pre-1934.
EL: Dunbar, pre-1934.

Turgenia Hoffm.

T. latifolia (L.) Hoffm.   Greater Bur-parsley
WL: Kirkliston, pre-1934.
ML: Slateford, Leith and Musselburgh, pre-1934.

Daucus L.

D. carota L.   Carrot
?Native. Status: scarce, often casual, sown in
'wild flower seed'. Formerly widespread on
waysides.
Waste ground, roadsides, railway embankments, arable
fields, dunes.

WL: rare, sown on the reclaimed bing near
Bo'ness; (Carriden, 1955).

ML: scarce, railway near Nether Longford;
Granton; Monktonhall; Bawsinch; Musselburgh;
Trinity.
EL: scarce, Longniddry; Drem; East Linton;
Tantallon; Hedderwick; West Barns.

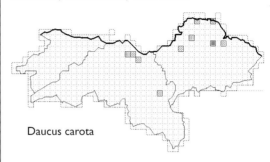

Daucus carota

GENTIANACEAE

Centaurium Hill

C. erythraea Rafn   Common Centaury
Native. Status: scarce.
Waste ground, roadsides, coastal habitats, rough
grassland, bings, bare ground, woodland paths.

WL: rare, Craigton; Philipstoun; slip road to
Dalmeny; good colonies at Bo'ness and on tracks
near Bedlormie.
ML: rare, Cousland; Arniston; Roslin; near
Easter Road; Newtongrange.
EL: occasional on or near the coast:
Prestongrange; Aberlady Bay; Gullane;
Yellowcraig; Tyninghame; Belhaven Bay; rare
along woodland paths: Saltoun Forest;
Gladsmuir; Bolton Muir Wood; (Yester,
pre-1863).

C. littorale (Turner ex Sm.) Gilmour
    Seaside Centaury
Native. Status: rare.
Dunes, coastal turf.

EL: rare, previously recorded in Gullane area,
1858, re-found in 1995, common over dunes
north of Gullane; (Longniddry, 1946).

C. pulchellum (Sw.) Druce   Lesser Centaury
Native. Status: rare.
Sandy places.

EL: rare, Aberlady Bay Reserve, 1994, previously
recorded from the adjacent Gullane Links,
pre-1894; (Yellowcraig, 1964).

Gentianella Moench

G. amarella (L.) Börner
                    Autumn Gentian, Felwort
ssp. amarella
Native. Status: rare.
Dunes, coastal turf, rough grassland.

WL: (Dalmeny, 1955).
ML: rare, Middleton Quarry.
EL: occasional along the coast: Aberlady Bay;
Gullane; Yellowcraig; Peffer Sands; Belhaven Bay;
Catcraig; rare inland: Bolton Muir Wood;
Kidlaw; (Gosford Links, 1844).

G. campestris (L.) Börner   Field Gentian
Native. Status: rare.
Coastal turf, dunes, rough grassland.

WL: rare, single colony at Hound Point; (South
Queensferry, 1954).
ML: rare, Fullarton Water; (Pentland Hills,
pre-1934).
EL: rare, Gullane, 1988; Belhaven Bay; (links to
the east of Cockenzie, pre-1824; Gosford, 1844;
Cockenzie, Gullane, North Berwick and Gifford,
pre-1934).

APOCYNACEAE

Vinca L.

V. minor L.   Lesser Periwinkle
Introduced. Status: quite widespread.
Woodland, scrub, hedgerow, waste ground.

WL: local.
ML: quite widespread.
EL: scattered sites, Pencaitland; Cuddie Wood;
Haddington; Gifford; Spott; Dunglass.

V. major L.   Greater Periwinkle
Introduced. Status: scarce.
Woodland, hedgerow, waste ground, dunes.

WL: rare, Linlithgow; (Kirkliston, 1956).
ML: rare, naturalised in woodland at Roslin and
New Hailes, 1998.
EL: established at a few sites, Seton Sands;
Longniddry Bents; Aberlady village; wood west
of Whitekirk.

SOLANACEAE

Lycium L.

L. barbarum L.   Duke of Argyll's Teaplant
Introduced. Status: rare, but long-lived.
Hedgerows, roadsides, waste ground.

WL: (Bathgate, 1958; Bo'ness, 1976).
ML: a rare alien, Seafield, 1950 to present;
(Balerno, 1952 to ?)
EL: rare, Saltcoats Castle; West Barns; Dunbar;
Skateraw; (Pefferside, 1962; Thorntonloch,
1968).

Atropa L.

A. belladonna L.   Deadly Nightshade
Introduced. Status: rare, possibly overlooked.
Waste ground.

WL: rare, a large colony on Philpstoun Bing.
ML: rare, Warriston; Carberry; Comely Bank;
(Borthwick, 1971; Granton, 1979).
EL: rare, Ormiston Hall; (Morham, pre-1837;
Prestonpans, pre-1894).

Hyoscyamus L.

H. niger L.   Henbane
Introduced. Status: rare and sporadic.
Coastal habitats, rough grassland.

WL: (Kirkliston, pre-1934).
ML: rare, Arthur's Seat; Cramond Island;
Liberton; Fisherrow; (Warriston, 1970).
EL: rare, usually coastal, Prestonpans; Soutra
Aisle excavations, 1996; Aberlady Bay; Luffness;
North Berwick; Tantallon; Auldhame; Sandy
Hirst, where it is established; Hedderwick; West
Barns; Broxmouth; (Longniddry, pre-1934;
Kilspindie, 1957).

Physalis L.

P. ixocarpa Brot. ex Hornem.   Tomatillo
Introduced. Status: rare, casual.
Waste ground.

ML: rare, Dalkeith, 1999.

Lycopersicon Mill.

L. esculentum Mill.   Tomato
Introduced. Status: rare, casual.
Beaches, dunes, waste ground, rubbish tips.

WL: rare, Bo'ness.
ML: rare, usually on the shoreline, sometimes ruderal.
EL: occasional on the tideline, Prestongrange; Prestonpans; Longniddry Bents; Aberlady Bay; John Muir Country Park.

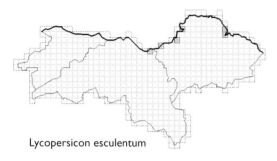

Lycopersicon esculentum

Solanum L.

S. nigrum L.   Black Nightshade
ssp. nigrum
Introduced. Status: rare.
Waste ground, roadsides, scrub, gardens, arable fields, dunes.

WL: (Bo'ness, 1970s).
ML: rare, Comely Bank; (Borthwick railway tip, 1963).
EL: rare, Cockenzie; Yellowcraig; North Berwick; Newbyth; Pefferside; (Longniddry Bents, pre-1824; Gullane, 1897).

S. physalifolium Rusby   Green Nightshade
var. **nitidibaccatum** (Bitter) Edmonds
Introduced. Status: rare.
Cultivated and waste ground, sandy soils.

ML: (a rare casual recorded in 1965 from a railway tip at Borthwick).
EL: rare, Aberlady Bay, 1991; Yellowcraig, known there since 1972.

S. dulcamara L.   Bittersweet, Woody Nightshade
Native. Status: quite widespread.
Waste ground, hedgerows, beside roads, railways, ditches and rivers, woodland, coastal habitats.

WL, ML: quite widespread in N., rare elsewhere.
EL: quite widespread in lowland areas.

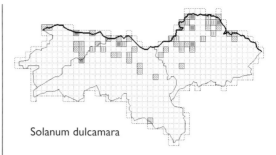

Solanum dulcamara

S. tuberosum L.   Potato
Introduced. Status: local, often casual.
Waste ground, cereal fields, rough grassland, roadsides, rubbish tips, quarries.

WL: occasional.
ML: rare.
EL: widely scattered sites.

S. rostratum Dunal   Buffalo-bur
ML: rubbish tip at Hailes Quarry, 1971.

Datura L.

D. stramonium L.   Thorn-apple
Introduced. Status: rare, casual.
Waste ground.

ML: rare, Leith Docks.
EL: rare, Gullane, 2001; two plants in a felled and replanted wood near Whitekirk, 1992.

CONVOLVULACEAE

Convolvulus L.

C. arvensis L.   Field Bindweed
Native. Status: local.
Roadsides, railway embankments, cereal fields and other cultivated land.

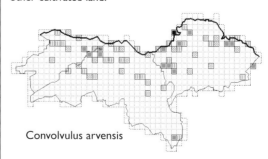

Convolvulus arvensis

WL: local, mainly in N.

ML: local, mainly in N., locally frequent.
EL: widely scattered distribution in lowland areas, locally frequent.

Calystegia R. Br.

C. soldanella (L.) R. Br.   Sea Bindweed
Native. Status: rare.
Sand-dunes.

EL: rare, near Dunbar, pre-1934, site re-found, 1982; (Aberlady Bay, 1958).

C. sepium agg.   Bindweed
Includes both native and introduced taxa. Status: widespread.
Roadsides, waste ground, riversides, railway embankments, hedgerows, scrub.

WL: widespread but not locally frequent.
ML: widespread in lowland areas, rare elsewhere.
EL: quite widespread and locally frequent in lowland areas, rare elsewhere.

Includes the following four taxa that were not usually separated in the survey.

C. sepium (L.) R. Br.
ssp. sepium   Hedge Bindweed
Native. Status: widespread, the commonest taxon.
Roadsides, waste ground, riversides, railway embankments, hedgerows, scrub.

WL: widespread but not locally frequent.
ML, EL: widespread in lowland areas, rare elsewhere.

C. pulchra Brummitt & Heywood
     Hairy Bindweed
?Introduced. Status: local.
Waste ground, hedgerows.

WL, ML: occasional.
EL: rare, Drem; Congalton; Gifford; Sandy's Mill; Beanston Mill; Knowes Mill; Ware Road, Tyninghame; (North Berwick, 1959; Overhailes, 1963; Longniddry station, 1972).

C. silvatica (Kit.) Griseb.   Large Bindweed
Introduced. Status: local.
Hedgerows, roadsides, waste ground.

WL: local.
ML: occasional, locally frequent.
EL: occasional in lowland areas, Prestongrange; Pencaitland; Aberlady village; West Barns.

ssp. disjuncta Brummitt
Introduced. Status: local.
Hedgerows, waste ground.

ML: in lowland areas only, occasional, Craigmillar Castle; Buckstone; Warriston; Millerhill; Midmar.
EL: (Whitekirk, 1955; North Berwick, 1958).

CUSCUTACEAE

Cuscuta L.

C. epithymum (L.) L.   Dodder
WL: appeared as a garden weed at Parkhead, 1962.
ML: on gorse, heather and thyme, pre-1894.
EL: on clover at Gleghornie, pre-1871; on heathland at Hedderwick Hill, 1878, and Lammer Law, pre-1934; on cowslips at Longniddry, 1911.

MENYANTHACEAE

Menyanthes L.

M. trifoliata L.   Bogbean
Native. Status: local (declining).
Marshes, ponds and other wet places.

WL: scarce, mainly in S., (Riccarton, Linlithgow and Philpstoun, pre-1934).
ML: local, mainly at reservoirs, locally frequent.
EL: rare, Gosford Pond; Aberlady Bay; Danskine Loch; Faseny Water; Caldercleuch; Black Loch; Aikengall.

Menyanthes trifoliata

Nymphoides Ség.

N. peltata Kuntze   Fringed Water-lily
Introduced. Status: rare, garden escape.
Lochs.

ML: established in Duddingston Loch, 1956 to present; Bawsinch.

POLEMONIACEAE

Polemonium L.

P. caeruleum L.   Jacob's Ladder
Introduced. Status: local, garden escape.
Waste ground, hedgerows, deciduous woodland, dunes.

WL: scarce, Broxburn; Philpstoun; Craigton;
Bedlormie; Fauldhouse; Gowanbank.
ML: occasional.
EL: abundant in an abandoned garden near
Haddington, 1993; otherwise rare, Keith;
Aberlady Bay; Dunglass; (North Berwick, 1958;
Blance, 1968).

BORAGINACEAE

Lithospermum L.

L. officinale L.   Common Gromwell
ML: fields and waste places at Buckstone, Leith and
Slateford, pre-1934; previously widespread.
EL: Dirleton, pre-1894.

L. arvense L.   Corn Gromwell, Field Gromwell
WL: South Queensferry, 1863; Kirkliston, pre-1934.
ML: in fields at Slateford, Leith and Musselburgh,
pre-1934; previously widespread.
EL: Dirleton, 1836; Whitekirk, 1866; Drem, pre-1934;
casual at West Barns, 1974.

Echium L.

E. vulgare L.   Viper's Bugloss
Native. Status: local.
Waste ground, coastal habitats, roadsides, quarries,
railway embankments, coal waste.

WL: rare, Bo'ness; north of Armadale;
(Craigiehall and Kirkliston, pre-1934; Carriden,
1955).
ML: occasional, Salisbury Crags.
EL: occasional, mainly coastal, Gosford Bay;
Aberlady Bay; Yellowcraig; Belhaven Bay.

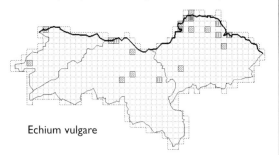

Echium vulgare

E. plantagineum L.   Purple Viper's Bugloss
Casual.
EL: West Barns, 1974.

Pulmonaria L.

P. officinalis L.   Lungwort
Introduced. Status: scarce.
Roadsides, waste ground, woodland, scrub.

WL: scarce, established at Almondell and Pepper
Wood.
ML: very occasional, Juniper Green; Carberry;
Hillend.
EL: rare, near East Linton, 1993; (ditch near
Gifford, 1964, but gone by 1970).

P. rubra Schott   Red Lungwort
Introduced. Status: rare.
Woodland.

ML: rare, Redhall Haugh.

P. 'Mawson's Blue'   Mawson's Lungwort
Introduced. Status: rare.
Shady places.

WL: rare, Almondell.

P. longifolia (Bastard) Boreau
                Narrow-leaved Lungwort
Introduced. Status: rare, garden escape.
Waste ground.

WL: rare, Bedlormie House.
ML: rare, Roslin Castle; (Balerno, 1941).

Symphytum L.

S. officinale L.   Common Comfrey
Native. Status: scarce, confused with *S. x
uplandicum* in survey. The following records are
accurate.
Beside roads, rivers and ditches, woodland, waste
ground, hedgerows, scrub.

WL: rare, Craigton Quarry; (Almondell,
Carlowrie, Dalmeny, Queensferry and Niddry,
pre-1934).
ML: local, Warriston; Inveresk; Lochend.
EL: rare, Ormiston; Fountainhall; Pencaitland;
Saltoun Forest; St. Germains; Peffer Burn;
Whittingehame; Deuchrie; (Gosford, 1903).

S. x uplandicum Nyman (*S. officinale* x
S. asperum)   Russian Comfrey
Introduced. Status: local, under-recorded.

Roadsides, waste ground, riversides, woodland, scrub, hedgerows.

WL, ML, EL: rare in upland areas, elsewhere quite widespread but not locally frequent.

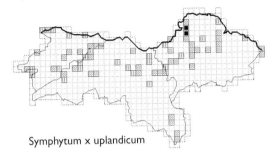

Symphytum x uplandicum

S. asperum Lepech.   Rough Comfrey
Introduced. Status: rare, garden escape.
Waste ground, woodland.

WL: (Carlowrie, pre-1934).
ML: rare, Marchbanks, 1999; (Warriston and The Inch, pre-1934).
EL: rare, Bilsdean where it has been known since 1903; (Wester Pencaitland, 1956).

S. asperum x S. caucasicum
Native. Status: rare, spontaneous hybrid.
Rough grassland.

ML: rare, King's Buildings, probably now extinct. This is the only known British and West European record for this hybrid.

S. tuberosum L.   Tuberous Comfrey
Native. Status: widespread.
Woodland, beside roads and rivers, scrub, hedgerows.

WL: widespread.
ML: rare in some upland areas, elsewhere quite widespread, locally abundant.
EL: widespread and locally frequent in lowland areas, rare in upland areas.

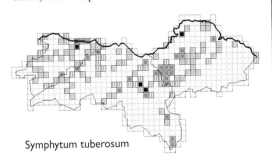

Symphytum tuberosum

S. 'Hidcote Blue' (*S. grandiflorum* x *?S. x uplandicum*)   Hidcote Comfrey
Introduced. Status: rare, garden escape.
Woodland.

EL: rare, Redmains, Pencaitland, 1994.

S. grandiflorum DC.   Creeping Comfrey
Introduced. Status: rare, garden escape.
Woodland edge.

EL: rare, Redmains, Pentcaitland, 1994.

S. orientale L.   White Comfrey
Introduced. Status: rare.
Roadsides, woodland, public parks, policies.

WL: rare, Midhope.
ML: rare, Craigmillar Castle, 1954 to present; Cockpen; Colinton; Calton Hill; Dean Village.
EL: rare, West Saltoun; Yester; (Winton House, 1975).

S. caucasicum M. Bieb.   Caucasian Comfrey
Introduced. Status: rare, garden escape.
Rough grassland.

ML: rare, King's Buildings alongside *S. asperum* x *S. caucasicum*.

Brunnera Steven

B. macrophylla (Adams) I. M. Johnst.
                     Great Forget-me-not
Introduced. Status: rare, garden throw-out.
Roadsides, scrub.

WL: rare, near East Bonhard Farm.
EL: (lane near Dirleton Church, 1963).

Anchusa L.

A. officinalis L.   Alkanet
Introduced. Status: rare, garden escape.
Rough grassland.

ML: (Leith and Musselburgh, pre-1934).
EL: rare, Gullane, one plant near path to Links, 1997; (Aberlady and Gullane, pre-1934).

A. arvensis (L.) M. Bieb.   Bugloss
Native. Status: local.
Fields of cereals and other crops, waste ground, rough grassland, roadsides, dry soils.

WL: scattered, mainly in N.
ML: local, Leith Docks; Niddrie; Roslin; (Howgate, 1973).
EL: widespread and locally frequent on sandy soil in N., rare elsewhere.

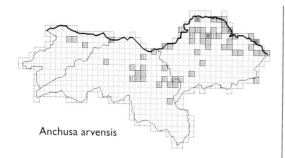

Anchusa arvensis

A. azurea Mill.   Garden Anchusa
ML: Borthwick and Davidson's Mains, *c.* 1959.
EL: Ormiston, 1887.

Pentaglottis Tausch

P. sempervirens (L.) Tausch ex L. H. Bailey
                                    Green Alkanet
Introduced. Status: local.
Roadsides, woodland, hedgerows, scrub, waste ground.

WL: local, mainly in N.
ML: widespread in N., locally frequent e.g. at
Hillend.
EL: occasional in lowland areas, locally frequent
in policies, Seton; Broxburn; Dunglass; in small
quantities when a garden escape.

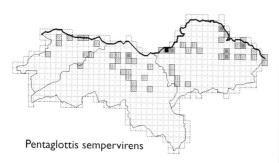

Pentaglottis sempervirens

Borago L.

B. officinalis L.   Borage
Introduced. Status: rare, casual.
Waste ground.

WL: rare, Hopetoun, 1986; (Linlithgow,
pre-1934).
ML: rare, Granton, 1982.
EL: rare, Gullane, 1999; Bara, 1996; (Gosford,
pre-1863; Canty Bay, 1932; North Berwick,
pre-1934).

Trachystemon D. Don

T. orientalis (L.) G. Don   Abraham-Isaac-Jacob
Introduced. Status: rare.
Damp woodland.

ML: rare, long established in some policy
woodlands, Dalkeith Estate; The Drum; Balerno;
Mortonhall, 2001.
EL: (Gifford, 1967).

Mertensia Roth

M. maritima (L.) Gray   Oysterplant
Native. Status: rare.
Beaches.

WL: (South Queensferry, 1823).
EL: rare, two plants, one flowering, were found
east of Bilsdean, 1989, but were washed away in
winter storms; (near North Berwick, 1865;
Skateraw, 1867; near Dunglass, pre-1894).

Mertensia was not listed by Martin in 1927 or in
1934, but was apparently growing on the foreshore near Fidra in 1901 where it was 'much
grazed by rabbits'.

Amsinckia Lehm.

A. micrantha Suksd.
                        Tarweed, Common Fiddleneck
Introduced. Status: local.
Waste ground, cereal fields, farmyards, rough grassland,
sandy soils, coastal scrub.

WL: rare, Parkley; by Union Canal near
Linlithgow; (Carriden, 1972).
ML: occasional, sometimes locally frequent but
not persisting, Leith; Niddrie; Loanhead; Balerno;
(Borthwick, 1950s).
EL: occasional in N. and increasing, sometimes
abundant on sandy ground and in fields near the
coast, Longniddry Bents; Aberlady; Gullane;
Drem; Haddington; Yellowcraig; Barns Ness;
Dunglass; (first record, Archerfield, 1954).

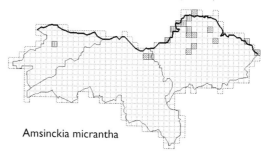

Amsinckia micrantha

Asperugo L.

A. procumbens L.   Madwort
WL: Kinneil, pre-1900.
ML: a rare casual, Slateford, Leith and Portobello,
pre-1934.
EL: Luffness, Gullane, Dirleton and Dunbar, pre-1934.
Introduced in fodder at the racecourse on Gullane
Common.

Myosotis L.

M. scorpioides L.   Water Forget-me-not
Native. Status: widespread.
In and beside rivers, ditches, ponds, lochs and the
canal, marshes.

WL, ML: widespread and locally frequent, except
in some upland areas.
EL: rare in NW., quite widespread elsewhere.

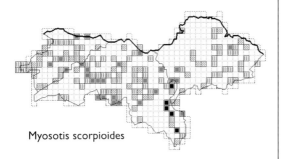

Myosotis scorpioides

M. secunda A. Murray
                Creeping Water Forget-me-not
Native. Status: widespread in upland areas, rare
elsewhere.
Marshes, in and beside upland streams and ditches.
Generally found in acid conditions.

WL: widespread in SW., rare elsewhere.
ML: widespread in upland areas, locally frequent
in SE.
EL: widespread in upland areas, especially in SE.

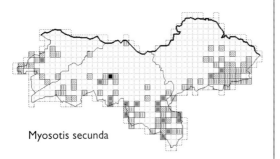

Myosotis secunda

M. stolonifera (DC.) J. Gay ex Leresche & Levier
                        Pale Water Forget-me-not
Native. Status: rare.
Marshes, ditches, streams.

ML: rare, Pentland Hills. This is the most
northerly station in Britain for this species.

M. laxa Lehm.   Tufted Water Forget-me-not
ssp. caespitosa (Schultz) Hyl. ex Nordh.
Native. Status: widespread in upland areas, local
elsewhere.
Marshes, in and beside upland streams, ditches and
ponds.

WL: widespread in the less acid areas of the SW.,
local elsewhere.
ML: widespread in upland areas, locally frequent
in SE.
EL: widespread in upland areas, rare in lowland
areas.

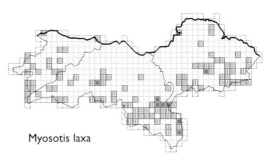

Myosotis laxa

M. sylvatica Hoffm.   Wood Forget-me-not
Native and often introduced. Status: local.
Woodland, roadsides.

WL: occasional.
ML: widely scattered sites.
EL: widely scattered distribution, mainly in estate
woodlands, locally frequent; rare and probably
native in Sheeppath Glen; (Cauld Burn, 1885).

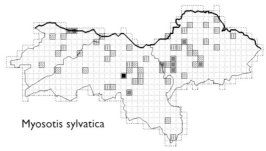

Myosotis sylvatica

M. arvensis (L.) Hill   Field Forget-me-not
Native. Status: common.
Arable fields, roadsides, waste ground, woodland, rough grassland.

WL: common.
ML, EL: local in some upland areas, common elsewhere.

var. **sylvestris** Schltdl.
Liable to be confused with *M. sylvatica*.

WL: Birkhill.
ML: Musselburgh
EL: Dirleton; North Berwick; Whitekirk; Dunglass.

M. ramosissima Rochel   Early Forget-me-not
Native. Status: local.
Dunes, coastal turf, cliffs, rough grassland, quarries, well-drained soils, bare dry areas.

WL: rare, Dalmeny; (Linlithgow, pre-1900).
ML: rare, Arthur's Seat; Blackford Hill; West Calder; Bawsinch; Bowshank; (Cramond Island, 1957).
EL: locally frequent along the coast, occasional elsewhere, mainly on basalt, Hummell Rocks; Yellowcraig; Hailes Castle; Black Loch.

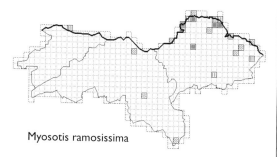

Myosotis ramosissima

M. discolor Pers.   Changing Forget-me-not
Native. Status: widespread.
Rough grassland, beside roads and rivers, intensive grassland, waste ground, marshes, heath grassland, quarries.

WL: widespread in SW., local in N.
ML: widespread in SE., local elsewhere.
EL: rare and apparently declining in NW., elsewhere widely scattered sites, mainly on rocky outcrops and in upland areas.

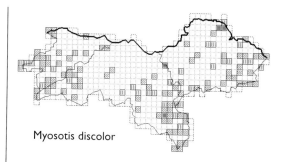

Myosotis discolor

Two subspecies have been recorded.

ssp. discolor
EL: Whitekirk Hill; Markle.

ssp. dubia (Arrond.) Blaise
EL: Rook Law; Elmscleugh.

Lappula Gilib.

L. **squarrosa** (Retz.) Dumort.   Bur Forget-me-not
WL: Philpstoun, pre-1906.

L. **echinata** Gilib.
ML: Leith Docks and Borthwick railway tip, 1950s.

Omphalodes Mill.

O. **verna** Moench   Blue-eyed Mary
Introduced. Status: rare, garden escape.
Waste ground, railway embankments.

WL: (Carlowrie pre-1927; Bo'ness and Linlithgow, pre-1934).
ML: (Craigmillar, pre-1927).
EL: rare escape, Skateraw, 1982; (railway near Biel, 1962), liable to be confused with *Brunnera macrophylla*.

Cynoglossum L.

C. **officinale** L.   Hound's Tongue
Native. Status: scarce.
Coastal scrub, dunes, meadows.

WL: (South Queensferry and Carriden, pre-1934).
ML: rare, Polton; (Roslin, pre-1934).
EL: scarce, scattered along the coast, Aberlady Bay; Yellowcraig; Pefferside; Tyninghame; Belhaven Bay; Catcraig; (Dunglass, 1972; Gosford, 1973).

LAMIACEAE

Stachys L.

S. officinalis (L.) Trevis.   Betony
Native. Status: rare, also in 'wild flower seed'.
Rough grassland, woodland, cultivated ground.

WL: rare, Almondell, 1982; (near Queensferry,
(Sibbald), 1684; Dundas Hill, 1894).
ML: (Colinton and Auchendinny, pre-1934).
EL: rare, Dry Burn, near Thurston, 1990s;
(Saltoun, pre-1934).

S. sylvatica L.   Hedge Woundwort
Native. Status: widespread.
Woodland, beside roads and rivers, scrub, waste
ground, hedgerows.

WL: widespread and locally frequent in N.,
scattered distribution in S., rare in far SW.
ML: rare in urban and upland areas, widespread
and locally frequent elsewhere.
EL: rare in upland areas and in intensively farmed
areas in N., widespread and locally abundant
elsewhere.

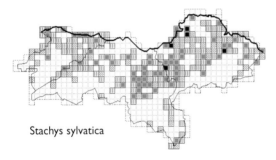

Stachys sylvatica

S. x ambigua Sm. (*S. sylvatica* x *S. palustris*)
                    Hybrid Woundwort
Native. Status: scarce, possibly under-recorded.
Beside rivers, ditches, canals and roads, scrub, waste
ground.

WL: rare, south-east of Broxburn; Burnhouse.
ML: rare, near Bell's Quarry; Blackford area.
EL: rare, Butterdean Wood; Colstoun; Garvald;
(North Berwick, pre-1863; Haddington, 1906;
Drem, 1907; Prestongrange, 1979).

S. palustris L.   Marsh Woundwort
Native. Status: widespread.
Beside rivers and ditches, marshes, roadsides, waste
ground.

WL: occasional in N., widespread in S., not
locally frequent.
ML: scattered distribution, seldom locally
frequent.
EL: scattered distribution in lowland areas,
usually in small quantities.

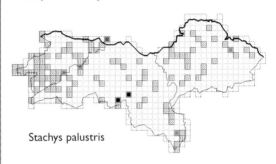

Stachys palustris

S. arvensis (L.) L.   Field Woundwort
Native. Status: rare, casual.
Cereal and other cultivated fields.

WL: rare, Broxburn, 1989; Wester Ochiltree,
1994.
ML: rare, Cockpen; Pencaitland, Dalkeith.
EL: (Whittingehame, 1872; Dunbar, pre-1894
and pre-1934).

S. byzantina K. Koch   Lamb's Lugs
ML: Craigmillar Quarry, 1907.

S. annua (L.) L.   Annual Yellow Woundwort
Casual.

ML: Leith Docks, sporadically 1889–1910; Slateford
rubbish tip, 1904.

Ballota L.

B. nigra L.   Black Horehound
Introduced. Status: scarce.
Roadsides, waste ground, rough grassland.

WL: rare, Hopetoun, last seen there in 1998;
(Dalmeny, 1836; Carriden, 1958; Linlithgow,
1977).
ML: local, Arthur's Seat; Calton Road Crags;
Levenhall; Craigmillar Castle; Fushiebridge.
EL: scattered distribution, Prestonpans;
Longniddry; Aberlady; Gullane; Whitekirk;
Tyninghame; Pressmennan; Thornton.

Leonurus L.

L. cardiaca L.   Motherwort
ML: Colinton, Musselburgh, Slateford and Leith, pre-1934.
EL: North Berwick, pre-1934.

Lamiastrum Heist. ex Fabr.

L. galeobdolon (L.) Ehrend. & Polatschek *s.l.*
Yellow Archangel
Introduced. Status: local.
Woodland, scrub, roadsides, waste ground.

WL: occasional, mainly in N.
ML: local.
EL: rare, Whittingehame; Belhaven; (Longniddry, Gosford and Dunglass, pre-1934).

The subspecies have not been consistently recorded, but modern records are usually ssp. *argentatum*.

ssp. montanum (Pers.) Ehrend. & Polatschek
Introduced. Status: local.
Woodland, scrub, riversides.

WL: rare, Pepper Wood; Threemiletown.
ML: (Liberton, nineteenth century).
EL: (Haddington, 1821).

ssp. argentatum (Smejkal) Stace
Introduced. Status: local, long-persisting garden escape.
Roadsides, waste ground.

WL: occasional in N., rare in S.
ML: local, Warriston, 2000; Boghall; Symington; cycle path at Ravelston.
EL: rare, Tranent; Ormiston; Dirleton; Colstoun; Morham.

Lamium L.

L. album L.
White Dead-nettle
Native. Status: common.
Roadsides, waste ground, hedgerows, riversides, scrub, woodland, rough grassland, cultivated fields.

WL, ML: occasional in S., widespread and locally frequent elsewhere.
EL: rare in upland areas, common elsewhere.

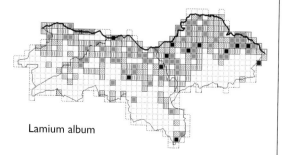
Lamium album

L. maculatum (L.) L.
Spotted Dead-nettle
Introduced. Status: local, naturalised.
Beside roads and rivers, hedgerows, woodland.

WL: occasional, well-naturalised and lacking spots, along the River Almond and Niddry Burn.
ML: local, River Almond, 1958 to present.
EL: rare, long-established at Longniddry Bents and West Barns; Longyester, 1994; (Wester Pencaitland, 1957; Kilspindie, 1964; Athelstaneford, 1973).

A garden form with white-blotched leaves was recorded in 1986 from Duddingston and Ratho (ML).

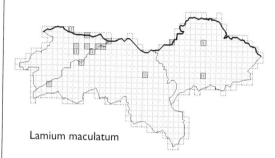
Lamium maculatum

L. purpureum L.
Red Dead-nettle
Native. Status: common.
Cultivated fields, roadsides, waste ground, rough grassland, gardens, farmyards.

WL: widespread, locally frequent in N.
ML: rare in upland areas, widespread and locally frequent elsewhere.
EL: occasional in upland areas, common elsewhere.

L. hybridum Vill.
Cut-leaved Dead-nettle
Native. Status: local, possibly under-recorded.
Arable fields, waste ground, roadsides, rough grassland.

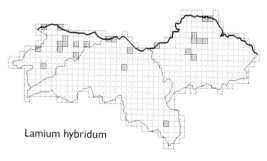
Lamium hybridum

WL: scattered, mainly in N.

ML: scarce, mainly in N., Borthwick; Arthur's Seat; Straiton.
EL: scattered distribution in lowland areas in N., Longniddry; Aberlady area; Clerkington; Athelstaneford; North Berwick; Dunbar. Cut-leaved variants of *L. purpureum* also occur.

L. confertum Fr.   Northern Dead-nettle
Native. Status: local.
Arable fields, waste ground.

WL: quite widespread in N., rare or absent elsewhere.
ML: rare, possibly overlooked, Fala; Stow; (Musselburgh, Buckstone and Duddingston, pre-1934).
EL: scattered distribution mainly in N. and centre, Prestonpans; Aberlady; Coates; North Berwick area; West Barns; Torness.

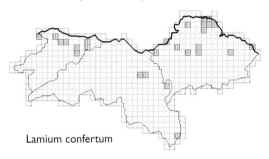

Lamium confertum

L. amplexicaule L.   Henbit Dead-nettle
Native. Status: widespread in N. and centre, rare in S.
Arable fields, waste ground, roadsides, gardens.

WL: quite widespread in N., but not locally frequent, rare or absent elsewhere.
ML: local, mainly in N. and centre, locally frequent in centre.
EL: rare in upland areas, widespread elsewhere.

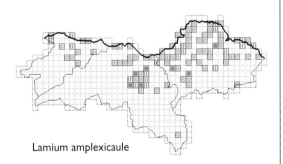

Lamium amplexicaule

Galeopsis L.

G. speciosa Mill.   Large-flowered Hemp-nettle
Native. Status: local.
Arable fields, roadsides, waste ground.

WL: scarce and casual, Kinneil; near Uphall; Port Edgar.
ML: local, widespread and locally abundant in the valleys of Gala Water and Heriot Water, scattered sites elsewhere.
EL: rare, Gilston; Coates; Markle; Monynut; (North Berwick, 1958; Broxmouth, 1965; East Saltoun, 1966; Bothwell, 1971).

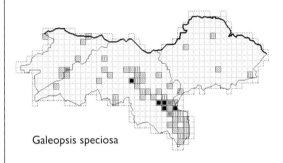

Galeopsis speciosa

G. tetrahit agg.   Hemp-nettle
Native. Status: common.
Arable fields, waste ground, rough grassland, roadsides, woodland.

WL: common.
ML: common in W. and in the valleys of Gala Water and Heriot Water, occasional elsewhere.
EL: local in upland areas and in intensively farmed arable land in N., widespread elsewhere.

Records include the following two taxa that can be separated only when in flower.

G. tetrahit L.   Common Hemp-nettle
Native. Status: common.
Arable fields, waste ground, rough grassland, woodland.

WL: common.
ML: common in W. and S.
EL: widespread except in intensively-farmed arable land and in upland areas.

G. bifida Boenn.   Bifid Hemp-nettle
Native. Status: widespread, probably almost as common as *G. tetrahit* and found in similar places.
Arable fields, roadsides, woodland.

WL: widespread.

ML: probably widespread.
EL: widely scattered distribution.

G. segetum Neck.   Downy Hemp-nettle
ML: fields at Lochend, 1863, G. McNab. This is the first
Scottish record for this species.

G. angustifolia Ehrh. ex Hoffm.   Red Hemp-nettle
WL: Kirkliston and Winchburgh, 1904.
ML: Newbattle, 1805; Inveresk, 1837.
EL: Elphinstone, 1882; Dunbar, pre-1894 and pre-1934.

G. ladanum L.   Broad-leaved Hemp-nettle
A rare casual often confused with *G. angustifolia.*

WL: Kirkliston, 1946.
ML: Dalkeith, Leith, Blackford Hill and Turnhouse, *c.*
1900.

Marrubium L.

M. vulgare L.   White Horehound
Introduced. Status: rare.
Waste places, mainly coastal.

WL: (Hopetoun, 1903).
ML: (Slateford, Leith and Inveresk, pre-1934).
EL: rare but thriving in small sporadic colonies:
Luffness; North Berwick; Broxmouth; Catcraig;
(Dunbar, 1882; Longniddry, Aberlady, Gullane
and Innerwick Castle, pre-1934; Thornton, 1972;
Ballencrieff, 1973).

Scutellaria L.

S. galericulata L.   Skullcap
Native. Status: rare, declining.
Beside lochs, ponds and streams.

WL: rare, Lochcote Marsh; Dundas.
ML: rare, Penicuik; (Mortonhall, 1949; Roslin,
1952).
EL: rare, doubtfully native, Pressmennan;
(Whittingehame, 1873; Dunbar, pre-1934;
Balgone, 1972).

Teucrium L.

T. scorodonia L.   Wood Sage
Native. Status: widespread.
Woodland, scrub, quarries, heath grassland, dry rocky
places.

WL: local, on rocky outcrops particularly on the
hills north of Bathgate.
ML: quite widespread, locally frequent.
EL: scattered distribution, locally frequent in
centre.

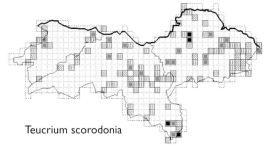

Teucrium scorodonia

T. chamaedrys L.   Wall Germander
Rare garden escape.
ML: a verge near Blackford Hill, 1965.

T. botrys L.   Cut-leaved Germander
ML: Leith, 1905.

Ajuga L.

A. reptans L.   Bugle
Native. Status: widespread.
Woodland, riversides, marshes, scrub.

WL: local, mainly in W.
ML: occasional in N. and W., quite widespread
elsewhere, locally frequent.
EL: rare in coastal areas, quite widespread
elsewhere, locally frequent in centre.

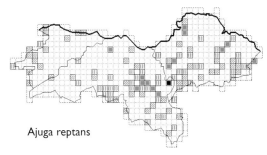

Ajuga reptans

Nepeta L.

N. x faassenii Bergmans ex Stearn (*N. racemosa*
Lam. x *N. nepetella* L.)   Garden Catmint
Introduced. Status: rare, garden throw-out.
Garden rubbish tips, roadsides, scrub.

WL: (Carribber, pre-1934).
ML: (Kingsknowe, 1902; Slateford rubbish tip,
1946; Roslin, 1961).
EL: rare, Aberlady Bay, 1988; Stoneypath Tower;
lane side of garden hedge, Dunglass.

N. cataria L.   Catmint
Rare garden escape.
ML: waste ground at Granton, 1979.

Glechoma L.

G. hederacea L.   Ground Ivy
Native. Status: widespread.
Woodland, scrub, roadsides, riversides, waste ground,
hedgerows.

WL: widespread, locally abundant in N.
ML: locally abundant in centre, locally frequent
in W., scattered distribution elsewhere.
EL: widespread and locally abundant in lowland
areas.

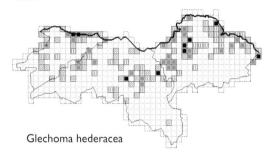

Glechoma hederacea

Prunella L.

P. vulgaris L.   Selfheal
Native. Status: common.
Roadsides, rough grassland, riversides, marshes,
woodland, heath grassland, waste ground.

WL, ML: common.
EL: occasional in N., common elsewhere.

Dracocephalum L.

D. parviflorum Nutt.   Dragon-head
Introduced. Status: rare, casual.
Waste ground, railway tip, roadsides.

ML: rare, Leith, 1982–3; (Borthwick railway tip,
1956; several old records from *c.* 1900).
EL: (near Gifford, 1906).

Melissa L.

M. officinalis L.   Balm
Introduced. Status: rare, garden escape.
Roadsides, waste ground.

ML: rare, Whitecraig, 1999; (Blackford Hill,
1977).

Clinopodium L.

C. vulgare L.   Wild Basil
?Native. Status: scarce.
Rough grassland, hedgerows, dunes.

WL: rare, near Woodcockdale.
ML: scarce, Roslin; Blackford Hill; Chalkieside;
Gorebridge; (previously much more widespread,
Dalhousie; Lasswade; Arthur's Seat; Colinton).
EL: rare, Longniddry railway walk, 1999;
Aberlady Bay; (Pressmennan, pre-1934; disused
railway near Pencaitland, 1973).

C. grandiflorum (L.) Stace   Greater Calamint
Introduced. Status: rare, garden escape or
throw-out.
Waste ground.

EL: rare, Dunbar; Skateraw.

C. acinos (L.) Kuntze   Basil Thyme
Native. Status: rare.
Arable fields and open habitats, calcareous grassland.

ML: (Arthur's Seat, pre-1934; Borthwick, 1971).
EL: rare, Yellowcraig, on a remnant of Dirleton
Common; (Aberlady, pre-1934).

Origanum L.

O. vulgare L.   Marjoram
?Native. Status: rare.
Hedgerows, dry grassland, waste ground, walls.

WL: rare, well-established on shingle banks on
the Dalmeny Estate; Bathgate; near Armadale;
(Carribber, pre-1934; South Queensferry, 1954;
Kirkliston, 1958).
ML: (Arniston, Roslin and Cramond, pre-1934;
Borthwick railway tip, 1963).
EL: rare, Aberlady Bay; Yellowcraig; (East
Linton, pre-1934; Archerfield, 1959).

Thymus L.

T. polytrichus A. Kern. ex Borbás   Wild Thyme
ssp. britannicus (Ronniger) Kerguélen
Native. Status: local.
Rough grassland, dry heath grassland, moorland,
streamsides, rocky outcrops, quarries, coastal habitats.

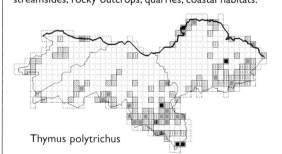

Thymus polytrichus

WL: local.
ML, EL: widespread in upland areas, occasional elsewhere.

T. vulgaris L.   Garden Thyme
Garden escape.

ML: Levenhall (1918).

Lycopus L.

L. europaeus L.   Gypsywort
Native. Status: scarce.
In and beside canals, lochs and reservoirs.

WL: scattered along the Union Canal.
ML: rare, Cobbinshaw Reservoir; Slateford Aqueduct.
EL: (Dunbar, pre-1934).

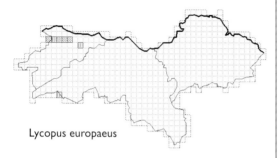

Lycopus europaeus

Mentha L.

M. arvensis L.   Corn Mint
Native. Status: scarce, declining.
Woodland, riversides, rough grassland, marsh, waste ground.

WL: scarce.
ML: rare, Gorebridge; (Cramond and Dalmahoy, 1956).
EL: rare, Keith Water; Laverocklaw; Luffness; Yellowcraig; Tyninghame; Dunglass; (Yester, 1882; East Saltoun, 1957; Bothwell Water, pre-1962; West Barns, 1973).

M. x verticillata L. (*M. arvensis* x *M. aquatica*)
                                                 Whorled Mint
Native. Status: scarce.
Marshes, beside ponds, rivers and lochs.

WL: rare, Dechmont; Kinneil; near Stoneyburn.
ML: scarce, Threipmuir; Almondell; Linhouse Water.
EL: rare, Yellowcraig, 1998; (Gullane, pre-1934; Longniddry, 1979).

M. x gracilis Sole (*M. arvensis* x *M. spicata*)
                        Bushy Mint, Ginger or Scotch Mint
Native. Status: scarce, possibly overlooked.
Beside ponds and lochs, waste ground.

WL: rare, Bathgate Hills; Bo'ness; Carribber; Tailend Moss; (Carlowrie, 1915).
ML: rare, Threipmuir Reservoir, 1999; (Craigmillar Quarry, 1905–13).

M. aquatica L.   Water Mint
Native. Status: widespread.
Marshes, in and beside rivers, ditches, ponds, lochs and canals.

WL: widespread and locally frequent in centre, occasional elsewhere.
ML: occasional in N., quite widespread elsewhere, locally frequent.
EL: occasional in W. and NW., widespread elsewhere, locally frequent.

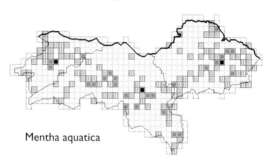

Mentha aquatica

M. x piperita L. (*M. aquatica* x *M. spicata*)
                                                 Peppermint
Native. Status: scarce, spontaneous hybrid or horticultural introduction.
In and beside rivers and ditches, derelict gardens.

WL: scarce, Blackfaulds; Bonnytoun; north of Almondell; south of Kirkliston; Livingston Reservoir.
ML: local, Bonnyrigg; Balerno; Musselburgh.
EL: rare, Elphinstone; Whiteadder Water area; (roadside, Pencaitland, 1969).

M. spicata L.   Spearmint
Introduced. Status: local.
In and beside rivers and ditches, marshes, roadsides, waste ground.

WL: local, mainly in S.
ML: quite widespread.
EL: scarce, scattered distribution.

M. x villosonervata Opiz (*M. spicata* x *M. longi-folia* (L.) Huds.)   Sharp-toothed Mint
Introduced. Status: rare, garden escape.
Waste ground.

ML: rare, Blackford Glen, since 1976; (Colinton; Linhouse Water).
EL: (Gullane, 1960).

M. x villosa Huds. (*M. spicata* x *M. suaveolens*)
                                                  Apple Mint
var. **alopecuroides** (Hull) Briq.
Introduced. Status: scarce, garden escape, may have been confused with *M. x rotundifolia* and some records here refer to that species.
Waste ground, marshes, beside rivers, ponds and canals.

WL: scarce, Bathgate; Bo'ness; Faucheldean; near Broxburn; near Torphichen.
ML: quite widespread.
EL: rare, Longniddry Bents; Aberlady; Saltcoats; North Berwick; Garvald; Papana Water; Hedderwick; (Gosford, 1967).

M. x rotundifolia (L.) Huds. (*M. longifolia* x *M. suaveolens*)   False Apple Mint
var. **webberi** (J. Fraser) Harley
Introduced. Status: rare, confused with *M. x villosa*.
Riversides, canal, railways, waste ground.

WL: (Kirkliston, 1917; Linlithgow Loch, 1970).
ML: (old records from Colinton, Slateford, Middleton, Craigmillar Quarry and Ratho).
EL: rare, Whittingehame; (North Berwick, 1976).

M. suaveolens Ehrh.   Round-leaved Mint
Introduced. Status: rare, escape.
Ditches, damp roadsides.

WL: rare, a number of records from around Broxburn.
ML: rare, Granton, probably now extinct.
EL: (River Tyne above East Linton, 1909).

M. x smithiana R. A. Graham (*M. arvensis* x *M. aquatica* x *M. spicata*)   Tall Mint
ML: Crichton, Musselburgh and elsewhere on the River Esk, pre-1934.

Rosmarinus L.

R. officinalis L.   Rosemary
Introduced. Status: rare, planted.
Public amenity areas.

WL: rare, Bo'ness.
ML: rare, Cramond car park.

Salvia L.

S. verbenaca L.   Wild Clary
WL: Linlithgow, pre-1934.
ML: Arthur's Seat, 1979.

S. viridis L.   Red-topped Sage, Annual Clary
ML: Leith Docks, 1895–1906; Slateford tip, 1905–6.
EL: Gullane, 1952.

S. verticillata L.   Whorled Clary
Casual, sometimes imported with ballast.
WL: Kirkliston, pre-1934.
ML: Leith Docks, Slateford tip, Straiton, Inveresk, Barnton Gate and Ratho, 1887–1914; Slateford Quarry, Gorebridge and Borthwick, pre-1934.
EL: Prestonpans and Seton Mains, pre-1934.

HIPPURIDACEAE

Hippuris L.

H. vulgaris L.   Mare's Tail
Native. Status: scarce.
Ponds, lochs, ditches, marshes, quarry pools.

WL: occasional, scattered throughout, but mainly in SW.
ML: scarce, Cobbinshaw; Duddingston Loch; Murieston; Livingston; (previously Portobello and Lochend Loch).
EL: rare, Aberlady Bay and four new sites: Kidlaw Reservoir, 1988; Latch Loch, 1988; Eaglescairnie Pool, 1996; Kippielaw Ponds, 1990.

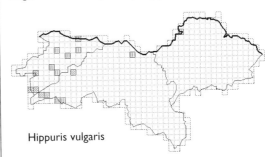

Hippuris vulgaris

CALLITRICHACEAE

Callitriche L.

C. hermaphroditica L. Autumnal Water Starwort
Native. Status: rare.
Ponds, lochs, canals, reservoirs.

WL: rare, Beecraigs Loch; Linlithgow Loch; (Dundas, 1905).

ML: rare, Duddingston and Dunsapie Lochs; Union Canal, west of Ratho; Quarrel Burn and Glencorse Reservoirs.
EL: rare, Stobshiel; Kidlaw and Donolly Reservoirs; (Spott Loch and Broxmouth, pre-1934, both possibly the same site).

C. stagnalis Scop.   Common Water Starwort
Native. Status: widespread.
In and beside rivers, ditches, ponds and lochs, marshes.

WL: widespread.
ML: widespread in upland areas, local elsewhere, locally frequent.
EL: quite widespread in E., local in W.

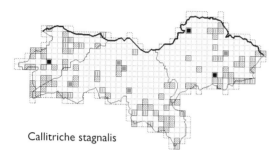

Callitriche stagnalis

C. platycarpa Kütz
                Various-leaved Water Starwort
Native. Status: rare.
Rivers and lochs.

ML: (Duddingston Loch, 1975).
EL: rare, Ormiston Hall; Stobshiel Reservoir; Papana Water; Johnscleugh; Black Loch.

C. hamulata Kütz ex W. D. J. Koch
                Intermediate Water Starwort
Native. Status: local.
Rivers, ponds, canals, lochs, ditches, reservoirs.

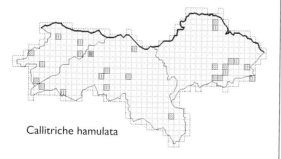

Callitriche hamulata

WL: local, in Union Canal and Lochcote Reservoir; rare elsewhere.

ML: local, Ratho; Dunsapie Loch; Threipmuir, Clubbiedean, Harperrig, North Esk and Rosebery Reservoirs.
EL: rare, mainly in upland areas in SE.; Whitekirk Covert; Dunbar Common; Faseny, Whiteadder and Monynut Waters.

PLANTAGINACEAE

Plantago L.

P. coronopus L.   Buck's-horn Plantain
Native. Status: common along the coast.
Coastal turf, cliffs, beaches, dunes, quarries, waste ground.

WL: occasional from Blackness to Dalmeny.
ML: local and locally frequent on coast.
EL: common along the coast.

P. maritima L.   Sea Plantain
Native. Status: widespread along the coast.
Cliffs, saltmarshes, coastal turf, beaches, dunes, waste ground.

WL: scattered along the coast, not locally frequent, Bo'ness; Blackness; Hopetoun; South Queensferry; Dalmeny.
ML: occasional along coast and on Cramond Island.
EL: widespread along the coast, very variable in quantity, locally abundant.

P. major L.   Greater Plantain, Rat's-tail Plantain
ssp. major
Native. Status: common.
Roadsides, waste ground, rough grassland, public parks, intensive grassland.

WL, ML, EL: common.

P. media L.   Hoary Plantain
Native. Status: scarce.
Roadsides, calcareous grassland, waste ground.

WL: rare, on restored bing at Bo'ness, possibly sown; (Overton, 1906).
ML: local, Newtongrange; Carrington.
EL: rare and declining, but thriving at one site on calcareous grassland near Gifford; (Pogbie, 1917; North Berwick, pre-1934; Gifford churchyard, 1960; Saltoun brickworks, 1966; Gosford House, 1974; Gosford Bay, 1968–81).

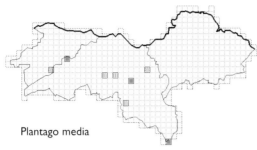

Plantago media

P. lanceolata L.   Ribwort Plantain
Native. Status: common.
Roadsides, rough grassland, waste ground.

WL, ML, EL: common.

P. afra L.   Glandular Plantain
Introduced. Status: rare, grain casual.
Waste ground.

ML: rare, Leith Docks, 1989.

P. lagopus L.   Hare's-foot Plantain
Introduced. Status: rare, grain alien.
Waste ground.

ML: rare, Leith Docks, 1989; (recorded on
several occasions between 1891 and 1905 from
Leith Docks and Slateford rubbish tip).

Littorella P. J. Bergius

L. uniflora (L.) Asch.   Shoreweed
Native. Status: local.
Shallow water at the edge of lochs and reservoirs.

WL: rare, Beecraigs; Bangour and Lochcote
Reservoirs; Parkley Craigs Quarry; (Philpstoun
Loch, 1825).
ML: local, at most reservoirs, locally frequent.
EL: rare, Stobshiel, Hopes, Thorters and
Whiteadder Reservoirs; (Gullane, pre-1934;
Aberlady Bay curling pond, pre-1977).

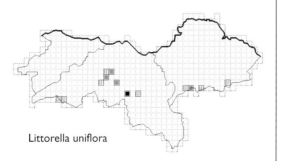

Littorella uniflora

BUDDLEJACEAE

Buddleja L.

B. davidii Franch.   Buddleia, Butterfly Bush
Introduced. Status: local, spreading.
Waste ground, roadsides, woodland, quarries, buildings.

WL: scattered sites.
ML: quite widespread in urban areas, rare
elsewhere.
EL: occasional, near habitation.

B. x weyeriana Weyer (*B. davidii* x *B. globosa*)
                    Weyer's Butterfly Bush
Introduced. Status: rare, planted.

WL: rare, by a track south of Muirend.

B. globosa Hope   Orange-ball Tree
Introduced. Status: rare, occasionally planted.
Public amenity areas, roadsides.

WL: rare, Kinneil.
ML: rare, riverbank, Gorgie, 1996.
EL: rare, path from Seafield pond to Belhaven.

OLEACEAE

Forsythia Vahl

F. x intermedia Zabel (*F. suspensa* (Thunb.) Vahl
x *F. viridissima* Lindl.)   Forsythia
Introduced. Status: rare, planted, spreading.
Public amenity areas, policies.

WL: rare, Uphall Station; Port Edgar; New
Burnshot.
ML: rare, riverbank, Murrayfield, 1994.
EL: rare, East Fortune, 1994; garden escape,
Whittingehame, 1992.

Jasminum L.

J. humile L. var. **revolutum** (Sims) Kobuski
EL: Tantallon Castle, 1913/1914.

Fraxinus L.

F. excelsior L.   Ash
Native. Status: common, sometimes planted.
Mixed and deciduous woodland, roadsides, hedgerows,
scrub.

WL: common.
ML, EL: rare in some upland areas, common
elsewhere.

forma **diversifolia** (*F. monophylla*) Single-leaf Ash, with only 1–3 leaflets was recorded from Musselburgh (**ML**) and from hedgerows in **EL**, 1998.

F. ornus L.   Manna or Flowering Ash
Introduced. Status: rare, planted.
Policies.

WL: rare, Newliston; Dundas.

F. angustifolia Vahl   Narrow-leaved Ash
Introduced. Status: rare, planted.
Policy woodland.

WL: rare, Hopetoun.

Syringa L.

S. vulgaris L.   Lilac
Introduced. Status: local, mainly planted.
Woodland, hedgerows, roadsides, public parks, policies.

WL, **ML**, **EL**: local, mainly in lowland areas, usually single trees.

Ligustrum L.

L. vulgare L.   Wild Privet
Introduced. Status: widespread in lowland areas.
Woodland, hedgerows, roadsides, scrub, disused railways.

WL: widespread.
ML, **EL**: widespread and locally frequent in lowland areas, rare or absent elsewhere.

L. ovalifolium Hassk.   Garden Privet
Introduced. Status: local, planted or garden escape.
Hedgerows, woodland, roadsides, scrub, public parks.

WL: quite widespread.
ML: occasional escape, mainly in lowland areas.
EL: occasional, probably always planted, persisting.

SCROPHULARIACEAE

Verbascum L.

V. virgatum Stokes   Twiggy Mullein
Introduced. Status: rare, casual.
Waste ground.

WL: rare, Craigie Quarry; bing near Uphall,

1987, this bing has subsequently been removed.
ML: local, Granton; Millerhill; Duddingston; Leith Docks; Fushiebridge.

V. x lemaitrei Boreau (*V. virgatum* x *V. thapsus*)
Introduced. Status: rare, casual.
Waste ground.

ML: rare, Duddingston, 1996–7, possibly only the second British record.

V. thapsus L.   Great Mullein
Native. Status: local.
Waste ground, bings, scree, quarries, railways, dry places.

WL: local.
ML: local, Warriston; Duddingston; Straiton Pond; (Stow, 1973).
EL: occasional, mainly in lowland areas, Spittal; Luffness quarry; Tyninghame.

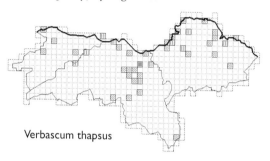

Verbascum thapsus

V. nigrum L.   Dark Mullein
Introduced. Status: rare.
Waste ground.

WL: (Broxburn, 1970; Cathlaw Estate, 1972).
ML: (Leith Docks and Borthwick, 1959, Roslin, 1969, Polton and Juniper Green, 1973).
EL: rare, Ormiston, 1992; (Gullane and links east of Cockenzie, pre-1934; East Saltoun, 1956; Garleton Hills, 1972).

V. lychnitis L.   White Mullein
Introduced. Status: rare, casual.
Waste ground.

ML: rare, Duddingston, appeared briefly with *V. x lemaitrei* in 1997.
EL: (Humbie churchyard, 1960).

Scrophularia L.

S. nodosa L.   Common Figwort
Native. Status: widespread.

Woodland, scrub, beside rivers and roads, waste ground.

WL, ML, EL: widespread but patchy distribution, seldom if ever locally frequent.

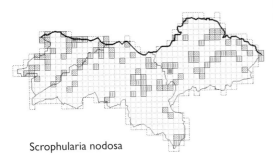

Scrophularia nodosa

S. auriculata L.　　　　　Water Figwort
?Native. Status: scarce.
In and beside rivers, lochs and ditches, marshes.

WL: rare, Hopetoun Estate; River Almond.
ML: occasional along the River Almond; rare elsewhere, Gogar and Figgate Burns.
EL: rare, at Newbyth, in or beside the loch, Peffer Burn and a new pond; beside a new pond at Tyninghame, apparently planted.

S. umbrosa Dumort.　　　　Green Figwort
Native. Status: scarce, but apparently increasing.
Beside rivers, ponds and lochs, woodland.

WL: (Dalmeny, 1836; Hopetoun, 1959).
ML: local, Cramond; Gogar; Newbattle; Musselburgh; Calder Wood; Rosewell.
EL: rare, Aberlady Bay, 1986–9; River Tyne at Haddington; Pressmennan Lake; ditch beside Ware Road, Tyninghame.

S. vernalis L.　　　　　Yellow Figwort
Introduced. Status: local, naturalised.
Woodland, coastal scrub, dunes, roadsides, waste ground.

WL: local, Craigton; Carribber; Hopetoun; Pepper Wood; plentiful at The Binns; (Carriden, 1970).
ML: local, Harvieston House; Dalkeith Country Park; (Oxenfoord, 1968; Juniper Green, 1972).
EL: scattered distribution along the coast, well established in some estate woodlands inland, Pencaitland; Gosford; Yellowcraig; Whitekirk; Tyninghame Links Wood.

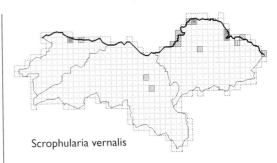

Scrophularia vernalis

Mimulus L.

M. moschatus Douglas ex Lindl.　　Musk
Introduced. Status: rare.
Riversides, marshes.

WL: rare, Dundas.
ML: (Boghall, 1949; Heriot, 1959).
EL: rare, Hopes Water below the reservoir, known there since 1957.

M. guttatus agg.　　　　Monkeyflower
Introduced. Status: local.
In and beside rivers and ditches.

WL: local, mainly central.
ML: local, seldom locally frequent.
EL: scarce.

Includes *M. guttatus*, *M. x robertsii* and *M. luteus* which are often confused.

M. guttatus DC.　　　　Monkeyflower
Introduced. Status: local.
Riversides.

WL: not distinguished from *M. x robertsii*.
ML: local, Liberton, 1996; (Braidburn; Cramond Bridge).
EL: scarce, Keith Water; Birns Water; River Tyne; Bara Loch; Papana Water; above Nunraw; Biel Water; (Dry Burn, 1962).

M. x robertsii Silverside (*M. guttatus* x *M. luteus* var. *rivularis*)　　Hybrid Monkeyflower
Introduced. Status: local. This is by far the most common taxon in the aggregate. Most or all records of *M. luteus*, and perhaps some of those of *M. guttatus*, are likely to refer to this hybrid which spreads independently of the parents. Spontaneous hybrids are unknown in the Lothians.
Riversides.

WL: not distinguished from *M. guttatus*.

ML: local, seldom locally frequent, Middleton; Stow; Morton; Kirknewton.

EL: scarce, Linn Dean; Humbie Water; Birns Water; Gifford Water; River Tyne; Faseny Water; Whiteadder Water; Sauchet Water; Spott Burn.

M. x burnetii hort. ex S. Arn. (*M. guttatus* x *M. cupreus* hort. ex Dombrain)

Coppery Monkeyflower

Introduced. Status: rare, garden escape.
In and beside rivers and ditches.

WL: rare, Dechmont, Glendevon.
ML: rare, Morton.
EL: rare, the ancient cv. **Duplex** has been recorded at three sites along Monynut Water.

M. luteus L.   Blood-drop-emlets
var. **rivularis** Lindl.
Introduced. Status: rare, garden escape.
Beside and in ditches and streams.

ML: rare, Fullarton Water.
EL: rare, Faseny Water.

M. x maculosus hort. ex T. Moore (*M. cupreus* hort. ex Dombrain x *M. luteus* var. *rivularis*)

Scottish Monkeyflower

Introduced. Status: rare, ancient garden hybrid.
Streamsides.

ML: rare, Luggate Water.
EL: rare, beside a stream in a deep, almost inaccessible, gully in a hill above Deuchrie.

M. x smithii Paxton (*M. luteus* var. *rivularis* x *M. variegatus* J. Saint-Hilaire)
Introduced. Status: rare.
In and beside streams.

EL: rare, by hill streams at Stoneypath (a large patch) and Deuchrie; and downstream in Biel Water at West Barns, 1988.

Limosella L.

L. aquatica L.   Mudwort
EL: Gullane Links, 1854; Drem, pre-1894.

Antirrhinum L.

A. majus L.   Snapdragon
Introduced. Status: local, garden escape, often established.
Waste ground, walls, roadsides, bings.

WL: rare urban casual, Bo'ness; (Bathgate,

Broxburn, Carriden, Kirkliston and Linlithgow, 1970s).
ML, EL: scattered distribution, mainly in urban areas.

Chaenorhinum (DC. ex Duby) Rchb.

C. minus (L.) Lange   Small Toadflax
Introduced. Status: local.
Current and disused railway tracks, waste ground, arable land.

WL: local, mainly in S.
ML: local, locally frequent in the valley of the Gala Water.
EL: rare, Prestongrange Mining Museum, 2001; Ormiston; south of Setonhill; North Berwick; (Haddington, 1888; Longniddry, 1895; Dunbar, 1959; Dunglass, 1966; Pencaitland, 1972).

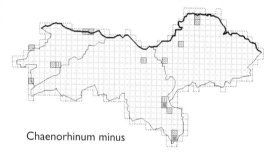

Chaenorhinum minus

Misopates Raf.

M. orontium (L.) Raf.   Weasel's Snout
Ruderal casual.
ML: Hailes rubbish tip, 1971; more frequent previously i.e. 1904–12.

Cymbalaria Hill

C. muralis P. Gaertn., B. Mey. & Scherb.
Ivy-leaved Toadflax
ssp. muralis
Introduced. Status: widespread.
Walls, pavements, hard ground.

WL, ML: widespread and locally frequent in N., occasional elsewhere.
EL: widespread near habitation in lowland areas.

C. pallida (Ten.) Wettst.   Italian Toadflax
Introduced. Status: rare, garden escape.
Walls.

WL: (Linlithgow, 1972).
ML: rare, Warriston, 1993; (Princes Street, Edinburgh, 1969).
EL: (Haddington, 1972).

C. hepaticifolia (Poir.) Wettst.
                    Corsican Toadflax
Introduced. Status: very local horticultural weed.
Gardens.

ML: very local, Royal Botanic Garden
Edinburgh, possibly overlooked elsewhere.

Kickxia Dumort.

K. elatine (L.) Dumort.   Sharp-leaved Fluellen
Introduced. Status: rare, casual.
Waste ground.

ML: on railway tip at Borthwick, 1954–87;
Currie, 2000.

Linaria L.

L. vulgaris Mill.   Common Toadflax
Native. Status: widespread.
Railway embankments, waste ground, roadsides.

WL: widespread, seldom frequent.
ML: widespread in N., locally frequent in urban
areas, local elsewhere.
EL: scattered distribution in lowland areas.
Almost all sites are beside, or within 1.5km, of a
present or former railway line.

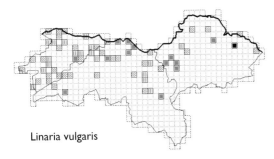

Linaria vulgaris

L. x sepium G. J. Allman (*L. vulgaris* x *L. repens*)
Native. Status: rare, spontaneous hybrid with
parents.
Railway embankment.

ML: rare, railway at Granton, 1988.

L. purpurea (L.) Mill.   Purple Toadflax
Introduced. Status: local.
Waste ground, roadsides, walls.

WL: local, mainly in N.
ML: common in urban areas in N., local
elsewhere.
EL: occasional, near habitation in lowland areas.

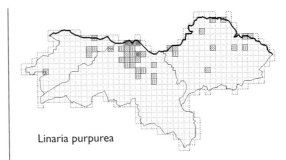

Linaria purpurea

L. repens (L.) Mill.   Pale Toadflax
Introduced. Status: rare.
Railway embankments, walls, roadsides.

WL: (Carriden, 1972).
ML: rare, Smeaton; Millerhill; (Levenhall, 1955;
Davidson's Mains, 1962; Braid Hills, 1967).
EL: (shore near Prestongrange, 1905).

L. maroccana Hook. f.   Annual Toadflax
A rare escape, not naturalising but frequently grown in
gardens.
ML: Braehead rubbish tip, 1978.

Collinsia Nutt.

C. heterophylla Buist ex Graham
                    Chinese-houses
Introduced. Status: rare, casual.
Gardens.

WL: rare, bird-seed alien recorded from
Threemiletown, 1990.
ML: (Leith Docks, 1906–7).

Digitalis L.

D. purpurea L.   Foxglove
Native. Status: common.
Woodland, scrub, beside roads and rivers, rough
grassland, dry heath, waste ground.

WL, ML: common.
EL: local in N., common elsewhere.

Erinus L.

E. alpinus L.   Fairy Foxglove
Introduced. Status: rare.
Walls.

WL: rare, Abercorn; Hopetoun; (Kirkliston,
1916).
ML: (Dalmahoy, 1915).

EL: rare, naturalised at Leaston and Morham; (Gifford, 1977).

Veronica L.

V. serpyllifolia L.　　　Thyme-leaved Speedwell
ssp. serpyllifolia
Native.　Status: common.
Rough grassland, roadsides, woodland, intensive grassland, heath grassland, waste ground, lawns.

WL: very widespread.
ML, EL: local in N., common elsewhere.

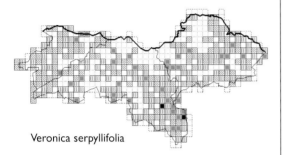

Veronica serpyllifolia

V. austriaca L.　　　Large Speedwell
ssp. teucrium (L.) D. A. Webb
Introduced.　Status: rare.
Open ground, dunes.

EL: rare, a small clump has persisted for several years in dunes near Gullane Point; (wall, East Linton, 1922).

V. officinalis L.　　　Heath Speedwell
Native.　Status: widespread.
Heath grassland, moorland, rough grassland, woodland, scrub, beside roads and rivers.

WL: widespread but patchy distribution.
ML: common in S., rare in N.
EL: widespread, locally frequent in S., patchy distribution in N.

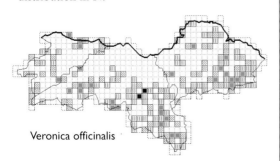

Veronica officinalis

V. chamaedrys L.　　　Germander Speedwell
Native.　Status: common.
Roadsides, woodland, rough grassland, scrub.

WL, ML, EL: common.

V. montana L.　　　Wood Speedwell
Native.　Status: local.
Old woodland, scrub, beside rivers and roads.

WL: scattered.
ML: quite widespread in E., local elsewhere.
EL: quite widespread in centre, virtually absent from N. and from upland areas.

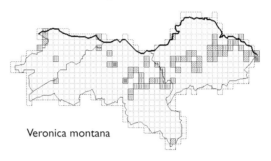

Veronica montana

V. scutellata L.　　　Marsh Speedwell
Native.　Status: local.
Marshes, in and beside ditches and streams, particularly in acid conditions.

WL: widespread in S. and W., rare or absent elsewhere.
ML, EL: quite widespread in upland areas, rare elsewhere.

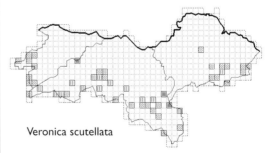

Veronica scutellata

V. beccabunga L.　　　Brooklime
Native.　Status: common.
In and beside rivers, ditches, ponds, lochs and canals, marshes.

WL: common.
ML: local in N., common elsewhere.
EL: rare in NW., widespread and locally frequent elsewhere.

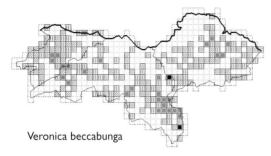

Veronica beccabunga

V. anagallis-aquatica L.   Blue Water Speedwell
Native. Status: local.
In and beside rivers and ditches, not in acid conditions.

WL: local, mainly in N.
ML: rare, Mortonhall, 1993; (Duddingston and Gogar, pre-1934; Addiewell, 1972).
EL: scattered distribution, Aberlady Bay; Stoneypath; Whiteadder Water; West Barns.

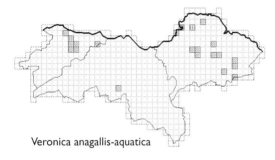

Veronica anagallis-aquatica

V. catenata Pennell   Pink Water Speedwell
Native. Status: rare.
Muddy places beside lochs and streams, marshes.

ML: (Duddingston Loch, 1977).
EL: rare, Aberlady Bay; (Yellowcraig, 1961).

V. arvensis L.   Wall Speedwell
Native. Status: widespread.
Arable fields, roadsides, waste ground, rough grassland, walls.

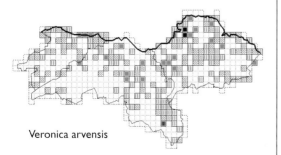

Veronica arvensis

WL: widespread.
ML: local in upland areas, widespread and locally frequent elsewhere.
EL: widespread and locally frequent.

V. peregrina L.   American Speedwell
Introduced. Status: rare, casual or garden weed.
Waste ground, gardens.

ML: (Slateford, pre-1927).
EL: rare, two or three plants on gravel path, Tyninghame House, 1992.

V. agrestis L.   Green Field Speedwell
Native. Status: local, possibly overlooked.
Arable fields, roadsides, waste ground, bare ground.

WL: local.
ML: local, mainly in SE., (Inveresk, 1956; Currie, 1959; Cramond, 1966).
EL: scarce, but wide distribution, Prestongrange; Markle; Woodhall; Oldhamstocks.

V. polita Fr.   Grey Field Speedwell
Native. Status: rare, possibly overlooked.
Arable fields, waste ground.

WL: rare, Westfield; Kirkliston; (South Queensferry, 1955).
ML: rare, Cockpen; Ormiston; (Portobello and Slateford, pre-1934; Leith Docks, 1955).
EL: rare, Longniddry; Gladsmuir; Kilspindie; East Saltoun; Whittingehame; (Canty Bay, 1932; Gullane, Gifford and Dunglass, 1955; Bothwell Water, pre-1962; Gosford, 1976).

V. persica Poir.   Common Field Speedwell
Introduced. Status: common.
Arable land especially cereal fields, roadsides, waste ground.

WL, ML, EL: rare in upland areas, common elsewhere.

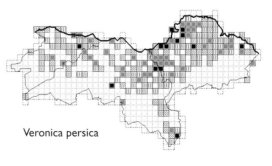

Veronica persica

V. filiformis Sm.   Slender Speedwell
Introduced. Status: widespread, a lawn weed.
Public parks, gardens, roadsides, riversides.

WL: widespread in N., occasional elsewhere.
ML: common in N., locally frequent in the valleys of Heriot Water and Gala Water, rare elsewhere. The first record in Scotland appears to have been at Swanston in 1937.
EL: widespread and locally frequent in lowland areas, rare in upland areas.

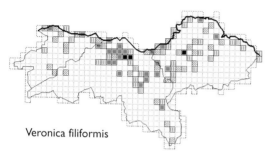

Veronica filiformis

V. hederifolia L.   Ivy-leaved Speedwell
Native. Status: widespread, mainly in lowland areas.
Arable land, especially cereal fields, roadsides, waste ground, woodland, gardens, rough grassland, coastal habitats.

WL, ML, EL: widespread in lowland areas.

Subspecies were not always distinguished in the survey.

ssp. hederifolia
Waste ground, sandy arable land, seldom in shady places.

ML: occasional.
EL: quite widespread in the northern half of the vice-county.

ssp. lucorum (Klett & Richt.) Hartl
On woodland banks and as a garden weed, rare on arable land; often in shady places.

ML: widespread.
EL: quite widespread in lowland areas but not recorded from eastern areas bordering the North Sea.

V. triphyllos L.   Fingered Speedwell
Casual.
ML: Slateford, pre-1927.

V. longifolia L.   Garden Speedwell
ML: Leith Docks, 1910.
EL: Dirleton, 1970.

Hebe Comm. ex Juss.

H. salicifolia (G. Forst.) Pennell   Koromiko
Introduced. Status: rare escape, naturalising.
Walls and rocks.

ML: rare, on rock, Roslin Castle; on wall, Trinity.
EL: rare, on wall, North Berwick, 1994.

Melampyrum L.

M. pratense L.   Common Cow-wheat
Native. Status: local.
Moor, heath grassland, scrub, woodland.

WL: (Linlithgow, pre-1934).
ML: local, mainly Pentland and Moorfoot Hills; Roslin; West Calder.
EL: rare, in the Lammermuir Hills: Hopes Reservoir area; Pressmennan; Elmscleugh; (Cauld Burn, 1885).

var hians Druce
EL: Monynut Woods, 1842, (recorded as *M. sylvaticum*).

Euphrasia L.

E. officinalis agg.   Eyebright
Native. Status: widespread.
Upland pasture, moorland, beside streams and roads, waste ground, coastal turf.

WL: widespread, locally frequent.
ML: widespread and locally frequent in upland areas, rare elsewhere.
EL: widespread in upland areas, scattered along the coast where it is locally abundant, rare elsewhere.

Includes the following thirteen taxa that were rarely identified in the survey.

E. arctica Lange ex Rostrup
ssp. borealis (F. Towns.) Yeo
Native. Status: apparently rare.
Hill pasture, grassy waste ground.

WL: rare, Bathgate; Broxburn.
ML: rare, Moorfoot Hills.
EL: rare, scattered distribution along the coast; inland in SW. at Linn Dean, Ormiston, Saltoun and Petersmuir Wood.

E. arctica x **E. nemorosa**
EL: beside the main path at Aberlady Bay, 1998.

E. x pratiuscula F. Towns. (*E. arctica* x *E. tetraquetra*)
EL: dunes, Gullane, 1998.

E. arctica x **E. confusa**
Native. Status: rare.
Hill pasture.

ML: rare, Moorfoot Hills; (Balerno, pre-1934).

E. x venusta F. Towns. (*E. arctica* x *E. scottica*)
Native. Status: rare.
Upland heath.

ML: rare, Moorfoot Hills near Stow, 1986.

E. tetraquetra (Bréb.) Arrond.
Native. Status: rare.
Coastal turf.

EL: rare, Aberlady Bay; (North Berwick, pre-1934; Catcraig, 1970s).
East coast plants, in general, lack the distinctiveness of populations on exposed cliffs in western Britain.

E. tetraquetra x **E. nemorosa**
EL: rare, in coastal turf at Aberlady Bay.

E. nemorosa (Pers.) Wallr.
Native. Status: local.
Dry heath grassland, woodland rides, roadsides, wet flushes, coastal turf.

WL: status unknown, Bo'ness area; Broxburn area; (Linlithgow, 1935).
ML: local, Hermiston; Cobbinshaw; Stow; Bowland; Blackhope Scar.
EL: rare, Aberlady Bay; Gullane; Colstoun; North Berwick; Dunglass; (Dunbar, pre-1934; Faseny Water, 1971).

E. nemorosa x **E. confusa**
Native. Status: local.
Dune grassland, railways, waste ground, heath grassland, open ground.

ML: rare, Millerhill, 1988.
EL: intermediate populations, presumably originally of hybrid origin, occur on the coast, Aberlady Bay; Yellowcraig; Belhaven Bay; also Rook Law in the Lammermuir Hills.

E. confusa Pugsley
Native. Status: local, undoubtedly under-recorded.
Short, dry, hill and coastal turf, waste ground.

WL: status unknown, Linlithgow; Torphichen.
ML: quite widespread in upland areas.
EL: current status unknown, Monynut Water, 1983; no confirmed recent records; (recorded along the coast and in upland areas in 1971, possibly lost to hybridisation at some sites).

E. confusa x **E. micrantha**
Native. Status: rare.
Dry upland pasture.
ML: rare, Stow, 1987; (Balerno, pre-1934).

E. micrantha Rchb.
Native. Status: rare.
Heather moorland.

ML: (Balerno Moor, 1935; Glencorse Reservoir, 1955).
EL: rare, Hopes Water; White Castle; Sheeppath Burn; (Faseny Water, 1971).

E. scottica Wettst.
Native. Status: rare, one known site.
Wet peaty flush.

EL: rare, Dunbar Common; perhaps overlooked elsewhere in the hills.

E. rostkoviana Hayne
WL: waste ground at Bo'ness, 1971.

E. micrantha x **E. nemorosa**
WL: Carribber, 1902.
ML: Balerno Moor, pre-1934.

Odontites Ludw.

O. vernus (Bellardi) Dumort.   Red Bartsia
Native. Status: widespread.
Roadsides, waste ground, rough grassland.

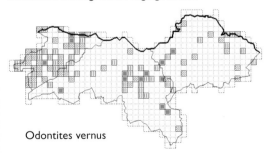

Odontites vernus

WL: widespread, locally frequent.
ML: local, quite widespread in E., locally frequent.
EL: scattered distribution throughout the vice-county.

Subspecies were not recorded in the survey.

ssp. vernus
EL: a verge at Dunbar, 1960; dunes at Gosford, 1970.

ssp. serotinus (Syme) Corb.
ML: Roslin, Penicuik and Turnhouse, pre-1934.
EL: Morham, 1909; Bolton Muir Wood, 1953; Gullane, 1957.

Parentucellia Viv.

P. viscosa (L.) Caruel — Yellow Bartsia
Introduced. Status: rare, casual.
Roadsides.

WL: rare, casual, sown in wildflower seed near Livingston, 1995.

Rhinanthus L.

R. minor L. — Yellow Rattle
Native. Status: widespread.
Rough grassland, beside roads and rivers, old bings, dune grassland.

WL: widespread.
ML: quite widespread, patchy distribution, locally abundant.
EL: occasional, Keith; Longniddry railway walk; Yellowcraig; Papana Water; Belhaven Bay; Harehead.

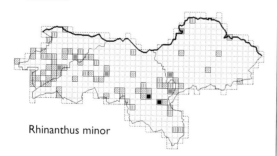
Rhinanthus minor

Subspecies were not recorded in the survey.

ssp. minor
On relatively dry and lime-rich soils.

Recorded with certainty only at Aberlady Bay (EL).

ssp. stenophyllus (Schur) O. Schwarz
Typically in damp, rather base-poor grassland, slag heaps.

ML: (Ratho, 1904; Straiton, 1968).
EL: rarely recorded, Aberlady Bay, 1998.

Pedicularis L.

P. palustris L. — Marsh Lousewort, Red Rattle
Native. Status: local.
Marshes, bogs, riversides.

WL: rare, north of Blackridge; Blawhorn Moss; Riccarton Hills; (Lochcote Marsh).
ML: local, entirely in upland areas.
EL: rare, Aberlady Bay; Linn Dean; a few sites in the Lammermuir Hills; (Gullane Loch, 1848; Longniddry, 1976).

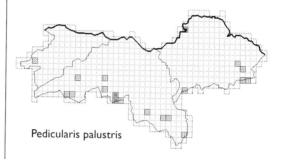
Pedicularis palustris

P. sylvatica L. — Lousewort
Native. Status: local.
Heath grassland, marshes, bogs.

WL: local, in S. only.
ML, EL: local, upland only.

Nemesia Vent.

N. strumosa Benth. — Nemesia
Garden escape.
ML: Braehead rubbish tip, 1978.

OROBANCHACEAE

Lathraea L.

L. squamaria L. — Toothwort
Native. Status: rare.
Damp woodlands on rich soils. Parasite on roots of elm and hazel.

ML: local, Dalkeith Estate; Lady Lothian Wood;

Cowpits; Arniston; Roslin; Inveresk; Gorebridge; (Temple, 1978).
EL: rare, Yester Estate; (Sandy's Mill, 1910; Humbie, 1957; Tyninghame, 1969).

L. clandestina L. Purple Toothwort
Introduced. Status: scarce, initially introduced when trees were planted, more or less naturalised. Estate grounds, riverside trees. Parasitic on roots of poplar and willow.

ML: rare, Roslin Glen; Bush Estate; Royal Botanic Garden Edinburgh; Auchendinny; Penicuik; River Esk at Musselburgh.
EL: rare, naturalised, parasitic on willow, (*Salix fragilis*). First recorded beside Birns Water near Spilmersford, 1956 and (upstream) by Keith Water, 1968; reached the River Tyne by 1971, and East Linton by 1999, via Samuelston, Clerkington, Haddington and Stevenson.

Orobanche L.

O. minor Sm. Common Broomrape
Native. Status: rare, possibly extinct.

EL: appeared in 1995 on a soil heap associated with open-cast mining near Tranent. The site has since been destroyed.

O. rapum-genistae Thuill. Greater Broomrape
ML: West Calder, 1841.

LENTIBULARIACEAE

Pinguicula L.

P. vulgaris L. Common Butterwort
Native. Status: local.
Damp heath grassland, marshes, bogs, beside streams and ditches.

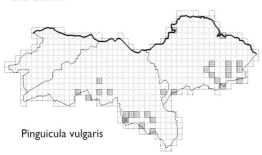

Pinguicula vulgaris

WL: rare, Cairnpapple, 1993; (Cockleroy, 1974).

ML: occasional in upland flushes, Moorfoot and Pentland Hills.
EL: rare along the coast, occasional in upland areas, Aberlady Bay; Yellowcraig; Papana, Bothwell and Monynut Waters.

Utricularia L.

U. vulgaris L. Greater Bladderwort
Native. Status: rare.
Ditches, ponds.

ML: rare, Bawsinch, planted, 1990; (Balerno, 1894).
EL: rare, Aberlady Bay; (Gullane Loch, 1848 and pre-1934).

U. minor L. Lesser Bladderwort
ML: Balerno, pre-1934.

CAMPANULACEAE

Campanula L.

C. lactiflora M. Bieb. Milky Bellflower
Introduced. Status: rare, garden escape, not naturalising.
Waste ground, roadside verges.

ML: rare, Craiglockhart Dell; Bawsinch; (Braid Hill Drive, 1970–1).

C. persicifolia L. Peach-leaved Bellflower
Introduced. Status: scarce, garden escape.
Rough grassland, grassy banks.

ML: scarce, Warriston; Balgreen; Cramond; Millerhill.
EL: rare, Aberlady; Luffness; Gullane Bents; (Longyester, 1975).

C. medium L. Canterbury Bells
Introduced. Status: rare, escape, not persisting.
Waste ground, rough grassland.

ML: rare, Pilton, 1997.

C. glomerata L. Clustered Bellflower
Native and introduced. Status: rare.
Dunes, waste ground, pavements.

ML: rare, Craigleith Hill, 1999.
EL: rare, Ferny Ness/Gosford; (Pencaitland, 1931; Gullane, 1957; formerly along the coast from Cockenzie to Gullane; also at North Berwick and Dunbar).

C. portenschlagiana Schult.   Adria Bellflower
Introduced. Status: rare, long-persisting garden escape.
Walls, pavements.

ML: rare, Warriston Gardens and Inverleith, 1985 to present.

C. latifolia L.   Giant Bellflower
Native. Status: local.
Old woodland, especially by rivers.

WL: quite widespread, mainly on the Dalmeny Estate and along the Almond and Avon Rivers.
ML: scattered distribution, mainly in lowland areas.
EL: scarce, Saltoun; Humbie; Butterdean; Colstoun; Papana Water; Woodhall Dean.

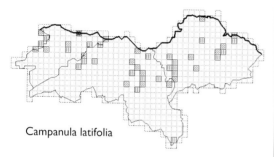

Campanula latifolia

C. trachelium L.   Nettle-leaved Bellflower
Introduced. Status: rare, garden escape.
Waste ground, walls.

WL: rare, Broxburn.
ML: rare, Warriston; Trinity; Blackhall; Balerno.
EL: (Luffness and North Berwick, pre-1934; garden relic at Longniddry, 1975).

C. rapunculoides L.   Creeping Bellflower
Introduced. Status: rare, garden escape and persistent weed.
Roadsides, railway embankments, hedgerows.

ML: rare, Easter Newton; Easter Road; (Granton, 1956; Redheughs, 1977).
EL: rare, Luffness; Yellowcraig; (East Linton, pre-1934; Innerwick, pre-1962; Gullane, 1972).

C. rotundifolia L.   Bluebell, Harebell
Native. Status: common.
Rough grassland, moors, beside roads and rivers, quarries, scrub, woodland, coastal habitats, dry banks, walls.

WL: widespread.
ML, EL: common in S., less widespread in N.

Distinctive, dwarf specimens from Arthur's Seat and Roslin (ML) and from the coast near Tantallon Castle (EL), 1907, have been determined as the north-European segregate, *C. groenlandica* L. These identifications must be regarded as very doubtful without supporting chromosome studies.

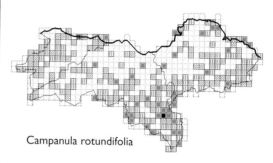

Campanula rotundifolia

Legousia Durande

L. hybrida (L.) Delarbre   Venus's Looking-glass
Introduced. Status: rare.
Cereal fields and near grain silos.

ML: rare, Leith Docks, 1981; (frequent *c.* 1900)
EL: rare, at the same site near Aberlady, 1993 and 1996; Tyninghame, 1996; (Luffness, Gullane and Dirleton, pre-1934; Yellowcraig, 1960).

L. speculum-veneris (L.) Chaix
    Large Venus's Looking-glass
Rare casual grain alien.
ML: Leith Docks, 1980.

Jasione L.

J. montana L.   Sheep's-bit
ML: Cousland Crags, 1871.

Lobelia L.

L. erinus L.   Garden Lobelia
Introduced. Status: rare, garden escape.
Pavement.

EL: rare, Gullane, 1997.

Pratia Gaudich.

P. angulata (G. Forst.) Hook. f.   Lawn Lobelia
Introduced. Status: rare, horticultural weed.
Gardens.

ML: widespread in lawns at Royal Botanic Garden Edinburgh and Malleny House.

RUBIACEAE

Sherardia L.

S. arvensis L.   Field Madder
Native. Status: local, possibly overlooked.
Sometimes sown with grass seed.
Rough grassland, quarries, public parks, roadsides, waste ground, scrub, dunes, rocky outcrops.

WL: rare, two sites near Cauldcoats Holdings; (Kirkliston and Turnhouse, pre-1934; Linlithgow, 1977).
ML: scattered distribution, Arthur's Seat; Blackford Hill; Fisherrow; Musselburgh; Middleton; Lothianburn; Roslin.
EL: scattered distribution, mainly in rocky places in NE., Gullane; Hailes; Tantallon; Stottencleugh.

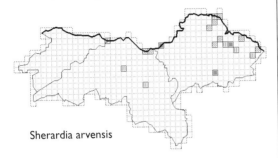

Sherardia arvensis

Asperula L.

A. taurina L.   Pink Woodruff
Introduced. Status: rare, garden escape.
Deciduous woodland.

WL: rare, Abercorn area; Hopetoun Estate; (Carlowrie, 1868).
ML: rare, Currie, 1996; (Arniston, Prestonhall and Little Vantage, pre-1934).
EL: rare, Redmains, Pencaitland; (road between Whittingehame and Stenton, 1848).

A. arvensis L.   Blue Woodruff
Grain alien.
ML: Ratho, 1881; Leith Docks, 1890–1904; Craigmillar Quarry, 1914; Musselburgh and Woodhouselee, pre-1934.

Galium L.

G. odoratum (L.) Scop.   Woodruff
Native. Status: local.
Woodland, roadsides.

WL: absent from S., scattered elsewhere, especially in old woodland along the River Avon.

ML: common in a relatively small area centred around Gorebridge, occasional and locally frequent elsewhere.
EL: quite widespread in centre, Humbie; Colstoun; Woodhall Dean; Dunglass; no recent records from NW. or SE.

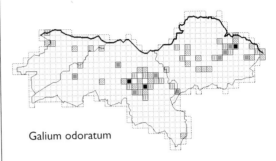

Galium odoratum

G. uliginosum L.   Fen Bedstraw
Native. Status: widespread.
Marshes, beside rivers and ditches, in acid to neutral soils

WL: occasional, mainly in S. and W.
ML: quite widespread, mainly in upland areas, locally frequent.
EL: occasional, mainly in upland areas, Linn Dean; Stobshiel Reservoir; Stoneypath; White Castle; Monynut Water.

G. palustre L.   Common Marsh Bedstraw
Native. Status: widespread.
Marshes, in and beside rivers and ditches, ponds, lochs and canal, bogs.

WL: widespread, especially in S. and W.
ML: occasional in N., common elsewhere.
EL: occasional in N., widespread and locally frequent in S.

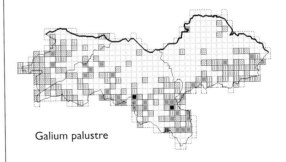

Galium palustre

The subspecies were rarely recorded during the survey.

ssp. palustre
EL: rarely recorded, Saltoun; Aberlady Bay; Latch Loch; White Castle; Dunbar Common; Sheeppath Burn.

ssp. elongatum (C. Presl.) Arcang.
ML: recorded recently, it is the most common, and probably the only, ssp. in ML.
EL: rarely recorded, Marl Loch, Aberlady Bay.

G. verum L.   Lady's Bedstraw
Native. Status: common.
Roadsides, rough grassland, heath grassland, woodland, scrub, quarries, coastal habitats, dry soils.

WL, ML, EL: common.

G. x pomeranicum Retz. (*G. verum* x *G. mollugo*)
Native. Status: rare, with parents.
Rough grassland, roadsides, hard ground.

ML: rare, Stow; Middleton.
EL: rare, Broxmouth.

G. mollugo L.   Hedge Bedstraw
Native. Status: scarce.
Scrub, roadsides, hedgerows, rough grassland.

WL: rare, near Dundas; south of Bathgate; (Kirkliston, pre-1934; Carriden, 1972).
ML: local, Monktonhall; Oxenfoord; Duddingston; Musselburgh; (Borthwick railway tip, 1960; Stow, 1967).
EL: rare, St Germains; Longniddry; Linn Dean; North Berwick; Pefferside; West Barns; Broxmouth; (Dirleton, 1836; East Linton, 1909; Archerfield, 1955).

ssp. erectum Syme
Native. Status: rare.
Dry calcareous grassland.

ML: rare, Ratho; Dalmahoy; (Portobello, pre-1934).

It is likely that some or all of the records for *G. mollugo* are this subspecies, but more research is needed.

G. sterneri Ehrend.   Limestone Bedstraw
Native. Status; rare.
Dry, base-rich grassland.

ML: rare, Glencorse; Loganlea; Gladhouse.
EL: ('Glen above Garvald', 1871, doubtful record, reputed herbarium specimen not found).

G. saxatile L.   Heath Bedstraw
Native. Status: common.
Moorland, heath grassland, mixed and coniferous woodland, beside roads and rivers, scrub, marshes, bogs.

WL: common, especially in S. and W.
ML, EL: local in lowland areas, very common in upland areas.

G. aparine L.   Sticky Willie, Goosegrass, Cleavers
Native. Status: common.
Roadsides, woodland, hedgerows, waste ground, scrub, cereal fields, rough grassland.

WL: common.
ML, EL: local in upland areas, very common in lowland areas.

G. tricornutum Dandy   Corn Cleavers
Grain alien.
ML: Ratho, 1881; sporadically from Leith Docks, 1881–1909; near Slateford, 1906; Craigmillar Castle quarry, 1914.

G. parisiense L.   Wall Bedstraw
ML: near Musselburgh, 1910.

Cruciata Mill.

C. laevipes Opiz   Crosswort
Native. Status: common.
Roadsides, open woodland, riversides, rough grassland, scrub, hedgerows.

WL: widespread and locally frequent in N., local elsewhere.
ML: local in urban and upland areas, common elsewhere.
EL: local in N. and S., common elsewhere.

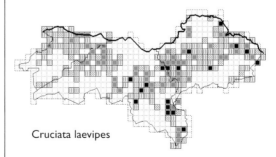

Cruciata laevipes

Crucianella L.

C. angustifolia L.
Rare birdseed alien.
ML: Slateford, 1906; Levenhall, 1911–12.

CAPRIFOLIACEAE

Sambucus L.

S. racemosa L.   Red-berried Elder
Introduced. Status: local.
Woodland, scrub.

WL: occasional.
ML: local, about 20 sites known.
EL: rare, Tranent; Ormiston; Fountainhall;
Humbie; Yester; Monynut Water; Stottencleugh.

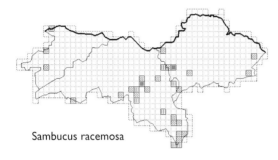
Sambucus racemosa

S. nigra L.   Elder
Native. Status: common.
Woodland, roadsides, hedgerows, scrub.

WL: common.
ML, EL: local in upland areas, common
elsewhere.

var. **laciniata** Miller   Cut-leaved Elder
Introduced. Status: rare, planted or escaped from
cultivation.
Woodland, roadsides, quarries.

WL: rare, Abercorn; Bo'ness; Midhope;
Swineburn Wood; Hopetoun; (Carlowrie).
ML: (Granton; Dalhousie).
EL: rare, Blinkbonny; Garleton Hills quarry;
Papple; (Athelstaneford; Tyninghame, 1866;
Pitcox, 1955; Broxmouth, 1973).

var. **viridis** Ait.
ML: Dalkeith Country Park; Inveresk; Lasswade.

S. ebulus L.   Danewort, Dwarf Elder
Introduced. Status: rare.
Roadsides, hedgerows, waste ground.

WL: (Bo'ness, pre-1934).
ML: (Moredun and Water of Leith, pre-1850).
EL: rare, Dirleton; (Pressmennan, pre-1934).

Viburnum L.

V. opulus L.   Guelder Rose
Introduced. Status: local, often planted.
Woodland, scrub, hedgerows, reclaimed bings.

WL, ML: local.
EL: occasionally planted in woods and hedges in
lowland areas; (said to grow wild in most of the
Lammermuir deans, 1884; Sheeppath Dean,
1967).

V. lantana L.   Wayfaring Tree
Introduced. Status: scarce, not naturalising.
Woodland, scrub, hedgerow, old bings.

WL: scarce, planted, Dalmeny; Dundas;
Hopetoun; near Broxburn; Westfield; Whitburn.
ML, EL: occasionally planted.

V. tinus L.   Laurustinus
Introduced. Status: rare, escape.
Policy woodland.

ML: rare, Newhailes.
EL: rare, Broxmouth; (Archerfield Estate, 1956;
Biel Estate, 1957).

Symphoricarpos Duhamel

S. albus (L.) S. F. Blake   Snowberry
var. **laevigatus** (Fernald) S. F. Blake
Introduced. Status: widespread, often planted
and forming thickets that can be a serious threat
to native plants.
Woodland, scrub, hedgerows, beside roads, rivers and
railways, public parks, policies.

WL: widespread.
ML, EL: widespread and locally frequent in
lowland areas, occasional elsewhere.

S. x chenaultii Rehder (*S. microphyllus* Kunth x
S. orbiculatus Moench)   Hybrid Coralberry
Introduced. Status: scarce, planted.
Roadsides, public amenity areas.

WL: scarce, Bathgate; Bo'ness; Dechmont; Port
Edgar.
ML: rare, Powderhall; established in
Craiglockhart Dell.

Linnaea L.

L. borealis L.   Twinflower
ML: Bavelaw, 1921.

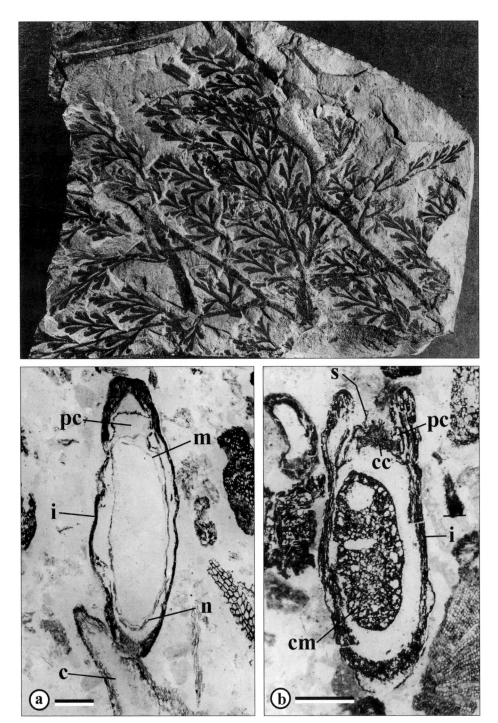

Plate 17 Top: *A frond of* Sphenopteris, *a seed fern (pteridosperm) from the Lower Carboniferous. Reproduced by courtesy of the National Museums of Scotland.* **Bottom right and left:** *Longitudinal sections of peels made by Albert Long from the Calciferous Sandstone, revealing anatomically preserved ovules of the Early Carboniferous seed fern* Stamnostoma huttonense. **a:** *ovule still attached to its branched cupule (c). The integument (i) of the ovule is thin and contains the nucellus (n), equivalent to the sporangial wall, and a single large megaspore (m). Distally in the nucellus is the pollen chamber (pc) that functioned to capture and retain pollen pre-fertilisation.* **b:** *ovule of the same species post pollination containing a multicellular megagametophyte (cm). At this stage, the pollen chamber is sealed by a parenchymatous mass called the central column (cc) that seals the apex of the pollen chamber at the base of a distally extending tube called the salpinx (s).* (a) *slide 788,* (b) *slide 1015, both from the A. G. Long slide collection in the Hancock Museum, Newcastle-upon-Tyne. Scale bars = 0.5 mm. Figure preparation by J. Hilton, National Museums of Scotland. Reproduced by courtesy of the National Museums of Scotland.*

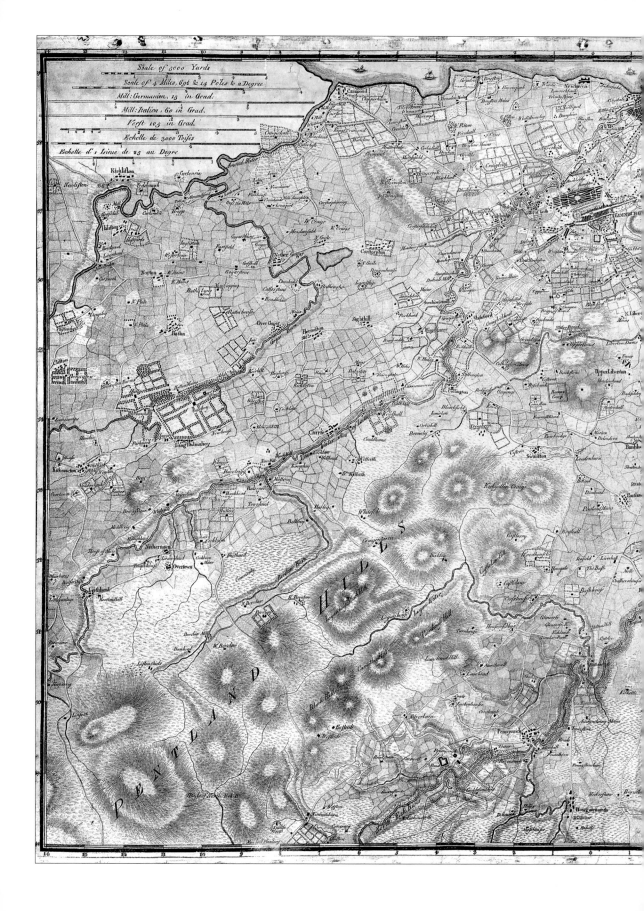

Scale of 3000 Yards
Scale of 4 Miles, 69¼ & 14 Poles to a Degree
Mill: Germanica, 15 in Grad.
Mill: Italica, 60 in Grad.
Verst 103 in Grad.
Echelle de 3000 Toises
Echelle d's Lieue de 25 au Degre

PEATLAND HILLS

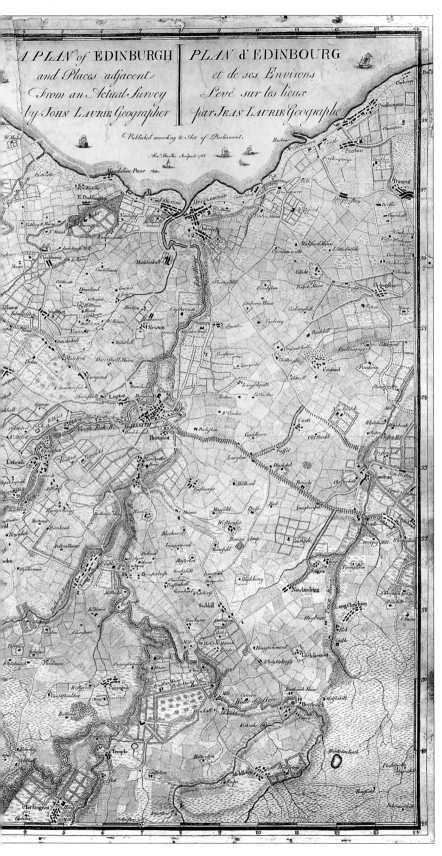

Plate 18 *This is a reproduction of part of John Laurie's map of the Lothians, 1766 edition. The extensive and mainly treeless arable land in the east is dissected by the ancient, but clearly surviving ravine and river valley woodlands which persist to the present time. Reproduced by kind permission of the Trustees of the National Library of Scotland.*

Plate 19 *Lichen:* Cladonia arbuscula, *a common, locally abundant lichen of acid heathland on moors and in the dune heath.* C. E. Jeffree.

Plate 20 *Microfungi (Chapter 4).* **Top:** *The common and widespread apple scab, the most serious disease of commercial apples, is caused by* Venturia inaequalis. **Bottom left:** *The aecia of* Gymnosporangium cornutum *appear on Rowan* (Sorbus aucuparia) *and related species. The aeciospores cannot re-infect this host, but cross over to Juniper* (Juniperus communis) *to complete the rust's life cycle. This photograph shows the telia (fruiting bodies of the final teliospore-producing stage) emerging from a juniper stem.* **Bottom right:** *Spores of* Alternaria *are charactistically multicellular, as shown in this light micrograph.* S. Helfer.

Plate 21 Top: *Common Earthball* (Scleroderma aurantium). **Bottom:** *The Morel* (Morchella hortensis) *recorded here at Liberton Hospital (see Chapter 5). Top: C. E. Jeffree; bottom: R. Watling.*

Plate 22 *Massed male gametophytes of the moss* Rhizomnium punctatum, *syn.* Mnium punctatum, *in 'flower'. C. E. Jeffree.*

Plate 23 *Great Horsetail* (Equisetum telmateia) *(Chapter 10) reaches its north-eastern limit in the Lothians.* **Top:** *The fertile shoots bearing cones emerge in April ahead of sterile stems,* **Bottom,** *which may reach 2m in height. A. F. Dyer.*

Plate 24: *Beech* (Fagus sylvatica), *shown here in mast, is one of several broadleaved tree species to have naturalised successfully in the Lothians.* C. E. Jeffree.

Plate 25 *Spring Beauty or Miner's Lettuce* (Claytonia perfoliata, *syn.* Montia perfoliata), *an introduction from North America, is common in coastal woodland at Aberlady, Yellowcraig and Tantallon. This edible species is now in demand as a gourmet salad vegetable. J. Muscott.*

Plate 26 *Bottle Sedge* (Carex rostrata) *is frequent in the margins of peaty lochs.*
C. E. Jeffree.

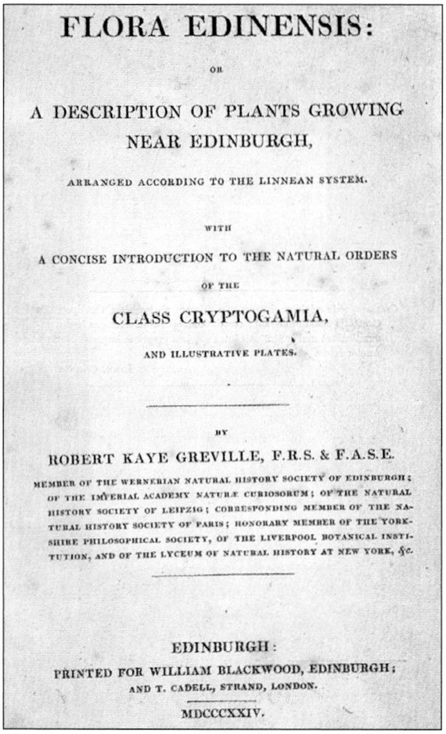

FLORA EDINENSIS:

OR

A DESCRIPTION OF PLANTS GROWING NEAR EDINBURGH,

ARRANGED ACCORDING TO THE LINNEAN SYSTEM.

WITH

A CONCISE INTRODUCTION TO THE NATURAL ORDERS OF THE CLASS CRYPTOGAMIA,

AND ILLUSTRATIVE PLATES.

BY

ROBERT KAYE GREVILLE, F.R.S. & F.A.S.E.

MEMBER OF THE WERNERIAN NATURAL HISTORY SOCIETY OF EDINBURGH; OF THE IMPERIAL ACADEMY NATURÆ CURIOSORUM; OF THE NATURAL HISTORY SOCIETY OF LEIPZIG; CORRESPONDING MEMBER OF THE NATURAL HISTORY SOCIETY OF PARIS; HONORARY MEMBER OF THE YORKSHIRE PHILOSOPHICAL SOCIETY, OF THE LIVERPOOL BOTANICAL INSTITUTION, AND OF THE LYCEUM OF NATURAL HISTORY AT NEW YORK, &c.

EDINBURGH:

PRINTED FOR WILLIAM BLACKWOOD, EDINBURGH; AND T. CADELL, STRAND, LONDON.

MDCCCXXIV.

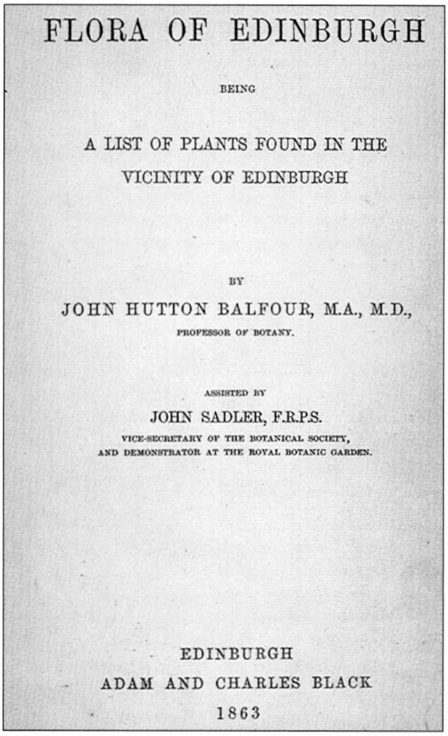

FLORA OF EDINBURGH

BEING

A LIST OF PLANTS FOUND IN THE VICINITY OF EDINBURGH

BY

JOHN HUTTON BALFOUR, M.A., M.D.,

PROFESSOR OF BOTANY.

ASSISTED BY

JOHN SADLER, F.R.P.S.

VICE-SECRETARY OF THE BOTANICAL SOCIETY, AND DEMONSTRATOR AT THE ROYAL BOTANIC GARDEN.

EDINBURGH

ADAM AND CHARLES BLACK

1863

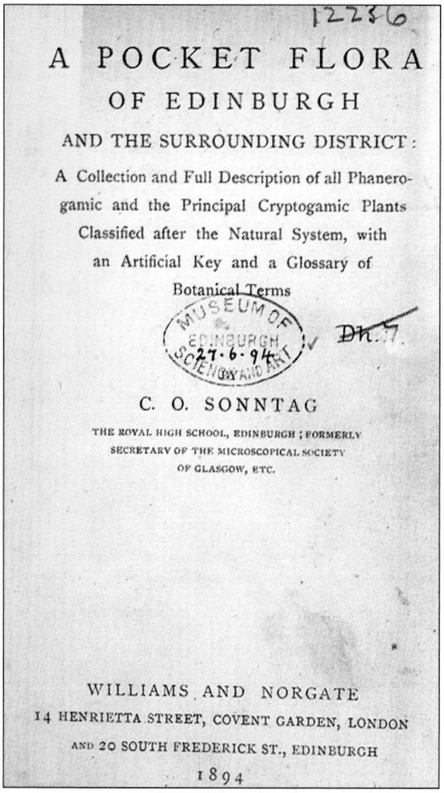

12256

A POCKET FLORA OF EDINBURGH AND THE SURROUNDING DISTRICT:

A Collection and Full Description of all Phanerogamic and the Principal Cryptogamic Plants Classified after the Natural System, with an Artificial Key and a Glossary of Botanical Terms

BY

C. O. SONNTAG

THE ROYAL HIGH SCHOOL, EDINBURGH; FORMERLY SECRETARY OF THE MICROSCOPICAL SOCIETY OF GLASGOW, ETC.

WILLIAMS AND NORGATE

14 HENRIETTA STREET, COVENT GARDEN, LONDON

AND 20 SOUTH FREDERICK ST., EDINBURGH

1894

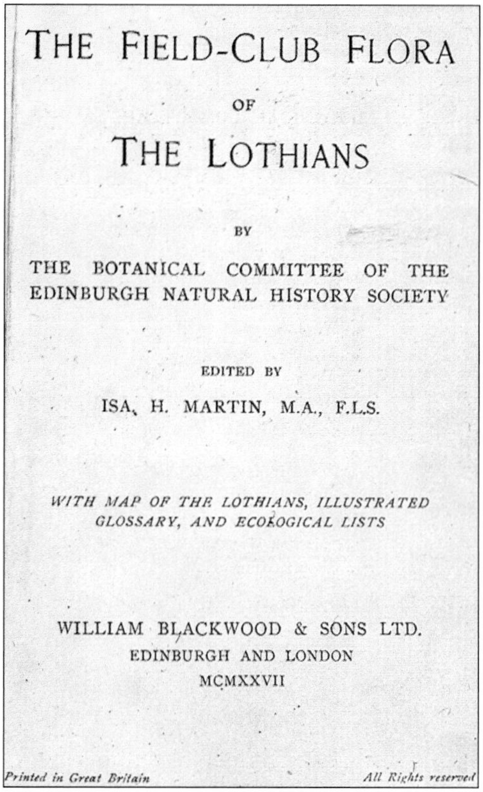

THE FIELD-CLUB FLORA OF THE LOTHIANS

BY

THE BOTANICAL COMMITTEE OF THE EDINBURGH NATURAL HISTORY SOCIETY

EDITED BY

ISA. H. MARTIN, M.A., F.L.S.

WITH MAP OF THE LOTHIANS, ILLUSTRATED GLOSSARY, AND ECOLOGICAL LISTS

WILLIAM BLACKWOOD & SONS LTD.

EDINBURGH AND LONDON

MCMXXVII

Printed in Great Britain *All Rights reserved*

Plate 27 *Title pages from Floras of the nineteenth and early twentieth centuries (Chapter 11).* **Top left:** *Greville's* Flora Edinensis *of 1824, the first Flora of Edinburgh and the Lothian area.* **Top right:** *Balfour and Sadler's* Flora of Edinburgh *(1863).* **Bottom left:** *Sonntag's* A Pocket Flora of Edinburgh and surrounding district *(1894).* **Bottom right:** *Martin's* Field-Club Flora of the Lothians *(1927). Photography (S. Clarke) by courtesy of the Royal Botanic Garden Edinburgh.*

Plate 28 *A turnip crop being grazed* in situ *by sheep is a common sight in the Lothians. C. E. Jeffree.*

Plate 29 *Managed heather moorland occupies a significant area of the Lothians. Degenerate phase heather* (Calluna vulgaris) (**Top**) *is of poor dietary value for grouse and sheep, but its decline offers opportunities for bryophytes, lichens and vascular plants to colonise.* **Bottom:** *Regular burning encourages vigorous and nutritious growth of a heather monoculture. The heather is burned in strips known as swiddens, the burn being controlled by beating. C. E. Jeffree.*

Plate 30 *Common Cottongrass* (Eriophorum angustifolium) *flowering in abundance on old spoil heaps near Fauldhouse. J. Muscott.*

Plate 31 *The view to the south-east from Arthur's Seat, overlooking the carr and marginal woodland of Duddingston Loch, with a golf course beyond. In the distance, patches of industrial, residential and arable land use and woodland provide a variety of plant habitats. C. E. Jeffree.*

Leycesteria Wall.

L. formosa Wall.   Himalayan Honeysuckle
Introduced. Status: local, escape, increasing.
Estate woodlands, waste ground, riversides.

WL: rare, Bo'ness; South Queensferry;
(Almondell, 1956)
ML: local, Danderhall; Cramond; Craiglockhart
Dell; (Craigmillar Castle quarry and Dreghorn,
1956; Carberry Tower, 1960).
EL: rare, Garleton.

Weigela Thunb.

W. florida (Bunge) A. DC.   Weigelia
Introduced. Status: rare.
Policies.

WL: rare, Dundas.

Lonicera L.

L. nitida E. H. Wilson   Wilson's Honeysuckle
Introduced. Status: scarce, planted or escaped.
Public amenity areas, policies, woodland, scrub,
hedgerow, waste ground.

WL: scarce, near Bo'ness; north of Linlithgow;
Champfleurie; Port Edgar; Dalmeny Estate.
ML: rare, Hermitage of Braid.
EL: rare, Luffness; north of Drem; Haddington;
Athelstaneford; Gifford; Whittingehame.

L. xylosteum L.   Fly Honeysuckle
Introduced. Status: rare, originally planted.
Open woodland.

WL: rare, planted at Hopetoun.
ML: rare, naturalised in Colinton Dell; Roslin;
(Arniston, 1960).
EL: (North Berwick and Tantallon, pre-1934;
Gosford, 1976).

L. periclymenum L.   Honeysuckle
Native. Status: widespread.
Woodland, hedgerows, roadsides, scrub.

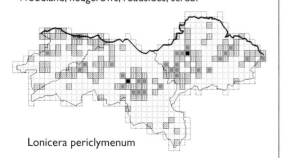

Lonicera periclymenum

WL: widespread.
ML, EL: widespread and locally frequent in
centre, local elsewhere.

L. caprifolium L.   Perfoliate Honeysuckle
Introduced. Status: rare.
Scrub, roadsides.

WL: rare, Dalmeny; Port Edgar; (Linlithgow,
pre-1934).
ML: rare, naturalised in Colinton Dell.
EL: (Tyninghame, 1868; East Linton, pre-1934,
possibly the same record).

ADOXACEAE

Adoxa L.

A. moschatellina L.   Moschatel, Townhall Clock
Native. Status: local.
Damp woodland on rich soils.

WL: scarce, scattered along the River Avon;
Dundas Estate; mouth of the River Almond.
ML: local, mainly in centre, locally frequent.
EL: local, mainly in centre, Fountainhall;
Humbie; Yester; East Linton; Biel; Dunglass.

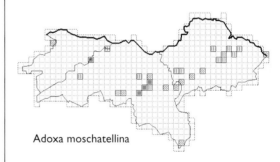

Adoxa moschatellina

VALERIANACEAE

Valerianella Mill.

V. locusta (L.) Laterr.
          Common Cornsalad, Lamb's Lettuce
Native. Status: scarce.
Dunes, cliffs, railway embankments, waste ground.

WL: rare, Torphichen; Dalmeny; (Hopetoun,
1956).
ML: scarce, Duddingston; Borthwick; Blackford
Hill.

EL: occasional along the coast, rare inland,
Aberlady Bay; Yellowcraig; Tyninghame; Barns
Ness; Longniddry railway walk; Luffness quarry;
Woodhall Burn; Broxburn; Aikengall.

var. **dunensis** D. E. Allen has been recorded in
EL.

V. rimosa Bastard   Broad-fruited Cornsalad
Native.
WL: near Linlithgow, 1866.

V. dentata (L.) Pollich   Narrow-fruited Cornsalad
Native.
Field weed.
ML: Crossgate Toll, three miles south of Musselburgh,
1809, (R. Maughan, new record for Scotland); Newhaven
and Granton, 1824; casual at Portobello and Leith,
c. 1910.
EL: Dirleton Common, 1835; Aberlady, 1836; Gullane,
1904; Yester, pre-1934.

V. eriocarpa Desv.   Hairy-fruited Cornsalad
WL: Bo'ness, pre-1934.
ML: Leith and Slateford, *c.* 1820.

Valeriana L.

V. officinalis L.   Common Valerian
Native. Status: local.
Marshes, waterside habitats, wet woodland.

WL: widespread.
ML: local, mainly in lowland areas.
EL: scattered distribution, mainly in lowland
areas, Aberlady Bay; Bolton Muir Wood;
Woodhall Dean.

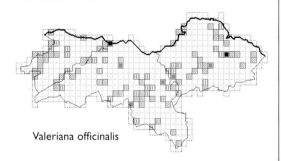

Valeriana officinalis

V. pyrenaica L.   Pyrenean Valerian
Introduced. Status: scarce, locally naturalised.
Damp, shady areas, watersides.

WL: scarce, Blackburn; Gowanbank; Carribber;
Midhope; Pepper Wood; River Almond west of
Livingston; (Fauldhouse, 1976).

ML: local, Penicuik Estate; Auchendinny;
Harburn; Roslin; Ratho; Inveresk; Colinton Dell.
EL: rare, Pencaitland; Gifford; Smeaton; (Bara
Wood and Thurston, 1969; Athelstaneford,
1973).

V. dioica L.   Marsh Valerian
Native. Status: rare.
Marshes, bogs, riversides.

WL: (Almondside, Linlithgow and Kirkliston,
pre-1934).
ML: rare, North Esk Reservoir, 1995; Longmuir
Moss, 2000; (Threipmuir, Dalkeith and Gogar,
pre-1934).
EL: rare, confined to upland areas, King's Inch;
Fennie Law; Danskine Loch; Papana Water;
Friardykes; Aikengall Water; (formerly coastal,
Luffness Links, 1906).

Centranthus Neck. ex Lam. & DC.

C. ruber (L.) DC.   Red Valerian
Introduced. Status: local, naturalising.
Walls, cliffs, roadsides, waste ground, railway
embankments.

WL: rare, Blackness; (Abercorn, 1970; Broxburn,
1973).
ML: rare, Roslin; Musselburgh; Edinburgh Castle
Rock; Salisbury Crags, 1998.
EL: scarce, in lowland areas, close to houses,
Athelstaneford; North Berwick seafront;
Belhaven; Dunbar; Thurston.

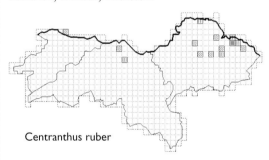

Centranthus ruber

DIPSACACEAE

Dipsacus L.

D. fullonum L.   Wild Teasel
Native. Status: local.
Waste ground, hedgerows, railway embankments,
dunes, bings.

WL: in N. only, scattered distribution.
ML: local, mainly in N.
EL: occasional in N., scattered distribution along or near the coast; inland at Tranent; Pencaitland; Haddington; East Linton.

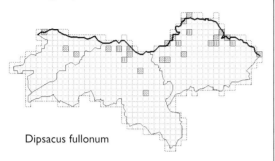

Dipsacus fullonum

D. sativus (L.) Honck. Fuller's Teasel
ML: Leith and Slateford, 1904 and 1908; Hailes Quarry tip, 1971.

D. pilosus L. Small Teasel
ML: Warriston, 1863; Duddingston, 1949 and 1953.

Cephalaria Schrad.

C. gigantea (Ledeb.) Bobrov Giant Scabious
Introduced. Status: rare.
Waste ground, roadsides, dunes.

WL: (Winchburgh, 1979).
ML: (Gorebridge, 1956; Buckstone sandpit, 1978).
EL: rare, path between Seafield Pond and Belhaven; (Gullane, 1957).

Knautia L.

K. arvensis (L.) Coult. Field Scabious
Native. Status: local.
Roadsides, railway embankments, rough grassland, cereal fields, dry banks.

WL: quite widespread, mainly in N., in small numbers.
ML: local, mainly in lowland areas, not locally frequent.
EL: occasional in lowland areas.

Succisa Haller

S. pratensis Moench Devil's-bit Scabious
Native. Status: widespread.
Rough grassland, roadsides, dry moorland, woodland, marshes, beside rivers and ditches, heath grassland.

WL: occasional in N., common in S.
ML: occasional in N., quite common elsewhere.
EL: widespread in S., frequent at a few sites elsewhere.

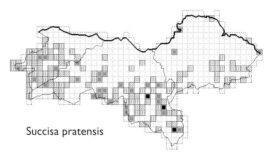

Succisa pratensis

Scabiosa L.

S. columbaria L. Small Scabious
Native. Status: rare.
Base-rich sea-cliffs, dry grassland.

WL: (Carriden, pre-1934).
ML: rare, Colinton, 1996.
EL: rare, North Berwick; Canty Bay; (Tantallon and Bass Rock, 1873).

ASTERACEAE

Echinops L.

E. exaltatus Schrad. Globe-thistle
Introduced. Status: rare, garden escape.
Roadsides, waste ground.

EL: local, Aberlady; old garden site near Spittal; well naturalised beside railway east of Drem.

E. bannaticus Rochel ex Schrad.
Blue Globe-thistle
Introduced. Status: rare, garden escape.
Rough grassland.

WL: rare, in an old garden at Overton.
ML: (Kinauld, 1973; Calton Hill, 1974; Portobello, 1979).
EL: (Gullane quarry, 1972).

E. sphaerocephalus L. Glandular Globe-thistle
WL: Carmelhill, 1972.

Carlina L.

C. vulgaris L. Carline Thistle
EL: Aikengall Dean, 1884; Sheeppath Dean, 1908, possibly the same site.

Arctium L.

A. minus agg.   Burdock
Native. Status: widespread.
Roadsides, woodland, waste ground, scrub, rough
grassland, cereal fields.

WL: widespread in NE., occasional in S.
ML: widespread in lowland areas, locally
frequent in E., occasional in upland areas.
EL: widespread and locally frequent in lowland
areas, occasional in upland areas.

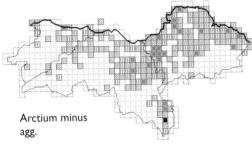

Arctium minus
agg.

Includes the following two taxa that were not
usually distinguished during the survey.

A minus (Hill) Bernh.   Lesser Burdock
Native. Status: rare.
Woodland, waste ground.

WL: (Bo'ness and Hopetoun, 1970s).
ML: rare, Granton; (Salisbury Crags, 1835).
EL: rarely recorded, but records are widely
scattered, Tranent; Gullane; Papana Water;
Tyninghame; Biel; (Skateraw, 1963; Bolton and
Dunglass, 1972; Yellowcraig, 1976).

A. nemorosum Lej.   Wood Burdock
Native. Status: the most common species of
Arctium in the Lothians, but not separated from
A. minus until recently, and not distinguished
from it in the survey.
Roadsides, woodland, waste ground, scrub, rough
grassland, cereal fields.

WL: probably widespread, Dyland Cotts;
(Bo'ness, 1970s).
ML: widespread in lowland areas, locally
frequent in E., occasional in upland areas.
EL: scattered distribution, mainly in lowland
areas.

Carduus L.

C. tenuiflorus Curtis   Slender or Seaside Thistle
Native. Status: local.

Coastal turf and scrub, waste ground, rough grassland,
roadsides, cliffs.

WL: rare, South Queensferry; (Hopetoun, 1978).
ML: scarce, mainly coastal, Cramond; locally
abundant on Inchmickery; Musselburgh; also
introduced inland, Borthwick; Canonmills.
EL: occasional along the coast, a few sites inland,
Ferny Ness; North Berwick; Whitekirk; Dunglass.

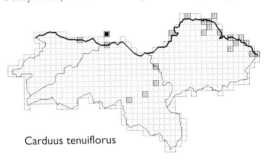

Carduus tenuiflorus

C. crispus L.   Welted Thistle
ssp. multiflorus (Gaudin) Gremli
Native. Status: local.
Waste ground, roadsides, scrub, rough grassland.

WL: widespread in N., rare in S.
ML: local, mainly in NE.
EL: quite widespread in N., rare elsewhere, often
only a single plant.

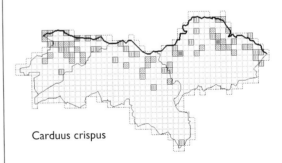

Carduus crispus

C. nutans L.   Musk Thistle
Native. Status: scarce.
Waste ground, roadsides, dunes, coastal turf.

WL: (Hopetoun, 1956).
ML: rare, Musselburgh; Cousland; near
Whitecraig; (Straiton, pre-1934; Duddingston,
1960; Crossgatehall, 1979).
EL: scarce, mainly along the NW coast from
Prestonpans to Yellowcraig; Keith; Drem;
Morham; (Dunbar, 1958; Catcraig, 1959).

C. x stangii H. Buek (*C. crispus* x *C. nutans*)
Native, spontaneous hybrid.
ML: found only once at Slateford quarry tip, 1907.
Presumably of spontaneous occurrence and probably the
only Scottish record.

Cirsium Mill.

C. vulgare (Savi) Ten.   Spear Thistle
Native. Status: very common.
Roadsides, rough grassland, waste and cultivated
ground, intensive grassland, riversides, scrub.

WL, ML, EL: very common.

C. heterophyllum (L.) Hill   Melancholy Thistle
Native, possibly also a garden escape. Status:
scarce.
Roadsides, waterside habitats, hill pastures.

WL: scarce, Bedlormie; Polkemmet Moor;
Armadale area; Breich Water; Philpstoun;
Stoneyburn; (Hopetoun, Kirkliston and
Livingston, pre-1934).
ML: scarce, Mid Calder; Gorebridge; Carlops;
Cockmuir; Breich Water; Howgate; Penicuik.
EL: rare and sporadic, Cuddie Wood; Aberlady
Bay; Woodhall Dean; (Drem and Longyester,
pre-1934; Sheeppath Dean, 1967; Gifford, 1972).

C. palustre (L.) Scop.   Marsh Thistle
Native. Status: common. Sometimes white-
flowered.
Marshes, beside rivers and ditches, wet parts of rough
grassland, woodland and roadsides.

WL, ML, EL: common in wet places, mainly in S.

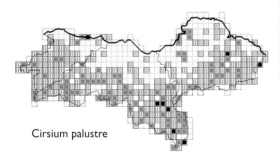

Cirsium palustre

C. arvense (L.) Scop.   Creeping Thistle
Native. Status: very common.
Roadsides, rough grassland, arable fields, intensive
grassland, waste ground, besides rivers and ditches,
mixed and coniferous woodland, scrub.

WL, ML, EL: very common.

Onopordum L.

O. acanthium L.   Cotton Thistle
Introduced. Status: rare, escape.
Waste ground, roadsides, railway embankments.

WL: rare, Kinneil, 1986; (South Queensferry,
1903; old garden, Linlithgow, 1960).
ML: rare, Sheriffhall, 1996; Duddingston, 1997;
(Leith, 1968).
EL: rare, sometimes persisting, Longniddry Bents;
Luffness/Gullane; West Barns/Dunbar;
(Prestonpans, 1976).

Silybum Adans.

S. marianum (L.) Gaertn.   Milk Thistle
Introduced. Status: rare, casual.
Waste ground.

WL: (Linlithgow, pre-1934).
ML: rare, Leith; (Ratho, 1973).
EL: rare, Luffness, 1983; two records from the
coast east of North Berwick, *c.* 1990; (Dunglass,
1875; North Berwick and Tantallon, pre-1934;
Auldhame, 1968).

Centaurea L.

C. scabiosa L.   Greater Knapweed
Native. Status: rare.
Railway embankment, cultivated ground, dunes, coastal
grassland.

WL: rare, north of Blackburn; (Linlithgow,
pre-1934).
ML: (Musselburgh, Howgate and Gogar,
pre-1934).
EL: scattered distribution along the coast,
Prestongrange; Ferny Ness; Kilspindie; Luffness;
Gullane; Yellowcraig; (Dunbar, 1963;
Thorntonloch, 1972).

C. montana L.   Perennial Cornflower
Introduced. Status: local, garden escape,
sometimes persisting.
Roadsides, waste ground.

WL: scattered records, usually a single plant.
ML: local, mainly in lowland areas.
EL: local, in lowland areas only, Elphinstone;
Spittal; Luffness; North Berwick; Innerwick
Castle.

C. cyanus L.   Cornflower
?Native/Introduced. Status: rare, old cornfield
weed, now usually a birdseed alien, a garden
escape, or a component of wildflower seed mixes.
Rubbish tips, cereal and other arable fields, roadsides.

WL: rare, recently sown but ephemeral at two
sites near Livingston.
ML: rare, Sighthill; Hermiston; Pilton; Gogar;
North Middleton.
EL: (nineteenth century records from Gullane,
Dirleton, Drem, Haddington, North Berwick and
Innerwick; potato field near Elphinstone, 1955;
rubbish tips at North Berwick, 1966, Gifford,
1970, and Pencaitland, 1973).

C. nigra L.   Common Knapweed, Hardheads
Native. Status: common.
Roadsides, rough grassland, riversides, waste ground,
scrub, railway embankments, woodland.

WL: very widespread, locally frequent. Rayed
forms at Drumshoreland, Dalmeny Village and
Livingston are suggestive of wildflower seed.
ML, EL: occasional in upland areas, common
elsewhere.

C. solstitialis L.   Yellow Star-thistle
ML: waste ground at Musselburgh, pre-1934; Borthwick,
1965.

C. diluta Aiton   Lesser Star-thistle
ML: waste ground and rubbish tips at Slateford and Leith
Docks, 1905 and 1909; Hailes, Levenhall and Granton,
1960s.

Mantisalca Cass.

M. salmantica (L.) Briq. & Cavill.
Rare grain/birdseed casual.
ML: Hailes rubbish tip, 1971.

Cichorium L.

C. intybus L.   Chicory
Introduced. Status: rare, casual, grown for pigs
in 1960s.
Roadsides, waste ground.

WL: (South Queensferry, pre-1934).
ML: rare, Musselburgh Lagoons, 1988; Temple,
1999; (Lauriston Castle and Cramond Bridge,
1967).
EL: rare, Prestonpans, 1984; Gullane rubbish tip,
1995; (Scoughall, 1969; Canty Bay, 1980; formerly quite widespread in lowland areas).

Lapsana L.

L. communis L.   Nipplewort
Native. Status: common.
Roadsides, waste ground, arable fields, woodland, scrub,
hedgerow.

WL: common in N., local in S.
ML, EL: widespread and locally frequent in
lowland areas, rare in upland areas.

Hypochaeris L.

H. radicata L.   Cat's-ear, Long-rooted Cat's-ear
Native. Status: widespread.
Roadsides, rough grassland, waste ground, scrub, bings.

WL: common.
ML, EL: quite widespread, somewhat patchy
distribution, locally frequent.

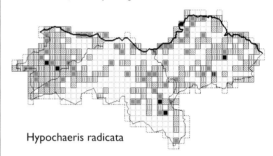

Hypochaeris radicata

Leontodon L.

L. autumnalis L.   Autumn Hawkbit
Native. Status: widespread, salt-tolerant,
frequent on some verges.
Roadsides, rough grassland, riversides, marshes, waste
ground, heath grassland, intensive grassland.

WL: widespread, locally frequent.
ML: widespread especially in upland areas,
locally frequent throughout, locally abundant in
SE.
EL: widespread especially in upland areas, seldom
locally frequent.

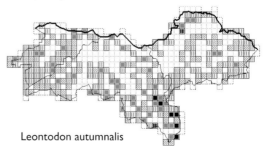

Leontodon autumnalis

var. **sordidus** Bab.
EL: beside Biel Water at West Barns.

L. hispidus L.   Rough Hawkbit
Native. Status: local.
Roadsides, rough grassland, mainly on calcareous soils, coastal grassland, disused railways.

WL: rare, old railway at Fauldhouse; Auchenard; Blackburn; a good colony, probably sown, on reclaimed bing at Bo'ness; (Stoneyburn and Almondell, 1972).
ML: local, mainly in S., Morton; (Stow and Heriot, 1967; Harburn, 1968; Addiewell, 1972; Cobbinshaw, 1977).
EL: local, Linn Dean; Pencaitland railway walk; Ferny Ness; North Berwick east cliff walk; Killmade Burn; Dunglass.

L. saxatilis Lam.   Lesser Hawkbit
Native. Status: scarce, introduced inland.
Coastal turf, dunes, waste ground.

WL: (Dalmeny, 1907).
ML: introduced at some sites; (Polton, 1883; Musselburgh, 1912; Levenhall, 1917; Leith, pre-1934).
EL: scarce, mainly along the coast, Cockenzie; Ormiston; Ferny Ness; Aberlady Bay; Yellowcraig; St Baldred's Cradle; West Barns; Dunbar; (Catcraig, 1962).

Picris L.

P. echioides L.   Bristly Oxtongue
Introduced. Status: rare.
Waste ground, disturbed ground.

ML: rare, garden weed, Corstorphine, 1995; (Leith and Murrayfield, pre-1934).
EL: rare, Prestongrange, where it is declining as the site becomes more grassy; (North Berwick, nineteenth century; Tynefield, 1864; East Linton, pre-1934; Scoughall, 1962).

P. hieracioides L.   Hawkweed Oxtongue
Introduced. Status: rare, casual.
Beside railway tracks, waste ground.

WL: (Carriden, 1972).
ML: (Leith Docks and Warriston, 1905).
EL: rare, Longniddry railway walk and nearby roadside; Drem; Kingston.

Scorzonera L.

S. hispanica L.   'Vegetable' Scorzonera
ML: grassland on Cramond Island, 1966.

Tragopogon L.

T. pratensis L.
          Goat's-beard, Jack-go-to-bed-at-noon
ssp. minor (Mill.) Wahlenb.
Native. Status: quite widespread.
Roadsides, waste ground, rough grassland.

WL: local.
ML: quite widespread in N. and E., rare elsewhere.
EL: quite widespread in lowland areas, rare or absent in upland areas.

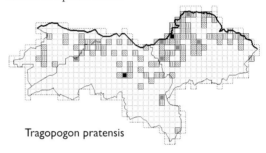

Tragopogon pratensis

ssp. pratensis
ML: Colinton, 1866; Hailes Quarry tip, 1902.
EL: Canty Bay, 1932; Cockenzie, Longniddry and Innerwick pre-1934.

T. porrifolius L.   Salsify
WL: railway siding in South Queensferry, 1907.
ML: Leith Docks, 1904.

Sonchus L.

S. arvensis L.
          Perennial Sow-thistle, Corn Sow-thistle
Native. Status: local.
Roadsides, waste ground, arable fields, beaches, rough grassland, dunes, coastal turf.

WL: quite widespread in N., locally frequent along the coast, occasional elsewhere.
ML: quite widespread in E., locally frequent in centre, rare elsewhere.
EL: quite widespread and locally frequent in lowland areas and along the coast, rare in upland areas.

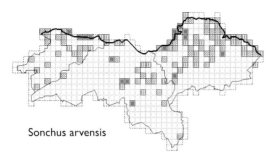

Sonchus arvensis

S. oleraceus L.   Smooth Sow-thistle
Native. Status: widespread.
Roadsides, waste ground, arable fields, walls, coastal habitats, gardens, hedgerows, farmyards.

WL, ML, EL: widespread and locally frequent in lowland areas, rare in upland areas.

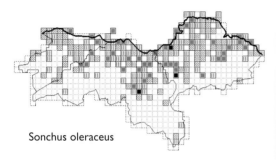

Sonchus oleraceus

S. asper (L.) Hill   Prickly Sow-thistle
Native. Status: widespread.
Roadsides, waste ground, arable fields, rough grassland, scrub, walls, railway embankments, farmyards.

WL: very widespread, seldom locally frequent.
ML, EL: widespread and locally frequent in lowland areas, occasional elsewhere.

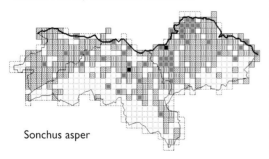

Sonchus asper

S. palustris L.   Marsh Sow-thistle
ML: Lochend, 1824; Murrayfield, pre-1934. The 1824 record was cited by Prof. Greville in his *Flora Edinensis* but was probably a wrong identification.

Lactuca L.

L. virosa L.   Great Lettuce
?Native. Status, rare.
Basalt crags, waste ground, railway tracks.

ML: rare, Leith Docks and Granton, 1997 to present; (Duddingston Loch, Lochend, Salisbury Crags and Leith, pre-1934).
EL: (Dunbar, pre-1894).

L. serriola L.   Prickly Lettuce
ML: Murieston, 1910.

Cicerbita Wallr.

C. macrophylla (Willd.) Wallr.
ssp. uralensis (Rouy) P. D. Sell
                    Common Blue Sow-thistle
Introduced. Status: local, garden escape, well-naturalised, spreading.
Roadsides, waste ground, hedgerows.

WL, ML: local, mainly in N.
EL: scarce, mainly in centre, north of Stobshiel; south of Gladsmuir; Drem; around Haddington; Whittingehame.

C. plumieri (L.) Kirschl.
                    Hairless Blue Sow-thistle
Introduced. Status: rare.
Woodland.

ML: rare, Corstorphine Hill, 1996.

Mycelis Cass.

M. muralis (L.) Dumort.   Wall Lettuce
Native. Status: local.
Damp woodland, walls, roadsides.

WL: local, in N., locally frequent on walls east of South Queensferry.
ML: local, mainly in N. and E.
EL: scarce, mainly in lowland areas, Gosford; Aberlady; Luffness; Yellowcraig; Donolly Reservoir; Woodhall Dean.

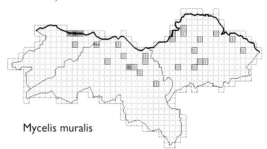

Mycelis muralis

Taraxacum F. H. Wigg.

T. officinale agg.   Dandelion

Native. Status: common.

Roadsides, waste ground, rough grassland, woodland, public parks, riversides, scrub, arable fields.

WL, ML, EL: common.

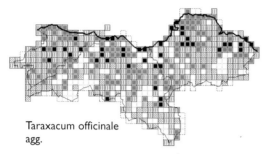

Taraxacum officinale agg.

Taraxacum (Dandelions) is a critical genus for several reasons. The characteristics of individual taxa vary considerably with the season and the environment, and dandelions are largely apomictic, producing clonal microspecies. In this survey, dandelions were mainly recorded under the catch-all name of *Taraxacum officinale* agg., which is almost a folk taxon, such is the frequency and confidence with which it is used in identification guides. In the most recent monograph on British material, Dudman and Richards (1997) point out that *T. officinale* Weber is a synonym of *T. campylodes* G. E. Haglund, a species from Swedish Lapland, not known from Great Britain and Ireland. Most common, wayside dandelions are referable to the *T. officinale* Wigg. group, here treated as in Stace (1997) as *Taraxacum* sect. *Ruderalia*. Other dandelion taxa are treated under the appropriate section. The general treatment followed in relation to names and authorities is that of Dudman and Richards (1997) except that within sections, the taxa have been put in alphabetical order for easy access as in Kent (1992). The records published here have been authenticated by national experts but, as these records represent only a small fraction of the dandelion populations in the Lothians, little can be deduced about the distribution of individual taxa. Dandelions are most surely identified during their main flowering and fruiting period, which in the Lothians is usually the first three weeks of May.

The following species have been recorded.

sect. **Erythrosperma** (Lindb. f.) Dahlst.

T. argutum Dahlst.
EL: dunes, Longniddry Bents, 1983; roadside verge, Waughton, 1991.

T. brachyglossum (Dahlst.) Raunk.
ML: heath, Arthur's Seat, 1971.
EL: dry grassland, Prestonpans, 1978; Longniddry Bents, 1983; Garleton Hills quarry, 1982; North Berwick golf course, 1974.

T. fulviforme Dahlst.
ML: Arthur's Seat, 1979; Penicuik Estate, 1989.

T. haworthianum Dudman & A. J. Richards
EL: Port Seton, 1973; Garleton Hills, 1982; Yellowcraig, 1982.

T. lacistophyllum (Dahlst.) Raunk.
ML: Blackford Quarry, 1971; Musselburgh, 1975.
EL: in dune turf, Prestonpans, 1973; Longniddry Bents, 1992; Aberlady Bay, 1998; North Berwick, 1983; on volcanic outcrops, Waughton, 1991; Hailes Castle, 1995.

T. oxoniense Dahlst.
ML: Blackford Quarry, 1973.
EL: Gullane Bents, 1985; Dirleton dunes, 1970; bank above stream, Sheeppath Burn, 1987.

T. proximiforme Soest
EL: Garleton Hills quarry, 1982.

T. rubicundum (Dahlst.) Dahlst.
EL: Dirleton dunes, 1970.

T. scoticum A. J. Richards
ML: Arthur's Seat, 1979.

sect. **Obliqua** (Dahlst.) Dahlst.

T. obliquum (Fr.) Dahlst.
EL: dune turf, north of Gullane, 1982; Yellowcraig, 1982; North Berwick golf course, 1974.

T. platyglossum Raunk.
EL: dune turf, Yellowcraig, 1982; North Berwick, 1912 and 1904.

sect. **Palustria** (Lindeb. f.) Dahlst.

T. palustre (Lyons) Symons
EL: Molat Hill, 1932. Site not located, possibly now lost. Dudman and Richards (1997) map it for the Haddington area.

sect. **Spectabilia** (Dahlst.) Dahlst.

T. faeroense (Dahlst.) Dahlst.
ML: quite common in wet hill pastures.
EL: White Castle, 1993; Dunbar Common, 1993; Sheeppath Burn, 1994.

sect. **Naevosa** M. P. Christ.

T. euryphyllum (Dahlst.) Hjelt
ML: heath, Linhouse Glen, 1995.
EL: roadsides, Garleton Hills, 1982; Markle Mains, 1994; verge of wooded lane, Spott, 1999; steep scrub-covered bank, Woodhall Burn, 1988.

T. maculosum A. J. Richards
WL: Hopetoun, Dalmeny.
ML: damp turf and wet rock faces, Linhouse Glen; Moorfoot Hills; Dreghorn; Carberry Tower.
EL: marshy places, North Berwick, 1974; Woodhall Dean, 1986, 1987, 1988.

T. pseudolarssonii A. J. Richards
ML: verges, Dalkeith, 1970.
EL: coastal habitats and roadsides, Longniddry Bents, 1983; Stoneypath, 1988; Elmscleugh, 1994; Bilsdean, 1989.

T. subnaevosum A. J. Richards
ML: hill pasture, Dewar Hill, 1984.
EL: waste ground, Prestonpans, 1979.

sect. **Celtica** A. J. Richards

T. bracteatum Dahlst.
WL: Dalmeny; South Queensferry, 1983.
ML: Barnton, 1986; Duddingston, 1996; Musselburgh, 1975.

T. britannicum Dahlst.
ML: canal bank, Ratho, 1978.

T. duplidentifrons Dahlst.
Mainly roadsides, also woodland.
WL: Linlithgow, 1987; South Queensferry, 1983.
ML: Roslin, 1897; Stow, 1987; Granton 1973; Inveresk, 1973.
EL: occasional/relatively common, Saltoun, 1983; St Germains, 1986; Markle Mains, 1994; Bass

Rock, 1985; Woodhall Dean, 1986; Oldhamstocks, 1981.

T. excellens Dahlst.
ML: Inveresk, 1992.

T. gelertii Raunk.
ML: East Calder, 1987; Stow, 1984; Musselburgh, 1992.
EL: verge of wooded lane, Spott, 1999.

T. landmarkii Dahlst.
WL: one record.
ML: Crosswood Burn, 1976.

T. nordstedtii Dahlst.
ML: Stow, 1984, 1986.
EL: mainly coastal, Cockenzie, 1992; Hanging Rocks north of Gullane, 1982; St Baldred's Cradle, 1979; Sandy Hirst, 1987; mixed woodland and roadside, Elmscleugh, 1994.

T. palustrisquameum A. J. Richards
EL: wet calcareous grassland, Aberlady Bay, 1994.

T. subbracteatum A. J. Richards
ML: Temple, 1997.

T. unguilobum Dahlst.
ML: occasional in dampish turf, Musselburgh, 1982; Stow, 1987.
EL: Prestonpans, 1978; Hanging Rocks north of Gullane, 1982; Garleton Hills, 1982; Whitekirk Hill, 1992; Woodhall Dean, 1987; Dunglass, 1989.

sect. **Hamata** H. Oellgaard

T. boekmanii Borgv.
Pastures, roadsides, waste ground.
WL: Wallhouse, 1984.
ML: Granton, 1979.
EL: Prestonpans, 1979; Tranent, 1999; St Germains, 1986.

T. hamatiforme Dahlst.
WL: in NE., 1980; South Queensferry, 1983.
ML: pastures, roadsides, waste ground, Inveresk, 1975; Hermitage of Braid, 1976.
EL: coastal waste ground and dunes, Prestonpans, 1979; Seton Sands, 1973; Longniddry Bents, 1983.

T. hamatum Raunk.
ML: in lawns and on verges, Edinburgh, 1976;
Penicuik, 1970.
EL: roadside verge, Ballencrieff, 1991;
Haddington area, 1982.

T. hamiferum Dahlst.
ML: verge, Musselburgh, 1992.
EL: verge of wooded lane, Spott, 1999.

T. marklundii Palmgr.
ML: refuse tips, Slateford and Blackford quarries,
1971.

T. prionum Hagend., Soest & Zevenb.
ML: verge, Carberry Tower, 1994.

T. pseudohamatum Dahlst.
ML: verges, Cammo, 1986; Stow, 1987;
Duddingston, 1996.
EL: roadsides and waste ground, Tranent, 1988;
Pencaitland, 1996; Ballencreiff crossroads, 1991.

T. quadrans H. Oellgaard
WL: South Queensferry, 1983.

T. sahlinianum Dudman & A. J. Richards
EL: North Berwick area (Dudman and Richards
1997).

T. subhamatum M. P. Christ.
ML: verges, East Calder, 1987; Arthur's Seat,
1990; Cramond, 1992; Musselburgh, 1992.
EL: roadside verges, Tranent, 1988; Waughton,
1991.

sect. **Ruderalia** Kirschner, H. Oellgaard &
Stepanek

T. aequilobum Dahlst.
EL: Keith area (Dudman and Richards 1997).

T. alatum H. Lindb.
ML: waste ground, roadsides, Borthwick, 1970;
Mortonhall, 1990.
EL: Prestonpans, 1978.

T. angustisquameum Dahlst. ex H. Lindb.
EL: Prestonpans, 1979.

T. cophocentrum Dahlst.
ML: waste ground, roadsides, Inveresk, 1975.
EL: rough grassland near shore, Gosford, 1978.

T. cordatum Palmgr.
ML: verges and waste ground, North Middleton,
1970; Leith Docks, 1971; Musselburgh, 1975;
Lasswade, 1976.

T. croceiflorum Dahlst.
WL: South Queensferry.
ML: riverside, Currie, 1982.
EL: dune grassland, Seton Sands, 1973; roadside
near Humbie, 1996.

T. cyanolepis Dahlst.
WL: South Queensferry, 1983.
EL: lush, grassy roadsides, Tranent, 1988;
Ormiston, 1992; Ballencreiff, 1991;
Athelstaneford, 1981.

T. dahlstedtii H. Lindb.
WL: in NE., 1982.
ML: seven widespread records from waste
ground and roadsides.
EL: wasteground, Prestonpans, 1979; dune,
Longniddry Bents, 1983; Gosford, 1979;
roadside, Pencaitland, 1996; North Berwick,
1980.

T. ekmanii Dahlst.
ML: Blackford quarry, 1971; Churchill, 1975;
Arthur's Seat, 1980.
EL: St Germains, 1986; roadside, Pencaitland,
1996.

T. exacutum Markl.
Probably introduced.
Waste ground, roadsides, stony places, disturbed
dunes, farmyard.

WL: Bo'ness; South Queensferry.
ML: five records in Edinburgh, 1990–6; Linhouse
Water, 1995.
EL: Cockenzie, 1992; Tranent, 1988; Longniddry
Bents, 1983; East Linton, 1995.

T. expallidiforme Dahlst.
Waste ground, roadsides.

WL: South Queensferry, 1983.
ML: North Middleton, Dreghorn and Borthwick,
1970; Blackford Hill, 1971; Granton, 1979.
EL: Prestongrange, 1985; Gifford, 1970.

T. exsertum Hagend., Soest & Zevenb.
ML: waste ground, roadsides, Leith Docks, 1979.
EL: waste ground, Prestonpans, 1979.

T. fagerstroemii Såltin
EL: waste ground, Tranent, 1988.

T. fasciatum Dahlst.
ML: waste ground, roadsides, Mid Calder and Mortonhall, 1970.
EL: roadside verges, Dunbar and Dunglass, 1970.

T. huelphersianum G. E. Haglund
ML: Leith Docks, 1996.

T. incisum H. Oellgaard
ML: railway bank, Trinity, 1999.

T. insigne Ekman ex M. P. Christ. & Wiinst.
WL: in NE., 1982.

T. interveniens G. E. Haglund
ML: waste ground, Granton, 1979.

T. laeticolor Dahlst.
EL: wall at roadside, Saltoun, 1999.

T. laticordatum Markl.
WL: in NE., 1980; South Queensferry, 1983.
ML: riversides, Penicuik, 1989; Cramond, 1992.
EL: roadsides and waste ground, Tranent, 1988; Pencaitland, 1996.

T. lepidum M. P. Christ.
EL: roadside verge, Tranent, 1988.

T. lingulatum Markl.
ML: waste ground, roadsides, Penicuik, 1970; Musselburgh, 1976; Stow, 1984.
EL: Garleton Hills quarry, 1981; roadside, Gifford, 1970; West Barns, 1981.

T. longisquameum H. Lindb.
ML: waste ground, Granton, 1979.

T. mimulum Dahlst. ex H. Lindb.
ML: ruderal, Leith Docks, 1996.

T. obliquilobum Dahlst.
EL: waste ground near shore, Prestonpans, 1971; golf course, North Berwick, 1974.

T. oblongatum Dahlst.
ML: canal bank, Hermiston, 1998.

T. obtusilobum Dahlst. ex G. E. Haglund
ML: grassy bank, Dunsapie Loch, 1979.

T. pachymerum G. E. Haglund
EL: roadside, Prestongrange, 1985.

T. pannucium Dahlst.
Waste ground, roadsides.
WL: one record, 1980.
EL: Prestonpans, 1978; Haddington and Garvald, 1970.

T. pectinatiforme H. Lindb.
WL: in NE., 1980.
EL: rough ground near shore, Prestonpans, 1978.

T. polyodon Dahlst.
One of the commonest waste ground microspecies, also found on roadsides, coastal grassland and rubbish tips.
WL: Dalmeny; South Queensferry, 1983.
ML: Mortonhall, 1970; Slateford rubbish tip, 1971; Granton, 1979; Leith, 1981; Stow, 1984; Gogar, 1985.
EL: Tranent, 1988, 1999; Pencaitland, 1996; Ballencrieff, 1991; Yellowcraig, 1994; Hailes, 1995; Woodhall ford, 1987.

T. sagittipotens Dahlst. & R. Ohlsen ex G. E. Haglund
ML: ruderal, Leith Docks, 1996.

T. sellandii Dahlst.
WL: in NE., 1982.
ML: riverside, Cramond, 1973.
EL: Drem, 1981; broad-leaved woodland, Tyninghame Estate, 1987.

T. subcyanolepis M. P. Christ.
EL: dunes, Longniddry Bents, 1983.

T. subexpallidum Dahlst.
ML: Musselburgh, 1975.
EL: Prestongrange, 1978.

T. subundulatum Dahlst.
ML: river bank, Inveresk, 1973. First Scottish record.

T. tenebricans (Dahlst.) Dahlst.
ML: roadsides, Stow, 1970.

T. vastisectum Markl. ex Puol.
EL: verge of wooded lane, Spott, 1999.

T. xanthostigma H. Lindb.
ML: canal bank, Slateford, 1997.

Crepis L.

C. paludosa (L.) Moench   Marsh Hawk's-beard
Native. Status: local.
Riversides, marshes, wet woodlands, ravines.

WL: local, mainly in SW.
ML: scattered throughout the southern half of the vice-county.
EL: local, mainly in the Keith/Humbie area and the Lammermuir Hills; White Castle; Woodhall Dean; Aikengall Water.

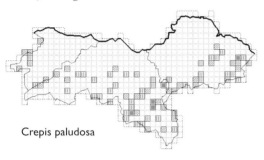

Crepis paludosa

C. capillaris (L.) Wallr.   Smooth Hawk's-beard
Native. Status: local.
Roadsides, waste ground, rough grassland, intensive grassland, public parks, coastal habitats, railway embankments.

WL: local, mainly in NE., locally frequent, Bo'ness; Port Edgar.
ML: local, mainly in lowland areas in E., locally frequent.
EL: local, mainly in lowland areas, locally frequent along the coast.

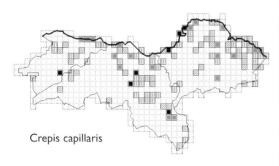

Crepis capillaris

C. mollis (Jacq.) Asch.   Northern Hawk's-beard
ML: Habbie's Howe, Carlops, 1847.

[**C. biennis** L.   Rough Hawk's-beard
EL: an introduction recorded pre-1934 from Aberlady, Luffness, Drem, North Berwick and Tyninghame, but not confirmed by specimens. Confusion with tall glandular plants of *C. capillaris* seems likely.]

C. setosa Haller f.   Bristly Hawk's-beard
Casual.
ML: Leith Docks, 1979.
EL: near Drem, 1847.

Pilosella Hill

P. officinarum F. W. Schultz. & Sch. Bip.
      Mouse-ear Hawkweed
Native. Status: widespread.
Short, rough grassland, roadsides, railway embankments, heath grassland, waste ground, rocky outcrops, moorland, scrub, bings, dunes.

WL: widespread, locally frequent.
ML: widespread and locally frequent in SE., local and locally frequent elsewhere.
EL: scattered distribution throughout the vice-county, locally frequent.

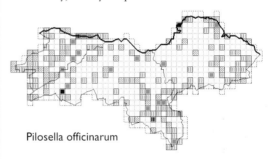

Pilosella officinarum

The following variants occur. These are sometimes recognised as subspecies, but are considered by Stace (1997) to be varieties.
ssp. **euronota** ML, EL.
ssp. **melanops** ML, EL.
ssp. **micradenia** ML, EL.
ssp. **officinarum** ML.
ssp. **tricholepia** EL.
ssp. **trichosoma** widespread, WL, ML, EL.

P. flagellaris (Willd.) P. D. Sell & C. West
      Shetland Mouse-ear Hawkweed
ssp. **flagellaris**
Introduced. Status: local.
Railway embankments, waste ground, dunes, bings.

WL: local, Kirkliston; (Dalmeny Station, 1954; Fauldhouse, 1976).

ML: local, Danderhall; Niddrie; Turnhouse;
(Canonmills, 1876; Seafield, 1967; Musselburgh,
1977).
EL: scarce, Prestongrange; Seton Sands; New
Town; Luffness quarry; Gullane; Garleton area;
Gifford; (Longniddry, 1895; Seton Halt, 1958;
Saltoun, 1963; Pencaitland, 1973).

First recorded in Britain from Granton (as
Hieracium stoloniferum) by Prof. J. H. Balfour in
1869.

P. praealta (Vill. ex Gochnat) F. W. Schultz. &
Sch. Bip.   Tall Mouse-ear Hawkweed
ssp. thaumasia (Peter) P. D. Sell
Introduced. Status: rare.
Reclaimed bing.

WL: rare, Bo'ness.

P. caespitosa (Dumort.) P. D. Sell & C. West
                              Yellow Fox-and-Cubs
ssp. colliniformis (Peter) P. D. Sell & C. West
Introduced. Status: rare, escape.
Waste ground.

ML: rare, Cockpen.
EL: (railway near Longniddry, pre-1968).

P. aurantiaca (L.) F. W. Schultz. & Sch. Bip.
                                      Fox-and-Cubs
Introduced. Status: rare, escape.
Waste ground, garden rubbish tips, railway
embankments.

WL: local.
EL: rare, Fountainhall; Aberlady Bay; Innerwick;
(Saltoun Station, 1969; Gullane Station, 1973).

ssp. aurantiaca
Introduced. Status: rare, escape.
Waste ground.

EL: rarely recorded, sea front North Berwick;
(Ormiston, 1877).

ssp. carpathicola (Nägeli & Peter) Soják
Introduced. Status: local, escape.
Waste ground, garden rubbish tips, railway
embankments.

WL: local.
ML: widespread, Granton; Morningside;
Gorebridge; (Borthwick, 1847; Balerno, 1954;
Davidson's Mains, 1962.
EL: rarely recorded, Hummell Rocks; North
Berwick Law; Clerkington; Gifford.

Hieracium L.

Hieracium spp.   Hawkweeds
In this genus, as in *Taraxacum*, apomixis occurs
producing many clonal microspecies.
Microspecies for which there are records have
been treated as in Kent (1992 and Supplement 1,
1996).

sect. **Sabauda** (Fries) Gremli

H. sabaudum L.
Native.
Roadsides and tracks, old railways, rough grassland.

WL: occasional, about a dozen records from
woodland and roadside verges.
ML: occasional in SE.
EL: rare, Prestongrange; Ormiston; Saltoun; New
Winton; Longniddry; Gladsmuir; Spott;
Broxmouth.

H. virgultorum Jordan
Introduced. Status: naturalised and spreading.
Roadsides, woodland, dunes, bings.

WL: Bathgate, 1983; Seafield, 1983; Greendykes
Bing, 1989.
ML: five records, including Leith, 1910.
EL: rare, Prestongrange; New Town; Pencaitland;
Hummell Rocks.

H. rigens Jordan
Native.

WL: Broxburn.
ML: Warriston; Holyrood Park; Calton Hill.
EL: rare, beside railway east of Drem, 1980.

H. salticola (Sudre) Sell & C. West
WL: Birkhill Mine, 1988; Faucheldean Bing;
Blackburn.
ML: Arthur's Seat; Musselburgh.

H. vagum Jordan
WL: Windyknowe, Bathgate, 1983.

sect. **Hieracioides** Dumort.

H. umbellatum L.
Native.

WL: Hound Point, Dalmeny.
ML: Figgate Burn.
EL: rare near the coast, Bass Rock, pre-1968;
Dunbar, pre-1934.

sect. **Foliosa** (Fries) Dahlst.

H. subumbellatiforme (Zahn) Roffey
WL: Carribber, 1970; Bridge Castle, south of
Westfield, 1989.

H. reticulatum Lindeb.
WL: by the River Avon, 1976.
ML: river-bank, Roslin.

H. strictiforme (Zahn) Roffey
WL: one record, 1938.

H. drummondii Pugsley
WL: River Avon at Linlithgow.

H. latobrigorum (Zahn) Roffey
WL: Drumshoreland Moor, 1955; Faucheldean
Bing; Rousland Farm, south of Kinneil, 1976.
ML: woodland, Roslin.

sect. **Tridentata** (Fries) Gremli

H. calcaricola (F. Hanb.) Roffey
WL: Windyknowe, Bathgate, 1983.
ML: Arthur's Seat.

H. scabrisetum (Zahn) Roffey
ML: verge, Blackford Glen.

H. lissolepium (Zahn) Roffey
ML: railway bank, Leith.

sect. **Prenanthoidea** Koch

H. prenanthoides Villars
ML: Moorfoot and Pentland Hills; Auchendinny.

sect. **Vulgata** (Griseb.) Willk. & Lange

H. cravioniense (F. Hanb.) Roffey
Native.

EL: rare, on rocks at Gullane Point, 1953.

H. vulgatum Fries
Native. Status: widespread and fairly common.
Walls/rocks, rough grassland, woodland, old railways,
dunes, waste ground, bings.

WL: widespread, the commonest microspecies.
ML: common.
EL: quite widespread and relatively common,
Prestongrange; Saltoun; Hummell Rocks; Hopes
Reservoir; Sheeppath Burn; Monynut Water.

var. **sejunctum** W. R. Linton
ML: Balerno, 1954.

H. rubiginosum F. Hanb.
ML: on rocks, Carlops; Granton; Burdiehouse;
Loganlea.

H. maculatum Smith
ML: walls, Dalhousie Castle.

sect. **Hieracium**

H. oistophyllum Pugsley
Native.

ML: on rocks, Carlops.
EL: rare, on conglomerate rocks in oakwood,
Woodhall Dean, 1987.

H. auratiflorum Pugsley
ML: Holyrood Park.

H. duriceps F. Hanb.
Native.

EL: rare, shady rocks beside stream, Sheeppath
Burn, 1987.

H. grandidens Dahlst.
Introduced.

ML: Borthwick; Cramond.
EL: rare on dunes north of Gullane, 1991 and
1999.

H. exotericum agg.
WL: railway viaduct at Linlithgow, 1988.
ML: on rocks, Carberry; Moorfoot Hills.
Earlier (1970s) records of '*H. exotericum* agg.'
may refer to *H. grandidens* or to other,
unrecognised, species.

sect. **Oreadea** (Fries) Dahlst

H. subrude (Arv.-Touv.) Arv.-Touv.
Native.

ML: on basalt rocks, Blackford and Dalmahoy
Hills; Arthur's Seat.
EL: rare on volcanic outcrops, south of North
Berwick, 1982.

H. caledonicum F. Hanb.
Native.

ML: on rocks, Arthur's Seat; Loganlea; Carlops.

EL: rare on rocky outcrops, south of North
Berwick, 1982; Tantallon, pre-1968.

H. argenteum Fries
ML: on rocks, Arthur's Seat.

H. dicella Sell & C. West
ML: on rocks, Arthur's Seat.

sect. **Cerinthoidea** Aug. Monnier

H. anglicum Fries
ML: near Ratho, 1870.
WL: Cockleroy.

Filago L.

F. vulgaris Lam.   Common Cudweed
Native. Status: rare, casual.
Rocky outcrops, waste ground.

WL: (Linlithgow, pre-1934).
ML: rare, Duddingston, 1998; (Arthur's Seat,
Lasswade and Slateford, pre-1934).
EL: rare and sporadic, Luffness quarry; outcrop
near Whitekirk; (path to Dunglass Dean, 1972;
old records from a few other sites).

F. minima (Sm.) Pers.   Small Cudweed
Native. Status: rare.
Sandy soils, rocky outcrops.

WL: rare, locally frequent on waste ground,
Bo'ness, and in an old sand-pit north of Couston.
ML: (Arthur's Seat, Dalmahoy and Blackford
Hill, pre-1934; Fairmilehead, 1956).
EL: rare and sporadic, Luffness quarry; Gullane
Hill; Hedderwick Hill; (old records from several
coastal sites).

Antennaria Gaertn.

A. dioica (L.) Gaertn.   Mountain Everlasting
Native. Status: rare.
Moorland, heath grassland.

WL: (Cockleroy, pre-1934).
ML: rare, Arthur's Seat; west of Kipps; south of
Gladhouse; Fairliehope; (Carlops, 1976).
EL: (Gullane area and Gifford area, some time
between 1950 and 1962).

Anaphalis DC.

A. margaritacea (L.) Benth.   Pearly Everlasting

Introduced. Status: rare, garden escape.
Waste ground.

WL: (formed 'a large colony' at Kinneil, 1957).
ML: rare, Leith Docks, 1904 and 1996.

Gnaphalium L.

G. sylvaticum L.   Heath Cudweed
Native. Status: scarce.
Watersides, waste ground, scrub, roadsides, railway
tracks, forestry tracks.

WL: rare, single plants on a bing by Drumtasie
Burn; north of Blackridge; north-west of
Torphichen; (Carribber, 1960).
ML: scarce, Bilston; Penicuik Estate; South
Melville; West Calder; Tynehead; Cobbinshaw.
EL: rare and sporadic, Gullane Hill; Yellowcraig;
Binning Wood; St Baldred's Cradle; Aikengall;
(several inland records, pre-1962).

G. uliginosum L.   Marsh or Wayside Cudweed
Native. Status: quite widespread.
Roadsides, waste ground, pastures, arable fields,
marshes, watersides, muddy places such as ruts in
paths.

WL: widespread in W. and S., occasional else-
where.
ML: quite widespread, mainly in E. and SW.
EL: local in Lammermuir Hills, rare elsewhere.

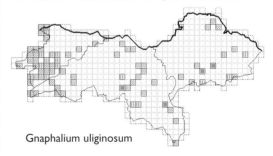

Gnaphalium uliginosum

G. luteoalbum L.   Jersey Cudweed
Introduced. Status: rare, casual, horticultural
weed.
Gardens.

ML: rare, Royal Botanic Garden Edinburgh,
1991.

Inula L.

I. helenium L.   Elecampane
Introduced. Status: rare, garden escape.
Watersides, roadsides, waste ground, railway
embankments, woodland.

WL: rare, established at South Queensferry, first recorded there in 1962.
ML: rare, Bawsinch, 1993; a large clump on Corstorphine Hill, 1999; (Vogrie, 1955).
EL: rare, Pencaitland; Hailes; (East Linton, 1973).

I. conyzae (Griess.) Meikle
Ploughman's Spikenard
Introduced. Status: rare, casual, possibly introduced with birdseed.
Waste ground, gardens.

ML: rare, garden weed, Barnton, 2000.
EL: (Dirleton, 1961).

Pulicaria Gaertn.

P. dysenterica (L.) Bernh.   Common Fleabane
Native. Status: rare.
Marsh, beside ponds and streams, dune scrub.

WL: (Bo'ness, pre-1934).
EL: rare, Longniddry Bents; Gosford House; Aberlady Bay where it has been known since 1908; (Gosford Links, 1849; Seton Sands, pre-1934).

Telekia Baumg.

T. speciosa (Schreb.) Baumg.   Yellow Ox-eye
Introduced. Status: rare, garden escape.
Coastal scrub, beside roads and ponds.

WL: rare, on derelict ground, Port Edgar, 1992.
ML: rare, Auchendinny; Oxenfoord; garden weed at Stow, 1986.
EL: rare, Haddington; Bara Loch.

Solidago L.

S. virgaurea L.   Goldenrod
Native. Status: scarce.
Heath grassland, woodland, ravines, rocks, riversides.

WL: rare, Carribber; Fauldhouse; Dalmeny near Snab Point; (Kinneil, 1971).
ML: scarce, scattered distribution, Moorfoot and Pentland Hills; Roslin Glen; Arniston; Middleton; River Almond; Torphin Quarry; The Howe.
EL: rare, mainly in ravines in the Lammermuir Hills; Priest Law; Killmade Burn; Woodhall Dean.

S. canadensis L.   Canadian Goldenrod
Introduced. Status: rare, garden escape, possibly confused with *S. gigantea*.
Waste ground, riversides, hedgerows, rough grassland.

WL: (Linlithgow, 1977; also some older records that may refer to *S. gigantea*).
ML: rare, disused railway cycle track at Trinity, 1999; beside Water of Leith near Gallery of Modern Art.
EL: rare, Haddington; West Barns; (Pencaitland, 1956).

S. gigantea Aiton   Early Goldenrod
Introduced. Status: rare, garden escape.
Roadsides, railway embankments, waste ground.

WL: widely scattered sites.
ML: rare, Fillyside, 2000; (Middleton, 1972; Ratho and Penicuik, 1978; Cramond Island, 1980).
EL: rare, Prestonpans; Longniddry; Gullane; Belhaven; (Haddington, 1971; Whitekirk, 1973).

Aster L.

A. novi-belgii agg.   Michaelmas Daisy
Introduced. Status: local, garden escape.
Waste ground, roadsides, railway embankments, canalsides, walls, dunes, beaches.

WL: scattered distribution.
ML, EL: local, mainly in lowland areas.

Includes the following six taxa that were not always distinguished in the survey.

A. novae-angliae L.   Hairy Michaelmas Daisy
Introduced. Status: rare.
Waste ground, walls.

ML: rare, Monktonhall Colliery, 1994; (Blackford Quarry, 1980).

A. x versicolor Willd. (*A. laevis* x *A. novi-belgii*)
Late Michaelmas Daisy
Introduced. Status: rare.
Roadsides, railway embankments, waste ground.

ML: occasional, Warriston; Murrayfield; Drylaw.

A. novi-belgii L.   Confused Michaelmas Daisy
Introduced. Status: rare.
Waste ground, dunes.

WL: rare, Uphall Station.
ML: scattered distribution, Loanhead; Warriston; (Penicuik Estate; West Calder; Musselburgh; Balerno; Leith).
EL: rarely recorded, Aberlady Bay; north of Drem; (Longniddry, 1968; Innerwick, 1970).

A. x salignus Willd. (*A. novi-belgii* x *A. lanceolatus*)   Common Michaelmas Daisy
Introduced. Status: local, garden escape.
Waste ground.

WL: scattered distribution.

A. lanceolatus Willd.
                    Narrow-leaved Michaelmas Daisy
Introduced. Status: rare.
Waste ground.

ML: (Slateford, 1919; Ratho, 1976).
EL: rarely recorded, Aberlady Bay.

A. laevis L.   Glaucous Michaelmas Daisy
ML: Levenhall rubbish tip, 1911; Balerno, 1945.
EL: shore at North Berwick, 1907.

A. tripolium L.   Sea Aster
Native. Status: local.
Saltmarsh, beaches, cliffs, sea walls, coastal turf.

WL: rare, Bo'ness; Blackness.
ML: (Leith Docks, 1908).
EL: local, locally frequent, Seton Sands; Aberlady Bay; Tyninghame Bay; Belhaven Bay; Dunbar; Catcraig area; (Prestongrange, pre-1871).

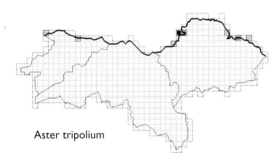

Aster tripolium

Erigeron L.

E. karvinskianus DC.   Mexican Fleabane
Introduced. Status: rare.
Walls.

EL: rare, recorded in 1967 as established and spreading on a wall on the Biel Estate.

E. acer L.   Blue Fleabane
Introduced. Status: rare, spreading.
Dunes, stony places.

ML: rare, on a cobbled street at Granton, 1998.
EL: rare, known at Aberlady Bay since 1998; at Yellowcraig since 1960, and at Belhaven Bay since 1993.

Conyza Less.

C. canadensis (L.) Cronquist   Canadian Fleabane
Introduced. Status: rare.
Waste ground.

ML: rare, one huge colony at Leith Docks, 1989 to present; car park at Roslin Glen Country Park, 1999; (Leith Docks, 1904-1905).

Bellis L.

B. perennis L.   Daisy
Native. Status: common.
Roadsides, short grassland, waste ground, public parks, riversides, gardens.

WL, ML, EL: common.

Tanacetum L.

T. parthenium (L.) Sch. Bip.   Feverfew
Introduced. Status: widespread.
Roadsides, waste ground, walls, gardens.

WL, ML, EL: widespread, mainly in lowland areas.

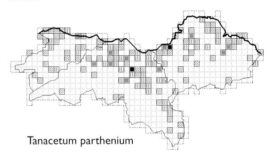

Tanacetum parthenium

T. macrophyllum (Waldst. & Kit.) Sch. Bip.
                    Rayed Tansy
Introduced. Status: rare.
Deciduous woodland.

WL: rare, Bedlormie; Blackness; Hopetoun; Port Edgar; South Queensferry; (Pepper Wood, 1960).
ML: rare, Cramond Bridge; Juniper Green; Blackford Glen, where it was planted about 10 years ago.

T. vulgare L.   Tansy
Native. Status: local.
Waste ground, roadsides, riversides, coastal habitats.

WL: widespread, not locally frequent.
ML: local, mainly in urban areas and river valleys.

EL: occasional in lowland areas, Cockenzie; Longniddry Bents; Dirleton; Haddington; West Barns.

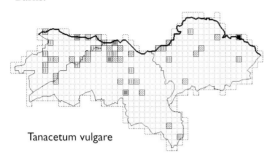

Tanacetum vulgare

T. balsamita L.   Costmary
ML: Slateford Quarry, 1904.
EL: Longniddry, 1942 and 1966–70, lost when the coast road was realigned.

Seriphidium (Besser ex Hook.) Fourr.

S. maritimum (L.) Polj.   Sea Wormwood
Native. Status: local.
Coastal habitats.

ML: (Leith Docks, 1905, 1906, 1910, 1927).
EL: local, rare on rocks at Gullane Point and North Berwick; quite frequent at the top of saltmarsh and beach at Tyninghame Bay; Belhaven Bay; Dunbar; Catcraig; (Luffness, 1812).

Artemisia L.

A. vulgaris L.   Mugwort
Native. Status: widespread.
Waste ground, railway embankments, roadsides, bings.

WL, ML, EL: rare in upland areas, widespread and locally frequent elsewhere.

A. absinthium L.   Wormwood
Introduced. Status: rare.
Waste ground, roadsides, railway embankments.

WL: rare, Bo'ness; Greendykes Bing; (Bathgate, 1973).
ML: rare, Warriston; Lochend; Smeaton; Leith Docks.
EL: rare, Thornton; (Prestongrange and Aberlady, pre-1934; West Barns, 1970; Thorntonloch, 1975).

A. campestris L.   Field Wormwood
Casual.
WL: South Queensferry, pre-1934.
ML: Borthwick railway tip, 1958; Leith, 1960.

Achillea L.

A. ptarmica L.   Sneezewort
Native. Status: quite widespread.
Marshes, beside roads, rivers and ditches, rough grassland.

WL: very widespread but seldom frequent in S., occasional in N.
ML: widespread and locally frequent in S., rare in N.
EL: occasional throughout SE., rare and declining elsewhere, Aberlady Bay; Gullane.

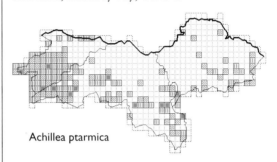

Achillea ptarmica

A. millefolium L.   Yarrow
Native. Status: common.
Roadsides, rough grassland, waste ground, riversides, intensive grassland, scrub, heath grassland.

WL, ML, EL: common.

Chamaemelum Mill.

C. mixtum (L.) All.
Introduced. Status: rare, casual.
Waste ground.

ML: rare, Leith Docks, 1989.

C. nobile (L.) All.   Chamomile
EL: Dirleton, 1959.

Anthemis L.

A. arvensis L.   Corn Chamomile
Native/Introduced. Status: rare, casual.
Rough grassland, canalside, waste ground.

WL: rare, possible record from the south side of Linlithgow, 1991; (north side of Linlithgow, pre-1934; Carriden, 1972).
ML: rare, Russell Road, 1988; (Lasswade, Roslin, Leith and Slateford, pre-1934; Borthwick railway tip, 1956; Leith Docks, 1958).
EL: (Aberlady, pre-1934; Dirleton area, 1959).

A. cotula L.   Stinking Chamomile
Introduced. Status: rare.
Waste ground, rubbish tips.

ML: rare, Russell Road; (Borthwick, 1959; Leith and Sighthill, 1963).
EL: (Gullane, 1844; Drem, pre-1894; Port Seton, Aberlady and Dirleton, pre-1934).

A. tinctoria L.   Yellow Chamomile
Introduced. Status: rare, casual.
Roadsides, gardens.

ML: rare, Blackford Quarry tip; (Cocklaw, 1973).
EL: rare, East Barns; (Longniddry, pre-1934).

Chrysanthemum L.

C. segetum L.   Corn Marigold
Introduced. Status: scarce, once a common and troublesome field-weed, now declining, casual.
Waste ground, arable fields.

WL: rare, north-west of Linlithgow; south-east of Philpstoun; a single plant near Blackness Castle, 1990; (South Queensferry, 1833; Kirkliston, 1848; Niddry, 1889; Abercorn, pre-1934; abundant in field near Blackness, 1972).
ML: (common in fields near Newbridge, 1945; Granton, 1975).
EL: occasional in arable fields in Gosford/Aberlady/Luffness/Kingston area; on waste ground at Pencaitland; Keith; West Barns; (field near Prestonpans, pre-1824).

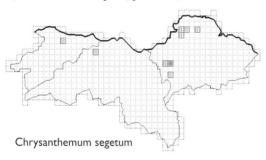

Chrysanthemum segetum

C. coronarium L.   Crown Daisy
Introduced. Status: rare, garden escape.
Waste ground.

ML: rare, Leith Docks and Seafield, 1989.

Leucanthemum Mill.

L. vulgare Lam.   Ox-eye Daisy, Moon Daisy
Native. Status: widespread.

Roadsides, rough grassland, waste ground, railway embankments, riversides, scrub, bings.

WL: very widespread, occasionally abundant.
ML: widespread and locally frequent, mainly in lowland areas.
EL: widespread and locally frequent in lowland areas, rare in upland areas.

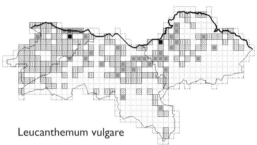

Leucanthemum vulgare

L. x superbum (Bergmans ex J. W. Ingram) D. H. Kent (*L. lacustre* (Brot.) Samp. x *L. maximum* (Raymond) DC.)   Shasta Daisy
Introduced. Status: rare, garden escape, sometimes persisting.
Roadsides, hedgerows, rough grassland.

WL: rare, Armadale; Carriden; north of Craigton; Craigiehall; Linlithgow; (Bathgate, 1973; Woodend, 1977).
ML: local, disused railway cycle track at Ravelston; Blackhall; Craiglockhart.
EL: rare, Prestongrange; Longniddry Bents; Aberlady Bay; Gullane; Humbie; Haddington area; West Barns; (North Berwick, 1972).

Matricaria L.

M. recutita L.   Scented Mayweed
Native. Status: scarce.
Arable fields, waste ground.

WL: rare, probably casual, Whitburn, 1995; M9 near Linlithgow, 1992; Bo'ness rubbish tip, 1998; (Kirkliston and Linlithgow, pre-1934; Carriden, 1972).
ML: rare, probably only casual, Leith Docks, 1989; Stow, 1996; Pilton, 1997; Granton, 1998; (Musselburgh, 1972; Cramond, 1974).
EL: local, south of Aberlady; Luffness; north of Drem; West Barns; (Longniddry, Haddington and North Berwick, pre-1934; Dirleton, 1972).

M. discoidea DC.   Pineappleweed
Introduced. Status: common.
Roadsides, waste ground, cereal fields and other cultivated ground, rough grassland, farmyards.

WL: common.

ML, EL: local in some upland areas, common elsewhere.

Tripleurospermum Sch. Bip.

The recording of the following two species was confused early in the survey, but the data presented here are accurate, though perhaps incomplete.

T. maritimum (L.) W. D. J. Koch   Sea Mayweed
Native. Status: local. Not always separated from *T. inodorum* and hence under-recorded.
Coastal sand, rocks and walls, waste ground near the sea.

WL: local, Bo'ness; Dalmeny.

ML: scarce, Cramond Island; Granton; Musselburgh.

EL: occasional along the beach, rarely somewhat inland, Gosford Bay; North Berwick; White Sands; Prestongrange Mining Museum, 1998; Gullane rubbish tip, 1997.

T. inodorum (L.) Sch. Bip.   Scentless Mayweed
Native. Status: common.
Roadsides, waste ground, cereal fields and other cultivated land, rough grassland, farmyards, coastal habitats.

WL: very widespread, locally frequent.

ML: rare in upland areas, quite widespread elsewhere, locally frequent in N.

EL: rare in upland areas, common elsewhere

Senecio L.

S. cineraria DC.   Silver Ragwort
Introduced. Status: rare, garden escape.
Coastal turf.

EL: rare, seafront at North Berwick.

S. x albescens Burb. & Colgan (*S. cineraria* x *S. jacobaea*)
Spontaneous hybrid. Status: rare.
Waste ground, coastal turf.

ML: rare, Craiglockhart, 2000.

EL: rare, isolated plant on the coast west of Yellowcraig, 1999; thriving population with parents on the seafront at North Berwick; one plant with *S. jacobaea* on the seawall at Dunbar, 1987. *S. cineraria* is grown in nearby gardens.

S. fluviatilis Wallr.
    Broad-leaved Ragwort, Saracen's Woundwort
Introduced. Status: rare, but increasing locally.
Woodland, roadsides, coastal scrub.

WL: rare, Bathgate; Craigton; Port Edgar; (Pepper Wood, 1868).

ML: rare, abundant at Cramond; Lanark Road; Roslin Glen.

EL: rare, Luffness; near Samuelston; Lennoxlove; (Dunbar, pre-1934; Wester Pencaitland 1956; Dirleton, 1958-68; below Stoneypath Tower, 1968; West Barns, 1973).

S. jacobaea L.   Common Ragwort
Native. Status: common.
Roadsides, rough grassland, waste ground, woodland, riversides, intensive grassland, scrub.

WL, ML, EL: common.

S. x ostenfeldii Druce (*S. jacobaea* x *S. aquaticus*)
Native. Status: rare, probably under-recorded.
Beside rivers and lochs, ditches.

WL: rare, near Westfield; (Linlithgow Loch, 1977).

ML: rare, Orchard Brae; (Linhouse Water, 1978).

S. aquaticus Hill   Marsh Ragwort
Native. Status: local.
Beside rivers and ditches, marshes.

WL: quite widespread in S. and W. where the terrain is wetter, rare elsewhere.

ML: quite widespread in SW., local and locally abundant elsewhere.

EL: rare, Keith Water; Woodhall Burn; (North Berwick and Tyninghame, pre-1934; Humbie, 1962; Garleton Hills, Whiteadder Water and Monynut Water, 1972).

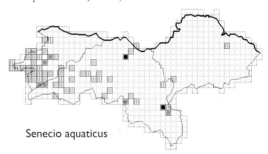

Senecio aquaticus

S. squalidus L.   Oxford Ragwort
Introduced. Status: local.
Waste ground, roadsides, rough grassland, walls, railway embankments, coastal habitats.

WL: occasional.

ML: widespread in N., locally abundant in urban areas, rare elsewhere.

EL: widespread and locally frequent in lowland areas, rare in upland areas, up to 6km from a railway line.

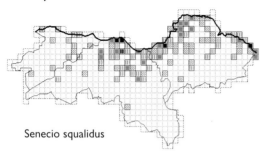

Senecio squalidus

S. x baxteri Druce (*S. squalidus* x *S. vulgaris*)
Native. Status: rare.
Waste ground.

ML: rare, Granton; Leith.
EL: (North Berwick area, 1972).

S. x subnebrodensis Simonk. (*S. squalidus* x *S. viscosus*)
Native. Status: rare.
Waste ground, railway lines.

ML: rare, Leith, 1982; (Seafield, 1959; Borthwick railway tip, 1962; Musselburgh, 1966).
EL: (Longniddry Station, 1970).

S. cambrensis Rosser   Welsh Groundsel
Native. Status: rare.
Waste ground.

ML: rare, Leith Docks, 1982; Carron Place, 1986.
This plant arose in Scotland from a stable cross between *S. squalidus* and *S. vulgaris*. The species also arose in Wales in 1948 and has since spread to neighbouring counties. Unfortunately the Leith population has not been recorded for several years in spite of diligent searches.

S. vulgaris L.   Groundsel
var. **vulgaris**
Native. Status: common.
Roadsides, cultivated ground, waste ground, rough grassland, farmyards.

WL: widespread but not locally frequent in S., common elsewhere.
ML: local in upland areas, common elsewhere.
EL: rare in upland areas, common elsewhere.

var. **hibernicus** Syme   Rayed Groundsel
Native. Status: widespread.

Roadsides, cultivated ground, waste ground, rough grassland, farmyards.

WL: quite widespread.
ML: common.
EL: scattered distribution in lowland areas.

S. sylvaticus L.   Heath Groundsel
Native. Status: local.
Quarries, rocky outcrops, rough grassland, scrub.

WL, ML: local.
EL: local, Garleton area; Whitekirk; Bothwell Water; Dunglass Dean.

S. viscosus L.   Sticky Groundsel
Native. Status: local.
Waste ground, roadsides, railway embankments, bings.

WL: quite widespread, seldom frequent.
ML: local, mainly in N., locally frequent.
EL: occasional, locally frequent.

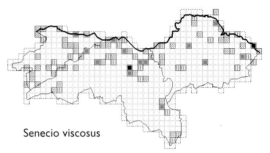

Senecio viscosus

S. vernalis Waldst. & Kit.   Eastern Groundsel
Grass-seed alien.
ML: waste ground, West Dock, Leith, 1974.

S. x viscidulus Scheele (*S. sylvaticus* x *S. viscosus*)
ML: on rocky outcrops on Arthur's Seat, 1967.

Sinacalia H. Rob. & Brettell

S. tangutica (Maxim.) B. Nord.   Chinese Ragwort
Introduced. Status: rare, garden escape.
Riversides, waste ground.

WL: rare, beside burn, Woodcockdale, 1984; (Linlithgow, 1962).
ML: (Cramond, 1903–75; Ravelston, 1961).
EL: rare, beside burn, Bilsdean; (Gifford, *c*. 1981).

Doronicum L.

D. pardalianches L.   Leopard's Bane
Introduced. Status: local, well naturalised.

Mixed and deciduous estate woodlands, wooded riversides and roadsides, scrub.

WL: local, locally frequent in N.
ML: local, locally frequent in river valleys.
EL: local, locally frequent in centre.

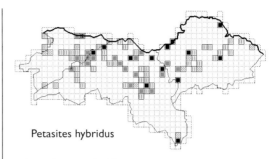

Petasites hybridus

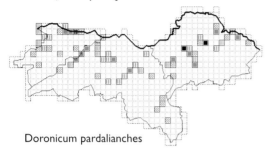

Doronicum pardalianches

D. plantagineum L.
Plantain-leaved Leopard's Bane
Introduced. Status: scarce, naturalised.
Policy woodland, grassy banks.

WL: rare, Kirkliston; Pepper Wood and disused railway line nearby; (Carriden, 1972; Hopetoun, 1977).
ML: local, Dalkeith Estate; Newhailes; Arniston Glen; Craiglockhart Dell; (Dalhousie Castle and Ravelston, 1975).
EL: rare, Macmerry; Whitekirk; Dunbar; (Whittingehame, 1871; Gifford area, 1973).

Tussilago L.

T. farfara L.   Colt's-foot
Native. Status: common.
Roadsides, waste ground, riversides, rough grassland.

WL: common.
ML, EL: local in some upland areas, common elsewhere.

Petasites Mill.

P. hybridus (L.) P. Gaertn., B. Mey. & Scherb.
Butterbur
Native. Status: local.
Damp places beside rivers and roads, on waste ground and in woodland and scrub.

WL: local, mainly in N., female plants at Hopetoun.
ML: local, mainly in lowland areas, locally abundant.
EL: local, mainly in centre, locally abundant. (Several colonies of female plants were recorded along Papana/Whittingehame/Biel Water from Garvald to West Barns, 1967 to 1981).

P. japonicus (Siebold & Zucc.) Maxim.
Giant or Creamy Butterbur
Introduced. Status: rare.
Riversides, woodland.

ML: rare, Braid Burn at Dreghorn and Blackford Glen, where it is spreading.

P. albus (L.) Gaertn.   White Butterbur
Introduced. Status: local.
Beside streams and ditches, damp woodland.

WL: local, well established in Pepper Wood and along parts of River Avon, with female plants near Brunton Burn.
ML: scarce, male plants at Cramond Bridge and at Bush and Carberry Estates; Bilston; Dalkeith Country Park.
EL: rare, Colstoun Water near Colstoun House and downstream beside River Tyne near Nether Hailes; abundant beside Gifford Water, spreading from Yester Estate; (small colony of female plants at Archerfield, 1971).

P. fragrans (Vill.) C. Presl   Winter Heliotrope
Introduced. Status: scarce.
Beside roads, rivers, ditches and railways, waste ground, woodland.

WL: rare, west of Blackburn; Carriden; East Bonhard; well established in Pepper Wood.
ML: local, Bilston Glen; Granton; (Pathhead, 1971; Stow, 1973).
EL: local, locally abundant, North Berwick Glen; Dryburn Bridge.

Calendula L.

C. officinalis L.   Pot Marigold
Introduced. Status: rare, garden escape.
Roadsides, waste ground, railway embankments, rubbish tips.

WL: rare, Bo'ness/Kinneil area; (Carriden and Broxburn, 1972).

ML: rare, Musselburgh; Granton.
EL: rare, near habitation.

Ambrosia L.

A. artemisiifolia L.   Ragweed
Introduced. Status: rare, casual, birdseed alien.
Waste ground.

ML: rare, Baberton; (Sighthill, 1968; Leith, 1978).

Guizotia Cass.

G. abyssinica (L. f.) Cass.   Niger
ML: Murieston, 1908.

Rudbeckia L.

R. laciniata L.   Coneflower
EL: banks of the River Tyne at East Linton, 1909.

Helianthus L.

H. annuus L.   Sunflower
Introduced. Status: rare.
Beach.

EL: rare, one plant on drift line, Aberlady Bay, 1982.

H. tuberosus L.   Jerusalem Artichoke
Introduced. Status: rare, long-persisting garden escape.
Waste ground, rubbish tips, marsh.

WL: rare, West Long, Livingston, 1994.
ML: rare, by Water of Leith, Canonmills and Slateford, 2000; (Hailes and Slateford Quarry tips and Leith Docks, *c.* 1907).
EL: rare, on rubbish tip, Aberlady Bay, 1989, 1991.

H. x laetiflorus Pers. (*H. rigidus* (Cass.) Desf. x *H. tuberosus*)
         Perennial Sunflower
WL: waste ground at Bo'ness, 1960.
EL: Prestonpans, 1980.

Galinsoga Ruiz & Pav.

G. quadriradiata Ruiz & Pav.   Shaggy Soldier
Introduced. Status: local, garden weed.
Waste ground, garden rubbish tips, municipal flower beds, pavements.

ML: local, corner of Bernard Terrace and St

Leonards; other sites in Edinburgh; Penicuik, 2000.

Bidens L.

B. cernua L.   Nodding Bur-marigold
Native. Status: rare.
Lochsides.

ML: rare, Duddingston, 1982; (Lochend, pre-1900).

B. tripartita L.   Trifid Bur-marigold
WL: Linlithgow Loch, pre-1934.
ML: Duddingston and Lochend, pre-1934.

Eupatorium L.

E. cannabinum L.   Hemp Agrimony
Native. Status: scarce.
Damp woodlands, waste ground, hedgerows, riversides, damp cliffs.

WL: (Linlithgow, pre-1934; River Almond at Craigiehall, 1962).
ML: rare, Crichton; (Roslin and Lasswade, pre-1934; planted at Bawsinch *c.* 1980).
EL: rare, on the North Sea coast, Seacliff; Bilsdean; Dunglass Dean.

BUTOMACEAE

Butomus L.

B. umbellatus L.   Flowering Rush
Introduced. Status: rare, planted and naturalising.
Ponds, lochs, marsh.

WL: planted in pond near Easter Inch; (Linlithgow Loch, pre-1934).
ML: rare, Carcant, 1976–present; Bawsinch, 1993; (Prestonhall, pre-1927; Carberry, 1962).
EL: (Pencaitland, 1954–67, introduced from there to Bara, 1967).

ALISMATACEAE

Sagittaria L.

S. sagittifolia L.   Arrowhead
Introduced. Status: rare, planted.
Ponds, lochs.

ML: rare, Bawsinch.

Baldellia Parl.

B. ranunculoides (L.) Parl.

Lesser Water Plantain

Native. Status: rare.
Marshes, ponds.

WL: (Winchburgh, 1867; Philpstoun, 1884).
ML: rare, Duddingston Loch; (Myreside, 1764; Hunter's Bog, 1824).
EL: rare, Aberlady Bay; (Dunbar, 1838).

Alisma L.

A. plantago-aquatica L.   Water Plantain
Native. Status: local.
Canal, lochs, reservoirs, ponds, rivers.

WL: quite widespread, at intervals along the Union Canal and scattered throughout the rest of the vice-county.
ML: local, Bush, Vogrie and Dalmahoy Estates; Threipmuir; Duddingston; Edgelaw; Slateford; Ratho.
EL: rare, Stobshiel Reservoir; Haddington; Danskine Loch; Newbyth; Tyninghame; Black Loch; (Wester Pencaitland, 1956; Aberlady Bay, 1957).

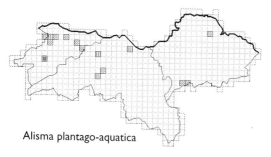

Alisma plantago-aquatica

A. lanceolatum With.

Narrow-leaved Water Plantain
Native. Status: rare.
Beside lochs, marshes.

ML: rare, Duddingston, 1988; (Lochend Loch, 1839; Duddingston Loch, 1969).

HYDROCHARITACEAE

Hydrocharis L.

H. morsus-ranae L.   Frogbit
Introduced. Status: rare.
Canal.

WL: rare, appeared in Union Canal, 1999, perhaps transferred from the Forth-Clyde Canal.

Stratiotes L.

S. aloides L.   Water Soldier
Introduced.
WL: reported to have become extinct by 1934 in Linlithgow Loch.
ML: reported to have become extinct by 1934 in Blackhall Quarry and Duddingston.
EL: pond in a wood between Haddington and Morham, 1909.

Elodea Michx.

E. canadensis Michx.   Canadian Waterweed
Introduced. Status: local.
Canal, lochs, ponds, rivers.

WL: local, widely scattered sites including the Union Canal.
ML: local, widely scattered sites, locally abundant.
EL: locally abundant at the few sites from which it is recorded, Ormiston Hall; Bara Loch; Donolly Reservoir; Pressmennan Lake; Pefferside; (earliest record, Dunglass, pre-1863).

The very rare male flowers were first recorded in Britain in 1880 from the Braid Hill marshes (**ML**) by the explorer, David Douglas.

E. nuttallii (Planch.) H. St. John

Nuttall's Waterweed
Introduced from aquaria and garden ponds.
Status: rare, but increasing.
Ponds, reservoirs, canal.

WL: rare, Livingston Reservoir; Union Canal.
ML: rare, Auchendinny, first recorded in 1987; Portobello; Lauriston Castle.

Lagarosiphon Harv.

L. major (Ridl.) Moss   Curly Waterweed
Introduced. Status: rare, probably planted.
Ponds.

ML: rare, Bawsinch, 1993; Vogrie Pond, 1994.
EL: rare, flooded quarry, Pencaitland, 1997.

JUNCAGINACEAE

Triglochin L.

T. palustre L.   Marsh Arrowgrass
Native. Status: local.
Marshes, beside streams and ditches.

WL: local, concentrations in S. and W.,
Fauldhouse Moor; Gowanbank area; scattered
elsewhere.
ML: local, mainly in centre and S., locally
frequent in SE.
EL: local, mainly in upland areas, rare on the
coast.

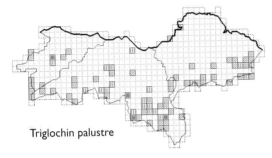

Triglochin palustre

T. maritimum L.   Sea Arrowgrass
Native. Status: local and decreasing.
Saltmarsh, dunes, coastal rocks.

WL: local, Grangemouth to Bo'ness; Blackness
area; Dalmeny.
ML: rare, Cramond Island.
EL: local along the coast, abundant at Aberlady
Bay and Heckies Hole, otherwise in small
numbers.

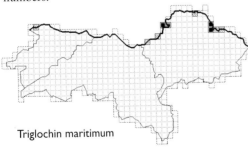

Triglochin maritimum

POTAMOGETONACEAE

Potamogeton L.

P. natans L.   Broad-leaved Pondweed
Native. Status: quite widespread..
Ponds, canal, rivers, lochs, reservoirs.

WL: widespread in Union Canal and other water
bodies.
ML: quite widespread.
EL: local, Aberlady Bay; Whiteadder Reservoir;
Catcraig.

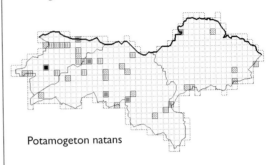

Potamogeton natans

P. polygonifolius Pourr.   Bog Pondweed
Native. Status: local.
In and beside ditches, streams and ponds, marshes,
bogs.

WL: local, concentrated in SW.
ML: local, mainly in S.
EL: rare, upland areas only, Linn Dean;
Sheeppath Burn; Whiteadder Water; Bothwell
Water; Monynut Water.

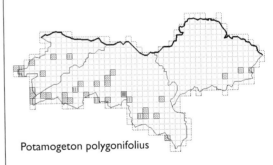

Potamogeton polygonifolius

P. coloratus Hornem.   Fen Pondweed
Native. Status: rare.
Ponds, ditches.

EL: rare, locally frequent with *P. natans* at
Aberlady Bay.

P. gramineus L.   Various-leaved Pondweed
Native. Status: local.
Lochs, reservoirs.

WL: rare, Bathgate Reservoir, 1982; (Linlithgow
Loch, pre-1934).
ML: local, mainly in SW., in almost all reservoirs.

P. x zizii W. D. J. Koch ex Roth (*P. gramineus* x *P. lucens*)   Long-leaved Pondweed
Native. Status; rare.
Lochs and reservoirs.

WL: (Linlithgow Loch, pre-1934).
ML: rare, Clubbiedean Reservoir, 1998.

P. x nitens Weber (*P. gramineus* x *P. perfoliatus*)
                Bright-leaved Pondweed
Native. Status: rare.
Canal.

WL: rare, Union Canal near Winchburgh, 1993.

P. alpinus Balb.   Red Pondweed
Native. Status: rare.
Reservoirs, ponds.

WL: (Bo'ness, pre-1934; near Livingston, 1968; near Knockhill, 1977).
ML: local, Cobbinshaw, Threipmuir, Clubbiedean and Corston Hill Reservoirs; Levenseat Quarry pool.
EL: (Gullane, pre-1863).

P. perfoliatus L.   Perfoliate Pondweed
Native. Status: rare.
Ponds, lochs, canal.

WL: (Union Canal west of Linlithgow, 1942).
ML: rare, in the Union Canal at Slateford, 2000; (Lochend and Duddingston, 1764; Union Canal at Sighthill, 1955, and Ashley Terrace, 1969).
EL: (Danskine, 1873; Haddington, pre-1934).

P. x cooperi (Fryer) Fryer (*P. perfoliatus* x *P. crispus*)   Cooper's Pondweed
Native. Status: rare.
Canal.

WL: (Union Canal, 1941).
ML: rare, locally abundant in the Union Canal at Ashley Terrace, 1987 and at Slateford, 1999; (Union Canal at Sighthill, 1955).

P. friesii Rupr.   Flat-stalked Pondweed
Native. Status: local.
Lochs, ponds, canal.

WL: (Union Canal and Linlithgow Loch, 1977).
ML: local, all along the Union Canal from Merchiston to Hermiston.

P. pusillus L.   Lesser Pondweed
Native. Status: local.
Ponds, lochs, reservoirs, canal.

WL: rare, possibly under-recorded because of confusion with *P. berchtoldii*, Union Canal near Linlithgow; Linlithgow Loch; Lochcote Reservoir; Beecraigs Loch.
ML: local, Blackford and Dalmahoy Ponds; Duddingston and Dunsapie Lochs; North Esk Reservoir; Roslin Sand Quarry.
EL: rare, Gosford Pond; Bara Loch; Scoughall; (Whiteadder Reservoir, 1972).

P. obtusifolius Mert. & W. D. J. Koch
                Blunt-leaved Pondweed
Native. Status: rare.
Reservoirs, lochs, canal.

WL: rare, Bangour Reservoir, 1993; (Kinneil, 1871; Newliston, 1901; Union Canal and Linlithgow Loch, pre-1934).
ML: rare, Threipmuir and Gladhouse Reservoirs; Howgate Pond.
EL: rare, frequent at Donolly Reservoir; Whiteadder Reservoir; (Gullane, pre-1934, this record may refer to a broad-leaved form of *P. berchtoldii*).

P. berchtoldii Fieber   Small Pondweed
Native. Status: scarce.
Ponds, marshes, lochs, reservoirs.

WL: a record exists from Linlithgow Loch in 1990, and a number of records exist from a survey in 1982 but no specimens are available and there is a possibility of confusion with *P. pusillus*; (Philpstoun Loch, 1825).
ML: scarce, Quarrel Burn; North Esk, Rosebery, Threipmuir and Gladhouse Reservoirs; (Edgelaw Reservoir, 1965).
EL: rare, Gladsmuir; Aberlady Bay; Luffness; near Kidlaw; Markle; Philip Burn; (Whiteadder Reservoir, 1972; Gosford House, 1978).

P. crispus L.   Curled Pondweed
Native. Status: local.
Lochs, ponds, reservoirs, rivers, canal.

WL: rare, Linlithgow Loch; Lochcote and Bangour Reservoirs; Almond Pools; Union Canal near Linlithgow.
ML: local, locally frequent, Union Canal; pond at Lauriston; Duddingston Loch; Threipmuir, Morton, Edgelaw and Bonaly Reservoirs; River Esk; Figgate Burn.
EL: scarce, wide distribution, River Tyne near Samuelston; Markle; Kidlaw; Pressmennan Lake; Biel Water at West Barns.

P. pectinatus L. Fennel Pondweed
Native. Status: scarce.
Ponds, lochs, rivers, canal.

WL: rare, Linlithgow Loch, (Union Canal, pre-1934).
ML: scarce, Union Canal; Inverleith and Craiglockhart Ponds; Lochend; Duddingston; River Almond at Gavieside.
EL: rare, mainly in new ponds or estate ponds, Gosford House; Haddington; North Berwick area; Markle; West Barns.

P. lucens L. Shining Pondweed
ML: Lochend and Duddingston, and the Union Canal at Slateford, 1961.

P. filiformis Pers. Slender-leaved Pondweed
ML: Mount Lothian Quarry pool, 1935.

Groenlandia J. Gay

G. densa (L.) Fourr. Opposite-leaved Pondweed
ML: Corstorphine and Duddingston, pre-1934.
EL: Gullane, 1884; Haddington, 1912.

RUPPIACEAE

Ruppia L.

R. maritima L. Beaked Tasselweed
Native. Status: rare.
Ponds.

ML: rare, Bawsinch, appeared in 1992 after *Utricularia vulgaris* was transferred there from the Marl Loch, Aberlady Bay.
EL: (in salt-water pools at Aberlady Bay and Gullane Links pre-1824; Tyninghame, pre-1894; all three sites, pre-1934).

ZANNICHELLIACEAE

Zannichellia L.

Z. palustris L. Horned Pondweed
Native. Status: rare.
Lochs, ponds, canal.

WL: rare, Linlithgow Loch; Glendevon Pond, 1982; Union Canal near Linlithgow, 1986; (Dundas Loch, 1905).
ML: rare, Duddingston Loch; Craiglockhart Pond; Union Canal; (Ravelrig, Braid Hill Marshes and Lochend, pre-1934).

EL: rare, Gosford House; Scoughall; Biel Water at West Barns; (near Luffness, pre-1824; Aberlady Bay, pre-1977).

ZOSTERACEAE

Zostera L.

Z. marina L. Eelgrass
Native. Status: rare, probably under-recorded.
Coastal, below low water mark.

WL: rare, west of Blackness and west of Cramond, 2000.
ML: (Leith and Granton, pre-1934).
EL: rare, Tyninghame Bay; (Gosford Bay and Aberlady Bay, 1970).

Z. angustifolia (Hornem.) Rchb. Narrow-leaved Eelgrass
Native. Status: rare.
Intertidal mud.

WL: rare, Carriden and Blackness areas, with *Z. noltei*, but in deeper water associated with stream outflows.
ML: (below low water mark, Leith and Granton, pre-1934).
EL: rare, Craigielaw Bay, 1991; Aberlady Bay, 1991, has declined here markedly since then due to algal bloom; Tyninghame Bay. Exposed at low tide more readily than *Z. marina*.

Z. noltei Hornem. Dwarf Eelgrass
Native. Status: rare.
Intertidal mud.

WL: local, mud-flats between Bo'ness and Hopetoun and west of Cramond, forming a turf in places.
ML: (Leith and Granton, pre-1934).
EL: rare, Craigielaw Bay; Kilspindie; Hummell Rocks; Tyninghame Bay.

ARACEAE

Acorus L.

A. calamus L. Sweet-flag
Introduced. Status: rare.
Beside lochs and ponds.

WL: (marsh in Kirkliston, 1869).
EL: rare, Pressmennan Lake.

Lysichiton Schott

L. americanus Hultén & H. St. John
American Skunk Cabbage
Introduced. Status: rare, probably planted.
Marshy ground in policies.

ML: rare, Bush Estate, 2000; (Carberry, 1976).
EL: rare, beside the lake, Smeaton, 1988.

Calla L.

C. palustris L.   Bog Arum
Introduced. Status: rare.
Beside ponds, wet woodland.

WL: rare, Linlithgow Golf Course.
ML: rare, introduced at Bawsinch *c.* 1980 or earlier.

Arum L.

A. maculatum L.   Lords and Ladies, Cuckoo Pint
?Native. Status: local.
Mixed and deciduous woodland, roadsides, scrub.

WL: local, mainly in N., associated with old estates.
ML: local, mainly, if not entirely, in N.
EL: scarce, mainly beside paths in old estate woodland, or by riverside walks.

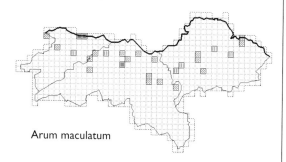
Arum maculatum

LEMNACEAE

Spirodela Schleid.

S. polyrhiza (L.) Schleid.   Greater Duckweed
Introduced. Status: rare.
Canal, ponds.

WL: rare, Union Canal near Woodcockdale, 1984.
ML: rare, Bawsinch, 1984; (Duddingston, pre-1894).

Lemna L.

L. gibba L.   Fat Duckweed
Native. Status: local.
Canal, lochs.

WL: locally abundant in Union Canal.
ML: local, Duddingston Loch, Union Canal.

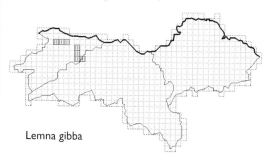
Lemna gibba

L. minor L.   Common Duckweed
Native. Status: local.
Ponds, ditches, rivers, canal, marshes, in still water.

WL: quite widespread, locally abundant.
ML: local, locally abundant.
EL: quite widespread, seldom locally frequent.

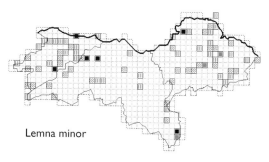
Lemna minor

L. trisulca L.   Ivy-leaved Duckweed
Native. Status: local.
Canal, ponds, lochs.

WL: frequent along Union Canal, rare elsewhere.
ML: occasional, Union Canal; Duddingston Loch, 1987.
EL: rare, Pencaitland; Elvingston; Gifford area; Aberlady Bay, 1982; (pond in a wood between Haddington and Morham, 1909; pool in Ogle Burn, Blackcastle Hill, 1969).

L. minuta Kunth   Least Duckweed
Introduced. Status: rare.
Still water in ditches and lochs.

WL: rare, Kinneil Loch, recorded for the first time in 1999.
ML: rare, ditch by Duddingston Loch, recorded for the first time in 1999; Bawsinch, 1999.

JUNCACEAE

Juncus L.

J. squarrosus L.   Heath Rush
Native. Status: widespread in upland areas.
Moorland, heath grassland, marshes, bogs, beside roads and rivers.

WL: widespread in wet acid areas in SW., local elsewhere, seldom frequent.
ML: in upland areas only, common.
EL: widespread and locally frequent in upland areas, local on heathland in lowland areas.

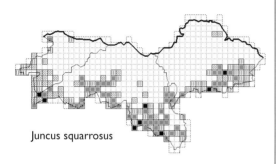

Juncus squarrosus

J. tenuis Willd.   Slender Rush
Introduced. Status: rare.
Beside roads, ditches, railway tracks and the canal, marshes, rough grassland.

WL: rare, towpath of Union Canal until recently; (Torphichen/Cairnpapple area *c.* 1970).
ML: rare, Drylaw; Roslin; Threipmuir; (Ratho, 1973).
EL: rare, Saltoun Forest; Gullane; (West Barns, 1959).

J. gerardii Loisel.   Saltmarsh Rush
Native. Status: local.
Saltmarshes, marshes, riversides.

WL: occasional along the coast.
ML: rare, Cramond Island.
EL: local along the coast, locally abundant at Aberlady Bay and Belhaven Bay.

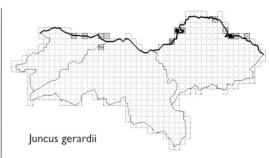

Juncus gerardii

J. bufonius L.   Toad Rush
Native. Status: widespread.
Marshes, beside roads, streams and ditches, waste ground, tracks, damp grassland.

WL: widespread, seldom in quantity.
ML: widespread and locally frequent in SE., local elsewhere.
EL: widespread, generally in small quantity.

J. ambiguus Guss.   Frog Rush
Native. Status: rare.
Beaches, coastal rocks.

EL: rare, Aberlady Bay, Barns Ness area.

J. articulatus L.   Jointed Rush
Native. Status: widespread.
Marshes, beside rivers and ditches.

WL: widespread, particularly in S. and W.
ML: common in SE., occasional in N., quite widespread elsewhere.
EL: quite widespread in centre and S., locally frequent, local in well-drained N.

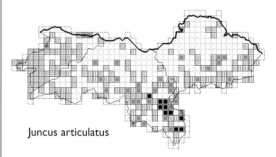

Juncus articulatus

J. x surrejanus Druce ex Stace & Lambinon (*J. articulatus* x *J. acutiflorus*)
Native. Status: rare, with both parents, possibly overlooked.

ML: rare, Threipmuir, 1998; (Colzium, 1838; Edinburgh, nineteenth century).

J. acutiflorus Ehrh. ex Hoffm.

Sharp-flowered Rush

Native. Status: widespread.
Marshes, beside rivers and ditches, rough grassland.

WL: widespread and locally frequent in S. and W., local elsewhere.
ML: mainly in S., common in SE., patchy distribution elsewhere, locally abundant.
EL: quite widespread in S. and E., locally frequent, rare in drier NW.

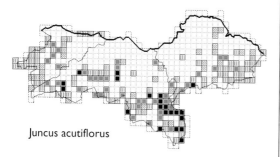

Juncus acutiflorus

J. bulbosus L.   Bulbous Rush
Native. Status: widespread in upland areas.
Marshes, bogs, beside ditches and streams, ponds.

WL: widespread in S. and SW., local elsewhere.
ML: quite widespread in S., locally frequent in SE., occasional elsewhere.
EL: quite widespread in upland areas, rare elsewhere.

J. inflexus L.   Hard Rush
Native. Status: local.
Marshes, beside ditches, streams, ponds and roads, usually on basic soils.

WL: quite widespread in N., rare in S. and W.
ML: local, locally frequent in the Pentland Hills.
EL: local, mainly in lowland and coastal areas.

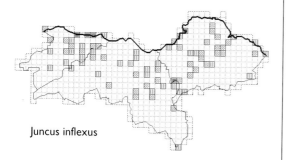

Juncus inflexus

J. effusus L.   Soft Rush
Native. Status: common.
Marshes, beside rivers and ditches, rough grassland, mixed and coniferous woodland, roadsides, scrub, moorland, bogs.

WL: common.
ML: widespread in N. except in built-up areas, common elsewhere.
EL: widespread in N., common elsewhere.

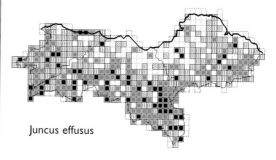

Juncus effusus

J. conglomeratus L.   Compact Rush
Native. Status: widespread.
Marshes, rough grassland, beside rivers, ditches and roads, moorland, mixed and coniferous woodland, scrub.

WL: common, particularly in SW.
ML: occasional in N., widespread and locally frequent in S.
EL: widespread in upland areas, local elsewhere.

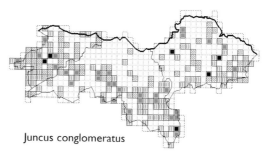

Juncus conglomeratus

J. compressus Jacq.   Round-fruited Rush
WL: South Queensferry, pre-1934; near Broxburn, 1956.
ML: Granton, Cramond and Threipmuir, pre-1934.
EL: Longniddry, 1871; Wester Pencaitland, 1957; Aberlady Bay, 1979.

J. subnodulosus Schrank   Blunt-flowered Rush
WL: Abercorn, pre-1934.
EL: Broxmouth, 1836.

J. maritimus Lam.   Sea Rush
WL: Blackness and Bo'ness, pre-1934.
EL: Luffness, pre-1934.

Luzula DC.

L. nivea (L.) DC.   Snow-white Woodrush
Introduced. Status: rare.
Gardens.

EL: rare, Belhaven, 1992; (Ormiston Hall gardens, 1911).

L. pilosa (L.) Willd.   Hairy Woodrush
Native. Status: local.
Moorland, heath grassland, woodland, riversides.

WL: rare, Carribber; Westfield; Woodcockdale; Barbauchlaw Burn; (Hopetoun, 1959).
ML: local, mainly in upland areas.
EL: quite widespread in upland areas, occasional in heathy woodland in lowland areas.

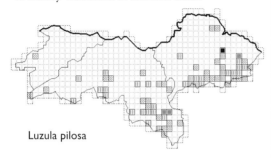

Luzula pilosa

L. sylvatica (Huds.) Gaudin   Great Woodrush
Native. Status: widespread.
Mixed and deciduous woodland, riversides, heath grassland.

WL: local, in old woodlands along River Avon; scattered elsewhere.
ML: widespread and locally frequent/abundant in river valleys and some upland areas, local elsewhere.
EL: rare in NW., quite widespread elsewhere, locally abundant at Humbie, Pressmennan, Woodhall Dean and Dunglass Dean.

L. luzuloides (Lam.) Dandy & Wilmott
                     White Woodrush
Introduced. Status; rare, fairly persistent escape.
Woodland, riversides.

WL: rare, Whitrigg Bing.
ML: rare, Dalhousie and Hawthornden Castles; (Penicuik, 1912; Lasswade, 1922; Cobbinshaw, 1978).

EL: rare, Gifford, escaped from Yester Estate; (Yester, 1913).

L. campestris (L.) DC.   Field Woodrush
Native. Status: widespread.
Rough grassland, roadsides, heath grassland, riversides, moorland, intensive grassland, public parks.

WL: widespread, locally frequent.
ML: widespread except in arable areas, common in SE.
EL: common, except in arable areas.

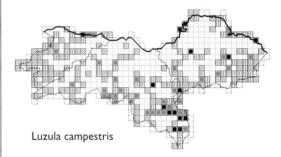

Luzula campestris

L. multiflora (Ehrh.) Lej.   Heath Woodrush
Native. Status: widespread.
Rough heath grassland, moorland, marshes, woodland, roadsides, riversides, scrub, bogs, bings.

WL, ML, EL: widespread and locally frequent in upland and boggy areas, local elsewhere.

ssp. congesta (Thuill.) Arcang. and **ssp. multiflora** occur widely but most recorders did not differentiate between them.

CYPERACEAE

Eriophorum L.

E. angustifolium Honck.   Common Cottongrass
Native. Status: local.
Bogs, moorland.

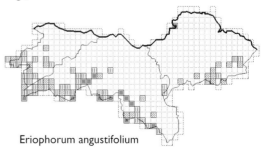

Eriophorum angustifolium

WL: in S. only, widespread, but not locally frequent.

ML: upland areas only, widespread and locally frequent.
EL: at Aberlady Bay; otherwise confined to the Lammermuir Hills where it is quite widespread but not locally frequent; (Balgone Estate, 1971).

E. latifolium Hoppe   Broad-leaved Cottongrass
Native. Status: rare.
Base-rich flushes.

ML: rare, Pentland Hills at Carlops, 1998; Harbour Hill, 2000; (Borthwick and Crichton Castles, 1871 and 1863).

E. vaginatum L.   Hare's-tail Cottongrass
Native. Status: local.
Bogs, moorland.

WL: in S. only, widespread, not locally frequent.
ML: confined to upland areas, common.
EL: rare at Aberlady Bay; widespread and locally frequent in the Lammermuir Hills.

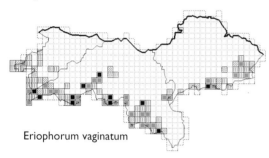
Eriophorum vaginatum

Trichophorum Pers.

T. cespitosum (L.) Hartm.   Deergrass
Native. Status: local.
Moorland, bogs, heath grassland.

WL: local, confined to the boggy areas of the S. and SW.
ML: local, confined to upland areas, locally abundant.
EL: local, now confined to upland areas, locally frequent; (Aberlady Bay, pre-1962).

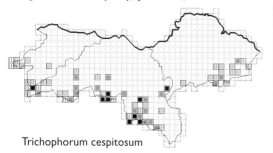
Trichophorum cespitosum

ssp. **cespitosum**
Moorland, raised bogs, in wetter areas than the following ssp.

ML: rare, Longmuir Moss, 2000; (Balerno, 1931).

ssp. **germanicum** (Palla) Hegi
Moorland, heath grassland.

ML: common in some upland areas.
EL: recorded from Sheeppath Burn, 1987.

nothossp. **foesteri** Swan (*ssp. germanicum* x *ssp. cespitosum*)
Moorland, raised bogs.

ML: rare, Longmuir Moss, 2000; (Threipmuir, 1834; Auchencorth Moss, 1870).

Eleocharis R. Br.

E. palustris (L.) Roem. & Schult.
                        Common Spike-rush
Native. Status: local.
Marshes, in and beside ponds, rivers, ditches, lochs and reservoirs.

WL, ML: local, widely scattered distribution, locally frequent.
EL: local, widely scattered distribution, locally frequent, Aberlady Bay; Stobshiel Reservoir; Whiteadder Reservoir.

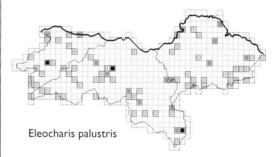
Eleocharis palustris

E. uniglumis (Link) Schult.   Slender Spike-rush
Native. Status: rare.
Marshes.

WL: rare, Blackness.
ML: (Pentland Hills, 1806).
EL: rare, Aberlady Bay; Gullane Bents; (Yellowcraig, 1974).

The erroneous records of *E. multicaulis* from the Lothians have been found to be *E. uniglumis*.

E. quinqueflora (Hartmann) O. Schwarz
                        Few-flowered Spike-rush

Native. Status: scarce.
Marshes, upland flushes, dune slacks, sandy tracks by the sea.

ML: scarce, Moorfoot and Pentland Hills; Mount Main; Fairliehope; Pate's Hill; Gutterford Burn; (Leith Links and Holyrood Park, pre-1806).
EL: rare, Aberlady Bay; Tyninghame; Stottencleugh.

E. acicularis (L.) Roem. & Schult.

Needle Spike-rush

Native. Status: rare.
Beside lochs and ponds.

WL: rare, Beecraigs Loch; (South Queensferry, 1840).
ML: Bawsinch Pond, 1984, destroyed when the pond was dug out.
EL: (Luffness, pre-1765).

Bolboschoenus (Asch.) Palla

B. maritimus (L.) Palla   Sea Club-rush
Native. Status: rare.
Saltmarshes, marshes, ponds and ditches in coastal locations.

WL: rare, Hopetoun, 1988; (Dalmeny, 1836; Blackness, 1867).
EL: occasional along the coast, Longniddry Bents; Aberlady Bay; Pefferside; Heckies Hole; Belhaven Bay; Barns Ness.

Scirpus L.

S. sylvaticus L.   Wood Club-rush
Native. Status: rare.
Damp woodland, riversides.

WL: rare, Carribber Glen; south-west of Blackness; wood near Balgornie, 1982; (Hopetoun, 1914).
ML: rare, Auchendinny; Howgate; Linburn; (River Esk at Lasswade, 1836; Cramond Bridge, 1863; Roslin, pre-1934).
EL: (Saltoun, pre-1934).

Schoenoplectus (Rchb.) Palla

S. lacustris (L.) Palla   Common Club-rush
Native. Status: rare.
Marshes, beside ditches, ponds and lochs.

WL: (recorded from Linlithgow Loch, 1980, but *S. tabernaemontani* was not differentiated at the time).

ML: (Duddingston Loch and Lochend, 1835; Cobbinshaw Loch, 1967).
EL: rare and doubtfully native, pond on Haddington Golf Course, 1996; Pressmennan Lake; (near Dunbar 1826).

S. tabernaemontani (C. C. Gmel.) Palla

Grey Club-rush

Native. Status: rare.
Marshes, ponds.

WL: rare, ponds south of Bathgate; Tailend Moss.
ML: (Lochend and Threipmuir, 1903).
EL: rare, Elvingston; Aberlady Bay; north of Drem; Dirleton; Balgone.

Isolepis R. Br.

I. setacea (L.) R. Br.   Bristle Club-rush
Native. Status: local.
Beside rivers and ditches, marshes, damp sandy ground.

WL, ML: local, quite widespread in S., not locally frequent.
EL: rare, now apparently confined to Lammermuir Hills, Newhall Burn; Faseny Water; Deuchrie; Monynut Edge; (Gullane Golf Course, 1976).

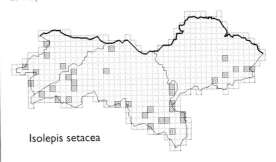

Isolepis setacea

Eleogiton Link

E. fluitans (L.) Link   Floating Club-rush
WL: Philpstoun Loch, 1871.
ML: Braid Hill Marshes, 1835; Balerno, pre-1934.

Blysmus Panz. ex Schult.

B. compressus (L.) Panz. ex Link   Flat Sedge
Native. Status: rare.
Base-rich marshes.

ML: (Balerno, Borthwick and Crichton, pre-1934).
EL: rare, recent records only from the

Lammermuir Hills, Whiteadder Water, 1982; (Faseny Water and Tavers Cleugh, 1977; older records from coastal and lowland locations, Aberlady Bay, 1847; Gullane 1887; Haddington, 1908; Broxmouth).

B. rufus (Huds.) Link Saltmarsh Flat Sedge
Native. Status: rare.
Saltmarsh.

WL: (east of South Queensferry, 1899).
EL: rare, Aberlady Bay; Heckies Hole; Belhaven Bay; Barns Ness; (Gosford, 1831; Prestonpans, pre-1871; Cockenzie, 1874; Thorntonloch, 1969).

Schoenus L.

S. nigricans L. Black Bog-rush
Native. Status: rare.
Dune-slacks.

EL: rare, Aberlady Bay; (Broxmouth, ?1758).

Carex L.

C. paniculata L. Greater Tussock Sedge
Native. Status: local.
Base-rich marshes.

ML: local, abundant at Crichton Castle; Longmuir Rig; Vogrie; Tynebank; Rosebery, Threipmuir and Edgelaw Reservoirs.
EL: rare, Keith Water; Aberlady Bay; River Tyne near Samuelston; Bara Loch; Bearford Burn; Garvald; Pressmennan Lake; Broxmouth; Innerwick; (Ogle Burn, 1967).

C. x boenninghausiana Weihe (*C. paniculata* x *C. remota*)
Native. Status: rare.
Base-rich marshes with parents.

ML: rare, Crichton Castle.

C. diandra Schrank Lesser Tussock Sedge
Native. Status: rare.
Marshes.

WL: (Dundas, pre-1863).
ML: rare, Rosebery and Crosswood Reservoirs; Longmuir Rig; (Ravelrig, pre-1863).
EL: rare, Aberlady Bay; (Longniddry area, 1890; Danskine, 1961).

C. otrubae Podp. False Fox Sedge
Native. Status: local.
Marshes, coastal turf, dunes, waste ground, canalside.

WL: scattered along the coast, rare inland,

Bo'ness; Blackness; South Queensferry; Dalmeny; Union Canal at Winchburgh.
ML: local in the valley of Gala Water, near Stow; Cockmuir; Polton; Cramond Island, 1957–94; (Musselburgh and Figgate Burn, 1878).
EL: occasional along the coast, locally frequent at Aberlady Bay; (inland at Lammer Law, pre-1962).

Records of *C. vulpina* in Martin (1934) refer to this species.

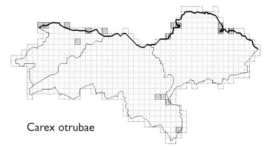

Carex otrubae

C. muricata L. Prickly Sedge
ssp. lamprocarpa Celak.
Native. Status: rare.
Rough grassland on rocky outcrops.

WL: (South Queensferry, pre-1934).
ML: rare, Arthur's Seat; Blackford Hill; Howdean; (Castle Rock, 1850; Dalkeith, 1863; Harburn, 1871).
EL: rare, North Berwick Law; Markle; Hairy Craig; Traprain Law; (Yester, pre-1863).

C. divulsa Stokes Grey Sedge
ssp. leersii (Kneuck.) W. Koch
Native. Status: rare.
Woodland, grassy embankments.

ML: rare, Newbattle.
EL: rare, Luffness.

C. arenaria L. Sand Sedge
Native. Status: local.
Beaches, dunes, coastal turf.

WL: rare, Blackness; Dalmeny.
ML: rare, Cramond Island; Musselburgh; (previously more widespread along the coast).
EL: widespread along the coast, locally frequent.

C. disticha Huds. Brown Sedge
Native. Status: local.
Marshes, beside canal and rivers.

WL: rare, burns near the centre of Bo'ness;

Beecraigs; Lochcote Reservoir; Union Canal near Winchburgh.
ML: quite widespread, rarely frequent; Longmuir Rig; Middleton; Carlops; Threipmuir; Fullarton Water; Holyrood Park; Ravensneuk.
EL: local, mainly in upland areas, frequent at Aberlady Bay.

C. maritima Gunnerus   Curved Sedge
Native. Status: rare.
Coastal sand.

EL: rare, near Gullane, 2000; (shore between Prestonpans and Longniddry, 1853; between Prestonpans and Cockenzie, 1867; Longniddry, pre-1934).

C. remota L.   Remote Sedge
Native. Status: local.
Marshy woodland, beside rivers, canal and ponds, scrub.

WL: rare, Carribber; Inveravon; Kinneil; Union Canal south of Broxburn.
ML: local, Crichton; West Calder; Roslin; Ratho; Penicuik.
EL: scarce, along north edge of Lammermuir Hills, Keith/Humbie/Saltoun; Colstoun; Gifford/Yester/Papana Water; Pressmennan; Woodhall Dean; Dunglass Dean.

C. ovalis Gooden.   Oval Sedge
Native. Status: widespread.
Marshes, beside rivers and ditches, rough grassland, roadsides, wet flushes in heath grassland.

WL: widespread and locally frequent in S., local in NW., rare in NE.
ML: widespread in SW. and SE., local elsewhere, locally abundant in SE.
EL: widespread in upland areas, local elsewhere, Gladsmuir; Humbie; Binning Wood; Thurston.

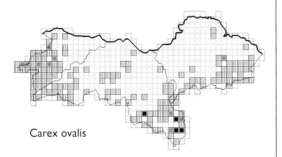

Carex ovalis

C. echinata Murray   Star Sedge
Native. Status: widespread.
Marshes, bogs, moorland, heath grassland, riversides.

WL: quite widespread in S., rare in N.
ML: widespread and locally frequent in upland areas, rare or absent elsewhere.
EL: widespread and locally frequent in upland areas, apparently now absent elsewhere; (Tyninghame, 1955; Gullane area, pre-1962).

C. dioica L.   Dioecious Sedge
Native. Status: local.
Marshes, heath grassland, bogs, usually in wet stony flushes.

ML: local, mainly in the Moorfoot and Pentland Hills.
EL: rare, Aberlady Bay; Lammer Law; Hopes Reservoir; Bothwell Water; Killmade Burn; Yearn Hope; (Faseny Water, 1972).

An old record from Gullane Loch (**EL**) (Greville, 1824) for *C. davalliana* refers to this species.

C. curta Gooden.   White Sedge
Native. Status: local.
Marshes, bogs, beside ditches and rivers.

WL: local, mainly, in S.
ML: local, mainly in upland areas, locally frequent.
EL: rare, Armet, Faseny, Whiteadder and Monynut Waters; Yearn Hope; (Spartleton, 1972; Sheeppath Dean and Burn Hope, pre-1980).

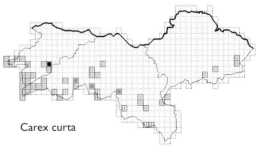

Carex curta

C. hirta L.   Hairy Sedge
Native. Status: local.
Marshes, beside rivers, canal and ponds, wet grassland.

WL: quite widespread, mainly in centre and N.
ML, EL: widely scattered distribution.

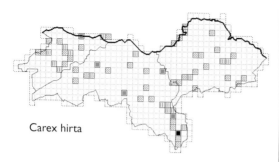

Carex hirta

C. acutiformis Ehrh.   Lesser Pond Sedge
Native. Status: scarce.
Marshes, riversides.

WL: (Carribber Glen, 1971).
ML: scarce, Duddingston; Threipmuir; Crichton;
Rosebery Reservoir; Vogrie; Boghall; Penicuik
Estate; Calder Wood.
EL: rare, Keith Water; Clerkington; Newmains;
Monynut Water; Aikengall Water; (Wester
Pencaitland, 1957; Yester, 1965).

C. riparia Curtis   Greater Pond Sedge
Native. Status: rare.
In and beside lochs, wet scrub.

ML: rare, Duddingston Loch; Bawsinch; Carberry
Tower; Currie Wood; Borthwick; (Arniston,
pre-1934).
EL: rare, Aberlady Bay; Danskine Loch; Garvald;
(Longniddry and Yester, pre-1934).

C. rostrata Stokes   Bottle Sedge
Native. Status: local.
Marshes, in and beside ponds, rivers, lochs and ditches,
in acid to neutral conditions.

WL: quite widespread, mainly in S.
ML: locally frequent and quite widespread in
upland areas, rare elsewhere.
EL: scarce, widely scattered distribution, Linn
Dean; Aberlady Bay; Donolly Reservoir;
Pressmennan Lake.

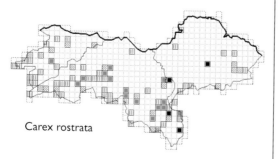

Carex rostrata

C. vesicaria L.   Bladder Sedge
Native. Status: rare.
Marshes, loch margins.

WL: (Linlithgow, pre-1934).
ML: (Blackhall, 1910; Pentland Hills,
Auchendinny and Threipmuir, pre-1934).
EL: rare, Elmscleugh Water; (Danskine Loch,
1972).

C. pendula Huds.   Pendulous Sedge
Native/Introduced. Status: local.
Woodland, beside rivers and ditches, scrub, waste
ground.

WL: rare, Inveravon; Carriden; Glendevon;
Hopetoun; Dundas; (Carribber Glen, 1971).
ML: local, Roslin; Carberry; Arniston; Kinauld;
Dreghorn; Trinity.
EL: scarce, Binning Wood; Pressmennan;
Bilsdean; Dunglass Dean.

C. sylvatica Huds.   Wood Sedge
Native. Status: local.
Mixed and deciduous woodland, gorges.

WL: rare, Carribber; Hopetoun; Polkemmet;
woodland by Union Canal north-west of
Winchburgh; (Linlithgow, pre-1934).
ML: local, Roslin; Crichton and Maggie Bower's
Glens; Cockpen; Murieston Water; Calder Wood;
Corstorphine Hill.
EL: scarce, widely scattered distribution,
Pencaitland; Humbie; Yester; Danskine;
Pressmennan; Tyninghame.

C. flacca Schreb.   Glaucous Sedge
Native. Status: widespread.
Marshes, rough grassland, beside rivers and ditches,
heath grassland, moorland, roadsides.

WL: widespread.
ML: widespread especially in upland areas,
locally frequent.
EL: quite widespread in upland areas and along
the coast, rare elsewhere.

C. panicea L.   Carnation Sedge
Native. Status: widespread.
Bogs, acid heaths and moors, hill flushes.

WL: widespread in S., absent from NE.
ML: quite widespread, mainly in upland areas,
Moorfoot and Pentland Hills; Holyrood Park.
EL: widespread in the Lammermuir Hills, rare in
lowland areas, mainly along the coast.

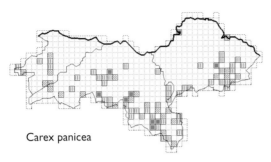

Carex panicea

C. laevigata Sm.   Smooth-stalked Sedge
Native. Status: rare.
Base-rich marshes, riversides, rock ledges in shady
ravines.

ML: rare, Borthwick; Roslin; (Arniston and
Tynehead, pre-1934).
EL: rare, Keith Water; Humbie; Gifford area;
Pressmennan; Bothwell area; Philip Burn;
(Haddington, pre-1934; Sheeppath Dean, 1966;
Woodhall Dean, 1967; Ogle Burn, 1970).

C. binervis Sm.   Green-ribbed Sedge
Native. Status: local.
Moorland, heath grassland, rough grassland.

WL: local, in hilly areas north and west of
Bathgate.
ML: quite widespread, mainly in upland areas,
seldom locally frequent.
EL: quite widespread in the Lammermuir Hills,
rare in lowland areas, in heathy woodland.

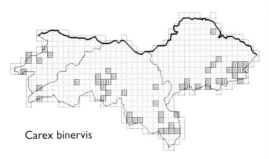

Carex binervis

C. distans. L.   Distant Sedge
Native. Status: rare.
Brackish marsh, coastal rocks.

WL: rare, Blackness; Dalmeny; (Bo'ness,
pre-1934).
ML: rare, Cramond Island.
EL: rare, Longniddry Bents; Aberlady Bay;
Belhaven Bay; Catcraig; (Fidra, 1969;
Tyninghame, 1972).

C. extensa Gooden.   Long-bracted Sedge
Native. Status: rare.
Saltmarshes, beaches.

EL: rare, Aberlady Bay; Tyninghame; Belhaven
Bay; east of Dunbar to Barns Ness;
(Thorntonloch, ?1909; North Berwick,
pre-1934).

C. hostiana DC.   Tawny Sedge
Native. Status: local.
Marshes, flushes in heath grassland.

WL: (South Queensferry, pre-1934; River Avon
near Brunton, 1977).
ML: local, mainly Moorfoot and Pentland Hills.
EL: rare, Aberlady Bay and upland areas, Linn
Dean; Stobshiel Reservoir; White Castle; Bothwell
Water; Heart Law; (Aikengall, 1967).

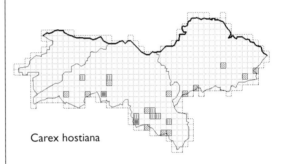

Carex hostiana

C. x fulva Gooden. (*C. hostiana* x *C. viridula*)
Native. Status: rare, where parents meet,
possibly overlooked.
Marshes, wet flushes.

ML: rare, near Gladhouse Reservoir, 1997.

C. viridula Michx.   Yellow Sedge

ssp. brachyrrhyncha (Celak.) B. Schmid
                    Long-stalked Yellow Sedge
Native. Status: local.
Coastal marshes, flushes in heath grassland.

WL: rare, Logie Water near Westfield, 1982; hills
north of Torphichen, 1983.
ML: local, widely scattered distribution, mainly
in upland areas.
EL: scarce in the Lammermuir Hills, rare in
lowland areas, in heathy woodland and at
Aberlady Bay.

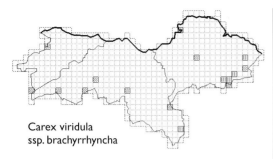

Carex viridula
ssp. brachyrrhyncha

ssp. oedocarpa (Andersson) B. Schmid

Common Yellow Sedge

Native. Status: widespread.

Marshes, beside ditches and rivers, rough grassland, wet flushes.

WL: widespread in S. and W., local elsewhere.
ML: quite widespread in upland areas, local elsewhere.
EL: widespread in upland areas.

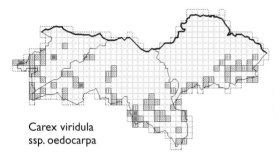

Carex viridula
ssp. oedocarpa

ssp. viridula   Small-fruited Yellow Sedge

Native. Status: rare.

Marshes, dune slacks.

EL: rare, coastal, Aberlady Bay; (Gullane, 1958; St Baldred's Cradle, 1973; Yellowcraig, 1980).

Records of *C. flava* in Martin (1934) refer to this species.

C. pallescens L.   Pale Sedge

Native. Status: scarce.

Moorland, heath grassland, damp rough grassland, riversides.

WL: rare, on the banks of the River Avon near Linlithgow, 1982.
ML: scarce, Threipmuir; Pentland Hills; Carlops; Gladhouse Mains; Rosebery Farm; Linhouse Water; Huntly Cot Hills, Moorfoot Hills.
EL: rare, Saltoun Forest; (Bothwell Water area, pre-1970).

C. caryophyllea Latourr.   Spring Sedge

Native. Status: local.

Rough grassland, heath grassland, coastal turf, volcanic outcrops.

WL: scarce, scattered records.
ML: local, mainly in upland areas, near Loganlea; Threipmuir; Cobbinshaw and Clubbiedean Reservoirs; Dalkeith Estate; Ladyside Burn; Murieston Water; Dewar Hill.
EL: local, mainly in upland and coastal areas.

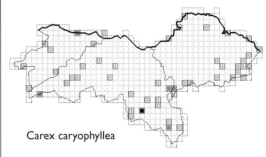

Carex caryophyllea

C. pilulifera L.   Pill Sedge

Native. Status: local.

Heath grassland, moorland, marshy rough grassland, riversides.

WL: local, mainly in S.
ML: local, in upland areas, widespread in SE.
EL: local, in upland areas, rare on lowland heaths.

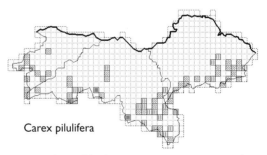

Carex pilulifera

C. aquatilis Wahlenb.   Water Sedge

Native. Status: scarce.

Marshes, in and beside lochs and rivers.

WL: rare, Blackridge; Fauldhouse; Barbauchlaw Burn, near Armadale; Westfield.
ML: scarce, Cobbinshaw; Threipmuir; Linhouse Water; South Livingston; (Carberry Estate, 1958).

C. acuta L.   Slender Tufted Sedge

Native. Status: rare.

Marshes, riversides.

WL: rare, River Almond near Blackburn;
(Linlithgow, pre-1840).
ML: rare, Breich Water at Breich.
EL: (Longniddry, 1871).

C. nigra (L.) Reichard   Black Sedge
Native. Status: widespread.
Marshes, bogs, heath grassland, rough grassland,
riversides, coniferous woodland.

WL: widespread in S., locally frequent in SW.,
scarce elsewhere.
ML: common in upland areas, local elsewhere.
EL: widespread in upland areas, local in lowland
areas, near Humbie, and along the coast.

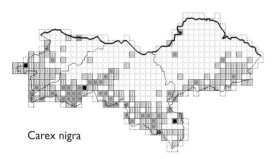

Carex nigra

C. pulicaris L.   Flea Sedge
Native. Status: local.
Marshes, heath grassland, beside rivers and ditches.

WL: rare, Fauldhouse Moor; Drumshoreland;
Riccarton Hills; Petershill; (Blawhorn, 1971;
Cockleroy, 1974).
ML: in upland areas only, local.
EL: quite widespread in upland areas, rare on the
coast, Aberlady Bay; (St Baldred's Cradle, 1866).

C. x pseudoaxillaris K. Richt. (*C. otrubae* x *C. remota*)
WL: Carribber and Drumshoreland, pre-1934.
ML: Musselburgh, Roslin and Balerno, pre-1934.
EL: Saltoun, pre-1934.

C. lasiocarpa Ehrh.   Slender Sedge
ML: Ravelrig and Duddingston, pre-1934.

C. strigosa Huds.   Thin-spiked Wood Sedge
ML: Arniston Woods, pre-1934, probably in error for *C.
sylvatica*.

C. depauperata Curtis ex With.   Starved Wood Sedge
ML: Bonaly Woods, late nineteenth century.

C. elata All.   Tufted Sedge
WL: Linlithgow Loch, pre-1934, probably in error for *C.
sylvatica*.
ML: Leith Docks, 1911.

POACEAE

Sasa Makino & Shibata

S. palmata (Burb.) E. G. Camus
       Broad-leaved Bamboo
Introduced. Status: rare but long-persisting
escape, occasionally planted.
Damp policy woodland.

WL: rare, Dundas Estate.
ML: rare, Silverknowes.
EL: rare, Dunglass Estate.

Nardus L.

N. stricta L.   Mat Grass
Native. Status: widespread.
Moorland, heath and other rough grassland, riversides,
bogs.

WL: widespread in S. and SW., rare or absent
elsewhere.
ML: very common in upland areas, rare or absent
in N.
EL: frequent in the Lammermuir Hills, rare
elsewhere.

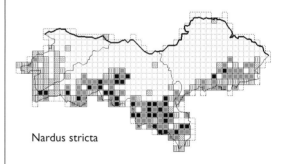

Nardus stricta

Milium L.

M. effusum L.   Wood Millet
Native. Status: local.
Mixed and deciduous woodland, wooded roadsides and
riversides.

WL: local, mainly in old woodlands and estates.
ML: local, Colinton Dell; Roslin Glen; Calder
Wood; Arniston Glen; Cramond; Old Craighall.
EL: rare, in old woodland and estates, Linn
Dean; Ormiston; Humbie; Yester; Yarrow.

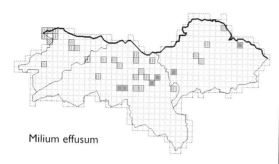

Milium effusum

Festuca L.

F. pratensis Huds.   Meadow Fescue
Native. Status: local, possibly under-recorded.
Roadsides, rough grassland.

WL: quite widespread in small quantities.
ML: scattered, mainly in lowland areas, locally frequent.
EL: local, mainly in lowland areas.

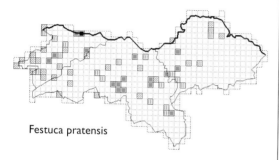

Festuca pratensis

F. arundinacea Schreb.   Tall Fescue
Native. Status: local.
Beside roads and rivers, rough grassland.

WL, ML, EL: scattered distribution, mainly in lowland areas and river valleys.

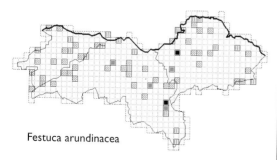

Festuca arundinacea

F. gigantea (L.) Vill.   Giant Fescue
Native. Status: local.
Mixed and deciduous woodland, wooded riversides.

WL: scattered along River Avon, local elsewhere.
ML: local, mainly in centre.
EL: scarce, mainly in estate woodlands.

F. altissima All.   Wood Fescue
Native. Status: rare.
Woodland, wooded ravines.

ML: rare, Roslin; Hawthornden, 1997; Edgelaw Reservoir Gorge, 1992; (Colinton, pre-1934).
EL: rare, Woodhall Dean.

F. heterophylla Lam.
        Grandmother's Hair, Various-leaved Fescue
Introduced. Status: rare, probably planted.
Estate woodland.

ML: rare, locally frequent at Lauriston Castle, Cramond, 1998.
EL: rare, Tyninghame Estate; (Yester, 1972).

F. arenaria Osbeck   Rush-leaved Fescue
Native. Status: rare.
Dunes.

WL: (South Queensferry, 1878).
EL: few sites, but frequent where it occurs, Aberlady Bay; Gullane; Yellowcraig; Belhaven Bay.

F. rubra L.   Red Fescue
Native. Status: common.
In many habitats, but most commonly found beside roads, in grassland and scrub, along riverbanks and in mixed woodland.

WL, ML: common.
EL: common, especially along the coast.

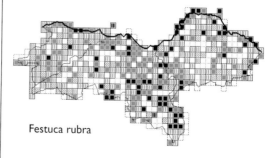

Festuca rubra

Subspecies of this variable grass were not systematically collected or recorded during the survey, but there are confirmed records for the following taxa:

ssp. juncea (Hack.) K. Richt.
Native.
Coastal rocks and grassland, rocky outcrops inland.

ML: Salisbury Crags; Granton.
EL: occasional along the coast, also on North Berwick Law.

ssp. litoralis (G. Mey.) Auquier
Native.
Coastal rocks.

EL: rarely recorded, The Leithies and coast east of North Berwick; (Bass Rock, 1974).

ssp. commutata Gaudin
Native.
Rough grassland, grassy waste ground.

ML: Granton, 1990.

ssp. megastachys Gaudin
Introduced.
Rough grassland, roadsides.

ML: Fountainhall, 1991.
EL: Prestongrange, pre-1988.

ssp. arctica (Hack.) Govor.
Native.
Wet upland flushes.
ML: near Loganlea Reservoir, 1956.

F. ovina L.　　　　　　　Sheep's Fescue
Native. Status: widepread.
Moorland, heath grassland, rough grassland, coniferous and mixed woodland, beside roads and rivers, quarries, waste ground, scrub.

WL: widespread and locally frequent in S., local in N.
ML: common in upland areas, local in lowland areas.
EL: common in upland areas, local and locally frequent in lowland areas, especially near the coast.

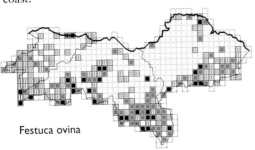

Festuca ovina

The subspecies of *F. ovina* were not recorded during the survey. However, **ssp. hirtula** (Hack. ex Travis) M. J. Wilk. is probably present in the Lothians.

F. vivipara (L.) Sm.　　　Viviparous Sheep's Fescue
Native. Status: rare.
Heath grassland.

ML: rare, Moorfoot and Pentland Hills; re-discovered on Arthur's Seat, 1990, last recorded there pre-1824.

F. filiformis Pourr.　　　Fine-leaved Sheep's Fescue
Native. Status: local, probably under-recorded.
Heath grassland, moorland, dunes, coastal turf, deciduous woodland.

WL: scarce.
ML: scarce, Moorfoot and Pentland Hills; Logan Waterfall; Habbie's Howe.
EL: scarce, scattered distribution, Aberlady Bay; Garleton Hills; White Castle; Sheeppath Burn.

x **Festulolium** Asch. & Graebn.

x **F. loliaceum** (Huds.) P. Fourn. (*Festuca pratensis* x *Lolium perenne*)　　　Hybrid Fescue
Native. Status: rare.
Roadsides, waste ground, cultivated ground.

WL: rare, a few sites where both parents occur.
ML: rare, Braid Hills; Fushiebridge.
EL: rare, Keith, 1993; Gifford, 1994; Garvald, 1991.

x **F. braunii** (K. Richt.) A. Camus (*Festuca pratensis* x *Lolium multiflorum*)
Native. Status; rare.
Waste ground.

WL: rare, Inveravon, 1992.

Lolium L.

L. perenne L.　　　　　Perennial Rye-grass
Native. Status: common, often sown.
Roadsides, rough and intensive grassland, waste ground, cereal and other arable fields, leys, public parks.

WL: common.
ML, EL: common, less so in some upland areas.

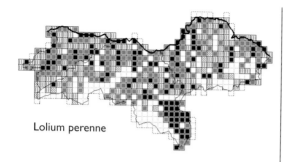

Lolium perenne

L. multiflorum Lam.   Italian Rye-grass
Introduced. Status: rare, agricultural relic.
Waste-places, field margins.

WL: local, small amounts at scattered locations,
except in S.
ML: rare, Loanhead; Niddrie; Kirknewton;
Holyrood Park; previously more widespread.
EL: widely scattered distribution, mainly in
lowland areas.

L. x boucheanum Kunth (*L. multiflorum* x *L.
perenne*)
Native. Status: rare, non-persistent agricultural
relic.
Waste ground, field margins, roadsides, farm yards.

ML: (Blackford Quarry, 1974).
EL: rare, Garleton, 1983; Drem, 1981 and 1982;
Sheriff Hall, 1982.

L. temulentum L.   Darnel
ML: frequent, pre-1934; on disturbed ground in Edinburgh,
1966.
EL: cornfields at Tranent and Haddington, pre-1871;
Innerwick, pre-1934; rubbish tip at North Berwick, 1961.

Vulpia C. C. Gmel.

V. bromoides (L.) Gray   Squirreltail Fescue
Native. Status: local, probably under-recorded.
Dry waste ground, roadsides, rubbish tips, volcanic
outcrops, railway embankments, allotments, dry
grassland.

WL: scarce, Bo'ness; Carriden; Craigton;
Drumshoreland; (South Queensferry, pre-1934).
ML: local, mainly in urban areas, Granton;
Midmar; Duddingston.
EL: scarce, scattered distribution, mainly in
lowland areas, railway cycle track, Ormiston;
Luffness quarry; Hedderwick Hill.

V. myuros (L.) C. C. Gmel.   Rat's Tail Fescue

Native. Status: rare.
Dry waste ground, roadsides, old railway lines.

WL: rare, Bo'ness; Armadale.
ML: rare, Leith; Granton; Monktonhall Colliery.
EL: rare, Prestongrange Mining Museum, 2001;
Tranent; West Saltoun; Pressmennan; Hedderwick
Hill; (Gifford, pre-1934; East Saltoun, 1956;
Thornton, 1972).

V. ciliata Dumortier   Bearded Fescue
ML: Musselburgh and Leith, pre-1934.

ssp. ambigua (Le Gall.) Stace & Auquier
ML: Comiston and sandy places at Leith, pre-1934.

Cynosurus L.

C. cristatus L.   Crested Dog's Tail
Native. Status: common.
Rough grassland, roadsides, intensive grassland,
riversides, marshes, heath grassland, waste ground,
scrub.

WL: common.
ML: local and locally frequent in N., common
elsewhere.
EL: local in N. and W., widespread and locally
frequent elsewhere.

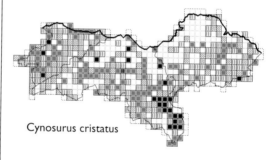

Cynosurus cristatus

C. echinatus L.   Rough Dog's Tail
Rare, ruderal casual.
WL: Linlithgow, pre-1934.
ML: Musselburgh, Leith and Slateford, pre-1934.

Puccinellia Parl.

P. maritima (Huds.) Parl.
           Common Saltmarsh Grass
Native. Status: local.
Saltmarshes, beaches, coastal turf within reach of the
tide.

WL: rare, Bo'ness; Blackness to Hopetoun;
Dalmeny.

ML: rare, locally frequent on the edge of coastal lagoons, Musselburgh.
EL: local along the coast, frequent in places, Aberlady Bay; The Leithies; Bass Rock; Belhaven to Barns Ness; (Longniddry area, pre-1962).

P. distans (Jacq.) Parl.   Reflexed Saltmarsh Grass
Native and introduced. Status: scarce, possibly increasing.
Waste ground, verges of salted roads, saltmarshes, sea cliffs, coastal rocks, piers.

WL: rare, Bo'ness; Blackness; near Cramond Bridge; by salted roads such as A90, A904 near South Queensferry, and a minor road north of Armadale; (South Queensferry, pre-1934).
ML: rare, Leith; Granton; Blackford Quarry.
EL: rare, Cockenzie; Dunbar area; (Prestonpans and Gullane, pre-1934; Aberlady Bay, pre-1962).

ssp. distans
ML: Leith Docks; Heriot; Musselburgh; Seafield; Roslin.
EL: Seafield; West Barns.

ssp. borealis (Holmb.) W. E. Hughes
ML: recorded from waste ground and dock areas.
EL: Hummell Rocks; Bass Rock; Tyne Estuary; (Fidra, 1957).

Briza L.

B. media L.   Quaking Grass
Native. Status: local.
Marshes, waterside habitats, roadsides, rough grassland, heath grassland. Usually on basic soils.

WL: scattered locations, mainly in hilly areas in W.
ML: quite widespread in upland areas, rare in N.
EL: local, confined to coastal and upland areas, Longniddry Bents; Aberlady Bay; Lammer Law; White Castle.

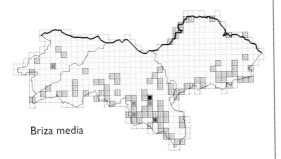
Briza media

B. maxima L.   Greater Quaking Grass
Rare casual.
ML: recorded sporadically from Leith Docks, 1891–1921.
EL: North Berwick, pre-1927.

Poa L.

P. annua L.   Annual Meadow Grass
Native. Status: common.
Roadsides, arable fields, rough grassland, waste ground, intensive grassland, leys, public parks, gardens.

WL, ML, EL: common.

P. trivialis L.   Rough Meadow Grass
Native. Status: common.
Roadsides, woodland, beside rivers and ditches, rough grassland, arable fields, intensive grassland.

WL, ML, EL: common.

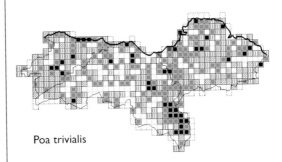
Poa trivialis

P. pratensis agg.   Smooth Meadow Grass
Native. Status: common.
Roadsides, rough grassland, mixed woodland, waste ground, beside rivers and ditches, scrub.

WL, ML, EL: common.

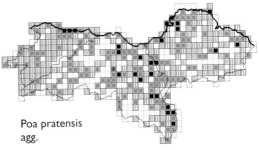
Poa pratensis agg.

Includes the following three taxa that were not usually distinguished during the survey.

382

P. humilis Ehrh. ex Hoffm.

Spreading Meadow Grass

Native. Status: probably widespread or even common, but distribution is uncertain due to confusion with *P. pratensis*.

Marshes, beside rivers and ditches, walls, rough grassland, roadsides.

WL: distribution uncertain, probably widespread.
ML: probably widespread, Fisherrow;
Meadowbank; Moorfoot Cleughs; Logan
Waterfall.
EL: widespread.

P. pratensis L.　　　　Smooth Meadow Grass
Native. Status: probably frequent, but possibly less widespread than *P. humilis*.

Roadsides, rough grassland, mixed woodland, waste ground, riversides, scrub.

WL: distribution uncertain.
ML: probably frequent.
EL: widespread, but apparently less frequent than *P. humilis*.

P. angustifolia L.

Narrow-leaved Meadow Grass

Native/Introduced. Status: rare, probably under-recorded.

Grassy banks, walls, waste ground, roadside verges.

ML: rare, Edinburgh Castle Rock above Johnston
Terrace; Sunnybank Place.
EL: rare, Prestongrange; West Mains near
Elphinstone; Longniddry/Gosford;
Brunt/Woodhall; (Innerwick area, 1980).

P. chaixii Vill.　　　Broad-leaved Meadow Grass
Introduced. Status: rare, escape.

Policies, coniferous woodland.

ML: very local, West Calder; Roslin Glen; Vogrie;
Glencorse; Hartwood House; (Leith, pre-1934)
EL: rare, Humbie; Yester; Tyninghame; (North
Berwick, 1911; Biel, 1971).

P. compressa L.　　　　Flattened Meadow Grass
Native. Status: rare, probably under-recorded.

Walls, rocks, roadsides, railways, dry waste ground.

WL: rare, Armadale; Bo'ness; near Easter Inch;
Whitrigg Bing; (Linlithgow, pre-1934).
ML: rare, Bonnyrigg; Loanhead; Wester Hailes
Quarry; Seafield; (Salisbury Crags, pre-1954).
EL: rare, Cockenzie; Waughton; East Linton;
(Longniddry and Gullane, pre-1934; Dirleton,
1968).

P. palustris L.　　　　Swamp Meadow Grass
Introduced. Status: rare.

Grassy places and rough ground, mainly ruderal.

WL: (South Queensferry, pre-1934.)
ML: rare, recorded from wet woodland at
Musselburgh, 1988, probably now extinct there;
(Dryden, Leith, Slateford and Murieston,
pre-1934).

P. nemoralis L.　　　　Wood Meadow Grass
Native. Status: quite widespread.

Mixed and deciduous woodland, scrub.

WL: quite widespread and locally frequent in N.
and E.
ML: widespread in lowland areas, locally
frequent.
EL: quite widespread in lowland areas, rare in
upland areas.

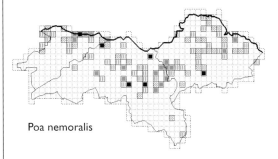

Poa nemoralis

P. imbecilla Sprèng.
var. **breviglumis** (Hook. f.) Cheeseman
Introduced. Status: rare, horticultural weed.

ML: a relict from the Royal Caledonian Grounds
in the nineteenth century at Inverleith, now a
weed at the Royal Botanic Garden Edinburgh.
Not yet recorded in the wild but possibly over-looked because of its similarity to delicate plants
of *Agrostis capillaris*.

P. bulbosa L.　　　　Bulbous Meadow Grass
Native and introduced. Status: rare.

Open ground on sandy soil, waste ground.

ML: (casual only, Musselburgh, Leith and
Slateford, pre-1934).
EL: rare, in natural vegetation on North Berwick
Golf Course. This site is well north of other
British localities and was not recorded until
1972.

Dactylis L.

D. glomerata L.   Cock's Foot
Native. Status: common.
Roadsides, rough grassland, waste ground, mixed woodland, scrub, arable fields, riversides, intensive grassland.

WL: common.
ML, EL: common, except in some upland areas.

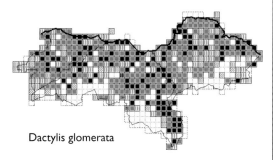

Dactylis glomerata

Catabrosa P. Beauv.

C. aquatica (L.) P. Beauv.   Water Whorl Grass
Native. Status: rare.
Ponds, lochs, streams.

WL: rare, stream near Easter Inch; pond at Ecclesmachen, 1988; (South Queensferry pre-1934; Linlithgow Loch, 1977).
ML: rare, Duddingston Loch; (Lochend Loch 1919; Leith and Slateford, pre-1934).
EL: (Gullane, pre-1863; North Berwick, 1882).

Catapodium Link

C. rigidum (L.) C. E. Hubb.   Fern Grass
Usually native. Status: rare.
Dunes, coastal turf, dry stony places, walls.

ML: rare, Musselburgh; Salisbury Crags; Eyre Place; (Arthur's Seat and Blackford Hill, 1902; an alien at Levenhall Quarry tip, 1912–16).
EL: scarce, mainly on or near the coast, Aberlady Bay; North Berwick area; Dunbar to East Barns; Dunglass; Bankton; Burn Hope; Oldhamstocks.

C. marinum (L.) C. E. Hubb.   Sea Fern Grass
Native. Status: rare.
Cliffs, coastal rocks, walls close to the sea.

ML: (west of Granton, 1863–1943).
EL: rare, Gullane Point; rocks to the east of North Berwick; Dunbar.

Parapholis C. E. Hubb.

P. strigosa (Dumort.) C. E. Hubb.   Hard Grass
Native. Status: rare.
Saltmarsh.

WL: (Blackness, 1836).
EL: rare, Aberlady Bay; Belhaven Bay; (Cockenzie and Gosford Links, 1831; Morrison's Haven, 1841; Longniddry shore, 1902).

Glyceria R. Br.

G. maxima (Hartm.) Holmb.   Reed Sweet Grass
Native. Status: local.
In and beside canals, rivers and ponds, marshes.

WL: local, mainly as a waterside dominant on the Union Canal; Linlithgow Loch; a few other ponds and streams.
ML: local, mainly in N., locally frequent/abundant, Union Canal; Water of Leith; Threipmuir Reservoir.
EL: rare, Aberlady Bay; Luffness; Dirleton; Hailes Castle; West Barns; (Gifford area, pre-1962).

var. **variegata** Boom & J. D. Ruys,
An introduction of garden origin.

EL: Pressmennan Lake, 1997.

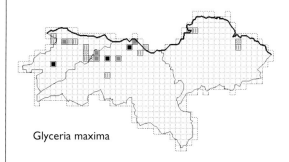

Glyceria maxima

G. fluitans (L.) R. Br.
           Floating Sweet Grass, Flote Grass
Native. Status: widespread, possibly over-recorded.
In and beside rivers, ditches and ponds, marshes.

WL: common in S., local in N.
ML: quite widespread in S., common in the valley of Gala Water, local in N.
EL: quite widespread in upland areas and around Humbie, rare elsewhere.

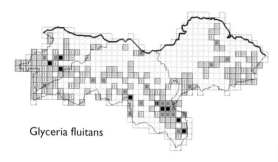

Glyceria fluitans

G. x pedicellata F. Towns. (*G. fluitans* x *G. notata*)   Hybrid Sweet Grass
Native. Status: rare.
Beside water.

ML: rare, Stow; Penicuik Estate; Burdiehouse Burn.
EL: rare, Longyester; Newmains, near Stenton.

G. declinata Bréb.
                    Small Sweet Grass, Small Flote Grass
Native. Status: quite widespread, sometimes confused with *G. fluitans*.
Marshes, in and beside streams and ditches.

WL, ML: quite widespread.
EL: quite widespread, Gosford House; Gifford; North Berwick Law; West Barns; Oldhamstocks area.

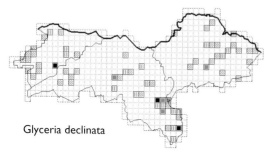

Glyceria declinata

G. notata Chevall.
                    Plicate Sweet Grass, Plicate Flote Grass
Native. Status: local, under-recorded because sometimes confused with *G. fluitans*.
In and beside streams, marshes.

WL: scattered distribution.
ML: local, mainly in SE.; (Duddington Loch, 1878; Bavelaw, 1902; Lochend, 1927).
EL: scattered distribution, mainly in centre, Keith Water; Humbie; Morham; Gifford; Garvald.

Melica L.

M. nutans L.   Northern or Mountain Melick
Native. Status: rare.
Woodland.

WL: rare, near Carriber; (Linlithgow, 1914; Woodcockdale, 1958).
ML: an unconfirmed record from Water of Leith at Balerno, 1989; (Penicuik Estate and Dalkeith, pre-1900; Roslin and Dryden, pre-1934).
EL: rare, Linn Dean; Sheeppath Dean; (Stottencleugh, 1885).

M. uniflora Retz.   Wood Melick
Native. Status: scarce.
Rocky outcrops in mixed and deciduous woodland, walls.

WL: rare, Carriber; Kinneil; River Almond near M8; (Linlithgow, 1905).
ML: rare, Bilston; Roslin and Gore Glens; Edgelaw Reservoir; Arniston; Lady Lothian Wood; (Penicuik, Tynehead and Colinton, pre-1934).
EL: rare, Linn Dean Water; Humbie; Papana Water; Pressmennan area; Woodhall Dean; Aikengall Water; (North Berwick, pre-1934; Sheeppath Dean, 1966).

Helictotrichon Besser ex Schult. & Schult. f.

H. pubescens (Huds.) Pilg.   Downy Oat-grass
Native. Status: local.
Rough grassland, public parks, riversides, usually on base-rich soils, often with *H. pratense*.

WL: rare, small patches, north of Blackridge; west of Westfield; Whitburn; Blackburn; Seafield.
ML: occasional, mainly in SE., Granton; Threipmuir, Harperrig and Morton Reservoirs; Carlops; Bowshank; Salisbury Crags; (Borthwick, Corstorphine and Hallyards, pre-1934).
EL: scattered distribution along the coast from west of Yellowcraig to east of Dunbar; inland at Linn Dean, Bothwell Water and Thurston.

H. pratense (L.) Besser   Meadow Oat-grass
Native. Status: local.
Basic grassland, quarries, rocky outcrops, dunes.

WL: rare, Riccarton Hills, 1982; Ochiltree Fort; Snab Point; old railway near Carlowrie; (on basic grassland at Broxburn, 1974; Winchburgh, 1974).
ML: scarce, Arthur's Seat; Dalmahoy; Middleton;

Pentland and Moorfoot Hills; (Braid Hills and Currie, pre-1934).
EL: occasional, mainly in NE., coast between Gullane and Bilsdean; Whitekirk Hill; Papana Water.

Arrhenatherum P. Beauv.

A. elatius (L.) P. Beauv. ex J. & C. Presl
False Oat-grass
Native. Status: widespread.
Roadsides, rough grassland, arable fields, waste ground, mixed woodland, scrub, riversides, hedgerows.

WL: common.
ML, EL: common except in some upland areas.

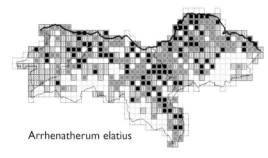

Arrhenatherum elatius

var. **bulbosum** (Willd.) St-Amans   Onion Couch
WL: Tailend Moss.
ML: (Dalhousie, 1836; Mortonhall, 1845; Granton, 1907; Musselburgh, 1914).
EL: Bilsdean, 1992; (Innerwick, pre-1934).

A. albumb (Vahl) Clayton
Wool casual.
ML: Levenhall, 1913–14.

Avena L.

A. strigosa Schreb.   Bristle Oat
Introduced. Status: rare.
Waste ground, coastal scrub/turf, rough grassland.

ML: rare, Smeaton, 1984; (cornfields at Meadowbank, 1824; later known mainly as a dockland casual, 1907–21).

A. fatua L.   Wild Oat
Probably introduced. Status: local, under-recorded in the survey.
Arable fields, field-margins and tracks, roadsides.

WL: local.
ML: scarce, Leith; Gorebridge; West Calder.

EL: quite widespread in lowland areas.

A. sativa L.   Oat
Introduced. Status: local, agricultural relic or escape, under-recorded in the survey.
Waste ground, cereal fields, roadsides.

WL: occasional.
ML: scarce, Granton; Leith Docks; Pilton; Braehead Quarry.
EL: rarely recorded, near Kingston and East Linton, early 1980s; Bilsdean, 1992.

Trisetum Pers.

T. flavescens (L.) P. Beauv.   Yellow Oat-grass
Native. Status: quite widespread.
Roadsides, rough and intensive grassland, quarries, usually on base-rich soils.

WL: quite widespread in N. and E., rare or absent in S.
ML: quite widespread in N. and E., locally frequent, rare elsewhere.
EL: quite widespread, locally frequent in lowland and coastal areas.

Koeleria Pers.

K. macrantha (Ledeb.) Schult.
Crested Hair-grass
Native. Status: local.
Dry rough grassland, quarries, rocky outcrops, dunes and other coastal habitats, on base-rich soils.

WL: rare, Binny Craig; Blackness; The Binns; Dalmeny; (Linlithgow, pre-1934; Carriden, 1956; Cockleroy, 1974).
ML: frequent in a few locations, Arthur's Seat; Dalmahoy; Blackford and Braid Hills; Cramond Island.
EL: scarce, mainly along the coast, Longniddry Bents; Aberlady Bay; Yellowcraig; North Berwick Law; Catcraig.

Deschampsia P. Beauv.

D. cespitosa (L.) P. Beauv.   Tufted Hair-grass
Native. Status: widespread.
Damp areas in mixed, coniferous and deciduous woodland, beside roads, rivers and ditches, rough grassland, marshes, scrub.

WL: widespread throughout, common in S.
ML, EL: common in upland areas, quite widespread in lowland areas.

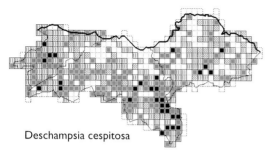

Deschampsia cespitosa

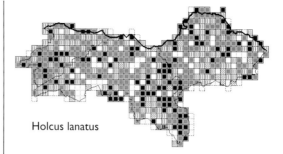

Holcus lanatus

ssp. cespitosa
Native. Status: the common spp. but rarely recorded in the survey.
Heath grassland, rough grassland, roadsides.

EL: Papana Water; White Castle; Dunbar Common.

ssp. parviflora (Thuill.) Dumort.
Native. Status: rare, possibly overlooked.
Woodland, hedgerows.

ML: frequent in Arniston Estate, 1993.
EL: Pressmennan Lake, 1992.

D. flexuosa (L.) Trin.   Wavy Hair-grass
Native. Status: common in upland areas.
Dry heath, moorland, bogs, beside roads and rivers, mixed, coniferous and deciduous woodland, rough grassland, scrub, on acid soils.

WL: scattered distribution in N., widespread and locally frequent in S.
ML, EL: common in upland areas, local elsewhere.

H. mollis L.   Creeping Soft Grass
Native. Status: widespread.
Mixed, coniferous and deciduous woodland, beside roads, rivers and ditches, scrub, rough grassland, cereal fields, waste ground.

WL: widespread, locally frequent.
ML: local in N., widespread elsewhere, locally frequent, common in SE.
EL: widespread except in some arable areas in N., locally frequent.

Two new varieties of *Holcus mollis* were described from Lothian material by Richard Parnell in his *Grasses of Scotland* (1842). A double-awned form, var. **biaristatus** was collected from Roslin Woods, (**ML**), and var. **parviflorus** was collected at Prestonpans, (**EL**). However, it has recently been shown that the diagnostic short lemmas in the latter were caused by an infestation with smut.

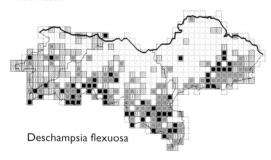

Deschampsia flexuosa

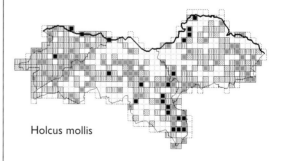

Holcus mollis

Holcus L.

H. lanatus L.   Yorkshire Fog
Native. Status: common.
Roadsides, watersides, rough grassland, woodland, waste ground, scrub, intensive grassland, leys, cereal fields.

WL, ML, EL: common.

Corynephorus P. Beauv.

C. canescens (L.) P. Beauv.   Grey Hair-grass
?Introduced. Status: rare.
Sandy clearing in conifer plantation.

EL: rare, Belhaven Bay, first recorded in 1986 and has increased considerably since then.

Aira L.

A. caryophyllea L.   Silver Hair-grass
Native. Status: local.
Waste ground, roadsides, quarries, bings, railway embankments, coastal turf, dry ground, shallow soils, rocky outcrops.

WL: scattered distribution.
ML: widespread in lowland areas, scarce in upland areas.
EL: scattered distribution, Ferny Ness; Gullane Point; White Castle; Stenton; Monynut Water.

ssp. multiculmis (Dumort.) Bonnier & Layens
ML: on coal waste at Monktonhall Colliery, Cockpen, and Bilston; in grassland on Arthur's Seat.
EL: roadside at Prestongrange, 1991; waste ground at Tranent, 1987; disused railway south of Macmerry, 1981.

A. praecox L.   Early Hair-grass
Native. Status: widespread.
Quarries, bings, waste ground, rough grassland, heath and moorland, beside roads, rivers and railways, coastal habitats, rocky outcrops.

WL, ML: widespread.
EL: quite widespread in upland areas, local elsewhere, Gosford Bay; Stobshiel Reservoir; White Castle; St Baldred's Cradle.

Anthoxanthum L.

A. odoratum L.   Sweet Vernal Grass
Native. Status: common.
Rough grassland, roadsides, moorland, heath grassland, watersides, woodland, intensive grassland, scrub, marshes.

WL: common.
ML: common in upland areas, local elsewhere, locally abundant.
EL: common in upland areas, local elsewhere, locally frequent.

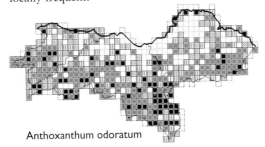
Anthoxanthum odoratum

Phalaris L.

P. arundinacea L.   Reed Canary Grass
Native. Status: widespread.
Riversides and other waterside habitats, marshes, mixed woodland.

WL: widespread.
ML: local, locally frequent/abundant.
EL: quite widespread in lowland areas, rare elsewhere.

var. picta L.   Gardener's Garters
Introduced. Status: rare, garden outcast.
Waste ground.

ML: (Cobbinshaw, 1958; Nine Mile Burn, 1977).
EL: Prestonpans, 1988.

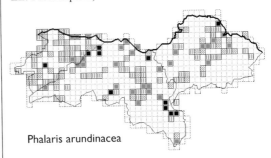
Phalaris arundinacea

P. canariensis L.   Canary Grass
Introduced. Status: rare, casual.
Roadsides, railway embankments, waste ground, rubbish tips.

WL: rare, Bo'ness; (Carriden, 1972; Bathgate and Broxburn, 1973).
ML: rare, Leith; Musselburgh; Borthwick; Penicuik; Seafield; Warriston.
EL: rare, near habitation, Tranent; Haddington; Dirleton; West Barns; (Drem, 1849; North Berwick, 1956; East Linton, 1957).

P. minor Retz.   Lesser Canary Grass
Introduced. Status: rare, casual, wool-alien.
Waste ground.

ML: rare, Leith, 1983 and pre-1934.

Agrostis L.

A. capillaris L.   Common Bent
Native. Status: common, frequently greatly dwarfed due to a fungal disease.
Rough grassland, roadsides, mixed woodland,

moorland, intensive grassland, riversides, waste ground, scrub.

WL, ML, EL: common.

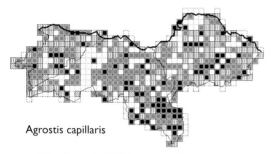

Agrostis capillaris

A. x bjoerkmanii Widén (*A. capillaris* x *A. gigantea*)
Native. Status: rare spontaneous hybrid growing with both parents.
Field margin.

ML: rare, Stow, 1988.

A. x murbeckii Fouill. ex P. Fourn. (*A. capillaris* x *A. stolonifera*)
Native. Status: rare, under-recorded.
Roadsides.

EL: rare, west of Elphinstone; Seton Sands; Drem.

A. gigantea Roth   Black Bent
Native. Status: local.
Roadsides, cereal fields, rough grassland, woodland.

WL: thinly scattered throughout.
ML: widely scattered, mainly in lowland areas, Holyrood; Dalkeith; Cramond; West Calder; New Hailes.
EL: thinly scattered, mainly in lowland areas, Tranent; Colstoun; Longyester; Tyninghame.

A. castellana Boissier & Reuter   Highland Bent
Introduced. Status: rare, possibly increasing because of planting on golf courses and other sports areas.
Roadsides, amenity and sports areas.

WL: rare, recorded from scrub near Drumcross, 1983.
ML: local, Penicuik Estate, 1999.

A. stolonifera L.   Creeping Bent
Native. Status: common.
Beside roads, rivers and ditches, marshes, rough grassland, waste ground, woodland, arable fields.

WL, ML, EL: common.

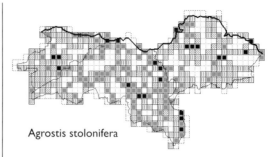

Agrostis stolonifera

A. canina agg.   Velvet Bent
Native. Status: local.
Heath and moorland, streamsides, marshes, rough grassland, woodland, acid soils.

WL: widespread in upland areas.
ML: widespread and locally frequent in S.
EL: local, mainly in upland areas.

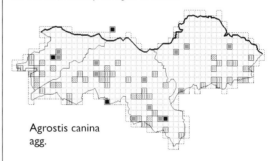

Agrostis canina agg.

Includes the following two taxa that were not always distinguished during the survey.

A. canina L.   Velvet Bent
Native. Status: local.
Heath and moorland, streamsides, marshes, rough grassland, bogs, damp woodland, acid soils.

WL: widespread and locally frequent in S., rare or absent in N.
ML: quite widespread in S. where it is locally frequent, rare or absent in N., Musselburgh; Auchencorth; West Calder.
EL: local, mainly in upland areas, White Castle; Sheeppath Burn; St Baldred's Cradle.

A. vinealis Schreb.   Brown Bent
Native. Status: local, probably under-recorded.
Dry heath and moorland, rough grassland.

WL: local, mainly in hilly areas in W.
ML: local, mainly in SE.
EL: local, mainly in upland areas, Garleton Hills; White Castle; Pressmennan; Whiteadder Water.

A. scabra Willd.   Rough Bent
Introduced. Status: rare, grain alien.
Waste ground.

ML: rare, Trinity, 1993; (regularly recorded from Leith Docks *c.* 1900).

Calamagrostis Adans.

C. epigejos (L.) Roth   Wood Small Reed
Native. Status: rare.
Damp areas, beside ditches and ponds.

WL: rare, Bo'ness; (Carriden, 1972).
ML: (one eighteenth-century record from Myreside Bog, an area now more or less drained).

C. canescens (F. H. Wigg.) Roth
     Purple Small Reed
Native. Status: rare.
Fens, open woodland.

ML: rare, south bank of River North Esk, Penicuik Estate, re-found in 1998 having previously been thought to be extinct.

Ammophila Host

A. arenaria (L.) Link   Marram Grass
Native. Status: local.
Dunes, beaches, coastal scrub/turf.

WL: now found only along Dalmeny foreshore; (Bo'ness area, 1955).
ML: local, Cramond; Cramond Island; Granton; Joppa.
EL: along much of the coast, locally abundant, enhanced by planting, since 1950, as part of coastal management.

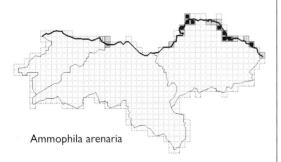

Ammophila arenaria

Lagurus L.

L. ovatus L.   Hare's Tail
Introduced. Status: rare, casual.
Pavements, waste ground.

ML: rare, George Street, half a block east of a flower shop, 1984; (the New Town, Edinburgh, 1970).

Apera Adans.

A. spica-venti (L.) P. Beauv.   Loose Silky Bent
Introduced. Status: rare.
Waste ground.

WL: rare, Glendevon Pond, 2000.
ML: rare, Warriston walkway, 1999; (Leith Docks, 1980; reported to have been more widespread pre-1927).

Mibora Adans.

M. minima (L.) Desv.   Early Sand-grass
Introduced. Status: rare.

EL: a subject of research by J. L. Knapp at Edinburgh in 1800, when this species was known only in Anglesey. Reputedly abundant along the shore at Gosford Bay, Gullane and Dirleton by the 1850s, and known to have been sown there. Now naturalised and thriving in dunes north of Gullane.

Polypogon Desf.

P. monspeliensis (L.) Desf.   Annual Beard-grass
Introduced. Status: rare, horticultural weed.
Gardens.

ML: rare, Old Pentland, 1991.

P. viridis (Gouan) Breistr.   Water Bent
ML: Musselburgh, 1913; Braehead Quarry, 1978.

Alopecurus L.

A. pratensis L.   Meadow Foxtail
Native. Status: common.
Roadsides, rough grassland, intensive grassland, beside rivers and ditches, cereal fields, woodland, waste ground.

WL: common.
ML: common except in some upland areas.
EL: widespread, locally frequent.

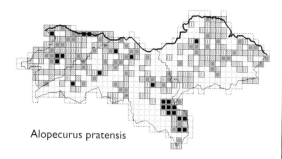

Alopecurus pratensis

A. x brachystylus Peterm. (*A. pratensis* x *A. geniculatus*)
Native. Status: rare.
Damp cereal field.

WL: rare, near Inveravon, 1992.

A. geniculatus L.   Marsh Foxtail
Native. Status: quite widespread.
Waterside habitats, roadsides, marshes, wet areas in rough and intensive grassland, leys, cereal fields and waste ground.

WL: widespread and locally frequent except in N. and E.
ML: rare in some upland areas and in NE., quite widespread elsewhere and locally frequent, common in SE.
EL: scattered distribution, but rare in drier N.

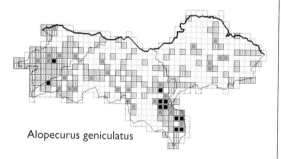

Alopecurus geniculatus

A. myosuroides Huds.   Black Grass
Introduced. Status: rare, casual, seldom persists for long.
Rough grassland, waste ground, railway embankments.

WL: (Linlithgow, pre-1934).
ML: rare, Newhaven, 1984; appeared at Granton after reseeding when the sea wall was dismantled in 1996; (Leith, 1965; Seafield, 1977).
EL: rare, edge of a root crop field, Stenton, 1993; (North Berwick, pre-1934; Thornton Station, 1970; Aberlady Station, 1971; West Barns, 1975).

Phleum L.

P. pratense agg.   Timothy
Native. Status: widespread.
Roadsides, rough and intensive grassland, leys, waste ground, cereal fields.

WL: widespread, locally frequent.
ML: common except in upland areas.
EL: widespread and locally frequent except in arable and upland areas.

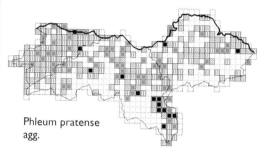

Phleum pratense agg.

Includes the following two taxa that were not always distinguished in the survey.

P. pratense L.   Timothy
Native. Status: widespread, often sown in grass-seed mixtures.
Roadsides, rough and intensive grassland, leys, waste ground, cereal fields.

WL: widespread, locally frequent.
ML: rare in upland areas, common elsewhere.
EL: local in upland areas and in arable land in N., widespread and locally frequent elsewhere.

P. bertolonii DC.   Smaller Cat's-tail
Native. Status: widespread, probably under-recorded.
Rough grassland, roadsides, waste ground.

WL, ML: widespread.
EL: widespread in lowland areas, often along the edges of golf courses; (recorded in upland areas, 1950s).

P. arenarium L.   Sand Cat's-tail
Native. Status: rare.
Coastal turf, dunes.

ML: (Cramond, *c.* 1800)
EL: rare, Aberlady Bay; Yellowcraig; formerly occasional along the coast; (Cockenzie, 1831; Prestonpans, pre-1839; Gosford Links, 1868; Tyninghame, pre-1871; Gullane, pre-1934; Dunglass, 1957).

Bromus L.

For the following species an interpretation of the genus *Bromus* has been used that differs from that in Stace (1997). Synonyms refer to the nomenclature used in Stace (1997).

B. commutatus Schrad.   Meadow Brome
Introduced. Status: rare, casual.
Waste ground, rubbish tips.

WL: (one record in E., *c.* 1950).
ML: rare, Granton, 1987; Leith Docks, 1977 and 1990; (Eskmouth, Newbattle, Duddingston, Leith and Slateford, pre-1934).
EL: (between North Berwick and Tantallon, 1844; Aberlady, pre-1934; Gullane area, 1955).

B. hordeaceus L.   Soft Brome
Native. Status: quite widespread.
Roadsides, rough grassland, waste ground, cereal fields, allotments.

WL: quite widespread.
ML: patchy distribution, mainly in lowland areas.
EL: quite widespread in lowland areas, locally frequent; rare or absent in upland areas.

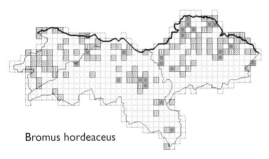

Bromus hordeaceus

ssp. thominei (Hardouin) Braun-Blanq.
EL: recorded from coastal turf at North Berwick and Catcraig; (Longniddry area and Dunbar, pre-1968).

ssp. ferronii (Mabille) P. M. Sm.
ML: Musselburgh, Portobello, and Leith, pre-1934, but the records are dubious.

B. x pseudothominei P. M. Sm. (*B. hordeaceus* x *B. lepidus*)   Lesser Soft Brome
Native, but also often introduced. Status: seemingly rare but much under-recorded. Widely sown as a grass seed contaminant.
Roadsides, rough and intensive grassland.

WL: (Linlithgow, 1962).

ML: (Goldenacre, 1953; Balerno and Comely Bank, 1954).
EL: rare, Seton Sands; East Fortune; Whitekirk Hill; Seacliff; Stoneypath; West Barns; Barns Ness; Skateraw.

B. lepidus Holmb.   Slender Soft Brome
Probably introduced in amenity grassland.
Status: rare, probably under-recorded.
Public parks, waste ground, grassland.

WL: (recorded in NW., 1956).
ML: rare, Arthur's Seat; (Malleny Mills, Comely Bank and Penicuik, pre-1955).
EL: rare, Craigielaw, 1989; (Haddington area, 1961; Aberlady Bay, 1963; Gullane, Bothwell Water and Monynut Water, 1972).

B. secalinus L.   Rye Brome
Introduced. Status: rare.
Waste ground, amenity grassland.

WL: (Kinneil, pre-1934).
ML: abundant in land sown with cheap grass seed, Balerno High School, 1983; (Musselburgh, Leith, Slateford and Roslin, pre-1934).

var. hirtus (F. Schultz) Asch. & Graebn.
EL: Tyninghame, 1960.

B. ramosus Huds. (*Bromopsis ramosa* (Huds.) Holub)   Hairy Brome
Native. Status: local.
Mixed and deciduous woodland, scrub, shady roadsides and watersides.

WL: quite widespread in N., rare or absent elsewhere.
ML, EL: local, locally frequent, mainly in centre.

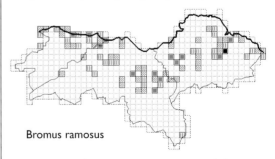

Bromus ramosus

B. erectus Huds. (*Bromopsis erecta* (Huds.) Fourr.)   Upright Brome
Native. Status: rare.
Roadsides, railway embankments, grassy slopes on basic soils, waste ground.

ML: rare, Blackford Hill; Granton; Arthur's Seat.
EL: rare, Gosford; West Saltoun; (Dunbar, pre-1934; Pencaitland, 1956).

B. inermis Leysser (*Bromopsis inermis* ssp. *inermis* (Leysser) Holub)   Hungarian Brome
Introduced. Status: rare.
Waste places, rough grassland.

ML: rare, Sighthill; Currie; Musselburgh Racecourse, 2001; (quarry tips, 1904–21; Fisherrow Harbour, 1976–8).

B. diandrus L. (*Anisantha diandra* (Roth) Tutin ex Tzvelev)   Great Brome
Introduced. Status: rare, usually casual.
Waste ground, rubbish tips.

ML: rare, Leith Docks, 1971–89; Rosewell, 1993; (Musselburgh, Portobello, Leith and Slateford, pre-1934).
EL: the record published in Watsonia, 1981, was redetermined as *B. rigidus*.

B. rigidus Roth (*Anisantha rigida* (Roth) Hyl.)
   Ripgut Brome
Introduced. Status: rare.
Rough, well-drained grassland, pathsides, waste ground.

ML: rare, Leith; (frequent on rubbish tips and at the docks, 1885–1922).
EL: rare, Prestongrange Mining Museum, 1998; Prestonpans; Seton Sands.

B. sterilis L. (*Anisantha sterilis* (L.) Nevski)
   Barren Brome
Native. Status: common in N., scarce in S.
Roadsides, cereal and other arable fields, waste ground, hedgerows, rough grassland.

WL: widespread in N., rare in S.
ML: common in N. and NE., scarce elsewhere.
EL: widespread and locally frequent in lowland areas, common along the coast, rare in upland areas.

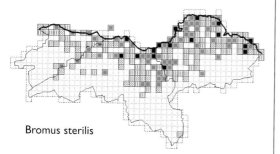

Bromus sterilis

B. tectorum L. (*Anisantha tectorum* (L.) Nevski)
   Drooping Brome
Introduced. Status: rare, casual.
Roadsides, rubbish tips.

WL: (Kirkliston, pre-1934).
ML: rare, locally abundant in 1990 alongside the Edinburgh City Bypass at Straiton; reseeded Lagoon, Musselburgh, 1996; (regularly found on rubbish tips at Levenhall, Portobello, Slateford, Craigmillar Castle and Leith Docks, 1883–1934).

B. carinatus Hook. & Arn. (*Ceratochloa carinata* (Hook. & Arn.) Tutin)   California Brome
Introduced. Status: rare.
Mixed and deciduous woodland, field margins, roadsides.

EL: rare, Saltoun; edge of shelter belt west of Yellowcraig, known since 1983; locally frequent at both sites.

B. arvensis L.   Field Brome
ML: Leith, Murrayfield and Slateford, pre-1934; Liberton, 1980.
EL: Innerwick, pre-1934.

B. racemosus L.   Smooth Brome
WL: historical record from Dalmeny.
ML: Musselburgh, Roslin and Craigmillar, pre-1934; Salisbury Crags, 1944.

B. interruptus (Hack.) Druce   Interrupted Brome
ML: garden weed, Leith, 1918–19.

B. catharticus Vahl (*Ceratochloa cathartica* (Vahl) Herter)
Rescue Brome
ML: Slateford quarry tip, 1971; waste ground at Leith, 1978.

Brachypodium P. Beauv.

B. pinnatum (L.) P. Beauv.   Tor-grass
Introduced. Status: rare.
Railway embankments, heath, coastal grassland.

ML: rare, Cobbinshaw.
EL: rare, Yellowcraig, known since 1977.

B. sylvaticum (Huds.) P. Beauv.   False Brome
Native. Status: local.
Old mixed and deciduous woodland, scrub, roadsides.

WL: local, scattered sites along the River Avon; Almondell; Hopetoun; South Queensferry; Dalmeny.
ML: local, mainly in N. and centre, locally frequent.
EL: local, mainly in centre, Elphinstone; Keith Glen; Garvald; Dunglass.

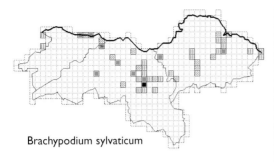

Brachypodium sylvaticum

Elymus L.

E. caninus (L.) L.   Bearded Couch
Native. Status: local.
Mixed woodland, wooded roadsides and riversides.

WL: local, mainly in old woodland along the
River Avon; Kinneil; Wester Shore.
ML: local, mainly in centre, locally frequent.
EL: occasional, mainly in centre, in old wood-
land, Keith/Humbie; Gifford; in woods beside
lower River Tyne; Dunglass Dean.

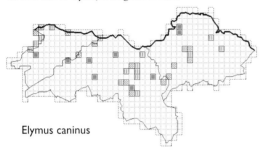

Elymus caninus

Elytrigia Desv.

E. repens (L.) Desv. ex Nevski   Common Couch
Native. Status: common.
Roadsides, cereal fields, waste ground, rough grassland,
hedgerows, potato and other vegetable fields, mixed
woodland, scrub, beaches, dunes.

WL: common.
ML, EL: common in lowland areas, rare in
upland areas.

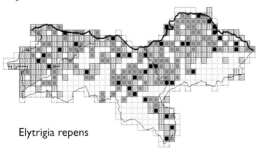

Elytrigia repens

var. **aristata** ined. occurs in farm and coastal
habitats and on roadsides in **EL** and could be
confused with *Elymus caninus*.

Coastal forms of *E. repens* have been mistaken
for *E. atherica* which does not occur here.

E. x laxa (Fr.) Kerguélen (*E. repens* x *E. juncea*)
Native. Status: rare.
Sandy beaches.

ML: (Granton and Musselburgh, pre-1934).
EL: rare, Cockenzie; Dunbar; (Seton, 1804;
Longniddry and Aberlady Bay, pre-1934).

E. juncea (L.) Nevski   Sand Couch
ssp. **boreoatlantica** (Simonet & Guin.) Hyl.
Native. Status: local.
Sandy beaches, dunes.

WL: rare, Carriden; Wester Shore; Dalmeny.
ML: rare, Musselburgh; Cramond; Seafield.
EL: scattered sites along the coast, Prestonpans;
Yellowcraig; Belhaven Bay; Barns Ness.

Leymus Hochst.

L. arenarius (L.) Hochst.   Lyme Grass
Native. Status: local.
Sandy beaches, dunes.

WL: rare, Dalmeny; (Carriden, 1964).
ML: local, Cramond to Musselburgh.
EL: common along the coast, enhanced by
planting, since 1950, as part of coastal
management.

Hordeum L.

H. vulgare L.   Six-rowed Barley
Introduced. Status: rare, escape.
Roadsides, vegetable crops, farmyards, intensive
grassland, coastal scrub.

WL, ML: local.
EL: rarely recorded, Craigleith Island; Drem;
Spott; Bilsdean.

H. distichon L.   Two-rowed Barley
Introduced. Status: rare, escape.
Fallow agricultural land; farmyards, roadsides.

EL: rarely recorded, Yellowcraig; Sheriff Hall;
Seacliff; Pefferside.

H. murinum L.   Wall Barley
ssp. murinum
Native. Status: local.
Roadsides, waste ground, public parks, rough grassland.

WL: scarce, Carriden; Grougfoot; Woodend; east of Queensferry; west of Cramond.
ML: widespread and locally frequent in urban areas, local elsewhere.
EL: quite widespread in N., rare or absent elsewhere.

ssp. glaucum (Steud.) Tzvelev
ML: Hailes Quarry, 1976.

H. jubatum L.   Foxtail Barley
Introduced. Status: scarce.
Roadsides (salted), waste ground, quarries, scrub, beaches.

WL: rare, verge of A799 near Tailend; verge of A803 at the west end of Linlithgow; (Bo'ness. 1970).
ML: rare, well-established at Musselburgh Lagoons, first recorded there in 1975; Fisherrow Harbour; verge of A68 Dalkeith to Sheriffmuir Roundabout, including roundabout.
EL: rare, Prestongrange; Prestonpans; Tranent area; Garvald; Seafield, West Barns.

H. secalinum Schreb.   Meadow Barley
Introduced. Status: rare, casual.
Rough grassland, waste ground.

ML: rare, Leith Docks, 1989; Trinity, 1995; (Hailes Quarry, 1971).

H. marinum Huds.   Sea Barley
WL: Bo'ness, pre-1934.
ML: shore at Musselburgh, pre-1934.
EL: coast near Tranent, pre-1934.

Secale L.

S. cereale L.   Rye
Agricultural relic.
ML: Leith Docks, 1883–1903; Slateford Quarry tip, 1903.
EL: sandy field at Gullane, 1902.

Triticum L.

T. aestivum L.   Bread Wheat
Introduced. Status: local, agricultural relic or escape.
Roadsides, farmyards, waste ground, rough grassland.

WL: rare, a patch on Foulshiels Bing.
ML: occasional/rare.
EL: rarely recorded, Prestongrange; Gosford Estate; Haddington; Sheriff Hall; West Barns; Bilsdean.

Danthonia DC.

D. decumbens (L.) DC.   Heath Grass
Native. Status: local.
Heath grassland, moorland, rough grassland, other well-drained acid soils, often on sloping ground.

WL: local, mainly in S.
ML: thinly widespread, mainly in upland areas.
EL: quite widespread in upland areas, occasional in lowland areas, especially along the coast.

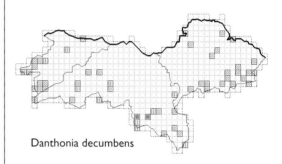

Danthonia decumbens

Cortaderia Stapf

C. selloana (Schult. & Schult. f.) Asch. & Graebn.   Pampas Grass
Introduced. Status: rare.
Policies.

EL: rare, Dunglass Estate.

Molinia Schrank

M. caerulea (L.) Moench   Purple Moor Grass
Native. Status: local.
Heath grassland, moorland, bogs, marshes, rough grassland, coniferous woodland, wet acid soils.

WL: widespread in S., rare elsewhere.
ML: in upland areas only, quite widespread, locally abundant.
EL: widespread and locally frequent in Lammermuir Hills; rare on coastal heath, Tyninghame; (Aberlady Bay, pre-1967).

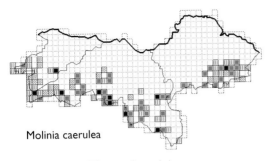

Molinia caerulea

Phragmites Adans.

P. australis (Cav.) Trin. ex Steud.   Common Reed
Native. Status: local.
In and beside lochs, ponds, rivers and ditches.

WL: quite scarce, not usually forming very large
stands, Kinneil area; Polkemmet Moor; west end
of Linlithgow Loch; Dalmeny.
ML: local, Duddingston Loch; Vogrie; Addistoun;
Loanhead; Leith Docks; Cobbinshaw Reservoir.
EL: local, Luffness; Danskine Loch; Pefferside;
Seafield Pond, West Barns.

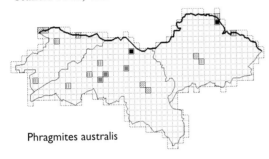

Phragmites australis

Spartina Schreb.

S. x townsendii H. & J. Groves (*S. maritima* (Curtis)
Fernald x *S. alterniflora* Loisel.)   Townsend's Cord-grass
WL: Bo'ness, 1921.

Panicum L.

P. miliaceum L.   Common Millet
Introduced. Status: rare, casual.

ML: rare, Trinity, 1996; (Leith Docks, 1903–10).
EL: rare, East Linton, 1999.

P. dichotomiflorum Michx.   Autumn Millet
ML: Borthwick railway tip, 1962.

Echinochloa P. Beauv.

E. crus-galli (L.) P. Beauv.   Cockspur
Ruderal casual, previously of regular occurrence.

ML: Musselburgh, Portobello, Slateford quarry tip and
Murieston, pre-1927; Hailes quarry tip, 1971; Leith Docks,
1976.

Setaria P. Beauv.

S. italica (L.) P. Beauv.   Foxtail Bristle-grass
Introduced. Status: rare, casual.
Waste ground.

ML: rare, Granton, 1982; Hamilton Place, 1991;
(Blackford Quarry, 1971).

S. viridis (L.) P. Beauv.   Green Bristle-grass
Ruderal casual of regular occurrence *c.* 1900.
ML: Leith Docks, Portobello and Slateford quarry tip,
pre-1927; Leith Docks, 1976 and 1979.

S. pumila (Poir.) Schult.   Yellow Bristle-grass
ML: Leith Docks, 1902–8, 1920 and 1976; Slateford, 1920;
Blackford and Hailes quarry tips, 1971.

Digitaria Haller

D. sanguinalis (L.) Scop.   Hairy Finger-grass
Rare grain casual.
ML: Leith Docks, 1904–6; Borthwick railway tip, 1962.

D. ciliaris (Retz.) Koeler   Tropical Finger-grass
Casual of wool, bird-seed, oilseed and cotton.
ML: Borthwick railway tip, 1962.

SPARGANIACEAE

Sparganium L.

S. erectum L.   Branched Bur-reed
Native. Status: quite widespread.
In and beside streams, ponds, ditches and lochs,
marshes.

WL: widespread, locally frequent.
ML: quite widespread.
EL: widely scattered sites, locally frequent at
Aberlady Bay; Pressmennan Lake; Stottencleugh.

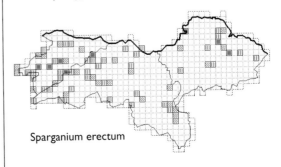

Sparganium erectum

Subspecies were not recorded during the survey.

ssp. microcarpum (Neuman) Domin.
ML: Duddingston Loch, 1969.

ssp. neglectum (Beeby) K. Richt.
Records confirmed from historic specimens.

WL: Kirkliston, 1857.
ML: Corstorphine Loch, 1857.

S. emersum Rehmann   Unbranched Bur-reed
Native. Status: local.
Streams, rivers, canal, ponds.

WL: local, Union Canal eastwards from
Broxburn; streams south of Blackridge;
occasional on the River Almond and on
Barbauchlaw Burn; Tailend Moss.
ML: local, Union Canal at Sighthill and Ratho;
Cobbinshaw Reservoir; West Murieston;
(Duddingston and Corstorphine, pre-1900).

S. angustifolium Michx.   Floating Bur-reed
Native. Status: rare.
In and beside lochs, ponds and slow-moving streams.

WL: (South Mains at Bangour, 1955).
ML: local, Pentland Hills; Threipmuir and
Cobbinshaw Reservoirs; Murieston Water; source
of Medwin Water; (Dalmahoy, pre-1934).

S. natans L.   Least Bur-reed
Native. Status: rare.
Peaty pools.

WL: rare, Drumshoreland, 1998 to present.
ML: rare, Cobbinshaw, 1996 to present;
(Pentland Hills and Ravelrig, pre-1934).

TYPHACEAE

Typha L.

T. latifolia L.   Bulrush, Reedmace
Native. Status: local.
In and beside ponds, lochs, streams and ditches.

WL: quite widespread.
ML: local, mainly in W.
EL: scarce, mostly beside estate lakes and new
ponds, and generally believed to have been
introduced, also appearing spontaneously,
Aberlady Bay; Donolly Reservoir; (West Barns,
1979).

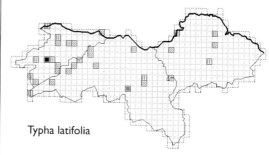

Typha latifolia

T. angustifolia L.   Lesser Bulrush
Introduced. Status: rare.
Pondside, lakeside.

ML: rare, planted at Bawsinch, *c.* 1984; Bush
Estate; (Dalkeith, 1865; Carberry, 1977).
EL: rare, Smeaton; (Pressmennan Lake, 1967).

LILIACEAE

Narthecium Huds.

N. ossifragum (L.) Huds.   Bog Asphodel
Native. Status: local.
Bogs.

WL: quite widespread in remnant bogs in S. and
W., absent elsewhere.
ML: local, confined to upland areas in S. and
SW., locally frequent in a very few places.
EL: rare, Danskine Loch; River Tyne below
Haddington, 1981; (Aikengall Moor, pre-1934;
Faseny Water, 1972).

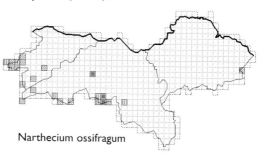

Narthecium ossifragum

Kniphofia Moench

K. uvaria (L.) Oken   Red-hot Poker
Introduced. Status: rare, garden escape.
Waste ground.

WL: rare, Bo'ness.
EL: rare, Canty Bay where it has spread from a
garden onto the dunes.

Gagea Salisb.

G. lutea (L.) Ker Gawl.
Yellow Star of Bethlehem
Native. Status: rare.
Wooded riverbanks.

WL: (Carlowrie Castle, 1950; Hopetoun House, *c*. 1980).
ML: rare, Inveresk; Lothian Bridge; Dalkeith Country Park; (Breich Water).

Erythronium L.

E. dens-canis L. Dog's-tooth Violet
Introduced. Status: rare, possibly planted.
Woodland.

ML: rare, on a riverbank at Gogar.

E. californicum Purdey
Introduced. Status: rare.
Woodland strip.

ML: rare, beside the Water of Leith, near Dean Village.

Tulipa L.

T. sylvestris L. Wild Tulip
Introduced. Status: rare.
Woodland.

WL: rare, Pepper Wood until late 1980s.
ML: (Arniston and Ravelston, 1927).
EL: rare, established in woodland in Tyninghame Estate; (bank of River Tyne near Sandy's Mill, 1972).

Lilium L.

L. martagon L. Martagon Lily
Introduced. Status: rare.
Old policy woodlands, roadside verges.

WL: rare, Bedlormie House.
ML: rare, Newhailes; (Slateford, 1960).
EL: rare, verge near cottages, Humbie area, 2000; (East Saltoun, 1957; Gifford area, 1972).

L. pyrenaicum Gouan Pyrenean Lily
Introduced. Status: rare.
Roadside verges.

WL: rare, verge near Overton.
EL: rare, verge north of Haddington, 1994.

Convallaria L.

C. majalis L. Lily of the Valley
Introduced. Status: local.
Policy woodland, rough grassland.

WL: rare, Hopetoun; Pepper Wood.
ML: local, Dalkeith, Arniston and Newhailes Estates.
EL: rare, Gosford Estate.

Polygonatum Mill.

P. x hybridum Brügger (*P. multiflorum* (L.) All. x *P. odoratum* (Mill.) Druce)
Garden Solomon's Seal
Introduced. Status: scarce, long-persisting garden escape.
Woodland, scrub, roadsides, waste ground.

WL: scarce, scattered records in N.
ML: rare, Roslin Castle; near Fala; Meggetland; New Hailes; Bawsinch; Threipmuir; Dreghorn; Trinity.
EL: scarce, on shady roadsides, Humbie Kirk; Longniddry Bents; Bolton Muir Wood.

Maianthemum Weber

M. bifolium (L.) F. W. Schmidt May Lily
Introduced. Status: rare.
Woodland.

WL: rare, Pepper Wood, since 1868, only one plant left.

Paris L.

P. quadrifolia L. Herb-Paris
Native. Status: rare.
Deciduous woodland.

ML: rare, thought to be extinct but re-found near Vogrie, 1989; (Arniston, Newbattle, Roslin and Currie, pre-1934).
EL: (Dunglass Dean, pre-1778; Pressmennan, pre-1934).

Ornithogalum L.

O. angustifolium Boreau Star of Bethlehem
Introduced. Status: local.
Roadsides, waste ground, rough grassland, riverbanks.

WL: rare, Hiltly area; Hopetoun area; near Craigton Quarry; west of Cramond; (Kirkliston, pre-1934; Pepper Wood, 1956).
ML: scarce, widely scattered, Whitecraig; near Newbridge; Melville Castle; Lasswade.

EL: scattered distribution in lowland areas, Seton Sands; Humbie to Leaston; Belhaven; East Barns.

O. nutans L.　　　　Drooping Star of Bethlehem
Introduced. Status: rare.
Roadsides.

EL: rare, Bolton; Drem.

Scilla L.

S. bifolia L.　　　　Alpine Squill
Introduced. Status: rare, garden escape.
Roadsides.

ML: rare, Balerno walkway.

S. bithynica Boiss.　　　　Turkish Squill
Introduced. Status: rare.
Riverside.

EL: beside River Tyne near Nether Hailes.

S. siberica Haw.　　　　Siberian Squill
Introduced. Status: rare, garden escape.
Woodland, public parks.

WL: rare, Dalmeny.
ML: rare, Warriston; Barnton.
EL: rare, Yellowcraig; Thornton Glen.

Hyacinthoides Heist. ex Fabr.

H. non-scripta (L.) Chouard ex Rothm.
　　　　　　　Wild Hyacinth, Bluebell
Native. Status: widespread, often mixed with the following two taxa when near habitation.
Mixed and deciduous woodland.

WL: local, locally abundant.
ML: occasional in upland areas, widespread elsewhere, locally abundant.
EL: over-recorded in the survey because of confusion with the hybrid. Occurs in old woods and related habitats, perhaps mainly in upland areas, Linn Dean; Woodhall Dean; Thornton.

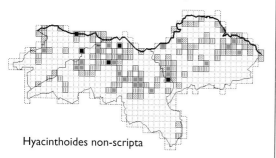

Hyacinthoides non-scripta

H. non-scripta x **H. hispanica**
Introduced and spontaneous hybrid. Status: widespread, often planted or escaped, under-recorded.
Woodland, coastal scrub, dunes.

WL: confused with *H. hispanica* but thought to be quite widespread in N.
ML: widespread in lowland areas.
EL: beside public paths and in lowland shelter belts, probably the common bluebell of lowland areas such as Pencaitland; Seton; Longniddry Bents; Luffness; Yellowcraig; Gifford. With both parents on the Clerkington and Tyninghame Estates.

H. hispanica (Mill.) Rothm.　　　Spanish Bluebell
Introduced. Status: local, garden escape, some records may have resulted from confusion with the hybrid.
Mixed woodland, roadsides, waste ground, dunes.

WL: confused with the hybrid.
ML: occasional.
EL: occasional in lowland areas, rarer than the hybrid, Seton Sands; Longniddry Bents; Dirleton (with hybrid); Bolton.

Chionodoxa Boiss.

C. forbesii Baker　　　　Glory of the Snow
Introduced. Status: rare, garden escape.
Woodland, roadsides, public parks.

ML: rare, Warriston; Roseburn.
EL: rare, roadside near car park, Yellowcraig.

C. sardensis Drude　　　Lesser Glory of the Snow
Introduced. Status: rare, garden escape.

EL: numerous plants on flood terrace of River Tyne below East Linton.

Muscari Mill.

M. armeniacum Leichtlin ex Baker
　　　　　　　Garden Grape Hyacinth
Introduced. Status: rare, garden escape or throw-out, also planted.
Roadsides, hedgerows, woodland.

WL: rare, near Cauldcoats Holdings; east of South Queensferry.
ML: rare, Warriston; Bawsinch; Stow.
EL: rare, Seton Sands; Luffness; Dirleton; Waughton; East Linton weir.

Allium L.

A. paradoxum (M. Bieb.) G. Don
Few-flowered Garlic or Leek
Introduced. Status: local.
Riversides, mixed and deciduous woodland, scrub, roadsides.

WL: local, invading many old woodlands, mainly in NE. First recorded in this vice-county around Binny Craig in 1866 and Carlowrie in 1907, it is now recorded from about thirty-five sites and continues to advance, competing successfully with native species.
ML: confined to lowland areas, quite widespread, locally abundant; (Moredon, 1842; Arniston, 1868).
EL: occasional, spreading to new sites, across lowland areas, generally in small quantity where it occurs, (earliest records, Saltoun and Winton, pre-1927).

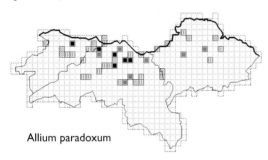

Allium paradoxum

A. ursinum L.   Ramsons
Native. Status: local.
Damp areas in mixed and deciduous woodland, riversides, scrub, roadsides.

WL: local in N. and E., locally abundant.
ML: quite widespread in lowland areas, locally abundant, occasional elsewhere.
EL: mainly in centre where it is widespread and locally abundant.

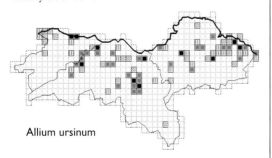

Allium ursinum

A. carinatum L.   Keeled Garlic
Introduced. Status: local.
Roadsides, riversides, waste ground.

WL: rare, Pepper Wood and nearby verge, 1982.
ML: local, on the banks of the River Esk and of the River Almond near Kirkliston; Gogar Bank; Saughton Park.
EL: (Aberlady, 1977).

A. scorodoprasum L.   Sand Leek
Introduced. Status: rare.
Roadsides.

ML: rare, Warriston walkway; Newhailes Estate; (Balerno, 1977).
EL: rare, Pencaitland; Colstoun.

A. vineale L.   Wild Onion, Crow Garlic
Native. Status: local.
Rocky outcrops, quarries, dry pasture, roadsides, railway embankments, mixed woodland, coastal grassland.

WL: (Craigie Hill, 1895; South Queensferry, pre-1934).
ML: local, Arthur's Seat; Edinburgh Castle Rock; Calton Road cliffs; Trinity; Ravelston.
EL: scattered distribution in lowland areas, Longniddry Bents; Hailes Castle; Whitekirk; West Barns.

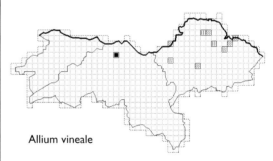

Allium vineale

A. oreophilum C. A. Mey.
Introduced. Status: rare.
Waste ground.

ML: rare, Granton Gas Works, 1995.

A. schoenoprasum L.   Chives
Garden relic.

ML: Musselburgh, 1965.

A. oleraceum L.   Field Garlic
WL: Kirkliston, 1906.
EL: Innerwick area, pre-1930.

Leucojum L.

L. aestivum L.   Summer Snowflake
Introduced. Status: rare, garden escape.
Public park.

ML: rare, naturalised in the grounds of the
National Gallery of Modern Art.

Triteleia Douglas ex Lindley

T. hyacinthina Greene
Introduced. Status: rare, garden escape.
Rough grassland.

ML: rare, Musselburgh Lagoons, 2000.

Galanthus L.

G. nivalis L.   Snowdrop
Introduced. Status: local, well naturalised.
Mixed woodland in estates and parks, scrub, roadsides.

WL: local, mainly in NE., abundant on Dalmeny
Estate, rare elsewhere.
ML: local, locally frequent.
EL: occasional on verges near houses, naturalised
in older woods.

Several hybrids and cultivars not listed here may
also be found.

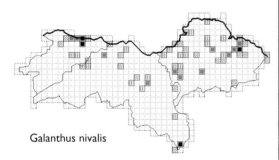

Galanthus nivalis

G. elwesii Hook. f.   Greater Snowdrop
Introduced. Status: rare, planted and escaped.
Coastal scrub, woodland, riversides.

EL: rare, Longniddry Bents, 1999; Greencraig,
1987; River Tyne flood terrace, East Linton,
1995.

Narcissus L.

N. poeticus L.   Pheasant's Eye Daffodil
ssp. **poeticus**
Introduced. Status: scarce, garden escape.

Mixed woodland, roadsides, waste ground, hedgerows,
scrub.

WL, ML: occasional.
EL: rarely recorded, scattered distribution in
lowland areas; (introduced on the Bass Rock
pre-1863).

N. x incomparabilis Mill. (*N. poeticus* x *N.
pseudonarcissus*)   Nonesuch Daffodil
Introduced. Status: rare.
Ditchside.

EL: rare, Longniddry Bents, 1990.

N. x medioluteus Mill. (*N. poeticus* x *N. tazetta*
L.)   Primrose Peerless
Introduced. Status: scarce.
Woodland, scrub, riverside, railway embankment.

ML: occasional, Warriston; Stockbridge.
EL: (Bass Rock, 1825 and 1831).

N. pseudonarcissus L.   Daffodil
ssp. **pseudonarcissus** cultivars
Introduced. Status: local, garden escapes, often
planted.
Roadsides, woodland, public parks, riversides,
hedgerows, waste ground.

WL: occasional escapes; also mass planting in
amenity areas, particularly around Livingston.
ML, EL: occasional escapes and some mass
plantings in amenity areas.

ssp. **major** (Curtis) Baker   Spanish Daffodil
Introduced. Status: rare, planted and persisting.
Woodland.

EL: rare, Cockenzie; Tyninghame Estate.

Asparagus L.

A. officinalis L.   Garden Asparagus
ssp. **officinalis**
Introduced. Status: rare, escape from cultivation.
Coastal grassland.

EL: rare, Aberlady Bay, 1988; (Gullane Links and
links near Gosford, pre-1934).

Ruscus L.

R. aculeatus L.   Butcher's Broom
Introduced. Status: scarce, surviving where
planted.
Policy woodland, scrub.

ML: rare, single plants at Craiglockhart Dell and Hermitage of Braid; Dalkeith Country Park, 1996; (Roslin, pre-1934; Colinton, 1956).
EL: scarce, Gosford House; Luffness Friary; Smeaton; Broxburn.

R. hypoglossum L.   Spineless Butcher's Broom
Introduced. Status: rare, garden escape.
Waste ground.

ML: rare, a quarry at Craigmillar, 1978 and 1997.

IRIDACEAE

Sisyrinchium L.

S. montanum Greene   American Blue-eyed Grass
EL: waste ground at West Barns, 1974.

Iris L.

I. germanica L.   Bearded Iris
Introduced. Status: rare.
Marsh, coastal scrub and grassland, dunes, mixed woodland.

ML: (one eighteenth-century record).
EL: rare, Seton Sands; Ferny Ness; Aberlady Bay; Yellowcraig; Belhaven Bay.

I. pseudacorus L.   Yellow Flag, Yellow Iris
Native. Status: widespread, sometimes planted.
Marshes, in and beside rivers, ponds, ditches and lochs.

WL: widespread.
ML: widespread, locally frequent in S.
EL: occasional, mainly in centre, locally frequent.

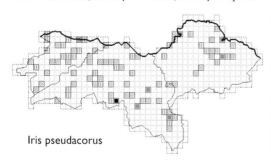

Iris pseudacorus

I. versicolor L.   Purple Iris
Introduced. Status: rare.
Beside ponds, damp places.

ML: rare, Inveresk, 1988.
EL: rare, Dunglass Pond.

I. foetidissima L.   Stinking Iris
Introduced. Status: rare.
Mixed woodland, scrub.

EL: rare, Gosford House; Smeaton; (in woodland on the shore below Seacliff, 1932).

Crocus L.

Large amenity plantings of a mixture of taxa occur in some places such as Linlithgow (**WL**) and Edinburgh (**ML**). These taxa remain largely undetermined.

C. vernus (L.) Hill   Spring Crocus
Introduced. Status: rare.
Coastal scrub, woodland clearings.

EL: Longniddry Bents car park, 1999; Luffness, 1983.

C. tommasinianus Herb.   Early Crocus
Introduced. Status: rare.
Coastal scrub.

EL: rare, Longniddry Bents, car park, 1999.

C. x stellaris Haw. (*C. angustifolius* Weston x *C. flavus* Weston)   Yellow Crocus
Introduced. Status: local, garden escape and in mass plantings of mixtures of *Crocus* species along verges in amenity areas.
Waste ground, scrub, roadsides, public parks and amenity areas.

WL: occasional escapes, also in amenity plantings.
ML: local, Stockbridge; Warriston; Duddingston; Melville Drive.
EL: rare, Pitcox; (Biel, 1967).

C. speciosus M. Bieb.   Bieberstein's Crocus
Introduced. Status: rare.
Coastal scrub, roadside.

EL: rare, Seton Sands, 1996; Gifford, 1994.

Crocosmia Planch.

C. paniculata (Klatt) Goldblatt   Aunt Eliza
Introduced. Status: rare, garden escape.
Dunes.

EL: rare, Belhaven Bay.

C. x crocosmiiflora (Lemoine) N. E. Br. (*C. pottsii* (Macnab ex Baker) N. E. Br. x *C. aurea* (Hook.) Planch.)   Montbretia
Introduced. Status: local, garden escape.
Waste ground, scrub, rough grassland, roadsides.

WL: local.
ML: widely scattered sites.
EL: scarce, Samuelston; Gullane Bents; East Bay, North Berwick; Dunbar.

ORCHIDACEAE

Cephalanthera Rich.

C. longifolia (L.) (Fritsch)   Narrow-leaved Helleborine
WL: Dalmeny, 1868
ML: in woods at Dalmahoy, 1894.

Epipactis Zinn

E. helleborine (L.) Crantz
                Broad-leaved Helleborine
Native. Status: scarce.
Deciduous and mixed woodland, scrub, car parks.

WL: rare, usually only one or two plants at each site, Belsyde; Torphichen; Bo'ness; Philpstoun Bing; Mid Tartraven; Carriber Mill; (Dalmeny, 1956; Almondell, 1969).
ML: scarce, Barnton; Cammo; Gogar; Polton; Lauriston Castle; Rosebank; Livingston; Millbank House.
EL: rare, Pencaitland; Garvald; Tyninghame; (Whittingehame, 1901; North Berwick and Dunglass, pre-1934; disused railway between Longniddry and Gullane, 1976).

E. youngiana A. J. Richards & A. F. Porter
                Young's Helleborine
Native. Status: rare.
Bings.

ML: rare, Gorebridge, 1994.

E. leptochila (Godfery) Godfery
                Narrow-lipped Helleborine
var. **dunensis** T. & T. A. Stephenson
Native. Status: rare, a recent arrival.
Woodland on old shale bings.

WL: rare, Philpstoun Bing; small bing near Mid Breich.
ML: rare, Gorebridge.

E. palustris (L.) Crantz.   Marsh Helleborine
ML: Lochend and Duddingston Lochs, Braid Hills, Roman Camp near Dalkeith, and Glencorse, pre-1900.
EL: Dunbar, pre-1835; Gullane, 1895.

Neottia Guett.

N. nidus-avis (L.) Rich.   Bird's Nest Orchid
Native. Status: rare.
Deciduous and mixed woodland, scrub.

WL: rare, Hopetoun; patch of woodland near Port Edgar, 1992, now built on; (Carriber, 1884; Craighall, 1894; Abercorn, 1903; Kirkliston and Almondell, pre-1934).
ML: rare, Pathhead; Dalhousie; Arniston Estate; Aikendean Wood.
EL: rare, New Town; Butterdean Wood; wooded roadside near Bolton; (Yester, 1902; Dunglass, pre-1934; Athelstaneford, 1937; Wester Pencaitland and Bara, 1957).

Listera R. Br.

L. ovata (L.) R. Br.   Common Twayblade
Native. Status: local.
Deciduous and mixed woodland, rough grassland, railway embankments, bings.

WL: local.
ML: local, mainly in S.
EL: scarce, in lowland areas, Pencaitland railway walk; Aberlady Bay; Yellowcraig; Catcraig.

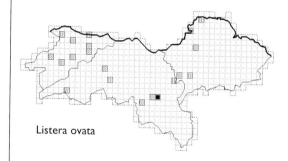

Listera ovata

L. cordata (L.) R. Br.   Lesser Twayblade
Native. Status: scarce.
Moorland, usually in *Sphagnum* under heather.

ML: local, upland areas only, especially the Moorfoot Hills.
EL: rare, upland areas only, Lammer Law; Killpallet Heights; Sheeppath Burn.

Goodyera R. Br.

G. repens (L.) R. Br.   Creeping Lady's Tresses
Native. Status: rare.
Coniferous woodland.

WL: (Barnbougle Castle until 1921).
ML: rare, pinewood near Stow; (Auchendinny, 1865; Fountainhall, 1906).
EL: rare, scattered distribution in woodland between Haddington and Gifford.

Corallorrhiza Ruppius ex Gagnebin

C. trifida Châtel   Coralroot Orchid
ML: in wet peaty places under trees or bushes, Ravelrig Bog, possibly up to 1840s.

Platanthera Rich.

P. chlorantha (Custer) Rchb.
        Greater Butterfly Orchid
Native. Status: scarce.
Unimproved grassland, waste ground, on somewhat basic soils.

WL: local, mainly in W.
ML: scarce, Addiewell; near West Calder; Breich; Linhouse Water; near Gladhouse and Threipmuir Reservoirs.

P. bifolia (L.) Rich.   Lesser Butterfly Orchid
WL: Uphall, 1868; Bathgate, 1898.
ML: Auchendinny, Threipmuir and Ravelrig, pre-1934.
EL: Gullane Links and Aikengall Moor, pre-1934.

Anacamptis Rich.

A. pyramidalis (L.) Rich.   Pyramidal Orchid
Native. Status: rare.
Dunes.

EL: rare, Aberlady Bay, 1998 to present; (Archerfield, 1908).

Pseudorchis Ség.

P. albida (L.) Á. & D. Löve   Small White Orchid
WL: Breich Water near Whitburn, pre-1934.
ML: West Calder, pre-1845.
EL: Longyester, sometime between 1950 and 1962.

Gymnadenia R. Br.

G. conopsea (L.) R. Br.   Fragrant Orchid
ssp. borealis (Druce) F. Rose
Native. Status: rare.

Limestone grassland, riversides, marshes.

ML: rare, Middleton; Linhouse Water; (Dalmahoy, Hallyards Castle, Tynehead, Currie, Brunstane and Auchendinny, pre-1900).
EL: rare, north-east side of Dod Hill, 1992; (Longyester, sometime between 1950 and 1962).

ssp. densiflora (Wahlenb.) E. G. Camus, Bergon & A. Camus
ML: Middleton, 1982 to present; (Borthwick, 1835).

x Dactylodenia (Gymnadenia x Dactylorhiza) Garay & H. R. Sweet

x D. st-quintinii (Godfery) J. Duvign. (*G. conopsea* x *D. fuchsii*)
Probably only occurs where parents grow together.
ML: Balerno, 1936.

Coeloglossum Hartm.

C. viride (L.) Hartm.   Frog Orchid
Native. Status: rare.
Rough grassland, wet pastures, dune turf.

WL: (South Queensferry, 1894).
ML: rare, near Gladhouse Reservoir, 1997, dwarf plants about 5cm tall; (Granton, Holyrood Park, Dalmahoy, Newbattle, Habbie's Howe, Currie and Ravelrig, pre-1900).
EL: rare, coastal locations only, thriving at Aberlady Bay; Gullane Hill and dunes; declining at Yellowcraig; (Seton Links, 1849; Canty Bay, 1932).

Dactylorhiza Necker ex Nevski

D. fuchsii (Druce) Soó
        Common Spotted Orchid
Native. Status: quite widespread.
Marshes, rough grassland, beside roads, rivers and ditches, waste ground, scrub, coniferous and mixed woodland, bings, neutral/basic soils.

WL: very widespread in S. and W., local in N. and E.
ML: quite widespread, mainly in centre, locally frequent.
EL: scattered distribution, mainly in centre, Aberlady Bay; Haddington railway walk; Catcraig.

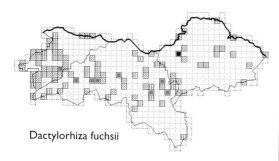

Dactylorhiza fuchsii

D. x venusta (T. & T. A. Stephenson) Soó (*D. fuchsii* x *D. purpurella*)
Native. Status: scarce, with parents.
Marshes, waste ground, grassy places.

WL: local, mainly in S.
ML: rare, Balerno, 1999; (Gorebridge, 1886).
EL: rare, Humbie, 1982; Aberlady Bay, 1998; Whittingehame, 1993; Dunglass, 1983.

Dactylorhiza x venusta

D. maculata (L.) Soó   Heath Spotted Orchid
ssp. ericetorum (E. F. Linton) P. F. Hunt & Summerh.
Native. Status: local.
Marshes, bogs, riversides, rough acid grassland.

WL: local, confined to S. and SW.
ML: local, mainly in upland areas, locally frequent.
EL: rare, mainly in Lammermuir Hills, also in heathy woodland near Gladsmuir.

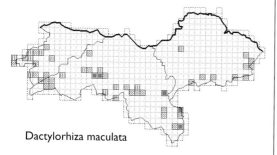

Dactylorhiza maculata

D. x formosa (T. & T. A. Stephenson) Soó (*D. maculata* x *D. purpurella*)
Native. Status: rare, with parents.
Marshes, rough grassland, birch woods.

ML: rare, near West Torphin, *c.* 1982.

D. incarnata (L.) Soó   Early Marsh Orchid
ssp. incarnata
Native. Status: rare.
Marshes, bogs, coastal habitats.

ML: rare, Temple; near Cousland; (Duddingston Loch, Ravelrig, Newbattle, Roslin and Craiglockhart Hill, pre-1900).
EL: rare, in coastal marshes, Aberlady Bay; Yellowcraig; Belhaven Bay; (inland at Danskine, 1868).

ssp. coccinea (Pugsley) Soó
EL: rare, recorded from dune marshes at Aberlady Bay and Gullane; (North Berwick area sometime between 1950 and 1962).

D. purpurella (T. & T. A. Stephenson) Soó
                Northern Marsh Orchid
Native. Status: local.
Marshes, rough grassland, flushes, old bings and quarries.

WL: widespread in S., local in N. and E.
ML: widely scattered distribution including Causewayside, 1992 (built on in 1993).
EL: scarce, mainly along the coast, Aberlady Bay; St Baldred's Cradle; Dunglass.

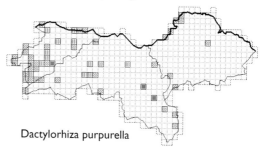

Dactylorhiza purpurella

Orchis L.

O. mascula (L.) L.   Early Purple Orchid
Native. Status: scarce and decreasing.
Mixed woodland, scrub, marshes, rough grassland, dunes.

WL: rare, Ochiltree Mill; (Binney Craig, 1868; Carribber, Hopetoun and Dalmeny, pre-1934).

ML: (Holyrood Park, Roslin Woods, Currie, Newbattle, Dalmahoy, Tynehead, Cramond Bridge, Ratho, Dalhousie, Lugton Wood, Penicuik and Colinton, pre-1900; Newtongrange, 1970; Silverknowes, 1972).
EL: scarce, Linn Dean Water; Hanging Rocks north of Gullane; Tantallon Castle; Newmains; Stenton; Dunbar;Aikengall Water; Bilsdean; (East Linton, 1863; Canty Bay, 1932; Deuchrie, 1970).

O. morio L.   Green-winged Orchid
Native. Status: rare.
Unimproved grassland.

ML: (in meadows, 1841).
EL: rare, in long-established lawn near Haddington.

14

Casual and Invasive Aliens

D. R. McKEAN

Non-native plants may be deliberately planted into the wild, an activity that is illegal, or they may arrive by accident. Some may become established and reproduce and these are said to be naturalised, while others such as some woody or herbaceous perennials, for example, rhubarb, persist for many years but do not spread, and these are classified as surviving. Others may persist for a short period only and, if the period is less than five years, they are known as casuals.

The first alien species were probably introduced into the Lothians by Neolithic people who cleared woodland and planted cereals. In the sixteenth century, the recording of plants began due to a slow accumulation of knowledge by clerics and monks. With increasing transport of raw materials by sea, more and more alien plants found their way here. Seafarers even as far back as the Vikings used ballast to stabilise their ships, and ballast heaps would have been a frequent sight around coastal towns. This continued to be so until the mid-1920s when iron ships were able to use water as ballast and the sand and gravel heaps were no longer needed (Macpherson 1997). Around 1870, a local amateur botanist, Alexander Craig-Christie, started taking the first serious interest in foreign plants, beginning with those found in the dock areas on old ballast heaps. He subsequently turned his attention to the grain and paper mills that had sprung up along the Rivers Esk, Tyne and Almond and the Water of Leith. Three other botanists later joined him, James Fraser, W. Wilson Evans, and then W. Edgar Evans. A son of Wilson Evans, Edgar Evans was a professional botanist at the Royal Botanic Garden Edinburgh and a former president of the Edinburgh Natural History Society. These enthusiasts collected at the known 'hot spots' such as Leith Docks, and quarry tips at Slateford, Levenhall, and Craigmillar Castle. Hundreds of records of alien plants were published from these sites and from various mills and dumps. Sometimes the collecting was sporadic and there were years in which no aliens were recorded.

By the late nineteenth century other kinds of transport were developing steadily with canals, railways and new roads being built and so aliens appeared more often and at more sites. Many of the old ballast heaps were later used to construct embankments and several aliens were unwittingly distributed mainly along railways.

Some of these plants became naturalised and still survive to this day as 'railway or ballast plants' for example Crown Vetch (*Securigera varia*), Melilots (*Melilotus* spp.), Wormwood (*Artemisia absinthum*), Lucerne or Alfalfa (*Medicago sativa*), Small Toadflax (*Chaenorhinum minus*), Wild Mignonette (*Reseda lutea*), Hoary Cress (*Lepidium draba*). These species and several more of similar origin are widely found throughout much of the Lothians. Wheat and oats arrived in bulk from Russia, Eastern Europe, Turkey and the Middle East and many aliens came in as impurities in these cargoes. Chick-pea Milk Vetch (*Astragalus cicer*) and Twiggy Spurge (*Euphorbia pseudovirgata*), for example, survived for many years near the mill at Catcune. Twiggy Spurge is still there and is also found near Haddington and at South Queensferry. In more recent times, grain imported from North America and seed imported for animal and poultry feed, and for cage birds and pheasants, were other vectors for bringing aliens to the Lothians. Many botanists, then as now, considered aliens of little importance and so the collecting and recording of them is inclined to wax and wane from year to year or even from decade to decade. After about 1930, the number of casuals found decreased dramatically and has continued to decrease since then. The main reason for the reduction in the number of these mainly ephemeral weeds has been mentioned already but other reasons in more recent times are that grain is harvested largely from fields treated with herbicides and that the docklands are sometimes sprayed with weedkiller. Many of the raw materials that were imported, for example, esparto grasses (*Stipa* spp.) from Spain and North Africa, for paper making, are no longer imported. Other commodities still being imported are subject to rigorous cleaning and are sealed in containers so that very few foreign propagules escape. It is possible that with the increasing popularity of organic farming many agricultural weeds will return. We botanists live in hope that herbicide-resistant genotypes will not come to dominate our agriculture. Our cereal and potato fields are already remarkably free of weeds.

From the 1950s onwards for many years, a group of expert amateurs, Ursula Duncan, Mary McCallum Webster, Betty Beattie, and Olga Stewart vied with each other to see who could find the most aliens. Their hunting grounds included Leith Docks, sites of old mills, disused railways, municipal rubbish tips and the railway tip at Borthwick Bank, near Gorebridge. Thanks to their efforts we have a good idea of what aliens have been seen until the late 1990s. Their records are included in the main text of the Flora. Aliens that were recorded in the Lothians before 1927 but have not been seen there since 1927 are listed at the end of this text.

Most aliens arrive in wild habitats as garden escapes or from birdseed. The latest to be found is the rhizomatous perennial Walthamstow Yellow-cress (*Rorippa armoracioides*) which seems to be sweeping the country. It has been found as a tenacious garden weed in Blackhall and on waste ground in Craigmillar and is probably elsewhere. Amenity areas and the verges of our dual carriageways are often planted with alien species such as Dwarf Mountain Pine (*Pinus mugo*), some *Cotoneaster* species, exotic brambles (*Rubus cockburnianus* and *R. armeniacus*), Japanese Rose (*Rosa rugosa*), Wilson's Honeysuckle (*Lonicera nitida*), Hybrid

Coralberry (*Symphoricarpos x chenaultii*) and also, unfortunately, the pernicious Snowberry (*Symphoricarpos albus*).

Horticulturalists have been responsible for the escape of many aliens into the countryside and a large number of these are aquatics or marsh plants. The Victorians planted many water plants such as Water Soldier (*Stratiotes aloides*) and Arrowhead (*Sagittaria sagittifolia*) but the latter eventually died out as it is not suited to the environment in Scotland. The non-native Water Soldier was known from as long ago as 1809 at Duddingston Loch and later from Corstorphine, Davidson's Mains, Blackhall, Linlithgow Loch (?1934) and near Haddington (1909). These records are especially interesting as they illustrate just how prevalent a practice it was at that time to plant exotics in the wild. A long-established example of the escape of imported plants is Canadian Waterweed (*Elodea canadensis*). This is dioecious and is normally found only as female plants. It is of local interest because Braid Hill marshes (now drained) was the first place in Britain (1879) where male plants were recorded (by the famous explorer-plant finder David Douglas!). The plant was noted from the Royal Botanic Garden Edinburgh pond as early as 1863 and soon spread to various localities such as the Water of Leith at Warriston, the Union Canal and Duddingston Loch. Druce (1932) states that the plant is now diminishing! This may have been so when he was writing but it is still abundant in all its old haunts and many more besides. In 1987, another North American water-weed, Nuttall's Waterweed (*Elodea nuttallii*), was found at Auchendinny and later near Livingston. This was reputed to be very invasive, so much so that it was believed to be supplanting Canadian Waterweed, in England at least. *E. nuttallii* has been reported from nine sites in the Glasgow area (Dickson *et al.* 2000), and it is believed to be spreading rapidly. In recent years, garden ponds and aquaria have become very popular and this has caused a boom in the sale of aquatic plants which has mixed blessings for the environment. On the whole it is beneficial to many forms of wildlife to have a garden pond as most of our natural ponds and wet areas have been drained. However, some introduced aquatic plants have escaped and cause serious competition with our native species. Two such plants that ecologists worry about are the little aquatic Water Fern (*Azolla filiculoides*) and Pigmy Weed (*Crassula helmsii*), from New Zealand. Water Fern carpets the Union Canal and pools at Bawsinch so densely that light is greatly reduced for the submerged plants, and even some floating plants such as Duckweeds (*Lemna* spp.) fail to compete. Pigmy Weed also now blankets the ponds at Bawsinch. Fringed Water-lily (*Nymphoides peltata*), which is sometimes mistaken for the Yellow Water-lily (*Nuphar lutea*), has escaped onto Duddingston Loch from the manse garden nearby. It appears not to be a nuisance as yet because it is probably at the northerly limit of its range and so does not spread too far before frost checks its growth.

Another group of garden marsh plants that have been with us for a long time are the Monkeyflowers (*Mimulus* spp.). They were recorded from the Pentland Hills in the mid-1800s and a few still grace our wet areas. The Hybrid Monkeyflower (*M. x robertsii*) would appear to be the most widespread of these but three other hybrids may also occur. Other ornamental water plants that are sometimes found include Sweet Flag (*Acorus calamus*), at Pressmennan Lake, and formerly in a marsh at Kirkliston, (possibly the marsh at Hallyards Castle, Midlothian), and the Flowering Rush (*Butomus umbellatus*), planted and then naturalising at Carcant near Heriot, at Bawsinch, near Pencaitland and formerly at Linlithgow Loch.

Probably the most notorious and expensive escapee is the Giant Hogweed (*Heracleum mantegazzianum*) with many city councils now spending tens of thousands of pounds trying to eradicate it, but to no avail. With thousands of its dormant seeds in the soil and just a few germinating at a time it is an impossible task to eradicate it. We should continue to try or the problem will become intractable. Japanese Knotweed (*Fallopia japonica*), Snowberry and Sycamore (*Acer pseudoplatanus*) are species that have long since escaped and become naturalised. Strenuous efforts are made to control them but often with limited effect.

As one might expect, it is the plants that grow in roughly similar climates to our own that are most likely to become pests. The Caucasus and mountainous areas of Europe, Asia and America which have relatively cool climates like ours are such areas where the following rampant herbs have come from: two species of Leopard's Bane (*Doronicum pardalianches* and *D. plantagineum*) (mid-Europe), Pyrenean Valerian (*Valeriana pyrenaica*) (Pyrenees), Common Blue Sow-thistle (*Cicerbita macrophylla*) (Urals). Fringecups (*Tellima grandiflora*) and Pick-a-back plant (*Tolmiea menziesii*), both from Western North America, are well known from policy woodlands and riverside sites such as those at Hopetoun, Carlourie (Pepper Wood), Roslin, Penicuik, Cramond, Dalkeith Palace, Arniston and Tyninghame.

Some incomers such as Indian Balsam (*Impatiens glandulifera*), Hybrid Spanish Bluebell (*Hyacinthoides non-scripta* x *H. hispanica*), Pineappleweed (*Matricaria discoidea*) and Butterfly Bush (*Buddleja davidii*) are so widespread in the Lothians that one could imagine that they are native. A plant of local interest is the New Zealand Willowherb (*Epilobium brunnescens*) which was first recorded in Britain by James Fraser as a garden weed in Craigmillar in 1905 and now can be found nationwide in open damp habitats including those in the Lothians. Interestingly, it is only in the last decade or so that some small creeping New Zealand willowherbs, for example, *E. brunnescens* and *E. pedunculare* have been found to be hybridising with our own native species, but none of these hybrids has been recorded in the Lothians so far. It may be worthwhile looking for them where the aliens occur among native populations. The American Willowherb (*E. ciliatum*), has now become probably the most widespread willowherb in the Lothians and in Britain. It too hybridises with native willowherbs such as the Broad-leaved Willowherb (*E. montanum*).

Long Extinct Casual Aliens

Aegilops biuncialis Vis. (*A. lorentii*) Wool casual.
ML: Leith Docks and tips at Slateford and Levenhall, 1904–11.

Aegilops crassa Boiss. (*Triticum crassum* var. *oligochaetum* Hack.)
The type specimen of the variety was described from Slateford (railway bank at Gorgie) in 1901.

Aegilops cylindrica Host
Grain casual.
ML: Leith Docks, 1906.

Aegilops geniculata Roth (*Triticum ovatum*)
ML: tips at Levenhall, Slateford, and Craigmillar Castle, and Leith Docks, 1904–13.
EL: Ormiston, 1883.

Aegilops markgrafii (Greuter) Hammer (*A. caudata*)
ML: Leith Docks, 1904, (not listed in Ryves *et al.* 1996), identified by P. H. Davis.

Aegilops neglecta Req. ex Bertol. (*A. triaristata*)
Grain alien.
ML: tips at Levenhall and Slateford, and at Leith Docks, intermittently 1883–1922.

Aegilops peregrina (Hack.) Maine & Weiller (*Triticum peregrinum*)
Grain alien.
ML: Leith Docks and Slateford. Type specimen from here in 1906.

Aegilops triuncialis L.
ML: tips at Levenhall and Slateford, and Leith Docks, 1903–12.

Aegilops ventricosa Tausch
Wool and ballast casual.
ML: very plentiful at tips at Slateford, Levenhall, Gorgie and Leith Docks.

Amberboa moschata (L.) DC. (*Centaurea moschata*)   Sweet Sultan
Garden escape.
ML: Slateford Quarry tip 1908, 1911.

Ambrosia trifida L.   Giant Ragweed
Oil-seed and grain casual.
ML: Leith Docks spasmodically from 1893 to 1921.

EL: Haddington, 1897.

Anacyclus clavatus (Desf.) Pers.
Wool and grain casual.
ML: Leith Docks 1885, 1895, 1905, 1908; Slateford Quarry tip, 1904; Duddingston 1903.

Anacyclus valentinus L.
Esparto grass and grain casual.
ML: Fillyside 1934; Leith Docks 1905–7; Musselburgh 1912–14; Slateford Quarry tip, 1906.

Anthemis orientalis agg.
ML: Leith Docks, 1906.

Anthemis pseudocotula Boiss.
ML: Leith Docks, 1921.

Anthemis ruthenica M. Bieb. (*A. maritima* auct.)
Bird-seed and grain casual.
ML: Leith Docks 1885, 1889, 1903, 1905.

Anthemis wiedmanniana Fischer & C. A. Mey.
ML: Leith Docks, 1885, 1905–6; Slateford Quarry tip, 1906.

Apera interrupta (L.) P. Beauv.   Dense Silky Bent
EL: in fields and by trackways on Dirleton Common, last seen, 1876.

Arabis glabra (L.) Bernh.   Tower Mustard
EL: North Berwick, pre-1894.

Argyranthemum frutescens (L.) Schultz-Bip.
                              Paris Daisy
Garden escape.
ML: Redhall Quarry (Slateford), 1907.

Aster linosyris (L.) Bernh. (*Crinitaria linosyris*)
                              Goldilocks Aster
ML: Davidson's Mains Railway Station, 1910.
This Red Data Book Species seems an amazing introduction!

Asteriscus aquaticus (L.) Less. (*Odontospermum aquaticum*)
Esparto grass and grain casual.
ML: Leith Docks, 1905–9; Slateford Quarry tip, 1906.

Baeria coronaria A. Gray.
ML: Leith Docks, 1904, (not listed in Clement and Foster 1994).

Bellis annua L.
ML: Leith Docks, 1920–1.

Bombycilaena erecta (L.) Smoljan.
ML: Levenhall Quarry tip, 1916–17.

Brachypodium distachyon (L.) P. Beauv.
Stiff Brome
Grain and esparto grass casual.
ML: Leith Docks, 1904.

Bromus madritensis ssp. rubens (L.) Domin
(*Anisantha rubens* (L.) Nevski)   Foxtail Brome
Grain and esparto grass casual.
ML: mainly at tips at Levenhall and Slateford;
Leith Docks and Juniper Green (mills),
intermittently 1874–1914.

Bromus marginatus Nees ex Steud. (*Ceratochloa
marginata* (Nees ex Steud.) B. D. Jacks.)
Western Brome
ML: Leith Docks, intermittently 1906–19.

Bromus scoparius L.
A wool and grain casual.
ML: Slateford Quarry tip and Leith Docks,
1903–6.

Bromus squarrosus L.
ML: tips at Slateford; Levenhall, Murieston and
at Leith Docks, 1903–12.

Callistephus chinensis (L.) Nees (*C. hortensis*)
China Aster
Garden escape.
ML: Leith Docks, 1903.

Carduus argentatus L.
Wool and grain casual.
ML: Fillyside (Seafield), 1934; Craigmillar Castle
Quarry tip, 1914; Slateford Quarry tip 1904–5;
Leith Docks, 1904.

Carduus hamulosus Ehrh.
ML: Leith Docks, 1888, 1905; Slateford Quarry
tip, 1907.

Carthamnus tinctorius L.   Safflower
Birdseed and grain casual.
ML: Slateford Quarry tip, 1904; Pilton tip, 1905.

Centaurea depressa M. Bieb.
ML: Leith Docks, 1904–7; Slateford Quarry tip,
1905.

Centaurea duriaei (Spach) Rouy
Grain casual.
ML: Slateford Quarry tip, 1904–6; Leith Docks,
1904, 1905. This North African species is not
listed in Clement and Foster (1994).

Centaurea hierapolitana Boiss.
ML: Portobello clay pit, 1912. This Turkish
species is not listed in Clement and Foster (1994).

Centaurea hyalolepis Boiss. (*C. pallescens* auct)
Grain casual.
ML: Leith Docks, sporadically from 1893–1906;
Liberton, 1909.

Centaurea iberica Trev. ex. Sprengel
Grain casual.
ML: Slateford Quarry tip, 1903–5; Leith Docks,
1905–6.

Centaurea jacea L.   Brown Knapweed
ML: Arniston woods 1883; near Edinburgh
1841; Leith Docks 1886.

Centaurea melitensis L.   Maltese Star-thistle
A grain, wool, linseed, bird-seed and esparto
grass casual.
ML: Leith Docks, sporadically from 1883–1904;
Slateford Quarry tip, 1904.

Centaurea nicaeensis All.
Vector unknown.
ML: Leith Docks, 1903, 1904, 1909; Slateford
Quarry tip, 1904, 1906.

Cephalaria syriaca (L.) Roem. & Schult.
ML: Slateford Quarry tip *c.* 1904.

Cichorium endivia L.   Endive
A wool and bird-seed casual and garden escape.
ML: near Edinburgh, 1884; Slateford Quarry tip,
1904–6; Murrayfield, 1904.

Cirsium eriophorum (L.) Scop.   Woolly Thistle
WL: seashore between Blackness and South
Queensferry, 1894.
ML: seashore at Leith, eighteenth century;
Oxenfoord Castle and Chesterhill, early
nineteenth century.

Cirsium rivulare (Jacq.) All. (*Cnicus rivularis*)
Garden escape.
ML: Craigmillar Quarry tip, 1906.

Cladanthus arabicus (L.) Cass.
ML: Slateford Quarry tip, 1904; Leith Docks, 1906.

Coleostephus myconis (L.) Reichb. f. (*Chrysanthemum myconis*).
A wool casual.
ML: Leith Docks, 1903, 1920–2.

Conyza bonariensis L.   Argentine Fleabane
ML: Leith Docks, 1887. (unconfirmed record).

Coreopsis tinctoria Nutt.   Garden Tickseed
Wool casual.
ML: Leith Docks, 1904, 1906.

Cotula coronopifolia L.   Buttonweed
Wool casual.
ML: Leith Docks, 1888.

Cotula squalida Hook f.   Leptinella
Garden escape and wool casual.
ML: Leith Docks, 1913; persistent lawn weed at Royal Botanic Garden for over twenty years.

Cotula tridentata (Del.) Benth. & Hook.
Grain casual.
ML: Leith Docks, 1980. Not listed in Clement and Foster (1994).

Crepis alpina L.
ML: Slateford Quarry tip, 1905–6.

Crepis aspera L.
ML: Slateford Quarry tip, 1904–5; Leith Docks, 1905, very rare, not listed in Clement and Foster (1994).

Crepis aurea (L.) Cass. (*Crepis foetida* ssp. *commutata*)
Grain casual.
ML: Leith Docks, 1889, 1904; Levenhall Quarry tip, 1912; Slateford Quarry tip, 1904, 1906.

Crepis foetida L.   Stinking Hawk's-beard
ML: Leith Docks, 1889; Slateford Quarry tip, 1906; Levenhall Quarry tip, 1912.

Crepis nicaeensis Balbis   French Hawk's-beard
Bird-seed and agricultural seed casual.
ML: Inveresk, 1904.

Crepis setosa Haller f.
EL: near Drem, 1847.

Crepis tectorum L.
Grain casual.
ML: Leith Docks, 1893, 1910.

Crepis vesicaria L. var. **taraxacifolia** (Thuill.) Thell. ex. Schinz & R. Keller.
ML: Leith Docks, 1911.

Crupinia vulgaris Cass.
ML: Levenhall Quarry tip, 1911–13.

Dianthis armeria L.   Depford Pink
EL: railway bank at Ormiston, 1882.

Digitaria ischaemum (Schreb. ex Schweigg.) Muhl.   Smooth Finger-grass
ML: Leith Docks, 1906.

Digitaria sanguinalis (L.) Scop.
        Hairy Finger-grass
ML: Leith Docks.

Eragrostis ?capillaris (L.) Nees
ML: Leith Docks, 1912.

Eragrostis cilianensis (All.) Vignolo ex Janch.
        Stink-grass
ML: Leith Docks, 1904, 1906.

Eremopyrum triticeum (Gaertn.) Nevski
ML: Leith Docks, 1905.

Fedia cornucopiae Gaertn.
ML: Leith Docks, 1906.

Filago pyramidata L. (*F. spathulata* auct.)
ML: Leith Docks, 1885; Slateford Quarry tip 1906; Levenhall Quarry tip, 1916–17.

Galactites tomentosa Moench (*G. galactites*)
Bird-seed casual.
ML: Leith Docks, 1904.

Guizotia abyssinica (L.f.) Coss.   Niger
Bird-seed, grain, oil-seed and wood casual.
ML: Leith Docks, 1906–7; Murieston, 1910.

Hedypnois cretica (L.) Dum. Cours.
Scaly Hawkbit
Wool, bird-seed and esparto grass casual.
ML: Leith Docks, 1889, 1907, 1921; Slateford
Quarry tip, 1904; Levenhall Quarry tip, 1916.

Helianthus angustifolius L. (incl. *H. giganteus*)
Swamp Sunflower
ML: Leith Docks, 1907–8.

Hemizonia kelloggii E. Greene
Grain casual.
ML: Leith Docks, 1893, 1906; Slateford Quarry
tip, 1902–6; mill near Currie, 1903.

Hordeum bulbosum L.
ML: Leith Docks, 1908, 1912.

Hordeum distichon L.
ML: Leith Docks, 1907.

Hypochaeris achyrophorus L. (*H. aetnensis*)
ML: Leith Docks, 1889, 1920.

Iberis amara L.   Wild Candytuft
EL: Dirleton Castle, before 1878.

Iris sibirica L.   Siberian Iris
ML: Redhall, *c.* 1770.

Koelpinia linearis Pallas
ML: Craigmillar Quarry, 1914. Not listed in
Clement and Foster 1994.

Lactuca saligna (L.)   Least Lettuce
ML: Leith Docks, 1910.

Lactuca sativa L.   Garden Lettuce
ML: Leith Docks, 1905.

Lactuca serriola L.   Prickly Lettuce
ML: Leith Docks, 1904, 1906, 1921; Slateford
Quarry tip, 1907; Murieston, 1910.

Lasthenia glabrata Lindley
Yellow-rayed Lasthenia
ML: Slateford Quarry tip, 1906.
Leith Docks, a few records, 1890–1909.

Leucanthemum lacustre (Brot.) Samp.
(*Chrysanthemum lacustre*)
Portuguese Swamp-daisy
Garden escape.
ML: Comiston Sandpit, 1915.

Lithospermum purpureocaeruleum L.
Purple Gromwell
WL: Pepper Wood.

Logfia heterantha (Rafin.) J. Holub (*Filago cupaniana*)
ML: Levenhall Quarry tip, 1916.

Madia elegans D. Don
ML: Leith Docks, 1906. Not listed in Clement
and Foster (1994).

Madia sativa Molina
ML: Portobello clay pit, 1910.

Notobasis syriaca (L.) Cass. (*Cnicus syriacus*)
Syrian Thistle
Grain casual.
ML: Slateford Quarry tip, 1904 and 1907; Leith
Docks, 1904, 1906 and 1907.

Onopordum tauricum Willd.
ML: Slateford Quarry tip, 1906.

Osteospermum glabrum (Lag.) Willk. (*Matricaria glabra*)
ML: Leith Docks, 1921. Not listed in Clement
and Foster (1994).

Panicum capillare L.   Witch-grass
ML: Leith Docks, 1903, 1911, 1921.

Picnomon acarna (L) Cass. (*Cnicus acarna*)
ML: near Levenhall Quarry tip, 1912.

Picris altissima Del.
Esparto grass and bird-seed casual.
ML: Leith Docks, 1905. Not listed in Clement
and Foster (1994).

Picris pilosa Del.
ML: near Levenhall Quarry tip, 1915.

Rhagadiolus stellatus (L.) Gaertn. (*R. edulis*)
Bird-seed, grain and wool casual.
ML: Leith Docks, 1889 sporadically to 1907;
Slateford Quarry tip, 1904.

Rhamnus cathartica L.   Buckthorn
EL: East Linton, 1857.

414

Rudbeckia laciniata L.   Coneflower
Double-flowered garden escapes.
ML: Leith Docks, 1906, 1908.
EL: East Linton, 1909.

Sanvitalia procumbens Lam.
ML: Leith Docks, 1904.

Scabiosa atropurpurea L.
ML: Leith Docks, 1906.

Scabiosa lucida Vill.
ML: Edinburgh, 1906.

Scabiosa prolifera L.
ML: Slateford Quarry tip, 1904; Leith Docks, 1904–5.

Schismus barbatus (L.) Thell.   Kelch-grass
ML: Levenhall.

Scolymus hispanicus L.   Golden Thistle
Bird-seed casual.
ML: Leith Docks 1893 and 1905.

Scolymus maculatus L.
Wool and bird-seed casual.
ML: Slateford Quarry tip, 1904.

Senecio aegyptius L.
Grain casual.
ML: Leith Docks, 1893, 1904.

Senecio doria L.   Golden Ragwort
Garden escape (confused with *S. fluviatilis*).
ML: (unspecified) wood near Edinburgh, 1867.

Senecio gallicus Chaix. (*S. subdentatus*)
                                        French Groundsel
ML: Leith Docks, 1908.

Setaria verticillata (L.) P. Beauv.
                                        Rough Bristle-grass
Esparto grass alien.
ML: Portobello clay pit, 1909.

Silphium perfoliatum L.   Cup-plant
ML: Craigmillar Castle Quarry tip, 1955.

Solidago graminifolia (L.) Salisb.
ML: Craigmillar and Slateford Quarry tips, 1907; Inveresk, 1809.

Tanacetum cinerariifolium (Trev.) Schultz-Bip.
                                        Dalmatian Pyrethrum
ML: Leith Docks, 1907.

Tragopogon crocifolius L.
ML: Leith Docks, 1904; Slateford Quarry tip, 1906.

Tragopogon hybridus L.   Slender Salsify
ML: Slateford Quarry tip, 1906; Leith Docks, 1904.

Tripleurospermum decipiens (Fischer & C. A. Mey.) Bornm. (*Matricaria decipiens*)
Grain and wool casual.
ML: Liberton, 1909; Leith Docks, 1890, 1909; Inveresk, 1903; Slateford Quarry tip, 1904–5; Portobello clay pit, 1910.

Triticum monococcum L.
ML: Leith Docks, 1909.

Triticum spelta L.
ML: Leith Docks, 1883; waste ground at Gorgie, 1903, 1908.

Triticum turgidum L.   Rivet Wheat
ML: Leith Docks, 1906, 1921.

Valerianella coronata DC.
ML: Leith Docks, sporadically 1890–1906; Craigmillar Castle Quarry tip, 1914.

Valerianella echinata DC.
ML: Slateford Quarry tip, 1906.

Valerianella microcarpum Lois
ML: Leith Docks, 1921–2.

Valerianella vesicaria Moench.
ML: Craigmillar Castle Quarry tip, 1914.

Verbena officinalis L.   Vervain
EL: Dunbar, pre-1894.

Volutaria lippii Cass.
ML: Leith Docks, 1910.

Wangenheimia lima (L.) Trin.
Esparto grass alien.
ML: Musselburgh and probably Levenhall, sporadically 1904–13.

Xanthium spinosum L.   Spring Cocklebur
Wool, grain and bird-seed casual.
ML: Leith, 1891, 1887, 1903–6; Canonmills, 1871.

Xanthium strumarium group.   Rough Cocklebur
Grain, wool and oil-seed casual.
ML: Leith Docks, 1906.

15

Notes from the Vice-counties

The Vice-county of West Lothian

J. MUSCOTT

When the vice-counties were set up in 1852 to facilitate biological recording, their boundaries more or less followed the county boundaries of the time. However, after the local government re-organisation of 1975, serious divergence set in and, as a result, the vice-county of West Lothian bears little relationship to the modern District of that name. Unlike the District, the vice-county includes a long stretch of coastline from the River Avon, which forms its western boundary, to the River Almond which forms the eastern boundary (apart from a curious diversion which excludes Pumpherston and much of Livingston New Town). The south-eastern boundary is Breich Water, and the south-western boundary follows a convoluted path similar to that of the modern District. The whole vice-county is very small, about 32,000 hectares (320 square kilometres).

The long coastline is home to some interesting maritime plants, though others, such as Scots Lovage (*Ligusticum scoticum*) and Sea Beet (*Beta vulgaris*) have been lost to industry and urban development. Extensive mudflats west of Hopetoun, and smaller ones just west of Cramond are home to a number of eelgrasses (*Zostera marina*, *Z. angustifolia* and *Z. noltei*, the latter forming an intertidal turf in places). All are scarce in the UK. The Dalmeny coastline, by contrast, has areas of calcareous sand, the only West Lothian site for plants such as Purple Milk Vetch (*Astragalus danicus*), Lesser Meadow Rue (*Thalictrum minus*), and Field Gentian (*Gentianella campestris*). Patches of saltmarsh are small and are scattered along the coast.

The highest point of the vice-county is Cairnpapple Hill (a prehistoric site including a burial chamber) at just over 300m. It is one of a number of small volcanic hills and crags to the north and west of Bathgate. This is the hilly part of West Lothian, and it consists largely of neutral grassland with rocky outcrops, though cultivation reaches quite high on some of the hills. The yellow form of Mountain Pansy (*Viola lutea*) can be found on some of the drier hillsides, and flushes and damp areas are quite rich in marsh plants. There is also high land to the south-west, but of a very different type – largely wet, acid moorland and bog. Much of it has been drained to produce rushy meadows, but small areas of natural 'moss' still remain, with an interesting selection of bog plants, including Round-leaved Sundew (*Drosera rotundifolia*), Bog Asphodel (*Narthecium ossifragum*) and Cranberry (*Vaccinium oxycoccus*). The largest surviving bog, Blawhorn Moss, is a Site of Special Scientific Interest (SSSI). Some secondary bogs have been formed as a result of mining subsidence.

There is one large natural water body, Linlithgow Loch, all that remains of a post-glacial lake, but now highly eutrophic, the edges choked by tall vegetation. There are old records of Water Soldier (*Stratiotes aloides*), and more recent ones of such local rarities as Trifid Bur-marigold (*Bidens tripartita*) and Water Purslane (*Lythrum portula*). Yellow Water-lily (*Nuphar lutea*) disappeared as recently as the late 1980s following an algal bloom. Greater botanical interest is currently provided by various reservoirs (Beecraigs, Lochcote, Bangour, Livingston) and smaller water bodies formed by quarrying and subsidence or on private estates. Other reservoirs have been drained, and Philpstoun Loch seems to have disappeared (under a bing?). The Union Canal is another substantial water body running right across the vice-county. It has been little used over the last forty years and so has developed an interesting flora, as has the towpath which passes through a variety of habitats. Among the interesting plants to be found in the canal are Gypsywort (*Lycopus europaeus*) and Tufted Loosestrife (*Lysimachia thyrsiflora*) (Plate 14). Neither plant is particularly conspicuous, and Tufted Loosestrife flowers only periodically. It is scarce in the UK, and a substantial part of the population seems to be in the Union and Forth-Clyde Canals. However, work is presently underway to make these two canals navigable and to re-link them. The subsequent expansion of canal traffic and developments along the towpath are bound to affect the water and towpath plants. The two boundary rivers and their tributaries also provide water habitats of varying interest, and it is along the Almond and the Avon (particularly the Avon) where most of the old woodland flora is to be found, usually in small gorges, several of which are SSSIs. There is a good deal of woodland elsewhere, but apart from some of the old estates (Hopetoun and Dalmeny in particular), it is usually poor for ground flora. Large conifer plantations occur at Beecraigs and on Polkemmet Moor in the southwest, with smaller plantations elsewhere. More recently, mixed woodland has been planted (at a highly unnatural density), and natural tree regeneration (mainly birch and willow) has taken place at a number of old industrial sites.

Major towns include the Royal and Ancient Burgh of Linlithgow, built beside the loch, and site of the royal palace, now in ruins, where Mary Queen of Scots was born. Bo'ness, its ancient rival on the coast, grew up around the coal and salt industries, later becoming a busy port exporting local products and importing timber for pit-props. All this has now been swept away, and for years the pit area, the docks and the lagoons which were built to take the remains of the coal bing (and later the local rubbish) have provided a habitat for ruderal plants and casuals. The old bing has now been converted into a wildflower meadow, but it seems likely that most of the area will eventually be built on, as Grangemouth expands across the river. South Queensferry, another coastal town, arose around the ferry started in the eleventh century by Queen Margaret, wife of Malcolm Canmore. Today it is probably more famous for the two bridges, road and rail, which cross the Forth at this point, and for the Hawes Inn which figures in *Kidnapped* by R. L. Stevenson. It seems to have been a good area for maritime plants in the past, but inevitably the less common ones have been lost. Bathgate is now the largest town in the vice-county, and it was from here that the Scottish oil industry took off in the mid-nineteenth century, before eventually moving to Grangemouth. There are small limestone outcrops to the north-east, and the area to the south, once a hunting ground known as Bathgate Moss, is now an interesting area, with lagoons and a golf course. During the oil boom, a number of new villages were founded and others expanded, most notably Broxburn which is now a substantial town and the home of Drambuie. Livingston New Town is the largest town in the area, but owing to a quirk of Victorian local government boundaries, only the western part, including the old village, is in the vice-county. Of particular interest is the fact that plants from Livingston formed the basis for the Edinburgh Physic Garden, which eventually developed into the Royal Botanic Garden Edinburgh. Sir Patrick Murray, Laird of Livingston at the beginning of the eighteenth century, was a great plant collector, and after his death, Sir Andrew Balfour and Sir Robert Sibbald, joint founders of the Physic Garden transported over 1,000 of his specimens to Edinburgh.

Agriculture in the area is mixed and relatively small scale, with rough grazing in the upland areas, and dairy and intensive arable farming elsewhere. Many of the old crop weeds have disappeared, and even poppies are now more often found on waste ground than at the edges of cornfields. Old meadow plants survive on roadside verges and embankments, little patches of unimproved grassland and old industrial sites. Greater Butterfly Orchid (*Platanthera chlorantha*) which is scarce in the Lothians can be found in wet grassland in a variety of such habitats.

Apart from the coal at Bo'ness which was mined from medieval times, exploitation of natural resources has been relatively recent, but has certainly left its mark on the landscape, the most obvious being the large red bings formed of oil-shale waste. Oil-rich shales, coal, limestone, sandstone, and fireclay were all laid down during the Carboniferous period, and all have been mined in West Lothian, but it is the oil-shale industry, which began in the 1850s and lasted for about 100 years which has made the biggest impact. Vast quantities of shale were required to produce relatively modest amounts of paraffin, and the mountains of waste were equally vast. They were dumped all over West Lothian with precious little respect for the environment, and while most of the old coal bings have been flattened, many of the larger shale bings still remain. Over the years, vegetation has gradually spread over them, and as agriculture has intensified round about, some have become wildlife refuges. The waste shale is dry and hard (like pottery shards), poor in all nutrients except calcium, and so inimical to the agricultural weeds which are usually the first colonisers of bare ground. The fifty-year old bings are still home to small plants such as Fairy Flax (*Linum catharticum*), Early Hair-grass (*Aira praecox*), Silver Hair-grass (*A. caryophyllea*), clovers and vetches, Wild Strawberry (*Fragaria vesca*) and Bird's-foot Trefoil (*Lotus corniculatus*). Some have become rough meadows (Plate 15) with a variety of grasses, and taller plants such as Common Knapweed (*Centaurea nigra*) and the semi-parasitic Yellow Rattle (*Rhinanthus minor*). Hawkweeds of various kinds (*Hieracium* and *Pilosella* spp.), Ox-eye Daisy (*Leucanthemum vulgare*), and Cat's-ear (*Hypochaeris radicata*) now grow on some of the drier slopes, while orchids have proved good colonisers in damper areas. Common Spotted Orchid (*Dactylorhiza fuchsii*), Northern Marsh Orchid (*Dactylorhiza purpurella*) and Twayblade (*Listera ovata*) are all to be found on bings and other brown-field sites. One of the most interesting colonisers of bings (including coal bings) has proved to be Stag's-horn Clubmoss (*Lycopodium clavatum*). Apparently this was once a lowland as well as a highland plant, and it has certainly shown a striking ability to make a come-back if conditions are right, growing profusely on one or two of the bings. Even more surprisingly, Alpine Clubmoss (*Diphasiastrum alpinum*) has been recorded on two shale bings, though it survives on only one. The sides of the bings are steep, but north-facing slopes are well colonised, even by trees, while many south-facing slopes remain bare. The latter have an almost Mediterranean micro-climate and native plants find it hard to cope. One plant which seems able to survive is Sticky Groundsel (*Senecio viscosus*) which is perhaps doubtfully native; another is Deadly Nightshade (*Atropa belladonna*) which has taken over the side of one bing and is definitely non-native this far north. An experiment with some Mediterranean species might be interesting!

The coal-mining industry produced much less waste than the shale oil industry, and most of the coal bings have been 'restored' – flattened and planted with trees (conifers in the early days, mixed plantings in more recent years) or sown with wildflower seed. But these areas too have their colonisers, Viper's Bugloss (*Echium vulgare*) seems to be particularly associated with coal waste, and the best colony of Common Wintergreen (*Pyrola minor*) is on a restored coal bing. Underground mining has now ceased in West Lothian (though a certain amount of opencast coal mining continues), but other legacies remain. Subsidence has produced a number of small ponds and secondary bogs, and prevents further industrial development on some sites. Water entering the old mine shafts produces pollution, which nowadays is countered by making lagoons and planting reeds – and, unlike new woodlands, ponds quickly become colonised by native species.

Following the collapse of the mining and other industries which had brought wealth to West Lothian there was a period of stagnation, when many brown-field sites gradually returned to nature. However, in recent years new industries have moved into both brown-field and green-field sites, and housing development continues around Livingston and elsewhere because the vice-county lies on major routes between Edinburgh and Glasgow and points west, and also between Edinburgh and the north. It is crossed not only by the Union Canal, but also by mainline railways, two motorways and a network of major and minor roads. Even the most remote corner is little more than a kilometre from a road of some sort. There are few substantial patches of 'wilderness', but plenty of little nooks and crannies where natural vegetation manages to survive and incoming species sometimes become established. However, without the industrial legacy of the oil-shale industry, West Lothian's botanical heritage would be much poorer.

Midlothian Specialities

D. R. McKEAN

Holyrood Park which encompasses Arthur's Seat and Salisbury Crags is a jewel as far as vascular plants are concerned. Nationally rare or locally scarce species such as Sticky Catchfly (*Lychnis viscaria*), Rock Whitebeam (*Sorbus rupicola*), Southern Polypody (*Polypodium cambricum*), Forked Spleenwort (*Asplenium septentrionale*) and Dropwort (*Filipendula vulgaris*), are found in its grassland and craggy areas. The lochs and surrounding marshes such as those in the Scottish Wildlife Trust (SWT) Bawsinch Reserve are home to other rare and local species: Rigid Hornwort (*Ceratophyllum demersum*), Nodding Bur-marigold (*Bidens cernua*), Lesser Water Plantain (*Baldellia ranunculoides*) and Various-leaved Water Starwort (*Callitriche platycarpa*). The last major survey of the Park carried out in 1975 listed almost 400 plant species that were either native or naturalising. The randomly selected square for this Flora surveyed only the Duddingston Loch and the Newington 1km squares, the latter including a small portion of Salisbury Crags.

A totally different habitat in Midlothian, but also an important one, is Red Moss, an SWT Reserve at Balerno. It is noted mainly for its raised bog and aquatic plant communities. The surrounding areas of moor, hill pasture and birchwood are interesting adjuncts to the reserve. The Ravelrig/Threipmuir area before it was largely drained was the home of several species that are now extinct in Midlothian, for instance, Narrow-leaved Helleborine (*Cephalanthera longifolia*), Coralroot (*Cardamine bulbifera*), Twinflower (*Linnaea borealis*), and Shepherd's Cress (*Teesdalia nudicaulis*). However, this is still a botanically rich site with the only extant colony of Bog Myrtle (*Myrica gale*) in the Lothians, and some species such as Northern Yellow-cress (*Rorippa islandica*), a rare moss (*Weissia rostellata*) and a rare liverwort (*Riccia canaliculata*) that are listed in the Red Data Book (2001). Also growing at Bavelaw is *Sphagnum magellanicum*, a moss of conservation merit because it is a good indicator for acid peat bogs. Both *R. canaliculata* and *S. magellanicum* are included in the City of Edinburgh's Biodiversity Action plans (2000).

Another SWT Reserve that is outstanding in Midlothian is Roslin Glen which is noted for its native woodland and varied ground flora. It is also a Site of Special Scientific Interest on account of having the type section of the Roslin Old Red Sandstone. The dominant trees in the wood are Oak (*Quercus* spp.), Wych Elm

(*Ulmus glabra*), Ash (*Fraxinus excelsior*) and Beech (*Fagus sylvatica*), with a good number of Scots Pine (*Pinus sylvestris*), Hazel (*Corylus avellana*), Holly (*Ilex aquifolium*) and a few planted Hornbeams (*Carpinus betulus*). By the river and in more open areas, Hawthorn (*Crataegus monogyna*), Goat Willow (*Salix caprea*), Alder (*Alnus glutinosa*), Elder (*Sambucus nigra*) and Aspen (*Populus tremula*) may be found, and round the castle are some venerable trees of Yew (*Taxus baccata*). The ground flora can be very beautiful in spring with sheets of Wild Hyacinth or Bluebell (*Hyacinthoides non-scripta*), Ramsons (*Allium ursinum*) and Wood Anemone (*Anemone nemorosa*). A tree parasite, Toothwort (*Lathraea squamaria*) is quite frequent on elm and hazel and in recent years Purple Toothwort (*Lathraea clandestina*) has been accidentally introduced with willows used for landscaping at the SWT car park at the foot of the glen. Giant Horsetail (*Equisetum telmateia*) is locally frequent in seeping wet ground and the locally rare Rough Horsetail (*Equisetum hyemale*) is just managing to survive by the path beneath the castle. The nationally scarce Wood Fescue (*Festuca altissima*) and Wood Melick (*Melica uniflora*) are found on shady slopes. Small sections of this wood are heathy with patches of heather (*Calluna vulgaris*) and Blaeberry (*Vaccinium myrtillus*) and some Common Cow-wheat (*Melampyrum pratense*). Unfortunately in the past many exotic conifers were planted and some of these such as Western Hemlock (*Tsuga heterophylla*) cast a very dense shade which is detrimental to the ground flora. Douglas Firs (*Pseudotsuga menziesii*), Spruces (*Picea* spp.) and Firs (*Abies* spp.) have also been planted throughout the area. Other notable woods in the vice-county are those at Colinton, Whitecraig, Gore Glen and Arniston, and in the Dalkeith and Penicuik Estates. The latter locality is the only Lothian site for Purple Small-reed (*Calamagrostis canescens*), refound in 1998 after not being seen in the Lothians for over 100 years.

The Union Canal is an especially important habitat because it helps to redress the balance of the loss of most of the ponds, lochs and marshes that were once widespread in the Lothians. As it is so important it has its own biodiversity plan. It is, of course, rich in aquatic plants and is home to some nationally scarce species such as Tufted Loosestrife (*Lysimachia thrysiflora*) (Plate 14) which has spread from the canal's feeder loch, Cobbinshaw Reservoir. Perfoliate Pondweed (*Potamogeton perfoliatus*) and Flat-stalked Pondweed (*P. friesii*) once thought to be local rarities have recently been found to grow near Meggetland, with *P. friesii* being found all the way along the canal to Ratho and probably beyond. This was only found out when a survey was done shortly after dredging operations. The plants had probably been there for years but their presence could not often be recorded because of the great tangle of smothering water weeds such as Blanket Weed (*Spirogyra* spp.) and Canadian Waterweed (*Elodea canadensis*). These weeds grew so thickly that it took a huge effort to drag a grapnel through them. Gypsywort (*Lycopus europaeus*) is another plant from the Cobbinshaw marshes that has recently spread to the Union Canal and has been found at the aqueduct at Slateford.

It was first believed that the bings around Lanarkshire and Stirlingshire were the only Scottish sites for the rarer helleborine orchids. However, a search of the Lothian coal/shale bings in the 1990s has confirmed that Narrow-lipped Helleborine (*Epipactis leptochila*), Young's Helleborine (*E. youngiana*) and possibly even the Green-flowered Helleborine (*E. phyllanthes*) occur. The last mentioned would be new to Scotland if it could be confirmed with a more adequate specimen. Differentiating these variable orchids is, however, a vexing task and other spoil heaps in Europe also seem to have their own helleborine variations.

There are various habitat and species biodiversity plans encompassing the Lothians and even a separate set of plans for the City of Edinburgh. Usually the species that have been selected for special protection by these plans are either nationally or locally rare or they may have been selected by the national biodiversity group, possibly as good indicators of a particular habitat. Just over thirty species of plants varying in form as widely as possible to cover as many differing groups as possible (from trees and mosses to lichens and fungi) are listed in the City's plans. Other parts of the Lothians are covered by similar plans but with different species lists. The Rio Summit Conference has come up with a winning idea in these action plans as they marry all manner of biologists, conservationists, the public and planners together. With their joint forces good conservation should result.

The East Lothian Coast

A. J. SILVERSIDE and E. H. JACKSON

East Lothian has the remains of a rich and varied coastline. The dune systems must once have been very fine, particularly the coast from Aberlady and Luffness to Gullane and Yellowcraig, the area including coastal marshes and the former sandy wastes of Dirleton Common. However, human pressures have taken their inevitable toll and many species have been lost. Such plants as Marsh Helleborine (*Epipactis palustris*), Mudwort (*Limosella aquatica*), Dense Silky Bent (*Apera interrupta*) and Sand Catchfly (*Silene conica*) have long since gone, while others are highly vulnerable. Nevertheless, the East Lothian coast still holds much of interest for the botanical visitor.

Not far out of Edinburgh, a sequence of public car parks at Longniddry Bents and Ferny Ness provides access to a strip of coastline with a lot of interest. There are no large dune systems or extensive marshes here, but the mix of scrub and grassland allows a diversity of species to survive. Motorised botanists may find they have actually parked on Spring Cinquefoil (*Potentilla neumanniana*). The open, rough grasslands are a delight, with colourful displays of Bloody Cranesbill (*Geranium sanguineum*) and Cowslips (*Primula veris*), the latter abundant and native here, though sometimes the results of well-intentioned plantings elsewhere on the coast. Clustered Bellflower (*Campanula glomerata*) survives in reasonable plenty in one area, though this species, very rare in Scotland, has gone from other local sites. The scrub areas are often dominated by Sea Buckthorn (*Hippophaë rhamnoides*) (Plate 16), one of the most characteristic of East Lothian coastal shrubs, though one that is not always welcome. It is said to be called locally the 'Baked-bean Bush' due to its orange berries. It is generally accepted as being native in the very east of the vice-county, but its status elsewhere is more uncertain, it having been planted in many places and then spread by the birds that feast on its berries in the winter months. The dense, spiny thickets can become a serious problem, but also can provide cover for wildlife and can protect more sensitive habitats from excess public pressure. Micro-organisms present in nodules on its roots take nitrogen from the atmosphere and fix it into organic compounds. The breakdown of the nodules enriches the soil. Nitrogen-loving plants grow under the buckthorn, notably Bur Chervil (*Anthriscus caucalis*), which is often abundant at Longniddry and elsewhere, while a very successful colonist from North America is Spring Beauty (*Claytonia perfoliata*) (Plate 25). The scrub at Longniddry is not all Sea Buckthorn. Wild roses form extensive thickets here, and with Burnet Rose (*Rosa pimpinellifolia*) and Sweet Briar (*R. rubiginosa*) both present and hybridising with other species, the result is an attractive mix of bushes of often uncertain parentage.

A little to the east, at Aberlady Bay, Gullane and Yellowcraig, the dunes become more extensive. Aberlady Bay is a Local Nature Reserve and one of the jewels of south-east Scotland, both botanically and for the rest of its wildlife. Although a number of species have been lost through past drainage, extensive coastal marshes and dune slacks still remain. Greater Bladderwort (*Utricularia vulgaris*), Marsh Stitchwort (*Stellaria palustris*) and Fen Pondweed (*Potamogeton coloratus*) are still to be found here. Of considerable specialist interest are rare hybrids between Horsetail (*Equisetum*) species, while the complex Marsh Orchid (*Dactylorhiza*) populations need further study. Some marsh plants, such as Red Rattle (*Pedicularis palustris*) and Dioecious Sedge (*Carex dioica*), are found at Aberlady but otherwise, in East Lothian, occur only in the Lammermuir Hills. A matter of concern, how-ever, is the loss of open, wet areas through recent invasion by more vigorous marsh vegetation. Smaller plants such as Brookweed (*Samolus valerandi*) and Fool's Watercress (*Apium nodiflorum*) are threatened or lost here. The latter, in its only East Lothian site and as a notable dwarf variant, should be looked for again and, if found, should be protected.

Eastwards at Gullane Bay and Yellowcraig the human influences are only too obvious. Gullane has been the site of a major reclamation scheme, reconstructing the dune landscape which was destroyed by military exercises in World War II. The long dune ridge here is entirely artificial. At Yellowcraig, we can only guess at the riches of the old Dirleton Common, now largely lost to agriculture and forestry. Basil Thyme (*Clinopodium acinos*) still survives precariously in just one tiny spot, while a rare, orange-yellow flowered dune dandelion, *Taraxacum obliquum*, is scattered over another limited area. A number of introductions are established in the dunes here, including alien species such as Pirri-pirri-bur (*Acaena novae-zelandiae*), plus species native further south in Britain, such as Blue Fleabane (*Erigeron acer*) and Tor-grass (*Brachypodium pinnatum*).

East from North Berwick to the vice-county boundary, the coastline becomes more rocky, often with sea cliffs. There are some small exposures of limestone and calcareous sandstone providing local refugia for such generally southern plants as Small Scabious (*Scabiosa columbaria*). Catcraig is an accessible site with extensive areas of limestone and here White Horehound (*Marrubium vulgare*) grows in some plenty.

Interrupting this rockier part of the coastline is the estuary of the Tyne, with much of the land and estuary now encompassed in the John Muir Country Park. Once again there are dunes, good here for many of the small annuals, while Henbane (*Hyoscyamus niger*), a characteristic if sporadic plant of the East Lothian coast, sometimes forms large populations on Sandy Hirst. Grey Hair-grass (*Corynephorus canescens*), a rare grass of the English coast, was found in one area of sandy turf in 1986 and has since become well established. Here too are some of the best areas of saltmarsh in the vice-county. Attractive, silvery clumps of Sea Wormwood (*Seriphidium maritimum*) are conspicuous on the margins of Belhaven Bay.

A remarkable aspect of East Lothian botany is, however, the paradox that some of the best East Lothian sites for 'coastal' plants are not on the coast at all. Outcrops of volcanic rocks (basalt and trachyte) locally develop a distinctive flora that may include Maiden Pink (*Dianthus deltoides*), Knotted Clover (*Trifolium striatum*), Heath Pearlwort (*Sagina subulata*), Hoary Cinquefoil (*Potentilla argentea*), Meadow Saxifrage (*Saxifraga granulata*), Annual Knawel (*Scleranthus annuus*), Hairy Whitlowgrass (*Erophila majuscula*), Blinks (*Montia fontana* ssp. *chondrosperma*), Changing Forget-me-not (*Myosotis discolor* ssp. *discolor*) and Prickly Sedge (*Carex muricata*). Associated with these are other species normally regarded as coastal but which here come well inland, including Slender Thistle (*Carduus tenuiflorus*) and small annuals such as Spring Vetch (*Vicia lathyroides*), Sea Mouse-ear Chickweed (*Cerastium diffusum*) and Early Forget-me-not (*Myosotis ramosissima*). Other notable East Lothian plants which show this same coastal/volcanic outcrop distribution are Purple Milk Vetch (*Astragalus danicus*), Spring Cinquefoil (*Potentilla neumanniana*) and Lesser Chickweed (*Stellaria pallida*), the last a frequent if much overlooked plant of sandy turf on the coast but only more recently recognised as a characteristic plant of the volcanic outcrops. It fluctuates markedly in abundance from one year to another. The off-shore island of Bass Rock is part of this same series of volcanic outcrops and might have been expected to be a particularly rich site for this assemblage of plants. However, here the high nitrogen and phosphate levels from the extensive sea-bird populations impose a very different flora. Areas are dominated by coarse grasses but spectacularly too by the introduced Tree Mallow (*Lavatera arborea*).

The East Lothian coast thus has both richness and variety, but also remains highly vulnerable. Perhaps most threatened of all are the strand line plants. The recent reappearance then disappearance again of Oyster Plant (*Mertensia maritima*) on the shingle between Bilsdean and Dunglass must be regarded as a natural phenomenon, but the recent introduction of beach-cleaning machines on and near popular beaches is a serious concern. Ray's Knotgrass (*Polygonum oxyspermum*) is a rare and sporadic plant on these beaches, but is of special significance in that East Scotland plants match the Scandinavian subspecies *oxyspermum*, rather than subspecies *raii* which occurs round the rest of Britain. Destruction of important strand line vegetation has meant that this plant may be close to not only local but also national extinction.

With the many and conflicting demands on East Lothian's coastline, careful and sensitive management is essential if the diversity of the flora is to be maintained. Introductions even of native species must be judicious and the dynamic nature of coastal microhabitats must be well understood.

16

The Phytogeography of Edinburgh and the Lothians

P. M. SMITH

Introduction

Phytogeography is an enormous, almost amoeboid subject, encompassing the study of the distribution of species (areography) and also of vegetation types; the evaluation of dispersal and vicariance explanations for disjunction; and floristic analysis. This chapter is a short phytogeographical review of the results of the Lothian survey in terms of floristic analysis only – of how the local flora relates to that of Britain and Europe as a whole.

Little has been written on this subject before, and it might be argued that the Lothian area is too small or even too dull to exhibit any meaningful phytogeographical correlation. It would be expected that within the broad climatic zones to which species are suited by physiological tolerance, they would occupy habitats – largely defined by soil and biotic factors – to which they are adapted. Since the habitat heterogeneity of the Lothians is greater than the climatic heterogeneity, Lothian plant distributions might therefore not be expected to contain much of a 'phytogeographical signal'. Nevertheless there are some interesting indications.

Like any location on earth, the Lothian area lies at the point of intersection of a number of climatic gradients. At this point, the flora will be a reflection of the tolerances evolved in local plants and the historical accidents that brought their ancestors here. The flora will be a variant of other floras nearby, just as the area is part of the Euro-Siberian floristic region.

Two climatic gradients are most relevant to the analysis of the Lothian flora. First, there is the gradient from the extreme oceanicity of the west of Scotland, to the more continental conditions of the east, and beyond into Eurasia. Oceanicity increases as mean winter and summer temperatures approach one another. Second, there is the broadly north-south temperature gradient. These two factors obviously interact to determine which plants characteristic of the north, south, east and west manage to survive in these three vice-counties.

Phytogeography and the Floras of NW Europe

After the initial surveys and evaluations of Watson (1873), a wave of enthusiasm for county Floras produced a flood of data on British plant distribution. Important summaries of the facts were given by Druce (1932) and in the BSBI *Atlas of the British Flora* (Perring and Walters 1962). European distributions are recorded in *Atlas Florae Europaeae* (Jalas and Suominen 1972–94).

Interpretation of floristic patterns in Britain as a whole needs to be updated, and it must be hoped that significant attention is paid to this following the publication of the new version of the BSBI *Atlas* (2002). Watson's six 'British Types' of floristic distribution/geographical association of species, were succeeded by the sixteen 'elements' of the British flora recognised in the work of J. R. Matthews of Aberdeen (Matthews 1937; 1955). This was a long overdue association of British distributions with those of Europe as a whole. Dahl's classification of floristic 'elements' (Dahl 1998) in NW Europe (summarised in Table 16.1) is a further step forward in that it focusses clearly on the two main climatic gradients. Dahl also connects the distributions with authoritative comment on the physiology of the plants involved. Agreement with Matthews' findings is considerable.

Table 16.1 Dahl's classification of NW European Floristic Categories (Dahl 1998) related to Scotland

Atlantic and Oceanic Element
        British-Atlantic sub-element
        West Scandinavian-Atlantic sub-element
        Scandinavian-Atlantic sub-element
        Baltic-Atlantic sub-element

Thermophilic Element
        Temperate sub-element
        Southern boreal sub-element
        Middle boreal sub-element
        Northern boreal sub-element

Boreal Element
        North European boreal sub-element
        Widespread boreal sub-element

Alpine Element
        Mid-alpine sub-element
        Low-alpine sub-element

Montane Element

Widespread Element

For our purposes, an analysis of the Lothian flora in terms of the ten relevant categories of Matthews, which are more precisely definitive than those of Dahl, seems appropriate. Some species belong to more than one Dahlian category – as might be expected where two gradients interact – but this does not make for ease of communication. A minor problem with both Matthews' and Dahl's classifications is that they refer to categories of distribution as 'elements', for example Matthews' sixteen 'elements of the British flora'. A floristic element is now more generally taken to mean a species characteristic of a particular phytogeographical area (phytochorion), and it is so used here. The categories of Matthews (Table 16.2) are here renamed 'British Floristic Units'.

Table 16.2 Native Lothian Plants in ten British Floristic Units (after Matthews 1955)

Mediterranean Floristic Unit
2 Lothian species (38 in Matthews' British list)

Catapodium marinum
Salvia verbenaca

Oceanic Southern Floristic Unit
11 Lothian species (82 in Matthews' British list)

Beta vulgaris ssp. *maritima*	*Limonium vulgare*
Carduus tenuiflorus	*Mibora minima* *
Carex extensa	*Parapholis strigosa*
Fumaria muralis ssp. *boraei*	*Phleum arenarium*
Hordeum marinum	*Ranunculus baudotii*
Ilex aquifolium	*Sagina maritima*

Oceanic West European Floristic Unit
17 Lothian species (87 in Matthews' British list)

Bromus hordeaceus ssp. *thominei*	*Ranunculus hederaceus*
Carex binervis	*Raphanus raphanistrum* ssp.
Conopodium majus	*maritimus*
Crambe maritima	*Salicornia dolichostachya*
Genista anglica	*S. ramosissima*
Hyacinthoides non-scripta	*Spartina x townsendii*
Lepidium heterophyllum	*Ulex europaeus*
Oenanthe crocata	*U. gallii*
Puccinellia maritima	*Vicia orobus*

Continental Southern Floristic Unit
15 Lothian species (129 in Matthews' British list)

Anacamptis pyramidalis	*Glaucium flavum*
Apium nodiflorum	*Lactuca virosa*
Arum maculatum	*Lotus pedunculatus*
Carex pendula	*Oenanthe lachenalii*

Catapodium rigidum
Dipsacus fullonum
Fumaria capreolata

Plantago coronopus
Salix purpurea
Torilis nodosa
Trifolium scabrum

Continental Floristic Unit
21 Lothian species (88 in Matthews' British list)

Allium vineale
Astragalus glycyphyllos
Corynephorus canescens
Cynoglossum officinale
Dianthus deltoides
Dipsacus pilosus * *
Gagea lutea
Galeopsis speciosa
Oenanthe aquatica
Orobanche minor

Potentilla argentea
Quercus petraea
Q. robur
Ranunculus lingua
Rosa rubiginosa
R. pimpinellifolia
Silene conica
Symphytum tuberosum
Teesdalia nudicaulis
Trifolium medium
Veronica montana

Continental Northern Floristic Unit
47 Lothian species (97 in Matthews' British list)

Alchemilla glabra
Andromeda polifolia
Angelica sylvestris
Astragalus danicus
Betula pubescens
Carex curta
C. diandra
C. dioica
C. disticha
C. echinata
C. lasiocarpa
C. pulicaris
Chrysosplenium alternifolium
Cicuta virosa
Circaea x intermedia
Cirsium heterophyllum
Coeloglossum viride
Crepis paludosa
Drosera anglica
D. rotundifolia
Eleocharis quinqueflora
Eriophorum angustifolium
E. latifolium
E. vaginatum

Galium uliginosum
Gentianella amarella
Hypericum hirsutum
Littorella uniflora
Menyanthes trifoliata
Meum athamanticum
Parnassia palustris
Pinguicula vulgaris
Pinus sylvestris * *
Potentilla palustris
Pyrola media
Rosa caesia
Sagina nodosa
Salix aurita
S. pentandra
Scrophularia umbrosa
Sedum villosum
Sparganium angustifolium
Stellaria palustris
Trichophorum cespitosum
Vaccinium myrtillus
Vicia sylvatica
Viola palustris

Northern Montane Floristic Unit

8 Lothian species (31 in Matthews' British list)

Antennaria dioica   *Rubus saxatilis*
Goodyera repens   *Salix phylicifolia*
Linnaea borealis   *Trientalis europaea*
Listera cordata   *Trollius europaeus*

Oceanic Northern Floristic Unit

13 Lothian species (23 in Matthews' British list)

Armeria maritima   *Honckenya peploides*
Atriplex glabriuscula   *Leymus arenarius*
Blysmus rufus   *Myrica gale*
Centaurium littorale   *Narthecium ossifragum*
Cochlearia danica   *Silene uniflora*
C. officinalis   *Thymus polytrichus* ssp. *britannicus*
                            Zostera angustifolia

Arctic-Subarctic Floristic Unit

5 Lothian species (28 in Matthews' British list)

Carex aquatilis   *Ligusticum scoticum*
Cochlearia micacea   *Mertensia maritima*
                            Rubus chamaemorus

Arctic-Alpine Floristic Unit

9 Lothian species (75 in Matthews' British list)

Carex maritima   *Oxyria digyna*
Empetrum nigrum   *Persicaria vivipara*
Epilobium alsinifolium   *Saxifraga hirculus*
Minuartia verna   *Sedum rosea*
                            Vaccinium vitis-idaea

---

* Naturalised
** Doubtfully native

- Mediterranean Unit: centred in the Mediterranean region with a northward extension into western France.
- Oceanic Southern Unit: centred essentially in SW Europe.
- Oceanic Western European species: almost exclusively to be found in Western Europe.
- Continental Southern elements: found chiefly in Southern and Central Europe, thinning out northwards.
- Continental Unit: species of Central Europe that thin out westwards, but which extend well into Russia.

- Continental Northern elements: Central and Northern Europe are their main centre, many being circumpolar. They are much less frequent to the south and become montane there.
- Northern Montane plants: species of Northern Europe, often circumboreal, becoming montane or sub-alpine in Central and Southern Europe.
- Oceanic Northern Unit: plants characteristic of NW Europe, some with a North American connection.
- Arctic-Subarctic species: exclusively northern in Europe, mostly centred in Scandinavia and extending into arctic-subarctic regions of Asia and North America.
- Arctic-Alpine Unit: circumpolar plants, found north of the tree boundary and in the mountains of Europe, Asia and North America.

Table 16.2 lists species native in the Lothians in their appropriate floristic units. Maps of the Lothian distribution of most of these species are to be found in the text of the vascular plant Flora (Chapter 13). Maps of some others, commented on here, are included in this account. Readers are reminded that, as explained in Chapter 12, each map reports the distribution of a taxon as recorded in the survey and does not necessarily include all the known locations for that taxon. For rare and scarce taxa, some or all of these locations are given in the Flora text.

Matthews included some 700 taxa in his categories, excluding a large number of British species because they are simply 'widespread', i.e. without a definable intra-British distribution. Thus *Calluna vulgaris* (Heather), though clearly a European Oceanic element as recognised by Dahl, was not classified in the ten categories of Matthews here reviewed, though it certainly thins out in central and eastern England. These 'widespread' taxa are also omitted from Table 16.2. In preparing Table 16.2, it has been necessary to amend Matthews' species names to match modern practice. Also, because modern concepts of native status are rather more rigorous, some taxa have been omitted. For instance, *Juncus tenuis* (Slender Rush) was included in Matthews' 'North American element', and so might have represented that category in the Lothians. However, no one would now regard it as a native British species.

A Floristic Analysis of the Lothian Flora

Mediterranean Unit

It is plain from Table 16.2 that species centred on the Mediterranean area are very poorly represented in the Lothian flora. One of the two is an old record. *Catapodium marinum* (Sea Fern-grass) however, demonstrates the fact that maritime Mediterranean plants have penetrated quite far north because of the uniformity, general equability and, critically, the continuity of the maritime environment.

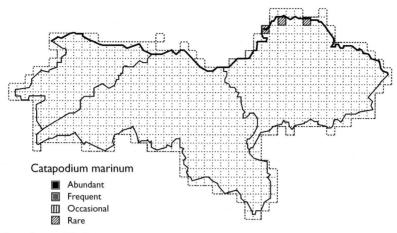

Catapodium marinum

- ■ Abundant
- ▦ Frequent
- ▥ Occasional
- ▨ Rare

Oceanic Southern Unit

Oceanic Southern species are more richly represented locally. Again, the maritime tendency and probable route of colonisation from the south are strongly indicated, as for instance in *Carduus tenuiflorus* (Slender Thistle) and *Sagina maritima* (Sea Pearlwort).

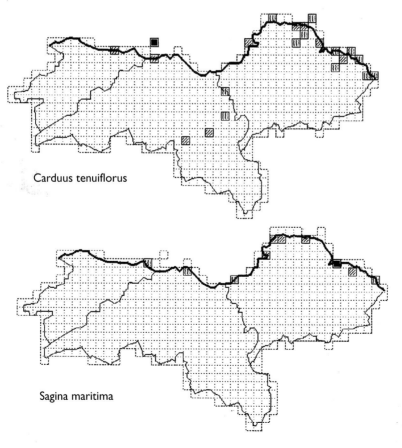

Carduus tenuiflorus

Sagina maritima

Fumaria muralis ssp. *boraei* (Common Ramping Fumitory) (see map in Flora) is a plant of arable and waste land, an opportunist for which a continuity of suitable habitats from the south and west will have existed probably from interglacial times. *Ilex aquifolium* (Holly), however, which is widespread in lowland areas of the Lothians, is clearly not merely clinging to a narrow ecological base: it is an Oceanic Southern species fully at home in a wide range of Lothian habitats.

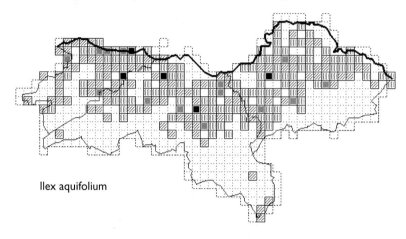

Ilex aquifolium

Oceanic West European Unit

Matthews' Oceanic West European species, of which six of the seventeen Lothian representatives are mapped in the Flora, often show some maritime tendencies in the Lothians. *Puccinellia maritima* (Sea Poa, or Common Saltmarsh Grass) is one

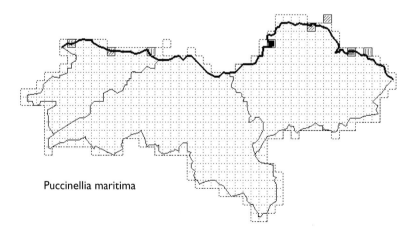

Puccinellia maritima

of them, and so too are *Oenanthe crocata* (Hemlock Water Dropwort), *Bromus hordeaceus* ssp. *thominei*, two *Salicornia* species (Glassworts), *Spartina x townsendii* (Townsend's Cord-grass), and *Raphanus raphanistrum* ssp. *maritimus* (Sea Radish). But in this Unit habitats widen to include inland, even upland wet places with *Ranunculus hederaceus* (Ivy-leaved Crowfoot) and *Carex binervis*

434

(Green-ribbed Sedge); woodland with *Hyacinthoides non-scripta* (Bluebell); and fleeting open places, with the opportunist *Lepidium heterophyllum* (Smith's Pepperwort). Two species, *Conopodium majus* (Pignut) and *Ulex europaeus* (Gorse) are common and widespread, clearly finding the Lothian climatic conditions generally undemanding. However, it is noticeable that, relative to populations further west, the local gorse suffers badly from frost scorch in a hard winter. Fitting into this floristic unit, though not mentioned in Matthews' lists, is *Anagallis tenella* (Bog Pimpernel), in Dahl's Oceanic British-Atlantic sub-element. In the Lothians it is a rare plant of bogs, not recorded for many years. *Asplenium adiantum-nigrum* (Black Spleenwort) and *Cerastium semidecandrum* (Little Mouse-ear Chickweed) from Dahl's Oceanic West Scandinavian-Atlantic sub-element, fall into this Matthewsian category: both show a markedly coastal distribution in the Lothian vice-counties.

Continental Southern Unit

Continental Southern plants are poorly represented in the Lothian flora. Often they have a scattered, patchy distribution such as that of *Salix purpurea* (Purple Willow), are sometimes scarce, and cling to the relatively mild conditions of the

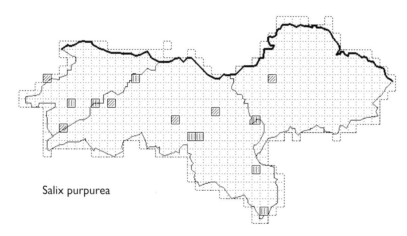

Salix purpurea

coast (*Torilis nodosa, Trifolium scabrum, Catapodium rigidum, Oenanthe lachenalii, Plantago coronopus, Anacamptis pyramidalis* etc.). The Great Lettuce (*Lactuca virosa*) barely creeps into the Lothians. Many of these plants seem to find the Lothians a challenging place in which to live.

Continental Unit

Continental plants fare somewhat better and some of them are widespread (e.g. *Rosa rubiginosa* – Sweet Briar), or very common (*Quercus robur* – Pedunculate Oak and *Q. petraea* – Sessile Oak, the latter being less common probably because of soil factors). *Allium vineale* (Crow Garlic), a common bulbous plant in southern England, perhaps shows its continentality, possibly an adaptation to hot, dry summers, by a pronounced East Lothian concentration. *Dianthus deltoides* (Maiden Pink) is here, but is scarce.

Continental Northern Unit

This Unit, of which many species are mapped, is well represented in the Lothians, having about half of Matthews' list. Many of these species are common and widespread members of the flora. Their 'northern-ness' partly reflects habitat preference, for they are mostly plants of wet uplands, of which there are many in the north. Unit members such as *Carices* (Sedges) and other Cyperaceae are characteristic of these habitats. Other species in the Unit have distributions exhibiting a western tendency (*Carex curta*, *C. echinata*, *Drosera rotundifolia* and the Cotton Grasses – *Eriophorum* spp.) while others have a more eastern (more continental) concentration, (*Carex dioica*, *Salix pentandra*, *Pinguicula vulgaris*). Some show a more generally upland pattern (*Crepis paludosa*, *Galium uliginosum*, *Salix aurita*). A few members have a scattered lowland distribution – inevitably therefore a more northern one in the Lothians. These include *Betula pubescens* (Downy Birch) and *Angelica sylvestris* (Angelica). Some species exhibit a marked maritime distribution (*Astragalus danicus*, *Stellaria palustris*, *Gentianella amarella*). Representing Dahl's Thermophilic Temperate sub-element, which overlaps with the Continental Northern Unit of Matthews, are *Bromus ramosus* (Hairy Brome) and *Brachypodium sylvaticum* (False Brome). Both (see maps in Flora) are scattered widely in the lowland areas. With these last two exceptions, there is little overall sign of 'continentality' and more of 'northern-ness' in this Unit, though this may simply arise from the high incidence of moor, bog and upland habitats across most of the southern part of the Lothians.

Northern Montane Unit

Northern Montane plants are not well represented in the Lothians, either in frequency or total number. *Listera cordata* (Twayblade) is one of them.

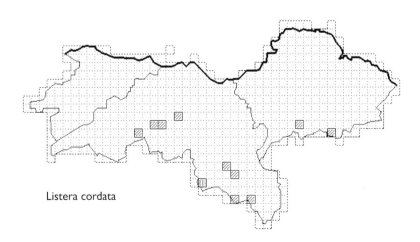

Listera cordata

Generally, they occur in a few upland squares only. *Linnaea borealis* (Twinflower) is represented by just one, old record, while *Salix phylicifolia* (Tea-leaved Willow) is scarce but more widespread. Dahl's Widespread Northern category overlaps with this Unit of Matthews. He exemplifies it by *Prunus padus* (Bird Cherry), *Alchemilla glabra* (Lady's Mantle) and *Nardus stricta* (Mat Grass). Bird Cherry is encountered locally throughout the lowland areas, with a minor concentration in West Lothian, though plantings confuse the interpretation. *A. glabra* is widespread in both lowland and upland areas, again with a stronger presence in the wetter, western third of the Lothians. Mat Grass is almost everywhere on high ground.

Oceanic Northern Unit

The Oceanic Northern Unit species show the distinctly maritime or bog distribution that would be expected, ten of the twelve species being coastal. Included in these ten are *Atriplex glabriuscula* (Babington's Orache), *Leymus arenarius* (Lyme Grass) and *Honckenya peploides* (Sea Sandwort).

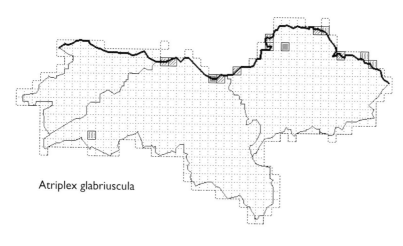

Atriplex glabriuscula

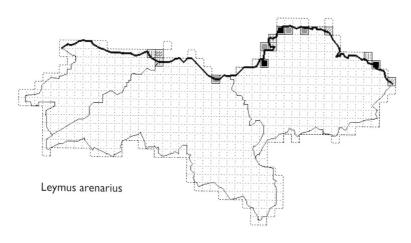

Leymus arenarius

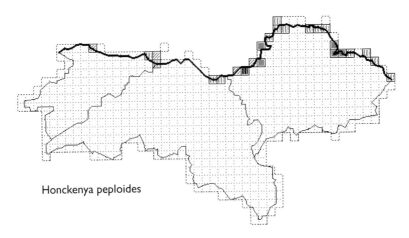

Honckenya peploides

Maps showing the distribution of other species in the Unit will be found in the Flora (Chapter 13). *Narthecium ossifragum* (Bog Asphodel) occurs mainly in wet, western, often upland sites. *Myrica gale* (Bog Myrtle) is a rare bog plant in the Lothians.

Arctic-Subarctic Unit

The five species of this Unit represent an outlier of a much more northerly group of plants. Two are maritime, for example, *Ligusticum scoticum* (Scots Lovage) while *Carex aquatilis* (Water Sedge) (see map), *Cochlearia micacea* (Mountain Scurvygrass) and *Rubus chamaemorus* (Cloudberry) occur in wet, western uplands.

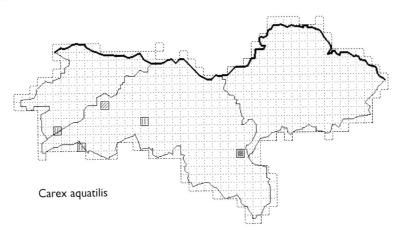

Carex aquatilis

Arctic-Alpine Unit

The Arctic-Alpine Floristic Unit has scant Lothian representation. Only *Empetrum nigrum* (Crowberry) is reasonably common, in the wet uplands of the south. The others are scarce. *Vaccinium vitis-idaea* (Cowberry), a plant which is commonplace to botanists familiar with the Scottish Highlands, is feebly scattered and local, mainly in the southern hills.

Conclusion

It can be concluded that, with due allowance for habitat incidence and distributions, the distributions of native Lothian plants fit well with the schemes of both Matthews and Dahl. The flora is predominantly an oceanic and generally northern one, with a very few outliers from the arctic. There is also a reasonable presence of southern and continental species, but many of these are scarce, and a significant number are maritime.

It must be remembered that this analysis is of a changing, perhaps now rapidly changing, phytogeographical scene. In a hundred or even in fifty years time, the floristic patterns may have altered quite markedly. We can begin to perceive what the change may be in Scotland, during a period of global warming (Watt *et al.* 1997). Crawford, Jeffree and Rees (2002) show that, while significant winter warming has taken place over the last century, summer warming has been significantly less. This seemingly inexorable, albeit irregular, diminution in the amplitude of annual temperature means that oceanicity is increasing – winter and summer temperatures are becoming less different. The likely ecological changes – forest retreat and bog growth – will certainly be accompanied by a territorial expansion and perhaps an enrichment of the oceanic floristic units here reviewed.

17

The Habitats, Distribution and Ecology of Plants in Edinburgh and the Lothians

P. M. SMITH

Introduction

The Botany of the Lothians survey produced a very large database of the distribution and frequency of plants in local habitats. It would take a space far beyond that available to review the findings completely in this chapter. There is material enough for further publication(s) and probably several higher degree theses. What can usefully be done here is to demonstrate the kind of information available and analyse some of it in a way that reveals both initial findings and the nature of the interpretations that are feasible. Those who might wish to extend the analysis will find that the data are lodged with the Botanical Society of Scotland (who have the copyright in them), the Scottish Wildlife Trust and Scottish Natural Heritage.

Habitats and frequency records have been mentioned earlier, in Chapter 12. It will be noted that the forty-one habitat categories necessarily reflect the high proportion of niches created or dominated by human intervention: such is the nature of the Lothian territory. The ecological mosaic is richer than it has ever been. The choice of habitats for recording, though far from being uninformed, was inevitably made *a priori* and, with hindsight, some alterations would have been made. Several apparently bizarre habitat records arose because of the incidence of microhabitats within otherwise very different environments – some examples are given in Chapter 12. No allowance was made for the explanation or indication of such microhabitats by recorders. Thus, waterside species such as *Ranunculus flammula* could be recorded from rough grassland because no microhabitat category of 'flush' had been defined. These anomalies are very few and do not impair the overall outcome. Were another survey to be undertaken, it could possibly be guided by the National Vegetation Classification Scheme (Rodwell 1991) which appeared well after the beginning of the present survey. However, it must be remembered that, in any venture involving recording by large numbers of people, simplicity is needed. What is practicable for use by professional ecologists may not be for the workers in a county Flora project.

Distribution of Habitats in Edinburgh and the Lothians

The maps (Figures 17.1 and 17.2) show the distribution of the main habitat categories in the three vice-counties. Moorland dominates on the higher ground to the south. Bog, rough grassland and heath are concentrated there. Woodland habitats are generally scattered, as are waterside categories. Conifer woodland is found on more upland sites than are mixed or deciduous woodland. 'Riversides' are more extensively recorded than rivers actually exist – many of these will refer to small burns, with which much of the Lothians is seamed. 'Ditchsides' seem more extensive than ditches: more plants grow beside water than in it. Climatic gradients imposed on the habitat distribution mean that the south and west are wetter than the north and east, and also that snow-lie and frost periods are generally lengthier to the south and east. The Firth of Forth acts as a notable climatic mitigation along the northern margin of the Lothians – the major maritime habitats obviously lie alongside it. Farmed land (arable and pasture – see 'Crop Plant Distribution' below) is scattered through the central and northern lowland, but there is a cereal farming area in East Lothian so extensive that it can give the impression of a 'prairie' (Plate 3). Major urban areas lie chiefly along the coast, from Bo'ness to Dunbar, Edinburgh being the largest. The major urban areas inland are Bathgate, Livingston, Penicuik, Dalkeith and Haddington. Base-rich soils are not a major feature of the Lothians, though limestone outcrops significantly near Dunbar. There are base-rich patches, rather than major zones, for example at Gifford and in Holyrood Park. The sea-coast and railway ballast comprise base-rich corridor habitats.

Further habitat continuity is provided by the road systems, though these are so extensive as to offer more a reticulum than a corridor opportunity for ruderals and hedgerow plants. Other notable corridor habitats are offered by the river valleys, which are still wooded for much of their length and by the Glasgow and Edinburgh Union Canal (Plate 14), running through West Lothian into central Edinburgh. The interruptions to this canal have recently been removed by the Millennium Link project.

The lowland area of the Lothians occupies most of the north and centre of the three vice-counties and in it are found the railways and the bulk of the waste ground, hedgerow and scrub habitats. A narrow, but prominent, south-eastern 'tongue', the valley of the Gala Water, extends into the Borders.

Plant Distribution Patterns and Ecology

Clear patterns of distribution are revealed by the results of the Botany of the Lothians survey. Most of these patterns can be correlated with habitat, land use

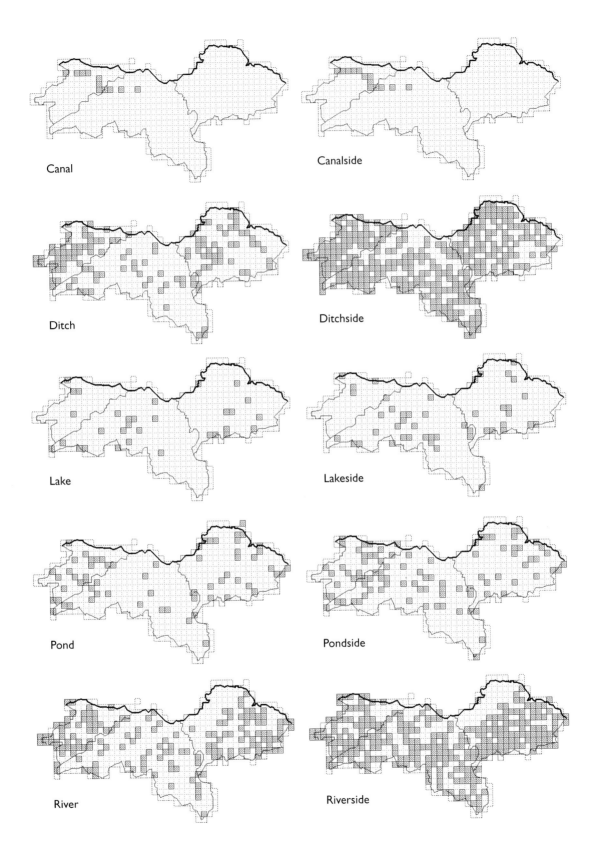

Canal

Canalside

Ditch

Ditchside

Lake

Lakeside

Pond

Pondside

River

Riverside

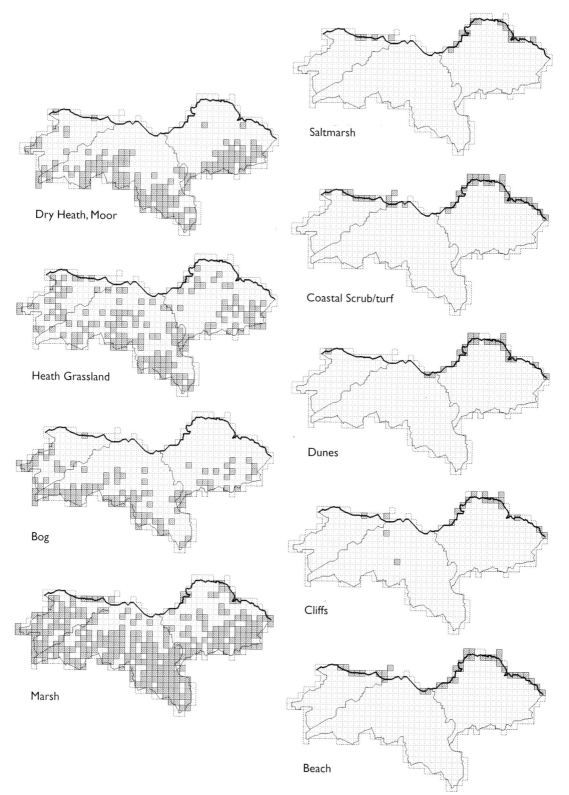

Dry Heath, Moor

Heath Grassland

Bog

Marsh

Saltmarsh

Coastal Scrub/turf

Dunes

Cliffs

Beach

Figure 17.1 (above and opposite) *Distributions of some of the habitats in the Lothians.*

and topography, and other major environmental factors. Most patterns are shown by a diversity of taxonomically unrelated species. The illustrations for this account are provided chiefly by the distribution maps printed in the Flora (Chapter 13), and appropriate reference should be made to them. In almost every case, many more examples of these distribution patterns exist than are mentioned here by way of definition and introduction. Only a few can be enumerated for each recognisable category of pattern. The main distribution patterns identified are summarised in Table 17.1, and these are exemplified and discussed below.

Table 17.1 The main types of plant distribution patterns in Edinburgh and the Lothians

Lowland Habitats	
General	**Coastal Habitats**
River valleys	Maritime
Lowland woodland	Near-maritime
Northern lowland	
	'Avoider Patterns'
Neutral and High-Base Substrates	Urban avoiders
	Arable avoiders
Waterside and Aquatic Habitats	
Canals, reservoirs, lochs and	**Widespread Patterns**
watersides	
Western, wet	**Curious Patterns**
Upland Habitats	**Crop Plant Distribution – 'Agricuse'**
General	
Upland, wet	**Arable and Ruderal Habitats**
Upland south-eastern	Weeds of agriculture
'Upland tendency'	Ruderals

Lowland Habitats

General

Numerous Lothian species show a general lowland distribution, whether they are common, widespread or less frequent there. In this group are many grasses with an agricultural connection – *Dactylis glomerata* (Cock's Foot) and *Phleum pratense* (Timothy) among them. *Elytrigia repens* (Common Couch), a common cereal and field margin weed, is another. Among shrubs and trees are *Ilex aquifolium* (Holly), *Fraxinus excelsior* (Ash), *Prunus spinosa* (Blackthorn), *P. avium* (Wild Cherry) and *Lonicera periclymenum* (Honeysuckle). Herbs showing similar patterns include *Vicia cracca* (Tufted Vetch), *Galeopsis tetrahit* (Common Hemp-nettle), *Bromus hordeaceus* (Soft Brome) and *Heracleum sphondylium* (Hogweed). The southern 'tongue' of lowland, extending into the Borders, is well shown by the distribution of many herbs, such as *Potentilla anserina* (Silverweed), *Pimpinella saxifraga* (Burnet Saxifrage), *Fumaria* spp. (Fumitories) and *Cruciata laevipes* (Crosswort). The Red-berried or Border Elder (*Sambucus racemosa*) is a shrub that has a distribution displaying this 'tongue'. *Symphytum tuberosum* (Tuberous Comfrey) and *Filipendula ulmaria* (Meadowsweet) have a lowland distribution prominently displaying what might be called the 'Pentland Gap', in which the Pentland Hills interrupt their lowland continuity.

Weedy or ruderal/roadside species with a wide lowland scatter include *Polygonum aviculare* (Knotgrass), *Chenopodium album* (Fat Hen), *Barbarea vulgaris* (Winter-cress) and *Convolvulus arvensis* (Field Bindweed).

River valleys

The generally wooded valleys in which Lothian rivers (Almond, Avon, Water of Leith, Esk and Tyne) run are corridor habitats for a good number of species, as their distributions attest. Such patterns are shown by *Geranium sylvaticum* (Wood Cranesbill), *Scrophularia nodosa* (Common Figwort), and *S. auriculata* (Water Figwort). *Mimulus* spp. (Monkeyflowers) (see Frontispiece) and *Valeriana officinalis* (Common Valerian) display this association, which is also exhibited, among introduced species, by *Doronicum pardalianches* (Leopard's Bane).

Lowland woodland

Several woodland-related patterns are discernible, apart from the river-valley distribution mentioned above. One is a scatter of woodland sites over a wide area, as with *Rumex sanguineus* (Wood Dock). Another is a probable association with the relic woodland of old estates, which are commoner in Midlothian and East Lothian than further to the west. Taxa in this category are numerous, but include some introduced or probably introduced species – *Vinca minor* (Lesser Periwinkle) and *Galanthus nivalis* (Snowdrop) and, perhaps *Arum maculatum* (Lords and Ladies). Among native grasses, *Poa nemoralis* (Wood Meadow Grass) and *Milium effusum* (Wood Millet) are probably in this group, as possibly also is *Festuca gigantea* (Giant Fescue). *Campanula latifolia* (Giant Bellflower) is a likely further example.

A special version of the woodland category is a group of distributions which centre on the southern or east-central part of Midlothian. These are plants of woodland on deep, richly humified soils where the habitat is probably long-established and comparatively undisturbed. The group can be exemplified by *Sanicula europaea* (Sanicle), *Adoxa moschatellina* (Moschatel, or Town Hall Clock) and *Circaea lutetiana* (Enchanter's Nightshade). *Carex pendula* (Pendulous Sedge) and *Galium odoratum* (Woodruff) probably also belong here.

Northern lowland

Lowland species with a generally more northerly and, to a degree, a sub-coastal pattern include *Dipsacus fullonum* (Teasel), *Carduus crispus* (Welted Thistle) and

Allium ursinum (Ramsons). Legumes figure notably in this group, examples being *Medicago lupulina* (Black Medick) and *Trifolium campestre* (Hop Trefoil). With urban habitats mostly located in the north of these vice-counties, it is not surprising that roadside, railway and ruderal species are well represented in this category. *Tragopogon pratense* (Goat's Beard, Jack-go-to-bed-at-noon), *Bromus sterilis* (Barren Brome), *Senecio squalidus* (Oxford Ragwort) and *Linaria vulgaris* (Common Toadflax) are instances.

Some northern lowland distributions occur with something of an inclination to the west, perhaps implying that the plants have some need for or tolerance of a higher precipitation/evaporation ratio. Examples include *Carex hirta* (Hairy Sedge), *Petasites hybridus* (Butterbur) and *Persicaria bistorta* (Common Bistort).

Northern lowland distributions with a cast to the north-east, perhaps suggesting a tolerance of, or need for, drier conditions, lighter soils and possibly more summer heat, are shown by *Urtica urens* (Small Nettle), *Sisymbrium officinale* (Hedge Mustard) and *Thlaspi arvense* (Field Pennycress). These may be partly determined in their distribution by their status as field weeds in the East Lothian 'prairie zone'. Perhaps, unlike some field weeds found more commonly elsewhere, they are, to a degree, indifferent to weed control regimes regularly practised in East Lothian. Other taxa seemingly associated with the warmer, drier conditions of the Lothian north-east are *Potentilla reptans* (Creeping Cinquefoil), *Malva sylvestris* (Common Mallow), *Silene vulgaris* (Bladder Campion) and *Allium vineale* (Crow Garlic). Three umbellifers seem to belong to this group: *Chaerophyllum temulum* (Rough Chervil), *Torilis japonica* (Upright Hedge Parsley) and *Conium maculatum* (Hemlock).

Neutral and High-Base Substrates

Lothian soils (see Chapter 1) are predominantly acidic. In a few places, where limestone or calcareous sandstones outcrop, there is the opportunity for calcicolous plants to establish. Other high-base niches are to be found on oil bings and on or near railway ballast, which is by no means all of local origin, and may be replenished periodically with new material still able to release minerals. There are volcanic outcrops. The occasionally shelly substrates near the sea also constitute possible niches for calcicoles.

The patterns of distribution do not generally provide clear evidence of association between high-base status and species incidence. An exception must be the coast but here there are more factors operating than the base status of the soils.

Evidence for a correlation between distribution and high-base status might be looked for in the patterns shown by such taxa as *Linum catharticum* (Fairy Flax), *Plantago media* (Hoary Plantain) *Primula veris* (Cowslip) and *Sanguisorba minor* ssp. *minor* (Salad Burnet). The evidence of overall correlation cannot be said to be overwhelming, but of course there are isolated associations that are recorded in the text of the Flora.

Waterside and Aquatic Habitats

Canals, reservoirs, lochs and watersides

Many species of water and watersides have a scattered distribution in the Lothians, as might be expected from the scattered incidence of the habitats themselves. However, there are some species that exhibit a marked association with particular aquatic habitats. Among these may be cited *Littorella uniflora* (Shoreweed) which is found in most of the southern reservoirs and reveals this in its distribution. So does *Alisma plantago-aquatica* (Water Plantain) though less markedly, but it also shows a notable correlation with the line of the Glasgow and Edinburgh Union Canal. *Lysimachia thyrsiflora* (Tufted Loosestrife) (Plate 14) is a striking associate of the canal, but its occurrences are rather discontinuous. Not so those of, for example, *Lycopus europaeus* (Gypsywort), *Glyceria maxima* (Reed Sweet Grass, or Water Manna) and *Lemna gibba* (Fat Duckweed). *L. minor* (Common Duckweed) has a much more scattered distribution, but also reveals a continuity along the canal.

Western, wet

Of wetland or marsh lowland taxa, there are a number in the Lothians that show a concentration towards the west, with their most notable occurrence in West Lothian. This is the wettest of the three vice-counties so it is not surprising that species with pronounced adaptations, physiological and/or morphological, to moister substrates should show this pattern. There are many mapped examples in the Flora, among them *Senecio aquaticus* (Water Ragwort), *Juncus articulatus* (Jointed Rush), *Achillea ptarmica* (Sneezewort) and *Hippuris vulgaris* (Mare's Tail). Others showing this association are *Lotus pedunculatus* (Greater Bird's-foot Trefoil), *Rumex longifolius* (Northern Dock), *Stachys palustris* (Marsh Woundwort), *Alopecurus geniculatus* (Marsh Foxtail) and *Gnaphalium uliginosum* (Marsh Cudweed).

A similar pattern but where a connection with moister habitats is not recorded, is shown by *Leucanthemum vulgare* (Oxeye Daisy) (Plate 15) and *Hypochaeris radicata* (Cat's Ear). These species may be showing a correlation with industrial dereliction. Easier to explain are some other 'western, wet' patterns shown by species with a generally upland distribution (see below).

Upland Habitats

General

Upland distributions are strikingly clear in the Lothians: they are of taxa well represented in the southern, largely upland area of the three vice-counties. Good examples, of many possible, include *Molinia caerulea* (Purple Moor Grass), *Empetrum nigrum* (Crowberry), *Nardus stricta* (Mat Grass) and *Juncus squarrosus* (Heath Rush).

Upland, wet

Many species showing upland concentrations in the Lothians may do so in whole or in part because they are plants of wetter places, rather than of uplands *per se*. As mentioned previously, some at least of these taxa might be expected to show a 'western, wet' pattern. This is clearly shown by very many taxa which are, overall, upland in distribution. Examples include *Eriophorum* spp. (Cottongrasses), *Callitriche stagnalis* (Common Water Starwort), *Succisa pratensis* (Devil's Bit Scabious), *Galium palustre* (Common Marsh Bedstraw) and *Potentilla palustris* (Marsh Cinquefoil). To be mentioned here also are *Isolepis setacea* (Bristle Club-rush) and several sedges (*Carex* spp.). *Vaccinium myrtillus* (Blaeberry), *Erica tetralix* (Cross-leaved Heath) and perhaps even *Calluna vulgaris* (Heather) show this tendency.

Upland south-eastern

In a mirror image of the previous pattern, are the distributions of upland species with a south-eastern concentration in the Lothians, though there are many fewer of these. *Erica cinerea* (Bell Heather) and *Vaccinium vitis-idaea* (Cowberry) show this pattern. Perhaps these eastern upland areas are marginally but significantly drier than the western upland areas. *Ajuga reptans* (Bugle), *Lysimachia nemorum* (Yellow Pimpernel), *Parnassia palustris* (Grass of Parnassus) and *Sedum villosum* (Hairy Stonecrop) seem to fall into this category, but are found in relatively moist habitats. *Potentilla sterilis* (Barren Strawberry) has a markedly south-eastern distribution, but is rather more scattered in lowland areas than are some others in this group.

'Upland tendency'

A distribution somewhat intermediate between upland and lowland is shown by species whose main incidence is more southerly than the central lowland area, but which do not occupy the highest ground. These species are here described as manifesting an 'upland tendency', possibly determined by the less intensive agriculture practised at intermediate altitudes. Among them are *Lathyrus linifolius* (Bitter Vetch), *Danthonia decumbens* (Heath Grass), *Potamogeton polygonifolius* (Bog Pondweed) and *Myosotis laxa* (Tufted Forget-me-not). In some cases the underlying factors may well be a requirement for water, or perhaps deep, moist soils, allied to a growth or reproductive constraint associated with late and early snow-lie, or frost. Other species seeming to show this pattern are *Pedicularis palustris* (Marsh Lousewort) and *Conopodium majus* (Pignut).

Coastal Habitats

Maritime

Various coastal habitats exist in the Lothians, ranging from cliffs (for example those of the Bass Rock) to the mudflats along the Firth of Forth – which figured so importantly in the seminal biosystematic work of J. W. Gregor. Dunes are well developed in parts of East Lothian. Species with adaptations to these facets of coastal diversity are found in increasing numbers from the west to the east, as estuarine conditions give way to fully maritime environments.

Many Lothian species exhibit a maritime distribution (see Chapter 16) – the main factors to which they are adapted including salt in the substrate and in the air (as sea spray), regular tidal immersion, high or at least neutral soil pH and a degree of exposure to high winds. Near the sea, most are low-growing herbs, such as *Atriplex* spp. (Orache), *Armeria maritima* (Thrift) and *Salsola kali* spp. *kali* (Prickly Saltwort). Others are wind-cut shrubs and small trees such as *Hippophae rhamnoides* (Sea Buckthorn) (Plate 16). In less exposed areas, and in the upper part of beaches, the larger herbs, for example *Oenanthe crocata* (Hemlock Water Dropwort), are to be seen. In regions of slack water, patches of saltmarsh are present, with *Glaux maritima* (Sea Milkwort), *Cochlearia* spp. (Scurvygrasses), *Triglochin maritima* (Sea Arrowgrass) and *Carex otrubae* (False Fox Sedge). Dune plants include *Carex arenaria* (Sand Sedge), *Leymus arenarius* (Lyme Grass) and *Ammophila arenaria* (Marram Grass).

Near-maritime

In addition to the long list of Lothian species shown by the maps to have a well-marked coastal distribution, there are some others that display a coastal and/or sub-coastal incidence. The mitigating effect of the sea on the absolute temperatures and lengths of frost-period and snow-lie are probably important factors determining the pattern of their occurrence. Perhaps the locally lighter, better-drained or higher pH soils are involved also. Species with such a pattern include *Anthyllis vulneraria* (Kidney Vetch, Lady's Fingers), *Conium maculatum* (Hemlock), *Daucus carota* (Carrot), *Vicia lathyroides* (Spring Vetch) and *Carduus tenuiflorus* (Slender Thistle).

'Avoider Patterns'

It is obvious that any plant with a propensity to grow and reproduce well in one set of conditions is likely to seem to avoid others: thus lowland species in the Lothians are not found in the uplands. But, beyond such major habitat differentiations there are a number of others where, within a generally suitable environment, patches occur where a species is seemingly excluded. Here the species seems to be avoiding some set of local factors that must be acting as performance constraints.

Urban avoiders

The Flora maps demonstrate that quite a number of species are scarce or absent in urban areas, which they surround and thus define by their pattern. *Vicia cracca* (Tufted Vetch) and *V. sepium* (Bush Vetch) show this distribution, as do *Filipendula ulmaria* (Meadowsweet) and *Moehringia trinervia* (Three-nerved Sandwort). The reasons for the exclusions will not be common to all the species excluded. *Primula vulgaris* (Primrose) is not found growing wild in, for instance, the City of Edinburgh, because its normal hedgebank and woodland habitats, with heavy, undisturbed soil, have largely been destroyed by building. Further, it would have been gathered by the populace. *Angelica sylvestris* (Angelica), otherwise a common lowland species, 'avoids' Edinburgh because, although there are streamsides there, they are highly accessible, much visited and disturbed, or else their banks are industrialised. The 'avoiders' do not therefore make a homogeneous group of species.

Arable avoiders

Intensive arable areas are subject to frequent ploughings and harrowings that furnish only brief and marginal opportunities for the growth of spontaneous vegetation. Grassland species will be excluded. For example, *Cynosurus cristatus* (Crested Dog's Tail) and *Anthoxanthum odoratum* (Sweet Vernal Grass) seem to avoid the cereal-growing areas of East Lothian as does *Veronica serpyllifolia* (Thyme-leaved Speedwell) which is generally found in permanent grassland. The East Lothian arable 'prairie' is a dry area as well as being farmed intensively: wetland and waterside habitats are sparse (Figure 17.1), hence species such as *Myosotis scorpioides* (Water Forget-me-not) and *Mentha aquatica* (Water Mint) are excluded.

A special case of arable avoidance may be revealed by such species as *Spergula arvensis* (Corn Spurrey). This is a common weed of agriculture, especially on light soils. Suitable conditions for it would seem to abound in the East Lothian arable area, yet it is scarcely known there now, though was formerly common. In West Lothian and Midlothian arable environments it is still frequent and familiar. Is the exclusion from East Lothian a result of sophisticated (or relentless) weedkiller regimes, that have not been, as yet, applied further west? *Persicaria maculosa* (Redshank) may be another example of this arable avoidance.

Widespread Patterns

Taxa with widespread patterns, apparently unconstrained by any facet of the Lothian environment, and nowhere near their tolerance limits, show few distributional features to connect with adaptation. Yet it would be naive to think that they have no interest because of this. It is wholly possible that they have diverse populations with large internal genetic variation – many available genotypes, many variants able to exploit all or most of the ecological opportunities that the Lothians provide, thus being among the most intriguing local taxa. Equally, they may have considerable phenotypic plasticity. Widespread upland and lowland types have already been exemplified. Some species are more catholic even than that, growing almost everywhere. One of them is *Sorbus aucuparia* (Rowan), though the natural distribution may be somewhat obscured by plantings. Others are *Lotus corniculatus* (Bird's-foot Trefoil) and *Agrostis capillaris* (Common Bent). Careful analysis of the Lothians survey habitat and frequency statistics might in such cases reveal ecologically significant facts and cryptic taxonomic differences, i.e. those which may indicate the presence of unrecognised taxa.

Curious Patterns

In many cases, several factors will be operating significantly to constrain or define the distribution of a species. Where, in one place one factor is particularly favourable, the species may succeed. In another place, it may also succeed because some other, potentially limiting factor is removed. Thus some species may occupy several distinctive niches. In such circumstances, the plant may present a curious distribution of which little or nothing can be offered by way of explanation. *Thymus polytrichus* ssp. *britannicus* (Wild Thyme) shows a polarised distribution in the Lothians – common in some dry, upland habitats, also some wetter ones, and showing additionally a pronounced maritime presence.

Another ecological puzzle is presented by *Galeopsis speciosa* (Large-flowered Hemp-nettle). A lowland, weedy, opportunist plant, it exhibits a notable concentration in the southern part of Midlothian, being comparatively unusual, though certainly not unknown, elsewhere.

Odontites vernus (Red Bartsia), a plant of grassland and roadsides, shows a concentration in central Midlothian, a good scatter through dry areas, but a prominently common incidence in the wetter, western area. The reasons are not readily apparent.

Crop Plant Distribution – 'Agricuse'

The maps in Figure 17.2 show the major categories of crop distribution recorded in the Botany of the Lothians survey. Farming activity is spread throughout the area. The type of farming falls into two distinct classes. Intensive grassland, including sown and managed pasture, is found throughout the area except on the higher ground and in built-up areas. Root and other vegetable crops show a major concentration in the northern part of East Lothian, with a lesser area in West Lothian. Cereal fields display this pattern more strikingly, with a large and continuous area in eastern Midlothian and most of the north of East Lothian, again with a lesser area in West Lothian. Cereal agriculture is best developed where the soils are lightest and the climate driest. These patterns appear to have obvious consequences for the distribution of wild plants – agricultural (mainly arable) weeds and the pasture grasses (native as well as sown) show marked correlated effects.

Arable and Ruderal Habitats

Weeds of agriculture

While *Urtica dioica* (Common Nettle), a common plant in the Lothians, abounds in and around pasture land, no doubt because of the high manurial (therefore phosphorus) content of the soil, and the mechanical/chemical anti-grazing defence of the plant, *Urtica urens* (Small Nettle) shows a marked association with areas of arable agriculture. Common segetal (cereal field) weeds, with a marked East Lothian representation, include *Bromus sterilis* (Barren Brome), *Avena fatua* (Wild Oat), and to some extent *Elytrigia repens* (Common Couch) though the latter abounds also on roadsides. Distributions that reveal a notable weediness in arable habitats are shown also by *Anchusa arvensis* (Bugloss), *Lamium confertum* (Northern Dead-nettle) and *L. amplexicaule* (Henbit Dead-nettle).

Ruderals

With so many roadsides and railways, and such a pace of demolition and rebuilding in urban areas, the Lothians are rich in scattered habitats for ruderals – offering fleeting opportunities for plants of short life-cycles and high reproductive capacity to colonise briefly. The sites are often stony and dry. The ruderal occupants are partly coincident with the weedy plants of dry, arable agriculture, and are found also as garden weeds and on walls. Generally, they are associated with human activities, and are least common in the undisturbed uplands and wetlands to the south. Prominent examples – a few of very many possible ones – are *Vicia hirsuta* (Hairy Tare), *Verbascum thapsus* (Great Mullein), *Euphorbia peplus* (Petty Spurge) and *Senecio squalidus* (Oxford Ragwort). Species found in both arable and ruderal habitats include *Aethusa cynapium* (Fool's Parsley) and *E. helioscopia* (Sun Spurge). *Linaria purpurea* (Purple Toadflax) shows a marked distributional correlation with urban areas, and seems to be increasing.

Some Lothian Autecological Portraits

Any or all of the distributional patterns just described, and others that may be discerned in the records, need to be accounted for more fully than has been possible above. The habitat statistics recorded in the survey would need to be considered, along with other information on the botany of the species concerned. To do this comprehensively would need a second volume of this book and is certainly beyond the aim of this chapter. Here an indication is given of some of the issues and possibilities, questions and answers that arise when the database is examined. Many other examples could have been used.

Ecological Differentiation and Adaptation within Genera

The discussion that follows – 'Three Buttercups' etc. – might have covered a large number of other species groups that show differentiation equally well. 'Three Heathers', 'Four Willows' and so on, have interesting tales to tell, and the reader is encouraged to look for them in the Flora maps and text.

Three buttercups

Ranunculus acris (Meadow Buttercup), *R. bulbosus* (Bulbous Buttercup) and *R. repens* (Creeping Buttercup) well illustrate the ecological differentiation arising

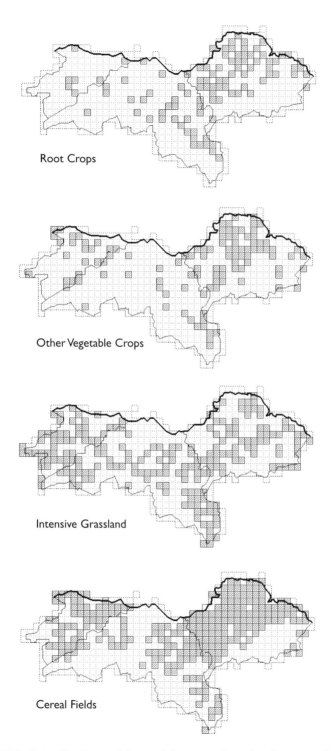

Figure 17.2 *Distributions of four of the agricultural habitats in the Lothians*

from the different adaptations of three well-recognised species in a genus. The Lothian results show *R. acris* to be common, but to be both an arable and an urban 'avoider', as well as thinning out on higher ground. It is at its most concentrated in lowland meadows, where its unpalatability to stock gives it advantage as a pasture weed. In life it evades the worst of the grazing pressure and, in death after cutting, because of its stem height, it has a notable presence in hay, in which its seeds can persist and spread. Hay is now less important as a crop than it was formerly.

A quite different distribution is shown by *R. bulbosus*, which is clearly a plant mainly of the coast and scattered northern and central grassland. In drier soils, it is at an advantage in having a bulbous base, enabling successful 'over-summering' survival. Its unpalatability probably reduces grazing, while Grime *et al.* (1988) point out that the worn or bare ground produced by grazers gives its locally dispersed seeds a good establishment opportunity. The Lothians distribution matches expectation very well.

R. repens is found almost everywhere in the Lothians. Though seeding less prolifically than the buttercups mentioned above, it spreads aggressively by stolons, colonising bare ground very rapidly and persisting well. It is not unpalatable, and Grime *et al.* suggest that its incidence implies past or present soil disturbance.

Analysis of the Lothian habitat records shows *R. acris* in thirty-five habitats, *R. bulbosus* in seventeen habitats, and *R. repens* in thirty-eight habitats. *R. repens* is much commoner as a weed of cereal fields and other arable crops than is *R. acris*. *R. bulbosus* is not recorded as a weed.

Two poppies

Two common poppies, *Papaver dubium* (Long-headed Poppy) and *P. rhoeas* (Corn Poppy), seem to show little distributional difference in the Lothians (see maps in Flora). Only slight ecological differentiation is revealed by the survey statistics. Each is recorded from a little over twenty habitats and the incidence in most of them is similar. However, the proportion of records in each kind of agricultural habitat is slightly greater for *P. rhoeas* than for *P. dubium*, while the proportion of records in ruderal habitats is slightly greater for *P. dubium* than for *P. rhoeas*. Both species are as common on roadsides and railway embankments as they are in cereal fields. Grime *et al.* (1988) suggest that as arable incidence is declining, ruderal incidence is increasing. This fits absolutely with Lothian observations. As the ecological differentiation wanes, perhaps the hybridisation (producing *P. x hungaricum*) will increase.

Three avens

Geum rivale (Water Avens) and *G. urbanum* (Wood Avens) hybridise when adjacent to produce *G. x intermedium*. The Lothian distributions (see Flora maps) neatly demonstrate the general spatial separation of the parents, with the hybrids chiefly in the zone of overlap. The habitat record shows a marked ecological differentiation between the parents. *G. urbanum* has its greatest incidence in mixed and deciduous woodland, hedgerows and scrub, though it is much recorded from road-sides. *G. rivale* is much commoner by watersides and in marshy ground. Though there is some roadside incidence shown in the habitat statistics, it is far less than that for *G. urbanum*, and probably involves roadsides near drainage ditches. The habitat particulars for the hybrid are intermediate between those of the parents. *G. urbanum* has some ruderal qualities, with a marked presence in farmland, gardens and waste ground.

Two bird's-foot trefoils

Lotus corniculatus (Bird's-foot Trefoil) and *L. pedunculatus* (Greater Bird's-foot Trefoil) show sharply distinctive Lothian distributions, the latter being largely absent from the north-central and eastern areas. The habitat details reveal clear evidence of significant ecological differentiation. *L. corniculatus* is in thirty-four habitats, many of them in dry grassland and dry ruderal (walls, quarries) sites. It has a maritime presence (beaches, cliffs and saltmarsh). By contrast, *L. pedunculatus* is a plant of moist places, waterside habitats collectively being commonest, and marshland being next. Unlike *L. corniculatus*, it has a presence in mixed woodland, clearly tolerating shade.

Four rushes

Juncus squarrosus (Heath Rush) has its upland, moorland distribution further illustrated by its habitat statistics. It was recorded from nineteen habitats, almost half the records being from dry heath/moor and heath grassland. Bogs, marshes and watersides make up most of the rest.

How different is the distribution of *J. bufonius* (Toad Rush), with a lowland scatter, thinner to the east, more continuous in the (wetter) west. This little annual (a taxonomic complex not unravelled in the survey) is also found in marshes and waterside habitats, but hardly at all in heath grassland and moors. Its main incidence is on roadsides, where it often occupies the drainage line on the road margin. It has a ruderal quality also, being found in waste ground and as an arable weed.

J. effusus (Soft Rush) and *J. conglomeratus* (Compact Rush) – see cover illustration – two very similar-looking species, are of informatively different distribution and ecology in the Lothians. *J. effusus* is widespread, except in urban or very intensively farmed areas, but reaches its commonest state on the wet moors and hills to the south, and the wet, western lowland areas. Bogs and marshes, ditch-sides, heath grassland, dry heath and moorland figure extensively in the recorded habitats for both species. *J. effusus* is less ecologically fastidious (thirty-two habitats compared with twenty-six for *J. conglomeratus*) but there is a huge overlap between the two species. The only two locational factors that seem to separate them are, in *J. conglomeratus*, a greater dependency on permanent moisture and, perhaps interacting with that, a typical incidence on slopes at higher altitude, where the water does not stand. The very small flowering time differences commonly reported for these two species are probably insufficient on their own to maintain their separateness.

Two sedges

The Botany of the Lothians survey results offer several instructive patterns of distribution and habitat statistics in *Carex* species. *C. curta* (White Sedge), for instance, has a south-western, upland distribution, implying a strong association with high precipitation, and is shown chiefly as a bog and marsh plant, with a secondary incidence along ditches and rivers. *C. nigra* (Common Sedge, Black Sedge), on the other hand, has a wider upland distribution with a much greater eastern presence. The habitat statistics show a considerable ecological overlap between the two species, but a large number of records for *C. nigra* from heath grassland, dry heath and moorland, and rough grassland. Drier habitats seem to suit the Black Sedge, and must account for its extensive distribution in the east.

Three sow-thistles

Interestingly different distributions are shown by three common sow-thistles (*Sonchus* spp.). *S. arvensis* (Perennial Sow-thistle, Corn Sow-thistle) shows a northern and north-eastern distribution, seemingly avoiding the area of greatest urbanisation. Habitat statistics show it to be in twenty-five habitats. While its highest incidence is on roadsides and waste ground, it has a significant presence in cereal fields and other arable crops, also on dunes and beaches. Apart from records in coastal scrub/turf, it has essentially no woodland presence.

    S. oleraceus (Smooth Sow-thistle) and *S. asper* (Prickly Sow-thistle) overlap with *S. arvensis* in cereal and other arable fields, but both have a high incidence in ruderal habitats. For *S. oleraceus* 80 per cent, and for *S. asper* 53 per cent, of records are from ruderal habitats. Four habitat features seem to separate them. *S. asper* has a much more marked presence, first, in woodland/scrub and, second, on various types of waterside. Third, it is more frequent in grassland (where its more notable mechanical defences may offer grazing resistance). Finally, it is much less characteristic of coastal habitats.

Ecological Amplitude within Species

Some widespread grasses

Festuca rubra (Red Fescue) is found almost everywhere in the Lothians. What do the habitat statistics reveal? It was recorded from thirty-five of the forty-one recognised habitats, a clear indication of its ecological catholicity. It was most commonly recorded on roadsides, then in descending order in rough grassland, waste ground, by rivers, in heath grassland and in intensive grassland. Lesser, but still frequent encounters with this grass are to be expected in habitats as diverse as mixed woodland, cereal fields, parks, gardens and coastal habitats. The ecological amplitude of this species must imply widely tolerant plastic phenotypes, great genetic heterogeneity, or both.

Agrostis capillaris (Common Bent) was found in thirty-three habitats, and much of what is said above for Red Fescue applies to it also. There is no clue in the habitat statistics that would explain the gap in its area to the south-west of Edinburgh. Its occurrence as a weed of cereals and vegetable crops is significantly greater than that of Red Fescue.

Lolium perenne (Rye-grass) is a much-sown grass as well as one of common native incidence. It is not surprising that so wide a distribution is shown. In thirty-three habitats, it clearly has a wide ecological amplitude. It is notable, however, that it is quite rarely found in the dry heath/moor and heath grassland categories – and its distribution does not extend to the higher upland areas in the south. A very similar pattern and habitat incidence is shown by *Dactylis glomerata* (Cock's Foot), a grass of very similar status.

Another very commonly encountered, widespread grass is *Anthoxanthum odoratum* (Sweet Vernal Grass). The distribution and habitat incidence (thirty-two habitats) of this reveal a wider ecological amplitude than that of most species in the Lothians. The statistics show numerous records from upland and moist grassland sites. Sweet Vernal Grass does not seem to avoid the higher altitudes as do Rye-grass and Cock's Foot, but unlike them and *Agrostis capillaris*, it has very little penetration into areas of arable cultivation. It seems to be a poor weed of agriculture, but an effective colonist more or less everywhere else.

The distributions and ecology of the familiar grasses show that, though they are with us apparently everywhere, significant ecological differentiations are present.

Ecological Tendency and Specialisation

Without showing an obvious predilection for an area or series of specialised habitats, some taxa nevertheless display a tendency towards this in their distribution, a kind of ecological half-heartedness. Undoubtedly, this will be because of limiting factors that inhibit a more patent association with a particular environment. Several examples of such species have been mentioned earlier in this chapter, and are mapped in the Flora.

Three umbellifers may be taken to illustrate what the habitat statistics add to the distribution, by way of explanation.

Torilis japonica (Upright Hedge Parsley) has some but not all the characters of what were above described as 'urban avoiders' and 'lowland north-eastern' species. Ecologically, it seems to have a good amplitude (thirty of the recognised habitats), doing quite well in mixed woodland/hedgerow/scrub habitats, roadsides, railway embankments, waste ground and similar places (76 per cent of all records). It is represented in cereal fields. Grassland of all kinds (closed communities), are almost completely denied to it. It seems to be restricted to areas where grazing or other attack is minimal, or to places such as in the hedgerow, where it can reproduce before being cut down, either by road verge control, or by the combine harvester. Another factor associated with it seems to be the dryness of soils – there are few waterside records. Relative freedom from disturbance or attack, allied to warm, well-drained soils, and its annual habit may be the principal factors encouraging it.

Conium maculatum (Hemlock) has a coastal/sub-coastal distribution in the north-east of the Lothians, and a small presence elsewhere. To what extent is it a coastal plant? What is the habitat basis for its north-eastern concentration? The statistics show that it is in twenty-five habitats, roadsides and waste ground being the commonest, so it is fundamentally ruderal. Hedgerow and farmyard incidences contribute to this pattern. There are few woodland or scrub records, and very few in rough grassland. Closed communities do not, therefore, seem to favour it. Two other indicators are: a significant presence as a marginal plant of cereal fields and vegetable crops, and also its status as a denizen of cliffs, coastal scrub/turf and dunes. This collectively gives it, in our area, a north-eastern distribution, for it is there that arable cultivation is most concentrated and coastal habitats are most diverse. Most coastal habitats (dunes, cliffs etc.) are open, so it is probably to be regarded as a ruderal with poor competitive ability in the closed communities (grassland and moorland) to the south and west, and low resistance to animal attack, despite its unpalatability. Yet so low is its incidence in ruderal situations to the south and west, that some climatic factor is probably also implicated – the warmer summers of East Lothian are the likeliest candidate. In this it resembles *Torilis japonica*.

Angelica sylvestris (Angelica) was recorded from twenty-eight habitats in the Lothians and the map shows its distribution to be predominantly southern and south-western. Habitat statistics reveal a concentration in waterside habitats (riversides and ditchsides) and marshes. Roadside habitats are probably associated with field drains. *A. sylvestris* seems to be a habitat specialist – moisture is the common feature in nearly all the survey records. The distribution tells the same story. It is not an upland plant, despite the precipitation and frequent watercourses there. It shuns areas of intensive arable cultivation – here the disturbance is great and the watercourses are fewer and more closely managed. It avoids conurbations where watercourses are polluted, or much visited or commercialised. Hence, surely, its greater continuity in West Lothian, where there is high precipitation, good development of streams and rivers, and where both agricultural and urban pressures form a mosaic rather than a continuum. There is no sense, as in *Torilis japonica* or *Conium maculatum*, of a continental phytogeography over-riding its habitat occupancy. It grows everywhere there are undisturbed lowland watersides.

Native and Introduced Species

It is no simple matter to determine what is native and what is introduced in any area (Webb 1985; Smith 1986) and a good deal more time and effort needs to be spent in examining the facts. 'Archaeophytes' and 'neophytes', defined on the basis of subfossils and historical records, may give indications related narrowly to the places with which the evidence specifically associates them. Such evidence is patchy and incomplete. Extrapolations are dangerous. What is native to one part of Britain may be introduced in another. Local studies are highly relevant. Populations of some taxa may derive in part from native incidence, in part from introduction. Native and introduced species have their own interesting histories: what matters more, ecologically, at any time, is what grows spontaneously, regardless of its origin. As concepts of native status change over time, there will be fashions for the presumption of introduction (perhaps as now), or *vice versa*. Hard facts are often scarce. An introduced plant is one that did not establish in an area by its own natural dispersal abilities. Generally, this casts doubt on the native status of species shown to have arrived almost anywhere during the last 400 years of extensive human migrations. In Britain, species shown to have been present in interglacial, periglacial and immediately post-glacial times are probably native. They are not certainly native, for they may have become extinct and been re-introduced. Most plants of the pre-Roman period are native, but they are not necessarily native everywhere in their British distribution. Most attributions of native status are uncontroversial, but what bearing do the Lothian distributions and habitat statistics have on less certain cases?

Is *Cytisus scoparius* (Broom) introduced or native, *locally*? Of nearly 500 records in twenty-six habitats, almost half are from niches created by human activity. The commonest occurrences are in these habitats, but there are also waterside, woodland and coastal records. Godwin (1975) notes that Broom is known as sub-fossil wood from 6,300 BP. It seems possible that it is a native, perhaps originally coastal plant that has greatly expanded following the ecological domination of humans, and their creation of open habitats inland.

Daucus carota (Carrot) is probably both native *and* introduced in the Lothians, and its distribution is probably changing. There are undoubtedly native populations near the coast. Helen Jackson, BSBI Recorder for East Lothian (personal communication), notes a change of habitat from the 'pastures' of Martin (1927, 1934) to the tips and dumps recorded in the 1950s by Betty Beattie (a previous Vice-County Recorder), to the cultivated fields and railway lines of today. Some of the latter carrot records may relate to the uninvestigated sightings of escaped, cultivated plants though, in this pre-packed age, few railway chefs will now be hurling surplus live carrots from passing trains. Undoubted introductions on roadsides follow the incorporation of carrot fruits into the 'wild plant mixtures' that are sown, confusingly but with the best of intentions, in new roadside seedings.

Aegopodium podagraria (Goutweed, Ground Elder) is another puzzle. Though often regarded as introduced, following Godwin's (1975) remark 'seldom seen in natural habitats such as the woodland habitats in Southern Sweden', Anderson (1967), writing locally, took a different view. The Lothian statistics certainly show the plant in plenty of habitats influenced or created by human activity, but it is also common in mixed and deciduous woodland, in hedgerow and scrub, as well as in waterside habitats. Though probably used as a potherb, having been found in association with Roman artefacts, Goutweed may have been misrepresented as a largely introduced plant.

Different considerations apply to the question of whether *Chamerion angusti-folium* (Rosebay Willowherb) (Plate 15) is native or introduced, locally. Recent British Floras come down on the side of native status, but in many places the evidence is quite equivocal. The plant was rare in the nineteenth century (Greville 1824) but became very common in the twentieth, posing an interesting ecological conundrum. Linton (1903) regarded early records as garden escapes. At the beginning of the twentieth century, Amphlett and Rea (1909) stated that it was a garden plant in Worcestershire, having first been recorded there in 1805. Considerable work was done in Edinburgh on the plant's status (Myerscough 1980). J. W. Heslop Harrison (1953) proposed the theory that introduction of new strains from North America had occurred, thus giving rise to a new, twentieth-century invasion. The previously-existing British plants may have been native only in mountain districts and on scree (Clapham *et al.* 1987). Higher incidence in the twentieth century is put down to increased soil disturbance and urban dereliction (Grime *et al.* 1988) or to woodland clearance (Clapham *et al.* 1987). Neither of these suggestions seems very plausible for the Lothians. Rosebay Willowherb is almost certainly introduced here, having been recorded in thirty-six diverse habitats, most occurrences being in open situations. The building of Edinburgh's New Town would have provided numerous disturbed sites and quarries in the eighteenth century. Yet Rosebay Willowherb went unrecorded, despite its considerable dispersal properties, and a local human society botanically alert well beyond the norm. The local woodland was largely cleared well before this (see Chapter 3) and cannot have been an ecological constraint or barrier probably for several thousand years. Now that there are simple, rapid molecular techniques for comparing the genetic affinities of different populations of plants, it would seem timely to re-examine Heslop Harrison's suggestion of 1953.

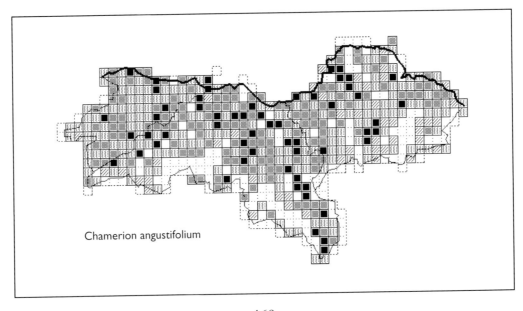

Chamerion angustifolium

Decline and Expansion: Losses and Gains

A scrutiny of the Flora will reveal that some species are known only from records of some antiquity, and perhaps also from very few or even single sites. Many of these are probably losses. But it is almost impossible to be sure when a species is lost from an area – how long does one wait before declaring it gone? Examples will be found in the Flora of rediscoveries after almost a century of apparent loss. It is similar with possible gains, where all that can be said is that a species has not previously been recorded. For plants that are inherently rare and perhaps in remote sites, discovery may be delayed for centuries. A decline may be very slow, and may eventually reverse, while some taxa may go as soon as they come (see *Orobanche minor*). The current nature of floristic change is thus more reasonably considered in terms of 'decline and expansion' rather than the more newsworthy, but possibly inaccurate, 'loss and gain'.

Some general features of the change in the flora since Greville's time (1824) have been commented upon in Chapter 3. The account following refers to changes since Martin's publications (1927, 1934). One general reason for the increase in the flora (of about 10 per cent overall since Martin) is that new niches, mainly created by human activity, have exceeded the number of those lost (largely wetland and sea-shore). Wetland and undisturbed coastal habitats have declined but not gone altogether. Another reason is that the passing of time has been an opportunity for more plant immigration, mainly of introduced species but also of some native species. It has also meant more time for search and discovery – the Lothians have never been scrutinised by botanists as thoroughly as they have been over the last twenty years.

Declining species fall into several categories. There are agricultural weeds whose specialised adaptations enabling them to contaminate crops are outrun by advances in agricultural hygiene (Smith 1986, Svensson 1986). *Agrostemma githago* (Corn Cockle), once a reasonably common segetal weed, is now an occasional casual – which is the intermediate fate of many such species perhaps before absolute loss. *Bromus secalinus* (Rye Brome) is another. A curious development with Corn Cockle is that it is now being sown on roadsides in 'wild seed mixtures'. *Ranunculus arvensis* (Corn Buttercup) is currently a 'near-loss', but *Chrysanthemum segetum* (Corn Marigold) still lingers with us, though Greville noted its near eclipse in 1824. It too is being sold in 'wildflower' seed mixtures. Other arable weeds that seem to be in decline are *Fumaria densiflora* (Dense-flowered Fumitory) and *Viola tricolor* ssp. *tricolor* (Wild Pansy).

Woodland species seemingly destined for lower incidence in future are *Pyrola minor* (Common Wintergreen), *P. media* (Intermediate Wintergreen) and *Trientalis europaea* (Chickweed Wintergreen) – see Flora text. *Rosa arvensis* (Field Rose) seems to be on its way out from hedgerow and scrub habitats. But we must remember that hedge loss has stabilised or ceased, while afforestation is now more likely than not.

Grassland species in decline include *Minuartia verna* (Spring Sandwort).

Saxifraga hypnoides (Mossy Saxifrage) seems to be going from wet upland rocks and streamsides. From dry grassland we may be seeing the eclipse and eventual loss of *Scleranthus annuus* (Annual Knawel) and *Filipendula vulgaris* (Dropwort). On rocks, walls and dunes, *Saxifraga tridactylites* (Rue-leaved Saxifrage) is seen less often.

Ulmus glabra (Wych Elm) has been in notable decline for some time in the Lothians, because of attack by Dutch Elm disease. The dead elms are a familiar feature of the Lothian landscape. Looking at the distribution map, however, we see encouraging frequency and width of distribution. The regenerative, suckering capacity of elm may be ensuring its continuance.

Coastal taxa now in small numbers and in decline include *Triglochin maritimum* (Sea Arrowgrass) in saltmarshes, *Viola hirta* (Hairy Violet), now confined to coastal dunes, and *Astragalus glycyphyllos* (Liquorice), last recorded from this type of habitat. *Rosa pimpinellifolia* (Burnet Rose) seems to be less frequent in coastal sites, but plantings obscure the situation overall. *Sisymbrium altissimum* (Tumbling Mustard, Tall Rocket), a plant of waste ground near the coast, is in decline as such habitats disappear, and *Orchis mascula* (Early Purple Orchid), a lime-lover, is now almost confined to a few coastal habitats.

Inland, *Lychnis viscaria* (Sticky Catchfly), on the verge of extinction at its first recorded British site (Arthur's Seat), has been the subject of a rescue attempt.

Waning wetland species include *Scutellaria galericulata* (Skullcap), *Mentha arvensis* (Corn Mint), *Stellaria palustris* (Marsh Stitchwort), *Anagallis tenella* (Bog Pimpernel) and *Samolus valerandi* (Brookweed). Even *Menyanthes trifoliata* (Bogbean) (Plate 11) is becoming scarcer. These changes are a continuation of the loss of wetland flora recorded steadily since 1824. It may now be, however, that with the enlightened modern approaches to conservation and, perhaps critically, with 'set-aside' policies in place, there will be reduced pressure to drain land for agricultural expansion. Survival of wetlands is more assured than at any time in the past. Moves to divert building development on to 'brownfield' sites may also help.

The Flora (Chapter 13) records expansions in many taxa, for instance *Phyllitis scolopendrium* (Hart's-tongue Fern) and *Scrophularia umbrosa* (Green Figwort), both native species. Most of the incomers are ruderals – weedy Oilseed Rape commands attention as a new, common and persistent roadside escape from fields. It is a likely permanent new ruderal, which flowers all through mild winters in Edinburgh. It was not recorded before 1990. *Epilobium brunnescens* (New Zealand Willowherb) is still expanding in upland areas while *Allium paradoxum* (Few-flowered Garlic or Leek) continues to spread in lowland areas. *Diplotaxis muralis* (Annual Wall-rocket) may also be expanding. *Heracleum mantegazzianum* (Giant Hogweed) (Plate 13) is relatively new, and certainly expanding, but the real position is obscured because it is so much attacked in speech and in deed by local authorities. It can produce a skin rash on contact. In the tradition of garden escapes such as *Veronica filiformis* (Slender Speedwell), now well established, we have records of expanding *Forsythia x intermedia* (Forsythia) and *Rosa rugosa* (Japanese

Rose). *Agrostis castellana* (Highland Bent) is more frequently encountered now than previously: it seems to have escaped from its role as a sown grass of golf courses. A new weed (1998) in the Royal Botanic Garden Edinburgh is *Crassula tillaea* (Mossy Stonecrop).

Erigeron acer (Blue Fleabane) is increasingly seen in stony habitats and on dunes. *Epipactis leptochila* (Narrow-lipped Helleborine) – a new native record for the Lothians – is colonising bings.

Among water or wetland newcomers or expansions, *Azolla filiculoides* (Water Fern) is notable. Also in this category are *Cardamine raphanifolia* (Greater Cuckoo Flower), *Crassula helmsii* (New Zealand Pigmyweed) and – new in 1999 – *Lemna minuta* (Least Duckweed) in ditches and lochs.

We must see these comings and goings of the flora as part of an inevitable change, not as indications of ecological disaster. We do not appreciate the significance of new plants (except the aggressive or poisonous ones) as easily as we see – and regret – the losses, because we know the incomers less well or not at all. And we cannot know until much later, the impact that they will have. In biological terms it matters not at all that much of what disappears is labelled by us, 'native', and much of what is new is 'introduced'. The Victorian establishment had a high regard for 'old money' and a tendency to sneer at 'trade'. We must take care not to be botanically snobbish, still less racist. We must venerate the old flora – it is part of our own past. We must delight in the present flora – it is our neighbour. We must welcome the new flora – it is part of our future. And we must know, and love, all three.

18

Land Use in the Lothians

G. RUSSELL

Introduction

There is very little land in Edinburgh and the Lothians that has not been subject to significant impacts from the activities of people. There are perhaps small areas in the steep-sided valleys of the Lammermuir Hills and on the sea cliffs where the vegetation is essentially natural, but elsewhere deliberate or inadvertent disturbance has changed the abundance and diversity of plant species from what it would have been had people not arrived. This does not mean that the flora has been impoverished. While some species have become rarer, others have spread and yet others have been introduced. The plant life of Edinburgh and the Lothians is characterised by change in space and time largely occasioned by the activities involved in using land.

A fundamental question concerns the factors affecting plant biodiversity. Without the impact of humans, plant biodiversity in the Lothians would depend on the soils and climate, and the ability of species to disperse into the region. The use of land by people changes things by altering the physical or the biological environment either continuously or intermittently. Both the above- and below-ground aspects of the physical environment can be modified. A few examples will illustrate these processes. The former type of change occurs after forest clearance when the light and thermal regimes experienced on the former forest floor are altered. The latter is a result of drainage, liming, fertiliser addition or ploughing. Atmospheric pollution can also have an impact. For example, the increased inputs of nitrogen associated with combustion of fossil fuels have contributed to a shift from heather moorland to grassland. The biological environment is altered by the encouragement or discouragement of animals, plants and micro-organisms. Examples include the extinction of some indigenous grazers and browsers such as the auroch and the wild ox in the Bronze Age and the European elk about 1,100 years ago, the reduction in numbers of predators of the remaining herbivores, the increase in grazing pressure by sheep, the introduction of new plant species into gardens, the chemical control of weeds in the second half of the twentieth century, and the advent of Dutch elm disease.

Land Use Change over the Centuries

About 13,000 years ago, the Lothians had emerged from the last ice age and the landscape was being colonised by plants. Hunter-gatherer communities were not far behind but their impact must have been small, at least until they began to use fire as a tool. Most of the Lothians may have been utilised from an early time, albeit at a low intensity. Recent excavations at Cramond, just west of Edinburgh, have revealed the remains of a temporary Mesolithic campsite dated from charred hazelnut shells to between 10,500 and 10,200 BP, the earliest record of a settlement in Scotland. About 5,000 years ago most of the area must have been covered with forest. It used to be believed that forest loss was largely due to clearance for agriculture. However, more recent work (Smout 2000) suggests that the major loss of forest had already occurred by 3,000 BP as a consequence of a deterioration in the climate, although further pressure did come from agriculture. A detailed map first published in 1610 (Pont 1664) (Plate 4) shows forty-eight enclosed areas in the Lothians, all relatively small and almost all marked as woodland. Many parts of the Lothians, particularly those areas underlain by poorly permeable Carboniferous drift, would formerly have been wetlands. However, these suffered a severe decline as soon as methods were developed for land drainage. The most rapid changes in land use occurred with the onset of industrialisation in the early nineteenth century, although coal mining started in the area around Tranent many centuries before. In the latter part of the twentieth century, urbanisation and the intensification of agriculture, particularly grassland, had a large impact on floral diversity.

Land Uses

There is a considerable amount of documentary and other evidence that can be used to look at the evolution of land use in the Lothians. Agricultural statistics have been collected for many years. However, comparisons over time are complicated by differences in methodology and by changes to region boundaries. A key issue is the spatial resolution of the data sets. For example, land marked as urban or industrial can contain a high proportion of vegetated land while the floristic diversity of arable fields bounded by hedges, ditches and streams is likely to be much greater than similarly mapped arable fields under intensive high-input management and bounded by fences.

Agriculture and Horticulture

From an agricultural point of view the Lothians can be divided into two parts, a low-lying belt bounded by the Firth of Forth averaging about 15 km in width, and a hillier upland part to the south. The potential for agriculture is shown by the Land Capability Class (Bibby *et al.* 1982) where Class 1 is suited to a wide range of crops, Class 4 provides extremely limited opportunities for cropping and Class 7 is unsuitable for agriculture. No land in the Lothians is classed as 7 for climatic or soil reasons, although urban and industrial areas have been included here for convenience (Table 18.1). The low ground corresponds to Classes 1 to 3 and the hill land to Classes 4 to 6. Note that the classification is with respect to the range of possible agricultural enterprises and not in terms of crop yield. Very high yields of cereals can be obtained in the Lothians (Plate 3, top) although the wet conditions often encountered in a late harvest can make the achievement of quality standards somewhat difficult. Although only a small proportion of the area is graded Class 1, it represents a significant proportion of all the Scottish Class 1 land. Cropping of one form or another would be possible on two-thirds of the Lothians (Classes 1 to 4) although not all the land is used for this purpose and only a narrow range of crops can be grown on Class 4 land. There is a strong demand for Class 1 and Class 2 land for building as it is well-drained and is relatively level. The only agricultural use for land in Classes 5 and 6 is grazing.

Table 18.1 *The proportion of the Lothians occupied by each Land Capability Class*

Class	1	2	3	4	5	6	unclassified
Percentage	2	15	32	15	20	6	10

Class 1 is suited to a wide range of crops. The only agricultural use for land in classes 5 and 6 is grazing. The unclassified category includes urban areas, industrial areas and water bodies.
Estimated from the land capability map of Bown and Shipley (1982).

Data from the June 1999 census show that 53 per cent of the total agricultural area was given over to cropping, including temporary grass and set-aside, 20 per cent was permanent pasture, generally for reasons of poor drainage or steepness, and 26 per cent was rough grazing. The corresponding figures for the period immediately before World War II were 45 per cent, 24 per cent and 32 per cent (Scola 1944). These figures exclude farm woodlands and other unfarmed areas on farms but do include important habitats such as verges, banks of water courses and field margins.

When the agricultural area is divided by farm type, cropping is the largest category (46 per cent), followed by cattle and sheep farms (38 per cent). These two categories are largely geographically separated as mentioned above. Mixed farms are less common (14 per cent). Horticulture and market gardening were formerly of great importance but occupy a very small area now.

Although there is a long history of settlement, it is not clear when the first farmers reached the Lothians. The Romans needed crops and animals to feed their troops and introduced new agricultural techniques. Many of the moors noted on Speed's map of 1610 (Pont 1664) would have been common grazings. Not all would be the typical wetland habitat by which we understand the term moor today as the ones in the drier east may have been more akin to dry heaths. Agricultural development remained small scale until after the Union of Parliaments in 1707 when trade improved and border raids stopped. There was an important sheep market at House o' Muir on the eastern slopes of the Pentland Hills between before 1658 and 1871, and Falkirk Tryst was an important staging post for cattle being driven south for export to England. The year 1760 marked the start of a rapid intensification of Scottish agriculture with new ideas and techniques being imported from England and the continent. A major development was the introduction of turnip cultivation, which revolutionised livestock husbandry by providing winter feed for cattle. In the eighteenth and early nineteenth centuries, agricultural changes led to enclosure with new settlements replacing the previous farm touns. By the middle of the nineteenth century, farming on the low ground parts of the Lothians was characterised by high inputs of labour and manure and high yields of cereals and vegetables. Changes in the twentieth century have included a reduction in the area of market gardens and dairying and an increase in cereals, particularly winter wheat, as a consequence of accession to the European Economic Community as it was then, and greater globalisation of markets for agricultural produce.

The effects of agriculture on the flora are diverse and depend on the nature of the land and the management practices adopted. In spite of the introduction of chemical herbicides, many weed species still depend on agriculture for their survival. Modern application techniques mean that farmers avoid spraying the field margins and use targeted, low doses to reduce competition from weeds without necessarily eradicating them. Agriculture has led to the alteration of the physical environment in several ways including drainage, addition of phosphate fertiliser, and liming to raise the pH. Much of the arable land was artificially under-drained in order to increase the amount of time during the year when the land could be grazed by animals or worked by machinery. Many of these drainage schemes date back to the nineteenth century and are beginning to collapse and be blocked with sediment. As the cost of under-drainage is now very high in relation to the value of produce, wetter areas have been put into permanent set-aside or are just not being cultivated. Ponds were also filled in and wetlands drained. Some hedges were removed, particularly in East Lothian, to produce larger fields. However, hedges have never been a major feature of the Lothians, mainly because of the low numbers of livestock on the low ground. Robert Louis Stevenson remarked of the south side of Edinburgh that 'The character of the neighbourhood is pretty strongly marked by a scarcity of hedge' (Stevenson 1913). Intensification of agriculture led to the decline of species associated with agriculture. However, there are many areas where the native flora can survive even in the most intensive farms. Contrary to what is often believed, the major impact of agriculture of the flora has not been on the arable farms but on livestock farms where pasture 'improvement', increased applications of nitrogen fertiliser and earlier cutting for silage rather than later cutting for hay has led to an impoverishment of the flora. Few plants are able to out-compete Perennial Ryegrass (*Lolium perenne*) at high levels of nitrogen nutrition, and cutting grass for silage in May prevents later-flowering species from reproducing. The ballad of Otterbourne includes the words 'It fell about the Lammas tide when the muir-men win their hay . . .'. Thus, in the fourteenth century hill pastures were cut in August after most species had flowered. Much more recently it was traditional to cut grass for hay in the week of the Royal Highland Show (mid June). However, silage cutting is now almost universal and often takes place in late May with one or even two subsequent cuts being taken.

Scola (1944) wrote that 'the major problems of to-day [the 1930s] were in being when Cobbett visited Scotland in 1832, and the main question ever since has been not so much concerned with the solution to problems of farming technique as in the adjustment of farming practice to the needs of the market'. After a few decades of prosperity, agriculture is passing once again through a difficult time of readjustment due to increasing costs and falling prices. On the other hand, the flora is likely to benefit from the consequent pressure to increase the efficiency with which fertilisers and crop protection chemicals are used, as well as from encouragement to promote habitat protection.

Forestry

Forest and woodland is largely made up of small units scattered throughout the region with a few larger blocks of coniferous plantation established on the peat soils of southern West Lothian in the 1960s, and more continuous cover in the valleys of the larger rivers. Unlike many other parts of Scotland, about 90 per cent of the woodland is in private ownership. About 10 per cent of the land area is now devoted to forestry of one sort or another compared with only 6 per cent in 1895. Conifers occupy 41 per cent of the forest area, broadleaves, predominantly Sycamore (*Acer pseudoplatanus*), 35 per cent, and mixed woodland a further 11 per cent with the balance made up of unplanted ground (Forestry Commission 1999). About one-third of the broadleaved woods were established before 1940 compared with fewer than 5 per cent of the coniferous plantations, reflecting the fashions and economic imperatives of the times. In recent years, the balance of new planting has swung back towards broadleaves. Many woods were not established primarily for timber production but rather for amenity or for shelter. Exposed promontories along coastal sand dunes were planted up for shelter, generally with Pine (*Pinus* spp.), by large estates. In recent years, community forestry has developed and the Central Scotland Woodland project has had an impact on amenity and conservation in West Lothian.

There are a few remnants of ancient woodlands, for instance at Roslin Glen, and Pressmennan, although most show signs of management in the past. Pollen analysis has shown that by 6,000 BP the forest cover had stabilised after the retreat of the last Ice Age. The wetter areas would have been covered in deciduous woodland but it is unclear whether there were also heaths in the drier areas. The tree line was probably close to the tops of the highest hills. By about 2,500 BP, the forest area was much smaller, largely for climatic reasons, as mentioned earlier, but also because of deliberate forest clearance and unplanned prevention of regeneration by grazing. Woodland would have been managed to conserve its value as a source of building timber while the soils of many parts of the forest were too damp for successful conversion to agricultural land. In 1511, when the warship the *Great Michael* was built at Newhaven by James IV, wood was brought from Fife and the rest of Scotland but there was no mention of locally sourced timber. This may, however, have been due to easier transportation by sea. There was re-establishment of woodland on a small scale from the beginning of the eighteenth century when farmers and landowners started planting woodlands for amenity, shelter and privacy. Some of the first tree planting in the region was carried out by the sixth Earl of Haddington who started planting trees on his Tyninghame Estate in 1707 (Ragg and Futty 1967). Many introduced species were included in the plantings. Old photographs of Edinburgh show remarkably little woodland and many of the woodlands in the vicinity of the city are of artificial origin in spite of their appearance of naturalness.

Woodlands in the Lothians are remarkably diverse with few large areas of even-aged, single species stands. However, even in dense coniferous plantations, there is bare ground in the rides between blocks or in areas where trees have died and where the original vegetation survives. The fungal flora of these woods is also of great interest.

Transport

Transport systems are essentially linear or reticulate and thus provide a large amount of edge habitat in relation to their area. Paths, roads, railways, motorways and, more recently, cycle paths often provide a range of habitats for plants as well as opportunities for dispersal. Disused railway lines in particular are home to a wide range of species. Cuttings and embankments expose bare soil and provide opportunities for colonisation. Roadworks in East Lothian are often accompanied by a flourishing of Common Poppies (*Papaver rhoeas*) as the seed bank is disturbed and brought to the surface. Embankments can provide areas of well-drained soil in an otherwise wet environment. Although motorway edges are usually landscaped with alien species, the indigenous vegetation soon invades. Floristically diverse road verges which used to be cut regularly throughout the year are now marked and cut in such a manner as to encourage floral diversity. The Union Canal (Plate 14), which was opened in 1822, is a contour canal extending from Edinburgh to Falkirk. It is now (2001) being re-opened to navigation and the new works have taken account of the need to protect and encourage the emergent vegetation. Ports and harbours, particularly Leith Docks, have been a rich source of aliens. Large quantities of Esparto Grass (*Stipa tenacissima*) were formerly imported through Granton Harbour for paper making but, although hardly a local family was reputed to be without a tortoise brought over by the crews, the climate was not congenial to the establishment of the North African plants whose seeds must have been widespread in the bales. Within the Lothians there are two Second World War air-fields and one major airport. The latter, with its close-cropped grass, is not noted for its floristic diversity although there are small areas near the perimeter fence that have benefited from protection. The region is crossed by several major gas and oil pipelines, but these have had only a very localised impact on vegetation.

Mineral Extraction

Much of the region is underlain by rocks of Carboniferous age including coal-bearing rocks and limestone. The monks of Newbattle Abbey near Dalkeith, or rather their tenants, started mining coal near Tranent in the thirteenth century. However, these early pits would have had little effect on the environment and it was not until the expansion of coal mining during the nineteenth century that large areas of land were covered by the coal bings on which the spoil was dumped. Although this material was often very acidic it did provide another habitat for plants and widespread subsidence caused by mine galleries close to the surface led to the development of marshy areas. There is now no deep mining in the region and many of the old bings have been removed and landscaped. Open-cast mining of coal still takes place but the landscape is now restored effectively once the coal has been won. The other major extractive industry was oil production from the shales of West Lothian and Midlothian. The shale was mined and then retorted to drive off the volatile oil. The resulting spent shale with its characteristic red colour was dumped on bings which, once salt had been leached, produced a much more congenial habitat for plants. The industry reached its peak at the beginning of the twentieth century and declined until the last mine closed in the 1950s. Many of these bings have now been removed to provide bottoming for roads. However, several floristically-rich bings still remain (Plate 15). Silver and lead were mined near Linlithgow in mediaeval times leaving behind cliffs which provide habitats for a range of plants, but there were few other mineral workings. Limestone is quarried near Dunbar for making cement and there is reputedly a localised increase in soil pH as a consequence. Most of the limestone bands are thin, however, with important but localised lime production for agriculture, for example at Burdiehouse where the lime kilns still remain.

Quarrying, once a major extractive industry, is now all but extinct. Huge quantities of stone were quarried to build the Georgian terraces of Edinburgh, and the prominent scarp of Salisbury Crags in Edinburgh includes a large former quarry. Some quarries still provide interesting habitats for plants but many have been used for landfill sites or otherwise filled in. Finally, commercial extraction of peat from Auchencorth Moss has had a profound effect on the associated plant species.

Although these industrial activities have produced habitats that are clearly not natural, they all have natural analogues elsewhere. The writer was struck many years ago by the resemblance between some areas of industrial dereliction in West Lothian and parts of Iceland.

Urbanisation

The structure of settlements remained largely unchanged from the early seventeenth until the late eighteenth century. The industrial revolution led to the development of towns and villages linked to industry. The advent of the railways and other means of mass transportation led to an expansion of the City of Edinburgh. In the late nineteenth century Robert Louis Stevenson noted that 'The builders have at length adventured beyond the toll [Morningside Station] which held them in respect so long, and proceed to career in these fresh pastures like a herd of colts turned loose' (Stevenson 1913). In the 1930s, houses and gardens accounted for only 2.7 per cent of the land area of the Lothians, although agriculturally unproductive land, presumably transport infrastructure and industrial sites, occupied another 4.0 per cent (Scola 1944). Outside Edinburgh, urbanisation proceeded slowly until the twin pressures of expensive housing in Edinburgh and cheap travel encouraged commuting. Another pole of growth was Livingston, a new town in West Lothian.

Although urbanisation has led to the loss of large areas of floristically interesting habitat, a surprisingly large proportion of the urban landscape of Edinburgh and the Lothians is still green, as can be seen from the air. The green belt round Edinburgh has largely been held, safeguarding a range of habitats. Derelict land and uncared-for walls and buildings soon sprout new vegetation and gardeners spend considerable time and effort in a continuing struggle to eradicate those members of the indigenous flora they call weeds. One consequence of the urbanisation was the need to supply water for domestic purposes and this led to the construction of a series of reservoirs, although water for the Lothians and Edinburgh now comes from outside their boundaries. There are still, however, more than twenty reservoirs in the region, many being used to provide compensation water for the local rivers. There is no consistent effect of reservoirs on floral diversity as some habitats are destroyed and others created. For example, Bavelaw Marsh, a nature reserve, is a consequence of the creation of Threipmuir Reservoir. In other cases, steeply shelving edges provide little opportunity for vegetation development although they have been found to provide an important habitat for rare mosses.

Industry

Industry in the Lothians has generally been on a small scale. Grain milling has taken place at Preston Mill in East Lothian since the sixth century and this was probably the earliest industry. Later developments using water power included extensive paper making on the River Esk and on the Water of Leith. These caused considerable pollution in the 1950s and 1960s. Salt production from sea water fuelled by coal from the local coalfields was an important industry, hence the place name Prestonpans. At the end of the seventeenth century, a small glassworks producing bottles was established at Morrison's Haven near Prestonpans although it did not thrive (Trotter 1796). It presumably used soda from the ashes of Common Glasswort (*Salicornia europaea*) (Plate 12, bottom left) and Prickly

Saltwort (*Salsola kali*), both of which now have a local distribution in the Lothians. The ashes were also used to manufacture soap. Small-scale iron processing took place at Cramond at the mouth of the River Almond from about 1750 although the workshops and docks are now derelict and overgrown. Shale oil refining was carried out at several locations in the region. There is relatively little heavy industry in the region now and the few factories are mainly on the outskirts of Edinburgh, at Livingston and at Bathgate. Power stations remain at Cockenzie (thermal) and Torness (nuclear). Impacts on vegetation have not always been deleterious as derelict sites have provided favourable habitats for plants to colonise, and sulphur deposition from coal burning power stations reduced soil deficiencies of this element, at least until cleaner fuels and technologies were introduced. Modern factories provide opportunities for sensitive landscaping and the protection of interesting habitats within their grounds.

Country Estates

The region includes a considerable number of country estates some on the lowland areas with gardens and estate woodlands and others being sporting estates in the Lammermuir, Moorfoot and Pentland Hills. As early as 1600 more than a hundred country seats were reputed to have been situated within 10km of Edinburgh. Many species, including trees such as Walnut (*Juglans regia*), were planted on the low ground estates and most of the oldest trees are found there. For example, there are 400-year-old Yews (*Taxus baccata*) at Malleny Garden in Balerno. Several other species of flowering plant such as Snowdrops (*Galanthus nivalis*) and Daffodils (*Narcissus pseudonarcissus*) have become naturalised in these estates. Upland estates are managed for grouse-shooting although bags have declined in recent years. Management involves regular burning of the heather (Plate 29, bottom) to produce a mosaic of different successional stages of vegetation. Well-managed muirburn in this part of Scotland tends not to lead to a diminution of floral diversity. There are fewer large private estates now but some, such as Almondell and Calder Wood, Dalkeith and Vogrie, have become country parks, which fulfil an important role in habitat protection as well as providing recreational opportunities.

The local monasteries had a major influence on the landscape with extensive estates of agricultural and forest land. The mediaeval church, being a multi-national organisation, was able to introduce new farming practices and plant species that were considered useful. In the fourteenth century, monks at Soutra on the upper slopes of the Lammermuir Hills ran Scotland's largest hospital. Recent excavations have identified the plants they cultivated for medical purposes. Plants have always been important for medicine. The Royal Botanic Garden Edinburgh started as the Edinburgh Physic Garden on a site now occupied by Waverley Railway Station.

Recreation

The recreational activity with the greatest impact on vegetation in this region is undoubtedly golf. The first mention of golf in the Lothians is in a document dated

1457. Considerable areas of land are now given over to golf courses, many of which are in floristically interesting coastal areas. Although some golf courses are heavily managed to preserve the fairways and greens in the optimum conditions for golf, there is an increasing trend to keep wild areas between the fairways. Indeed, there is a Scottish Golf Course Wildlife Group, and Scottish Natural Heritage funds two wildlife advisers to work with golf clubs. Without golf courses, the amount of green space in Edinburgh would be considerably less. Traditional public parks are often not very interesting floristically. Other land used for recreation includes the horse racing course at Musselburgh, the dry-ski slope at Hillend and numerous allotments. All provide opportunities for conservation management.

Military Works

Many battles have been fought in the Lothians. However, the main military impact now comes from an army training ground in the Pentland Hills. The numerous fortifications and other works from the two world wars have now largely become derelict and many have been removed. Some of them have provided specialised habitats such as walls on which crevice plants and lithophytes have been able to grow.

Nature Reserves

Edinburgh and the Lothians are well provided with protected areas although not all are managed specifically for vegetation. The only significant habitat not represented in a nature reserve is probably heather moorland. It is considered not in need of this form of protection although it is mentioned in the local Biodiversity Action Plans. The Scottish Wildlife Trust alone has nineteen reserves in the region, covering approximately 280ha, and there are Local Nature Reserves and wildlife sites such as Aberlady Bay in East Lothian and Linlithgow Loch in West Lothian. Holyrood Park, close to the centre of Edinburgh, contains the floristically interesting cliffs of Arthur's Seat and Salisbury Crags. There are many Sites of Special Scientific Interest and the Scottish Wildlife Trust has identified a further 270 interesting sites. Of course, many plants flourish in areas without any statutory protection. Grime (1972) identified the need to establish refugia for plant and animal life threatened by the destruction of habitats and the intensification of agriculture. He argued that there were numerous opportunities in the design and maintenance of forest plantations and reservoir catchments as well as the landscaping of urban areas for this to happen. In fact, this has often happened inadvertently in the Lothians with interesting pockets of biodiversity in areas of industrial dereliction or beside transport corridors.

The Future

Several factors relating to land use will influence floral diversity in the Lothians in the future. The support mechanisms for agriculture are changing as a consequence of the Agenda 2000 reforms of the Common Agricultural Policy of the European Union with a shift in emphasis away from production subsidies. This is likely to have a beneficial effect on the flora of the Lothians, particularly as government support will be progressively targeted at achieving landscape and environmental goals. The Water Framework Directive is also likely to have a beneficial impact by encouraging the development of buffer strips on agricultural land next to streams and rivers and the conservation of wetlands. On the other hand, there is no sign of a reduction in the rate of urbanisation or in the amount of commuter traffic in spite of policies to reduce car use. This is leading to a gradual loss of habitats, although many vulnerable areas are protected.

Acknowledgements

Many thanks to the Forestry Commission Woodland Survey section for providing statistics on forest area and to everyone else who provided information.

19

Ethnobotany of the Lothians

G. E. KENICER and T. M. DARWIN

Introduction

Since prehistoric times, Lothian has been an area of great movements of people, with plants and ideas about plant use following in the wake of hunter-gatherers, invading armies, migrants, travellers and traders to produce a dynamic and varied ethnobotanical history for the area. The earliest inhabitants were nomadic hunter-gatherers, reliant on wild plants for the majority of their basic needs. The Neolithic saw the introduction of agriculture – a whole new means by which supplies of useful plants could be ensured. Following this, the Bronze Age (second and first millennia BC) and Iron Age (*c.* 700 BC to early centuries AD) allowed the further refinement of the tools and processes used in agriculture, forestry and the exploitation of wild resources.

The Iron Age Votadini are the first inhabitants of the area for whom we have a contemporary name. They were succeeded by influxes of Mediterranean Romans, Germanic Northumbrians, Scandinavian Norse and Normans, all of whom brought innovations in plant use from their original homelands. By the late Middle Ages and subsequent 'Age of Empire', Lothian had established itself as an important international centre of trade and medicine, in which plants played a major role. Until very recently, transport by land or sea was slow and often difficult, with no refrigeration, so that growing or harvesting plants locally was an option worth considering whenever possible. Plants that were numerous enough, however, could be collected directly from the wild, and appear in many cases to have been used on an intuitively sustainable basis.

Today, there is a significant revival of interest in collecting and growing wild species for food, herbal remedies, cosmetic products and crafts. Collecting from the wild raises some serious conservation issues, as habitats are often under pressure from development and pollution in the Lothians. However some small scale, sustainable harvesting of wild plants can and should be compatible with protection of both individual species and habitats. It is, after all, what people have been doing since they first roamed this diverse and fertile area. Furthermore, the possibilities opened up by enlightened management of 'semi-wild' populations and the current push towards diversification in farming allow new opportunities for development of some of the Lothians' more useful plants.

This chapter will examine a few of the wild plants which the many and varied inhabitants of the Lothians have employed over the centuries, as well as some of those that have been deliberately introduced and cultivated for economic and medicinal purposes.

Commercially Important Plants

The commercial and domestic economy of the Lothians has been largely reliant on plants throughout the history of the area. Agriculture, importation from elsewhere and collection from wild or managed, 'semi-wild' populations have all been important sources of the raw materials for plant-based industries and crafts.

Agriculture and Imports

Although the bulk of Lothian's agricultural production has always been geared towards the provision of food, other introduced crops such as Flax (*Linum usitatissimum*) and Hemp (*Cannabis sativa*) were important in the production of textiles and cordage from the arrival of the Anglo-Saxons until the late 1700s at least (Bede 1990; Arnott 1998; Simmons 1998; Dumayne-Peaty 1999; Statistical Accounts). By the end of this period, however, most of the fibres were being imported from elsewhere for processing.

The Lothians have long been a major processing centre for plant materials imported from the rest of Scotland and further afield (Figure 19.1). During the eighteenth and nineteenth centuries, a huge variety of plants entered through Leith for processing in the numerous facilities in and around the port and Edinburgh. From the corners of the Empire, mahogany and other timbers, sugar, cotton, coffee and tobacco were all brought in to be processed by tradesmen including cabinet and coach makers, weavers and snuff-millers (Arnott 1998). 'Esparto-grasses' (a wide range of species) were imported from Spain for paper-making and in the 1800s, 'cutch' (the heartwood of *Acacia catechu*), came from The Raj in large quantities to supply Edinburgh's leather tanning industry and for use as a waterproofing for fishing nets. Cutch had replaced the traditional use of native Scottish oak bark as the principal commercial tanning agent, as it caused less reddening of the leather. In addition, it was actually cheaper to bring in cutch from abroad than it was to carry the oak bark overland from the Atlantic woodlands of Scotland's west coast.

Dyes

The use of dyestuffs in the Lothians is particularly interesting in terms of movements of plants. Given the appropriate mordant and sufficient skill almost any colour of fabric dye can be produced from a variety of Scottish native plant material including leaves, catkins, fruit, blossom, bark, roots, fungi and lichens.

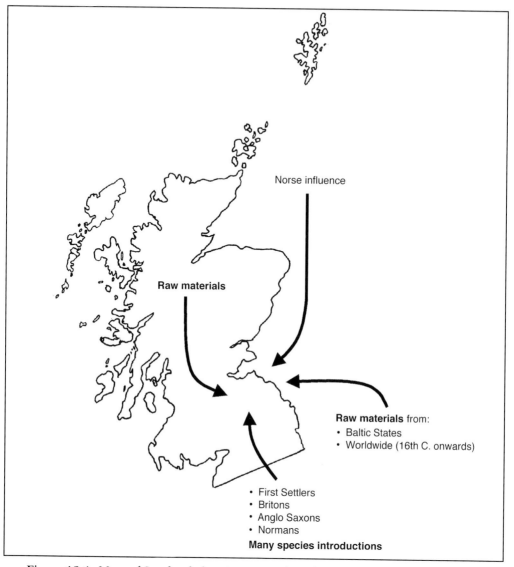

Norse influence

Raw materials

Raw materials from:
• Baltic States
• Worldwide (16th C. onwards)

• First Settlers
• Britons
• Anglo Saxons
• Normans
Many species introductions

Figure 19.1 *Map of Scotland showing routes by which plant materials reached the Lothians.*

However, the brightest, most vibrant and colour-fast shades of red and blue that are so appealing to the human eye can be most easily and consistently produced with non-native species such as Madder (*Rubia tinctorum*) and Woad (*Isatis tinctoria*) (Plate 16). Caesar famously reported that the ancient Britons were painted or tattooed with woad (Caesar 5/14, 1982), but the plant has not yet appeared in archaeological records in Scotland (Dickson and Dickson 2000).

Madder and woad can both be grown in Scotland, but generally produce poorer crops than when they are cultivated in warmer climates. Woad does well in sunny locations with rich soil, though the first-year leaves of madder can only be harvested once or twice a year in the Lothians (compared with up to four times elsewhere). Dyeing was largely a domestic affair, using locally gathered plant material, until the seventeenth century. One of the earliest commercial cloth weaving mills operated from 1640 until 1713 in Haddington, using locally grown woad. In 1757, *The Scots Magazine* reported that the Edinburgh Society was awarding prizes for the greatest quantity of madder, not under ten pounds weight, dressed and cured for the market, and another for the greatest weight of woad, not under fifty pounds weight. The Statistical Accounts (1/6) imply that woad was still being grown in 1791 but only in sufficient quantities to supply local requirements.

Although not using locally grown plant material, it is interesting to note that eighteenth-century Edinburgh played a significant role in the development of another important dye, known as cudbear. The recipe for this was patented in 1758 by brothers George and Cuthbert Gordon; the latter was a Leith merchant who began manufacturing the dye in a local factory and gave his name to the product. The principal ingredient was *Ochrolechia tartarea*, in combination with a number of other lichens – some texts suggest members of the genus *Cladonia* (particularly *C. pyxidata*) and others such as *Lecanora calcarea*. The commercial development of cudbear caused great excitement because it could be used as a substitute for expensive indigo (from tropical *Indigofera* spp.) and cochineal to produce shades ranging from blue and purple to pink and red. The business failed in Edinburgh (Arnott 1998), but moved to Glasgow with great success, later expanding to textile towns in England. However, the supply of lichens from Scottish sources was soon exhausted, and *O. tartarea* and a substitute (*Roccella* spp.), had to be imported from Scandinavia (Johnson 1993).

Wild Plants in the Domestic Economy

Underpinning large-scale agriculture and trade in plant-based goods and raw materials, was the use of locally sourced wild plants. Wild-collected plants were often significant on a quasi-industrial or 'home-industry' scale. There are a number of archaeological sites dating from the Bronze Age to Roman times that include charcoal remains (from most of the significant native tree species), but this implies little beyond their potential use as fuel (Hall and Tomlinson 1996). It is not really until the recent historical period that we get an indication of the extent of wild plant use in manufacturing or the domestic industries.

Willows – Basketry and Medicine

The traditional utilisation of willows (*Salix* spp.) is well known. The First Statistical Account for Uphall (1/6) confirms this, stating that, 'The great willow [*S. alba*] thrives wonderfully and is an useful tree for many country purposes'. R. K. Greville's interesting, but often vague *Flora Edinensis* (1824) records their presence in managed 'osier grounds'. He mentions Osier (*S. viminalis*), Bay Willow

(*S. pentandra*), White Willow (*S. alba*) and the rather unlikely Dark Leaved Willow (*S. myrsinifolia*) as having been used extensively in basketry and the production of wattle hurdles. Indeed, all of these species and a great many hybrids and colour variants are still used in the Lothians today by an expanding group of basketmakers, some growing their own willow, who produce both practical, traditional and more artistic, contemporary pieces. To weave in with the willow, or as an alternative, they gather a wide range of materials from gardens, hedgerows and woodland.

Historically, the bark of willow was processed in order to extract the 'bitter principle', used as a pain-killer and febrifuge, in much the same way as aspirin, its modern derivative. A chemical works at Beaverbank processed a lot of willow bark in the late nineteenth century, as a substitute for 'Peruvian bark' taken from *Cinchona* species, which yields quinine, another febrifuge. A letter held in Kew Garden's economic botany archives, written in 1909 by John Cowper, secretary of the Board of Agriculture mentions that the inert strips of willow bark left over from the process were recycled as ties for vegetables (information courtesy of Flora Celtica / Edinburgh Development Consultants).

Foods

Although wild-collected foods are becoming more and more popular as 'alternative' foodstuffs, they were certainly an important part of the everyday diet for the poorer people of the Lothians until the start of the 1900s at least. Robert Sibbald's 1709 *Provision for the Poor in time of Dearth and Scarcity*, a sort of handbook of edible wild plants for the philanthropic landowner, was written in Edinburgh. The book would have had as valid a relevance to the poor of Sibbald's local area as any other impoverished part of the country – had 'the poor' been able to read it!

Later writers, including Greville (1824), name several edible species presumably used in the area, including Elder (*Sambucus nigra*) berries ('in great request for making wine' – although they are inedible raw), the introduced Horse-radish (*Armoracia rusticana* – 'in constant use for the table') and Watercress (*Rorippa nasturtium-aquaticum*) – 'esteemed as a salad'. He also refers to Nettles (*Urtica dioica*) as having been gathered by the country people to make nettle kail (broth).

The Second Statistical Account of Cramond includes a list of species, some of which were used 'in the culinary arts'. Sea Beet (*Beta vulgaris* ssp. *maritima*), Sea Kale (*Crambe maritima*), Wild Carrot (*Daucus carota*) and a number of seaweeds (discussed below) are all mentioned (although wild carrot may have been used as a flavouring rather than a food).

Many of these historical food references are interesting as the plants may have been employed throughout their Scottish ranges, and the practices were certainly well known to the academics of the day (primarily through John Lightfoot's seminal and widely referenced *Flora Scotica*). However, the texts are seldom explicit about the extent of their use in Lothian. If the practices were ubiquitous at the time, it might not have been considered significant to specify the areas in which they were or were not used in Scotland.

Although the use of wild plants as food is on a reduced scale nowadays, it is undergoing a resurgence. For instance, growing numbers have become experts at home wine-making and a great many people collect blackberries from the hedgerows every autumn. However, a relatively recent case of (non-fatal) poisoning in Edinburgh by Deadly Nightshade (*Atropa belladonna*) highlights some of the dangers of wild food collecting and the importance of accurate identification (David Chamberlain pers. comm.).

Medicinal Plants

Edinburgh's role as a military stronghold, centre of population, trading port, traveller's staging post and university city produced a high demand for medical treatment and also many opportunities for the introduction of remedies from elsewhere. Medicinal plants are mostly used dried or as concentrated tinctures, so they travel particularly well. However, being able to grow or harvest a plant locally would of course always have been preferable to relying on the vagaries of imports, especially during the many turbulent periods of war and invasion to which the Lothians have been subject.

Soutra monastery, situated about seventeen miles south of Edinburgh on an important route (now the A68), was founded in 1165. Like most religious houses it had an influential role as a place of healing, with monks skilled in herbal lore, preparing their own remedies using plant material from the local area and imported from further afield. Although its importance declined in the mid-1400s, the hospital continued to care for travellers and the local community until around 1650. Archaeological excavations led by Brian Moffat have produced interesting information on the medicinal plants being used at the hospital (Moffat 1989). Most significant in the pollen record are Opium Poppy (*Papaver somniferum*), flax, hemp and Tormentil (*Potentilla erecta*). Although Moffat concludes that the flax and hemp were most likely used for their fibres, to make textiles and cordage, the seeds of poppy, flax and hemp all yield oils which would have had medicinal uses in ointments, salves and poultices. Robert Sibbald, in his *Scotia Illustra* of 1684, recommended poppies as a narcotic and soporific, hemp seed and sap for worms, diarrhoea and other fluxes, and flax for tumours and growths. It has been queried whether poppies grown in the temperate climate of the Lothians would have been pharmacologically potent. However, a two-year experiment conducted 'in the neighbourhood of Edinburgh' in 1819–20 by John Young, Fellow of the Royal College of Surgeons in Edinburgh, is said to have produced 'opium, superior in quality to the best Turkey opium'. One acre of poppies yielded 1,000 lbs of seed, which gave 375 lbs of oil (Young 1820)!

The Edinburgh Botanic Garden's Medical Background
From the seventeenth century, we have a fascinating insight into the knowledge of medicinal plants that had by then accumulated, from the records of the Physic

Garden that was to become the Royal Botanic Garden Edinburgh. Two local physicians, Andrew Balfour and Robert Sibbald, had completed their medical training at some of Europe's finest universities. Sibbald in particular was a keen naturalist, and states in his autobiography that during his studies he had become convinced that 'the simplest method of Physic was the best, and these that the country affoorded [sic] came nearest to our temper, and agreed best with us'.

When he came to practice in Edinburgh, Sibbald was appalled at the ignorance of the surgeons and apothecaries. In 1670, therefore, he teamed up with Balfour to establish, near Holyrood Abbey, a Physic Garden 'planted with all sorts of Phisical [sic], Medicinal and other Herbs' for the study of botany and herbal medicine. It contained between 800 and 900 plants from all over the world, including many indigenous Scottish species. In 1675 a larger piece of land attached to Trinity Hospital was acquired, below Calton Hill on a site now covered by Waverley Station (where a plaque commemorates it); this Physic Garden contained over 2,000 plants. This was the beginning of what eventually evolved into the Royal Botanic Garden Edinburgh, used by generations of Scottish medical students.

The Garden was managed by James Sutherland, who was appointed to the first Chair of Botany at the University of Edinburgh in 1676, and eventually became the first 'Botanist to the King in Scotland'. In 1683, Sutherland published the *Hortus Medicus Edinburgensis – A Catalogue of the Plants in the Physical Garden at Edinburgh*. This provides a very useful record of the plants being cultivated for medicinal use at that time. Plants and seeds were imported from 'all places', mainly by travelling nobility and medical students. In 1699, Sutherland had to replant 300 different species to replace those eaten by sheep that had strayed across the low dyke around the garden. The Garden moved to two more sites before finally coming to rest in its present location 'in the countryside' at Inverleith in 1820.

John Duncan, of the pharmaceutical firm Duncan, Flockhart and Co., had a pleasure garden on the south side of Edinburgh, where he is said to have grown a great variety of medicinal plants (Glode Guyer 1921). In 1915, his firm acquired a plot of neglected land at Warriston, near the Royal Botanic Garden Edinburgh, where they grew medicinal plants for their own laboratories. After a great deal of work to clear the plot and improve the soil, they established a 'Pharma-Farm' where, under carefully controlled conditions, they grew commercial crops including Aconite (*Aconitum* spp.), Belladonna or Deadly Nightshade (*Atropa belladonna*), Foxglove (*Digitalis purpurea*), Henbane (*Hyoscyamus niger*), Poppies (*Papaver* spp.) and Valerian (*Valeriana dioica*), with smaller plots of experimental crops such as Broom (*Cytisus scoparius*), Calendula (*Calendula officinalis*), Chamomile (*Anthemis* sp.), Dandelion (*Taraxacum* spp.), Elder (*Sambucus nigra*), Juniper (*Juniperus communis*) and Male Fern (*Dryopteris filix-mas*).

Some of these plants did very well in the Edinburgh soil and climate; belladonna root, for example, in two years attained a harvestable size that required three years growth in other locations, with alkaloidal values – a measure of potency – of up to

0.78 per cent, comparing very favourably with the US Pharmacopoeia standard of 0.45 per cent. Male Fern rootstocks were used to treat tapeworms and other intestinal parasites, the manufacturing process having been worked out by a Dr Christison of Edinburgh. Scottish supplies – more potent than those from the Continent – became very important when the First World War interrupted imports from Germany. The rose garden, where 2,000 plants (*Rosa* cultivars) were grown for pollen to treat hay fever, was said to be the most attractive part of the Pharma-Farm for visitors (Glode Guyer 1921). Timothy grass (*Phleum pratense*) was grown for the same purpose.

Henbane, a powerful narcotic, was used as an anaesthetic throughout Europe from early times and may have a very lengthy history of use in the Lothians for a 'non-native' species. Twenty-two seeds of henbane were found by archaeologists excavating a hearth on a midden at Edinburgh Castle dating from AD 100–300, but it is not known whether the plant was being used or was present by chance (Dickson and Dickson 2000). Sutherland grew henbane in the Physic Garden; 'a fickle plant to cultivate', it 'took kindly' to the site at Warriston in the early twentieth century (Glode Guyer 1921).

During the Second World War, large quantities of *Sphagnum* moss were collected for field hospitals by Girl Guides and others from sites such as Red Moss near Balerno; the dried moss is highly absorbent and its antibiotic qualities make it an excellent wound dressing. Today, although Napiers Herbalists (established 1860) in Edinburgh do not get their plant material supplies from the Lothians, they certainly continue the well-established tradition of providing herbal medicines and advice to people in the area. By way of deference to their origins, the Royal College of Physicians in Edinburgh has a small physic garden.

Folklore

South Queensferry's annual burry man parade revolves around a costume made from the strongly-hooked fruits of Lesser Burdock (*Arctium minus*) (Plate 16). The origins of this custom are uncertain; since 1687 it has formed part of the annual Ferry Fair, but local people believe it to be an ancient tradition. It may derive from a Pagan fertility ritual to ensure success for the fishing and harvest or to ensnare evil influences and malevolent spirits. Other, more sinister uses of plants are mentioned in local folklore, for example, the infamous North Berwick witches (late sixteenth century) are said to have met the devil himself bearing a 'baton of fir' (pine) in the Duddingston area.

Use of Aquatic and Marine Plants

The First Statistical Account (from the 1790s) discusses the commercial significance of reeds (*Phragmites australis*) taken from Duddingston Loch (Plates 5 and 31). The reeds were 'employed by weavers to supply their looms' (acting as guides for the warp-threads) and also served as 'a most valuable thatch for any species of houses, which by the strength and hardness of the fibres resists the attacks of sparrows, mice and common vermin which infest and deface straw roofs'. John Lightfoot and his travelling companion, Thomas Pennant, both mention Duddingston Loch as the source of an alga (possibly a species of *Cladophora*) from which 'a coarse kind of paper' was made. Lightfoot also adds that it was used as a stuffing and wadding for garments (Lightfoot 1777; Simmons 1998). Eelgrass (*Zostera marina*) was also used as a stuffing material, e.g. for mattresses, the Second Statistical Account for Cramond describing it as 'an object of interest in our manufactures'.

Commercial Use of Seaweeds

As recently as the 1970s, the larger members of the Fucaceae ('wracks') and the Laminariaceae were brought up from the shores in considerable quantities to fertilise the East Lothian farmlands. The Statistical Accounts mention the use of storm-cast 'sea ware' in all the coastal parishes from Cockburnspath to Dirleton (Statistical Accounts 1/3, 1/5, 1/13). In some areas, cuttings were made in the coastal dunes in order to facilitate the transport of the valuable commodity inland. Barley fertilised with seaweed was said to be particularly popular with brewers and, in the middle of the eighteenth century, commanded a higher price than that grown on dung, lime or bone meal – the other common fertilisers of the time (Statistical Accounts 1/13, 2/2).

The production of kelp (ash from burned brown seaweeds) also occurred in some of the eastern coastal parishes, albeit on a far smaller scale than on the West Coast, the Hebrides and in the Northern Isles, where this was one of the major industries of the eighteenth and early nineteenth centuries. Twelve tons were produced in Dirleton parish in one year in the 1740s (Statistical Accounts 1/3) and 'a small quantity' in North Berwick around the same time. It appears that (following the practice elsewhere) (Statistical Accounts 1/5) higher quality weed was harvested at low water for the production of kelp, which was then taken to Edinburgh for processing, the lower quality weed being used as manure. This local kelp would be added to the huge quantities coming into Leith from more northerly production centres, and used in Edinburgh's thriving glass and soap-making industries and in the bleaching of linen. Although Edinburgh was a major processing centre, kelp was never more than a minor industry on the coast of the Forth and by the mid-nineteenth century had died out altogether in the area (Statistical Accounts 2/2).

Seaweeds as Food and Medicine

As in many coastal towns in nineteenth-century Scotland, a number of seaweed species were eaten, hawked in the streets of Edinburgh to the cry of 'wha'll buy dulse and tang?' (Landsborough 1849). In this context, 'dulse' might refer to several filmy seaweeds, including Dulse (*Palmaria palmata*), Sea Lettuce (*Ulva lactuca*),

Sloke (*Porphyra umbilicalis*) and 'tang', large, brown species such as *Laminaria digitata*, *L. saccharina* and *Alaria esculenta*. All of these species are known to have been eaten elsewhere in Scotland at the time and most are mentioned in the Second Statistical Account for Cramond (2/1) amongst a list of seaweeds 'some of which were useful in the arts, in medicine and for culinary or domestic purposes'. Also included in this list is Carrageen (*Chondrus crispus*), which a footnote says was 'bleached and prepared [and] sold under the name of Irish moss' and used as both food and medicine.

Although the text is not explicit about carrageen having been used in the Lothians, it is probably safe to assume that what was standard practice elsewhere in the country applied to the Lothians as well. However, as discussed in the section on foodstuffs above, the well-educated, Edinburgh elite who recorded much of our information might have been unaware of the extent to which seaweeds were employed in the local area. Similar confusion exists over the role that many species of seaweed played in traditional medicine in the Lothians. Dulse (along with similar filmy seaweeds) is recorded from elsewhere in coastal Scotland as having been used as a vermifuge, cold compress and impromptu sticking plaster (making use of the plant's content of mucilaginous phycocolloid chemicals). It is, however, one of the few seaweeds with a direct reference to its medicinal use in the Lothians. Martin Martin, writing in the late 1600s talks of an obstetric miracle:

A large handful of the sea-plant dulse . . . takes away the after-birth with great ease and safety; this remedy is to be repeated until it produce the desired effect, though some hours may be intermitted: the fresher the dulse is, the operation is the stronger: for if it is above two or three days old, little is to be expected from it in this case. This plant seldom or never fails of success, though the patient had been delivered several days before; and of this I have lately seen an extraordinary instance at Edinburgh in Scotland, when the patient was given over as dead. (Martin 1994)

Today, derivatives of seaweeds from the Forth may prove to be a timely source of new antibiotics as resistance amongst pathogenic bacteria and fungi increases. Research into this, being carried out in the Lothians today, links to a long history of study into the distribution and properties of seaweed, carried out over a thirty-year period in the mid-1900s. Almost all trace of the Scottish Seaweed Research Association (SSRA, based at Inveresk) is gone now, but towards the end of the Second World War, when national resources were at a premium, the SSRA was Britain's premier research facility for work on macroalgae (Kenicer *et al.* 2000).

Table 19.1 Summary of uses of some native, naturalised and cultivated taxa mentioned in the text (arranged taxonomically). Most have explicit references to their use in the Lothians, although almost all were also in use elsewhere as well. It is essential to note that this is not a comprehensive list of all plants used in the Lothians – there is still plenty out there to discover!

Taxonomic name	Common name	Family	Notes
Cladophora sp.?			Green alga used for papermaking (eighteenth century).
Ulva lactuca	Sea lettuce	Ulvaceae	Green seaweed, edible.
Porphyra umbilicalis	Sloke	Bangiaceae	Red seaweed, edible.
Palmaria palmata	Dulse	Palmariaceae	Red seaweed, used as a foodstuff and (seventeenth century) in medicine – notably obstetrics.
Fucus spp. and *Ascophyllum nodosum*	Wracks	Fucaceae	Brown seaweeds, used as manure (up to twentieth century) and in the production of 'kelp' ash (eighteenth century).
Laminaria digitata and *L. hyperborea*	'Kelps'	Laminariaceae	Brown seaweeds, used as manure (up to twentieth century) and in the production of 'kelp' ash (eighteenth century).
Laminaria saccharina	Sugar Wrack	Laminariaceae	Brown seaweed, edible.
Sphagnum spp.	Bogmosses	Sphagnaceae	Highly absorbent and with anti-microbial properties. Used as a wound dressing.
Dryopteris filix-mas	Male Fern	Dryopteridaceae	Used to kill off intestinal parasites.
Juniperus communis	Juniper	Cupressaceae	Abortifacient, allegedly used by a handmaiden of Mary, Queen of Scots. Flavouring for gin. Modern 'Edinburgh gin' is made with imported berries.
Aconitum napellus	Monk's-hood	Ranunculaceae	Medicinal herb, extremely toxic, but long used in medicine – now employed in homeopathy.
Helleborus foetidus	Stinking Hellebore	Ranunculaceae	Toxic – historically used as a strongly purgative medicine – recorded in second Statistical Account of Cramond as a useful introduction.

Species	Common name	Family	Notes
Thalictrum minus	Lesser Meadow Rue	Ranunculaceae	Medicinal plant used as a purgative and in menstrual problems.
Papaver somniferum	Opium Poppy	Papaveraceae	Used as a sedative and possibly even as a simple anaesthetic.
Cannabis sativa	Hemp	Cannabaceae	One of the principal historical fibre plants in Scotland. Grown in southern Scotland from sixth century AD to late nineteenth century.
Urtica dioica	Nettle	Urticaceae	Historically used throughout Scotland by the poor for nettle broth and as a fibre source.
Beta vulgaris ssp. *maritima*	Sea Beet	Chenopodiaceae	Young leaves eaten boiled, cultivars grown as sugar source (sugar beet) and fodder (mangold).
Salix spp.	Willows	Salicaceae	Pliant twigs mainstay of basketry. 'Bitters' (containing salicin) extracted on a commercial scale as painkiller / febrifuge.
Cochlearia spp.	Scurvygrass	Brassicaceae	Leaves widely used in the treatment of scurvy. Mentioned in the second Statistical Account of Cramond amongst a list of useful plants.
Crambe maritima	Sea Kale	Brassicaceae	Boiled and eaten as a vegetable. Mentioned in the second Statistical Account of Cramond amongst a list of useful plants.
Rorippa nasturtium-aquaticum	Watercress	Brassicaceae	Historically collected for use in salads.
Vaccinium myrtillus	Blaeberry	Ericaceae	Widely eaten in upland Scotland in the past. Thought to be good in combating diarrhoea and improving eyesight. Plant stems used in contemporary basketry.
Sedum album	White Stonecrop	Crassulaceae	Found on walls throughout Britain, where it may have been planted for use medicinally. Sap used as a soothing agent for headaches, burns, mouth ulcers and venereal diseases. Mentioned in list of useful plants in second Statistical Account for Cramond.

Scientific name	Common name	Family	Notes
Potentilla erecta	Tormentil	Rosaceae	Astringent – used in tanning leather elsewhere in Scotland, but principally in medicine in the Lothians. Used from the Enlightenment onwards in the treatment of diarrhoea and other complaints of the bowel.
Rubus spp.	Brambles	Rosaceae	Fruits still widely collected during autumn – pliant stems used in contemporary basketry.
Cytisus scoparius	Broom	Fabaceae	Pliant twigs used in contemporary basketry. Cultivated for medicine by Duncan Flockhart & Co., Inverleith (early twentieth century).
Hippophae rhamnoides	Sea Buckthorn	Eleagnaceae	Mentioned in second Statistical Account of Cramond – berries said to be edible.
Daphne laureola	Spurge Laurel	Thymeliaceae	Bark used medicinally to raise blisters.
Cornus sanguinea	Dogwood	Cornaceae	Pliant twigs used in contemporary basketry.
Linum usitatissimum	Flax	Linaceae	Cultivated on a large scale historically, but lessened in importance as a crop in the Lothians during eighteenth century (fibre imported from elsewhere).
Aegopodium podagraria	Ground Elder, Goutweed	Apiaceae	Possibly a monastic medicinal plant used in the treatment of gout.
Myrrhis odorata	Sweet Cicely	Apiaceae	Widespread and successful umbellifer with a distinctive aniseed smell – probably used as a pot herb at one time.
Atropa belladonna	Deadly Nightshade	Solanaceae	Long-time medicinal plant (source of atropine and other tropane alkaloids). Toxic – some cases of poisonings in Lothians.
Hyoscyamus niger	Henbane	Solanaceae	Long-time medicinal plant – found in Fife from the Neolithic as a possible ritual hallucinogen. Toxic, containing similar alkaloids to those in *Atropa belladonna*.
Menyanthes trifoliata	Bogbean	Menyanthaceae	Collected from Duddingston Loch for medicinal use as a general tonic.

Scientific name	Common name	Family	Notes
Borago officinalis	Borage	Boraginaceae	Cultivated historically for medicinal use as a tonic and in a wide range of complaints. Currently thought to offer scope as an anti-cancer drug. Being re-introduced to Lothian on a small agricultural scale. Also grown ornamentally.
Symphytum spp.	White Comfrey	Boraginaceae	Mucilage from many species of comfrey used historically in the treatment of broken bones (helps a good 'set'). Also used as a 'green manure'.
Origanum vulgare	Wild Marjoram	Lamiaceae	This is the flavouring herb oregano. Widely used in modern herbal medicine.
Digitalis pupurea	Foxglove	Scrophulariaceae	Long used throughout Europe as a treatment for dropsy and other complaints caused by heart irregularities (contains digitalin and digitoxin).
Rubia tinctorium	Madder	Rubiaceae	Cultivated around Haddington (eighteenth century and earlier) as a source of red dye.
Sambucus nigra	Elder	Caprifoliaceae	Berries used in wine-making.
Valeriana officinalis	Common Valerian	Valerianaceae	Hugely popular as a herbal medicine today in stress relief. Formerly cultivated in physic gardens.
Taraxacum spp.	Dandelions	Asteraceae	Dandelions were used medicinally as a diuretic (to promote urination) and in the production of a number of 'tonic drinks', e.g. dandelion and burdock. Eaten as a salad.
Tussilago farfara	Colt's-foot	Asteraceae	'Dried and mixed with tobacco by the country people' (Greville 1824).
Matricaria recutita	Scented Mayweed (Chamomile)	Asteraceae	Very widely used in herbal medicine and as a calming tea today.
Artemisia spp.	Wormwoods	Asteraceae	Cultivated in physic gardens for use in combating worms and other intestinal parasites (hence the name).

Lactuca virosa	Great Lettuce	Asteraceae	Cultivated in physic gardens for use in a similar manner to opium – as a sedative and narcotic (although far less powerful).
Tanacetum parthenium	Feverfew	Asteraceae	Used historically in the treatments for worms.
Zostera marina	Eelgrass	Zosteraceae	Used to stuff matresses (probably a more northerly practice) and to pack brittle glassware.
Juncus effusus and J. conglomeratus	Rushes	Juncaceae	Pliant leaves used in contemporary basketry. Historically used to weave many functional items, including mats.
Ammophila arenaria	Marram	Poaceae	Long known as stabiliser for dune ecosystems – encouraged as such for the past two centuries at least (mentioned in second Statistical Account of Cramond).
Phleum pratense	Timothy	Poaceae	Cultivated by Duncan Flockhart & Co. at Inverleith as a source of pollen for the treatment of hay fever.
Phragmites australis	Common Reed	Poaceae	Historically used for thatching throughout its range. Collected from Duddingston Loch.

Conclusion

Despite the tendency towards urbanisation and development in the Lothians and their relatively small size amongst the Scottish botanical vice-counties, they have a diverse past and present use of plants. We have restricted ourselves to explicit references to plant use in the Lothians; much has been omitted due to constraints of space. There must be a host of plant uses that have never actually been recorded for the area, yet were a part of everyday life for the many different peoples who have lived here. Although by no means comprehensive, we hope that the information provided gives an insight into this diversity and provides starting points for those who are interested to investigate further. Current trends in farming and interest in the uses of native plants suggest there is potential for development of both historical and innovative means of using Lothian's botanical resources in a responsible manner.

Acknowledgements

Thanks are due to Kim Howell and Douglas McKean for valuable comments on the text.

References

Adam, J. C. (1870), 'Mosses of West Lothian (VC 84)', *Transactions of the Botanical Society of Edinburgh*, 10: 251–3.

Adam, J. C. (1917), 'Mosses of West Lothian (VC 84)', *Transactions of the Botanical Society of Edinburgh*, 27: 123–34.

Alesius, A. (1550), in Munster, S., *Cosmographica*, Basle.

Amphlett, J. and Rea, C. (1909), *The Botany of Worcestershire*, Birmingham: Cornish Brothers.

Anderson, M. L. (1967), *A History of Scottish Forestry*, Edinburgh: Oliver and Boyd.

Anon. (1976), *John Muir Country Park: Descriptive Management Plan*, East Lothian District Council.

Anon. (1977), *Aberlady Bay Local Nature Reserve: Descriptive Management Plan*, East Lothian District Council.

Arnolds, E. (1992), 'The analysis and classification of fungal communities with special reference to macrofungi' in W. Winterhoff, *Fungi in Vegetation Science*, Dordrecht: Kluwer, pp. 7–47.

Arnott, H. (1998), *The History of Edinburgh* (reprint of 1799 edition), Edinburgh: Westport Books.

ASH Consulting Group (1998), *The Lothians Landscape Character Assessment*, Scottish Natural Heritage Review No. 91, Perth: SNH.

Avery, B. W. (1990), *Soils of the British Isles*, Wallingford: CABI.

Baas-Becking, L. G. M. (1934), *Geobiologie of inleiding tot de milieukunde*, Den Haag: Van Stockum, quoted from Gams, W. (1992), 'The analysis of communities of saprophytic microfungi with special reference to soil fungi' in W. Winterhoff, *Fungi in Vegetation Science*, Dordrecht: Kluwer, pp. 183–223.

Balfour, J. H. (1849), *Manual of Botany*, Glasgow: Richard Griffin & Co.

Balfour, J. H. and Sadler, J. (1863), *Flora of Edinburgh*, Edinburgh: Adam and Charles Black, 2nd edition, 1871.

Barnard, P. D. and Long, A. G. (1973), 'On the structure of a petrified stem and some associated seeds from the Lower Carboniferous rocks of East Lothian, Scotland', *Transactions of the Royal Society of Edinburgh*, 69: 91–112.

Bede, the Venerable, Saint, 673–735 (1990), *Ecclesiastical History of the English People*, tr. L. Sherley-Price, revised by R. E. Latham, introduction and additional translation D. H. Farmer, rev. edition, Harmondsworth: Penguin Books Ltd.

Bell, W. and Sadler, J. (1869), 'Mosses collected on Excursions round Edinburgh in 1869', *Transactions of the Botanical Society of Edinburgh*, 27: 251–3.

Bell, W. and Sadler, J. (1870), 'Notice on the Grimmias collected on Arthur's Seat near Edinburgh', *Transactions of the Botanical Society of Edinburgh*, 10: 432–5.

Berkeley, M. J. and C. E. Broome (1879), 'Notices of Britsh Fungi, n.1796', *Annales and Magazine of Natural History*, 5 Series III: 202–15.

Bibby, J. S., Douglas, H. A., Thomasson, A. J. and Robertson, J. S. (1982), *Land Capability Classification for Agriculture*, Aberdeen: The Macaulay Institute for Soil Research.

Blockeel, T. L. and Long, D. G. (1998), *A Check-List and Census Catalogue of British and Irish Bryophytes*, Cardiff: British Bryological Society.

Boece, H. (1938), *The Chronicles of Scotland* as translated into Scots by John Bellenden, 1531, eds R. W. Chambers and E. C. Batho, Edinburgh and London: William Blackwood & Sons.

Bolton, J. (1785), *Filices Britannicae*, Leeds: John Binns.

Bown, C. J. and Shipley, B. M. (1982), *Soil and Land Capability for Agriculture: South-East Scotland*, Aberdeen: Macaulay Institute for Soil Research.

Braithwaite, M. E. and Long, D. G. (1990), *The Botanist in Berwickshire*, The Berwickshire Naturalists Club.

Braun, U. (1987), *A monograph of the Erysiphales (powdery mildews)*, Beihefte zur Nova Hedwigia 89, Berlin: Gebrueder Borntraeger.

Brown, P. H. (1891), *Early Travellers in Scotland*, Edinburgh: David Douglas.

Brown, R. (ms) (1793), *Plants in the Edinburgh Botanic Gardens or found in the neighbourhood* [copies at the British Museum in London and the Royal Botanic Garden Edinburgh Herbarium].

Bryant, J. A., Stace, C. A. and Stewart, N. F., 'Checklist of the charophytes of the British Isles', in preparation.

Cadbury, D. A., Hawkes, J. G. and Readett, R. C. (1971), *A Computer-mapped Flora. A Study of the County of Warwickshire*, London: Academic Press.

Cadell, H. M. (1913), *The Story of the Forth*, Glasgow: McLehose.

Caesar, J. (100 BC–44 BC) (1951), *The Conquest of Gaul (De Bello Gallico)*, ed. J. F. Gardner, trans. S. A. Handford, rev. edition, Harmondsworth: Penguin Books Ltd.

Chalmers, G. (1887), *Caledonia or A historical and topographical account of North Britain*, Paisley: Alexander Gardner.

Chandler, T. J. and Gregory, S. (eds) (1976), *The Climate of the British Isles*, New York: Longman.

Charlesworth, J. K. (1957), *The Quaternary Era, with special reference to its glaciation*, London: Edward Arnold.

Church, J. M., Coppins, B. J., Gilbert, O. L., James, P. W. and Stewart, N. F. (1996), *Red Data Books of Britain and Ireland: Lichens. Volume 1: Britain*, Peterborough: Joint Nature Conservation Committee.

Clapham, A. R., Tutin, T. G. and Moore, D. M. (1987), *Flora of the British Isles*, Cambridge: Cambridge University Press.

Clement, E. J. and Foster, M. C. (1994), *Alien Plants of the British Isles*, London: Botanical Society of the British Isles.

Colledge, D. (1980), *Lothian*, Edinburgh: Holmes McDougall.

Coppins, B. J. (1978), 'A glimpse of the past and present lichen flora of Edinburgh', *Transactions of the Botanical Society of Edinburgh*, 42 (Suppl.): 19–35.

Craig, G. Y. (ed.) (1983), *The Geology of Scotland*, 2nd edition, Edinburgh: Scottish Academic Press.

Craig, G. Y. and Duff, P. M. (1975), *Geology of the Lothians and South-East Scotland*, Edinburgh: Scottish Academic Press.

Crawford, R. M. M., Jeffree, C. E., and Rees, W. G. (2002), Paludification and forest retreat, *Annals of Botany* (in press).

Dahl, E. (1998), *The Phytogeography of Northern Europe*, Cambridge: Cambridge University Press.

References

Daiches, D. (1978), *Edinburgh*, London: Hamish Hamilton.

Darwin, T. (1996), *The Scots Herbal*, Edinburgh: Mercat.

Dept. of the Environment (DoE) (1996), *Review of the Potential Effects of Climate Change in the United Kingdom*, London: HMSO.

Dickson, C. and Dickson, J. H. (2000), *Plants and People in Ancient Scotland*, Stroud, Gloucestershire: Tempus.

Dickson, J. H., Macpherson, P. and Watson, K. (2000), *The Changing Flora of Glasgow*, Edinburgh: Edinburgh University Press.

Dobson, F. S. (2000), *Lichens: An Illustrated Guide to the British and Irish Species*, Slough: The Richmond Publishing Co.

Druce, G. C. (1932), *The Comital Flora of the British Isles*, Arbroath: Buncle & Co.

Dudman, A. A. and Richards, A. J. (1997), *Dandelions of Great Britain and Ireland*, London: Botanical Society of the British Isles.

Dumayne-Peaty, L. (1999), 'Late Holocene human impact on the vegetation of south-eastern Scotland: a pollen diagram from Dogden moss, Berwickshire', *Review of Paleobotany and Palynology*, 105: 121–41.

Dyer, A. (1995), 'Temperature as a factor in determining the distribution of *Asplenium septentrionale* in Britain', *Pteridologist*, 2: 288–90.

Edees, E. S. and Newton, A. (1988), *Brambles of the British Isles*, London: The Ray Society.

Edinburgh Biodiversity Partnership (2000), *The Edinburgh Biodiversity Action Plan*, Edinburgh: City of Edinburgh Council.

Ellis, M. B. and Ellis, J. P. (1988), *Microfungi on miscellaneous substrates: an identification handbook*, London: Croom Helm Ltd.

Ellis, M. B. and Ellis, J. P. (1996), 'Fungi and Slime Moulds in Suffolk', Supplement 3, *Transactions of the Suffolk Naturalists' Society*, 32: 165.

Ellis, M. B. and Ellis, J. P. (1997), *Microfungi on land plants: an identification handbook*, second enlarged edition, Slough: Richmond Publishing.

Evans, W. (1905), 'On the Ricciae of the Edinburgh District', *Transactions of the Botanical Society of Edinburgh*, 23: 285–7.

Evans, W. (1917), 'Some Moss Records for Selkirk, Peebles and the Lothians', *Transactions of the Botanical Society Edinburgh*, 27: 138–56.

Fletcher, H. R. and Brown, W. H. (1970), *The Royal Botanic Garden Edinburgh 1670–1970*, Edinburgh: HMSO.

Floate, M. J. S., Eadie, J., Black, J. S. and Nicholson, I. A. (1973), 'Improvement of *Nardus* dominant hill pasture by grazing control and surface treatment and its economic assessment', in *Hill Pasture Improvement and its Economic Utilisation*, Colloq. Proc. No. 3, Potassium Inst. Ltd, Edinburgh, 1972, pp. 33–9, cited in *Science and Hill Farming: HFRO 1954–1979*, Penicuik: HFRO.

Ford, M. A. and Younie, D. (1996), 'Greening the countryside: towards the creation of species-rich grasslands on former farmland', *Environmental & Food Science Research Report 1995*, Edinburgh: Scottish Agricultural College.

Forestry Commission (1999), *National Inventory of Woodland and Trees. Scotland – Lothian Region. Part 1 – Woodlands of 2 hectares and over*, Edinburgh: Forestry Commission.

Fowler, R. C. A. (1967), *The present condition of shelter-belts on farms in the vicinity of the Pentland Hills, Midlothian, Scotland*, M.Sc. Thesis, University of Edinburgh.

Francis, P. E. (1981), *The Climate of the Agricultural Areas of Scotland*, Climatological Memorandum No. 108, Bracknell: Meteorological Office.

Fraser, D. (1976), *Edinburgh in Old Times*, Montrose: Standard Press.

Furley, P. A., 'Soil determinants of and soil responses to variations in plant communities in the coastal sand dunes of East Lothian', in preparation.

Gams, W. (1992), 'The analysis of communities of saprophytic microfungi with special reference to soil fungi', in W. Winterhoff, *Fungi in Vegetation Science*, Dordrecht: Kluwer, pp. 183–223.

Geikie, J. (1866), 'On the buried forests and peat mosses of Scotland, and the changes of climate they indicate', *Transactions of the Royal Society of Edinburgh*, 24: 363–84.

Gilbert, O. L. (2000), *Lichens*, London: HarperCollins.

Glode Guyer, R. (1921), 'Cultivation of Medicinal Plants in Scotland – Past and Present', *The Pharmaceutical Journal and Pharmacist*, Feb.–Mar. 1921: 146–9, 168–71, 190–2.

Godwin, H. (1975), *History of the British Flora*, Cambridge: Cambridge University Press.

Gordon, J. E. (ed.) (1994), *Scotland's Soils: Research Issues in Developing a Soil Sustainability Strategy*, Scottish Natural Heritage Review No. 13, Edinburgh: SNH.

Gordon, J. E. and Sutherland, D. G. (1993), *Quaternary of Scotland*, 1st edition, London: Chapman & Hall.

Graham, G. G. and Primavesi, A. L. (1993), *Roses of Great Britain and Ireland*, London: Botanical Society of the British Isles.

Gravesen, S., Frisvad, J. C. and Samson, R. A. (1994), *Microfungi*, Copenhagen: Munksgaard.

Greville, R. K. (1824), *Flora Edinensis or A description of plants growing near Edinburgh*, Edinburgh: William Blackwood; London: T. Cadell.

Greville, R. K. (1828), *Scottish Cryptogamic Flora*, Edinburgh: Maclachlan and Stewart.

Grierson, S. (1986), *The Colour Cauldron*, Perth: Mill Books.

Grime, J. P. (1972), 'The creative approach to nature conservation', in F. J. Ebling and G. W. Heath (eds), *The Future of Man*, London: Academic Press, pp. 47–54.

Grime, J. P., Hodgson, J. G. and Hunt, R. (1988), *Comparative Plant Ecology*, London: Unwin Hyman.

Grove, J. M. (1988), *The Little Ice Age*, London: Methuen.

Hall and Tomlinson (1996) [http://intarch.ac.uk/journal/issue1/tomlinson/toc.html]

Hanson, W. S. (1997), 'The Roman presence: Brief Interludes', in K. J. Edwards and I. B. M. Ralston (eds), *Scotland: Environment and Archaeology 8000 BC–AD 1000*, Chichester: John Wiley.

Harrison, J. (1997), 'Central and South Scotland', in D. Wheeler, and J. Mayes (eds), *Regional Climates of the British Isles*, London: Routledge.

Hawksworth, D. L. (1991), 'The fungal dimension of biodiversity; magnitude, significance and conservation', *Mycological Research*, 95: 641–55.

Hawksworth, D. L. (1997), in M. B. Ellis and J. P. Ellis, *Microfungi on land plants: an identification handbook*, 2nd enlarged edition, Slough: Richmond Publishing, pp. i–vi.

Hawksworth, D. L., Kirk, P. M., Sutton, B. C. and Pegler, D. N. (1995), *Ainsworth and Bisby's Dictionary of the Fungi*, 8th edition, Wallingford: CABI.

Hay, R. A. (1700), *Genealogie of the Sainteclaires of Rosslyn*, privately printed, 1835.

Helfer, S. (1993), 'Rust Fungi – A Conservationist's Dilemma', in D. N. Pegler, L. Boddy, B. Ing and P. M. Kirk (eds), *Fungi of Europe: Investigation, Recording and Conservation*, London: Royal Botanic Gardens Kew, pp. 287–94.

Henderson, D. M. (2000), *A Checklist of the Rust Fungi of the British Isles*, Richmond, Surrey: British Mycological Society.

References

Heslop Harrison, J. W. (1953), 'The present position of the rosebay willowherb', *Vasculum*, 38: 25.

Hirsch, G. and Braun, U. (1992), 'Communities of parasitic microfungi', in W. Winterhoff, *Fungi in Vegetation Science*, Dordrecht: Kluwer, pp. 225–50.

Holliday, P. (1989), *A Dictionary of Plant Pathology*, Cambridge: Cambridge University Press.

Hooker, W. J. (1821), *Flora Scotica*, Edinburgh: Archibald Constable & Co.; London: Hurst Robinson & Co.

Hope, J. (1765), 'List of plants in the neighbourhood of Edinburgh, collected in flower', *Notes from the Royal Botanic Garden Edinburgh* (1907), 4: 18.

Hulme, M. and Barrow, E. (1997), *Climates of the British Isles*, London: Routledge.

Hunter, T. (1883), *Woods, Forests and Estates of Perthshire*, Perth: Henderson, Robertson and Hunter.

Hutton, J. (1795), *Theory of the Earth*, Edinburgh: Printed for Messrs Cadell Junior and Davies, London: and William Creech, Edinburgh, 1795–1899.

Ingram, D. and Robertson, N. (1999), *Plant Disease*, London: HarperCollins.

Ingrouille, M. (1995), *Historical Ecology of the British Flora*, London: Chapman and Hall.

Jalas, J. and Suominen, J. (1972–94), *Atlas Florae Europaeae, 1–10*, Helsinki: Societas Biologica Fennica Vanamo.

Johnson, W. T. (1995), *Cudbear Dye and its Discovery by Cuthbert Gordon (1730–1810)*, Livingston: Officina.

Jüelich, W. (ed.) (1994), *Colour atlas of micromycetes*, Stuttgart: Gustav Fischer Verlag.

Kendrick, B. (1992), *The Fifth Kingdom*, 2nd edition, Newburyport, MA: Focus Texts.

Kenicer, G., Bridgewater, S. and Milliken, W. (2000), 'The ebb and flow of Scottish seaweed use', *Botanical Journal of Scotland*, 52: 119–48.

Kent, D. H. (1992), *List of vascular plants of the British Isles*, London: Botanical Society of the British Isles.

Kent, D. H. (1996), *List of vascular plants of the British Isles, Supplement 1*, London: Botanical Society of the British Isles.

King, J. A., Smith, K. A. and Pyatt, D. G. (1986), 'Water and oxygen regimes under conifer plantations and native vegetation on upland peaty gley soils and deep peat soils', *Journal of Soil Science*, 37: 485–97.

Kirby, R. P. (1997), 'The Aberlady Bay coastal landforms and vegetation communities', *Scottish Geographical Magazine*, 113: 121–6.

Kirk, P. M., Cannon, P. F., David, J. C. and Stalpers, J. A. (eds) (2001), *Ainsworth and Bisby's Dictionary of the Fungi*, 9th edition, Wallingford: CABI.

Kurtzman, C. P. and Sugiyama, J. (2001), 'Ascomycetous yeasts and yeastlike taxa', in D. J. McLaughlin, E. G. McLaughlin and P. A. Lemke (eds), *The Mycota* Vol. 7A, Berlin: Springer, pp. 179–200.

Landsborough, D. (1849), *A popular history of British Sea-weeds*, London: Reeve, Benham and Reeve.

Last, F. T. and Watling, R. (1998), 'First record of *Laccaria fraterna* in Britain', *The Mycologist*, 12: 152.

Lawson, P. (1851), *The Vegetable Products of Scotland*, Edinburgh: Lawson and Sons.

Lightfoot, J. (1777), *Flora Scotica*, 2 vols, London: B. White.

Linnaeus, C. von (1938), The *Critica Botanica* of Linnaeus, trans. by A. Hort, rev. by M. L. Green, London: The Ray Society.

Linton, W. R. (1903), *Flora of Derbyshire*, London: Bemrose.

Long, A. G. (1979), 'Observations on the Lower Carboniferous genus *Pitus withami*', *Transactions of the Royal Society of Edinburgh*, 70: 111–27.

Long, A. G. (1961), '*Tristichia ovensi* gen. et sp. nov., – a protostelic Lower Carboniferous pteridosperm from Berwickshire and East Lothian, with an account of some associated seeds and cupules', *Transactions of the Royal Society of Edinburgh*, 64: 477–89.

Lothian Regional Council (1995), *A Guide to Lothian*, Edinburgh: Lothian Regional Council.

Mackie, J. D. (1964), *A History of Scotland*, Harmondsworth: Penguin.

Macpherson, P. (1997), 'Plants from ships ballast', *B.S.B.I. News*, 74: 51.

Maitland, W. (1753), *History of Edinburgh*, Edinburgh: Hamilton, Balfour and Neill.

Martin, I. H. (1927), *The Field-Club Flora of the Lothians*, Edinburgh: William Blackwood and Sons, 2nd edition, 1934.

Martin, M. (1994), *A Description of the Western Isles of Scotland circa 1695. Including a voyage to St. Kilda* (reprint edited by D. J. Macleod), Edinburgh: Birlinn.

Matthews, J. R. (1937), Geographical relationships of the British flora, *Journal of Ecology*, 25: 1–90.

Matthews, J. R. (1955), *Origin and Distribution of the British Flora*, London: Hutchinson.

Maugham, R. (1811), 'A List of the rarer plants observed in the neighbourhood of Edinburgh', *Wernerian Natural History Society*, vol. 1.

McAdam, A. D. and Clarkson, E. N. K. (eds) (1996), *Lothian Geology: An Excursion Guide*, Edinburgh: Edinburgh Geological Society.

McAdam, A. D., Clarkson, E. N. K. and Stone, P. (1992), *Scottish Borders Geology: an Excursion Guide*, Edinburgh: Edinburgh Geological Society: Scottish Academic Press.

McAndrew, J. (1912), 'Notes on some Mosses from the three Lothians', *Scottish Botanical Review*, 1912: 202–5.

McKean, D. R. (1989), *A Checklist of the Flowering Plants and Ferns of Midlothian*, Edinburgh: Botanical Society of Edinburgh.

Meteorological Office (1982), *The Climate of Great Britain: Edinburgh, the Lothian Region and Stirling*, Climatological Memorandum 115, Bracknell: Meteorological Office.

Meteorological Office (1989), *The Climate of Scotland: Some Facts and Figures*, Edinburgh: HMSO.

Miller, H. (1841), *The Old Red Sandstone, or, New Walks in a Old Field*, Edinburgh: Johnstone.

Miller, H. (1849), *Foot-prints of the Creator, or, The Asterolepis of Stromness*, Edinburgh: Johnstone and Hunter.

Moffat, B., Thomson, B. S. and Fulton, J. (1989), *The third report on researches into the medieval hospital at Soutra, Lothian/Borders region, Scotland*, Edinburgh: SHARP.

Moore, D. and More, A. G. (1866), *Contributions towards a Cybele Hibernica*, Dublin: Hodges, Smith; London: Van Voorst.

Moore, J. A. (1986), *Charophytes of Great Britain and Ireland*, B.S.B.I. Handbook No. 5, London: Botanical Society of the British Isles.

Moss Exchange Club (1907), *A Census Catalogue of British Mosses*, York: Moss Exchange Club.

Muscott, J. (1989), *A Checklist of the Flowering Plants and Ferns of West Lothian*, Edinburgh: Botanical Society of Edinburgh.

Myerscough, P. J. (1980), 'Biological Flora of the British Isles: *Epilobium angustifolium*', *Journal of Ecology*, 68: 1047–74.

Newey, W. W. (1967), 'Pollen analyses from South-east Scotland', *Transactions of the Botanical Society of Edinburgh*, 40: 424–34.

Newey, W. W. (1970), 'Pollen analysis of late Weichselian deposits at Corstorphine', Edinburgh, *New Phytologist*, 69: 1167–77.

Nortcliff, S. (1988), 'Soil formation and characteristics of soil profiles', in A. Wild (ed.), *Russell's Soil Conditions & Plant Growth*, 11th edition, Harlow, Essex: Longman, pp. 168–212.

Ogilvie, A. G. (1951), 'South-Eastern Scotland: the region and its parts', in *Scientific Survey of South-Eastern Scotland*, Edinburgh: British Association for the Advancement of Science, pp. 11–35.

Page, C. N. (1988), *A Natural History of Britain's Ferns*, London: Collins.

Parnell, R. (1842), *The Grasses of Scotland*, Edinburgh and London: Blackwood & Sons.

Parry, M. L. (1975), 'Secular climatic change and marginal land', *Transactions of the Institute of British Geographers*, 64: 1–13.

Peach, B. N., Clough, C. T., Hinxman, L. W., Grant Wilson, J. S., Crampton, C. B., Maufe, H. B. and Bailey, E. B. (eds) (1910), *The Geology of the Neighbourhood of Edinburgh*, 2nd edition, Memoirs of the Geological Survey of Great Britain (Scotland), Edinburgh: HMSO.

Perring, F. H. and Sell, P. D. (1968), *Critical Supplement to the Atlas of the British Flora*, London: Thomas Nelson and Sons.

Perring, F. H. and Walters, S. M. (1962), *Atlas of the British Flora*, London: Thomas Nelson and Sons for the Botanical Society of the British Isles, 2nd edition, 1976.

Piggott, S. (1955), *British Prehistory*, London: Oxford University Press.

Pont, T. (1636), *The Mercator-Hondius Atlas*. Facsimile of 1633 edition, 1968, Amsterdam: T.O.T.

Pont, T. (1664), 'A New Description of the Shyres Lothian and Linlitquo', in J. Blaue (ed.), *Geographiae Blavianae*, vol. 5, Amsterdam: Joan Blaeu.

References

Purvis, O. W. (2000), *Lichens* [The Life Series], London: The Natural History Museum.

Ragg, J. M. (1960), *The Soils of the Country around Kelso and Lauder*, Edinburgh: HMSO.

Ragg, J. M. (1973), 'Factors in soil formation', in J. Tivy (ed.), *The Organic Resources of Scotland*, Edinburgh: Oliver & Boyd, pp. 38–50.

Ragg, J. M. and Dent, D. L. (1969), *The Soils of the Bush Estate, Midlothian*, Aberdeen: Macaulay Institute for Soil Research.

Ragg, J. M. and Futty, D. W. (1967), *The Soils of the Country round Haddington and Eyemouth*, Memoirs of the Soil Survey of Great Britain, Edinburgh: HMSO.

Ramsay, S. and Dickson, J. H. (1997), 'Vegetational History of Central Scotland', *Botanical Journal of Scotland*, 49: 141–50.

Redfern, D. B. (1977), 'Dutch elm disease in Scotland', *Scottish Forestry*, 31: 105–9.

Rich, T. C. G. and Smith, P. A. (1996), 'Botanical recording, distribution maps and species frequency', *Watsonia*, 21: 155–67.

Richardson, M. J. (1970), 'Studies on *Russula emetica* and other agarics in a Scots pine plantation', *Transactions of the British Mycological Society*, 55: 217–29.

Richardson, M. J. (1998), 'New and interesting records of coprophilous fungi', *Botanical Journal of Scotland*, 50: 161–76.

Rodwell, J. S. (ed.) (1991–2000), *British Plant Communities* vols 1–5, Cambridge: UK Joint Nature Conservation Committee/Cambridge University Press.

Ryves, T. B., Clement, E. J. and Foster, M. C. (1996), *Alien Plants of the British Isles*, London: Botanical Society of the British Isles.

Sadler, J. (1868), 'Notice on some rare British mosses recently collected near Edinburgh', *Transactions of the Botanical Society of Edinburgh*, 9: 163.

Scola, P. M. (1944), 'The Lothians', in L. Dudley Stamp (ed.), *The Land of Britain* parts 16–18, London: Geographical Publications Ltd., pp. 125–214.

Sibbald, R. (ms) (1684), *Catalogue of Plants growing in King's Park* [in Royal Botanic Garden Edinburgh Herbarium].

Sibbald, R. (1684), *Scotia Illustrata*, Edinburgh: ex officina typographica J. Kniblo, J. Solingensis and J. Colmarii.

Sibbald, R. (1699), *Provision for the poor in time of Dearth and Scarcity*, Edinburgh: printed by James Watson.

Silverside, A. J. (2001), *Fungi of the Wicken Fen*. Online database. http://www-biol.paisley.ac.uk/research/Asilverside/Wickenfungi.html

Silverside, A. J. and Jackson, E. H. (1988), *A Checklist of the Flowering Plants and Ferns of East Lothian*, Edinburgh: Botanical Society of Edinburgh.

Simmons, A. (ed.) (1998), *A Tour in Scotland and voyage to the Hebrides 1772, by Thomas Pennant*, Edinburgh: Birlinn.

Simpson, N. D. (1960), *A Bibliographical index of the British Flora*, Bournemouth: published privately.

Sissons, J. B. (1958), 'The deglaciation of part of East Lothian', *Transactions of the Institute of British Geographers*, 25: 59–77.

Sissons, J. B. (1962), 'A re-interpretation of the literature on late glacial shorelines in Scotland with particular reference to the Forth area', *Transactions of the Edinburgh Geological Society*, 19: 83–99.

Sissons, J. B. (1963), 'Scottish raised shoreline heights with particular reference to the Forth Valley', *Geografiska Annaler*, 45 (2–3): 180–5.

Sissons, J. B. (1967), *The Evolution of Scotland's Scenery*, Edinburgh: Oliver & Boyd.

Sissons, J. B. (1971), 'The geomorphology of central Edinburgh', *Scottish Geographical Magazine*, 87: 185–96.

Smeaton, O. (1904), *Edinburgh and its Story*, London: J. M. Dent.

Smith A. N. (1995), 'The excavation of Neolithic, Bronze Age and Early Historic features near Ratho, Edinburgh', *Proceedings of the Society of Antiquaries of Edinburgh*, 125: 69–138.

Smith, P. M. (1986), 'Native or Introduced? Problems in the taxonomy and plant geography of some widely introduced annual brome-grasses', *Proceedings of the Royal Society of Edinburgh*, 89B: 273–81.

Smout, T. C. (2000), *Nature Contested*, Edinburgh: Edinburgh University Press.

Sonntag, C. O. (1894), *A Pocket Flora of Edinburgh and the Surrounding District*, London and Edinburgh: Williams and Norgate.

South, G. R. and Tittley, I. (1986), *A checklist and distributional index of the benthic marine algae of the North Atlantic Ocean*, St. Andrews, New Brunswick, and London: Huntsman Marine Laboratory and British Museum (Natural History).

Stace, C. (1997), *New Flora of the British Isles*, 2nd edition, Cambridge: Cambridge University Press.

Statistical Accounts of Scotland (1998), First 1791–99; Second 1845: 1/3 (Dirleton), 1/5 (North Berwick), 1/6 (Uphall), 1/10 (Athelstaneford), 1/13 (Cockburnspath), 1/18 (Duddingston), 2/1 (Cramond), 2/2 (Dirleton), Edinburgh: Westport Books. Also available as electronic resource [Edinburgh]: EDINA [2001].

Stenton, H. (1953), 'The soil fungi of Wicken Fen', *Transactions of the British Mycological Society*, 36: 304–14.

Stevenson, J. (1879), *Mycologia Scotica*, Edinburgh: Ballantyne Press.

Stevenson, R. L. [1878] (1913), *Edinburgh*, new edition, London: Seeley, Service & Co. Ltd.

Stewart, A., Pearman, D. A. and Preston, C. D. (1994), *Scarce Plants in Britain*, Peterborough: Joint Nature Conservation Committee.

Sutherland, J. (1683), *Hortus Medicus Edinburgensis*, Edinburgh: printed by the heir of Andrew Anderson.

Svensson, R. and Wigren, R. (1986), 'A survey of the history, biology and preservation of some retreating synanthropic plants,' *Acta Universitatis Upsaliensis Symbolae Botanicae Upsaliensis*, 25: 1–73.

The London Catalogue of British Plants, 11th edition, London: Bell.

Thompson, D. A. (1984), *Ploughing of Forest Soils*, Forestry Commission Leaflet No. 71, London: HMSO.

Tindall, F. (1998), *Memoirs and Confessions of a County Planning Officer*, Ford, Midlothian: The Pantile Press.

Traill, G. W. (1886), 'The marine algae of Joppa', *Transactions and Proceedings of the Botanical Society of Edinburgh*, 16: 395–402.

Traill, G. W. (1890), 'The marine algae of the Dunbar coast', *Transactions and Proceedings of the Botanical Society of Edinburgh*, 18: 274–99.

Tranter, N. (1979), *Portrait of the Lothians*, London: Robert Hale.

Triebel, D. (2001), *Microfungi exsiccati* – Online-version. Botanische Staatssammlung München: http://www.botanik.biologie.uni-muenchen.de/botsamml/arnoldia/mifuind.html – München.

Trotter, J. (1796), 'The Parish of Prestonpans', in Sir J. Sinclair (ed.), *The Statistical Account of Scotland: Drawn Up from the Communications of the Ministers of the Different Parishes* vol. 17, Edinburgh: William Creech, pp. 61–88.

Tulloch, W. and Walton, H. S. (1958), *The Geology of the Midlothian Coalfield*, Edinburgh: HMSO.

United Kingdom Review Group on Impacts of Atmospheric Nitrogen, 1994, *Impacts of Nitrogen Deposition on Terrestrial Ecosystems*, London: Department of Environment.

Various, (1845), *The new statistical account of Scotland, Vol II, Linlithgow, Haddington, Berwick*, Edinburgh: William Blackwood & Sons.

Walker, F. T. (1945/46), *Seaweed Survey of Scotland – Fucaceae*, SSRA Report No. 80, Musselburgh: SSRI.

Watling, R. (1986), '150 Years of Paddockstools: A history of agaric ecology and floristics in Scotland', *Transactions of the Botanical Society of Edinburgh*, 45: 1–42.

Watling, R. (1997), 'Secrets and Treasures in Edinburgh Gardens', *Mycologist*, 11: 62–4.

Watling, R. (2001), 'Larger Fungi', in D. Hawksworth (ed.), *The Changing Wildlife of Great Britain and Ireland*, London: Taylor & Francis.

Watson, E. V. (1981), *British Mosses and Liverworts*, 3rd edition, Cambridge: Cambridge University Press.

Watson, H. C. (1847), *Cybele Britannica*, London: Longmans.

Watson, H. C. (1873), *Topographical Botany*, Thames Ditton: privately published.

Watts, A., Carey, P. D. and Eversham, B. C. (1997), Implications of climate change for biodiversity, in L. V. Fleming, A. C. Newton, J. A. Vickery, and M. B. Usher (eds), *Biodiversity in Scotland: Status, Trends and Initiatives*, Edinburgh: HMSO.

Webb, D. A. (1985), 'What are the criteria for presuming native status?' *Watsonia*, 25: 231–6.

Wellman, C. H. and Richardson, J. B. (1993), 'Terrestrial plant microfossils from Silurian inliers of the Midland Valley of Scotland', *Palaeontology*, 36 (1): 155–93.

Whitfield, W. A. D. and Furley, P. A. (1971), *The Relationship between Soil Patterns and Slope Form in the Ettrick Association, South-East Scotland*. Special Publ. 3. London: Institute of British Geographers.

Whittington, G. and Edwards, K. J. (1997) 'Climate Change', in Edwards, K. J. and

Ralston, I. B. M. (eds), *Scotland: Environment and Archaeology, 8000 BC–AD 1000*, Chichester: Wiley.

Wigginton, W. J. (1999), *British Red Data Books. 1. Vascular Plants*, Peterborough: Joint Nature Conservation Committee.

Wilkinson, M. (1980), 'Benthic estuarine algae and their environment: A review', in J. H. Price, D. E. G. Irvine and W. F. Farnham (eds), *The Shore Environment: Methods and Ecosystems*, vol. 2, London: Academic Press, pp. 425–86.

Wilkinson, M. (1992), 'Aspects of intertidal ecology with reference to conservation of Scottish seashores', *Proceedings of the Royal Society of Edinburgh*, 100B: 77–93.

Wilkinson, M. and Rendall, D. A. (1985), 'The role of benthic algae in estuarine pollution assessment', in J. Wilson and W. Halcrow (eds), *Estuarine Management and Quality Assessment*, New York: Plenum, pp. 71–81.

Wilkinson, M. and Slater, E. M. (1997), 'Estuarine algal communities in two sub-estuaries of the Forth with different water quality', *Botanical Journal of Scotland*, 49: 387–96.

Wilkinson, M. and Tittley, I. (1979), 'The marine algae of Elie, Scotland: a reassessment', *Botanica Marina*, 22: 249–56.

Wilkinson, M., Scanlan, C. M. and Tittley I. (1987), 'The attached algal flora of the estuary and Firth of Forth', *Proceedings of the Royal Society of Edinburgh*, 93B: 343–54.

Wilkinson, M., Telfer, T. C. and Grundy, S. (1995), 'Geographical variations in the distribution of macroalgae in estuaries', *Netherlands Journal of Aquatic Ecology*, 29: 359–68.

Winterhoff, W. (1992), 'Introduction', in W. Winterhoff, *Fungi in Vegetation Science*, Dordrecht: Kluwer, pp. 1–5.

Wood, N. (1989), *Scottish Place Names*, Edinburgh: Chambers.

Woodforde, J. (1824), *A Catalogue of the indigenous phenogamic* [sic] *plants growing in the neighbourhood of Edinburgh*, Edinburgh: John Carfrae; London: Longman *et al.*

Young, J. (1820), 'On the preparation of opium in Great Britain', *Edinburgh Philosophical Journal*, 1820 I–II: 258–70.

List of Contributors to the Botany of the Lothians Project

This list gratefully records the names of those who contributed to the Botany of the Lothians survey as square recorders, or who helped in other ways to complete the record now published. Some contributions were large, others were slight – no distinction has been made. With such a long-running project and such a large and diverse company of contributors, errors and omissions are likely to have occurred in our records. For these we apologise.

Adamson, D.
Aglen, P. M.
Aiken, M.
Aitken, M.
Alexander, L.
Ambler, R. P.
Anderson, G.
Anderson, R.
Andrews
Argent, G.
Averis, B.
Bailey, M.
Bain, D. R.
Bareau, I.
Barker, R. W.
Battey, N.
Bell, C. M.
Bennell, A. P.
Bignell, J.
Black, R.
Blackmore, S.
Bland, V.
Boardman
Bogue, J.
Bonner, N.
Borland, I. J.
Brandt, N.

Broome, A.
Brown, N.
Bruce, Mr & Mrs W.
Calder, M. A.
Campbell, A.
Campbell, C.
Campbell, E.
Cant, M.
Carlyle, J.
Carroll, S.
Caulton, E.
Chamberlain, D. F.
Chambers, K.
Chess, C.
Christie, D.
Clark, N. W.
Clarke, S.
Clarkson, M.
Clayton, H.
Clement, R.
Clinton, P. H.
Cochrane, M. A.
Cochrane, M. P.
Corbett, N.
Cormack, J-A.
Cottrell, J. E
Couper, C. J.

Cranston, T.
Crawford, C.
Crawford, G.
Crossmann, R.
Cruikshank, M.
Cullen, J.
Cunningham, J. M.
Cunningham, M.
Curtis, A.
Dagg, E. A.
Dale, B.
Dale, J.
Davidson, C.
Davidson, E.
Davidson, M.
Denham, J.
Dennis, G.
Denton, D.
Dickson, G.
Dinwoodie, W.
Dixon, C.
Dixon, R. O. D.
Donald, D.
Doroszenko, A.
Dyer, A. F.
Edgar, C.
Edwards, I.

Elliot, M.
Ellis, N.
Elston, D.
England, F. J.
Eunson, S.
Fairnie, S.
Fairweather, A. D.
Faithfull, M. A.
Fenton, J.
Fife, W. A. A.
Ford, M. S.
Forrest, G.
Forrest, L.
Forster, R. H.
Fraile, C.
Fraser, A.
Galt, R.
Garner, D.
Garrod, R.
Garry, J.
Gent, J.
Gill, W. D.
Gillespie, A.
Gillespie, E.
Glass, E.
Godfrey, A.
Godfrey, C.
Godfrey, R.
Gordon, C.
Gordon, H.
Gordon, P.
Gough, F. J.
Gowers, B.
Gowler, Z.
Grace, J.
Grant, R.
Green, N.
Gregor, J.
Habgood, D.
Hadfield, P.
Hall, R. D.
Hamilton, J.
Hardy, A. L.
Harper, G. H.
Harper, P. C.
Harvie, B.

Hastie, V.
Hay, M.
Hay, W.
Hendry, S.
Hipkin, A.
Hobbs, A.
Hobson, B.
Hobson, P. M.
Hutchon, C.
Innes, E.
Jack, Dr & Mrs M.
Jackson, C.
Jackson, E. H.
Jackson, G.
Jackson, J.
Jamie, E.
Jeffree, C. E.
Johnston, C.
Johnstone, C.
Johnstone, F.
Keeling, S.
Keightley, P.
Kemp, R. F. O.
Kirk, F. A. and E. Kirk
Klayman, F.
Knees, S. G.
Kungu, E. M.
Langstaffe
Last, P.
Laverock, M.
Learmonth, R. W. C.
Leavy, P.
Leckie, S.
Legg, C. J.
Lindsey, S.
Linklater, A.
Litteljohn, M. S. R.
Little, M.
Long, D. G.
Longstaff, M.
Lumsden, B.
Luxmoor, S.
Lyburn, J.
Macintosh, F.
Mackay, K.
Mackay, S. J.

Mackenzie, K. J.
MacKintosh, J.
MacLeod, I.
Mann, D. G.
Manning, A. W. G.
Marland, A.
Marrian, M.
Marshall, W.
Maxwell, S.
McConnell, J.
McCraw, J.
McDougall, G.
McGilvray, A.
McInnes, D.
McKean, D. R.
McMillan, K. A.
McNab, W.
McPherson, J.
McVicar, M.
McWilliam, A.
Meek, E.
Meikle, M.
Meissner, S.
Miedzybrodzka, M.
Mill, R. R.
Millar, I. A.
Miller, A.
Miller, H.
Miller, J. I.
Mitchell, F.
Mitchell, M.
Mitchell, R. R.
Moeller, M.
Moffat, B.
Morrison, D.
Morss, L.
Mountford, Mrs
Muir, I.
Muirhead, V.
Munro, D. S.
Murdoch, G.
Muscott, J.
Mutch, A.
Nairn, I.
Needham, C. D.
Neild, P.

Nicholson, D.
Nimmo, M.
Nisbet, M. E.
Outterson, J.
Page, C. N.
Pankhurst, R. J.
Paterson, D.
Paterson, J.
Penny, M.
Perks, M.
Peters, J. C.
Peters, M.
Phillips, K. B.
Phillips, S.
Pitcairn, C. E. R.
Proctor, J.
Rae, D.
Raeburn, B.
Raeburn, J.
Ramage, P.
Rangeley, A.
Ratter, J. A.
Rawcliffe, C. P.
Rebane, N.
Ressle, C.
Richards, A. J.
Richardson, P.
Richardson, S. A.
Riley, D.
Ritchie, P.
Robertson, F. W.
Robertson, M.
Robertson, S. C.
Roe, E.
Ross, M.

Rowe, J.
Rutherford, A.
Rutherford, A.
Sales, F.
Salt, C.
Sandham, J.
Saville, R. E.
Schofield, A.
Schwarz, E.
Scott, C.
Scott, M. M.
Seaton, D. H.
Selfridge, M.
Silverside, A. J.
Sinclair, A.
Sinclair, I.
Sinclair, R. C.
Sladdin, S.
Slee, D.
Smith, D.
Smith, E.
Smith, J. M.
Smith, P. M.
Stace, H.
Staples, K.
Steele, M. J.
Stevenson, K. R.
Stevenson, T. S.
Stewart, N. F.
Stewart, O. M.
Stickle, A.
Stirling, A. McG.
Stobie, M.
Stone, C.
Strachan, I.

Struthers, C.
Sumner, B. E. H.
Swift, N.
Taylor, G.
Taylor, J.
Tebble, M. O.
Thom, H.
Thomas, P.
Thompson, I. A.
Thornton, J.
Tulett, A. J.
Upton, P.
Walker, R.
Wallace
Wardrop, V.
Watkins, J.
Watson, K.
Watt, E.
Weatherhead, R.
Webb, C
Weitzner, M.
West, R.
White, M.
Williams, A. E.
Wilson, A.
Winham, J.
Woods, J.
Woods, P. J. B.
Worsdall, B.
Wright, J.
Yeomann, M. M.
Young, S.
Younger, A.

Students (U o E)

Gazetteer

Six-figure references are given for sites outwith the City of Edinburgh where the name of the site is not on the 1:50,000 Ordnance Survey map of the area. Two grid references are given when two survey squares were located in an area identified by a single name.

Some of the more important locations are marked on the map of the vice-counties on page 517.

Abercorn **WL** NT0879
Aberlady **EL** NT4679
Aberlady Bay **EL** NT4681
Aberlady Bay Curling Pond **EL** NT4680
Aberlady Mains **EL** NT4779
Aberlady Point **EL** NT4580
Aberlady Station **EL** NT4779
Addiewell **ML** NS9962
Aikengall area **EL** NT7170
Aikengall Water area **EL** NT7071
Ainslie Park **ML** NT2376
Aitkendean Wood **ML** NT3161
Allermuir Hill **ML** NT2266
Almond Pools **WL/ML** NT022664
Almondell Country Park **WL/ML** NT0969
Archerfield (Estate) **EL** NT5084
Armadale **WL** NS9368
Armet Water **EL** NT4455
Arniston **ML** NT3259
Arthur's Seat **ML** NT2772
Athelstaneford **EL** NT5377
Auchencorth (Moss) **ML** NT2055
Auchendinny **ML** NT2562
Auchenhard **WL** NS9963
Auldhame **EL** NT5984
Baadsmill **ML** NT0059
Balerno **ML** NT1666

Balfour Monument **EL** NT5772
Balgone **EL** NT5682
Balgornie **WL** NS9365
Ballencrieff area **EL** NT4878
Ballencrieff Crossroads **EL** NT4877
Bangly Hill **EL** NT4875
Bangour (old hospital) **WL** NT0271
Bangour Reservoir **WL** NT0171
Bankton **EL** NT3973
Bara/Baro **EL** NT5669
Barbauchlaw Burn **WL** NS9471
Barns Ness **EL** NT7277
Barnton **ML** NT1975
Bass Rock **EL** NT6087
Bathgate **WL** NS9769
Bathgate Hills **WL** NS9870
Bavelaw **ML** NT1462
Bawdy Moss **ML** NT0655
Bawsinch **ML** NT2872
Beanston Mill **EL** NT5575
Bearford Burn **EL** NT5473
Bedlormie House **WL** NS8867
Beecraigs Country Park **WL** NT0074
Beeslack **ML** NT2461
Belhaven **EL** NT6678
Belhaven Bay **EL** NT6578
Bell, The **EL** NT6763
Bells Quarry **ML** NT0465

Belsyde **WL** NS9775

Biel (Estate) **EL** NT6375

Biel Water/Burn **EL** NT6476

Bielmill **EL** NT6275

Bilsdean **EL** NT7672

Bilston Burn **ML** NT2664

Binning Wood **EL** NT6080

Binns, The **WL** NT0578

Binny Craig **WL** NT0473

Birdsmill **WL** NT1071

Birkhill **WL** NS9679

Birns Water **EL** NT4567

Black Loch **EL** NT6673

Blackburn **WL** NS9865

Blackcastle Hill **EL** NT7171

Blackfaulds **WL** NS9172

Blackford Glen/Quarry **ML** NT2670

Blackford Hill **ML** NT2570

Blackhall **ML** NT0562

Blackhill **ML** NT1863

Blackhope Scar **ML** NT3148

Blackness **WL** NT0580

Blackridge **WL** NS8967

Bladdering Cleugh **EL** NT6969

Blaik Law **EL** NT6671

Blaikie Heugh **EL** NT575730

Blance **EL** NT4868

Blawhorn Moss **WL** NS8868

Bleak Law **EL** NT6066

Blinkbonny **EL** NT4772

'Blinkie Burn Moor' **EL** NT5262

Boghall **ML** NT2465

Bolton Crossroads **EL** NT5068

Bo'mains **WL** NS9979

Bonaly **ML** NT2167

Bo'ness **WL** NT0081

Bonnyrigg **ML** NT3165

Bonhard Farm **WL** NT020793

Bonnington **ML** NT2675

Bonnytoun Farm **WL** NT0078

Boonslie **EL** NT6670

Borthwick (railway tip) **ML** NT3760

Bothwell **EL** NT6864

Bothwell Water **EL** NT6766

Bowden Hill **WL** NS9774

Bowerhouse/Bourhouse **EL** NT6676

Bowshank **ML** NT4541

Braid Hills **ML** NT2569

'Braid marshes' **ML** NT2569
(now golf courses)

Breich **ML** NS9660

Breich Water **WL** NT0165

Bridgend **WL** NT0475

Broad Law **ML** NT3453

Broxburn **EL** NT6977

Broxburn **WL** NT0772

Broxmouth **EL** NT6977

Brunt, The **EL** NT6873

Brunton Burn **WL** NS9473

Buckstone **ML** NT2469

Burdiehouse **ML** NT2767

Burn Hope **EL** NT6969

Bush Estate **ML** NT2463

Butterdean Wood **EL** NT4572

Caerketton **ML** NT2366

Cairnpapple Hill **WL** NS9871

Cairntow **ML** NT2872

Caldercleugh **EL** NT6666

Calder Wood **ML** NT0766

Calton Hill **ML** NT2674

Calton Road **ML** NT2673

Cameron Toll **ML** NT2771

Camilty Hill **ML** NT0459

Cammo Country Park **ML** NT1774

Canonmills **ML** NT2575

Canty Bay **EL** NT5885

Carberry **ML** NT3669

Carcant **ML** NT3652

Carlops **ML** NT1656

Carlowrie Estate **WL** NT1474

Carmelhill Loch **WL** NT1075

Carnethy Hill **ML** NT2061

Carperstane **EL** NT5582

Carribber Glen **WL** NS9675

Carribber Mill **WL** NS9574

Carriden **WL** NT0281

Carrington **ML** NT3160

Carrington Barns **ML** NT3261

Catcraig **EL** NT7177

Cauld Burn (1885) **EL** NT6472

Cauldcoats Holdings **WL** NT0479

Cauldhall Moor **ML** NT2758

Chalkieside **ML** NT3768
Champfleurie **WL** NT0376
Clerkington **EL** NT5072
Clubbiedean Reservoir **ML** NT1966
Coates **EL** NT4775
Cobbinshaw Reservoir **ML** NT0157
Cockenzie (Power Station) **EL** NT3975
Cocklaw **ML** NT1668
Cockleroy Hill **WL** NS9874
Cockmuir **ML** NT2654
Cockpen **ML** NT3263
Colinton **ML** NT2168
Colinton Dell **ML** NT2169
Colstoun (House/Water) **EL** NT5171
Colstoun Wood **EL** NT5370
Colzium **ML** NT0858
Comiston (House) **ML** NT2468
Congalton **EL** NT5480
Corston Hill **ML** NT0963
Corstorphine Hill **ML** NT2074
Cousland **ML** NT3868
Couston **WL** NS9571
Cowieslinn **ML** NT2351
Cowpits **ML** NT3470
Craigcrook Castle **ML** NT2174
Craighall **ML** NT3370
Craigie Hill/Quarry **WL** NT1576
Craigiehall **WL/ML** NT1675
Craigielaw **EL** NT4579
Craigielaw Bay **EL** NT4479
Craigleith **ML** NT2274
Craigleith Island **EL** NT5587
Craiglockhart Dell **ML** NT2270
Craiglockhart Hills **ML** NT2370
Craigmillar **ML** NT2871
Craigmillar Castle Quarry **ML** NT2870
Craigton Quarry **WL** NT075770
Cramond **ML** NT1976
Cramond Bridge **WL/ML** NT1775
Cramond Island **ML** NT1978
Crichton (Castle/Kirk) **ML** NT3861
Crossgatehall **ML** NT3669
Crosswood Reservoir **ML** NT0557
Crowhill **EL** NT7374
Cuddie Wood **EL** NT4671
Currie **ML** NT1867

Dalhousie **ML** NT3254
Dalkeith Country Park **ML** NT3368
Dalmahoy **ML** NT1366
Dalmeny Estate **WL** NT1578
Dalmeny Village **WL** NT1477
Dalry **ML** NT2372
Danderhall **ML** NT3169
Danskine (Loch) **EL** NT5667
Darent House **EL** NT5866
Davidsons Mains **WL** NT2075
Dean Village **ML** NT2473
Deans, Livingston **WL** NT0268
Dechmont **WL** NT0470
Deuchrie **EL** NT6271
Dewar Burn **ML** NT3448
Dirleton (Castle) **EL** NT5183
Dirleton Common **EL** NT5085
Dirleton Station **EL** NT5282
Dod Hill **EL** NT7268
Donolly Reservoir **EL** NT5768
Dreghorn **ML** NT2268
Drem **EL** NT5179
Drem Pool **EL** NT5080
Drum, The **ML** NT2770
Drumcross **WL** NT0070
Drumshoreland **WL** NT0670; NT0770
Dry Burn **EL** NT7174
Dryburn Bridge **EL** NT7275
Dryden Tower **ML** NT2764
Duddingston **ML** NT2972
Duddingston Loch **ML** NT2872
Duddingston Wood **WL** NT1077
Dunbar **EL** NT6778
Dunbar Castle **EL** NT6779
Dundas Estate **WL** NT1176
Dundreich Plateau **ML** NT2848
Dunglass (Estate) **EL** NT7671
Dunglass Dean **EL** NT7772
Dunsapie Loch **ML** NT2873
Duntarvie **WL** NT0876
Eaglescairnie Pool **EL** NT5169
East Barns **EL** NT7176
East Cairn Hill **ML** NT1259
East Calder **ML** NT0867
East Craigs **ML** NT1873
East Fortune **EL** NT5579

East Lammermuir Deans area **EL** NT6969
East Linton **EL** NT5977
East Saltoun **EL** NT4767
Eastcraigs Hill **WL** NS9068
Easter Inch **WL** NS9966; NT0066
Eastside Heights **ML** NT3545
Ecclesmachen **WL** NT0573
Edgelaw Reservoir **ML** NT3058
Edgelaw **ML** NT2958
Elf Loch **ML** NT2569
Elmscleugh Area **EL** NT4767
Elmscleugh Water **EL** NT7072
Elphinstone **EL** NT3970
Elvingston **EL** NT4674
Emly Bank **ML** NT2947
Fairliehope **ML** NT1556
Fala **ML** NT4360
Fala Flow/Moor **ML** NT4258
Faseny Water **EL** NT6162
Faucheldean Bing **WL** NT085742
Fauldhouse **WL** NS9360
Fauldhouse Moor **WL** NS9161; NT9261
Fawnspark **WL** NT0675
Fennie Law/Burn **EL** NT5662
Ferny Ness **EL** NT4477
Fidra **EL** NT5186
Figgate Burn **ML** NT2973
Fisherrow **ML** NT3373
Five Sisters Bing **WL** NT0164
Foulshiels Bing **WL** NS9763
Flotterstone **ML** NT232631
Ford **ML** NT3864
Fountainhall **ML** NT4349
Fountainhall (plantation) **EL** NT4267
Frances Craig **EL** NT6381
Friardyke **EL** NT6668
Fullarton Water **ML** NT2857
Fushiebridge **ML** NT3460
Galabraes **WL** NS9870
Garleton area **EL** NT4976
Garleton Hills (quarry) **EL** NT5076
Garvald **EL** NT5870
Garvald Punks **ML** NT3247
Gavieside **ML** NT0265
Gifford **EL** NT5368

Gilston **EL** NT4456
Gladhouse **ML** NT2954
Gladhouse Reservoir **ML** NT2953
Gladsmuir **EL** NT4573
Gladsmuir **ML** NS9256
Glencorse Reservoir **ML** NT2163
Glendevon Pond **WL** NT007750
Glenhornie **EL** NT5983
Gogar **ML** NT1772
Gogarburn **ML** NT1671
Gore Glen **ML** NT3461
Gorebridge **ML** NT3461
Gosford House/Lake/Pond/Estate **EL** NT4578
Gowanbank **WL** NS9171
Granton **ML** NT2376
Greencraig **EL** NT4479
Greendykes Bing **WL** NT0873
Greendykes Quarry **WL** NT076733
Grougfoot **WL** NT0278
Gullane (Station) **EL** NT4882
Gullane Bay/Bents **EL** NT4783
Gullane Hill **EL** NT4782
Gullane Links **EL** NT4682
Gullane Point **EL** NT4683
Gullane rubbish tip **EL** NT4783
Gunpowder Trail, Roslin **ML** NT273624
Gutterford Burn **ML** NT1558
Habbie's Howe **ML** NT1756
Haddington **EL** NT5173
Hagbrae **ML** NT3762
Hailes area **EL** NT5776
Hailes Castle **EL** NT5775
Hailes Quarry (tip) **ML** NT2070
Hairy Craig **EL** NT5775
Half Loaf Pond **WL** NS960669
Hallheriot **ML** NT3852
Hallyards **ML** NT1273
Hanging Rocks **EL** NT4985
Harburn **ML** NT0460
Hare Burn/Moss **ML** NT2056
Hare Hill **ML** NT1762
Hare Moss Wood **WL** NS9265
Harehead **EL** NT6963
Harestone **EL** NT5662
Harlaw Reservoir **ML** NT1764

Harperrig Reservoir **ML** NT0961
Harthill **WL** NS9164
Harvieston House **ML** NT3460
Hawthornden Castle **ML** NT2963
Heart Law **EL** NT7166
Heckies Hole **EL** NT6379
Hedderwick area/Hill **EL** NT6478
Herdmanston **EL** NT4770
Heriot **ML** NT4053
Heriot Water **ML** NT3650
Hermand **ML** NT0263
Hermiston **ML** NT1770
Hermitage of Braid **ML** NT2570
Hillend **ML** NT2566
Hillhouse **WL** NT0075
Hiltly Crags **WL** NT003752
Hirendean Castle **ML** NT2951
Holyrood Park **ML** NT2773
Hope Hills **EL** NT5661
Hopes area/Water **EL** NT5562
Hopes Reservoir **EL** NT5462
Hopetoun Estate **WL** NT0979
Hopetoun Wood **WL** NT0877
Hoprig **EL** NT4574
Hound Point **WL** NT1579
Howe, The **ML** NT1862
Howgate **ML** NT2458
Howgate Pond **ML** NT2457
Humbie area **EL** NT4664
Humbie Kirk **EL** NT4663
Hummell Rocks **EL** NT4683
Huntington **EL** NT4874
Hunter's Bog **ML** NT2773
Inchmickery **ML** NT2080
Ingliston **ML** NT1372
Innerwick **EL** NT7274
Innerwick Castle **EL** NT7373
Inveravon **WL** NS9579
Inveresk **ML** NT3472
Inverleith **ML** NT2475
Jerusalem **EL** NT4770
John Muir Country Park **EL** NT6479
Johnscleugh **EL** NT6366
Juniper Green **ML** NT2068
Keith Glen **EL** NT4564
Keith Marischal **EL** NT4464

Keith Water **EL** NT6463
Kidlaw **EL** NT5063
Kidlaw Reservoir **EL** NT5163
Killmade Burn **EL** NT6662
Killpallet **EL** NT6260
Killpallet Heights **EL** NT6060
Kilspindie **EL** NT4580
Kinauld **ML** NT1767
King's Inch **EL** NT4557
Kingsknowe **ML** NT2170
Kinleith **ML** NT1866
Kinneil **WL** NS9880
Kinneil Kerse **WL** NS9680
Kippielaw **EL** NT5875
Kipps, The **ML** NT3049
Kips, The **ML** NT1860
Kirkliston **WL** NT1274
Kirknewton **ML** NT1167
Knock Hill **WL** NS990711
Knowes Mill **EL** NT6078
Lady Lothian Wood **ML** NT3265
Ladyside **ML** NT3650
Ladyside Height **ML** NT3647
Lamb **EL** NT5386
Lammer Law **EL** NT 5261
Lampinsdub **WL** NT0775
Lasswade **ML** NT3165
Latch Loch **EL** NT5263
Laughing Law **EL** NT7364
Lauriston Castle **ML** NT2076
Laverocklaw **EL** NT4775
Lawhead **EL** NT6079
Lawrie's Pen **EL** NT7078
Leadburn **ML** NT2355
Leaston **EL** NT4863
Leith Docks **ML** NT2776
Leithies, The **EL** NT5785
Lennox Tower **ML** NT1767
Lennoxlove **EL** NT5172
Levenhall **ML** NT23673
Liberton **ML** NT2768
Ling Hope **EL** NT7069
Linhouse Glen **ML** NT0663
Linkfield Car Park **EL** NT6578
Linkhouse Wood **EL** NT5284
Links Wood **EL** NT6380

Linlithgow **WL** NT0077
Linlithgow Bridge **WL** NS9877
Linn Dean (Water) **EL** NT4659; NT4759
Little Cathpair **ML** NT4546
Little Spott **EL** NT6574
Little Vantage **ML** NT1063
Livingston Reservoir **WL** NT032678
Livingston Village **WL/ML** NT0466
Lizzie Brice's Roundabout **ML** NT0666
Loanhead **ML** NT2865
Lochcote Marsh **WL** NS977742
Lochcote Reservoir **WL** NS9773
Lochend Pond **ML** NT2774
Logan Burn **ML** NT1761
Logan Cottage **ML** NT2063
Logan's Waterfall **ML** NT1861
Loganlea Reservoir **ML** NTI962
Longmuir Moss **ML** NT4651
Longmuir Plantation **WL** NT0273
Longniddry (Station) **EL** NT4476
Longniddry Bents **EL** NT4376
Longniddry Dean **EL** NT4375
Longridge **WL** NS9562
Longyester **EL** NT5463
Lothian Edge **EL** NT6471
Lothianbridge **ML** NT3264
Lothianburn **ML** NT2467
Luffness area **EL** NT4880
Luffness Friary **EL** NT4780
Luffness Quarry **EL** NT4781
Luggate Burn **EL** NT6074
Luggate Water **ML** NT4443
Macmerry **EL** NT4372
Maggie Bowers Glen **ML** NT3761
Malleny Mills **ML** NT1666
Markle (quarry/area) **EL** NT5777
Markle Mains **EL** NT5677
Marl Loch **EL** NT4681
Mavisbank **ML** NT2865
Meadowbank **ML** NT2874
Meadows Yard **ML** NT2975
Medwin Water **ML** NT0754
Meggetland **ML** NT2271
Meikle Says Law **EL** NT5861
Melville Castle **ML** NT3066
Mid Tartraven **WL** NT0072

Middleton **ML** NT3557
Midhope **WL** NT0778
Midmar **ML** NT2750
Millerhill **ML** NT3269
Mochries Craig Hill **WL** NT028742
Monktonhall (colliery) **ML** NT3370
Monynut (Wood/Water) **EL** NT7066
Moredun **ML** NT2770
Morham **EL** NT5572
Morningside **ML** NT2470
Morrison's Haven **EL** NT3773
Morton Reservoir **ML** NT0763
Mortonhall **ML** NT2668
Moss Law **EL** NT6065
Mount Main **ML** NT3848
Muiravonside **WL** NS9675
Muirend **WL** NT0970
Muirhouse **WL** NS9979
Mungoswells **EL** NT4978
Murieston **ML** NT0463
Musselburgh **ML** NT3472
Myreside **ML** NT2370
Nether Brotherstone **EL** NT4354
Nether Hailes **EL** NT5675
Nether Monynut **EL** NT7264
Netherton **ML** NT2357
New Burnshot **WL** NT1776
New Town **EL** NT4470
New Winton **EL** NT4271
Newbattle **ML** NT3365
Newbridge **ML** NT1272
Newbyth **EL** NT5880
Newhailes **ML** NT3272
Newhall Burn **EL** NT4966
Newhall, River North Esk **ML** NT1756
Newhaven **ML** NT2577
Newlands Burn **EL** NT5766
Newliston Estate **WL** NT1173
Newton **ML** NT3369
Newtongrange **ML** NT3364
Niddrie **ML** NT3071
Niddry **WL** NT0974
Niddry Burn **WL** NT1074
Nine Mile Burn **ML** NT1757
North Berwick **EL** NT5585
North Berwick Law **EL** NT5584

North Berwick rubbish tip **EL** NT5785
North Esk Reservoir **ML** NT1558
North Middleton **ML** NT3559
Nunraw **EL** NT5970
Ochiltree Hill (Fort) **WL** NT030741
Ochiltree Mill **WL** NT033739
Ogle Burn **EL** NT7272
Oldhamstocks (Burn/Water) **EL** NT7470
Ormiston **EL** NT4169
Ormiston **ML** NT0966
Ormiston Hall **EL** NT4167
Overhailes **EL** NT5776
Overton **WL** NT1074
Oxenfoord Castle **ML** NT3965
Papana Water **EL** NT5868
Papple **EL** NT5972
Pardovan House **WL** NT0477
Park Farm **WL** NT029771
Parkley Craigs **WL** NT0175
Pate's Hill **ML** NS9959
Pathhead **ML** NT3964
Peffer Burn **EL** NT5080
Peffer Burn, Newbyth **EL** NT5980
Peffer Sands **EL** NT6282
Pefferside **EL** NT6182
Pencaitland **EL** NT4468
Pencraig Hall **EL** NT5776
Penicuik **ML** NT2360
Penicuik House/Woods **ML** NT2159
Pepper Wood **WL** NT142745
Petershill (former Reservoir) **WL**
    NS985696
Petersmuir Wood **EL** NT4866
Philip Burn **EL** NT7163
Philpstoun Bing **WL** NT0576
Pilmuir (House) **EL** NT4869
Pinkerton area **EL** NT6975
Pitcox **EL** NT6475
Pogbie **EL** NT4660
Polkemmet Country Park **WL** NS9264
Polkemmet Moor **WL** NS9162; NS9262
Polton **ML** NT2964
Pomathorn **ML** NT2459
Port Edgar **WL** NT1178
Port Seton **EL** NT4075
Portobello **ML** NT3074

Powflats **WL** NT0871
Pressmennan (Lake/Wood)) **EL** NT6273
Preston (Castle) **EL** NT3974
Prestonfield **ML** NT2771
Prestonhall **ML** NT3965
Prestonpans marina **EL** NT3974
Priest Law **EL** NT5162
Priestlaw Hill **EL** NT6562
Pringles Green **ML** NT3744
Quarrel Reservoir **ML** NT1858
Ratho **ML** NT1370
Ravelrig **ML** NT1566
Ravelston **ML** NT2174
Ravensheugh (Sands) **EL** NT6281
Red Moss **ML** NT1663
Redden Grain **EL** NT546612
Redhall **ML** NT2170
'Redheughs' **ML** NT1870
Redhouse Castle **EL** NT4677
Redmains **EL** NT4469
Renton Hall **EL** NT5471
Riccarton Hills **WL** NT0173
River South Esk **ML** NT3157
River Tyne **EL** NT5274
Rockville Heughs **EL** NT5581
Rook Law **EL** NT6367
Rosebank **ML** NT0366
Rosebery Reservoir **ML** NT3050
Rosewell **ML** NT2862
Roslin **ML** NT2763
Rough Moss **ML** NT3348
Royal Botanic Garden Edinburgh **ML**
    NT1975
Ryal **WL** NT0871
St Baldred's Cradle **EL** NT6381
St Germains **EL** NT4274
St. Marks Park **ML** NT2575
Saltcoats Castle **EL** NT4881
Saltcoats **EL** NT4882
Saltoun area **EL** NT4667
Saltoun brickworks **EL** NT4868
Saltoun Forest/Wood/Big Wood **EL**
    NT4666
Saltoun Hall **EL** NT4668
Saltoun Station **EL** NT4566
Samuelston **EL** NT4870

Sandy Hirst **EL** NT6379
Sandy Knowe **EL** NT4985
Sandy's Mill **EL** NT5575
Sauchet Water **EL** NT6172
Scald Law **ML** NT1961
Scoughall **EL** NT6183
Seacliff **EL** NT6084
Seafield **WL** NT0066
Seafield **ML** NT2875
Seafield Pond **EL** NT6578
Selm Muir Wood **ML** NT0864
Seton (Halt) **EL** NT4174
Seton Dean/Glen/Mains **EL** NT4275
Seton House **EL** 4175
Seton Sands **EL** NT4276
Setonhill Car Park **EL** NT4675
Sheeppath Burn **EL** NT6870
Sheeppath Dean/Glen **EL** NT7070
Sheriff Hall **EL** NT5681
Sheriffhall **ML** NT3167
Sighthill **ML** NT1920
Silverknowes **ML** NT2077
Skateraw (Harbour) **EL** NT7375
Skid Hill Quarry **EL** NT5076
Slateford (quarry tip) **ML** NT2271
Smeaton **ML** NT3569
Smeaton House (Lake/Wood) **EL** NT5978
Snab Point **WL** NT1877
South Queensferry **WL** NT1278
Soutra **EL** NT4559
Soutra Aisle **EL** NT4558
Spartleton **EL** NT6565
Spilmersford **EL** NT4569
Spittal **EL** NT4677
Spott (Burn) **EL** NT6775
Spott Mill **EL** NT6574
Standing Stones Farm **EL** NT5773
Star Wood **EL** NT5968
Stenhouse **ML** NT2171
Stenton **EL** NT6274
Stevenson **EL** NT5474
Stobshiel Reservoir **EL** NT5062
Stockbridge **ML** NT2474
Stoneyburn **WL** NS9862
Stoneypath **EL** NT6171
Stoneypath Tower **EL** NT5971

Stottencleugh **EL** NT7270
Stow **ML** NT4644
Straiton **ML** NT2766
Swanston **ML** NT2467
Swineburn Wood **WL** NT0976; NT1075
Tailend Moss **WL** NT0067; NT0167
Tantallon (Castle) **EL** NT5985
Tavers Cleugh **EL** NT6266
Temple **ML** NT3158
Thornton **EL** NT7373
Thornton Station **EL** NT7474
Thorntonloch **EL** NT7574
Thorters Reservoir **EL** NT6069
Threemiletown **WL** NT0575
Threipmuir **ML** NT1663
Thurston area **EL** NT7174
Thurston Mains (Burn) **EL** NT7173
Torduff Reservoir **ML** NT2067
Torness **EL** NT7475
Torphichen **WL** NS9672
Torphin **ML** NT2067
Totleywells **WL** NT0976
Townhead **EL** NT5568
Tranent **EL** NT4072
Traprain Law **EL** NT5874
Trotter's Bridge **ML** NT324601
Turnhouse **ML** NT1674
Tweeddale Burn **ML** NT2752
Tyne Estuary **EL** NT6279
Tynefield **EL** NT6277
Tynehead **ML** NT3959
Tyninghame (House) **EL** NT6179
Tyninghame Bay/Estuary **EL** NT6378
Tyninghame Links Wood **EL** NT6381
Uphall **WL** NT0671
Uphall Station **WL** NT0670
Upper Dye Water **EL** NT5660
Vogrie **ML** NT3762
Vogrie Country Park **ML** NT3763
Wallhouse **WL** NS9572
Wamphray **EL** NT5683
Ware Road, Tyninghame **EL** NT6178
Warriston **ML** NT2575
Waughton area/Castle **EL** NT5680
Waughton Crossroads **EL** NT5681
West Barns **EL** NT6578

West Calder **ML** NT0163
West Saltoun **EL** NT4667
West Steel **EL** NT686700
West Torphin **ML** NT0360
Wester Ochiltree **WL** NT0374
Wester Pencaitland **EL** NT4368
Wester Shore **WL** NT0679; NT0779
Westfield **WL** NS9372
Whitburn **WL** NS9464
White Castle **EL** NT6168
White Sands **EL** NT7077
Whiteadder Reservoir **EL** NT6563
Whiteadder Water **EL** NT6365
Whitecraig **ML** NT0754
Whitekirk (Hill) **EL** NT5981
Whitekirk Covert **EL** NT5982
Whitelaw Cleugh **ML** NT3655

Whitrigg Bing **WL** NS968645
Whittingehame (Water) **EL** NT6073
Wide Hope **EL** NT7068
Winchburgh **WL** NT0874
Winton (House) **EL** NT4369
Wolf Cleugh **ML** NT3447
Woodcockdale **WL** NS 9776
Woodend **WL** NT0877
Woodhall Burn/Ford **EL** NT6873
Woodhall Dean **EL** NT6872
Woodmuir (plantation) **ML** NS9658
Wyndford Crags **WL** NT065732
Yarrow **EL** NT6172
Yearn Hope **EL** NT7269
Yellowcraig **EL** NT5185
Yester (Estate/House) **EL** NT54

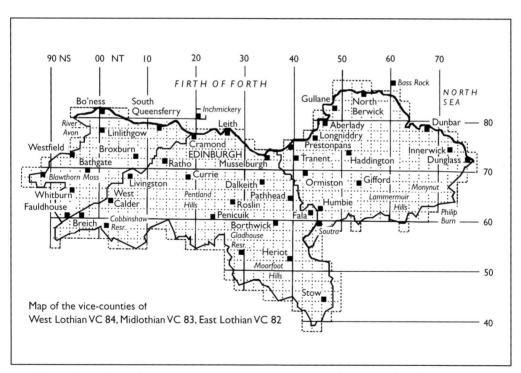

Map of the three vice-counties showing some of the places listed in the gazetteer.

List of Synonyms

Synonyms	Names used in Flora
Acaena anserinifolia auct. non (J. R. & G. Forst.) Druce	Acaena novae-zelandiae
Aconitum anglicum Stapf	Aconitum napellus
Agrimonia odorata (Gouan) Mill.	Agrimonia procera
Agropyron caninum (L.) Beauv.	Elymus caninus
Agropyron junceiforme (Á. & D. Löve) Á. & D. Löve	Elytrigia juncea ssp. boreoatlantica
Agropyron pungens auct. non (Pers.) Roem. & Schult.	Elytrigia atherica
Agropyron repens (L.) Beauv.	Elytrigia repens
Agrostis tenuis Sibth.	Agrostis capillaris
Alchemilla minor auct. non Huds.	Alchemilla glaucescens
Alchemilla vestita (Buser) Raunk.	Alchemilla filicaulis ssp. vestita
Amelanchier laevis auct. non Wieg.	Amelanchier lamarckii
Antirrhinum orontium L.	Misopates orontium
Aphanes microcarpa auct. non (Boiss. & Reut.) Rothm.	Aphanes australis
Arctium vulgare auct. non (Hill) Druce	Arctium nemorosum
Artemesia maritima L.	Seriphidium maritimum
Atriplex hastata auct. non L.	Atriplex prostrata
Bromus mollis L.	Bromus hordeaceus ssp. hordeaceus
Bromus thominei auct. non Hardouin	Bromus x pseudothominei
Bromus thominei Hardouin	Bromus hordeaceus ssp. thominei
Calamintha grandiflora (L.) Moench	Clinopodium grandiflorum
Callitriche intermedia Hoffm.	Callitriche hamulata
Callitriche polymorpha Fr.	Callitriche platycarpa
Calystegia sepium ssp. *pulchra* (Brummitt & Heywood) Tutin nom. inval.	Calystegia pulchra
Calystegia sepium ssp. *silvatica* (Kit.) Batt.	Calystegia silvatica
Carex demissa Hornem.	Carex viridula ssp. oedocarpa
Carex pairaei F. W. Schultz	Carex muricata ssp. lamprocarpa
Carex polyphylla Kar. & Kir.	Carex divulsa ssp. leersii
Carex serotina Mérat	Carex viridula ssp. viridula
Cedrus libanensis Mirbel	Cedrus libani
Chamaenerion angustifolium (L.) Scop.	Chamerion angustifolium

Cheiranthus cheiri L.	Erysimum cheiri
Chrysanthemum leucanthemum L.	Leucanthemum vulgare
Cornus stolonifera Michx.	Cornus sericea
Coronilla varia L.	Securigera varia
Corydalis claviculata (L.) DC.	Ceratocapnos claviculata
Corydalis lutea (L.) DC.	Pseudofumaria lutea
Dactylorchis fuchsii (Druce) Verm.	Dactylorhiza fuchsii
Dactylorchis purpurella (T. & T. A. Stephenson) Verm.	Dactylorhiza purpurella
Delphinium ajacis L. sec. J. Gay	Consolida ajacis
Delphinium consolida L.	Consolida regalis
Delphinium orientale J. Gay	Consolida orientalis
Desmazeria marina (L.) Druce	Catapodium marinum
Desmazeria rigida (L.) Tutin	Catapodium rigidum
Dipsacus fullonum ssp. *sativus* (L.) Thell.	Dipsacus sativus
Dryopteris abbreviata auct. non DC.	Dryopteris oreades
Dryopteris austriaca Woyn. ex Schinz & Thell. non Jacq.	Dryopteris dilatata
Elymus arenarius L.	Leymus arenarius
Endymion hispanica (Mill.) Chouard	Hyacinthoides hispanica
Endymion nonscriptus (L.) Garcke	Hyacinthoides non-scripta
Epilobium adenocaulon Hausskn.	Epilobium ciliatum
Epilobium pedunculare auct. non A. Cunn.	Epilobium brunnescens
Erigeron mucronatus DC.	Erigeron karvinskianus
Eruca sativa Mill.	Eruca vesicaria
Euphrasia occidentalis Wettst.	Euphrasia tetraquetra
Festuca juncifolia St.-Amans	Festuca arenaria
Filago germanica L. non Huds.	Filago vulgaris
Fumaria micrantha Lag.	Fumaria densiflora
Galeobdolon luteum Huds.	Lamiastrum galeobdolon
Galium cruciata (L.) Scop.	Cruciata laevipes
Glyceria plicata (Fr.) Fr.	Glyceria notata
Helianthemum chamaecistus Mill.	Helianthemum nummularium
Hieracium aurantiacum L.	Pilosella aurantiaca
Hieracium caespitosum Dumort.	Pilosella caespitosa ssp. colliniformis
Hieracium flagellare Willd.	Pilosella flagellaris ssp. flagellaris
Hieracium pilosella L.	Pilosella officinarum
Inula conyza DC.	Inula conyzae
Kochia scoparia (L.) Schrad.	Bassia scoparia
Lamium molucellifolium auct. non (Schumach.) Fr.	Lamium confertum
Larix leptolepis (Siebold & Zucc.) Endl.	Larix kaempferi
Larix x eurolepis A. Henry nom. illeg.	Larix x marschlinsii
Lathyrus montanus Bernh.	Lathyrus linifolius
Lemna polyrhiza L.	Spirodela polyrhiza

Leontodon taraxacoides (Vill.) Mérat nom. illeg.	Leontodon saxatilis
Lepidium smithii Hook.	Lepidium heterophyllum
Leucorchis albida (L.) E. Mey.	Pseudorchis albida
Lotus uliginosus Schkuhr	Lotus pedunculatus
Lycium halimifolium Mill.	Lycium barbarum
Lycopodium alpinum L.	Diphasiastrum alpinum
Lycopodium selago L.	Huperzia selago
Lycopsis arvensis L.	Anchusa arvensis
Malus sylvestris ssp. *mitis* (Wallr.) Mansf.	Malus domestica
Matricaria maritima L.	Tripleurospermum maritimum
Matricaria maritima ssp. *inodora* (L.) Clapham	Tripleurospermum inodorum
Matricaria matricarioides (Less) Porter nom. illeg.	Matricaria discoidea
Medicago falcata L.	Medicago sativa ssp. falcata
Medicago hispida Gaertn. nom. illeg.	Medicago polymorpha
Medicago x varia Martyn	Medicago sativa ssp. varia
Melandrium album (Mill.) Garcke	Silene latifolia ssp. alba
Mentha x cordifolia auct. ?non Opiz ex Fresen.	Mentha x villosa
Montia perfoliata (Donn ex Willd.) Howell	Claytonia perfoliata
Montia sibirica (L.) Howell	Claytonia sibirica
Myosotis brevifolia Salmon	Myosotis stolonifera
Myosotis hispida Schlecht.	Myosotis ramosissima
Narcissus hispanicus Gouan	Narcissus pseudonarcissus ssp. major
Narcissus majalis Curtis	Narcissus poeticus ssp. poeticus
Nasturtium microphyllum (Boenn.) Rchb.	Rorippa microphylla
Nasturtium officinale W. T. Aiton	Rorippa nasturtium-aquaticum
Naumburgia thrysiflora (L.) Rchb.	Lysimachia thrysiflora
Odontites verna (Bell.) Dum.	Odontites vernus
Oenothera erythrosepala Borbás	Oenothera glazioviana
Orchis fuchsii Druce	Dactylorhiza fuchsii
Orchis purpurella T. & T. A. Stephenson	Dactylorhiza purpurella
Ornithogalum umbellatum auct. non L.	Ornithogalum angustifolium
Oxalis floribunda auct. non Lehm.	Oxalis articulata
Oxycoccus microcarpus (Aiton) Pursh	Vaccinium oxycoccus ssp. microcarpum
Oxycoccus palustris Pers.	Vaccinium oxycoccus
Papaver orientale auct. non L.	Papaver pseudoorientale
Parietaria diffusa Mert. & W. D. J. Koch	Parietaria judaica
Peplis portula L.	Lythrum portula
Phleum nodosum auct. non L.	Phleum bertolonii
Phragmites communis Trin.	Phragmites australis
Plantago psyllium L. 1762 non 1753	Plantago afra
Platanus x hybrida Brot.	Platanus x hispanica
Poa subcaerulea Sm.	Poa humilis
Polygonum amphibium L.	Persicaria amphibia

Polygonum amplexicaule D. Don	Persicaria amplexicaulis
Polygonum aubertii L. Henry	Fallopia baldschuanica
Polygonum bistorta L.	Persicaria bistorta
Polygonum campanulatum Hook. f.	Persicaria campanulata
Polygonum convolvulus L.	Fallopia convolvulus
Polygonum cuspidatum Sielbold & Zucc.	Fallopia japonica
Polygonum hydropiper L.	Persicaria hydropiper
Polygonum lapathifolium L.	Persicaria lapathifolia
Polygonum persicaria L.	Persicaria maculosa
Polygonum polystachyum Wall. ex Meisn.	Persicaria wallichii
Polygonum raii Bab.	Polygonum oxyspermum
Polygonum sachalinense	
F. Schmidt ex Maxim.	Fallopia sachalinensis
Polygonum viviparum L.	Persicaria vivipara
Polypodium vulgare ssp. *prionodes*	
(Asch.) Rothm.	Polypodium interjectum
Polypodium vulgare ssp. *serrulatum*	
F. W. Schultz ex Arcang.	Polypodium cambricum
Polypogon semiverticillatus (Forssk.) Hyl.	Polypogon viridis
Populus canescens (Aiton) Sm.	Populus x canescens
Populus gileadensis Rouleau	Populus x jackii
Potentilla tabernaemontani Asch. nom. illeg.	Potentilla neumanniana
Poterium sanguisorba L.	Sanguisorba minor ssp. minor
Prunus amygdalus Batsch	Prunus dulcis
Pseudotsuga taxifolia (Poir.) Britt.	Pseudotsuga menziesii
Ranunculus lenormandii F. W. Schultz	Ranunculus omiophyllus
Raphanus maritimus Sm.	Raphanus raphanistrum ssp. maritimus
Reynoutria japonica Houtt.	Fallopia japonica
Reynoutria sachalinensis	
(F. Schmidt ex Maxim.) Nakai	Fallopia sachalinensis
Reynoutria x bohemica Chrtek & Chrtková	Fallopia x bohemica
Rheum rhabarbarum auct. non L.	Rheum x hybridum
Rhynchosinapis cheiranthos (Vill.) Dandy	Coincya monensis ssp. cheiranthos
Rhynchosinapis monensis (L.) Dandy	Coincya monensis ssp. monensis
Rorippa islandica auct. non (Oeder) Borbás	Rorippa palustris
Rosa coriifolia Fr.	Rosa caesia ssp. caesia
Rosa villosa auct. non L.	Rosa mollis
Rumex alpinus L. 1759 non 1753	Rumex pseudoalpinus
Salicornia perennis Mill.	Sarcocornia perennis
Salix atrocinerea Brot.	Salix cinerea ssp. oleifolia
Salix nigricans Sm.	Salix myrsinifolia
Salix viminalis x *Salix aurita*	Salix x fruticosa
Salvia horminoides Pourr.	Salvia verbenaca
Sarothamnus scoparius (L.) W. D. J. Koch	Cytisus scoparius
Scirpus maritimus L.	Bolboschoenus maritimus
Scirpus setaceus L.	Isolepis setacea

Scrophularia aquatica auct. non L.	Scrophularia auriculata
Sedum reflexum L.	Sedum rupestre
Senecio tanguticus Maxim.	Sinacalia tangutica
Setaria glauca auct. non (L.) P. Beauv.	Setaria pumila
Sieglingia decumbens (L.) Bernh.	Danthonia decumbens
Silene alba (Mill.) E. H. L. Krause nom. illeg.	Silene latifolia ssp. alba
Silene maritima With.	Silene uniflora
Silene x intermedia (Schur) Philp nom. illeg. non (Lange) Bocq.	Silene x hampeana
Solanum sarachoides auct. non. Sendtn.	Solanum physalifolium
Sparganium minimum Wallr.	Sparganium natans
Spergularia marginata Kitt. nom. illeg.	Spergularia media
Spergularia salina J. & C. Presl.	Spergularia marina
Stellaria alsine Grimm nom. inval.	Stellaria uliginosa
Symphoricarpos rivularis Suksd.	Symphoricarpos albus
Thelypteris dryopteris (L.) Sloss.	Gymnocarpium dryopteris
Thelypteris oreopteris (Ehrh.) Sloss.	Oreopteris limbosperma
Thelypteris phegopteris (L.) Sloss.	Phegopteris connectilis
Thymus drucei Ronn.	Thymus polytrichus ssp. britannicus
Thymus praecox auct. non Opiz	Thymus polytrichus
Tilia x vulgaris Hayne	Tilia x europaea
Tillaea muscosa L.	Crassula tillaea
Trigonella ornithopodioides (L.) DC.	Trifolium ornithopodioides
Tripleurospermum maritimum ssp. *inodorum* (L.) Hyl. ex Vaar.	Tripleurospermum inodorum
Viscaria vulgaris Bernh.	Lychnis viscaria

Index of Bryophyte Flora

Index of Flora (Latin Names)

Index of Flora (Common Names)

General Index

544